Jet Aircraft Power Systems

JET AIRCRAFT

THIRD EDITION

POWER SYSTEMS

JACK V. CASAMASSA

Russell Associates, Inc.

RALPH D. BENT

Northrop Institute of Technology

McGRAW-HILL BOOK COMPANY

New York London

St. Louis Toronto

San Francisco Sydney

JET AIRCRAFT POWER SYSTEMS

Library of Congress Catalog Card Number: 64-21627

10199

Preface

Even as changes and improvements are made in gas-turbine engines, so must changes and improvements be made in textbooks describing these engines. For this reason, the third edition of *Jet Aircraft Power Systems* has been written to bring new information on gas-turbine engines to the student of aircraft powerplants and to reflect the experience gained in the use of gas-turbine engines for aircraft during the time that such engines have been employed in regular commercial service.

Many improvements have been made during the past few years in the performance and dependability of gas-turbine powerplants. The advent of the large jet engines producing thrusts of 15,000 to 30,000 lb has led to the development of aircraft with unprecedented performance. The fanjet or turbofan engine has increased the efficiency of jet propulsion to the point where fuel consumption is no longer considered a serious problem, and the range of jet airliners is now such that routes of more than 7,000 miles can be flown without refueling. The large engines have made it possible to develop the supersonic transport which will be in service within a few years.

The use of gas-turbine powerplants in civil transport aircraft has made it essential that all possible information relating to the construction and operation of such powerplants be made available to nonmilitary personnel. To aid in accomplishing this objective, this edition includes descriptions of engines installed in our most modern commercial aircraft to prepare the student for employment in the jet engine field.

The new material for this edition has been drawn from many sources, the chief ones being engineers, manufacturers, and operators of gas-turbine engines. From the large amount of information gathered, the material was condensed and arranged for easy study. Like the previous editions, this text contains three major sections and the material contained in each section is arranged to emphasize the operational aspect of actual gas-turbine engines.

The first section deals with the history, theory, nomenclature, operation, types, and systems, pertaining to jet-propulsion devices. The second section shows how the basic principles, parts, and systems are applied to specific gas-turbine engines, including turbojet, turbofan, and turboprop types. The last section gives information on operation, maintenance, and overhaul together with background data to broaden overall understanding of the gas-turbine field.

Even though it would take many volumes to describe all types of gas-turbine engines and their associated systems, it is believed that this book contains the basic information needed by the technician, mechanic, or instructor to provide a clear understanding of the construction and operation of gas-turbine and other reaction engines.

Jack V. Casamassa
Ralph D. Bent

Acknowledgments

The authors acknowledge with appreciation the generous contributions of photographs, drawings, and technical information from the following organizations:

Aero Publishers, Inc., Fallbrook, California
Aircraft Porous Media, Inc., Glen Cove, New York
AiResearch Manufacturing Division, The Garrett Corporation, Los Angeles, California
AiResearch Manufacturing Division of Arizona, Phoenix, Arizona
Allison Division, General Motors Corporation, Indianapolis
Bendix Products Division, Bendix Corporation, South Bend, Indiana
Bendix Utica Division, Bendix Corporation, Utica, New York
The Boeing Company, Transport Division, Renton, Washington
Continental Air Lines, Los Angeles, California
Continental Aviation and Engineering Corporation, Gas Turbine Division, Detroit, Michigan
Delavan Manufacturing Company, West Des Moines, Iowa
Douglas Aircraft Company, Long Beach, California
Douglas Aircraft Company, Santa Monica, California
Esso Oil Company
Fairchild Stratos, Hagerstown, Maryland
General Dynamics-Convair, San Diego, California
General Electric Company, Flight Propulsion Division, Cincinnati, Ohio
General Electric Company, Small Engine Division, West Lynn, Massachusetts
Hamilton Standard Division, United Aircraft Corporation, Windsor Locks, Connecticut
Humble Oil Company
Lockheed California Company, Burbank, California
Lycoming Division, Avco Corporation, Williamsport, Pennsylvania
Marquardt Aircraft Company, Van Nuys, California
McDonnell Aircraft Company, St. Louis, Missouri
North American Aviation, Inc., Los Angeles, California
Northrop Institute of Technology, Inglewood, California
Pratt and Whitney Aircraft, East Hartford, Connecticut
Republic Aviation Corporation, Farmingdale, New York
Rocketdyne Division, North American Aviation, Inc., Canoga Park, California
Rolls-Royce, Ltd., Montreal, Canada, and Derby, England
Ryan Aeronautical Company, San Diego, California
United States Air Force, Washington, D.C.

Contents

CHAPTER 1

History and Development of Gas Turbines

Before World War II, all aircraft were powered by piston-type reciprocating engines. These engines are used to rotate a propeller, thus causing thrust which drives the aircraft through the air. The propeller acts somewhat like a large screw, and it is often called an "airscrew." This analogy is not entirely accurate, but it aids in understanding why a propeller creates thrust. In comparing a rotating propeller with jet propulsion, it is better to consider the effect of the propeller on the air which flows past it during operation.

Everyone who has watched an aircraft in operation on the ground and in the air is familiar with the "slipstream" caused by the propeller. This slipstream is merely a large volume of air which has been set in motion by the rotating propeller. On a conventional aircraft, the slipstream flows from the propeller toward the rear of the aircraft. It is this acceleration of the mass of air toward the rear of the aircraft which creates the thrust.

The turbojet engine performs a function similar to that of the reciprocating engine with the propeller. The difference in the operation of the two systems lies in the mass and velocity of the air or gases affected. The propeller hurls a comparatively large volume of air rearward at a relatively slow velocity, while the turbojet engine takes in a smaller volume or mass of air, expands it with burning fuel, and blasts it to the rear at a very high velocity. The larger the mass of air, and the more it is accelerated, the greater will be the thrust.

The aeolipile

Although the gas turbine or jet engine seems to be a product of very recent times, the principle of jet propulsion has been known for over two thousand years. Figure 1 · 1 illustrates a steam-jet machine called the "aeolipile." This device is said to have been invented by Hero of Alexandria about 150 B.C. It is not known who drew the sketch of the aeolipile shown in Fig. 1 · 1, but it is reported that the original aeolipile was somewhat different in shape, perhaps like a cylinder.

An unconfirmed story relates that the priests of Egypt were fond of amazing their worshippers with "supernatural" events and illusions. To accomplish one of their projects, they commissioned Hero to devise a method whereby the temple doors could be opened without the aid of visible power. The aeolipile was the result. It is not known whether the operation was successful, but it is an interesting sidelight in the history of jet propulsion.

An examination of the aeolipile shows that the energy for operation was supplied by the burning of fuel. The heat from the burning fuel was used to convert water to steam in a boiler. Attached to the head of this boiler are two tubular uprights which also form trunnions for a hollow sphere. Two nozzles projecting from the sphere discharge the steam at right angles to the axis of rotation. This is our first example of jet reaction, and while this may not have been able to produce any practical results as far as the production of power is concerned, still it did

Figure 1 · 1 Hero's aeolipile.

demonstrate the principle of what is today the latest trend in power plants.

The horseless carriage

Figure 1 • 2 illustrates a design for the first horseless carriage. This machine was produced by Sir Isaac Newton in 1680, but there is no record of its having operated. In theory, the jet of steam escaping from the nozzle at the rear of the boiler would cause a reaction which would propel the vehicle forward. It is apparent that thrust would be created, but it is doubtful that it would be sufficient to move the carriage. Even though the horseless carriage may not have operated successfully, it still provides a good illustration of Newton's third law of motion: **to every action, there is an equal and opposite reaction.**

Branca's application

Figure 1 • 3 illustrates what is possibly the first actual turbine. This turbine design is credited to the Italian engineer Giovanni Branca and is dated 1629, about 51 years before Newton's horseless carriage. A study of the illustration clearly shows how the turbine operates. Fuel is burned to heat water in a boiler, thus producing steam. The steam is expelled through a jet to impinge upon buckets around the periphery of a wheel. This causes the wheel to rotate and drive a stamp mill through a 90° reduction-gear arrangement.

It is interesting to note that the examples of Figs. 1 • 1 and 1 • 2 illustrate pure reaction, while Fig. 1 • 3 illustrates the first concept of the impulse wheel. It is also worthy of note that this is our first indication of the gas turbine's doing useful work.

Figure 1 • 2 Newton's horseless carriage.

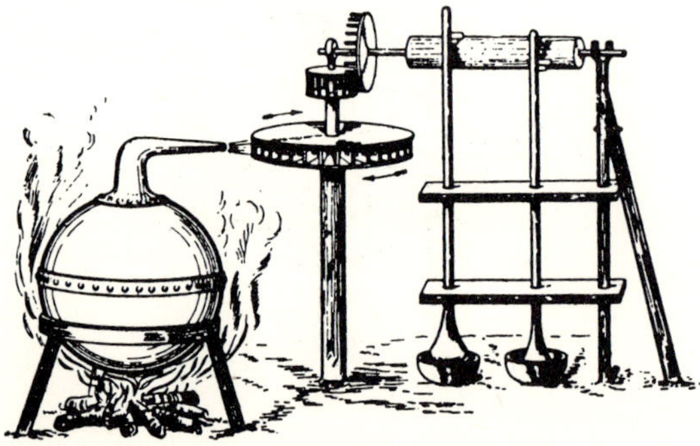

Figure 1 • 3 Branca's turbine.

Early rocket development

In our consideration of reaction devices, we must not overlook the fact that solid-fuel rockets have been in use to some extent for many centuries, possibly as early as A.D. 1100.

In records of a battle which took place in China in A.D. 1232, there is evidence that rockets were used as weapons. These rockets used a type of gunpowder as a fuel. From that time on, gunpowder and rockets were used in China.

The first gas turbine in the United States

What is believed to be the first gas turbine built in the United States was begun in 1902 at Cornell University under the supervision of Sanford A. Moss and was used to obtain information upon which he based his doctor's thesis. Numerous gas turbines were built in Europe between 1905 and 1920 for the purpose of driving air compressors. The earliest application of the gas turbine for aircraft use (in the United States) was to drive a supercharger for reciprocating engines. Dr. Moss supervised the design of the first machine of this type, which was built by the General Electric Company and was assembled in May, 1918. The usefulness of this type of gas turbine during World War II is well known. In 1963, turbosuperchargers were still in use on the Pratt & Whitney R-4360 engines which powered the Boeing Stratocruiser.

The turbosupercharger has emphasized the development of materials capable of operating at high temperatures. This emphasis has been a very important factor in making possible the gas turbine as a power plant for aircraft.

The value of the gas turbine used to drive the turbosupercharger was dramatically demonstrated in 1928 when a world's record for altitude was established by Major R. R. Schroeder in a supercharged Wright Apache airplane.

Development of the turbojet engine

There has been considerable controversy regarding who or what nation should have credit for the development of the first successful turbojet engine. In this text we shall make no attempt to give credit to any individual or nation, since experiments and development have continued concurrently in many nations and by many individuals.

The first jet-propelled flight was made on Aug. 27, 1939, in an aircraft produced by the Heinkel Aircraft Company in Germany. The plane was

Figure 1 · 4 The Heinkel He 178—first jet airplane to fly.

designated He 178, research aircraft, and was not intended to have any military application. This airplane is illustrated in Fig. 1 · 4. The power plant was an early Heinkel turbojet, the HeS 3B, with a thrust of 880 to 1,100 lb. It was an engine of the centrifugal-flow type and was mounted in the fuselage, with the air intake in the nose of the aircraft.

During World War II, German jet development reached the point where some 1,300 aircraft had been fitted with the Jumo 004 engine by the end of the war. Most of these aircraft were Me 262s. The BMW 003 was scheduled for similar production and was to have been installed in the He 162, Arado 234, Me 262, Junkers 287, and Me 1101. Thus it appears that the German development was somewhat ahead of that of other nations, since the Germans were the first to produce jet aircraft suitable for practical military operations.

British development

In 1926, Dr. A. A. Griffith of the British Royal Aircraft Establishment developed a theory of turbine design based upon flow past airfoils rather than through passages. In October of that year he presented his proposals to a meeting of representatives of the British Air Ministry and the Aeronautical Research Committee. These representatives were favorably impressed and recommended preliminary investigations. In 1927 two separate experiments were started.

In 1928, while Dr. Griffith was engaged in developing an axial-flow compressor and turbine with power output transferred to a propeller, Sir Frank Whittle, then a cadet at the Royal Air Force College in England, prepared a thesis describing the possibilities of jet propulsion. Eighteen months later he conceived the idea of using the gas turbine for jet propulsion, and in January, 1930, he applied for his first patent for an engine which used a compressor, burner, and turbine to produce jet thrust. Figure 1 · 5 is the sketch which illustrated his idea. The numbers on the illustration show (1) air inlet, (2) compressor, (3) combustion chamber, (4) turbine wheel, and (5) exhaust.

From 1930 to 1935, Whittle presented his idea for a jet propulsion engine to the Air Ministry and to a number of manufacturing firms, but he had no success. In May, 1935, he was approached by two former RAF officers, R. D. Williams and J. C. B. Tinling, who suggested that they should try to get something started. This eventually led to the formation of Power Jets, Ltd., in March, 1936. The president of the Air Council was a party to the agreement through which this was accomplished, and the Air Ministry was a shareholder from the start,

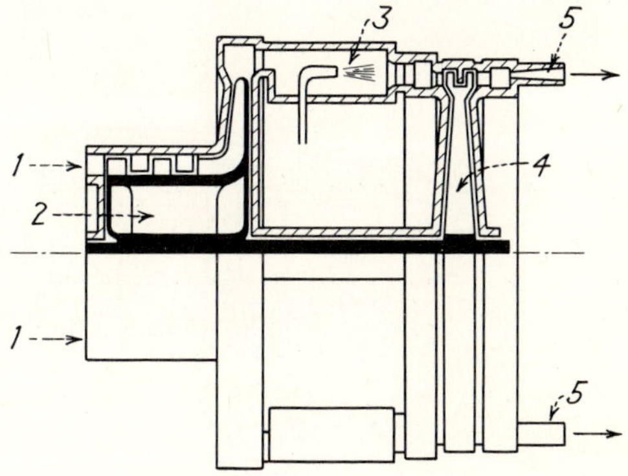

Figure 1 · 5 Whittle's patent sketch of basic turbine.

in that a portion of the shares allotted to the president was held in trust for the Ministry. Thus, with some official and financial backing, it was possible to proceed with the development of a turbojet engine.

The engine to be built was to have a centrifugal compressor with bilateral intakes, driven by a single-stage turbine which was directly coupled to the compressor. Combustion was to take place in a single combustion chamber through which the working fluid passed from the compressor to the turbine. The compressor was required to have a pressure ratio of 4:1, which was considerably higher than any which had been attained previously. Finally, it was necessary that the engine produce more than 3,000 shaft horsepower (shp) out of the single-stage turbine wheel, which was about 16½ in. in outside diameter.

The engine was built by the British Thomson-Houston Company for Power Jets, Ltd. As with any new engine development, many problems of combustion and structural design had to be solved. Testing of the first model indicated that there was a good chance for success; hence additional help was gained from the Air Ministry and other interested parties.

In the summer of 1939, the Air Ministry ceased to regard the development as a matter of long-term research and came to accept the fact that they had the basis for a practicable aircraft power plant. As a result, Power Jets received a contract for a flight engine. A short time afterward, a contract was placed with the Gloster Aircraft Company for the manufacture of an experimental airplane to Specification E28/39. Thereafter, Power Jets and the Gloster Aircraft Company worked in close collaboration. The Air Ministry also agreed to purchase the experimental engine and to cover all test running by contract, thus assuring support for additional development.

Power Jets subcontracted the drawing and manufacture of the flight engine to the British Thomson-Houston Company. The design team now included a few engineers recruited by Power Jets, as well as British Thomson-Houston engineers, and Whittle. The first flight engine was known as the W1.

Flight trials with the W1 engine began on May 14, 1941. The engine had been cleared for 10 hr of flying, and the program of flight testing laid down was completed in 14 days without special incident. The test pilot was Flight-Lieutenant P. E. G. Sayer.

The successful flight trials of the E28/39 aircraft with the W1 engine, shown in Fig. 1·6, gave considerable impetus to the program of jet-engine development. The Ministry of Aircraft Production decided to plan for the production of the W2B engine and the Meteor aircraft. Many skeptics were converted, and the firms already engaged in turbine engine work increased their activities considerably. Other firms which had practically ignored the gas turbine suddenly evinced a lively interest.

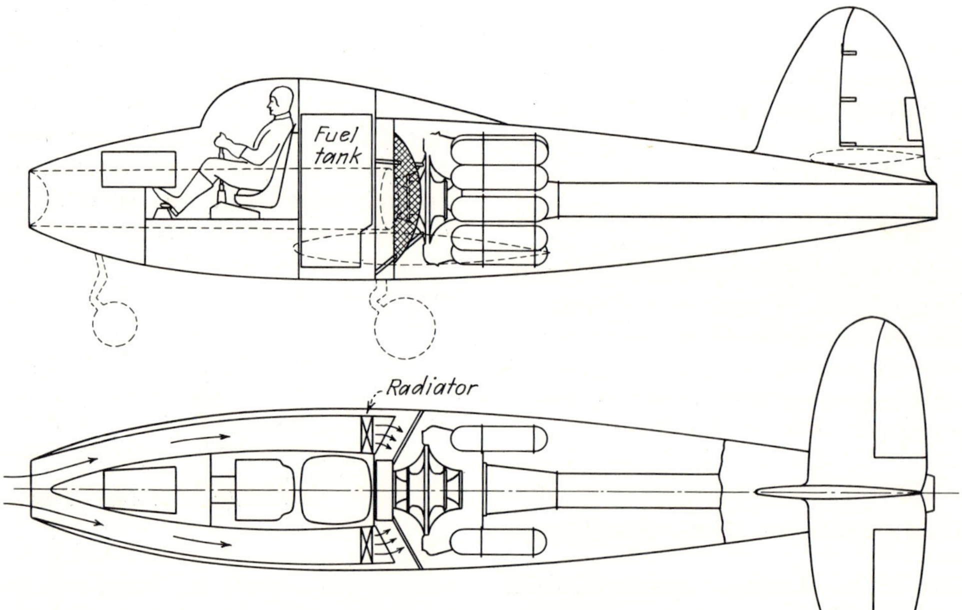

Figure 1·6 Fuselage arrangement of the E28/39 experimental airplane.

Early German development

Figure 1 · 7 shows a cross section of one of the first German turbojet engines to be brought to the United States for analysis and test. This is the Jumo 109-004B engine designed and built by the Junkers Aircraft Corporation of Dessau, Germany. This very interesting design employs an eight-stage axial-flow compressor, six individual can-type combustion chambers, a single-stage turbine, and a variable-area tail cone utilizing an adjustable bullet to vary the exhaust nozzle area. The compressor construction is interesting because of the disk design and the method of assembly. The design of the burners is also of interest because of the method employed for flame propagation.

Because of the lack of proper materials, more than 7 per cent of the air taken in was bled for cooling purposes. Even with all this air used for cooling, the engine had a service life of only about 10 hr. Just after the Germans capitulated, however, it was discovered that new alloys had been developed that would have made it possible to attain a service life of about 150 hr.

Near the end of the war, one of the most interesting of the German jet projects was the 109-011 A-O. This engine was also known as the Heinkel Hirth 011 and was just about ready for production. Many advanced features of compressor development, ice prevention, and turbine-nozzle-blade cooling were in evidence in this design.

The reason for the intensive development of turbojets in England and in Germany during the later stages of the war was quite simple. It was very desirable to develop a power plant capable of producing an extremely high power per pound of installed weight to enable fighters to climb to bomber altitudes in a very short time and to provide a sufficient margin of speed over bombers to make the fighters effective. In addition, the Germans faced the necessity of using fuels of relatively low grade which could be made available under the circumstances at that time. These immediate objectives were certainly attained, but it was done at a great sacrifice of engine durability and specific fuel consumption, compared with a reciprocating engine.

Development and manufacture in the United States

During World War II England and the United States had a reciprocal agreement whereby there was to be a constant interchange of ideas in order to expedite the development and production of military projects. Under this agreement, the first Whittle engine was brought to this country on Oct. 1, 1941.

The General Electric Company (GE) was awarded the development contract to produce an engine conforming to American standards and suitable for flight purposes. General Electric had a long history in the development of turbosuperchargers and air compressors, and since the aviation gas-turbine engine was frequently called a "glorified supercharger," it was logical to award the company this project.

Bell Aircraft was selected to construct the first American jet airplane. Only one year was allotted to build the engine and develop the aircraft. One year and three weeks later (1942) the Bell XP-59A, powered by the GE I-A turbojet engine, made a successful test flight. When all the complexities of a project such as this are realized, it may be called one of the outstanding engineering feats of the year. During normal times of peace, at least 5 years would be needed to develop an airplane-engine combination of this type. The performance of this first jet-powered airplane in the United States was comparable to that of military fighters of that day.

It is only fair to point out that our prewar and wartime reciprocating-engine manufacturers were completely occupied with producing and developing reciprocating engines with which to win the war. In spite of rapid jet-engine progress, it was common knowledge that the war would have to be won with conventional reciprocating engines. Nevertheless, General Electric emerged from the war as an aviation gas-turbine manufacturer. Since the war, this company has developed many jet engines of centrifugal-flow and axial-flow types. One of the most famous of the GE engines is the J47, which served so well during the Korean War. This engine was used to power the North American F-86 Sabrejet fighter, which has probably seen more actual combat service than any other jet-powered fighter. The J47 engine was followed by the J73, the J79 (CJ805), and the J85 (CJ610). One of the most recent General Electric turbojet engines is the J93. Six of these 30,000-lb thrust engines are used to power the North American Valkyrie supersonic reconnaissance bomber. The J47, CJ805, and CJ610 engines are described more fully later in this text.

Under Navy sponsorship, the Westinghouse Electric Corporation was given a contract to develop an all-American design of turbojet engine. This was successfully accomplished, and in 1944 an airplane using the Westinghouse engine made a test flight for the first time. An improved and larger model of this engine was flown in 1946,

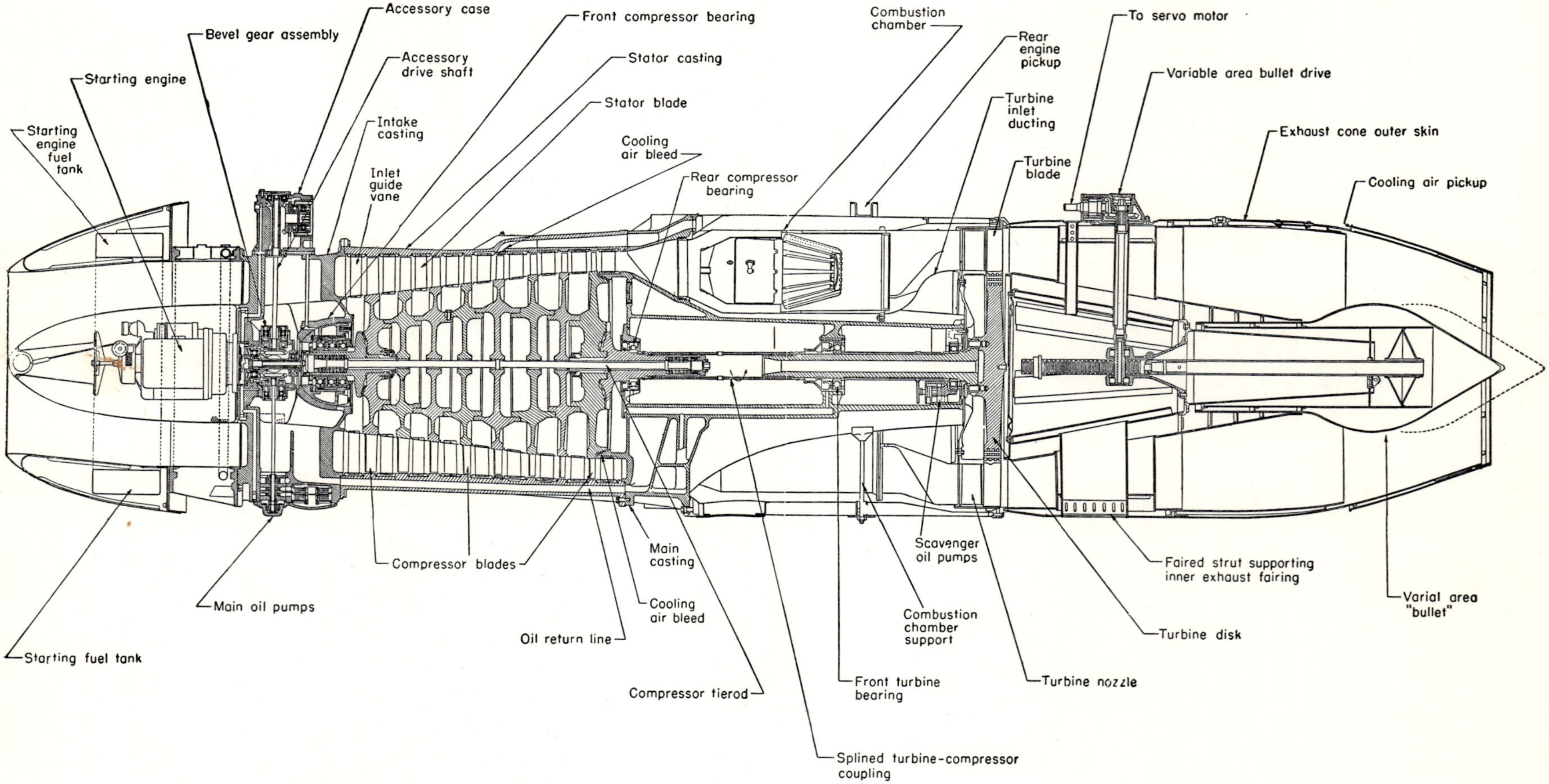

Figure 1 · 7   The Jumo 109-004B turbojet engine.

(a)

(b)

(c)

(d)

Figure **1 · 8** *Military jet aircraft. (a) North American RB-70 Valkyrie. (b) Lockheed F-104 Starfighter. (c) Convair F-106 Delta Dart. (d) North American A-3J Vigilante. (e) Republic F-105 Thunderchief.*

(e)

Figure 1·9 Turbojet and turboprop airliners.

and until recently, Westinghouse continued to develop and produce excellent turbojet engines for the Navy.

Principal features of the Westinghouse engine were the axial-flow compressor and the annular combustion chamber. Thus, these engines resembled the German designs more than they did the British types. Among well-known engines produced by Westinghouse were the J34, J40, and the J46.

Soon after World War II, the Allison Division of General Motors Corporation took over the production and development of the J33 and the J35 engines which had previously been manufactured by the General Electric Company. Allison continued to improve these engines and also to develop new models. One powerful engine designed and produced by Allison was the J71, which is in the 10,000-lb class. This engine was used to power long-range missiles and a number of fighter aircraft for both the Air Force and the Navy.

Allison has also been very active in the production of turboprop engines. Outstanding models of these turboprops are the T38, T40, T54, and T56. The commercial model of the T56 is designated the Model 501-D13 and 501-D15. The T56 is used in several military aircraft, and the 501 powers the Lockheed Electra airliner.

Another leading manufacturer of aircraft engines in the jet-engine field is Pratt & Whitney Aircraft, a division of the United Aircraft Corporation. Pratt & Whitney started in jet-engine manufacture by producing the J42 and the J48 jet engines under a licensing agreement from Rolls-Royce, Ltd., of Derby, England. The Rolls-Royce engines of similar design are the NENE and the TAY, respectively.

Outstanding among the turbine engines designed and produced by Pratt & Whitney are the J57 (JT-3) and the J75 (JT-4) turbojet engines. Both of these engines are used in the Boeing 707 and 720 jet airliners and in the Douglas DC-8 airliner. Among other turbojet and turbofan engines produced by Pratt & Whitney are the JT8, JT10, JT11, and JT12, and their military models. The T34 turboprop engine is used to power the Douglas C-133 Cargomaster military transport airplane. For helicopters, Pratt & Whitney developed the JFTD12 turboshaft engine from the JT12 turbojet.

The Wright Aeronautical Division of the Curtiss-Wright Corporation entered the jet-engine field by producing the J65 engine, which is a modified version of the British Armstrong Siddeley Sapphire engine. This engine was used by the Navy in the Douglas A4D Skyhawk, the Grumman F11F Tiger, and the North American FJ-4 Fury. Another engine produced under license from a British firm was the J67, a split-compressor engine similar to the Bristol Olympus. This engine was rated in the 10,000-lb-thrust class and resembles the Pratt & Whitney J57.

One of the most important accomplishments in gas-turbine engine design in recent years is the development of the fan-bypass principle utilizing either a forward fan or an aft fan. Engines of this type produce a greater thrust per pound of fuel consumed than does a pure turbojet engine; likewise, the thrust/weight ratio is superior. The use of these engines has made the operation of commercial jet airliners more economical; hence, excellent service can be provided by the airlines at a reasonable cost. The Pratt & Whitney JT8D and JT3D and the General Electric CJ805-23 turbofan engines are currently used commercially.

Among gas-turbine engine manufacturers, which we have not yet specifically mentioned, are the AiResearch Manufacturing Division of the Garrett Corporation, the Boeing Company, the Continental Aviation and Engineering Company, the Lycoming Division of the Avco Corporation, and the Solar Aircraft Company. Some of the engines manufactured by these companies are used as primary power sources for aircraft, and others are used as auxiliary power units for starting engines and for other purposes.

Because of the great strides which have been made over the past few years in the development of gas-turbine engines, our military services are now equipped with Mach-2-plus aircraft such as the Convair F-106 Delta Dart, Lockheed F-104 Starfighter, North American A3J Vigilante, North American RB-70 Valkyrie, and Republic F-105 Thunderchief. Photographs of these aircraft are shown in Fig. 1 · 8.

Commercial aviation has reaped great benefits from the advancements in turbine engine design, and today the airways of the United States and many other countries are traversed continually by such airliners as the Boeing 707, 720, and 727; Convair 880 and 990; Douglas DC-8; Fairchild F-27; Lockheed Electra; and Vickers Viscount. As a result of the commercial use of gas-turbine-powered aircraft, there are more people traveling by this means today than by any other form of commercial transportation. Turbojet and turboprop airliners are shown in Fig. 1 · 9.

CHAPTER 2

Theory of Jet Propulsion

Before starting a study of the various types of jet and turbine devices, it is well to become familiar with the reasons why such devices operate to produce thrust. In general, any device which causes acceleration of a mass (weight), whether it be air, gas, or a solid, will produce thrust. It is our purpose in this chapter to explain the reasons why jet thrust occurs under the proper conditions.

Application of Newton's laws of motion

The theory of jet propulsion may be considered the reaction of a **mass** when its velocity is changed.

The term "mass" is similar to "weight," but it is actually a different quantity. "Mass" refers to the quantity of matter, while "weight" refers to the pull of gravity on that quantity of matter. At sea level, under standard conditions, 1 lb of mass will have a weight of 1 lb. In engineering, mass is often measured in **slugs.** One slug is equal to a mass of approximately 32.2 lb at sea level. As will be shown later, an understanding of the term "mass" will aid in the development of the jet-thrust formula.

Newton's laws of motion relating to jet thrust may be stated as follows:

Second law: **Force is proportional to the product of mass and acceleration.**

Third law: **For every action, there is an equal and opposite reaction.**

An analogy that may be applied to explain the above principles is the recoil of a pistol. Although it may seem that a pistol recoils because of the push of the expelled gases on the surrounding air, this is not true. A pistol recoils in accordance with Newton's laws. It is interesting to note that whether a pistol is fired through water, through ordinary atmosphere, or through a vacuum, the recoil is the same in all cases. This, of course, disputes the commonly accepted theory that expelled gases push upon the surrounding air. Still another fact that should be noted is that the thrust is equal against the bullet and the pistol. The bullet flies off at a high velocity because it has much less mass or inertia than the pistol. The effect of the recoil is also lessened because the pistol is normally held in the hand. Figure 2 · 1 shows how the recoil of a pistol may be compared to the operation of a jet engine.

Figure 2 · 2 will help the student to understand the jet thrust principle. If the cubical container on the left is completely enclosed and a pressure of 100 psi exists inside the container, then equal force is exerted against each side of the container. No thrust results, because forces in the opposite directions cancel one another. Now, if we make an opening with an area of 1 sq in. in one side of the container and feed sufficient air into the container to maintain 100 psi, there will be 100 lb more push against the solid sides than there will be against the side with the opening. Thus we have produced a thrust of 100 lb, and the container will tend to move in a direction opposite the side with the opening. In a like manner, the net thrust of a jet engine or rocket will

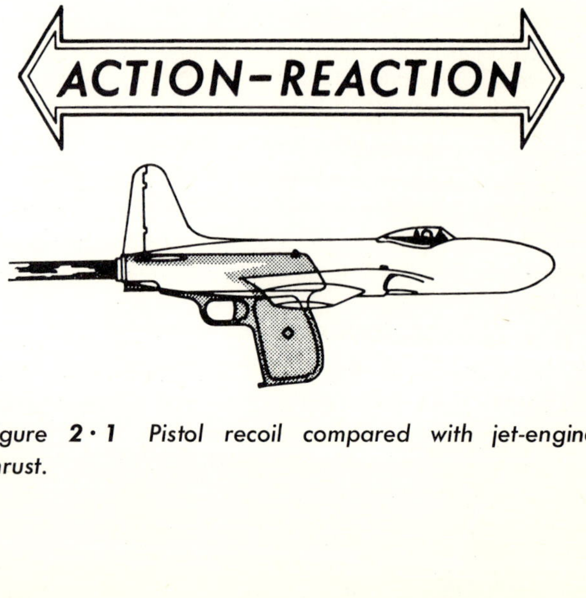

Figure 2 · 1 Pistol recoil compared with jet-engine thrust.

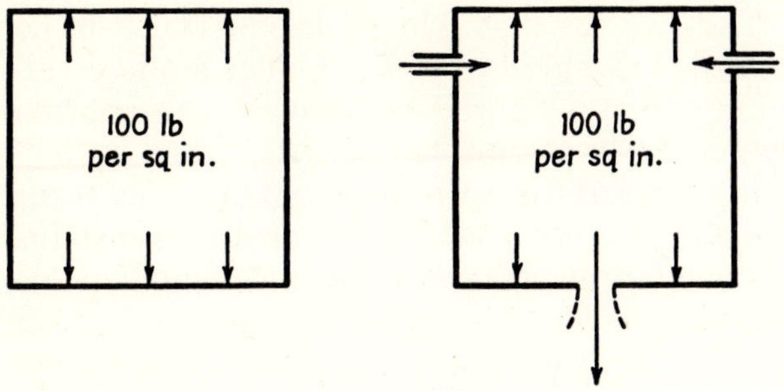

Figure 2 · 2 Principle of jet thrust.

be equal to the difference between the total thrusts acting in each direction within the unit.

Formula for jet thrust

The high-velocity jet from a jet engine may be considered as a continuous recoil imparting force against the airplane in which the engine is installed, thereby producing thrust. Refer again to Fig. 2 · 1. The formula for thrust may be derived from Newton's second law, which states that force is proportional to the product of mass and acceleration. This law is expressed in the formula

$$F = M \times A$$

where F = force, lb
M = mass, slugs
A = acceleration, ft per sec per sec

Since a mass of 1 slug has a weight of approximately 32.2 lb at sea level, we may state that mass (in pounds) is equal to M (in slugs) multiplied by 32.2 (the acceleration of gravity, usually expressed by the symbol g).

Since the acceleration of gravity is 32.2 ft per sec per sec, we arrive at the conclusion that a force of 1 lb will accelerate a mass of 1 lb 32.2 ft per sec per sec. (Feet per second per second is often expressed "ft per sec^2.") Also, a force of 10 lb will accelerate a mass of 10 lb 32.2 ft per sec per sec. In other words, if a force, equal in pounds to the mass of an object, is applied to the object, the object will be accelerated 32.2 ft per sec per sec. This is assuming that there is no friction or other resistance to overcome.

We can also reason that if an object is accelerated 322 ft per sec per sec, the force applied must be equal in pounds to ten times the mass of the object. This is because, in this case, the acceleration is ten times 32.2.

It is now apparent that the ratio of the force (in pounds) is to the mass (in pounds) as the acceleration in feet per second per second is to 32.2. Using m to represent the mass in pounds, the formula may be expressed thus:

$$\frac{F}{m} = \frac{A}{g}$$

Then $$F = \frac{mA}{g}$$

If we compute the force necessary to accelerate a mass of 50 lb 100 ft per sec per sec, we would require the same force to accelerate **50 lb per sec** 100 ft per sec. For example, by the foregoing formula,

$$F = \frac{50 \times 100}{32.2}$$

Then $$F = 155 \text{ lb.}$$

Thus, if the velocity of 50 lb of mass per second is increased by 100 ft per sec, the resulting thrust is 155 lb.

To compute the force necessary to accomplish a certain change per second in the velocity of a given mass, we arrange the formula as follows:

$$F = \frac{m_s(V_2 - V_1)}{g}$$

where F = force, lb
m_s = mass flow, lb per sec
$V_2 - V_1$ = change in velocity, i.e., difference between first velocity and second velocity
g = acceleration of gravity, or 32.2

For an example, if we wish to determine the force necessary to change the velocity of 100 lb of mass airflow per second from 600 ft per sec to 800 ft per sec, we apply the formula as follows:

$$F = \frac{100(800 - 600)}{32.2}$$

$$F = \frac{20{,}000}{32.2} = 621 \text{ lb}$$

Since the jet engine is a device which accelerates a mass of fuel and air, we can use the foregoing formula to determine jet thrust. If we know the mass airflow per second, and the difference in the velocity of the air from the intake of the engine to the exhaust nozzle, it is easy to compute the force necessary to produce the change in velocity. According to Newton's third law, for every action, there is an equal and opposite reaction. Therefore the jet thrust of the engine must be equal to the force required to accelerate the air mass through the engine. Then, using T as the symbol for thrust in pounds, our formula becomes

$$T = \frac{m_s(V_2 - V_1)}{g}$$

or $$T = \frac{m_s}{g}(V_2 - V_1)$$

It is interesting to note from the foregoing examples that the thrust of a gas-turbine engine can be increased either by increasing the mass flow of air through the engine or by increasing the jet velocity. Increases in mass flow may be obtained by utilizing water-injection systems and by the use of an afterburner. The afterburner also increases jet velocity. Jet velocity may also be increased by reducing the area of the jet nozzle, but this method is limited because reducing the nozzle area will result in higher temperatures and pressures within the engine. Methods for thrust augmentation are discussed in Chap. 12.

From the formula for jet thrust, it is apparent that, if the velocity of the jet remained constant with respect to the airplane, then as the speed of the airplane increased, the jet thrust would decrease. This is because V_1 would increase in value. This does not present a serious problem, however, because as airplane speed increases, more air enters the engine and jet velocity increases. The resultant net thrust is almost constant with increased airspeed.

Thrust compared with horsepower

Very often, direct comparisons between the reciprocating engine and the gas-turbine engine are attempted. Since the reciprocating engine is rated in brake horsepower (bhp) and the gas-turbine engine is rated in pounds of thrust (static, 59°F, 29.92 in. Hg), no direct comparison can be made. However, since the reciprocating engine-propeller combination receives its thrust from the propeller, a comparison can be made by converting the horsepower developed by the reciprocating engine to thrust.

If desired, the thrust of a gas-turbine engine may be converted into thrust horsepower (thp). It will be necessary, however, to take into consideration the speed of the airplane. This conversion may be accomplished by the use of the formula

$$\text{thp} = \frac{\text{thrust} \times \text{aircraft speed (mph)}}{375}$$

where 375 = mile-pounds per hour, derived from the basic horsepower formula:

$$1 \text{ hp} = 33{,}000 \text{ ft-lb per min}$$
$$33{,}000 \times 60 = 1{,}980{,}000 \text{ ft-lb per hr}$$
$$\frac{1{,}980{,}000}{5{,}280} = 375 \text{ mile-lb per hr}$$

Thus we see that 1 hp equals 33,000 ft-lb per min or 375 mile-lb per hr. Under static conditions, thrust is figured as equivalent to approximately 2.6 lb per hp.

If a gas turbine is producing 4,000 lb of thrust and the aircraft in which the engine is installed is traveling at 500 mph, the thrust horsepower will be

$$\frac{4{,}000 \times 500}{375} = 5{,}333.3 \text{ thp}$$

It can be seen that it is necessary to calculate the horsepower for each speed of an airplane, since the horsepower varies with speed. Therefore, it is not very practical to try to rate or compare the output of a jet engine on a horsepower basis.

If we plot the thrust curves from a reciprocating engine and a jet engine, as shown in Fig. 2·3, we shall note a number of interesting things about the relative performance of the two engines. For example, at flight speeds below approximately 375 mph, the conventional engine will outperform the jet engine, since the conventional engine produces higher thrust in this speed range. This means that the conventional engine will have better take-off and initial-climb characteristics. Continued development of jet engines with a higher static thrust is overcoming this disadvantage of the jet engine to a considerable extent. Turboprop engines have been developed which are as effective for takeoff as the reciprocating engines.

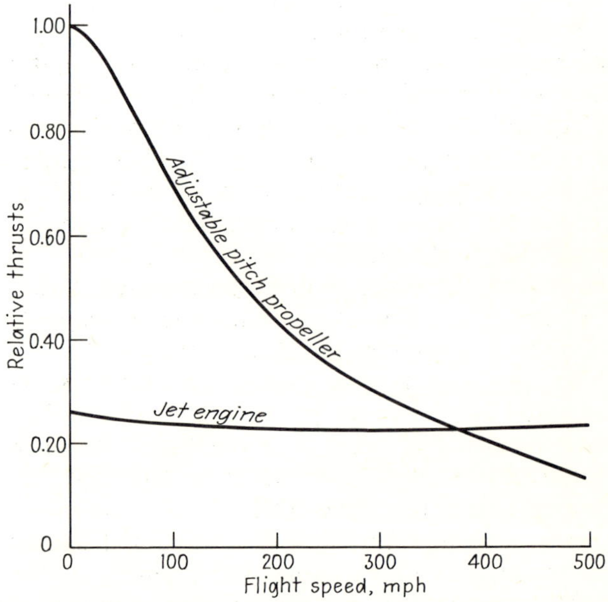

Figure 2·3 Thrust vs. speed for propeller-driven and jet-propelled aircraft.

Figure 2 • 4 is a plot of conventional-engine thrust, jet-engine thrust, and airplane drag which provides a good comparison between a conventional engine and a jet engine. To the left of the line *AA* on the chart, the propeller-driven plane will outperform the jet-propelled one. As was mentioned before, the propeller develops more thrust at the lower speeds. Between the speeds represented by lines *AA* and *BB,* the jet engine is more effective than the conventional engine because propeller thrust continues to drop as airspeed increases. The jet-engine thrust, during the increase in speed, remains almost constant. The line *BB* represents the maximum speed attainable by the conventional engine with a propeller, and the line *CC* is the maximum airspeed attainable by the jet engine. At this time, drag becomes equal to thrust.

The Brayton cycle

The Brayton cycle may be used to show the pressure and volume relationships affecting the air which is utilized in a jet engine. Figure 2 • 5 shows the curves for the Brayton cycle on a pressure-volume chart. Point 0 is the condition of the ambient air before it enters the compressor, and point 1 indicates the condition of this air at the inlet to the impeller after it has entered the ram inlet and has been diffused to reduce its velocity. The velocity at this point is almost zero with respect to the airplane. Point 2 represents the condition at the inlet to the combustion chamber. The curve from 2 to 3 shows what occurs in the combustion chamber of the engine. In this area the fuel is burned, thus adding heat to the air. Volume and velocity are increased, and pressure drops slightly.

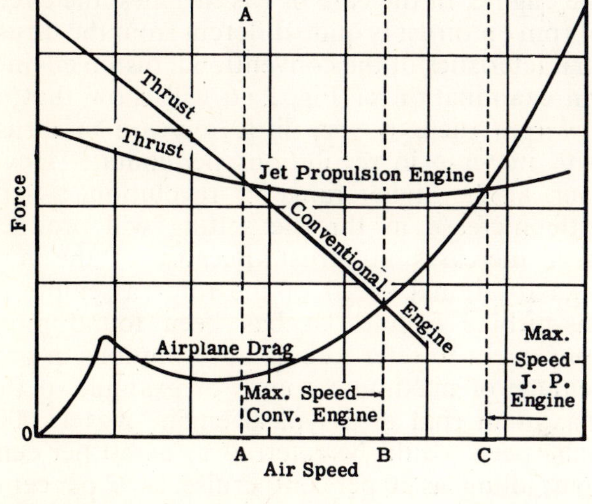

Figure 2 • 4 Comparison of conventional-engine performance with jet-engine performance.

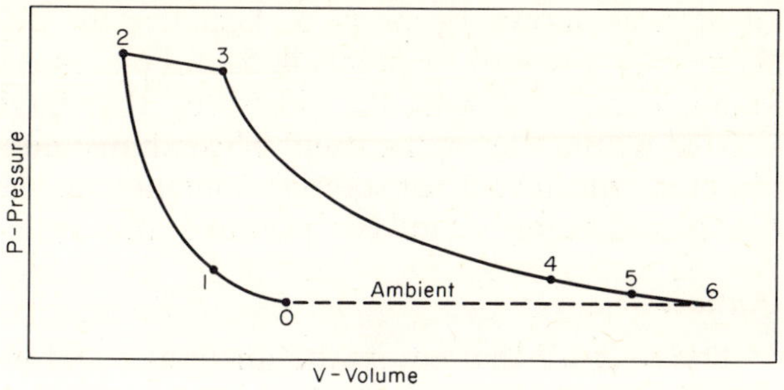

Figure 2 • 5 Brayton cycle, P-V diagram.

At point 3 a substantial pressure drop begins as the air expands through the turbine. Point 4 is the condition after the air has passed through the turbine. The curve from 4 to 5 indicates the continued expansion as the air flows to the nozzle outlet. Point 5 shows the pressure and volume relationships at the nozzle where the velocity is still supersonic. At point 6 the air has left the engine and is again at ambient pressure. The distance from point 0 to point 6 represents the difference in the volume of the air as the result of the heat which has been added because of the burning of fuel in the engine.

Application of the Brayton cycle

If we examine the curves of Fig. 2 • 5 and compare the indications to actual conditions in a specific engine, we shall understand how the cycle operates under these circumstances.

Air enters the intake of the engine at 350 mph (ram-air velocity) at a flow of 85 lb per sec. This is comparable to point 1 in the *P-V* diagram of Fig. 2 • 5. The air is compressed by the centrifugal impeller to a pressure of 48 psi. This is represented by point 2 on the *P-V* curves.

With the high pressure-ratio engines, point 2 on the chart might indicate pressures as high as 170 psi. The temperature of the compressed air is elevated to more than 650°F.

The curve from point 2 to point 3 in the diagram of Fig. 2 • 5 represents conditions through the combustion chamber. The volume of the gases increases, while the pressure drops from 48 psi to 44 psi. This is to be expected, since the combustion area is not an enclosed chamber such as the cylinder on the reciprocating engine.

Directly to the rear of the turbine wheel, there is a substantial pressure drop after expansion through the turbine. This is represented by point 4 on the diagram. At the nozzle throat, the velocity is increased and the pressure drops further, as shown by point 5. Usually, the differences in pressure at points 4, 5, and 6 are very small, compared with the pressures that have existed within the engine itself. The exhaust nozzle area will have a substantial influence on the pressure difference between points 5 and 6.

Airflow

The only air that enters the air intake and induction system of the reciprocating engine is that necessary for combustion in the cylinders. The turbojet, however, swallows all the air the reciprocating engine uses for combustion and cooling, plus the air handled by the propeller to create thrust. The quantity of this air assumes huge proportions. To appreciate the airflow of a turbojet engine, the following comparison is cited. The Pratt & Whitney Wasp Major engine (3,500 hp) consumes approximately 23,000 lb of air per hour, while the Pratt & Whitney J57 engine consumes as much as 600,000 lb of air per hour. For purposes of visualization, this is approximately the volume of two standard railway tank cars of air per second.

Considerable power is required to drive a compressor at its rated speed so that it will deliver the amount of air needed by the turbojet engine. It may require considerably more than 100 hp to deliver each pound of air per second. This depends upon the design of the compressor and inlet. Air weighs 0.076 lb per cu ft; hence it takes more than 13 cu ft of air to make 1 lb at standard sea-level pressure.

One pound of air per second through a turbojet engine may develop into 50 lb of thrust or more. In a turbojet engine rated at 4,000 lb of thrust, the turbine wheel must extract over 8,000 hp from the energy stream flowing through it and deliver this torque power to the rotor shaft of the compressor. For example, suppose the engine rating is 5,000 lb of thrust, and each 50 lb of thrust requires 1 lb of airflow per second. Therefore, the required airflow is

$$5{,}000 \div 50 = 100 \text{ lb per sec}$$

If delivering 1 lb of air per second requires 100 hp, then the required horsepower is

$$100 \times 100 = 10{,}000 \text{ hp}$$

The engine used in the above example is a low-pressure type and is comparatively low in power. Some of the modern engines require as much as 30,000 to 40,000 hp to drive their compressors.

It should be realized that, in order to test the compressors when out of the engines, the same amount of power is required as that developed by the turbine wheel. This means that the engine manufacturers, as far as compressors are concerned, must have testing facilities capable of delivering the required horsepower. This is a considerable undertaking, which entails the acquisition of huge and costly equipment.

On some airplanes, the air intakes, under static conditions on the ground and without the benefit of ram air, cannot supply a sufficient quantity of air to the compressors because of their size. Thus, when the compressor operates at high speeds, it tends to evacuate the air in the engine compartment. When this condition arose initially, it was noticed that the aircraft developed **duct rumble.** To remedy this condition, it was found necessary to add a spring-loaded door or opening in the engine compartment to admit an additional quantity of air. This, in effect, increased the area of air intake when needed on the ground. The door opens when the higher atmospheric pressure outside the engine compartment pushes the door inward toward the lower pressure in the engine compartment. When the airplane attains forward speed in its take-off run, ram air pressurizes the engine compartment and the "derumble" door is closed against a stop. The door opens only when engine compartment air pressure is lower than ambient air pressure. This, of course, occurs only when the compressor requires more air than the intakes can furnish.

Factors affecting thrust

Engine Revolutions per Minute. One of the most obvious factors that affects the thrust output is the speed of the rotor assembly, generally referred to as revolutions per minute (rpm) of the engine. In the case of gas turbines, the effect of rpm on thrust is quite different from the thrust characteristics of the conventional piston engine. An examination of Fig. 2 · 6 will show that at low rpm there is very little increase in thrust with increase in revolutions per minute. However, at the higher range of revolution, a very little increase in throttle setting will produce large increases in thrust. Because of the very high rotational speed of the rotor assembly of gas-turbine engines, it has been found more convenient to refer to engine revolutions in percentage of maximum rpm. For example, if the maximum rpm of a typical engine was 13,000, half speed would be referred to as 50 per cent rpm; idling, as 30 per cent; cruise, as 92 per cent; take-off, as 100 per cent; etc.

Referring to Fig. 2 · 6 again, it will be noticed

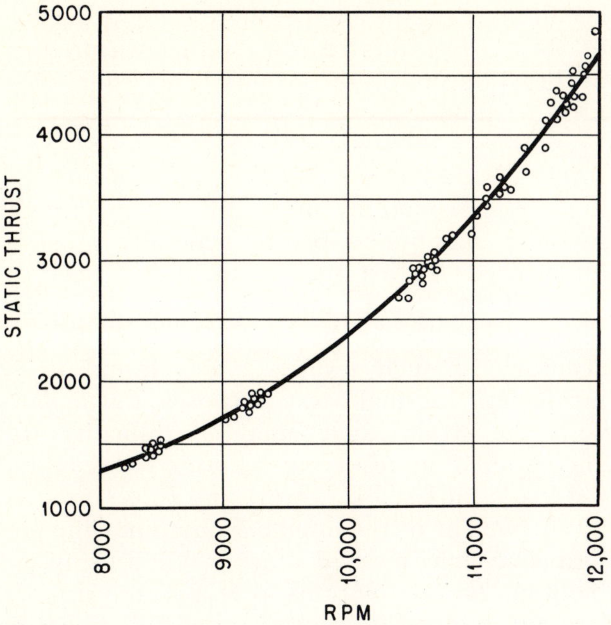

Figure 2·6 Increase of thrust with rpm.

that with an increase in revolutions from 8,000 to 9,000 per min, an increase of approximately 350 lb of thrust is realized. However, when the revolutions are increased from 11,000 to 12,000 per min, the thrust increase is approximately 1,500 lb. At the lower settings, fuel consumption is high for the amount of thrust produced. For this reason, gas turbines are operated near their maximum rpm. Cruise is generally 85 to 100 per cent of maximum, as compared to the 40 to 70 per cent generally associated with conventional piston-propeller engines.

Airspeed. With respect to airspeed, turbojet engines are referred to as **constant-thrust machines.** As was explained in the beginning of this chapter, a study of the thrust formula shows that if the velocity of the jet remains constant for any airplane speed, then, as the airplane's speed increases, the thrust would decrease in direct proportion. Because of the effect of ram-air intake, however, the mass flow and the velocity of the jet also increase with airplane speed, so that the resultant net thrust is practically constant with increased airspeed. Figure 2·7 illustrates this point by showing how the thrust curve dips slightly and then climbs again with an increase in airspeed, resulting in almost constant thrust for all practical purposes. It will also be noted that this holds true for different altitudes.

Altitude. Figure 2·7 shows that there is an appreciable thrust loss with increases in altitude. Any change in barometric pressure will, therefore, affect the performance of the engine thrustwise. In tests conducted on the Avro C102 Jetliner, it was found that the thrust loss at 30,000 ft was approximately 40 per cent. It must be realized, however, that the drag at this altitude is reduced considerably, being less than 25 per cent. With a thrust variation of only 5 per cent from 200 to 500 mph, it becomes apparent that the turbojet-powered aircraft must be flown at high altitude and relatively high speed to attain its highest degree of efficiency.

Temperature. Atmospheric temperature is another factor that affects the thrust of a gas-turbine engine. With increase in air temperature, less thrust is developed by the engine. This means that in tropical operation longer runways are necessary for take-off purposes. Fortunately, systems of thrust augmentation (described in Chap. 12) have been developed to overcome this deficiency. At a temperature of 110°F, a thrust loss of 16 per cent was experienced in the flight tests of the Avro C102 Jetliner.

On large airplanes that require considerable heating and ventilating airflow, air bled off the compressor for these purposes will result in a thrust loss. Additional air is also used internally in the engine for bearing and turbine-wheel cooling. However, this air is returned to the main engine air stream at the turbine exhaust.

Performance losses at full engine rpm are incurred with fixed-area exhaust nozzles because tailpipe temperature varies with speed and altitude. With the exhaust nozzle limiting the tailpipe temperature on the ground under static con-

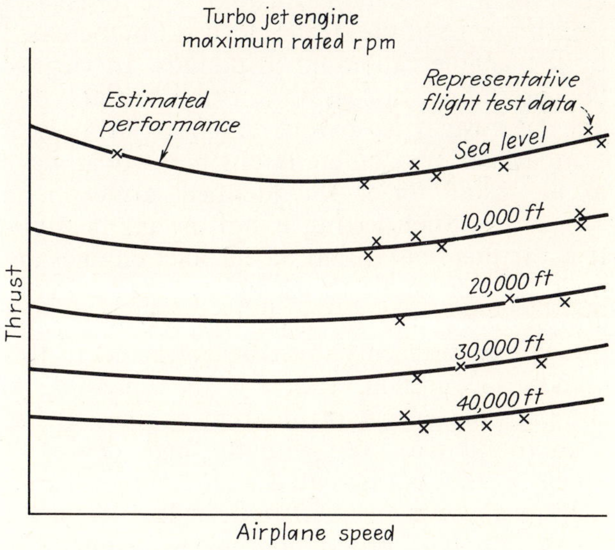

Figure 2·7 Effect of airspeed and altitude on thrust.

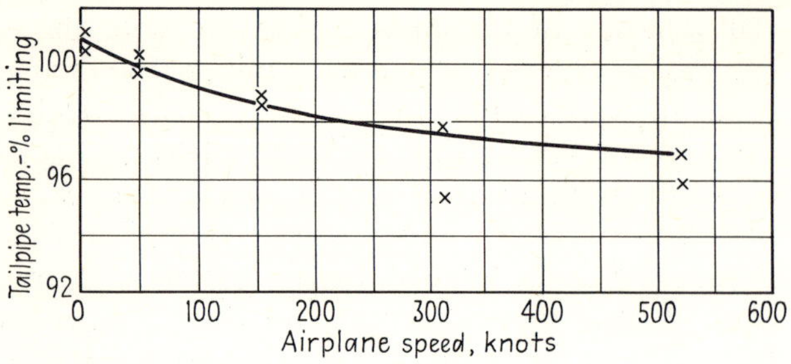

Figure 2·8 Variation of tailpipe temperature with airspeed.

ditions, a drop of 50 to 60°F will accompany a speed increase to climbing speed or high-speed level flight. This temperature drop is illustrated in Fig. 2·8. It results, of course, in a loss of thrust.

Another cause of performance loss is the sharp increase in tailpipe temperature at high altitude, usually above 30,000 ft. To prevent the temperature from exceeding the limiting value, the fuel flow must be reduced. The fuel-control units included with most turbojet fuel systems automatically control fuel flow to prevent excessive temperatures. This reduction in fuel flow will result in a loss of thrust.

Thrust measurement

Static Thrust Measurement. Under static conditions on the ground, the thrust output of a turbojet engine (either mounted on a test stand or installed in the airplane) can be determined readily by direct mechanical means. This measurement may be accomplished through strain gauges, thrust-balancing pistons, dynamometers, or spring scales, all with an accuracy suitable for powerplant evaluation. All that is required is a measurement of the straightforward pull created by the engine calibrated in pounds. In view of the preceding discussion on the factors that affect thrust, it is obvious that for purposes of comparison two different engines would have to be tested under the identical atmospheric conditions. Such testing is not always possible. It is for this reason that all turbojet engines are rated at their static thrust under standard conditions (59°F and 29.92 in. Hg). Therefore, all thrust measurements must be corrected to this standard condition, and it is then possible to compare the output of different engines. It has been found that even with the high degree of precision that is associated with turbojet manufacture, there is a considerable difference in the thrust output of production engines, sometimes as much as 13 per cent. Also, as operating hours accumulate on the engine, there is a gradual loss of thrust. For this reason, static measurement of thrust output on the ground is necessary from time to time.

Flight Measurement. Measurement of thrust in flight is considerably more difficult in that the methods mentioned before generally must be abandoned. Thrust measurement has been attempted by strain gauges or thrust-balancing pistons installed at the engine mounts, but the accuracy of such test data was found to be insufficient for flight-test purposes. Particularly with the strain-gauge method, difficulty was encountered in resolving measured forces into the loads imposed by thrust and those imposed by the weight of the engine. Sometimes the **performance-chart method** is used. When using this method, the revolutions per minute, altitude, free-air temperature, and airspeed must be known. By consulting a chart prepared by the engine manufacturer, it is possible to ascertain the approximate thrust in flight. Because of the differences of thrust on production engines, however, the standard chart is not very useful. The system is based on an average engine, and it is not accurate enough for test-flight purposes.

A satisfactory in-flight technique is based on the measurement of the flow conditions in the exhaust stream. To accomplish this, gas pressures and temperatures are sensed by an exhaust-gas survey rake placed across the tailpipe nozzle as shown in Fig. 2·9 and the thrust output of the engine is calculated from the data thus obtained. Errors are calibrated out by applying suitable coefficients (the ratio of calculated thrust to actual thrust) that are determined during static ground tests. The instrumentation is neither extensive nor unreasonably elaborate.

In order to calculate thrust by the method just described, two components of thrust must be evaluated: (1) the gross thrust output, and (2) the ram drag. The difference in these two compo-

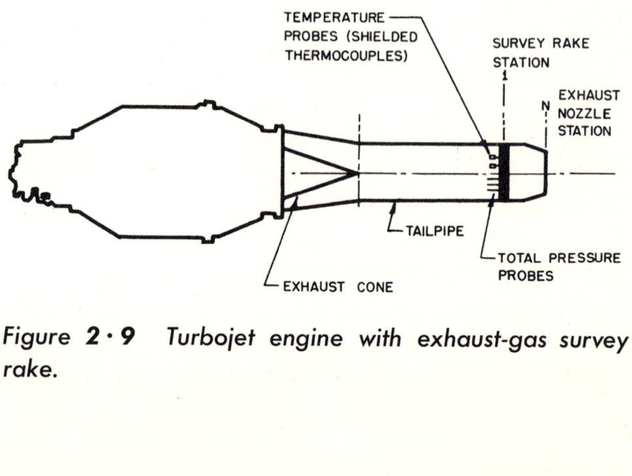

Figure 2·9 Turbojet engine with exhaust-gas survey rake.

nents is equal in magnitude to the net change in momentum of the engine induction air as it passes through the airplane.

Algebraically, the net thrust may be expressed

$$F_N = F_G - F_R$$

where F_N = net thrust of engine
F_G = gross thrust of engine
F_R = ram drag (component of gross thrust required to accelerate inlet air to airplane velocity)

The quantities that must be known in order to evaluate the gross thrust are:

1. Stagnation or total pressure of the gas
2. Stagnation temperature of the gas
3. Static pressure of the gas
4. Area of the exhaust nozzle

Items 1 and 2 are obtained by sampling the exhaust gases with a survey rake placed in the tailpipe as shown in Fig. 2 · 9. The static pressure at the tailpipe exit is set equal to the ambient air pressure, and the nozzle area is measured directly.

The gross thrust at the nozzle can be calculated, since all four variables are known.

The ram drag of the engine induction air is equal to the gross thrust times the ratio of airplane velocity to exhaust-gas velocity.

$$F_R = F_G\left(\frac{v_{\text{airplane}}}{v_{\text{exhaust gas}}}\right)$$

The assumption is made here that the mass of the induction air is equal to the mass of the exhaust gases. Actually, fuel is added to the flow, approximately one part in sixty, but this generally balances the air bled from the compressor for cabin ventilation, cooling, and compressor leakage.

Satisfactory calibrations may be readily obtained from static thrust measurements on the ground. An over-all probable error in net thrust determination of 1½ per cent over the normal operating range of the airplane is believed to be reasonable for this method. Of course, an individual calibration factor for each test powerplant installation must be established, since engines of the same series give varying thrust performance under identical operating conditions.

Compressibility phenomena

The subject of compressibility phenomena is of considerable interest to persons who work with jet engines and high-speed aircraft. The term "compressibility" refers to compression waves generated in air at sonic or supersonic speeds. Near-sonic and sonic airflow velocities through the engine itself make the understanding of compressibility and sonic speeds a virtual "must." It should be emphasized, however, that our knowledge of the subject is still somewhat incomplete.

Definition. First of all, we must determine what is meant by the term **compressibility phenomena.** A great many effects, seemingly unrelated, are included in this term. At relatively low airspeeds, the flow of gases (such as air) is quite similar to the flow of an incompressible liquid, such as water. Thus, pressure and velocity relationships for the flow of air around an airfoil can be calculated with satisfactory accuracy using Bernoulli's equation and other relationships as developed for the flow of an incompressible fluid through a pipe. However, when air at the surface of an airfoil flows at a speed that is close to the speed of sound (761 mph at sea level, with standard conditions), its behavior changes in that its flow characteristics cannot be predicted by the use of such simple relationships. When these flow changes occur, their effect on the airplane may be disturbance of its trim, a sudden change in control forces, and perhaps buffeting of the tail surfaces.

Effect of Speed and Sound. Since these changes in flow occur when the local airspeed around the wing or other parts of the airplane is approximately equal to the speed of sound, it is important to know how the speed of sound changes with altitude and air temperature and what its relationship is to the pilot's indicated airspeed. Table 2 · 1 shows these relationships for several altitudes.

At sea level, or any other altitude, the speed of sound varies with temperature, becoming

Table 2 · 1 Speed of sound in relation to altitude and temperature

Altitude, ft	Standard temperature, °F	Speed of sound, mph true airspeed	Speed of sound, mph calibrated airspeed
0	59.0	761	761
5,000	41.2	748	703
10,000	23.3	735	649
20,000	−12.4	707	544
30,000	−47.8	678	447

greater as the temperature rises. The actual temperature at a given altitude will change from day to day, and the true speed of sound will change correspondingly. However, since the temperature enters into the correction for both the pressure altitude to true altitude and the speed of sound, it is fortunately true that the indicated speed of sound is constant for a given pressure altitude. This means that if the altimeter is set at the standard pressure when the airplane is on the ground, the correct calibrated airspeed of the speed of sound is given in the fourth column of the table for the instrument reading of altitude.

Significance of Mach Number. In a discussion of compressibility phenomena, it is necessary to define another term, the **Mach number,** which was named for Mach, the Austrian physicist. This number is equal to the velocity of the air relative to the airplane, divided by the speed of sound. It may also be defined as the ratio of airspeed to the speed of sound. For example, if an airplane is flying with an airspeed of 600 mph and the speed of sound under the ambient conditions is 750 mph, then the Mach number is equal to 600 divided by 750. The result is a Mach number of 0.8.

At sea level, a speed of 381 mph would be a Mach number of 0.5, and 761 mph would be a Mach number of 1.0.

Influence on Structural Design. Compressibility effects may cause changes in the distribution of air loads on doors, cowling, and enclosures that may result in failures of these parts during high-speed dives, or supersonic flight in any attitude. Engineering departments attempt to predict these changes in distribution, and they have made substantial progress since supersonic flight has become commonplace.

All the mysteries of the "sonic barrier" have not yet been solved; however, it is routine with military aircraft to exceed the speed of sound. Although the exact figures are classified for military security, it has been reported that fighter aircraft have exceeded Mach 2.0 by a substantial amount and that rocket vehicles have traveled several times as fast.

Having conquered the "sonic barrier," the problem today is the "thermal barrier." The heat generated by the friction of air on the surface of an airplane can become so great that it will weaken or destroy the material of which the aircraft is made. Refrigeration systems have been developed to keep the interior of the cockpit cool, but these systems are necessarily limited in their capacity.

Compressor stall

During the past several years, compressor ratios for gas-turbine engines have been increased from about 4.5:1 to more than 12.5:1. These high pressure ratios have resulted in engines with excellent specific-weight and specific-fuel-consumption characteristics. Today, gas-turbine engines are capable of producing more than twice the power of early engines with the same consumption of fuel.

The improvements in engine performance have been accompanied by problems, one of the principal ones being **compressor stall.** Compressor stall may be encountered in any gas-turbine engine when certain conditions are present; however, the high-compression engines must operate in an area much nearer to stall than the low-pressure engines. For this reason, stall is encountered more frequently during the operation of the high-pressure engines.

To anyone reasonably familiar with the theory of aircraft flight, stall is well understood. It occurs when the smooth airflow over an airfoil (wing or control surface) breaks away as the result of an excessive angle of attack. The **angle of attack** is the angle formed by the direction of airflow relative to the chord line of the airfoil, as shown in Fig. 2 · 10. When the angle of attack becomes too great, the air will break away from the surface of the airfoil, and the airfoil will lose a substantial portion of its lift. This effect is shown in Fig. 2 · 11.

The blades of the compressor in an axial-flow turbine engine are actually small airfoils, and they have a shape in cross section similar to that of an airplane wing. When the angle of attack becomes too great on these blades, a stall results.

It may appear that the angle of attack for a compressor blade could not change, because the blades are installed with a fixed angle in the compressor disk. This would be true if the compressor did not rotate; however, the transverse

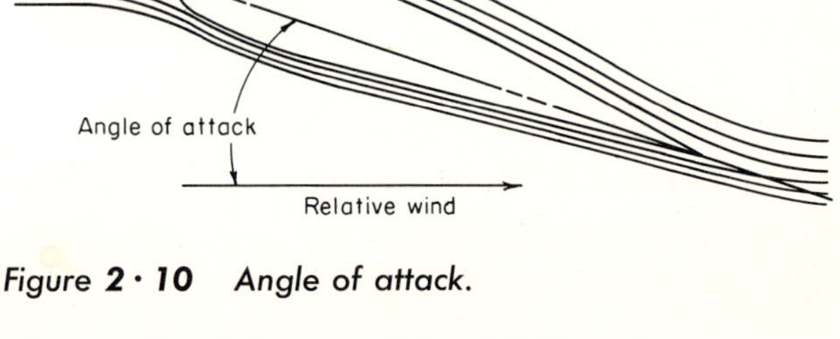

Figure 2 · 10 Angle of attack.

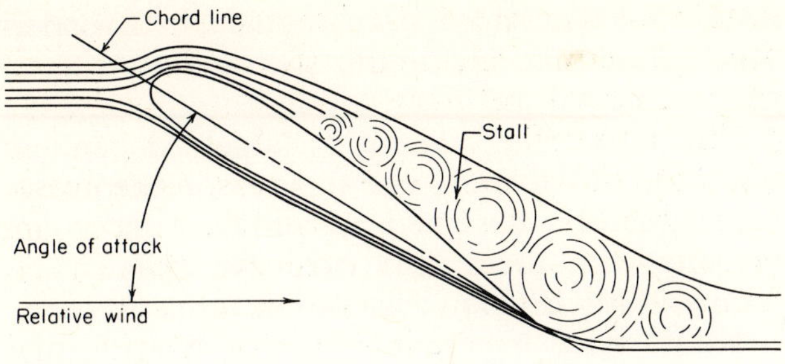

Figure 2 · 11 Schematic diagram illustrating stall.

velocity of the blades due to rotation has a substantial effect on the angle of attack. The angle of attack varies according to the inlet airspeed and the speed of rotation, as shown in Figs. 2 · 12 and 2 · 13.

In the diagram, *AC* represents the linear velocity of the compressor blade due to rotation. The line *BC* represents the velocity and direction of the air entering the compressor. The resultant is *AB*, which is the direction of airflow relative to the chord of the blade. Thus the angle of attack is established. If the inlet airspeed decreases, the line *BC* will become shorter and the angle *BAC* will be smaller. This will result in a larger angle of attack, as shown in Fig. 2 · 13. Thus we can see that a high compressor rpm with a low inlet airspeed will cause a high angle of attack and a possible stall. For any particular engine, the combinations of compressor rpm and inlet airspeed which will cause stall are known.

The degrees of compressor stall vary from a low engine rumble, resulting from partial stall,

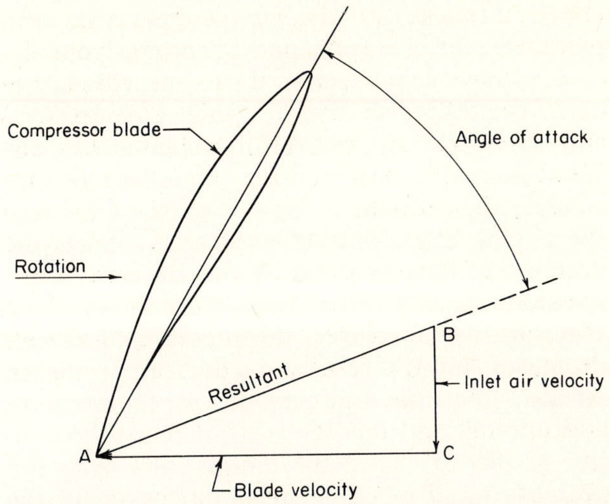

Figure 2 · 13 Angle of attack on compressor blade—low inlet-air velocity.

to a violent explosion, resulting from a complete stall. Explosive compressor stalls are frightening in their intensity, especially to a pilot who is sitting above the inlet air duct of the engine. These stalls usually occur without damage to the engine, but they can and do cause damage under some circumstances.

Partial stalls will often cause **surge.** Since a stall brings a reduction in the airflow through the engine, the fuel control will reduce fuel flow. This decreases the back pressure on the compressor and permits an increase in airflow which eliminates the stall. The fuel control will restore the fuel flow, and back pressure will increase. If the back pressure becomes too great, the stall will occur again, and the cycle may repeat until the pilot increases airspeed by "nosing down" his airplane or reduces fuel flow with the throttle.

Compressor stalls are most likely to occur during acceleration and at very high altitudes. Fuel-control units are designed to schedule fuel flow to avoid stalls, but under certain critical conditions, the stall may occur.

The design features of modern jet engines have virtually eliminated compressor stall through the use of variable inlet guide vanes, variable stator blades, or twin-spool compressors. The danger of compressor stall is also greatly reduced by the action of the automatic fuel control units which regulate fuel flow to compensate for changing conditions of operation.

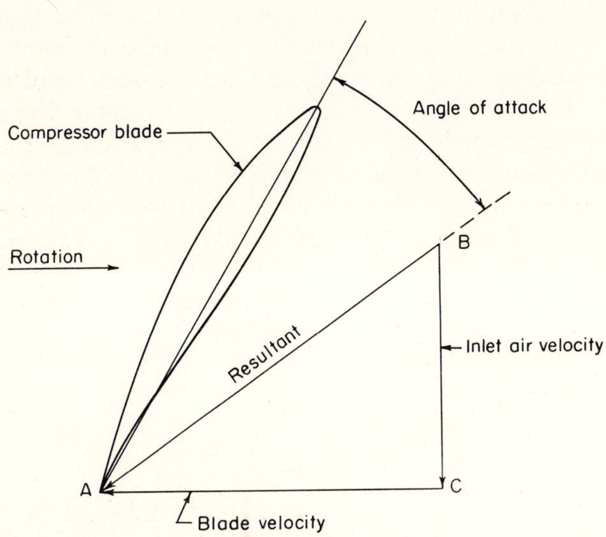

Figure 2 · 12 Angle of attack on compressor blade—high inlet-air velocity.

Summary of factors affecting thrust

The most significant factors affecting the thrust of a turbojet engine are airspeed, ram effect, altitude, air density (temperature and pressure), turbine rpm, and jet-nozzle velocity.

As airspeed increases, thrust decreases because the difference between inlet airspeed (V_a) and jet-nozzle airspeed (V_j) decreases. As airspeed increases, however, the ram effect also increases, thus forcing a greater mass of air into the engine inlet. This effect largely offsets the decrease of thrust caused by the increase in airspeed.

As altitude increases, the pressure of the air decreases, and this results in a decrease in thrust. At the same time, however, the air temperature is dropping, and this tends to increase the density of the air. Since the temperature does not decrease rapidly enough to compensate for the decrease in pressure, a decrease in thrust takes place. This decrease in thrust does not present a serious problem because the drag on the airplane decreases with altitude and it takes less thrust to maintain a given airspeed.

Since air density determines the mass of air that is used by a jet engine, the effects of temperature and pressure must be taken into consideration in any computation of engine thrust. This is particularly important during the takeoff of a jet airplane because it will make a substantial difference in the takeoff distance. It must be remembered that **with a given throttle setting engine thrust will decrease as air temperature increases and increase as air pressure increases.** This principle may also be expressed as follows: At a given throttle setting, engine thrust will increase as air density increases.

Turbine rpm controls the amount of air being used by the jet engine to a large extent, with the greatest effect on thrust occurring near maximum rpm. Therefore, if an engine will not operate at its design maximum speed, the thrust of the engine will be greatly reduced.

The jet-nozzle velocity is largely dependent upon the nozzle area. For this reason the nozzle area must be designed to provide maximum velocity without causing excessive exhaust gas temperature (EGT). The nozzle area is sometimes adjusted by the insertion of small fittings called "mice." When an engine is equipped with an afterburner, the jet nozzle is provided with "eyelids" or an iris construction to permit a variation in nozzle area.

STUDY QUESTIONS

1. What laws explain the theory of jet propulsion?
2. What is the basic formula for thrust?
3. What is the thrust horsepower of a gas-turbine engine which produces 4,000 lb of thrust and is moving at a velocity of 375 mph? What would the thrust horsepower be at 750 mph?
4. If a gas-turbine engine develops approximately 10,000 lb of static thrust, what is the approximate horsepower required to drive the compressor?
5. What are the principal factors affecting thrust in a turbojet engine?
6. During the static test of a turbojet engine, how may thrust be measured?
7. What is meant by "Mach number"?
8. Explain "compressor stall."
9. What conditions are likely to bring about compressor stall?
10. What engine-design features have been developed to combat stall?

CHAPTER 3

The Aviation Gas Turbine

Since the advent of the gas-turbine engine as an aircraft power plant, literally hundreds of different models and types of aircraft gas-turbine engines have been designed and built. Many of these have been used successfully in operational aircraft; however, the number accepted for service is considerably less than the number rejected. This ratio is due to a tendency of the engine-development program to exceed the time period for effective use, and it does not mean that the rejected engines were poorly designed.

From the many designs of turbine engines which have been built, four types have evolved as being the most practical: the centrifugal type; the axial type with single rotor; the axial type with dual rotor or split compressor; and the axial type with a bypass or a fan.

Comparison of types of turbojets

Figure 3 · 1 is a cutaway view of the Allison J33-A-35 engine. The construction of this engine is typical of **centrifugal-flow** turbojets. This particular engine has a double-sided compressor rotor, which means that air is admitted to both sides of the compressor.

Another centrifugal engine, similar in configuration to that shown in Fig. 3 · 1, is illustrated in Fig. 3 · 2. This is a drawing showing the airflow through a Pratt & Whitney JT6B engine.

Figure 3 · 3 illustrates the construction of the Pratt & Whitney J57 **twin-spool, axial-flow** turbojet engine. This type of engine has been used extensively in military fighter aircraft.

The configuration of an axial, **split-compressor** engine is shown in Fig. 3 · 4. Note that the two sections of the compressor are independent of each other. The forward, low-pressure section is driven by the two rear turbines, which are joined to the same shaft. The rear or high-pressure compressor rotor is driven by the front turbine.

The split-compressor design illustrated in Fig. 3 · 4 is used in the Bristol Olympus, a British engine, and the Pratt & Whitney J57, or JT3, engine. This design has proved very successful and is being used in many modern aircraft.

Typical of **bypass** engines is the Conway bypass engine manufactured by Rolls-Royce. A schematic diagram of the arrangement of this engine is shown in Fig. 3 · 5. Test flights with this engine indicate a remarkable specific fuel consumption and an excellent power/weight ratio.

It will be noted from a study of the foregoing illustrations that a basic difference between the centrifugal-type engine and the axial-type engine is in the airflow through the engine. The airflow through the centrifugal engine follows a rather tortuous route as compared with the airflow through the axial-flow engine. Violent changes in direction of airflow remove some of the energy from the air, and this energy is lost.

Major components of turbine engines

Because of the many types of turbine engines, it is not possible to list all the major components and have the list apply to all engines. There are several components common to most turbine engines, however, and a knowledge of these will be helpful in developing a further understanding of aviation gas-turbine engines.

The operation of any gas-turbine engine requires that provision be made for three principal functions: (1) the compression of air, (2) the expansion of the air by burning fuel, and (3) the extraction of power from the jet stream of the engine for driving the compressor and accessories. Thus, we may say that a gas-turbine engine comprises three main sections: the **compressor section,** the **combustion section,** and the **turbine section.**

In addition to these three main sections, there are also components which serve to provide transition from one main functional section to another. The following is a list of all the major components, arranged as they would appear from the front of the turbine to the rear:

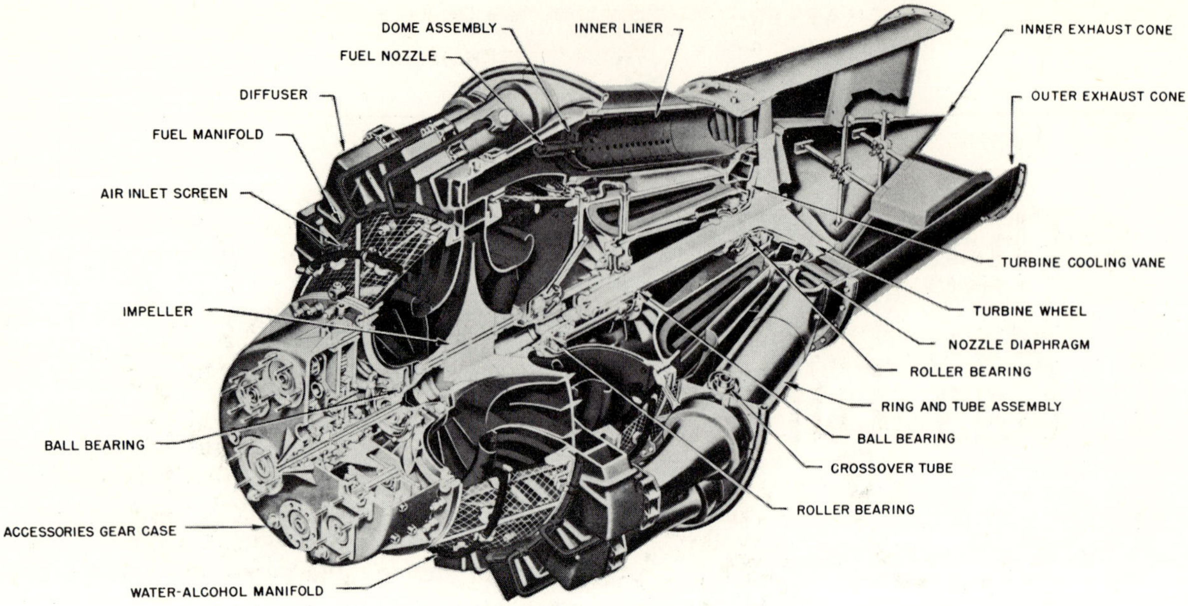

Figure 3 · 1 Centrifugal-flow turbojet.

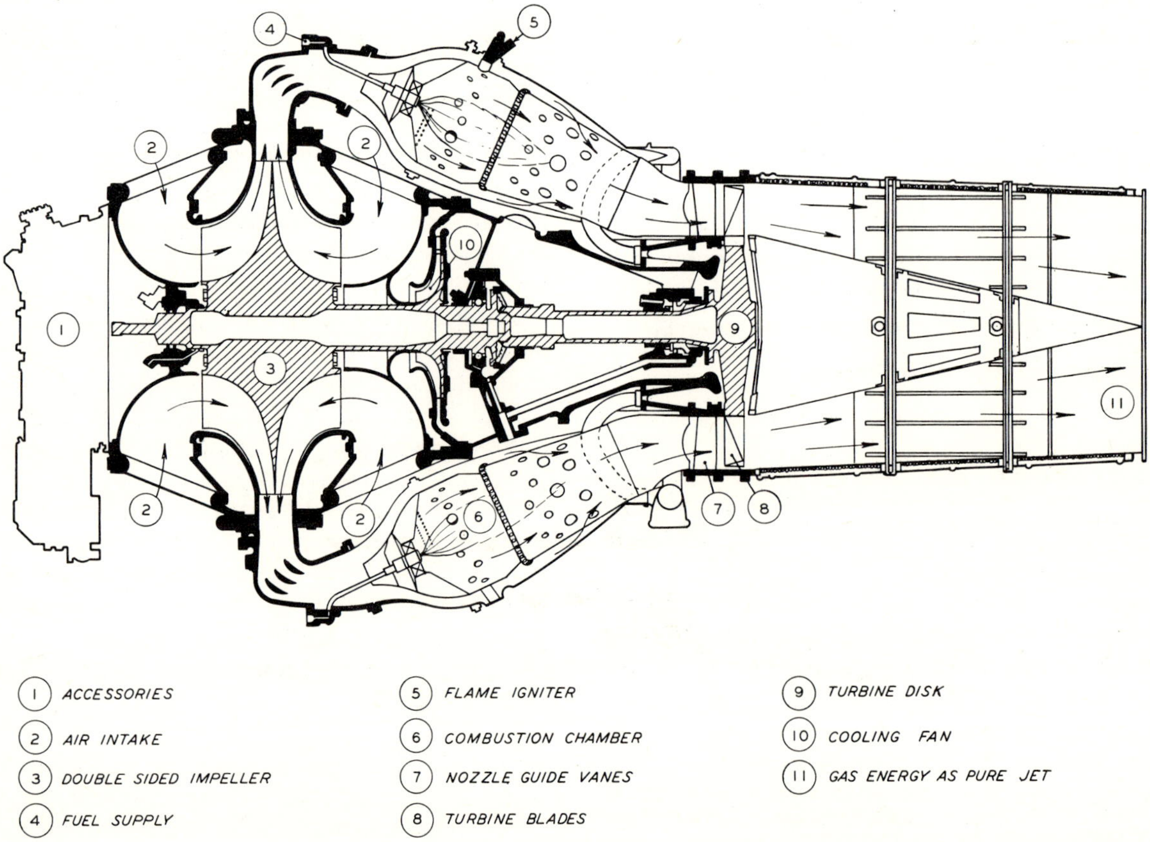

Figure 3 · 2 Airflow through the Pratt & Whitney JT6B engine.

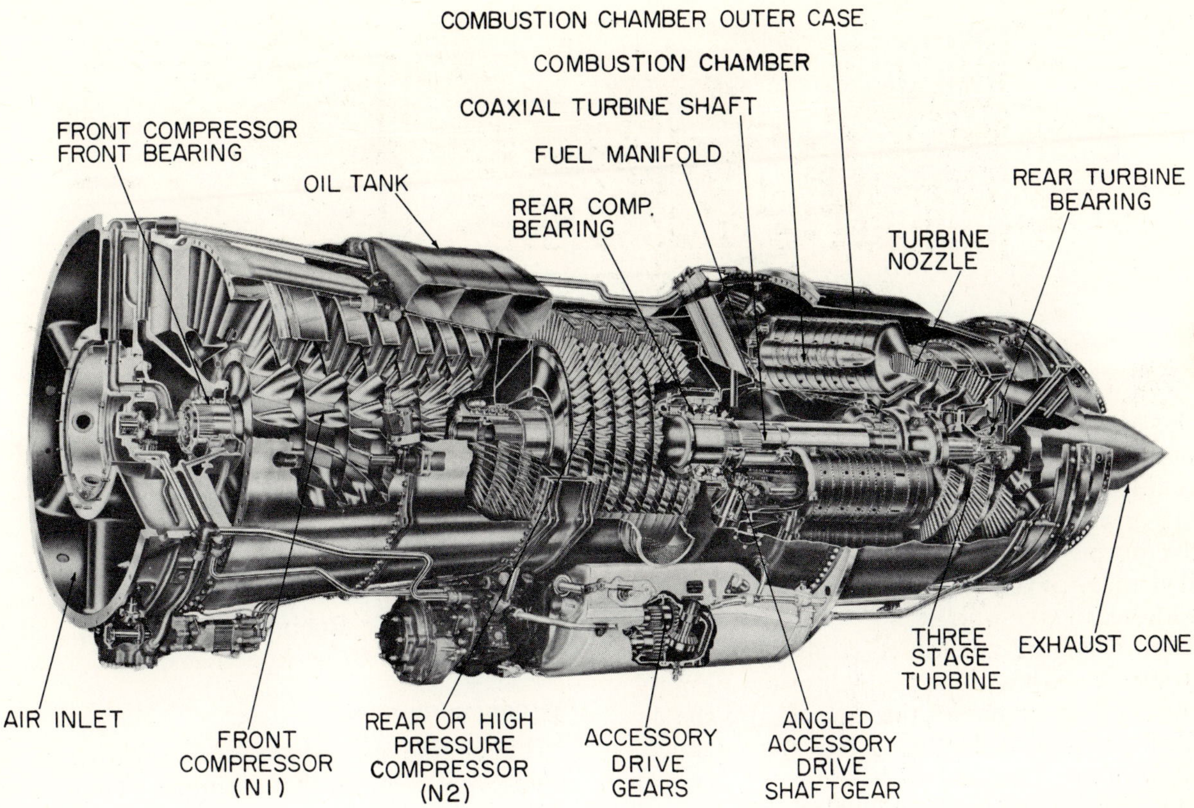

Figure 3·3 Pratt & Whitney J57 turbojet engine.

1. The inlet duct and guide vanes
2. The compressor
3. The diffuser, with or without air adaptors
4. The combustion chamber or chambers
5. The nozzle diaphragm
6. The turbine wheel or wheels
7. The exhaust cone
8. The afterburner (if the engine is so equipped)
9. The accessory section (which may be located at the front of the engine or farther to the rear)

The foregoing listing does not necessarily apply to all engines, but it will serve as a basic list for the purposes of study.

THE INLET DUCT AND GUIDE VANES

The aircraft inlet duct

The design of the inlet duct for a gas-turbine engine is an important factor in the performance

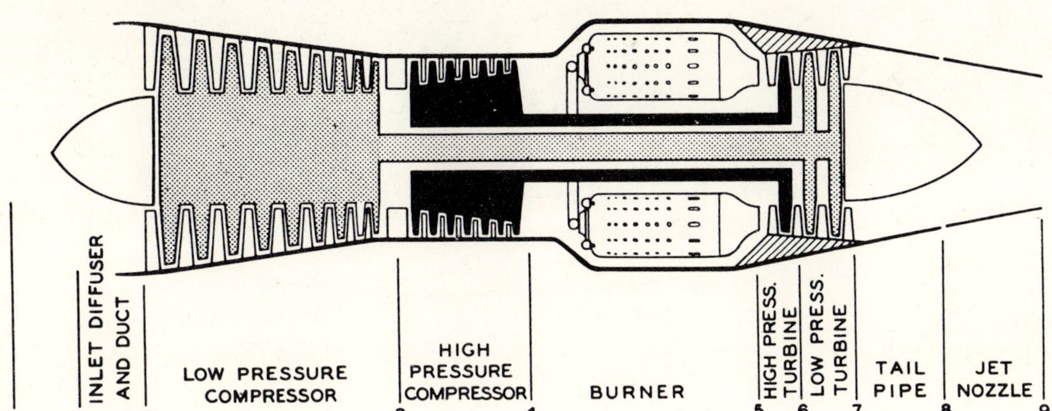

Figure 3·4 Arrangement of a split-compressor, axial-flow engine.

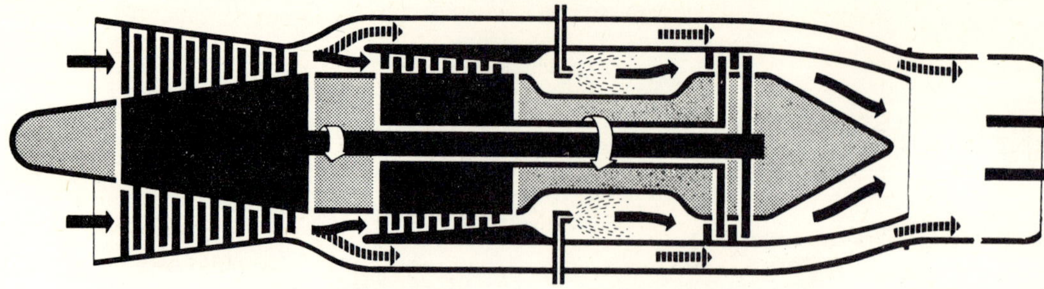

Figure 3·5 Arrangement of an axial-flow bypass engine.

of the engine. It is the responsibility of the airframe manufacturer to construct the type of inlet ducting that will produce the best results for all aircraft speeds, attitudes, and altitudes. It is the purpose of the ducting to supply a controlled mass of air to the engine inlet under all conditions of operation.

Aircraft-engine inlet ducts are designed with a variety of shapes, taking into consideration the position of the engine in the aircraft and the design of the aircraft. In the case of the Lockheed F-80 airplane, the inlet duct was split, with an air entrance at the leading edge of each wing at the root. The two ducts join together and feed into a large plenum chamber surrounding the forward portion of the engine. This arrangement is shown in Fig. 3·6.

Supersonic fighter aircraft such as the Lockheed F-104 and Republic F-105 have air-inlet ducts located on each side of the fuselage and designed to eliminate the "choking" created by shock waves.

The simplest type of ducting is found with engine mounted in pods beneath the wing of a transport such as the Douglas DC-8 shown in Fig. 3·7. This type of engine arrangement is also employed on the Boeing B-52 bomber and the Boeing Model 707 airline transport.

The engine air inlet

The engine air inlets for a centrifugal-type engine are comparatively simple. They consist of large screened openings with curved passages leading to the guide vanes near the center of the impeller. The inlets are clearly visible in the illustration of Fig. 3·6.

The arrangement of the air inlet for a centrifugal engine has some very important advantages. The plenum chamber surrounding the front of the engine serves to collect dirt, rocks, and other

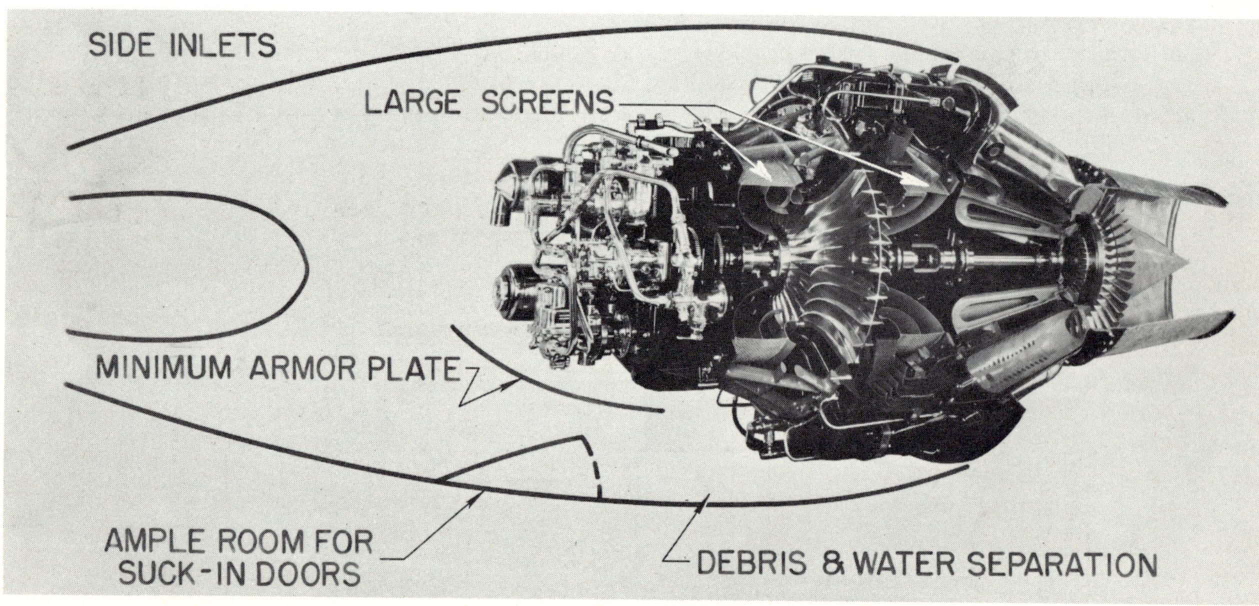

Figure 3·6 Schematic diagram of inlet ducts and plenum chamber used on Lockheed F-80 airplane.

Figure 3·7 Douglas DC-8 showing simplified engine installation.

foreign matter, thus preventing them from entering the engine and causing damage. In addition, because of the location of the inlet screens, they are practically ice-free even under the most severe conditions.

The air inlet of the axial-flow engine has been a problem to engineers from the beginning. Many engines of comparatively early design had the engine-driven accessories grouped in the center of the air inlet, forward of the front frame and bearing support. The accessories and accessory-drive gearing limited the area available for inlet air, thus decreasing the effectiveness of the compressor. With engines of later design, such as the Westinghouse J40 and J46 and the Pratt & Whitney J57, the main accessory gearing is located outside the air-inlet area. This arrangement provides a maximum of air-inlet area, which helps to produce a greater pressure ratio.

The front section of an axial-flow engine which forms the air inlet has a variety of names, according to the desire of the manufacturer. It may be called the **front frame,** the **front frame and accessory section,** the **air-inlet assembly,** the **front bearing support and shroud assembly,** the **compressor front bearing support,** or any other term descriptive of its function. In any event, it serves a variety of purposes. Usually, the outer shell of the front frame is joined to the center portion by means of struts. These are often called **compressor struts.** The entire section may be cast in one piece of aluminum alloy or magnesium alloy. The guide vanes, which direct air to the first stage of the compressor, are housed in the front frame or adjacent to it.

Some modern axial-flow turbine engines have variable-pitch inlet vanes mounted in the inlet section. The purpose of these vanes is to make corrections necessary to prevent compressor stall and surge during acceleration as the engine passes through critical speed ranges. The high pressure ratios of many present-day engines tend to bring about stall and surge unless fuel flow and airflow are carefully controlled. Variable inlet vanes and variable stator blades are quite effective in reducing the incidence of stall.

Foreign-object damage

One of the major problems encountered in the operation of axial-flow engines is foreign-object damage. Rocks drawn into the air inlet during taxiing cause considerable damage because they nick or scratch the compressor and turbine blades as they pass through the engine. Nicks and scratches may not be immediately apparent, but they can lead to fatigue failure with the result that the engine may throw a blade in flight.

This could result in loss of the aircraft or serious damage to the engine.

To prevent foreign-object damage, the air inlets on the engine are screened as shown in Fig. 3 · 8. These screens are effective in removing large objects from the air stream, but they will not prevent small rocks from entering the engine. Small rocks, sand, and gravel can do a great amount of damage to the engine.

Air inlet icing

The air screen at the inlet of an axial-flow engine is subject to icing, with the result that the engine may stop. The air-inlet sections of the jet engines installed on modern jet airliners are not equipped with air screens; thus the problem of screen icing has been eliminated. The engine nose cowling, nose dome, and inlet guide vanes are subject to icing, however, and it is necessary to incorporate provisions in the engine and in the engine nose cowling to prevent the formation of ice.

Jet engine anti-icing systems usually make use of the high-temperature air from the diffuser section of the engine. Since the air has been compressed to about 190 psi, the temperature of the air rises several hundred degrees, and the hot air thus produced is ducted through the inlet guide vanes, nose dome, and nose cowling.

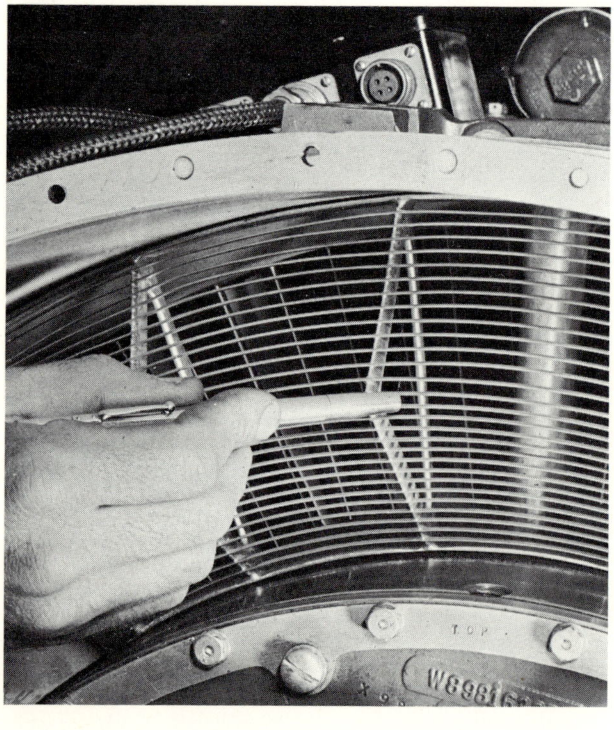

Figure 3 · 8 Air inlet screens.

COMPRESSORS

Basic types

Three basic types of air compressors are used in gas-turbine engines: the centrifugal type, the axial single-rotor type, and the axial split-compressor type. Each of these compressors has qualities which are desirable.

The centrifugal-type compressor produces a high-pressure rise in a single stage, it is easy to manufacture, it is durable, it will operate under a great number of adverse conditions such as ice and sand, and it is comparatively low in cost.

The axial-flow compressor is more efficient than the centrifugal-flow compressor, and it will handle a larger volume of air for a given diameter. It also provides for higher pressure ratios than are possible with the centrifugal compressor.

The split, or two-spool, compressor provides for very high pressure ratios which lead to excellent specific fuel consumption. Typical of the pressure ratios obtainable with the split compressor is a ratio of 13:1. Turbojet engines with this pressure ratio have attained a specific fuel consumption of approximately 0.75, which is exceptionally good for engines of this type. Turbojet engines equipped with forward or aft fans are now capable of obtaining sfc ratings of 0.53 lb per lb thrust per hr. This is an efficiency comparable to that of the best reciprocating engines.

Centrifugal-flow design characteristics

In the case of the centrifugal compressor, some engines use a single-sided, or single-entry, compressor. This type is a little more efficient than the double-sided type, but its use requires a larger diameter for the same result. Most gas turbines of the centrifugal type utilize the double-sided impeller in order to reduce the diameter of the engine.

One important condition to be considered in the diameter of a centrifugal compressor is the linear speed of the outer periphery (outside rim) of the impeller. If this speed becomes as great as the speed of sound, much energy will be lost because of shock waves, and severe vibration may occur. Therefore it is advantageous to use an impeller of smaller diameter so that the outer periphery will not attain the speed of sound.

Figure 3 · 9 is a drawing of a typical double-sided centrifugal impeller assembly. Air enters the compressor through circumferential inlets on the front and back of the impeller. The inlets are

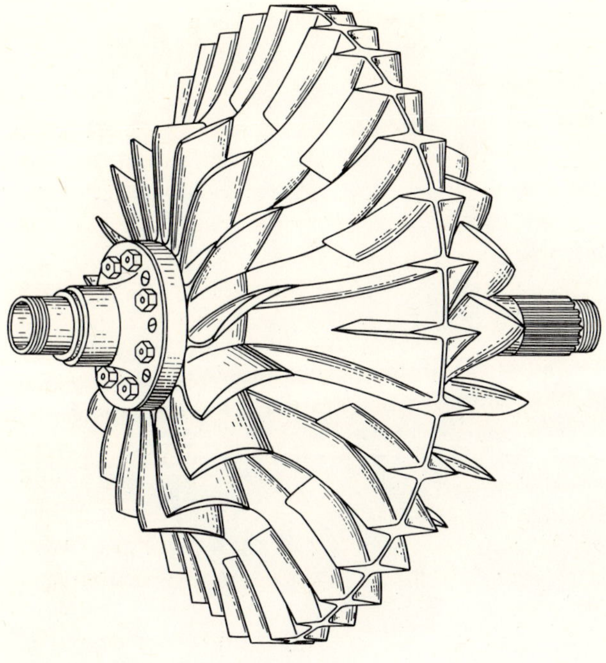

Figure 3 · 9 A double-sided centrifugal impeller assembly.

screened to prevent particles of dirt and other foreign matter from entering the air intake. The curved guide blades on each side of the impeller permit the high-velocity air to enter with a minimum of shock.

Single-stage centrifugal compressors are capable of delivering a 4:1 to 5:1 pressure ratio to the engine. Some engines are equipped with more than one centrifugal stage and a higher pressure ratio is thus attainable.

Construction of the centrifugal compressor

The impeller of a centrifugal compressor is usually constructed of a heat-treated aluminum alloy forging. The impeller is assembled together with the guide vanes and two stub shafts. Figure 3 · 10 is an exploded view of a double-sided impeller assembly. The two stub shafts provide bearing support for mounting the impeller in the engine. A ball bearing is mounted on the front shaft, and this positions the impeller both axially and radially. The ball bearing also serves to absorb axial thrust loads. The rear stub shaft is supported by a roller bearing and is splined to the turbine-shaft coupling.

The precision with which an impeller must be constructed is illustrated by Fig. 3 · 11. Note that the impeller vanes and the outer casting must fit with a minimum of clearance to prevent the loss of air pressure.

Still another feature which must be attained in the construction of an impeller is precise balance. The method for balancing is shown in Fig. 3 · 12. The semicircular cutouts in the web between the vanes represent the removal of weight from the heavy side of the impeller. In order to attain an accurate balance, the impeller is tested on a dynamic balancing machine. This machine rotates the impeller at a very high speed, thus multiplying the effect of unbalance many times. Weight is removed from the impeller until it produces a minimum of vibration while rotating at speeds greater than the normal operating speed.

Axial-flow design characteristics

As shown in Fig. 3 · 13, an axial compressor is made up of many stages. The typical compressor illustrated consists of 11 bladed disks shrunk on a steel shaft and enclosed in a cast-aluminum outer casing. This is called an "11-stage" compressor because of the 11 rows of compressor blades. Each row of blades constitutes a stage.

The number of stages in an axial compressor varies with engine design. The General Electric J47 engine has a 12-stage compressor, while the GE CJ805 engine has a 17-stage compressor.

The compressor illustrated in Fig. 3 · 13 is a

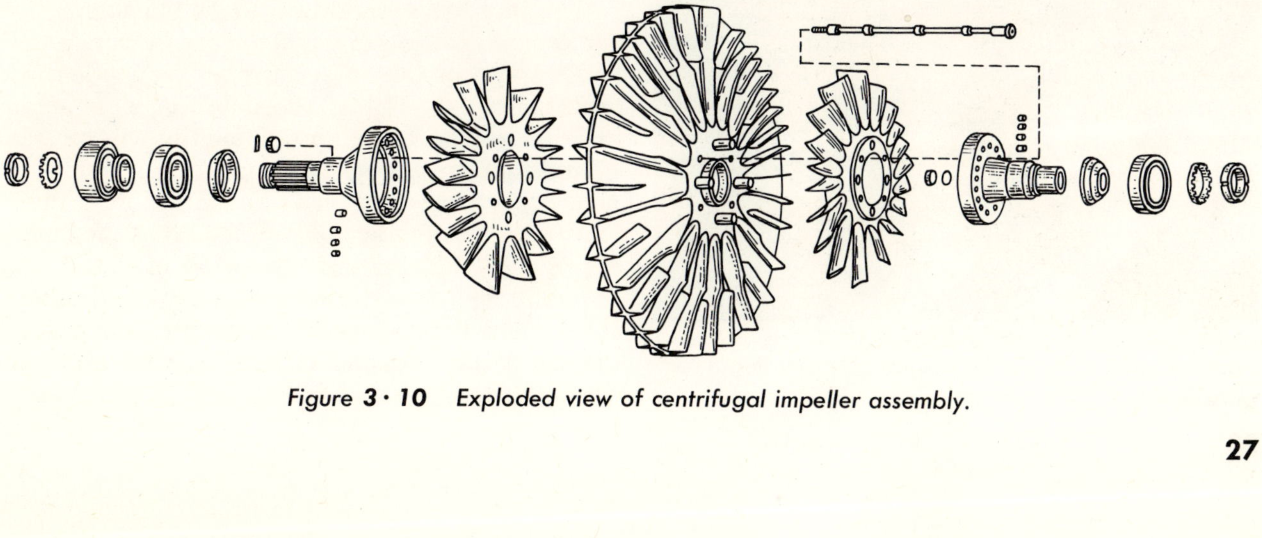

Figure 3 · 10 Exploded view of centrifugal impeller assembly.

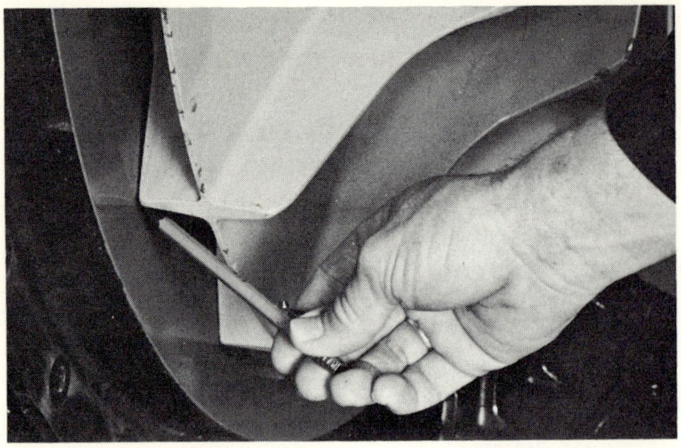

Figure 3 · 11 Minimum clearance required to prevent loss of compressed air.

Figure 3 · 13 Compressor rotor with half of case removed.

component of the Allison J35 engine. The compressor rotor consists of 10 forged aluminum disks and one steel disk. The steel disk is a heat-treated forging directly coupled to the turbine shaft by means of a splined coupling. The rotor blades, forged and coined to size, are attached to the disk rims with trapezoidally shaped dovetails. The outer compressor casing consists of two aluminum-alloy castings. Each casting forms a half cylinder, and the two halves, bolted together through flanges along the edges, form the cylinder which is the compressor casing. Grooves are milled into the inner surface of the casing to receive the **stator** blades. The stator blades fit between the rows of compressor blades to redirect the air from one stage to the next. They are constructed in the same manner as the rotor blades and have a similar shape. The shape of the rotor and stator blades is the same as that of an airfoil.

The compressor rotor for a Westinghouse J34 engine is shown in Fig. 3 · 14. This rotor is designed for use with shrouded stator blades, illustrated in Fig. 3 · 15. The shroud rings attached to the inner ends of the stator blades fit into the spaces between the rotor disks, thus sealing in the air. When the stator blades are not shrouded, the areas between the rotor disks are filled with spacer rings.

Figure 3 · 12 Cut outs for balancing a centrifugal compressor rotor.

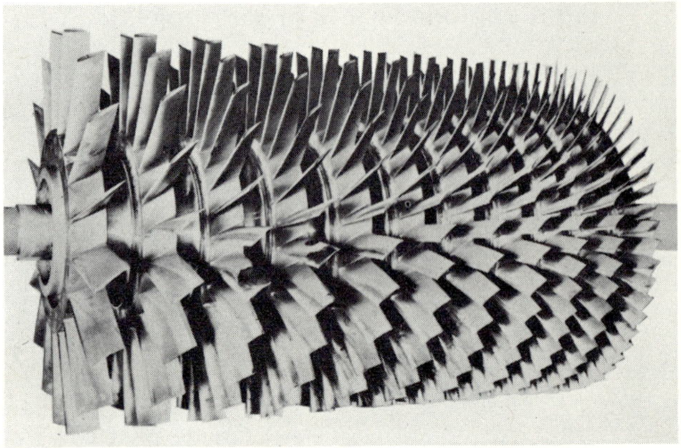

Figure 3 · 14 Compressor rotor assembly for axial-flow engine.

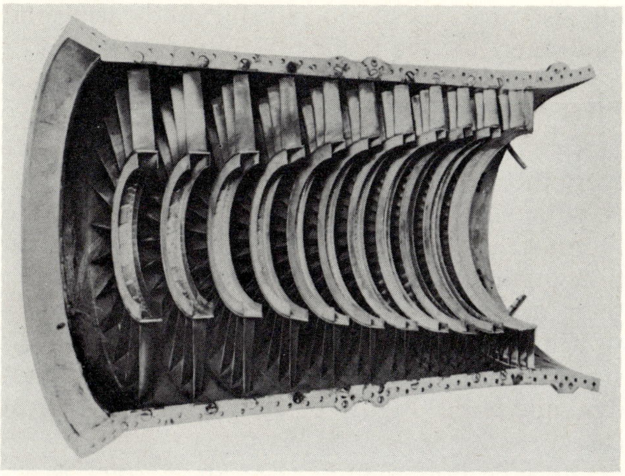

Figure 3·15 One section of compressor-stator assembly.

The General Electric CJ805 engine utilizes a 17-stage compressor. The compressor rotor consists of 17 Timken alloy disks with steel blades mounted by dovetails on the rims. The disks are bolted together with spacers and torque cones at the rear to form a single unit. The construction of this unit is shown in the section describing the CJ805 engine.

The stator for the CJ805 engine consists of an outer casing, 19 rows of vanes, and the actuating linkage for the front six rows of variable vanes. The variable stator vanes provide for optimum performance of the compressor over a wide range of operating conditions. The variable vanes are regulated through the fuel control unit.

The two-spool compressor

A typical two-spool compressor consists of a low-pressure rotor, designated N_1, and a high-pressure spool designated N_2. This type of compressor arrangement permits very high pressure ratios of from 9:1 to 12:1. Two-spool compressors are used in both turbojet and turboprop engines.

A two-spool or "split" compressor provides several advantages. Usually, only the high-pressure rotor is governed for speed, and the low-pressure rotor is free to operate at its own best speed. For this reason, the low-pressure rotor will increase speed as the compressor load decreases in the lower-density air of high altitudes. The increase in low-pressure rotor speed increases airflow, thus reducing the loss that would occur with a single-rotor engine.

The two compressor rotors in a two-spool system are mounted on separate shafts. The shaft through which the front rotor is driven passes through the hollow shaft of the rear rotor. Each rotor has a front and rear bearing. The rear bearing for each rotor is a heavy ball bearing assembly which absorbs axial thrust loads in addition to the radial loads. The arrangement of a two-spool compressor is shown in Fig. 3·16.

Since the two-spool engine is designed for high power and high pressure ratios, most of the

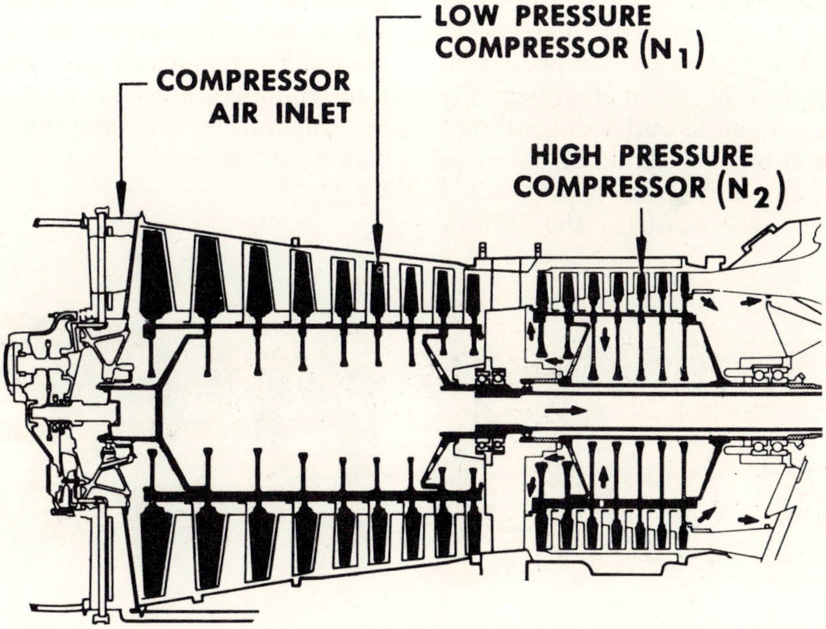

Figure 3·16 Arrangement of a two-spool compressor.

parts are constructed of alloy steel instead of aluminum alloys. The compressor rotors consist of steel disks held together by through bolts and separated by spacer rings. At each end is a steel hub and shaft for bearing support. The compressor blades are constructed of forged alloy steel and are mounted on the rims of the disks.

Between the two rotors of a two-spool compressor is a supporting structure for rotor bearings and guide vanes. The vanes are designed to give proper direction to the airflow from the low-pressure section to the high-pressure section.

The stator blades are shrouded and arranged in a manner similar to that shown in Fig. 3 · 15. The stator rings may be separate units stacked together, or they may be mounted together in the outer case. The stator blades are usually of the same shape as the adjacent rotor blades.

THE DIFFUSER

The function of the diffuser assembly is to direct air from the compressor to the combustion chambers and to change air pressure and velocity as required for best fuel combustion.

The air discharge from a centrifugal impeller enters equally spaced diffuser passages, and at the end of each is a Wirt type of elbow containing four vanes which turn the air 90° into the compressor discharge. The diffuser has a boxed type of single casting, with elbows and turning vanes cast integrally. Figure 3 · 17 illustrates a typical diffuser for a centrifugal-flow engine.

The diffuser for an axial-flow engine serves to carry the air from the compressor to the combustion chambers. For an engine equipped with individual "can"-type combustion chambers, the diffuser must have a separate outlet shaped to fit the inlet of each combustion chamber. In some axial-flow engines, the diffuser section is called the **midframe.** It not only contains the diffuser but also provides support for the midbearing and mountings for the fuel-nozzle assemblies.

Air adapters

Air adapters on a centrifugal-flow engine carry the air from the diffuser to the combustion chambers. They also provide attachment for the fuel nozzles, domes or end caps of the combustion chambers, and the igniter plugs. Because of their diverging shape, the air adapters aid in slowing the air velocity and increasing the pressure, as is desirable at this point of the thermodynamic cycle. Figure 3 · 18 illustrates a typical cast air adapter.

On axial-flow engines, the air adapter is actually the outlet of the diffuser section. Usually this portion of the assembly is not even named as the air adapter.

COMBUSTION CHAMBERS

Function

The purpose of the combustion chambers in a turbine engine is to expand the air passing through the engine by burning fuel in the air stream. The heat released by the burning fuel adds energy to the air in the form of velocity. The acceleration of the fuel-air mixture imparts positive thrust to the combustion-chamber structure.

Roughly one-fourth the air entering the combustion chamber area is burned with the fuel. The balance of the air serves to keep the temperature of the heated gases down to a level which will not damage the turbine nozzle and blades. For this reason the design of the combustion chambers must be such that adequate fuel combustion is accomplished and that proper cooling is attained.

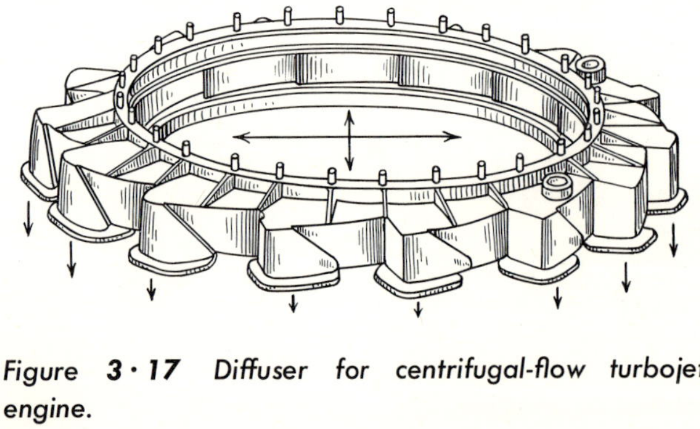

Figure **3 · 17** *Diffuser for centrifugal-flow turbojet engine.*

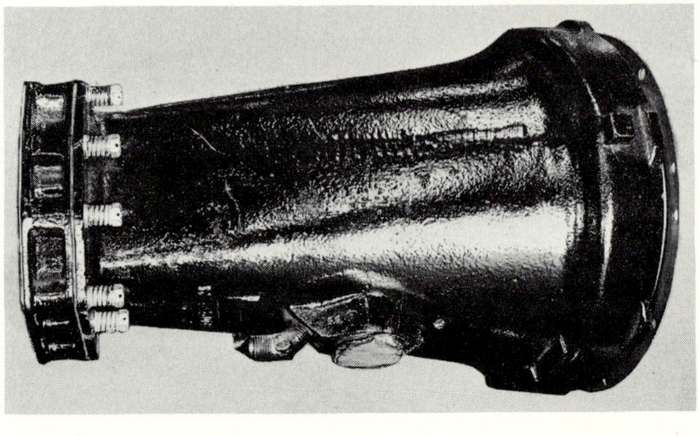

Figure **3 · 18** *Cast air adapter.*

Design of individual combustion chambers

The design of individual or "can"-type combustion chambers may seem to be rather simple, as illustrated in Fig. 3 · 19; however, there are a number of interrelated features which are quite complex in their operation. The hole-and-louver arrangement, the fuel-injection characteristics, the degree of uniformity of the air supply, and the nature and location of the igniter mutually affect one another. It has required a long program of testing and development to design combustion chambers which would produce the highest performance.

Some of the elements of airflow and combustion in the "can"-type combustor are illustrated in Fig. 3 · 20. This combustion chamber, or combustor, is typical of the individual or "can" type. One of the most important characteristics of this combustor is the flame stabilization or anchoring, that is, the smooth continuous ignition process which occurs in the dome end. It is obvious that any deficiencies of operation occurring in this flame-initiating zone must be compensated for elsewhere at many times the cost. In order to anchor the flame and secure complete combustion, it is necessary to produce a strong reverse flow of heated air and gas, as shown in Fig. 3 · 20. This backflow results in the ignition eddy which is called the **tore.** This characteristic flow pattern is visualized with some difficulty. To establish its existence beyond any doubt, various tests have been made with pressure probes, along with other tests such as fuel injection into the first ring of holes in the flame tube instead of injection at the normal fuel-nozzle position. These tests have proved conclusively that the flow obtained is truly that represented in Fig. 3 · 20.

Understanding of the backflow may be accomplished best from a study of the end view of the combustion chamber shown at the right in Fig. 3 · 20. It is obvious that the eight strong jets of air, impacting at the center as shown, must produce a substantial axial flow, both out of and into the plane of the page. At the dome end, this axial flow will be turned radially outward by the dome and will emerge through the eight substantially unobstructed longitudinal passages formed by the "watermelon slice" spaces between the longitudinal rows of air inlets.

It has been found that the strength of this backward flow may be controlled by various design factors. This is important, since having approximately one-fourth of the total airflow going into the tore gives the best results for gas-turbine operation.

One other important point may be seen from a study of this end view. Since the hot burning

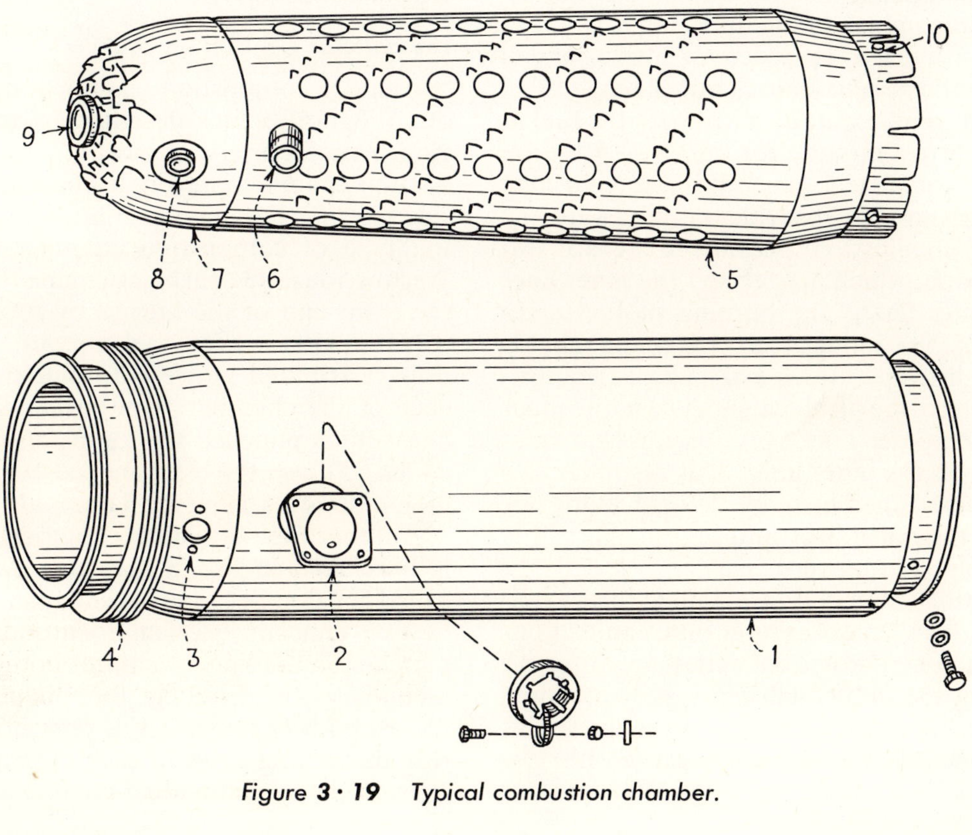

Figure 3 · 19 Typical combustion chamber.

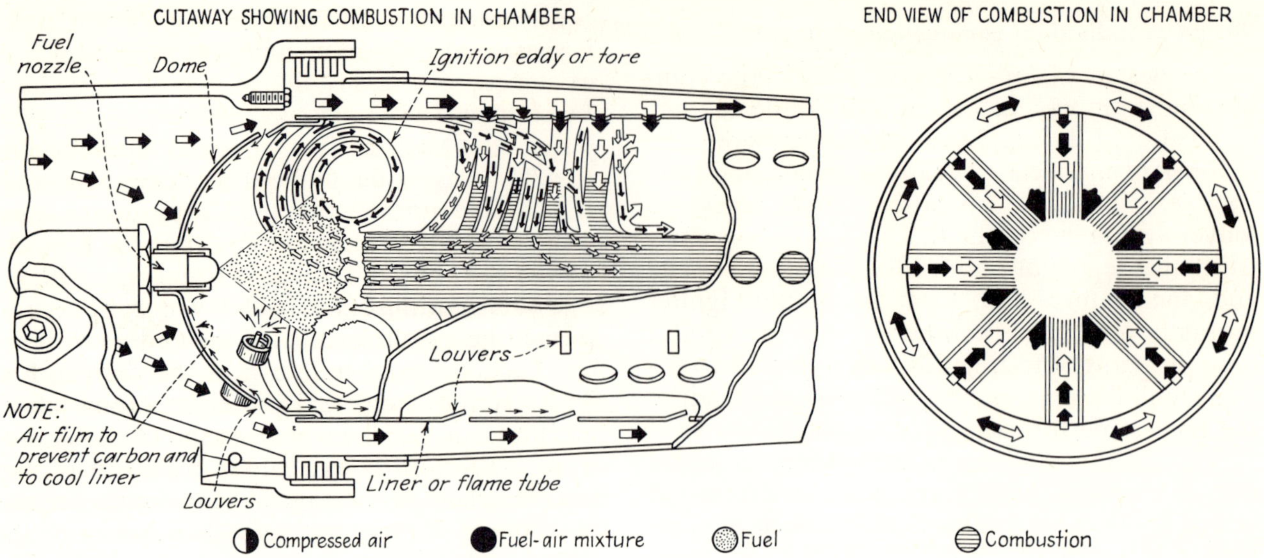

Figure 3 · 20 Pattern of burning fuel and air in a "can"-type combustion chamber.

gases pass in close proximity to the eight incoming air jets, a certain portion of the hot gas is entrained by these relatively cool jets, with the result that extremely rapid preheating of the incoming combustion air takes place. This has a most important effect in obtaining good ignition and combustion characteristics.

Cooling of combustion chambers

As was mentioned before, the quantity of air entering the combustion chamber area is about four times as much as is needed for combustion of the fuel. In the combustion-chamber flame tube, only enough primary air is mixed with the fuel to provide a correct mixture for burning. After the burning is completed, secondary air joins the burning mixture to provide cooling. In the "can"-type combustion chamber there are two principal parts, which are termed the **inner liner** and the **outer liner.** The burning of fuel takes place inside the inner liner. Even though extremely high temperatures exist where the burning takes place, the air flowing between the inner liner and the outer liner prevents the heat from penetrating to the outer liner. This secondary air also cools the inner liner by flowing along the outside of the inner liner and entering the liner through holes and louvers.

Combustion chambers are joined near their forward ends by cross-ignition tube connections into which inner crossover tubes are inserted, thus linking the individual liners and spreading combustion from one chamber to the next. This inner connection is necessary because only two spark plugs or igniters are generally used per engine, being placed diametrically opposite each other. Upon initial ignition, the flame travels from the combustion chambers that house these spark plugs or igniters to the other combustion chambers, one at a time. This would not be possible without the inner connectors. One of the advantages of having individual combustion chambers is that replacement of combustion chambers and liners is possible without complete engine disassembly.

The annular combustion chamber

Annular combustion chambers have a variety of configurations as designed by individual engine manufacturers. Basically, an annular combustion chamber is a single chamber completely surrounding the engine. Figure 3 · 21 illustrates the type of combustion chamber used on the Westinghouse J34 turbojet engine. Fuel is fed to the front end of the burner by means of a ring of 24 equally spaced fuel nozzles. The flame is thus distributed entirely around the circumference of the chamber. The burner outer liner or "basket" is punched with uniform rows of holes which diffuse the compressed air and aid in maintaining an efficient flame pattern.

Another type of annular combustion chamber is illustrated in Fig. 3 · 22. This design may be called a "double annular chamber." It consists of two concentric circular shrouds with holes punched in the liners to admit both primary and secondary air. Circular fuel manifolds supply fuel to both chambers, thus providing a uniform ring of burning gases in each burner space.

Even though individual combustion chambers

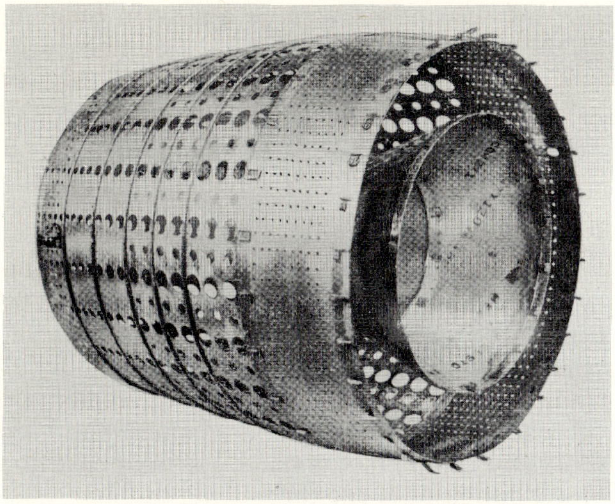

Figure 3 · 21 Annular combustion chamber for Westinghouse J34 engine.

are cheaper to develop and maintain, the annular chamber has important advantages. A larger volume is available in proportion to the inner exposed area. As a result, friction between the gas and the burner surfaces is minimized. Improved annular pressure equalization and reduced weight add to the efficiency of this type of burner.

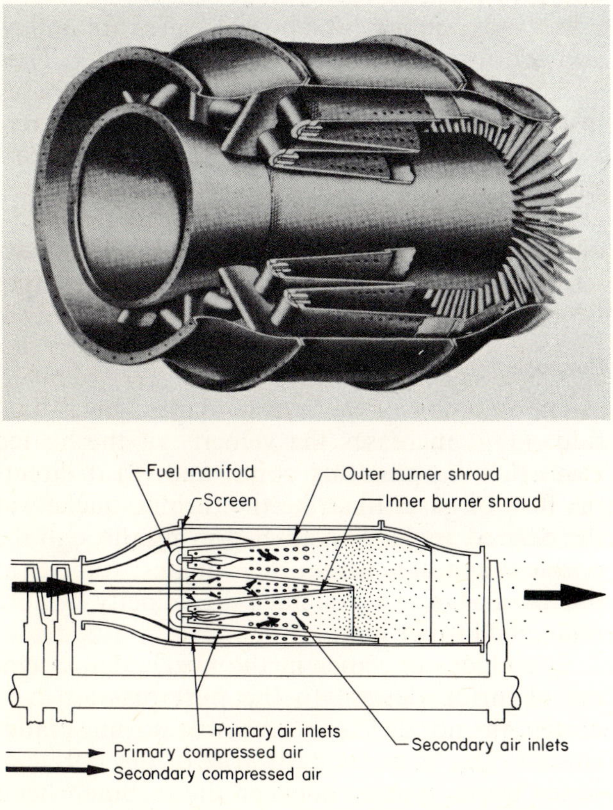

Figure 3 · 22 Double annular combustion chamber.

The cannular combustion chamber

The cannular or "can-annular" combustion chamber is a combination of the "can" type and the annular type. Figure 3 · 23 illustrates a cannular combustion chamber (burner) for a modern turbine engine. This burner consists of eight combustion chambers or "cans" arranged evenly around the circumference of an annular chamber. Each separate burner is in the form of a small annular chamber, in that there is an inner liner through which air flows to cool the inside of the chamber. Both the inner liner and the outer liner are provided with holes which permit the entry of primary and secondary air.

Each combustion chamber is supplied with fuel through six dual fuel nozzles mounted at the forward end of the chamber. The nozzles are attached to the fuel manifold in eight clusters of six nozzles each. Each cluster of nozzles serves one burner. Primary air for the combustion of fuel enters the chambers through guide vanes arranged circumferentially around the fuel nozzles. The vanes cause the air to flow with a swirling motion which aids in good mixing of fuel and even burning. This design also helps to bring about complete burning of the fuel before the fuel-air mixture leaves the chamber.

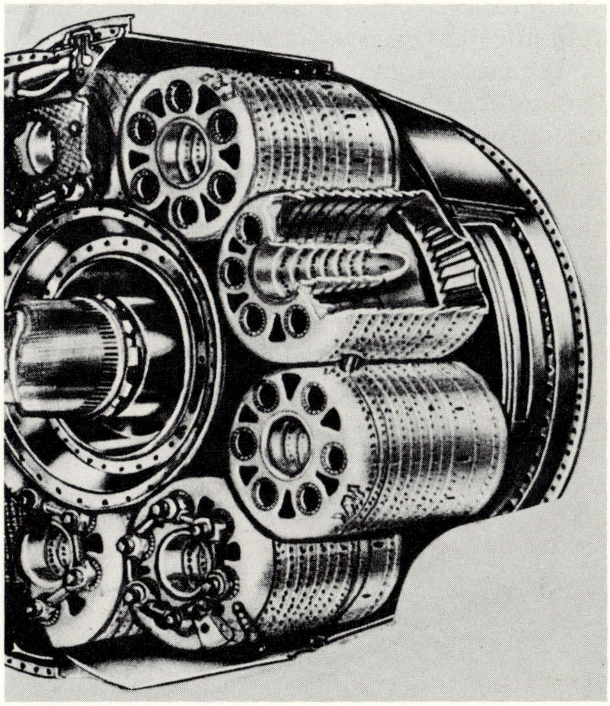

Figure 3 · 23 A cannular combustion chamber.

Fuel nozzles

The design and arrangement of fuel nozzles have an important effect on the operation of any combustion chamber. Many different types of fuel nozzles are employed in modern turbine engines, but we shall consider only a few of the most commonly used types.

An early type of fuel nozzle, called the **simplex nozzle,** is shown in Fig. 3 · 24. This is a fixed-orifice nozzle, and it was considered suitable for an engine which operates at a nearly constant load. At low pressures, the arc of spray from the orifice of the simplex nozzle is approximately 70 to 80°. This is desirable for starting. The larger the spray pattern, the more easily the atomized fuel can enter the vicinity of the spark plug. At higher pressures, the arc narrows and the flow is extended further into the combustion chamber. It is desirable to have the arc narrow as much as possible at high pressures to keep the flame narrow and centered. This prevents burning of the inner liner and spark-plug electrodes.

To overcome the limitations of the simplex fuel nozzle, various types of dual or duplex nozzles have been designed. A diagram of a **dual orifice** nozzle is shown in Fig. 3 · 25. The basic feature of a dual orifice nozzle is that it is a nozzle within a nozzle. Essentially, it is two concentric nozzles within one body: an inner low-capacity nozzle called the **primary stage** and an outer annular nozzle called the **secondary stage.** This design provides good atomization in the low-flow range, a wide-flow range, and a uniform spray angle over the entire range.

The mechanism of atomization in the primary stage of the dual orifice nozzle is the same as that of the simplex nozzle, and it operates initially with only the primary stage providing flow. Secondary fuel flow is admitted to the nozzle after the fluid pressure of the primary system has increased to the extent that the primary spray can assist the breakup of the secondary spray during the phase when secondary pressure is less than 10 psig. The primary stage assists in breaking up the secondary flow until the secondary pressure reaches a value that will provide atomization without assistance from the primary spray. The secondary stage supplies most of the fuel in the higher power ranges.

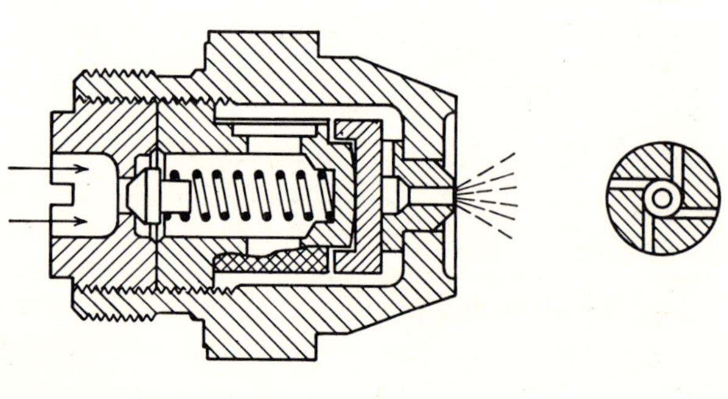

Figure 3 · 24 Simplex fuel nozzle.

Transition liners

On some engines, transition liners are used to direct air from the combustion chambers to the turbine nozzle. In a "can"-type axial-flow engine, the transition liner is mounted in the aft frame section. One end of the liner receives the aft end of the combustion chamber and the other end of the liner is shaped to adapt the flow of the hot gases to the turbine nozzle.

TURBINE NOZZLE AND NOZZLE DIAPHRAGM

Description and location

A typical nozzle diaphragm is illustrated in Fig. 3 · 26. This diaphragm consists of a group of nozzle vanes welded between two shroud rings. In the typical nozzle diaphragm, the inner and outer bands contain punched holes to receive the ends of the nozzle vanes. The nozzle vanes are usually constructed of high-temperature alloy, and they must be highly heat-resistant.

In many engines, the nozzle vanes are hollow and are formed from stainless-steel sheet. They are then welded and ground smooth before being installed between the shroud rings. When there is more than one turbine wheel, additional nozzle diaphragms are installed to direct the heated gases from one wheel to the next. Second-, third-, and fourth-stage nozzle vanes are often constructed of solid steel alloy. These may be either forged or precision-cast.

Purpose

The function of the nozzle diaphragm is twofold: (1) it increases the velocity of the heated gases flowing past this point, and (2) it directs the flow of gases to strike the turbine buckets at the desired angle. The gases flowing through the nozzle diaphragm attain their highest velocity at this point. On some engines the velocity is more than 2,000 ft per min.

The blades or vanes in the nozzle diaphragm are of airfoil design. In the pure reaction-type design, the nozzle blades resemble turbine blades closely. This is particularly true of the nozzle or stator blades (vanes) between the turbine wheels in multistage turbine assemblies.

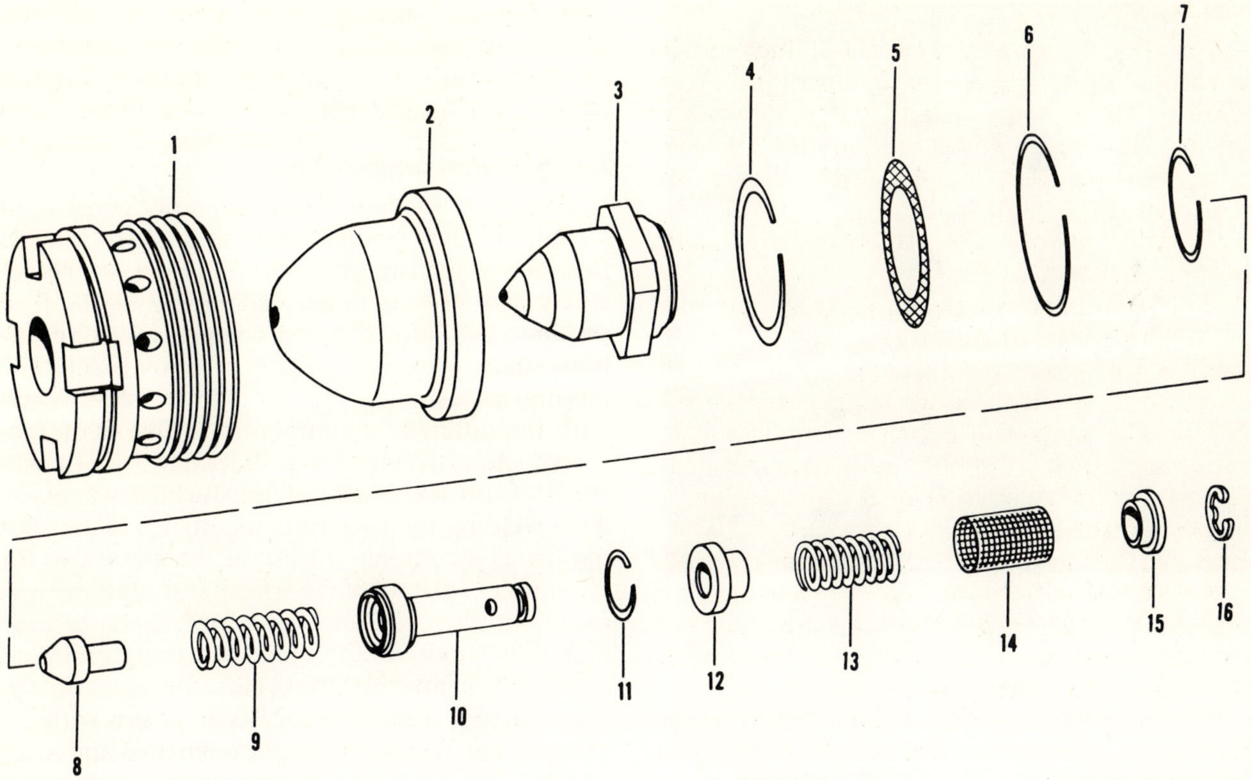

1. Retaining nut
2. Nozzle body
3. Insert
4. Insert snapring
5. Secondary strainer
6. Secondary strainer outer snapring
7. Secondary strainer inner snapring
8. Primary plug
9. Spring
10. Spring seat
11. Spring seat snapring
12. Lower primary strainer ferrule
13. Spring
14. Primary strainer
15. Upper primary strainer ferrule
16. Primary strainer snapring

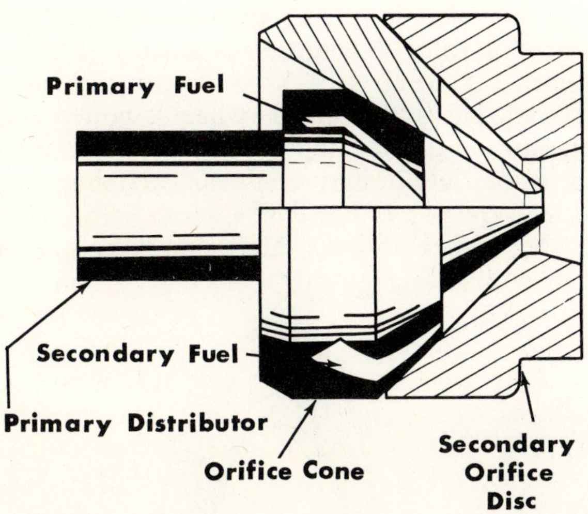

Figure 3 · 25 Two types of dual orifice fuel nozzles.

TURBINE WHEELS

Essential requisites

Among the most seriously stressed components of a gas-turbine engine is the turbine wheel. The buckets of the wheel are subjected to high centrifugal stresses and to a fluctuating stream of hot gases which raise their temperature and may, at the same time, introduce vibratory stresses. The disk of the wheel is subjected to heat by conduction from the buckets and may, in some cases, be directly heated by the combustion gases. Temperatures of over 1500°F are not uncommon near the rim of the first-stage turbine disk. Where there is more than one turbine disk in an engine, the second- and third-stage disks do not encounter such high temperatures. Turbine disks are often cooled by means of air bled from the compressor section and directed through various passages to flow around the turbine disk.

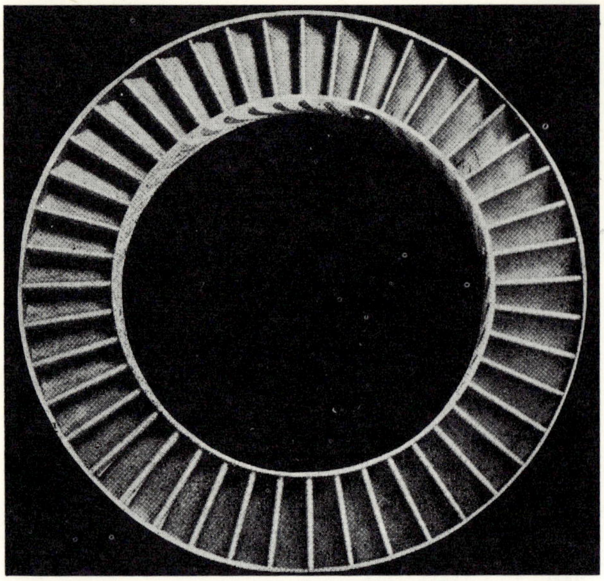

Figure 3 · 26 *A nozzle diaphragm.*

In most turbine disks, a high temperature gradient exists between the center of the disk and the rim. This gradient may be as much as 500°F or more. The differential expansion caused by this high temperature gradient introduces a system of thermal stresses that combine with the centrifugal stresses caused by rotation, thereby producing severe conditions of operation.

The gas-turbine industry has accomplished miracles in the development of new alloys which will withstand these conditions. The fact that turbine wheels and buckets (blades) will operate without failure for more than 4,000 hr of flight indicates the remarkable materials which have been developed.

Turbine-wheel components

The usual turbine-wheel assembly consists of a shaft, a wheel or disk, and the buckets (blades) fastened to the rim of the disk. A typical turbine-wheel assembly is illustrated in Fig. 3 · 27. The **turbine shaft,** also called a **compressor-drive turbine shaft,** is usually made of alloy steel. The method of fastening the wheel to the shaft varies with the different manufacturers. One manufacturer fastens the shaft to a short butt, which protrudes from the center of the turbine wheel, by flash welding the two units together. Flash welding forms a complete fusing of the metals, so for all practical purposes the wheel and shaft are one unit. Another manufacturer utilizes a turbine disk that is more in the shape of a doughnut than a solid disk and electric-welds the shaft to the outer turbine disk. As many as 26 passes with the arc rod are necessary to join the two units together. The front end of the shaft is splined to fit into the compressor rotor hub or into a coupling which joins the turbine shaft to the compressor rotor. A coupling of this type is illustrated in Fig. 3 · 28.

In some cases, the turbine wheel is bolted to a flange on the turbine shaft. This method is used with some late-design engines, including the Pratt & Whitney JT3 and JT4.

Turbine-wheel blanks are constructed of a

1. Turbine wheel
2. Buckets
3. No. 4 bearing journal
4. No. 4 bearing
5. No. 3 bearing journal
6. No. 3 bearing
7. Turbine shaft
8. Main drive spline to fit compressor rotor hub spline

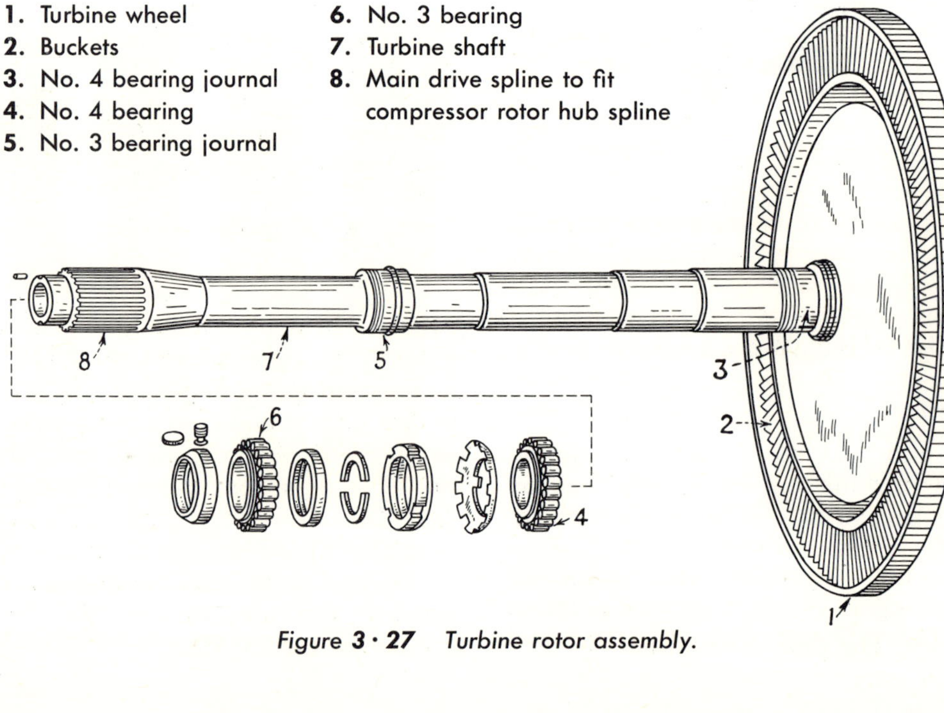

Figure 3 · 27 *Turbine rotor assembly.*

Figure 3 · 28 Turbine shaft-to-compressor coupling.

high-quality steel alloy such as Timken 17-22-A (S). The original wheel blank passes through a number of severe inspections before final machining and grinding, and the finished wheel is also inspected thoroughly. Dynamic balancing and spin testing are two of the typical procedures used to determine the quality and serviceability of a turbine wheel.

Turbine buckets must be designed to withstand a variety of severe conditions and forces such as heat (more than 1600°F), centrifugal force (more than 30,000 psi), thermostress, varying atmosphere, corrosion, vibration, and differential expansion of materials. To achieve reliability, efficiency, and long life for buckets, the following points must be controlled carefully during manufacture: (1) grain structure of the material, (2) aerodynamic performance, (3) surface finish, and (4) producibility.

The method for fastening the buckets to the turbine wheel rim should be designed for (1) strength, (2) low stress concentration, (3) simplicity, and (4) ease of replacement in the field.

Turbine buckets are manufactured by a variety of processes including forging, investment casting, and machining from the bar. In all cases it is necessary to perform final finishing and polishing operations.

Turbine-bucket attachment

Although turbine buckets have been attached to turbine-wheel rims by a variety of methods, it appears that the "fir-tree" attachment is the most satisfactory and the most widely used. One of the reasons for the design of the fir-tree attachment is the temperature differential between the wheel rim and the bucket. With early engines, when the turbine buckets were fitted solidly into the rim of the wheel, turbine buckets and wheels were often subject to failure. Radial cracks often developed at the rim of the wheel to compensate for a large amount of expansion and contraction.

The **fir-tree** attachment, illustrated in Fig. 3 · 29, consists of a series of grooves or notches broached on the root of the bucket and in the rim of the wheel. Sufficient clearance is allowed to permit a movement of the bucket tip approximately 0.050 in. when the wheel is cold. During operation, the expansion of the metal causes the

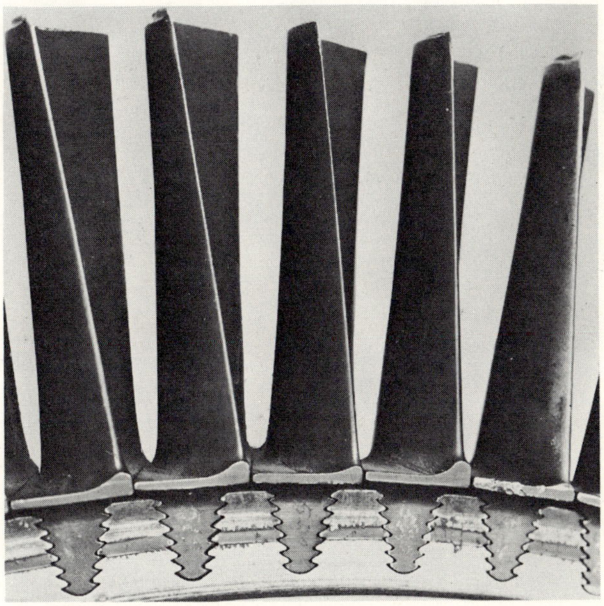

Figure 3 · 29 "Fir-tree" attachment for turbine blades.

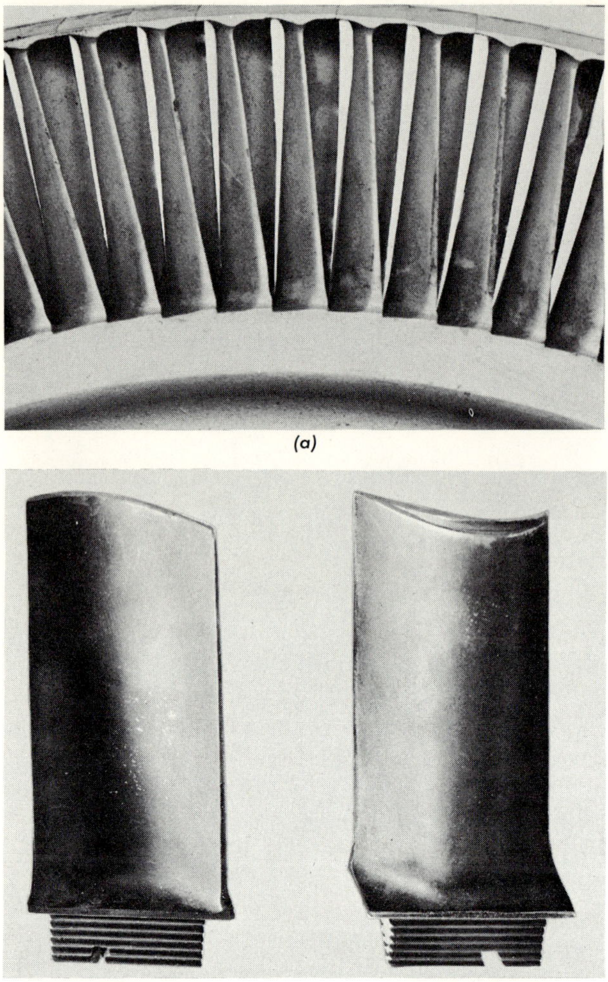

(a)

(b)

Figure 3 · 30 (a) Shrouded and (b) unshrouded turbine buckets.

Figure 3 · 31 Cutaway view of an exhaust cone.

bucket root to fit tightly in the rim of the wheel. Various methods are used to keep the turbine buckets from sliding out of the groove. In one method, a hole is drilled outward through the shoulder which forms the rim of the wheel. The outer end of the hole enters the bottom of the groove in which the bucket fits and is aligned with a hole in the root of the bucket. When the bucket is installed, a steel pin is driven through the hole in the rim and into the hole in the root of the bucket. This effectively locks the bucket into place. Another method for securing the bucket in the wheel rim is to provide a hole through the wheel rim and parallel with the groove. The hole is located so that it touches the root of the bucket. A rivet is placed in this hole and headed on each end. The heads prevent the bucket from sliding out of the groove.

Turbine blades are made in two principal configurations: **shrouded** and **unshrouded.** The shrouded bucket is provided with a shroud section at the outer end, so that when the buckets are installed, they form a continuous rim or shroud around the outer ends of the blades. With the unshrouded blades, it is necessary to provide a closely fitting **shroud ring** to prevent the escape of the hot gases which drive the turbine. Figure 3 · 30 illustrates both types of turbine buckets.

Some consideration has been given the possibility of using hollow steel or nickel-alloy blades for the turbine wheels, and some experimentation has been accomplished along this line. Information thus far released has not indicated that hollow turbine buckets have been used to any appreciable extent.

THE EXHAUST CONE

The exhaust cone is located directly behind the turbine wheel, and its main function is to collect discharge gases from the turbine wheel and expel them at the correct velocity. The exhaust cone illustrated in Fig. 3 · 31 consists of a stainless-steel outer shell and central cone supported from the shell by four streamlined struts. The function of the streamlined struts or fins is to straighten out the airflow from approximately 45° to an axial direction. Air flowing through this section decreases in velocity and increases in pressure.

The outer surface of the exhaust cone is insulated in most installations, and many different types of insulations are used. A typical arrangement consists of four layers of aluminum foil, each separated from the next by a layer of bronze screening. The insulation reduces the heat losses that would normally escape through the exhaust cone. The insulation also protects adjacent aircraft structures and equipment from damage caused by heat.

When the gas-turbine engine is delivered by the manufacturer to the ultimate consumer, the exhaust cone is the terminating component of the basic engine. In order to operate the engine and obtain the required performance, however, it is necessary to use a tailpipe and exhaust nozzle. The length of the exhaust pipe or tailpipe varies with each airplane installation, and therefore, by necessity, the pipe must be manufactured by the air-frame manufacturer.

THE THRUST REVERSER

Jet engines installed in jet airliners are equipped with thrust reversers to provide a braking action after the airplane has landed. The thrust reverser blocks gas flow to the rear and directs it forward to produce reverse thrust up to 5,000 lb or more. The reverse thrust is produced when the air baffle doors or "clamshells" are moved into the gas stream by actuating cylinders controlled by the reverse-thrust lever in the cockpit. Figure 3 · 32 is a drawing of a typical thrust-reverser unit designed for use with the General Electric CJ805 jet engine.

ACCESSORIES SECTION

Locations

The location of engine-driven accessories on a modern gas-turbine engine is substantially different from some of the early arrangements. Early-type engines usually had the accessory drive section located at the nose of the engine, where it had the effect of limiting the available area for air intake. This was particularly true on axial-flow engines.

Present designs place the main accessory drive section at the bottom of the engine, and fitted closely against the case. The Pratt & Whitney JT3 engine takes advantage of the "wasp waist" to nest the accessories snugly against the engine near the bottom center, under the high-pressure compressor section. The accessory drive is geared to the shaft of the high-pressure compressor.

Engine-driven accessories

Engine-driven accessories include the accessories needed for engine operation and also those required for the operation of the aircraft. The following is a list of accessories commonly driven by the engine and mounted on the accessory section:

1. Engine starter (connected to the engine during starting)

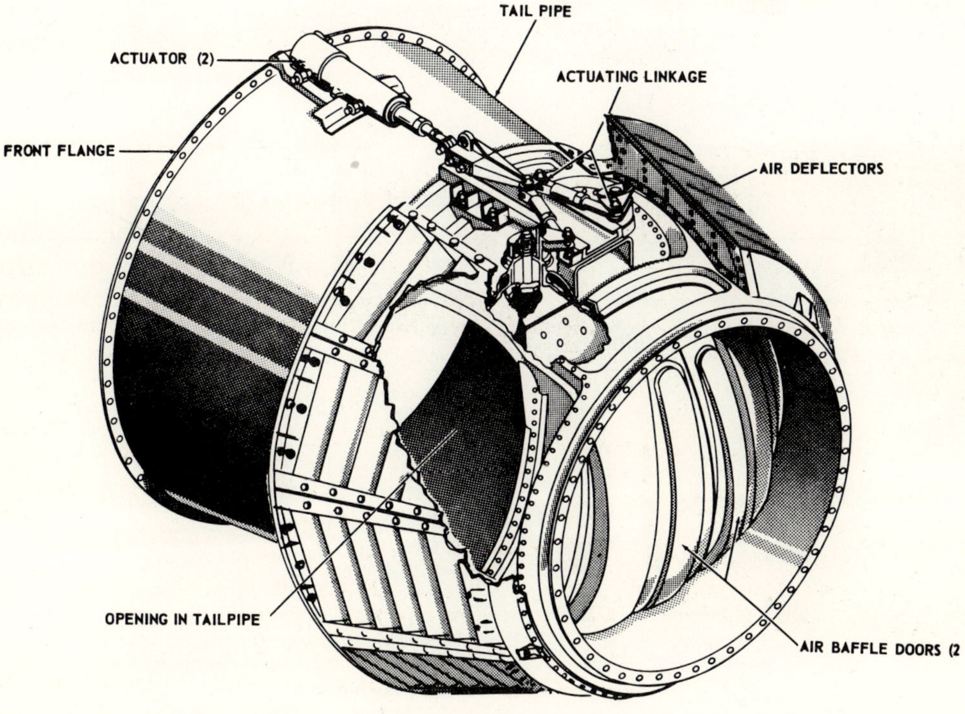

Figure 3 · 32 Thrust-reverser unit. (General Electric Co.)

2. Generator
3. Fuel pump
4. Emergency fuel pump
5. Afterburner fuel pump
6. Tachometer generator
7. Fuel-control unit
8. Air-bleed governor
9. Oil pump and scavenge pump
10. Hydraulic pump

Some gas-turbine engines have but one accessory power section, while others may have more. For example, one modern engine drives most of the accessories from a power takeoff at the bottom of the engine but also has an auxiliary accessory drive at the front of the engine. Two or three of the smaller accessories are driven from this front accessory section.

Development of accessory systems

Figure 3 · 33 illustrates the development of accessory systems and drives on turbine engines.

Engine *A* in the illustration was a booster-type engine; that is, it contained no accessories whatsoever for the benefit of the airplane. The engine accessories were the minimum needed for engine operation. These involved starting, ignition, lubrication, and the fuel and control system. In this case, the total weight of the accessories amounted to 11.6 per cent of the gross engine weight.

In engine *B*, drives were added for generator and hydraulic or vacuum pump for aircraft services, as well as an emergency fuel system for the engine. This development resulted in a gearbox redesign. At the same time, it was found possible to save some weight in the lubricating system. The resulting over-all accessory weight came out to be 12.9 per cent of the gross engine weight. It is clear that going from engine *A* to engine *B* involved considerable change in frontal area.

Engine *C* performs the same functions as engine *B*, but the horsepower to the aircraft accessory drive is increased 50 per cent and a manifold is added to provide compressor-air bleed for aircraft service. A fuel-drain system is added, and several refinements are made in the engine accessories. The total accessory weight increases

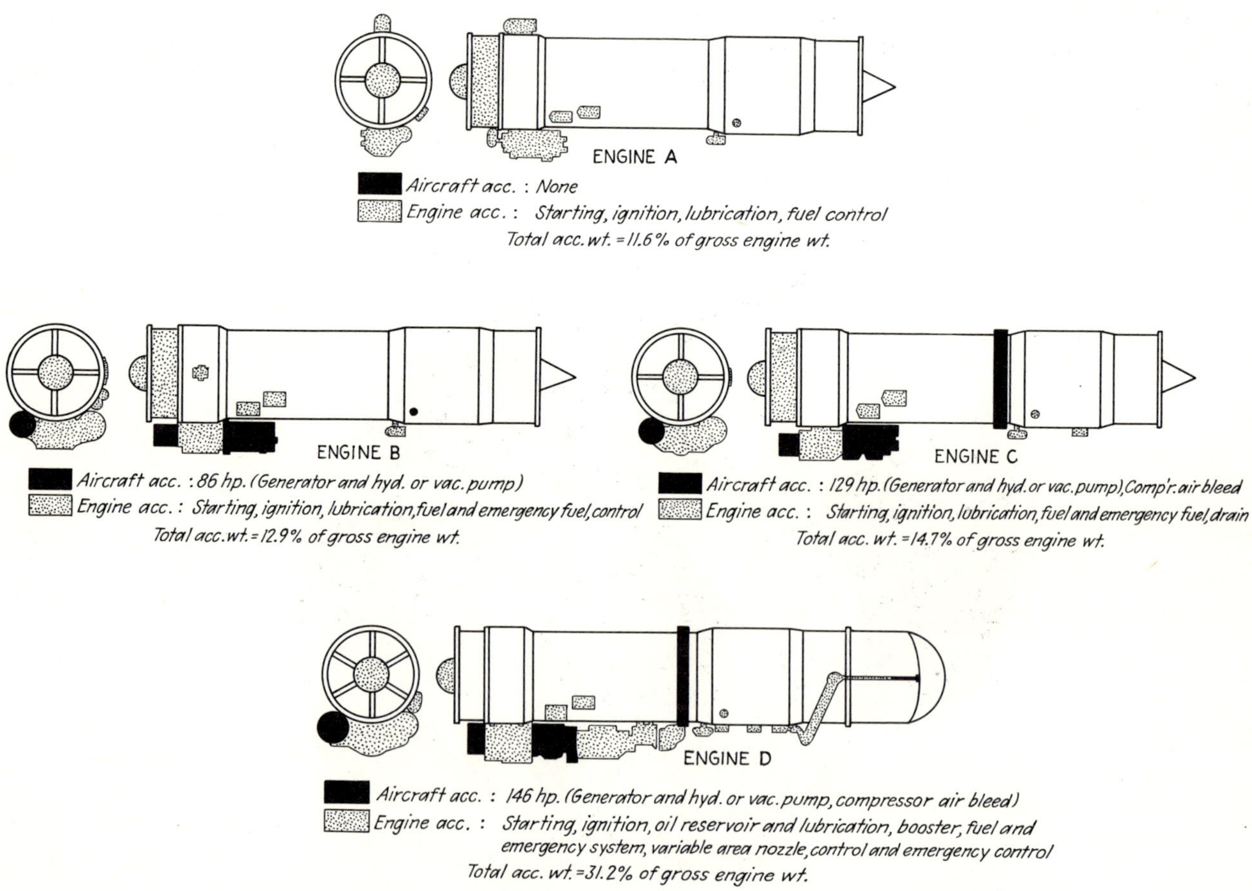

Figure 3 · 33 Development of accessory systems for turbine engines.

to 14.7 per cent of the over-all engine weight.

Engine *D* represents a study of an advanced engine, which is still more useful from an aircraft standpoint. Compared with engine *C*, engine *D* furnishes 13 per cent more aircraft accessory power. In this study, the engine is provided with anti-icing equipment, an integral oil reservoir, a fuel-booster pump, and a complete emergency control system. The engine is equipped with a variable-area exhaust nozzle and its control. As the illustration shows, the added features do not come free but involve a considerable addition to the total volume and weight. For this study, the accessory weight totaled 31.2 per cent of the gross engine weight, which is a very high percentage.

It is doubtful that the system shown for engine *D* will actually be used, because auxiliary units have been developed which serve the same purposes as many of the accessories shown. Small gas and air turbines are in common use for supplying aircraft services; hence it is not necessary to call upon the main power plant for power to supply all the auxiliary aircraft systems.

STUDY QUESTIONS

1. What are the two principal types of gas-turbine engines?
2. Compare the airflow through the centrifugal- and the axial-flow engines.
3. What are the major components of a gas-turbine engine?
4. Describe the construction of the two types of compressors.
5. What method is used to reduce air-inlet icing?
6. What are the advantages of a centrifugal compressor?
7. Describe the advantages of the two-spool compressor.
8. What is the function of the diffuser?
9. Describe three combustion-chamber designs.
10. How are combustion chambers cooled?
11. What is the function of the nozzle diaphragm?
12. How is a turbine wheel constructed?
13. Explain the action of a thrust reverser.
14. What methods may be used to drive engine accessories?
15. What is the best location for engine accessories?

CHAPTER 4

Rockets

Even though the rocket principle has been understood for many years, it is only recently that rockets have been adapted as powerplants for aircraft. Since rockets represent a form of jet propulsion, they should be discussed in this text as one of the types of reaction powerplants.

Because of the many restrictions relating to the details of specific rocket applications, only rocket principles and general methods of rocket operation can be discussed. Military security forbids the release of exact systems currently in use; also, it is not possible to give exact figures relating to rocket performance. Notwithstanding, we can obtain a good understanding of the operation of typical rockets and rocket systems, and with this knowledge we are able to undertake tasks pertaining to modern rocket operations with very little further orientation.

Types of rocket powerplants

The rocket powerplant produces its thrust in the same manner as the pulsejet and ramjet except for one outstanding difference. The rocket carries its own oxygen supply and does not require atmospheric oxygen; thus the rocket powerplant is not limited to travel in the earth's atmosphere.

The propellants employed in a rocket engine may be solid, two liquids (fuel plus an oxidizer), or materials containing an adequate supply of available oxygen in their chemical composition (monopropellants). Rocket engines generally are divided into two categories: solid and liquid. Figure 4 · 1 shows schematic illustrations of both types of engines. The rocket engine at the top of the illustration is a solid-propellant type, and the rocket engine at the bottom of the illustration is a liquid-propellant type.

LIQUID-PROPELLANT ROCKET POWERPLANTS

Principles of operation

The liquid-propellant rocket engine, as shown schematically in Fig. 4 · 1, consists of a combustion chamber, a nozzle, and an injector system with control valves. The propellant gases are generated in the combustion chamber at pressures governed by the chemical characteristics of the fuel and oxidizer, their rate of consumption, and the physical characteristics of the combustion chamber and nozzle. The gases are ejected into the atmosphere through the nozzle with supersonic velocity. The function of the nozzle is to convert the pressure of the propellant gases into **kinetic energy.** (Kinetic energy is the energy possessed by a mass because of its motion.) The thrust is generated by the reaction developed from the discharge of the high-velocity propellant gases through the nozzle.

The combustion chamber and nozzle of a liquid-propellant rocket engine may be cooled or uncooled. The uncooled chamber and nozzle are utilized normally for short duration (less than 30 sec) and are constructed of an outside metal casing for strength and an inner ceramic liner to resist the high temperature of combustion. The cooled rocket engine normally consists of an inner and an outer metal casing, constructed so that a cooling liquid can be circulated between the walls in sufficient volume and velocity to cool the inner walls of the chamber.

Several other methods have been developed for cooling rocket engines. A cooling liquid may be forced from the injector in such a manner that the liquid flows along the inside of the inner wall, thus cooling the chamber. Cooling liquid also may be injected at different points along the chamber and nozzle walls to form a protective fluid barrier on the inner surface. The cooling liquid generally is heated rapidly and reacts with the propellants. Another method for cooling the chamber and nozzle is to construct the inner wall of a porous metal so that the cooling liquid may enter through the wall into the chamber and nozzle at all points. In most systems, either the fuel or the oxidizer is employed as the cooling liquid. One of the propellants is circulated for

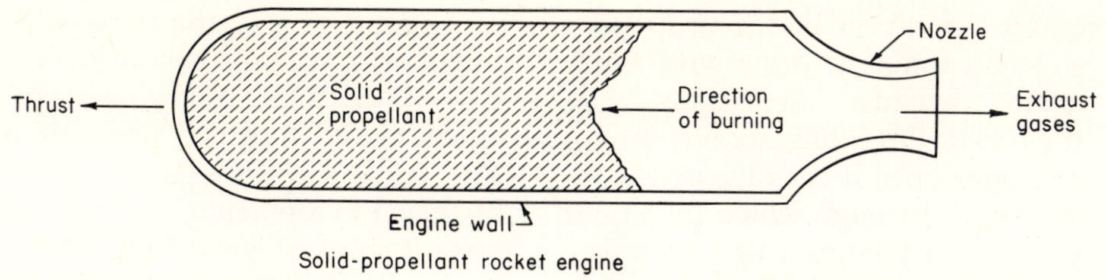

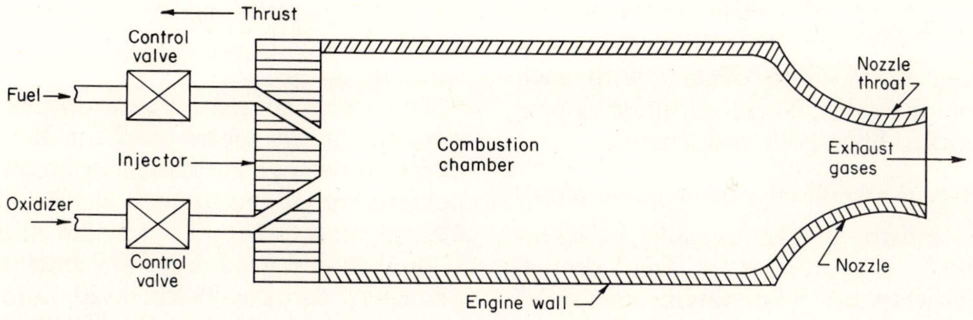

Figure 4 · 1 Typical solid- and liquid-propellant rocket engines with uncooled nozzles and combustion chambers.

cooling purposes; then it is injected into the chamber through the injector, through cooling holes, or through the porous walls of the chamber and nozzle. When one of the propellants is circulated through the engine walls for cooling and then is injected into the chamber, the system is called a **regeneratively cooled** rocket system. As the propellant passes along the inner wall of the chamber and nozzle, heat is transferred to the liquid, thus cooling the metal wall and heating the propellant. This heat is returned to the combustion chamber as the propellant is injected into the chamber, thus increasing the efficiency of the engine. The rocket powerplant shown schematically in Fig. 4 · 2 utilizes a regeneratively cooled rocket engine with fuel as the cooling liquid. The selection of the fuel or the oxidizer for the cooling liquid depends upon the volume of liquid required and the physical and chemical properties of the liquid, particularly the cooling ability of the liquid.

Figure 4 · 2 illustrates the essential components of a bipropellant liquid rocket powerplant. The propellants are removed from the supply tanks for use in the engine. The pressure of the propellants is raised by some suitable pressurizing means, such as pumps, and then forced into the combustion chamber against the combustion-chamber pressure. The pressurizing means may also be an inert gas held in a separate container at high pressure. Combustion-chamber pressures in liquid rocket engines normally vary from 300 to 800 psia, while solid-propellant chamber pressures may vary from 1,500 to 3,000 psia. The pressure normally is determined by the propellants employed and the physical configuration of the engine. The liquid-propellant rocket powerplant can be started and stopped at

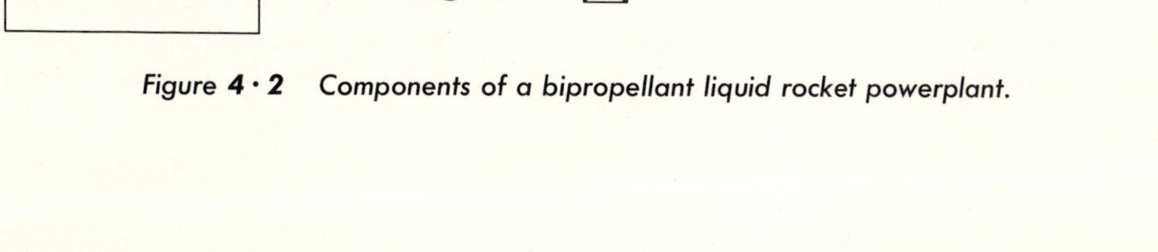

Figure 4 · 2 Components of a bipropellant liquid rocket powerplant.

will by the remote operation of the propellant control valves. This cannot be done with solid-propellant systems, because when they have once started, the fuel in the combustion chamber must continue to burn until it is exhausted.

The injection system through which the liquid propellants flow into the combustion chamber usually is designed to atomize the liquids as they enter in many pairs of impinging streams. To start a rocket engine, an electric igniter may be necessary for some propellants; however, propellants which ignite upon contact with each other are more desirable. These self-igniting propellants are called **hypergolic** propellants.

Pressurized liquid-propellant rocket powerplants

Figure 4 · 3 illustrates schematically the essential components of a pressurized liquid-propellant rocket powerplant. The system consists of a rocket engine, a remote-operated propellant control valve, fuel and oxidizer tanks, a remote-operated gas valve and pressure regulator, and a high-pressure inert-gas storage tank. This system is relatively simple compared to other pressurizing means; however, the duration of powerplant operation is limited by the fact that, for operations beyond the range of 30 to 45 sec, the weight of the tanks becomes excessive. In such a system, the fuel and oxidizer tanks must be able to withstand the feed pressure that is required. The pressurized system may be simplified further by replacing the remote-operated propellant control valve and the remote-operated gas-pressurizing valve with blow-out disks; however, no control for stopping and restarting the powerplant then exists. Nitrogen, helium, and occasionally air are employed as inert pressurizing gases.

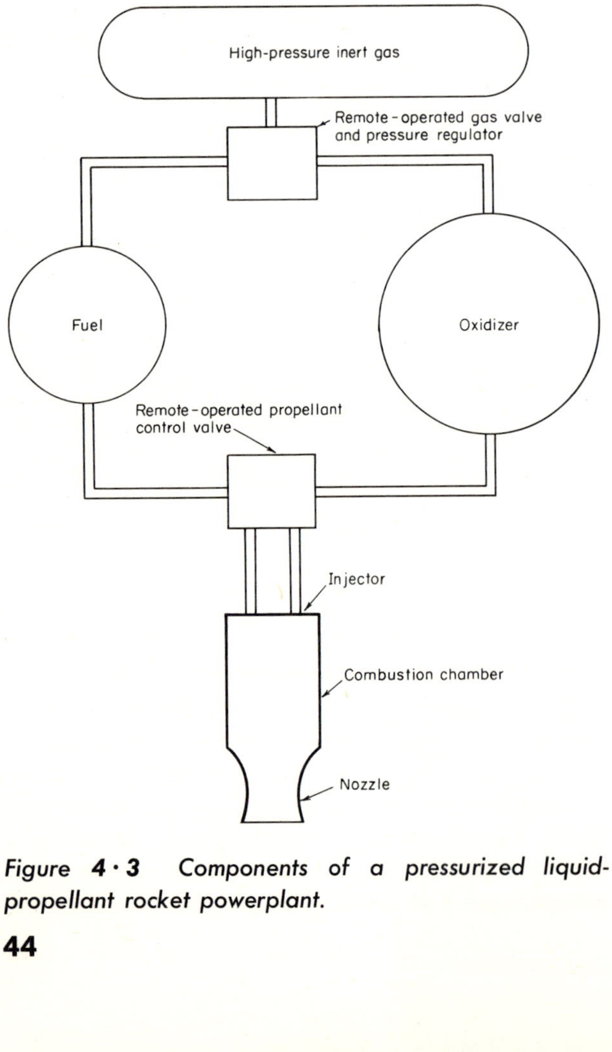

Figure ***4 · 3*** *Components of a pressurized liquid-propellant rocket powerplant.*

The pressurized liquid-propellant rocket powerplant has been used for RATO (rocket-assist takeoff) powerplants for aircraft. A rocket powerplant of this type is the Aerojet General Corporation YLR-13-AJ-5 developed for the United States Air Force. The unit was designed primarily for the B-45, B-29, and B-50. The pressurized liquid-propellant rocket powerplant also has been employed on several rocket-propelled test sleds in use by the USAF at Edwards Air Force Base and Holloman Air Force Base. One such sled employed for the high-speed testing of parachutes is shown in Fig. 4 · 4. The pressurized rocket powerplant for this sled generates 50,000 lb of thrust for over 4.5 sec duration and utilizes liquid oxygen and ethyl alcohol for propellants. The sled has reached velocities of over 800 mph on the 10,000-ft precision track at Edwards Air Force Base.

The weight limitations imposed by long-duration pressurized liquid-propellant rocket systems stimulated the development of pumping systems to raise the pressure of the propellants for injection into the combustion chambers.

Turbopump liquid-propellant rocket powerplants

Figure 4 · 5 illustrates schematically the essential components of a turbopump liquid-propellant rocket powerplant. The system consists of a rocket engine, a remote-controlled propellant control valve, fuel and oxidizer tanks, fuel and oxidizer pumps, and a gas turbine to drive the pumps. The fuel and oxidizer pumps are connected to a common shaft, which is driven by the gas turbine. The fuel and oxidizer pumps are required to withstand a relatively low pressure range of 25 to 60 psia, thereby ensuring positive suction feed pressure to the propellant pumps. The duration of a turbopump rocket powerplant is limited only by the amount of propellants and gases for the turbine. The powerplant may be started and stopped at will by simultaneous remote control of the propellant control valve and the gas supply to the turbine. The turbine ex-

Figure 4 · 4 Rocket-powered test sled.

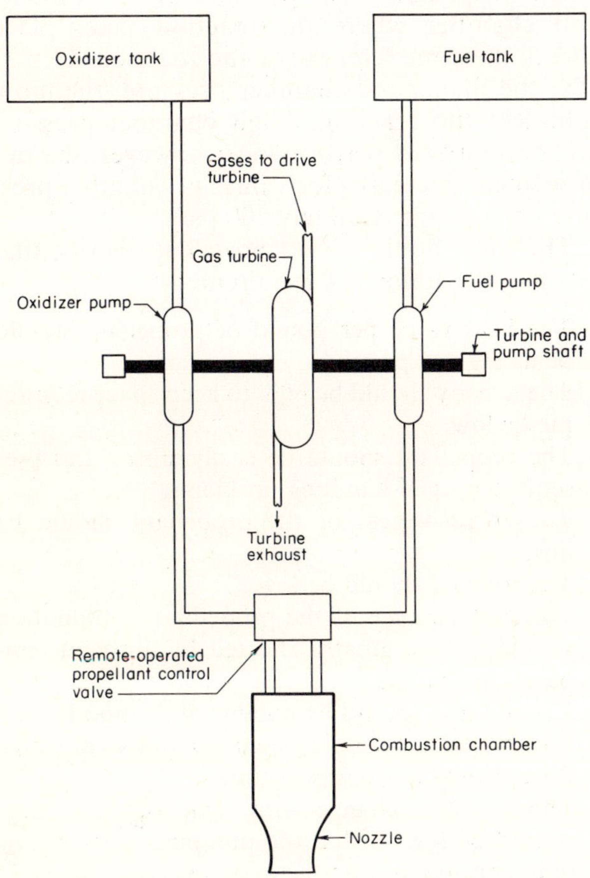

Figure 4 · 5 Components of a turbopump liquid-propellant rocket powerplant.

haust gases may be employed as a high-velocity jet stream for thrust augmentation, or in missile application they may be applied as a control for yaw and roll of the missile.

The design and construction of the propellant pumps, the turbine, and the connecting drive system usually are dependent upon the physical and chemical characteristics of the propellants and upon the operating conditions and configuration of the rocket engine. The chemical combination of the propellants determines the ratio of fuel to oxidizer, while the operating conditions and configuration of the rocket engine determine the total amount of propellants required and the pressure at which the propellants must be delivered to the combustion chamber. In some cases, different speeds may be required for the fuel and oxidizer pumps, and in this case a gear system may be used between the turbine and the pumps. In most applications, the pumps are of a centrifugal type.

Generation of gases for turbopump drive

Two different methods for generating high-temperature gases for the turbopump drive are illustrated in Fig. 4 · 6. At the top of the figure is a schematic illustration of a gas generator utilizing liquid rocket propellants to generate a high-temperature gas. The propellants react in a gas generator to form a hot gas similar to the re-

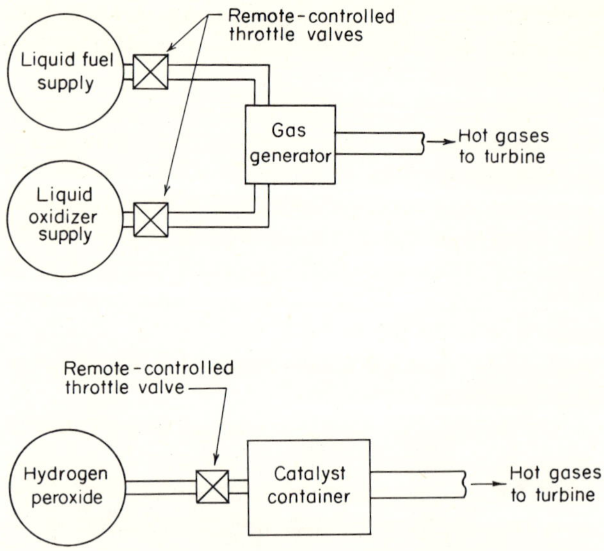

Figure 4·6 High-temperature gas-generating systems for operation of turbopumps.

action in the combustion chamber of the rocket engine. Normally, the fuel and oxidizer are diluted, or the mixture ratio is fuel-rich or oxidizer-rich to prevent excessive temperatures. At the bottom of the figure is a schematic illustration of a hydrogen peroxide gas generator. Concentrated hydrogen peroxide is reacted with a solid catalyst such as calcium or potassium permanganate, usually in the form of stones impregnated with the catalyst, or in the form of a coated screen, to produce high-temperature gases. The catalyst does not combine chemically with the hydrogen peroxide, but it does cause the hydrogen peroxide to break down into water and oxygen. The water is released in the form of steam. The temperature of the steam can be controlled by the strength of the hydrogen peroxide and the design of the catalyst container.

Throttle valves are employed to control the liquid flow to the reactant chamber to ensure a steady flow of hot gases for the turbine. In applications where multiple-rocket engines employ the same turbopump system, the throttle valve may be used to vary the flow of hot gases to the turbine, thus varying the pumping rate of the propellant pumps as different chambers are started and stopped.

Some liquid-propellant rocket powerplants have been experimentally tested employing a bipropellant gas generator firing directly into the fuel tank and another gas generator firing directly into the oxidizer tank to generate the pressure necessary to force the propellants into the combustion chamber. In other applications, solid-propellant charges have been utilized to generate high-temperature gases to pressurize the fuel and oxidizer tanks.

One unique turbopump drive system for a rocket powerplant employed in an aircraft for thrust augmentation utilizes an air bleed from the turbojet engine to drive the turbopump turbine. The rocket engine in this application is used primarily for thrust augmentation on takeoff.

Liquid propellants

The energy developed in the rocket engine for propulsion purposes is derived from the thermochemical energy of the propellants. Their chemical reaction, called "combustion," is accompanied by the evolution of large quantities of gases at high temperature. Since the fuels commonly employed are hydrocarbons, the products of combustion usually contain carbon dioxide (CO_2), carbon monoxide (CO), and water (H_2O) in the form of steam, as the principal constituents. The temperature attained by the reaction and the composition of the reaction products are influenced to a large extent by the mixture ratio of the propellants. The pressure in the combustion chamber where the reaction takes place affects the completeness of the reaction. Generally, the higher the chamber pressure, the more complete the reaction. High chamber pressure favors improved performance; however, the improvement decreases for values of chamber pressure above approximately 300 psia.

The ideal liquid rocket propellant is one that satisfies the following requirements:

1. The heat value per pound of propellant should be as high as possible.
2. The density should be high to keep space requirements low.
3. The propellant should be easily stored and present no special handling problems.
4. The corrosiveness of the propellant should be low.
5. The toxicity should be low.
6. The performance of the propellant combination should not be greatly affected by ambient temperature changes.
7. The ignition should be reliable and smooth.
8. The propellant should be stable and storageable for reasonable lengths of time.
9. The viscosity change with temperature change should be low so that the pumping work at low temperatures will not be excessive.

None of the known liquid rocket propellants satisfies all these requirements. Table 4·1 lists

Table 4 · 1 Common liquid rocket-propellant combinations and formulas

Oxidizer	Fuel
Liquid oxygen (O_2)	Ethyl alcohol (C_2H_5OH) Gasoline (C_8H_{18}) Kerosene Hydrazine (N_2H_4) Liquid hydrogen (H_2) Ammonia (NH_3)
Concentrated fuming nitric acid (red and white) (HNO_3)	Gasoline (C_8H_{18}) Hydrazine (N_2H_4) Kerosene Furfuryl alcohol ($C_4H_3OCH_2OH$) Aniline ($C_6H_5NH_2$)
Hydrogen peroxide (H_2O_2)	Ethyl alcohol (C_2H_5OH) Hydrazine (N_2H_4) Ethylene diamine ($C_2H_4N_2H_4$)

some of the more commonly used fuel and oxidizer combinations, together with their chemical formulas.

Liquid oxygen presents several problems, such as handling because of its extremely low temperature (−298°F) and the possibility of explosion upon impact with many hydrocarbons. The storage of liquid oxygen at this low temperature is difficult. Pumping problems are many because of the low temperature and the high vapor pressure. Liquid oxygen has been used in rocket engines for over 10 years, and techniques for the design of equipment and for handling have been largely solved.

Nitric acid is less hazardous than liquid oxygen or hydrogen peroxide, as far as storageability is concerned. The extreme corrosiveness of the liquid virtually requires that stainless steel (or, in some applications, certain aluminum alloys) be used exclusively for storage vessels. Many problems are involved in the selection of materials for valve seats, packing, etc. Most of the problems have been solved, and several reliable rocket powerplants are in use with nitric acid as the oxidizer.

Hydrogen peroxide, in concentrations of 80 to 90 per cent, was used extensively by the Germans during World War II, and work has continued in the United States since the end of the war. The liquid presents problems of storage and handling because of its unstable characteristics and its toxicity. Hydrogen peroxide has been adopted for use with a number of systems for generation of gas to drive the pump turbine. The British have successfully developed a unit employing hydrogen peroxide as the oxidizer. Of the three oxidizers mentioned, hydrogen peroxide is used the least.

Some testing has been accomplished in the liquid-propellant field utilizing monopropellants. A **monopropellant** is a liquid that requires no auxiliary material (oxidizer) for bringing about the release of the thermochemical energy. To this class of materials belong such explosives as nitroglycerine, trinitrotoluene, and nitromethane. For a liquid to be a satisfactory monopropellant for a rocket powerplant, it must satisfy the practical considerations mentioned previously. It must be stable under all conditions but must decompose completely once it has been introduced into the combustion chamber. These two requirements are normally conflicting and greatly restrict the choice of liquid monopropellants. Several monopropellants which have been tested to a certain degree of success are nitromethane, propyl nitrate, and hydrogen peroxide.

Multiple combustion chambers

It has been found in practice that the rocket engine is highly inefficient when throttled. Usually a rocket engine is designed for, and will satisfactorily operate only over, a small thrust range. Some rocket powerplants may be specifically adjusted for different thrust ranges up to approximately 50 per cent of the total thrust range; however, the adjustments must be made before each different operation and may require the substitution of some parts such as the injector. The rocket engine at present is considered a constant-thrust apparatus. Some testing and research work has been done in an attempt to develop a variable-thrust rocket powerplant, but no satisfactory results have been attained.

To solve the problem of obtaining different thrust levels from a liquid-propellant rocket powerplant, the rocket industry has turned to the use of multiple combustion chambers as one possibility. Figure 4 · 7 illustrates one application of a multiple-combustion-chamber liquid-propellant rocket powerplant. This powerplant is employed by the USAF at Edwards Air Force Base to propel a test sled on a 10,000-ft precision track. The system utilizes red fuming nitric acid as an oxidizer and a mixture of aniline and furfuryl alcohol as a fuel. Nitrogen is used as the pressurizing medium to force the propellants into the combustion chambers. Figure 4 · 8 shows the combustion chambers and associated parts of the sled powerplant. The engine utilizes five different combustion chambers, each rated at approximately 4,500 lb of thrust. Two of the

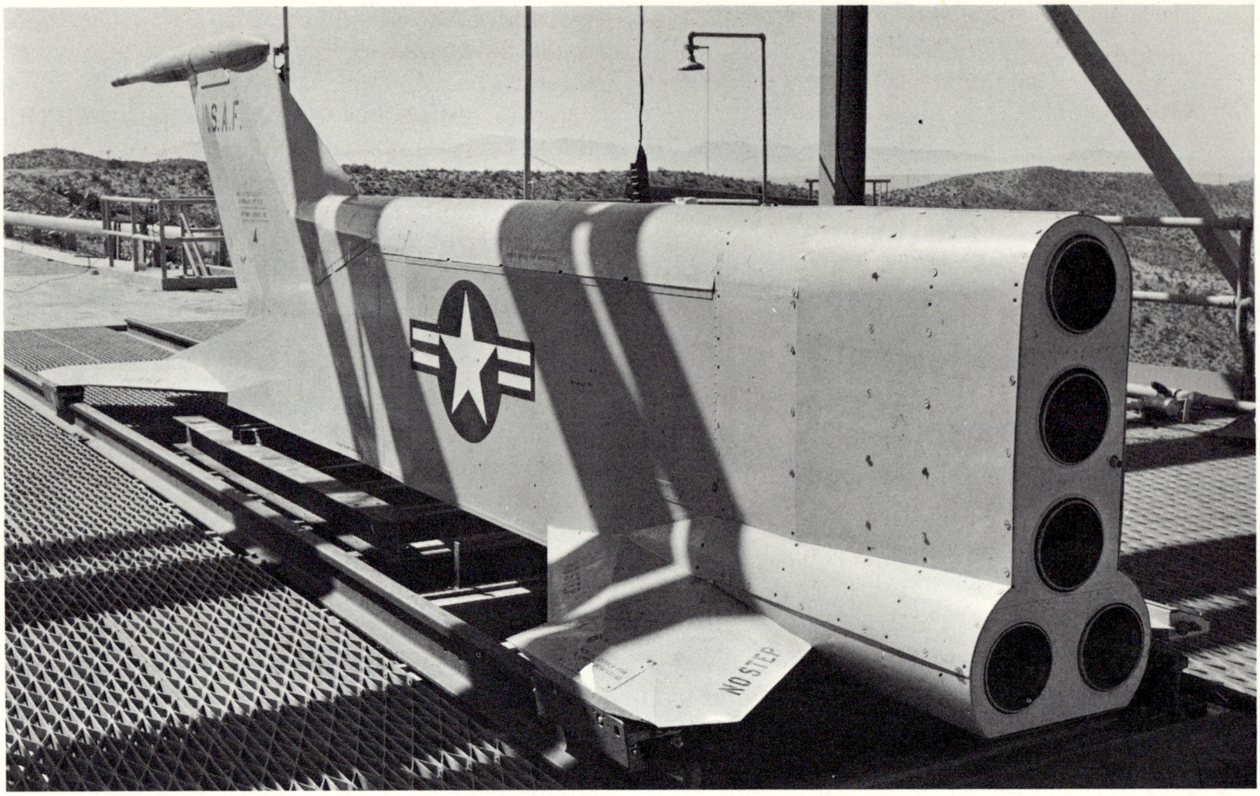

Figure 4·7 Multiple combustion chambers on rocket-propelled sled.

Figure 4·8 Multiple-chamber rocket engine for a test sled.

chambers are designed to operate for approximately 5 sec, and the remaining chambers are designed to operate for about 8 sec. All chambers operate upon the start of the powerplant operation. Any combination of chambers and any duration of operation up to maximum time available from the propellant supply may be selected for operation. The variable thrust is required to permit extensive variation in the acceleration and maximum velocity of the vehicle. The sled has attained a maximum velocity of over 825 mph on the 10,000-ft track.

Another application of the multiple-combustion-chamber idea is illustrated in Fig. 4·9. In this instance the purpose is to multiply thrust. This is a photograph showing the cluster of Rocketdyne H1 engines used as booster power for the Saturn I space research vehicle. This cluster of eight engines produces a thrust of 1,500,000 lb.

The H1 engine is an outgrowth of earlier engines used for the Thor and Jupiter IRBMs (intermediate-range ballistic missiles) and was chosen because of its availability, reliability, and simplicity. Each of the eight H1 engines comprising the Saturn I booster system operates independently. Each has its own thrust chamber, gas generator, and turbopump. The propellants used are liquid oxygen and a kerosene fuel designated RP-1 which are fed to all engines from common sources.

The four outer engines of the booster cluster are swiveled, and their movements control the direction of the vehicle's flight. Each of the outer engines is mounted on a gimbal bearing and is moved by hydraulic arms actuated by the vehicle's guidance system. The four inboard engines are mounted in a fixed position.

The H1 engine has a thrust chamber formed of thin light-weight metal tubes through which fuel is forced at high speed before it enters the combustion chamber to be burned. In this way the fuel serves to cool the thrust chamber and prevent it from melting under the extremely high temperatures of combustion.

Fuel consumption and efficiency

The efficiency of a rocket engine is determined by the **specific impulse** of the engine which, in effect, is the number of pounds of thrust developed for every pound of total propellant consumed per second. Multiplying the weight of the fuel consumed per second by 32.2 ft per sec per sec, we obtain the **effective jet velocity,** which is the average velocity of the escaping gases. At the present time, velocities of 7,000 ft per sec have been obtained repeatedly, and under certain special conditions, jet velocities as high as 8,500 ft per sec have been indicated. The fuel-consumption rate at a jet velocity of 7,000 ft per sec would be approximately 0.0044 lb per sec per pound of thrust. To compete with the conventional reciprocating engine in miles per pound of propellant used, the rocket engine would have to travel at speeds in the neighborhood of 3,000 mph. It is apparent that the rocket engine is an inefficient propulsion device at speeds comparable to those currently being attained by fighter-type aircraft.

Application of liquid-propellant rocket engines in aircraft

Liquid-propellant rocket engines have a definite and proved place in the aircraft propulsion field. In this country, the main application has been to power research aircraft beginning with the X-1 series and continuing to the Dyna Soar X-20. This is an airplane type vehicle by which astronauts may return from space flights and reenter the earth's atmosphere under the positive control of a pilot.

Liquid-propellant rockets have also been applied to aircraft for the purpose of rocket-assisted takeoff. They can be mounted as fixed installations in the aircraft or in such a manner that they can be dropped by parachute after use. Reliable liquid-propellant rocket engines have been developed that operate without difficulty for continuous periods of several minutes. These

Figure 4 · 9 Cluster of Rocketdyne H1 rocket engines.

engines are light in weight and normally are cooled by circulating one of the propellants, usually the fuel, around the combustion chamber and nozzle. One advantage of the liquid-propellant rocket engine over the solid-propellant rocket engine is that the takeoff may be refused by stopping the liquid-propellant engine. Once the solid-propellant engine is started, the aircraft is committed to takeoff.

In addition to thrust augmentation for an aircraft during takeoff, the liquid-propellant rocket powerplant may be operated in flight to obtain "superperformance." This superperformance for short periods of time is particularly applicable to the interceptor type of fighter aircraft.

Several different types of liquid-propellant rocket engines have been employed as powerplants for aircraft. During World War II the Germans developed and tested a rocket-propelled interceptor airplane designated the ME-163 using concentrated hydrogen peroxide as the oxidizer and a mixture of hydrazine hydrate plus methyl alcohol as the fuel. The airplane was successful in defense against high-flying bombers; however, they were too few in number to affect the outcome of the war.

The X-15 research airplane, designed and built by North American Aviation, Inc., and operated by NASA, utilizes a multichamber rocket engine arrangement for propulsion and at present holds the altitude and speed records for manned aircraft. This airplane has been described as the most effective research aircraft ever built because its use has yielded an abundance of vital information relative to flight in the outer atmosphere and the reentry of manned vehicles from space. When operating in the upper regions of the atmosphere, the X-15 cannot be controlled aerodynamically; hence, small reaction jets are installed at points where they can apply corrective forces. These jets are operated through the normal cockpit controls of the airplane after the aerodynamic control surfaces become ineffective.

Application of liquid-propellant rocket engines in missiles

The liquid-propellant rocket powerplant is almost ideally matched for ballistic-missile application and, to a certain degree, for other types of missiles. The powerplant requirements for a missile have two important and stringent conditions:

(1) the powerplant must not require oxygen from the atmosphere, since the powered phase of the missile flight normally is required to continue to extreme altitudes; and (2) the missile must have tremendous power, particularly at the start of the flight. The rocket powerplant meets both these requirements and is the only practical powerplant that will meet them. A general idea of the tremendous power requirements for missiles and spacecraft may be obtained from the fact that powerplants for such vehicles produce thrusts of from a few thousand pounds up to 1,500,000 lb. The Saturn V booster for the Apollo moon explorer requires five of the 1,500,000-lb of thrust F1 rocket engines.

One of the most spectacular military applications of the rocket principle was the German V-2, which first began to rain down upon England in the late summer of 1944. A simplified drawing of this missile is shown in Fig. 4 · 10. The V-2 was powered by a rocket engine utilizing liquid oxygen and ethyl alcohol. The liquid oxygen and alcohol were carried in tanks, located amidships, which contained about 7,500 lb of alcohol and 11,000 lb of oxygen. A turbine, located just below the liquid-oxygen tank, was driven by superheated steam produced by reacting hydro-

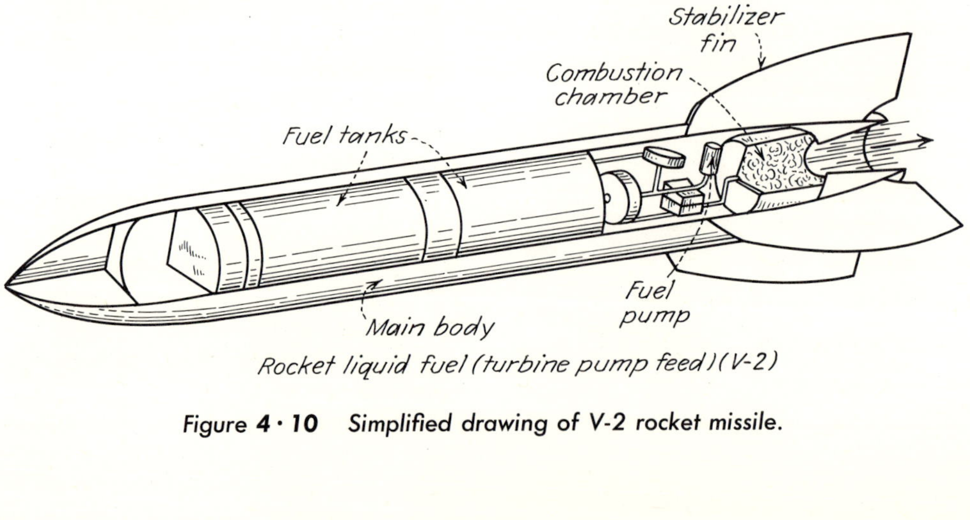

Figure 4 · 10 Simplified drawing of V-2 rocket missile.

gen peroxide with calcium permanganate. The turbine drove the pumps that forced the liquid oxygen and alcohol into the combustion chamber.

Electric ignition, remotely controlled from the launching platform, started the combustion of the prestage which constituted approximately 10 per cent of the full thrust of 52,000 lb. After several seconds of prestage combustion, main-stage operation was remotely started. The path of the projectile as it left the launching platform was almost vertical. Gyroscopic devices, located just behind the warhead, operated the control surfaces and caused the rocket to curve from the vertical toward the target. The range of the rocket was controlled by presetting an automatic fuel cutoff. By controlling the length of time during which the fuel burned, the height of the trajectory was determined. Thus the range was calculated in much the same manner that the student in sophomore mathematics determines the course of a mythical projectile that flies along a parabolic path through resistanceless space. The V-2, because of its tremendous starting thrust, reached a maximum speed of approximately 4,000 mph and, accordingly, passed through the earth's atmosphere and into virtually empty space in the few seconds that elapsed before the fuel was exhausted. Since the rocket was pointing upward at about 45° when the fuel was cut off, it continued along a parabolic trajectory, achieving a maximum altitude of approximately 68 miles. Then, under the influence of gravity, it rushed toward the earth at a speed that may have exceeded 1 mile per sec. The V-2 traversed the entire course (some 200 ground miles maximum) in approximately 5 min. As the rocket descended through the atmosphere from its maximum elevation, it was slowed down to about 2,000 mph by the resistance of the air. At the same time, it was heated in much the same manner as a meteorite that enters the earth's atmosphere. Observers who survived the arrival of a V-2 in their immediate vicinity have noted that it appeared to glow a dull red. Some trajectory and flight characteristics are listed below:

Initial acceleration	1 g
Duration of powered flight	71 sec
Velocity at end of powered flight	5,300 ft per sec
Height at end of powered flight	22 miles
Maximum height reached	68 miles
Velocity at top of trajectory	3,900 ft per sec
Time to reach top of trajectory	193 sec
Horizontal distance traversed	200 miles

Research and development on liquid-propellant rocket engines have been conducted continually in this country and in other countries since World War II. Literally hundreds of engine-vehicle combinations have been tested operationally for weapon systems, space research, and space travel. Some combinations of engines and vehicles, such as Thor, Jupiter, Redstone, Atlas, and Titan, have been used for both military applications and space research. The Atlas was used extensively in the Mercury project for the orbiting of astronauts in space capsules.

At the present time the principal strategic missiles in operation for the United States are the Atlas, Titan, Polaris, and Minuteman. Of these, the Atlas and Titan are equipped with liquid-propellant engines. Although solid-propellant rocket engines have replaced liquid-propellant engines in many applications because of the comparative simplicity of the solid-propellant engine, it is certain that liquid-propellant engines will continue in use wherever flexibility of thrust control is necessary.

Liquid-propellant engines for space flight

We shall consider briefly the engines employed for the Saturn-Apollo project as examples of liquid-propellant engines used for space flight. The engines that power the Saturn V booster and space-vehicle combination are shown in Fig. 4 · 11. Three of the four engines pictured, the H1, J2, and F1, were developed and built by the Rocketdyne Division of North American Aviation, Inc. The F1 is used in a cluster of five to provide 7,500,000 lb of first-stage thrust. The hydrogen-fueled J2 powers the second stage in a cluster of five engines and is used singly in the third stage. The RL10 is produced by the Pratt & Whitney Division of United Aircraft Corporation and is used in the upper stage of Saturn I.

As mentioned previously, the Rocketdyne F1 engine delivers a thrust of 1,500,000 lb. Its propellants are liquid oxygen (LOX) and kerosene (RP-1). The engine is similar in basic design to many of the previously developed smaller engines. The fuel and oxidizer are pumped to the combustion chamber by means of a turbopump which utilizes the rocket propellants for fuel. The thrust-chamber design provides a tubular wall for regenerative cooling, an extension skirt, a double-inlet oxidizer dome, and a flat-face injector. The wall construction of the thrust chamber is seen in Fig. 4 · 12.

The combustion chamber is designed to flow 2 tons per sec of liquid oxygen and 1 ton per sec of fuel and to burn these propellants at a pres-

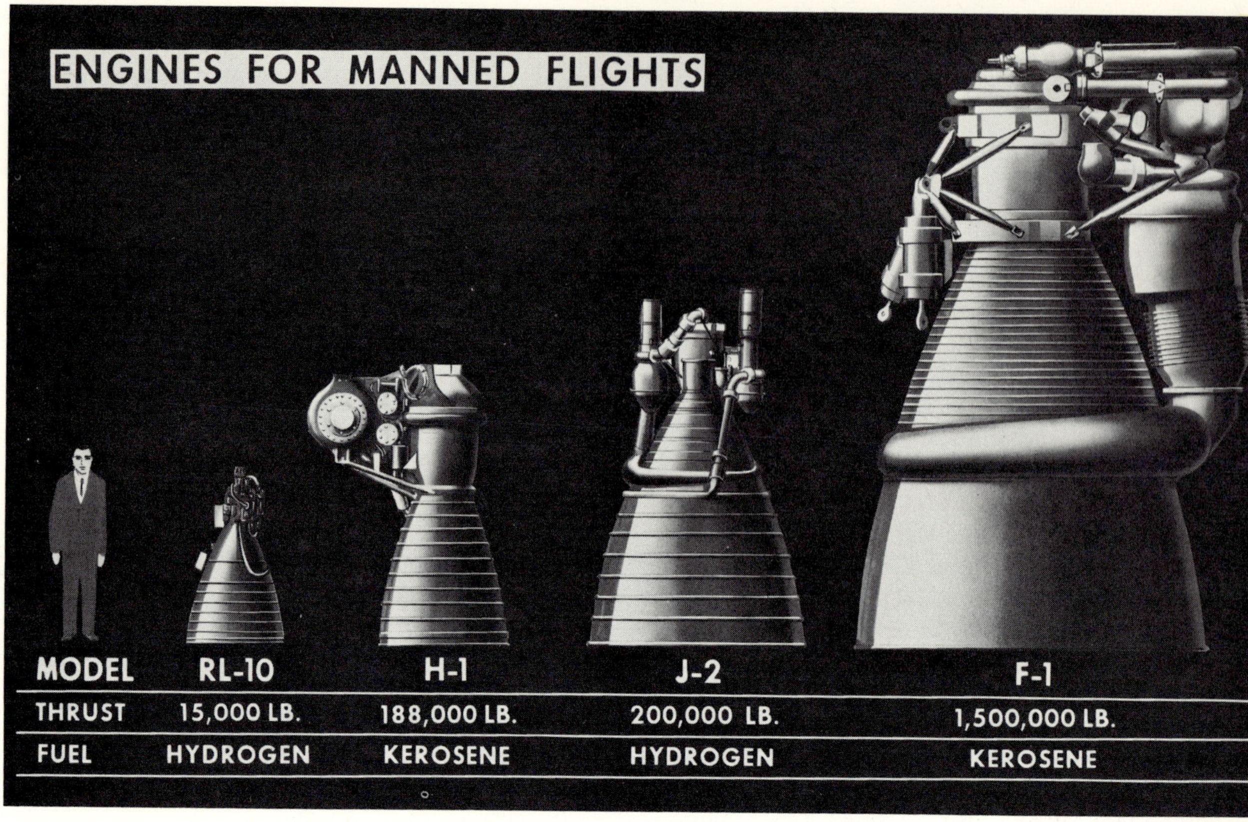

Figure 4·11 Rocket engines for Saturn V.

Figure 4·12 Wall construction of F1 thrust chamber.

sure of approximately 1,150 psi at the injector face. The fuel-cooled portion of the thrust chamber extends to a 10:1 expansion area ratio. Its approximate measurements are:

Combustion chamber diameter	40 in.
Nozzle exit diameter	9½ ft
Length	11 ft

To meet the requirements of expansion area ratios up to 16:1, a skirt is attached to the thrust chamber. This skirt is cooled by turbine exhaust gases from the turbine which drives the turbopumps flowing through the double walls at temperatures of 480 to 650°C. Although these temperatures may seem high, they are much cooler than the main exhaust of the engine.

The material used for the tubular walls of the thrust chamber is a high-nickel alloy. The strength property of this alloy permits design of thinner wall section tubes, resulting in reduced weight. The thinner walls also provide a lower gas side-wall temperature even though their thermal conductivity is lower than that of some other materials considered. The tubes are hydraulically formed and furnace-brazed.

The thrust-chamber liquid-oxygen dome is

provided with a dual inlet for improved liquid-oxygen distribution to the injector. The dome material of high-nickel alloy was selected for high strength because the dome is the focal point for the transmission of the thrust through the gimbal bearing to the vehicle.

The turbopump consists of a gas-turbine engine and two direct-driven centrifugal pumps. The fuel pump is placed between the turbine and the oxygen pump to separate the elements having the greatest temperature extremes. The turbine design is two-stage and velocity-compounded and consists of two rotating impulse wheels separated by a set of stationary impulse stators. A light, strong aluminum-alloy casting was selected for the volutes, impellers, and inlets which have operating pressures as high as 2,000 psia. For the turbine wheels and manifold a nickel alloy having high strength at elevated temperatures is used to provide adequate strength for high operating pressures and temperatures. The turbopump assembly weighs approximately 2,500 lb and delivers 60,000 hp. To do this, the turbine consumes about 2.5 per cent of the total engine propellants.

The feed of propellants to the engine is controlled by two poppet-type fuel valves and two poppet-type oxidizer valves. These valves are designed to minimize sealing areas of each of the valves and to provide valving in locations that allow a better distribution of propellant feed to the thrust chamber. The actuators for the main propellant valves and the gas generator valves utilize engine fuel as a hydraulic fluid. This system avoids the installation of a pneumatic system in the engine and makes use of a fluid which is already available in the engine at the required pressures.

The other major control components for the engine include dual-linked valves, gas-generator controls, and a four-way solenoid valve. The valves were selected for the gas generator to provide fast and repeatable actuation while minimizing the pressure drop and valve-seal forces.

The four-way solenoid valve controls the sequencing of the engine system by regulating the flow of fuel to the actuation mechanism of the main propellant and gas-generator valve. This valve and the spark-ignition system for the gas generators are the only engine components requiring electric energy.

The starting of the engine is automatically sequenced, beginning with the gas-generator spark igniter and followed by the electric actuation of the four-way solenoid valve. Movement of this valve puts into motion the **pressure ladder** of a start-sequencing system whereby each resulting action is linked to a necessary prior event. After the gas generator starts at a preset pressure level, an ignition source in the form of a **hypergolic** fluid is introduced into the thrust chamber, followed by the opening of fuel valves and the accomplishment of main-propellant ignition in the thrust chamber. The hypergolic fluid automatically ignites when it comes in contact with liquid oxygen.

Liquid rocket formulas

In computing the values for **thrust, specific impulse,** and **effective exhaust velocity,** we employ formulas through which the unknown quantities can be determined when other quantities are known. For example, if the pressure within a rocket thrust chamber and the area of the nozzle throat are known, the thrust produced by the rocket engine can be obtained from the **general thrust formula.**

$$F = P_c A_t$$

where F = force, lb
P_c = pressure inside the combustion chamber
A_t = area of nozzle throat, sq in.

Because the thrust nozzle or expansion nozzle has a decided effect on the thrust of a rocket engine, we must provide a factor which includes this effect. For this purpose a **thrust coefficient** C_f is employed with the preceding formula. The formula then becomes

$$F = P_c A_t C_f$$

To arrive at a more precise measurement of rocket thrust, the basic formula for **momentum thrust** is used. That is,

$$F = ma$$

or force is equal to the product of mass and acceleration in appropriate units.

For the rocket engine we use $\dot{m}$ to represent the propellant mass flow in **slugs** (1 slug = 32.2 lb at sea level) per second and v_e to represent acceleration. Actually, v_e is the theoretical exhaust velocity, but it also represents acceleration because the propellant velocity is zero in the propellant tanks. The formula then becomes

$$F = \dot{m} v_e$$

In addition to the foregoing formula we must include another factor which affects thrust values. This is the difference in exit pressure and ambient pressure. The force exerted by this difference in pressure is dependent upon the area of the nozzle exit (A_e). The formula then becomes

$$F_2 = (p_e - p_a)A_e$$

where p_e = exit pressure
p_a = ambient pressure
A_e = exit area, sq in.

The **fundamental thrust equation** then becomes

$$F = \dot{m}v_e(p_e - p_a)A_e$$

This equation is also expressed

$$F = \frac{\dot{W}}{g} v_e(p_e - p_{am})A_e$$

In the latter formula $\dot{W}$ (pronounced W dot) represents propellant mass flow in **pounds per second.** When this value is divided by g (32.2 or acceleration of gravity), the mass in slugs ($\dot{m}$) is obtained.

As mentioned previously, the performance of a rocket engine is conveniently expressed as **specific impulse** in seconds. The actual quantity represented by specific impulse I_s is pounds of thrust produced for each pound of propellant consumed per second. The formula is simply

$$I_s = \frac{F}{\dot{W}}$$

where I_s = specific impulse
F = thrust in lb
$\dot{W}$ = weight of fuel consumed per second in lb

From the preceding formula **effective exhaust velocity** is derived. This is expressed

$$c = I_s g$$

It is possible to derive and express many additional formulas for rocket performance; however, for the purpose of this text, the foregoing are considered adequate.

SOLID-PROPELLANT ROCKETS

Construction

As mentioned previously, solid-propellant rockets have replaced liquid-propellant rocket engines in a variety of applications. They are used extensively to power rocket projectiles, guided missiles, and spacecraft boosters, and also as boosters for aircraft. A typical solid-propellant rocket engine or booster is illustrated in Fig. 4 · 13. As shown in the drawing, the complete assembly includes the **chamber, liner, igniter assembly, spring, propellant grain,** and **nozzle.** The entire mass of the solid propellant is called the "grain." The grain contains both oxidizer and fuel and is burned so that the propellant is consumed in a direction normal to (perpendicular to) the burning surface. The grain material, size, geometrical form, and exposed burning surface are the determining factors influencing chamber pressure, thrust, and duration.

Solid propellants

There are two principal types of solid propellants: the composite propellant, consisting of an inorganic oxidizer such as potassium perchlorate or sodium nitrate dispersed in a mixture of organic plastic, and the double-base propellant, consisting of the colloid of nitroglycerin-nitrocellulose.

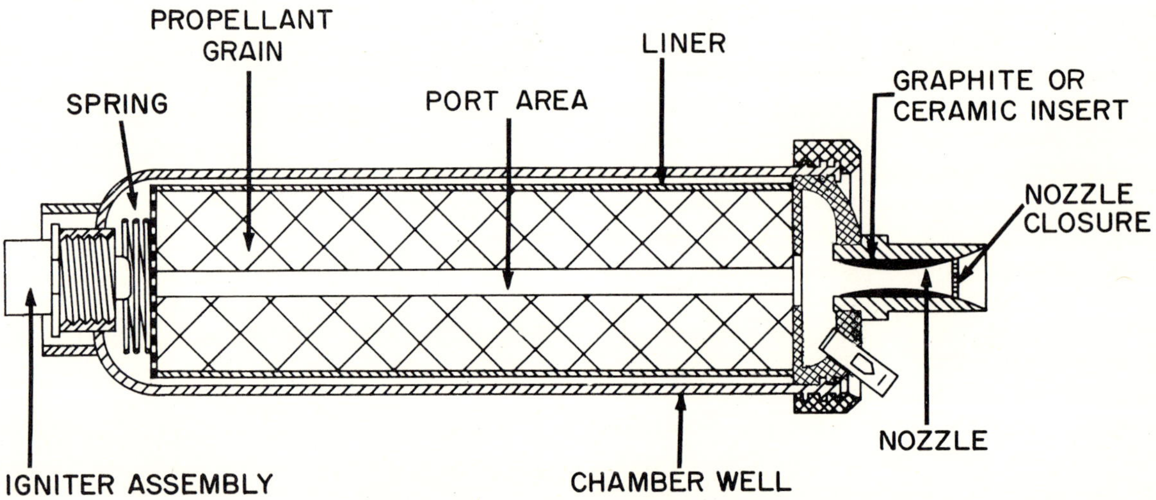

Figure 4 · 13 Solid-propellant rocket engine.

Oxidizers and monopropellant fuels are derived from several sources. Some of the more common oxidizer crystals used in composite propellants are from the perchlorate and inorganic nitrate families as indicated previously. These crystals are dispersed in fuels which have a base of asphalt-oil, plastic, or rubber. Some of the common double-base propellant compositions are from the organic nitrate family.

In addition to the chemicals which make up the major ingredients, a small percentage of additives are used to modify the burning rate, increase stability, and alter the physical properties. The use of additives makes possible the preparation of homologous propellants whose ballistic and physical properties vary over a wide range.

The grain configuration for solid-propellant rocket engines varies according to the required thrust-time program for the engine. The rate of burning is controlled by the configuration of the grain as shown in Fig. 4 · 14. Each of these forms produces a different burning characteristic.

Jet-assist takeoff (JATO)

The use of added thrust supplied by a jet (rocket) motor for a limited period of time has been termed **jet assistance** or **rocket assistance.** Thus the expression JATO or RATO was coined. The term is used to describe the application of rocket power to help in launching airplanes or other types of aircraft.

Model designation

Undoubtedly the most successful of takeoff assistance units for aircraft is the JATO solid-propellant rocket developed by Aerojet General Corp., of Azusa, Calif. A typical model of this unit is designated as Aerojet Model 14AS-1000D-4. The model number may be interpreted as follows:

14	duration of thrust, in seconds, at 60°F
A	Aerojet General Corporation
S	solid-propellant charge
1000	thrust, in pounds, at 60°F
D	basic modification of original design
4 or 5	minor change in D modification

General description

A JATO rocket unit, Model 14AS-1000D-4, is illustrated in Fig. 4 · 15. The unit is of basically simple construction, and since it contains no moving parts, it requires no lubrication or maintenance procedures. Essentially, the unit consists of a solid-propellant cartridge charge contained within a cylindrical steel chamber assembly to which attachment fittings are welded. The forward end of the chamber is closed by a cap which is equipped with a ring stand, and the aft end contains a nozzle assembly. Adjacent to the aft end of the chamber assembly are two bosses; one contains the igniter assembly, and the other contains the safety pressure-release assembly.

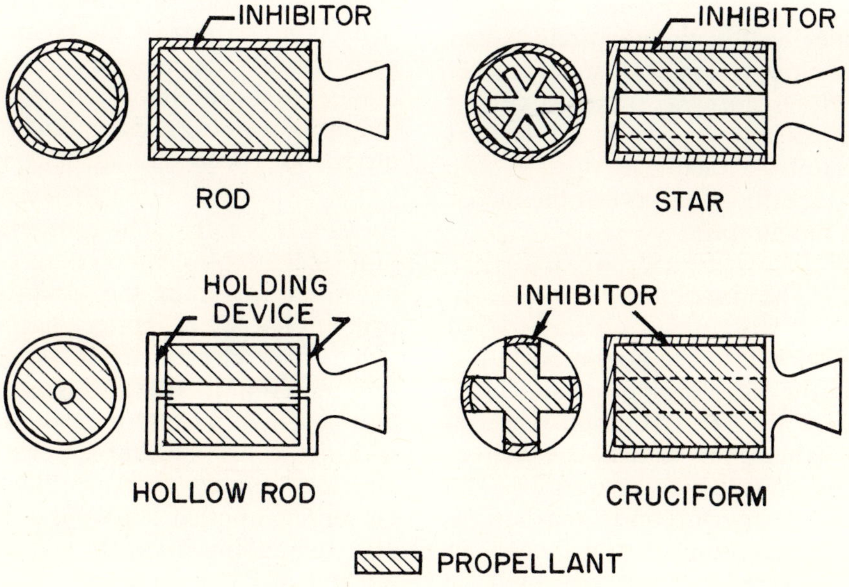

Figure 4 · 14 Grain configurations.

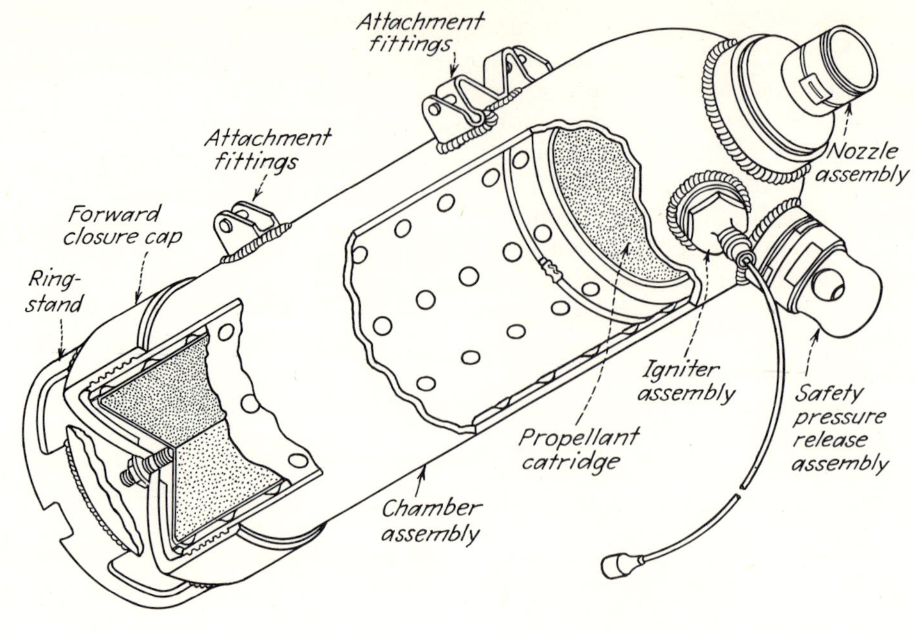

Figure 4 · 15 A JATO rocket unit.

Detailed description

Chamber Assembly. Although the maximum operating chamber pressure of a JATO rocket unit approximates 2,650 psi at 130°F, chambers of these motors are designed to withstand pressures of 5,500 psi or more in order to provide a substantial safety factor. The chamber assembly consists of a two-piece pressure vessel with a chamber body and a forward closure cap. The chamber body is fabricated from seamless steel tubing and is threaded at its forward end to engage the closure cap. The aft end of the chamber body is drawn into a hemispherical shape, and three threaded bosses are welded into the openings in the chamber wall in the positions shown in Fig. 4 · 15. The closure cap consists of a steel dished head to which the ring stand is welded to facilitate handling and to provide for vertical storage of the unit. Attachment fittings are welded to the chamber body to permit the motor to be attached to the aircraft.

Nozzle Assembly. The nozzle assembly is constructed of steel and is threaded on its forward end to permit insertion into the nozzle boss. The throat section of the nozzle is protected from the erosive action of the exhaust gases by a carbon insert, thus enabling this critical component to withstand many consecutive operations. The diameter of the nozzle aperture in the carbon insert is determined and preset at the factory so that the thrust delivered by the unit will be within the specified limits. A thin copper burst diaphragm, which fragments at the start of rocket operation, is provided in the nozzle assembly to seal the motor prior to its use and also to serve as an aid to prompt and full ignition. This seal assures that there will be no fuel contamination.

Safety Pressure-release Assembly. JATO rockets are equipped with a safety pressure-release assembly which consists of a copper burst diaphragm in an adapter threaded into the boss provided for it in the chamber wall. The diaphragm is designed to rupture at a pressure between 3,200 and 3,600 psi. This is low enough so that there is no danger of the chamber's bursting before the diaphragm ruptures. Also, it is sufficiently high so that there is no danger of the diaphragm's rupturing unless the propellant charge has been damaged.

Igniter Assembly. The igniter assembly consists of a steel adapter, which is screwed into the boss provided for it in the chamber body wall; a primer charge; an igniter charge; and an electric cable. The electric cable is a single-lead wire terminated at its free end in a snap-on connector and at the other end in a nichrome wire section within the primer charge. The standard aircraft electrical supply source of 12 to 24 volts direct current is applied across the cable and ground, and, depending upon the value of the voltage, a current of 25 or 50 amp is drawn by the igniter for 0.032 to 0.110 sec. This current fuses the

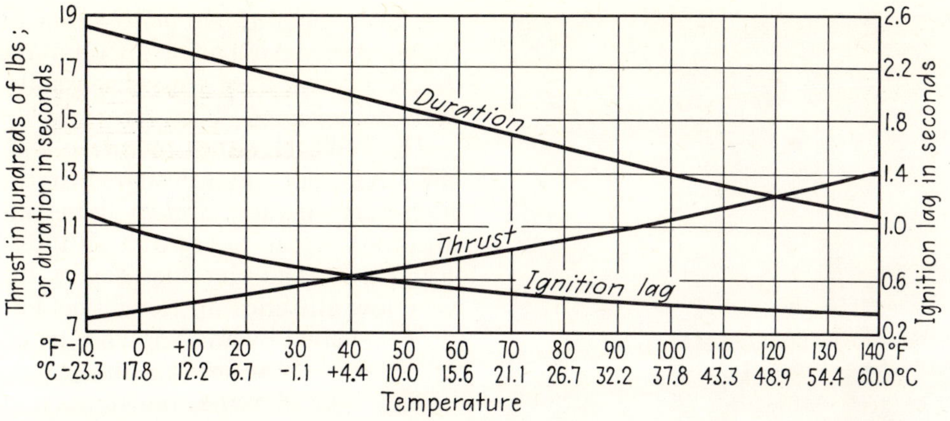

Figure 4 · 16 Effect of temperature upon performance of JATO rocket engine.

nichrome wire and causes the primer charge to burn, thus raising the temperature of the igniter charge to its autoignition value of 575°F. Although ignition of the primer charge causes the propellant to burn, such burning is below the normal rate, and so appreciable gas pressure is generated within the chamber. It is the burning of the igniter charge that raises the pressure and temperature within the chamber to the point at which the propellant commences to burn normally. When this occurs, the rocket is in operation.

Propellant. Standard rocket propellant for the Aerojet Model 14AS-1000D-4 rocket contains an oxidizer and a fuel blended in the proportions necessary to provide complete combustion. The oxidizer is potassium perchloride ($KClO_4$), which is a slightly hygroscopic, white, crystalline solid. A mixture of two asphalts and a paraffin-base motor oil is used as the fuel. The propellant resembles tar or asphalt and has a specific impulse of 165 lb per pound of propellant per sec. The autoignition temperature is in excess of 750°F.

Figure 4 · 17 A Nike Hercules missile battery.

Figure **4 · 18** *Nike Hercules missile.*

Nominal performance

Aerojet Model 14AS-1000D-4 rockets are designed to deliver a thrust of 1,000 lb for a duration of 14 sec when operated at an ambient temperature of 60°F. The total impulse, which represents the power derived from operation of the rocket, is obtained by multiplying thrust by duration and is, therefore, 14,000 lb per sec.

As indicated in Fig. 4 · 16, duration and thrust vary with temperature. However, operation of the JATO rocket at any temperature within the range of 0 to 130°F will provide a total impulse which is a close approximation of the nominal total impulse stated.

MISSILES POWERED BY SOLID-PROPELLANT ROCKET ENGINES

The Nike Hercules

Typical of missile applications of solid-propellant rocket engines is the Nike Hercules air defense missile manufactured by the Douglas Aircraft Company. The Nike Hercules missile has been operational with units of the United States Army Air Defense Command since 1958. A battery of these missiles ready for firing is shown in Fig. 4 · 17.

The Nike Hercules missile is the successor to the Nike Ajax and is designed for defense against high-performance aircraft flying at a variety of altitudes. It can operate at altitudes in excess of 150,000 ft. The Hercules covers the range from very low altitudes up to ceilings far greater than those reached by any known type of manned aircraft presently in use for military purposes.

The Nike Hercules booster is a cluster of four solid-propellant rocket engines, shown in Fig. 14 · 18, and the sustainer engine for the second stage is a single solid-propellant rocket. The weight of the missile at launch is 10,000 lb.

The Nike Zeus

The Nike Zeus missile, shown in Fig. 4 · 19, is a surface-to-air type designed to intercept and destroy enemy ICBMs before they reach their intended targets. The Zeus carries a nuclear warhead, thus enabling it to destroy an enemy missile even though it does not make physical contact.

The power system for the Nike Zeus missile

Figure **4 · 19** *Nike Zeus missile.*

Figure 4 · 20 Launch sequence for the Nike Zeus missile.

consists of three stages of solid-propellant rockets which produce an eventual warhead speed of many times the speed of sound. This speed is essential in the interception of the hypersonic ICBM warheads.

The Zeus is controlled by automatic command guidance. When the acquisition radar locates a target, the missile is fired. The target track radars then take control and pinpoint the target, furnishing continuous, precise trajectory information to the computers to determine the intercept point. The missile is automatically guided to arrive at this point simultaneously with the target. The launch sequence for the Nike Zeus missile is shown in Fig. 4 · 20.

STUDY QUESTIONS

1. Rocket engines are divided into two general categories. Describe each briefly.
2. What action generates thrust in a rocket engine?
3. How may a rocket combustion chamber and nozzle be cooled?
4. What is meant by "regenerative cooling"?
5. What are the essential components of a bipropellant liquid rocket powerplant?
6. Describe two methods by which liquid rocket propellants may be forced into the rocket combustion chamber.
7. What pressures exist in the combustion chamber of a liquid-propellant rocket when firing?
8. What is meant by "hypergolic" propellants?
9. What method is used to drive the turbopump in a rocket propellant system?
10. Explain the function of the two components of a bipropellant fuel mixture.
11. What is a monopropellant?
12. What is the specific impulse and the effective exhaust velocity for the Rocketdyne F1 engine? (Use formulas.)
13. Describe a solid-fuel rocket, naming the principal components.
14. What is the **grain** of a solid-propellant rocket?
15. What is the principal advantage of solid-propellant rockets for missiles such as the Nike Hercules and Zeus?

CHAPTER 5

Athodyds—Ramjets and Pulsejets

Although the greater portion of this text deals with aviation gas turbines, there is another group of jet-propulsion powerplants which play an important part in the operation of aircraft, especially aircraft of the guided-missile type. This group of jet powerplants is called the **athodyd** group. The word "athodyd" is derived from the words "AeroTHermODYnamic Duct."

Briefly, the athodyds differ from the gas-turbine group in that the turbine wheel and compressor assembly are omitted. In the case of the **ramjet** engine, compression is supplied merely by ram pressure as the engine is moved forward at a very high velocity. In the **pulsejet,** air intake is obtained by reducing the pressure in the combustion chamber momentarily to induce a fresh charge of fuel-air mixture to enter the combustion chamber.

THE RAMJET

General description

The ramjet engine, shown in Fig. 5 · 1, is an air-breathing engine which operates on the same principle as the turbojet engine. Its basic operating cycle is similar to that of the turbojet. It compresses the incoming air by ram pressure, adds heat energy at a high pressure, converts the heat energy to velocity, and produces thrust. By converting the kinetic energy of the incoming air into pressure, the ramjet is able to operate without a mechanical compressor. Therefore, the engine requires no moving parts and is mechanically the simplest type of jet engine which has been devised. Since it depends on the velocity of the incoming air for the needed compression, the ramjet will not operate statically, that is, when it is not moving. For this reason, it requires a turbojet or rocket assist to accelerate it to operating speed.

At supersonic speeds the ramjet engine is capable of producing very high thrust with high efficiency. This characteristic makes it quite useful on high-speed aircraft and missiles, where its great power and low weight make flight possible in regions where it would be impossible with any other powerplant except the rocket. Ramjets have also been used at subsonic speeds where their low cost and light weight could be used to advantage.

At supersonic speeds the power required for aircraft flight becomes enormous, and the amount of power that can be produced per pound of engine weight becomes vital. The ramjet produces more than 100 hp per pound of engine weight at supersonic speeds. It is exceeded in this respect only by the rocket, and it is much more economical in the amount of fuel burned. For this reason, ramjets are used where boost up to "takeover" speed can be provided by rockets or turbojet engines. Guided missiles and high-supersonic-speed piloted aircraft are typical ramjet applications.

Principles of operation

The ramjet consists of a diffuser, fuel injector, flame holder, combustion chamber, and exit nozzle. These components are shown in the drawing of Fig. 5 · 2. The air taken in by the diffuser is compressed in two stages. The external compression, as shown in Fig. 5 · 3, takes place because the bulk of the approaching engine forces the air to change its course. Further compression is accomplished in the diverging section of the ramjet diffuser. Fuel is injected into and mixed with the air in the diffuser. The flame holder provides a low-velocity region favorable to flame propagation, and the fuel-air mixture burns in this region. The burning mixture recirculates within this sheltered area and ignites the fresh charge

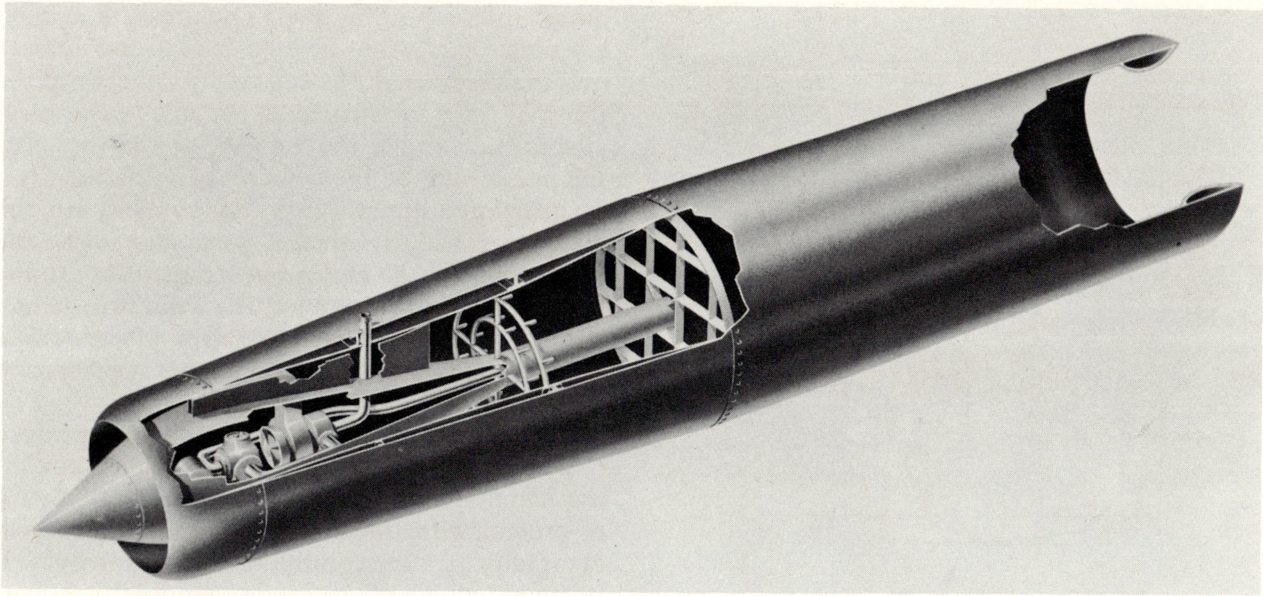

Figure 5 · 1 Cutaway view of a supersonic ramjet.

as it passes the edge of the flame holder. The burning gases then pass through the combustion chamber, increasing in temperature and therefore in volume. Because the volume of the air increases, it must speed up to get out of the way of the fresh charge following behind it, and a further increase in velocity occurs as the air is squeezed out through the exit nozzle. The thrust produced by the engine is proportional to this increase in velocity.

The behavior of the pressure, velocity, and temperature of the gases going through the engine is indicated in the lower part of Fig. 5 · 3. The external compression is seen to be the most important pressure rise available. The further increase in pressure accomplished within the diffuser is also important, since it is this pressure which bears on the ramjet structure and actually transmits the thrust force from the air to the ramjet. Through the diffusion process, the total

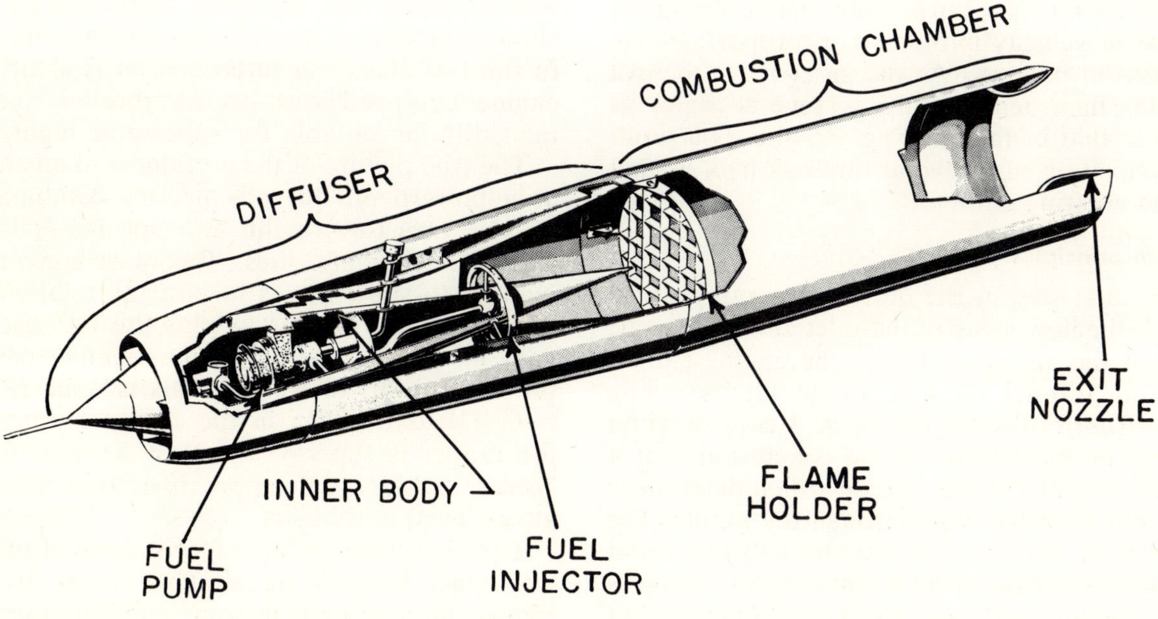

Figure 5 · 2 Supersonic ramjet, showing components.

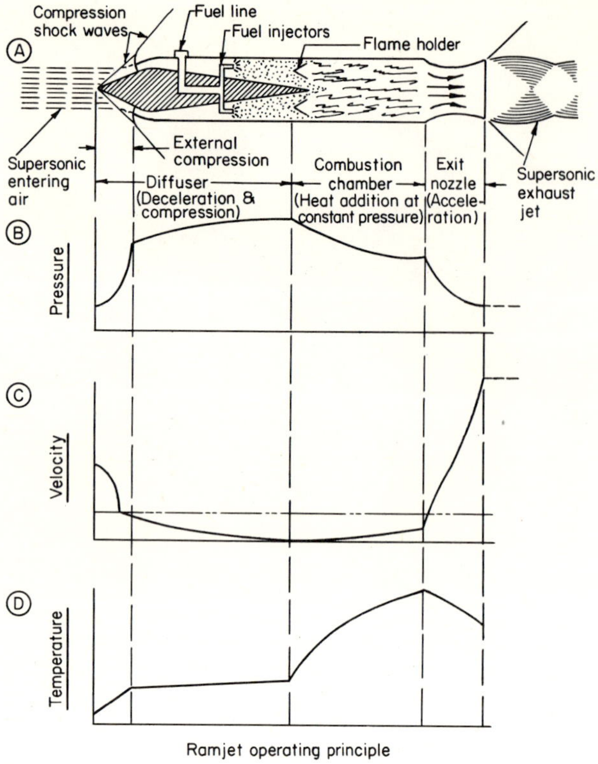

Figure 5 · 3 Operating conditions within a supersonic ramjet.

amount of energy in the air remains constant, and therefore the energy which causes the pressure rise is obtained by transforming the kinetic energy of the incoming air into pressure. This can be seen by noting the relationship of pressure and velocity as seen in Fig. 5 · 3.

In the combustion chamber, heat energy is added as the fuel burns, and the resulting increase in velocity is more than proportional to the loss in pressure. As the gases leave the exit nozzle, their temperature is several times as high as that of the entering air, and as a result the velocity is also several times as high as that of the entering air.

Design principles

The first step in the design of a ramjet is to match the flow areas of the inlet and exit nozzle so that the air which enters the engine can all flow out through the exit nozzle without piling up in front of the diffuser inlet. A consideration of this problem leads to the conclusion that it is the area of the exit nozzle which determines how much air will flow through the ramjet. The amount of air that will pass through the throat of the exit nozzle is determined by its temperature, pressure, and velocity. The temperature of the air can be determined from data on the heat added by the burning fuel. The pressure may be calculated, knowing the efficiency of the diffuser and the pressure drops caused by the drag of the internal engine structure. In the case of a supersonic ramjet, the velocity at the throat of the nozzle will be the velocity of sound and can be calculated directly from the temperature, the change in velocity being proportional to the square root of the change in temperature. In the case of a subsonic ramjet, the same can be accomplished by a trial and error process. The inlet size is then determined so that it will scoop up the required amount of air.

When the relative sizes of the exit and entrance have been established, the design of the remaining components is completed with the objective of providing the most efficient operation of each component for the conditions under which it will be operated. The fuel injectors and flame holder present a possibility of aerodynamic drag losses and are made as streamlined as possible. In the case of a supersonic ramjet, the contours of the diffuser inlet are highly important, and the design of this component is usually developed by a series of aerodynamic tests. The combustion system of ramjets is nearly always developed experimentally, since no analytical method has yet been devised.

Applications

The place of the ramjet in the aircraft powerplant picture is determined by its thrust-to-weight ratio and its fuel consumption in comparison to other available engines. Such a comparison is shown in Figs. 5 · 4 and 5 · 5. It will be noted that the term "turbo-ram-jet" is shown for one of the engines being compared in the two charts. A **turbo-ram-jet** is a turbojet engine equipped with an afterburner and an inlet diffuser suitable for supersonic flight.

The true picture of the usefulness of an engine as a powerplant depends upon a combination of the thrust-to-weight and specific-fuel-consumption characteristics. The most convenient manner to show this is to plot the total weight of the engine installation plus the fuel used to gain an understanding of the relative cost of power in terms of weight. This is done in Fig. 5 · 6. The data given in the chart indicate that the ramjet is superior to other powerplants at speeds of Mach 2.0 (approximately 1,500 mph at sea level) and higher.

The usefulness of a ramjet is limited mainly by its lack of static thrust. This means that the ramjet must be used in some combination with another form of propulsion capable of bringing

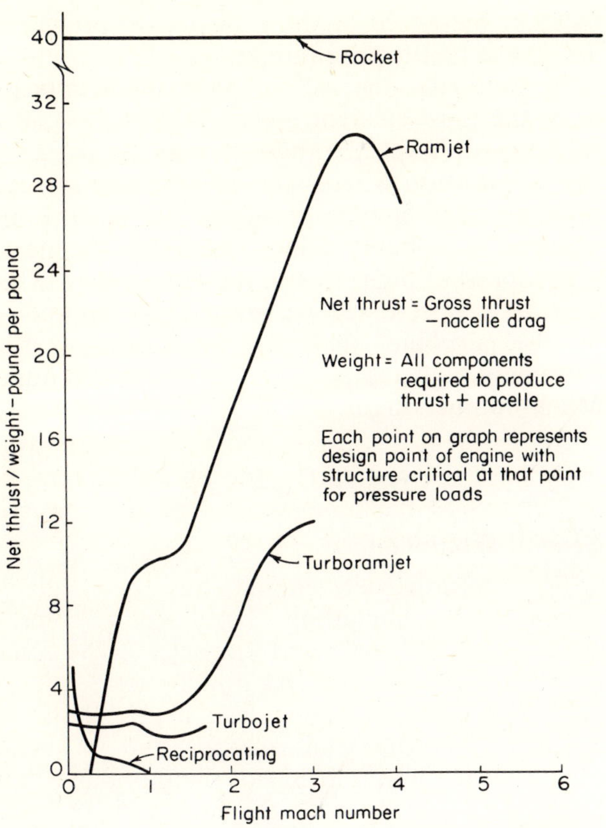

Figure 5 · 4 Specific weight in relation to speed.

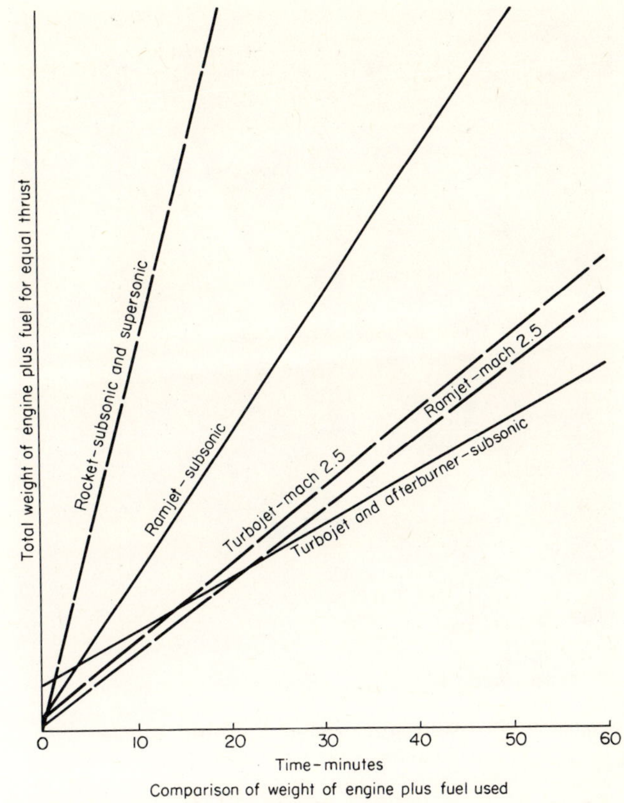

Figure 5 · 6 Comparison of total engine-system weight.

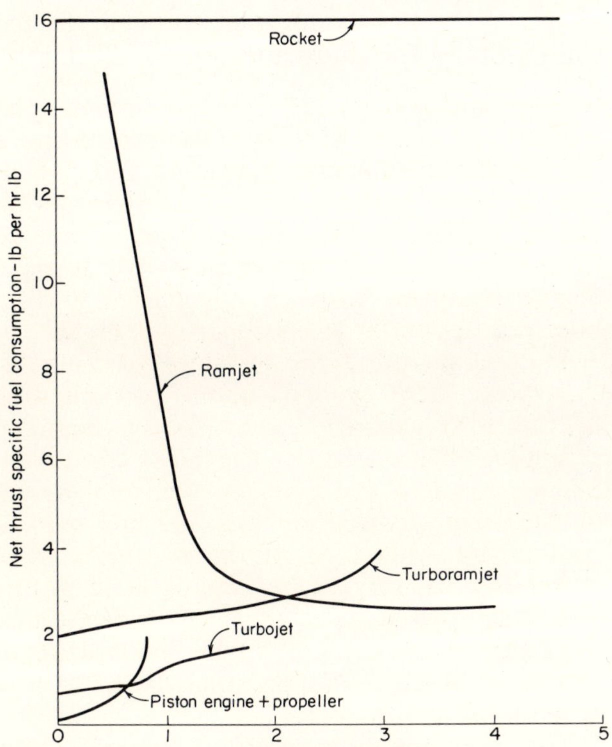

Figure 5 · 5 Specific fuel consumption in relation to speed.

it up to speed from rest. A turbojet or rocket engine is usually used to accomplish this. For many missile applications the rocket engine can be a simple solid-propellant unit which is jettisoned when it is burned out. A typical supersonic ramjet-powered missile is shown in Fig. 5 · 7. In this case, the booster unit is a rocket in the missile fuselage.

The subsonic ramjets have a rather limited field of application but have been successfully used on target drones, as shown in Fig. 5 · 8, and on helicopters, as shown in Fig. 5 · 9. This last illustration shows the tip of a helicopter rotor blade in which the ramjet is built as a part of the blade. The opening at the leading edge of the blade is the air inlet for the subsonic ramjet. The advantage of the ramjet engine as used in the target drone and the helicopter is the low cost.

Piloted ramjet

During early tests of ramjet engines, two of the 30-in. Marquardt C30-1.0 ramjets were mounted on the wing tips of a Lockheed F-80 Shooting Star jet fighter. This arrangement is shown in Fig. 5 · 10. At speeds of over 250 mph the ramjet can be ignited.

The effect is a bit startling to an observer on

Figure 5 · 7 Supersonic ramjet-powered missile.

the ground. The ramjets take hold with a sound resembling that of a large cannon, and the Shooting Star proceeds to live up to its name by shooting flames some 40 to 50 ft to the rear. The flames practically disappear, however, as speed is increased. The purpose of the project was not to increase the speed of the F-80 but to provide a flying test bed for the ramjet.

In summary, the ramjet powerplant appears to be the most efficient powerplant in the 1,500- to 2,500-mph speed range. It may be used for any application where static thrust can be supplied by another type of engine. Its upper limit of speed is probably coincidental with the maximum speed of flight within the atmosphere. This maximum speed will be limited by the ability of the airplane structure to withstand the extremely high temperatures generated in high-Mach-number flight.

THE PULSEJET

General description

The pulsating-jet engine has been given several names, including "buzz engine," "resojet," "aeroresonator," and "pulsejet." The commonly acceptable term has now become "pulsejet." The pulsejet engine was first scrutinized seriously by American military and technical men after the initial firings of the German robot bomb, or V-1 weapon, in the late spring of 1944. This weapon was commonly called the "buzz bomb" because of the loud buzzing noise made during operation. A drawing of the V-1 arrangement is shown in Fig. 5 · 11. Captured models of the V-1 robot bomb were brought to the United States for study and test. A substantial

Figure 5 · 8 Subsonic ramjet target drone.

Figure 5 · 9 Subsonic ramjet helicopter engine.

number were built in this country and used in research for guided-missile development.

As a general type, the pulsejet engine may be considered as an intermittent, compressorless, aerodynamic powerplant, with few or none of the mechanical features of conventional aviation powerplants. In its simplest form, the operation of the pulsejet depends only on the properties of atmospheric air, a fuel, a shaped tube, and some type of flow-check valve, and not on the interposition of pistons, impellers, blades, or other mechanical parts whose geometry and motion are controllable.

The pulsejet engine differs from other types of air-breathing engines in that the airflow through it is intermittent. Unlike the ramjet, it can produce static thrust. Experimental evidence has shown that the pulsejet is capable of producing higher thrust statically than when in motion. At low speeds, up to 400 mph, it produces acceptable performance. It is comparatively simple and requires very little maintenance. A typical pulsejet engine is shown in Fig. 5 · 12.

Operating principle

The pulsejet operates on a four-phase cycle as shown in Fig. 5 · 13. To start the cycle, air is forced into the engine through the flapper valves

Figure 5 · 10 Testing Marquardt ramjets on Lockheed F-80.

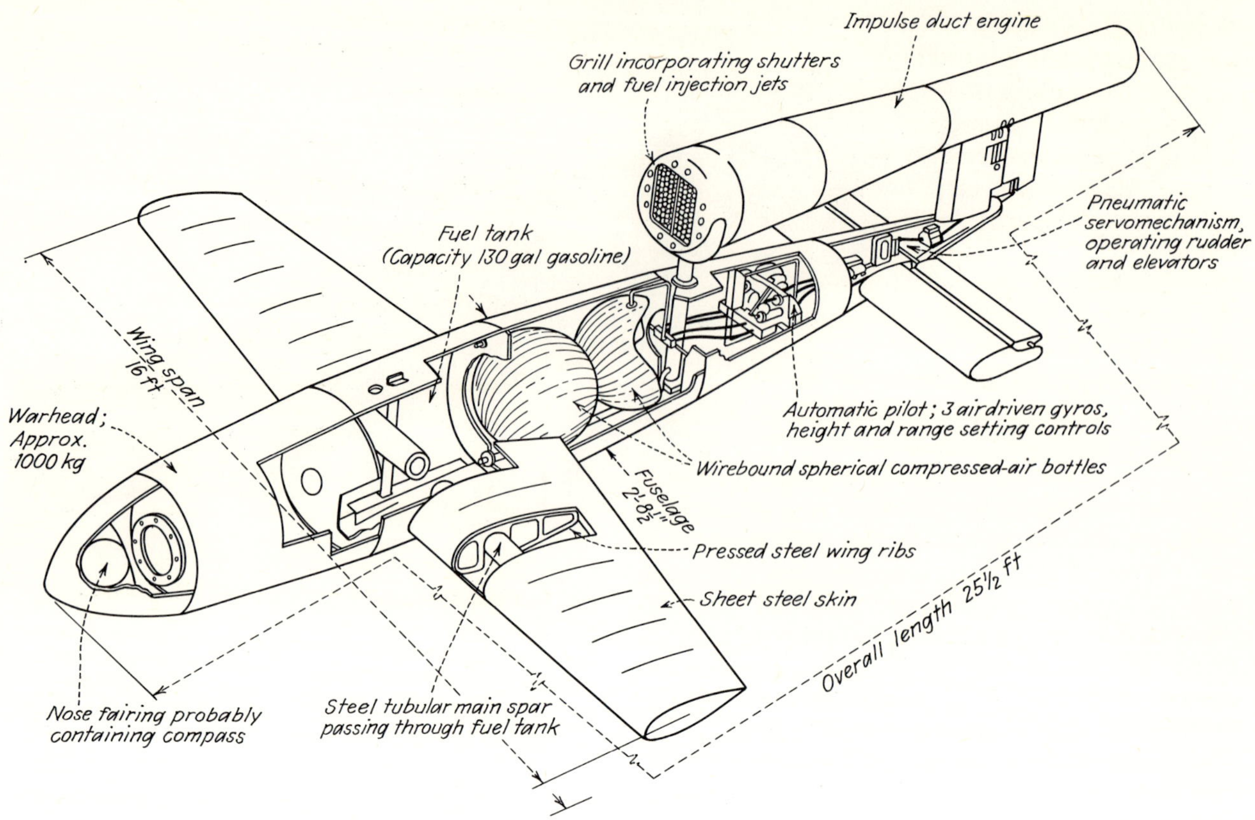

Figure 5·11 *The German V-1 "buzz bomb," powered by a pulsejet.*

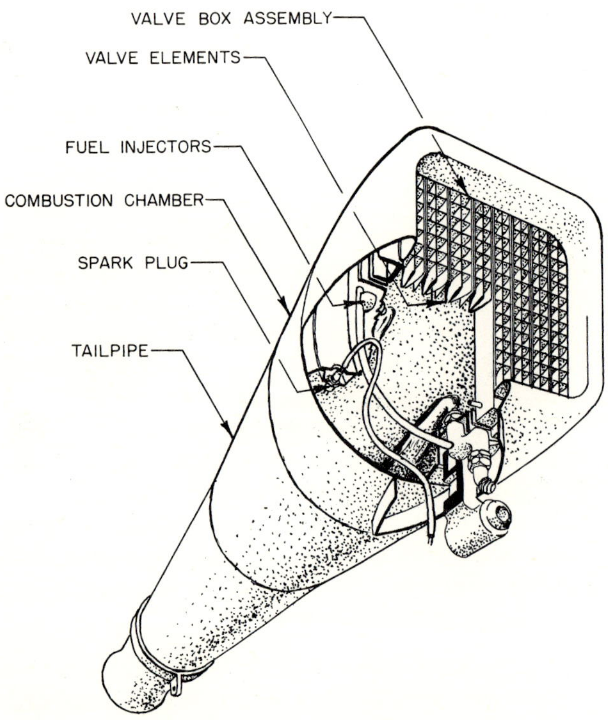

Figure 5·12 *Typical pulsejet engine.*

in the inlet. In practice this may be done by blowing compressed air through the valve box or by the motion of the engine through the air. The air that enters the engine passes by the fuel injector and is mixed with fuel. The cycle is now at the stage indicated in part *A* of Fig. 5·13. When the fuel-air mixture reaches the proper proportion to burn, it is ignited by a spark plug. The burning takes place with explosive force, thus causing a very rapid rise in pressure. The increase in pressure forces the flapper valves shut and propels the charge of burned gases out of the tail pipe, as in *B* of the figure. The momentum of the gas leaving the tailpipe causes the air to continue to flow out even after the pressure within the engine has reached atmospheric pressure. The pressure within the engine is therefore evacuated to below atmospheric pressure, part *C* in the figure. After the pressure has reached its lowest point, atmospheric pressure (and ram pressure, if the engine is in flight) forces air into the engine through the flapper valves. At the same time, air will also be drawn in through the tailpipe, since the pressure within the tailpipe is low and has nothing to prevent the entry of air. At this point, part *D* in the figure, the engine is ready to begin another cycle.

Once engine operation has become established, the spark plug is no longer necessary. The reignition between each cycle is accomplished when the fresh charge of fuel and air is ignited by some residual flame which is left over from the preceding cycle. The airflow which reenters the tailpipe is important from both the engine-operation and thrust standpoints. Experiments have shown that the amount of air which flows into the tailpipe can be several times as much as that which flows into the inlet. This mass of air increases the thrust of the engine by providing additional mass for the explosion pressure to work on. It also increases the pressure within the engine at the beginning of each explosion cycle, resulting in a more efficient burning process. Reentry of air into the tailpipe is made more difficult as the airspeed surrounding the engine increases. The thrust of the engine, therefore, tends to decrease with speed. As the speed increases, the amount of reentering airflow decreases to the point where the internal pressure is eventually too low to support combustion and the engine will no longer operate. Various methods of increasing the limiting speeds of the pulsejet have been tried with some success. At speeds above 500 mph, the fuel efficiency of the pulsejet falls below that of a ramjet, however, and its thrust-to-weight ratio is not so good.

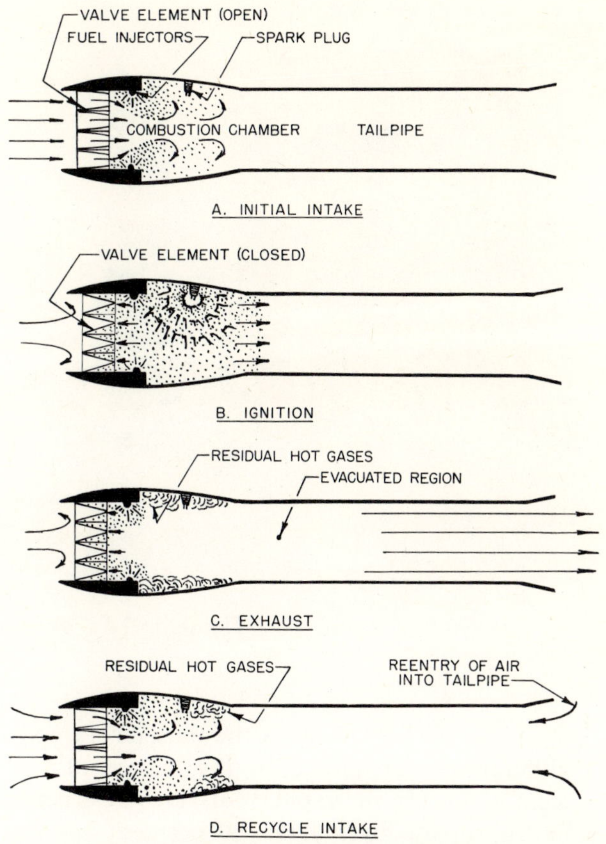

Figure 5 · 13 Pulsejet-engine operating cycle.

Characteristics of the pulsejet

The chief advantages of the pulsejet are its simplicity, light weight, low cost, and good zero-speed thrust characteristic. Its particular disadvantages are its 400- to 500-mph operating speed limit, rather limited altitude range, and somewhat unpredictable valve life. These valves must be made of comparatively light sheet-metal elements, as shown in Fig. 5 · 14, so that they provide a minimum of interference with the inflow of air. As a result, they are sensitive to the high temperatures and high vibratory stresses to which they are subjected in operation. Although considerable progress has been made in developing comparatively long-lived valves, the actual operating life in practice is still not consistent or predictable.

One interesting, and sometimes objectionable, feature of the pulsejet engine is the sound it makes when in operation. This sound is a series of loud reports caused by the firing of the individual charges of fuel and air in the combustion chamber. The frequency of the reports depends upon the length of the engine from the inlet valves to the end of the tailpipe and upon the temperature of the gases within the engine. The resulting sound is a continuous, loud, and vibratory note that can usually be heard for several miles.

Fuel studies

Many tests have been performed to study the effects on performance of various fuels with widely differing properties. Reciprocating-engine fuels have been tested in a range from 100 per cent cetane to fuels with a 120 performance rating. Kerosene, benzene, methyl and ethyl ether, propane, and a variety of related compounds have been tried. It is reported that, in one instance, nitroglycerine was added to gasoline. Generally, tests were made to determine what effects could be expected with fuels having different properties. A few significant facts have been developed:

1. The pulsejet engine will run on almost any common fuel with little variation in performance.
2. Physical and chemical properties of fuels affecting the operation of diesel and Otto-cycle engines

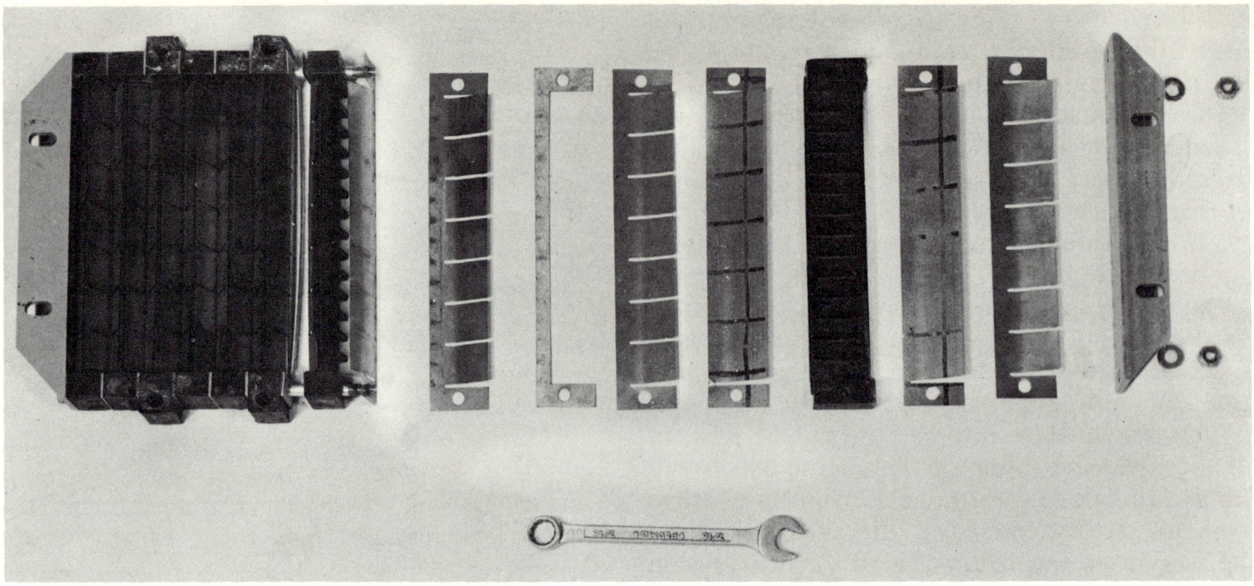

Figure 5 · 14 Valve assembly of a pulsejet engine.

have no marked effect on the pulsejet. Principal differences were noted in the degree of body heating and in the rapidity of valve destruction.

3. Nitropropane gives about 20 per cent improvement on thrust with about a 100 per cent in specific fuel consumption. Favorable comparisons at altitude, expected on the basis of the oxygen content in the fuel, have not been realized.
4. Changing from liquid to gaseous fuels is possible with minor changes in the fuel injectors.

Generally, the fuel studies indicate that liquid fuels are the most satisfactory because of the ease of handling and storing, and because of the lower cost and availability.

Applications

The most successful use of the pulsejet engines was on the German V-1 "buzz bomb," mentioned earlier in this chapter. This application ideally suited the characteristics of the engine. Another outstanding application was on the American Helicopter Company's Jet Jeep helicopter. In this latter application, the light weight of the engine made it possible to design a very light and simple helicopter. Other successful uses include target drones in the 200- to 400-knot range. One of these is shown in Fig. 5 · 15. Light pulsejets such as that shown in Fig. 5 · 16 may also be used as an auxiliary powerplant for sailplanes.

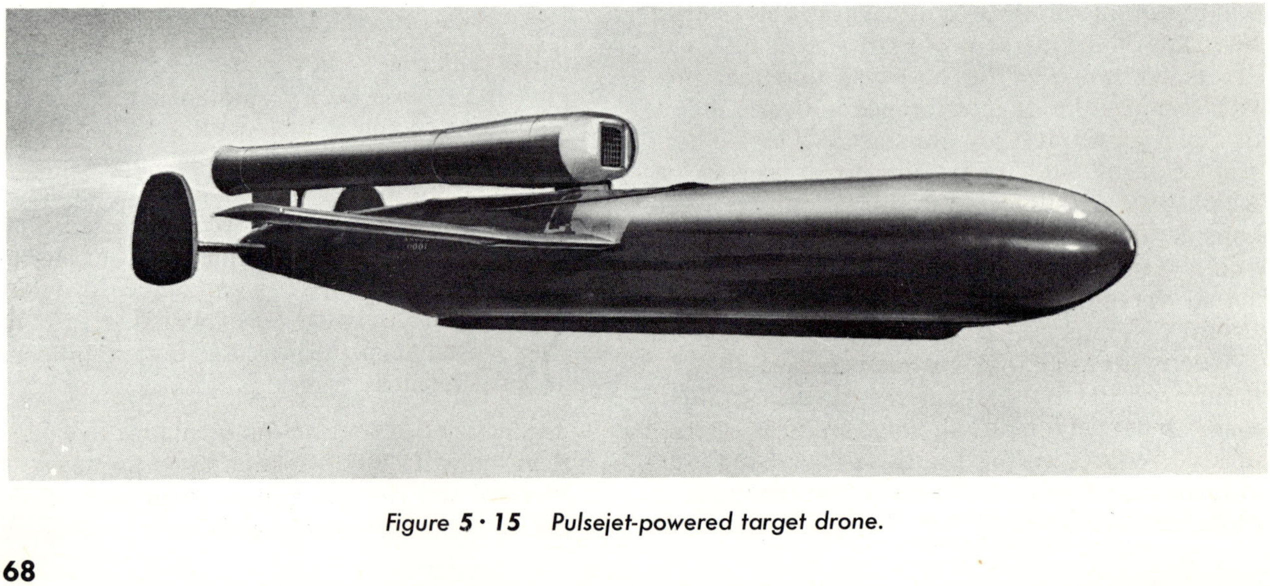

Figure 5 · 15 Pulsejet-powered target drone.

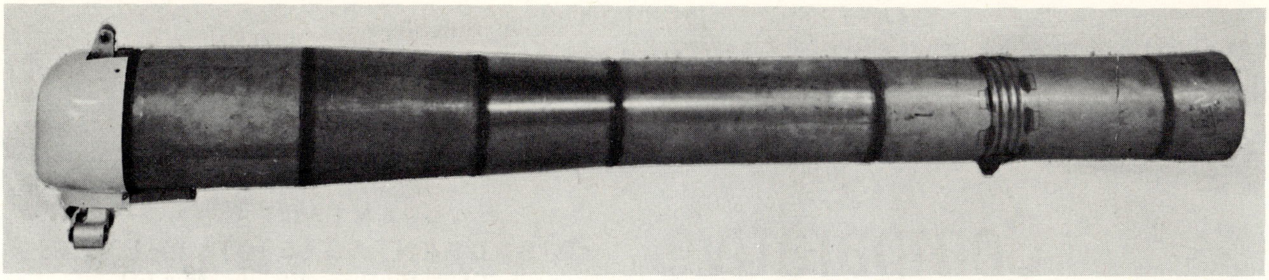

Figure 5 · 16 Typical lightweight pulsejet.

STUDY QUESTIONS

1. What is the meaning of the term "athodyd"?
2. Compare the operation of the ramjet with that of the turbojet.
3. Why will the ramjet not operate statically?
4. Discuss the specific weight of a ramjet at supersonic speeds.
5. Describe the airflow through the ramjet engine.
6. At what speeds is the ramjet superior to the turbojet?
7. What are the two principal limiting factors affecting the use of the ramjet?
8. Compare the pulsejet with the ramjet engine.
9. Explain the operating principle of the pulsejet engine.
10. What are the chief advantages of the pulsejet engine?

CHAPTER 6

Comparative Advantages and Disadvantages of Gas-turbine Engines

In describing the various types of turbine engines in the preceding chapters, it has been necessary to discuss some of their advantages and disadvantages briefly. In following chapters, it will also be necessary to make comparisons of the characteristics peculiar to the different types of engines. In order to avoid confusion, this chapter has been prepared to consolidate all the information available relating to advantages and disadvantages of turbine engines so that a true evaluation can be made.

Since the successful development of the first gas-turbine engine, many opinions have been expressed regarding the relative merits of the various powerplants. Comparing the performance of the gas-turbine engine with that of the reciprocating engine has been a favorite topic of discussion. It is hoped that some light will be shed on this subject by the facts, tables, and charts in this chapter, because it is obvious as we look ahead that gas-turbine engines are taking over the job of supplying power for all modern high-performance aircraft, both military and commercial.

Gas-turbine engines to drive propellers are common for all types of aircraft and pure jet engines of comparatively low power are now being employed on private and business-type aircraft.

ADVANTAGES AND DISADVANTAGES OF THE GAS TURBINE AS A MAIN POWERPLANT

In its present forms, the gas turbine, either propeller-drive or pure jet, has numerous attractive features that have been mentioned frequently. These may be reviewed briefly as follows:

Advantages

1. FREEDOM FROM VIBRATION. This permits lighter propeller sections and mounting structure. Vibration is reduced by the elimination of reciprocating parts such as connecting rods and pistons.
2. SIMPLICITY OF CONTROL. Only one lever is required for controlling the speed and power of the unit.
3. NO RADIATORS OR OTHER COOLING SURFACES. Such surfaces add weight and drag. Very small coolers for lubricating oil are used on large jet engines and turboprop engines.
4. NEGLIGIBLE COOLING AIR REQUIRED. Conventional engines require from five to eight times as much air for cooling as is required for power production. The acceleration of this air to airplane speed represents an appreciable loss of power, particularly in climb, even though much of this may be recovered later by the use of carefully designed radiators.
5. NO SPARK PLUGS REQUIRED EXCEPT FOR STARTING. After combustion is once established, it is self-supporting.
6. NO CARBURETORS. Hence there is no carburetor icing and no mixture controls. There is some question as to this advantage, since large gas turbines do require very complex fuel-control units. The automatic features of these units compensate for their complexity, however.
7. AVAILABLE SUPPLY OF COMPRESSED AIR. This air is used for driving cabin superchargers and small turbines and for anti-icing purposes.
8. DECREASED FIRE HAZARD. Fuels used for gas turbines are usually less volatile than the high-octane fuels used in reciprocating engines.
9. LOWER SPECIFIC WEIGHT. A gas turbine may

develop several times as much power as a reciprocating engine of the same weight.

10. LOWER OIL CONSUMPTION.

Disadvantages

1. HIGH SPECIFIC FUEL CONSUMPTION AT LOW AIRSPEEDS. This applies chiefly to pure jet engines. Turboprop engines have performance comparable to reciprocating engines in some instances, since they have attained specific fuel consumption as low as 0.40 lb per hp per hr.
2. INEFFICIENT OPERATION AT LOW POWER LEVELS.
3. SLOW ACCELERATION FROM MINIMUM TO MAXIMUM POWER LEVEL. This condition applies chiefly to turbojet engines. Turboprop and turbofan engines are able to accelerate quite rapidly.
4. HIGH STARTING POWER REQUIREMENTS. Starting large gas-turbine engines has been a problem in the past; however, starters have been developed within the past few years which make it relatively simple.
5. HIGH COST OF MANUFACTURE. Although the gas turbine is much simpler in operation than the piston engine, the special materials and manufacturing processes needed make the cost of the gas turbine much higher.
6. SUSCEPTIBILITY TO DAMAGE BY FOREIGN MATERIAL. Such material is readily drawn into the air inlet.

GAS-TURBINE ENGINE DEVELOPMENT

It must be remembered that, during the short history of the gas-turbine engine, improvements have been made which are almost fantastic in their scope. Today's gas turbines are the most efficient and reliable aviation powerplants ever manufactured. A few years back, the use of turbojet engines for commercial transport aircraft was considered to be something which might occur in the remote future. Today such use is commonplace.

In view of the many improvements which have been made in turbine engines, the advantages and disadvantages listed in the foregoing section of this chapter may be temporary in nature. Scientists and engineers are constantly at work overcoming the disadvantages; hence the modern aviation gas turbine is one of the most efficient powerplants in existence.

Applications of powerplants

Table 6 · 1 illustrates how the various types of powerplants are used and how the turbine, rocket, and athodyd engines compare with reciprocating types. It will be noted that the speed range of the turbine engines and other reaction types is in the higher bracket. An examination of the table explains why gas turbines and reaction engines of the rocket type have replaced reciprocating engines for aerospace vehicles which require high performance in speed, altitude, and range.

BASIC DIFFERENCES—GAS TURBINES VS. RECIPROCATING ENGINES

The basic differences between the gas-turbine engine and the reciprocating engine may be classified into five main groups.

These five classifications may be discussed in more detail by reviewing the advantages, disadvantages, and neutral characteristics of the aviation gas turbine.

Aerodynamic

Advantages. Smaller nacelles possible; negligible cooling power required; high-speed jet a more efficient propulsive means than propeller at high flight speeds.

Disadvantages. A high-speed jet is a less efficient propulsive means than a propeller at lower flight speeds and during takeoff. The development of turboprop and turbofan engines has made it possible to combine the advantages of the turbine engine with the efficiency of the propeller for lower speeds and for takeoff.

Weight

Advantages. Turbine engines are considerably lighter than reciprocating engines for the same power output. Turbojet engines have been developed with a weight/power ratio of less than 0.13 lb per lb of thrust. Turboprop engines have attained 0.39 lb per equivalent shaft horsepower (eshp) in comparison with reciprocating engines which usually have a weight/power ratio of approximately 1.0 lb per hp or more.

Fuel consumption

Neutral Characteristics. Best specific fuel consumption occurs near maximum output. (Best specific fuel consumption of reciprocating engines occurs at about one-half maximum power.) At a given flight speed, specific fuel consumption of the turbojet engine tends to decrease with altitude. The specific fuel consumption for turboprop engines is comparable to that of the best

Table 6 · 1 Powerplant applications

Engine type	Use	Speed range	Advantages	Limitations
Piston	Long-range moderate-speed transports and bombers	0–350	Low cruising fuel consumption, reliability, long service life	Operates at relatively low altitudes
	Moderate-range, training aircraft	0–400	Versatility, reliability	Low speed, medium altitude
	Small private aircraft, executive aircraft	0–250	Low cost, reliability	Low altitude, short range
Turboprop	Medium- and long-range transports and bombers	0–subsonic	High sea-level power, low fuel consumption, low specific weight, long range	Speed limited by propeller efficiency
Turbojet	All fighters, interceptors; most bombers and transports; includes commercial airliners	0–supersonic	High speed, high altitude, smooth performance	Low thrust horsepower at low speeds
Rocket	Missiles and high-speed interceptors	Unlimited	High speed, low weight; ability to operate outside atmosphere	Extremely high fuel-load requirements
Pulsejet	Missiles, targets, experimental craft	0–1,500	Simple design, cheap manufacture	High fuel consumption, noise, vibration, limited speed
Ramjet	Pilotless aircraft, missiles	0–3,000	Simple design, cheap manufacture, high speed	Requires assistance from other power source to reach operating speed

reciprocating engines. It is likely that continued development will produce a turboprop engine with much better specific fuel consumption than any reciprocating engine. Turbojet engines have been developed to a point where specific fuel consumption is excellent, and at operating speeds and altitudes, they are more efficient than reciprocating engines. High compression ratios of 12:1 and better have brought about specific fuel-consumption figures of less than 0.70 lb per pound of static thrust.

Disadvantages. Best fuel consumption is in general poorer for the turbojet engine; however, this disadvantage appears to be decreasing rapidly.

Output

Neutral Characteristics. Operation of the gas-turbine engine at varying altitudes is somewhere between that of an unsupercharged and a supercharged reciprocating engine. That is, the turbine engine is more adaptable to varying altitudes than the unsupercharged engine and perhaps a little less adaptable than the supercharged reciprocating engine. Here again, constant improvements are being made in the gas-turbine engine, and it is expected that the turbine engine will be superior to any reciprocating engine in this respect.

General

Advantages. Low powerplant vibration, relatively constant speed over a wide range of output.

Disadvantages. High engine speed (advantageous for generator drive).

PERFORMANCE COMPARISON

A comparison of the performance of aircraft powered by reciprocating, turboprop, and turbojet engines indicates that for a cruising altitude of 35,000 ft:

1. Aircraft with top speeds below 335 mph achieve their maximum range when powered by reciprocating engines. Late developments in turboprop engines may change this figure downward.
2. Aircraft with top speeds above 610 mph achieve their maximum range when powered by turbojet engines.
3. Aircraft with speeds between those specified in the foregoing statements in general achieve their maximum range when powered by turboprop or turbofan engines.

The acceptance by commercial airlines of both turboprop and turbojet powerplants for air liners is good evidence that the gas-turbine engine has been developed to a point where it exceeds the performance of the reciprocating engine in both efficiency and economy.

AXIAL FLOW VS. CENTRIFUGAL FLOW

Within the classification of turbojet engines, some engines have advantages over others. At the present stage of development of these engines, the axial-flow type has the following advantages over the centrifugal-flow type:

1. LOWER SPECIFIC FUEL CONSUMPTION. This advantage is quite important, since turbine-powered aircraft are now designed to fly vast distances without refueling. The lower fuel consumption is accomplished with the axial-flow engine because the axial compressor makes possible a higher pressure ratio. This is particularly true of the turbofan engines.
2. SMALLER DIAMETER OR FRONTAL AREA. This characteristic makes the axial-flow engine more suitable for wing installation.

The following are some of the advantages of the centrifugal-flow engine over the axial-flow type:

1. SIMPLE MANUFACTURE AND FEWER PARTS. This reduces initial cost and maintenance.
2. LOWER WEIGHT. A centrifugal engine with an equivalent compression ratio may be lower in weight for the same amount of thrust.
3. EASIER INSTALLATION AND REMOVAL. No close-fitting ducts to engine are necessary.
4. MORE EFFECTIVE WATER INJECTION. On this type of engine the water can be injected directly into the compressor, whereas in an axial-flow engine, water is injected into the combustion chamber.
5. FASTER ACCELERATION OF THE ROTOR SECTION.

The choice between the axial-flow and centrifugal-flow engine at the present time is a matter of conjecture. Each engine, as the preceding items show, has its own particular advantages. It is true, however, that the advantages of higher compressor ratios possible with the axial-flow engines will make this type of engine more desirable, especially for high-performance aircraft.

Development of turbofan and other bypass engines has progressed to such an extent that these types may well become the leading turbine engines. Engines of these types are now in production and seem to be receiving very favorable consideration because of their outstanding performance and efficiency. As has been previously explained, these types of engines increase air-mass flow by feeding additional air into the jet stream directly to the rear of the turbine. This results in thrust augmentation because a greater mass of air is accelerated than would be the case with the simple jet engine.

SUMMARY

A great variety of propulsion systems can be developed around the gas turbine. The combinations are almost limitless. The gas turbine, built up of a number of relatively simple components, each of which is susceptible of individual development, allows great latitude to the design engineer in creating an engine for optimum performance for a given application. In the future, more successful models of gas-turbine engines may be expected than there were types of reciprocating engines before World War II.

The gas-turbine engine has demonstrated its adaptability to the many powerplant requirements of modern industry. These engines have found a place not only in aviation but also in many industrial applications. For example, gas-turbine engines are now being used in automobiles and trucks and for a variety of other purposes where effective shaft horsepower is needed.

Considering the great progress made in the improvement of gas-turbine engines during recent years, it is conceivable that these engines will be used to supply nearly all power produced as a result of the burning of fuel.

STUDY QUESTIONS

1. List the advantages of the gas-turbine engine as a main powerplant.
2. What are the basic differences between the reciprocating engine and the turbojet engine?
3. Which type of turbojet engine has the lower specific fuel consumption, the axial-flow type or the centrifugal-flow type?
4. Discuss the fuel consumption of the turbofan engine.
5. What applications are best suited to the characteristics of the turboprop engine?
6. Discuss the indications for future use of gas-turbine engines.

CHAPTER 7

Gas-turbine Fuels

Many variables are involved in the selection of suitable fuels for gas-turbine engines. Military requirements differ somewhat from commercial ones because of unlike operating conditions. Although a gas-turbine engine may operate satisfactorily on a variety of fuels, certain fuels are more suitable than others. In this chapter we shall discuss some of the requirements of gas-turbine fuels, but we shall make no attempt to define the ideal fuel, for the simple reason that it has not yet been produced.

Fuel consumption

The amounts of fuel and air consumed by the gas-turbine engine in developing its great power are astounding. A turbojet engine developing 4,000 lb of thrust, equivalent to 4,000 hp at 375 mph, may require more than 4,000,000 cu ft of air in an hour. More efficient engines may require as little as 2,500,000 cu ft of air per hour to produce 4,000 lb of thrust, but in either case, the volume is tremendous. At these rates of air consumption, a typical five-room house could be exhausted in less than 10 sec.

From 15 to 20 barrels of fuel may be burned each hour—enough fuel, if it were gasoline, to drive an automobile over 10,000 miles at 60 mph, or, if heating fuel, to heat an average house for over six months of winter.

Heat is released in the combustion chambers of the turbojet engine at the rate of about 20 million Btu per hr per cubic foot of combustion zone, which may be compared with a rate of 1 to 2 million Btu per hr per cubic foot in the case of industrial furnaces. The letters Btu stand for **British thermal unit** and represent a unit of heat. One Btu is the heat energy required to raise the temperature of 1 lb of water 1°F. Another unit of heat energy is called the **calorie.** One calorie is the heat energy required to raise the temperature of 1 g of water 1°C.

Developed fuels

The following developed fuels have been used to operate aviation gas-turbine engines:

Military designation	*Remarks*
MIL-F-5572	Aviation-grade gasoline
AN-F-32a (JP-1)	Light, wax-free kerosene; not often used
AN-F-34 (JP-2) (canceled)	Kerosene "blend"; could be used only for ground running; no longer used
AN-F-58a (JP-3)	Blend of gasoline, kerosene, and fuel oil
MIL-F-5624a (JP-4)	Characteristics between kerosene and aviation gasoline; has proved to be a very good fuel
JP-5	Similar to JP-1; not commonly used

Other fuels have also been used, but those listed have been the most active.

Properties of fuels

Historically, airline fuel specifications have been essentially the same as those used by the military services. They have differed to a minor degree, but these differences have not placed an undue burden on the producers or users. There is a question as to whether this will be true of fuels of aviation gas turbines.

The selection of military fuels involves consideration of the petroleum industry's productive capacity in the light of other military and essential civilian wartime requirements. A study of this problem leads to the conclusion that the only substantial source of material which may be made available in wartime by denying the civilian economy is that derived from motor gasoline, which represents close to 50 per cent of the output of the industry. Some of this gasoline is too volatile for use in turbine fuel. A **volatile** fuel is one which vaporizes easily.

The end result of various compromises between performance and availability requirements is a product boiling between 150 and 550°F, known as JP-4. Performance requirements of future high-speed combat aircraft indicate that the fuel should be of lower volatility, which is contrary to the direction desired for maximum wartime availability.

Early turbine enthusiasts who claimed that we now have an engine capable of burning any fuel may be said to have been somewhat optimistic. Nevertheless, the turbine engine does demonstrate a capacity to digest a wider range of fuels than the reciprocating engine. This characteristic, therefore, provides an opportunity to utilize fuels other than those specified by the military, if such are desired.

In analyzing this problem, the difference in objectives of the military and the air-transport industry must be kept clearly in mind. Whereas the military must strive for performance and wartime availability, the objectives of the airlines are performance, economy, and safety.

The problems that would exist during conditions of war are too complex for analysis without defining the type of military operation and logistic requirements involved. The discussion in this chapter considers the problem only in the light of our peacetime economy.

The petroleum refiner finds it advantageous to consider fuel characteristics in two distinct categories: (1) the physical properties relating to volatility and viscosity; and (2) the chemical properties relating to stability, corrosion characteristics, and residues. Obviously, the combination of these groups of properties must be considered when dealing with combustion phenomena. One of the most restrictive requirements is a very simple one: the fuel must remain liquid until it reaches the combustor nozzle or vaporizer. There is more to this requirement than meets the eye. Under the extreme conditions to which it is subjected in the fuel system, the fuel must not vaporize. Vaporization will result in fuel loss, vapor lock, or interruption of control-system functioning. The fuel must not form solid materials which plug fuel nozzles, and it must remain pumpable. Having met these requirements, the fuel must then be capable of being burned efficiently, not only at the design point but over a wide range of operating conditions imposed on the engine. Furthermore, it must not cause corrosion or excessive deposits in the combustor or turbine sections.

In addition to the foregoing requirements, gas-turbine fuel must be available in the quantities required and at the locations desired at a price which will permit a healthy economic development of the air-line industry. Finally, the fuel must provide the highest degree of safety possible, based on present knowledge of fuels and fuel systems.

In Table 7 · 1 are shown the general properties of a number of fuels now produced. Fuels com-

Table 7 · 1 Properties of gas-turbine fuels

Test	100/130 Aviation gasoline	JP-4	Kerosene	JP-1 Low-pour-point kerosene	JP-5 Low-pour-point kerosene	Premium diesel	Diesel	Bunker C
Gravity, °API	71.5	50.5	42.5	43.5	43.0	42.5	33.6	10.5
Specific gravity	0.697	0.777	0.813	0.808	0.810	0.813	0.857	0.986
Weight, lb per gal	5.8	6.48	6.78	6.74	6.76	6.78	7.14	8.30
Distillation, °F:								
Initial boiling point	102	135	352	325	355	340	395	
50%	207	313	422	370	410	435	530	
End point	324	484	514	450	500	560	676	
Flash point, °F	...	...	135	120	140	130	158	225
Viscosity, at 100°F, centistokes	0.6	0.9	1.59	1.6	1.6	1.71	3.0	700
Freezing point, °F	< -76	< -76	...	< -76	−45			
Pour point, °F	...	...	−35	...	...	−20	5	25
Cloud point, °F	...	...	−30	...	...	...	15	
Calorific value, Btu per lb	19,025	18,730	18,570	18,590	18,580	18,570	18,350	17,560
Btu per gal $\times 10^3$	112	121	126	125	126	126	131	146

plying with military specifications are generally in distribution systems to military installations and military contractors. The other fuels are in general distribution throughout the United States and, in most instances, throughout the world. Because of their characteristics, aviation gasoline and "bunker C" may be ruled out, leaving for consideration JP-4, kerosene, special low-pour-point kerosene as typified by JP-1, premium diesel fuel, and regular diesel fuel. These and any special fuels lying within this framework represent possible commercial-aircraft turbine fuels.

With these concepts in mind, we shall discuss in some detail several of the more important variables of petroleum and their relationship to aircraft operation. These variables fall into three major categories: performance, availability, and economics.

Performance

Calorific Value. Calorific value of a fuel represents its heat energy for a given weight. Calorific value per pound of fuel decreases with increasing molecular weight and increasing carbon/hydrogen ratio. The reverse is true in respect to calorific value per unit volume. Fuel **density** (weight per unit volume) is for all practical purposes a good index of calorific value, as shown in Fig. 7 · 1.

In the chart of Fig. 7 · 1, the density (gravity) is noted in degrees API. This is a unit which has been developed over many years by the petroleum industry and is accepted as standard by the American Petroleum Institute. The formula for converting specific gravity to degrees API is as follows:

$$1°\text{API} = \frac{141.5}{\text{specific gravity at } 60°\text{F}} - 131.5$$

Specific gravity is the ratio of the weight of a substance to an equal volume of water at 60°F.

The density of fuel is influenced not only by fuel type but also by temperature, as shown in Fig. 7 · 2.

Many analyses have been made of weight vs. volume limitation of turbine-powered aircraft. Although the question is quite controversial, there appears to be some agreement that short-range aircraft will tend toward the volumetric limit, while long-range aircraft will be weight-limited.

If it is correct to assume that a preponderance of the operations will result in a weight-limited condition, then the most desirable fuels for maximum range or payload will be volatile paraffinic hydrocarbons. For the reverse condition, heavy aromatics are desired, assuming that they can be utilized efficiently.

Front-end Volatility. This fuel characteristic influences a number of performance features. Figure 7 · 3 illustrates the ASTM (American Society for Testing Materials) distillation curves for types of fuels under consideration. For starting on the ground or air relighting, a high volatility is desired. However, this requirement is

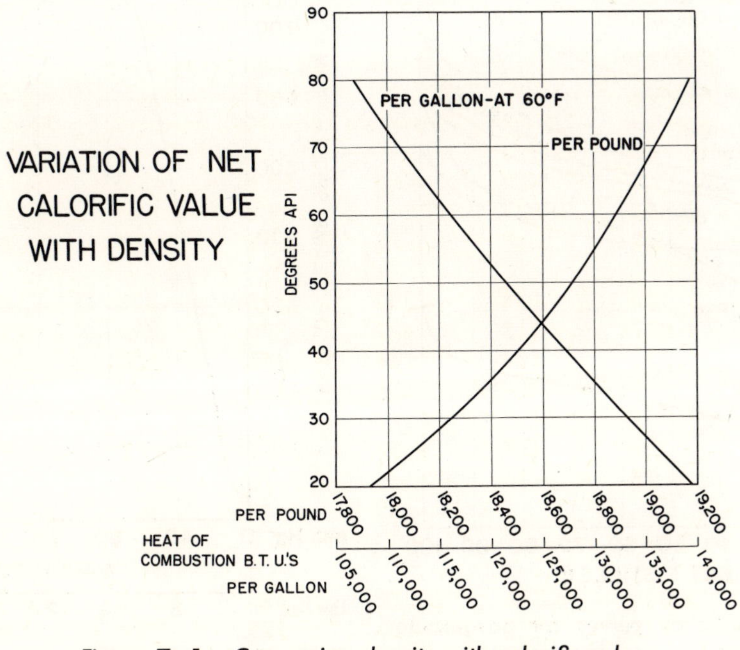

Figure 7 · 1 Comparing density with calorific value.

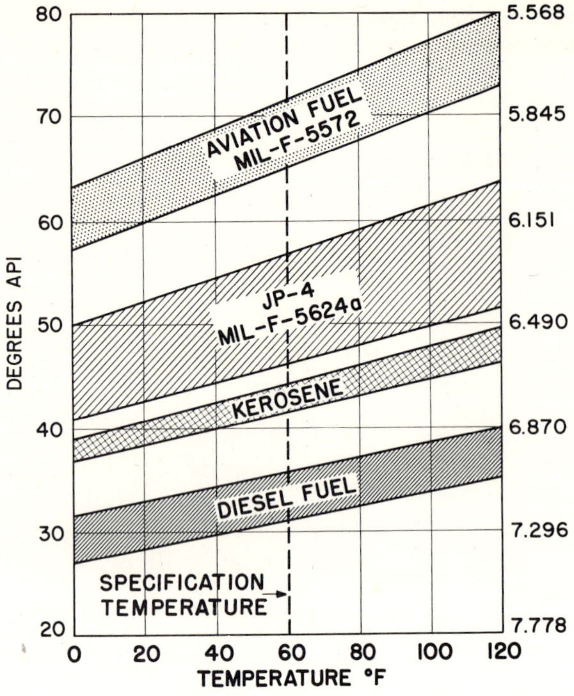

Figure 7·2 Comparison of density with temperature.

opposed by a number of others. For the purpose of this discussion, it is assumed that ignition systems are designed to provide satisfactory ignition of fuels within the framework under consideration.

Fuel loss, by evaporation or "slugging," and vapor lock are not new problems to the aircraft industry. These factors are directly related to fuel volatility. When the boiling tendency of gasoline and the foaming tendency of kerosene were combined in the form of JP-3 fuel, the results were catastrophic, with slugging losses as high as 90 per cent reported. Available information indicates that front-end volatility equal to or less than that of JP-4 fuel should be satisfactory from the vapor-lock and vapor-loss aspect in subsonic aircraft at the altitudes contemplated.

The influence of front-end volatility on hazards or safety is quite significant. Fuel in a liquid state will not burn. It may be vaporized by increasing temperature or reducing pressure. It may also be atomized with a nozzle or by imparting to it sufficient impact energy to form a mist or fog, which, like the vapor, may be quite easily ignited, when combined with adequate quantities of air. NACA tests clearly showed this fact, thus indicating that low volatility does not guarantee complete absence of fire hazards. Another factor to be considered is the autoignition temperature, which for kerosene or diesel fuel is somewhat lower than for gasoline.

Still another factor relating to crash fires is flame speed or propagation. Figure 7·4 shows the influence of volatility on rate of flame speed.

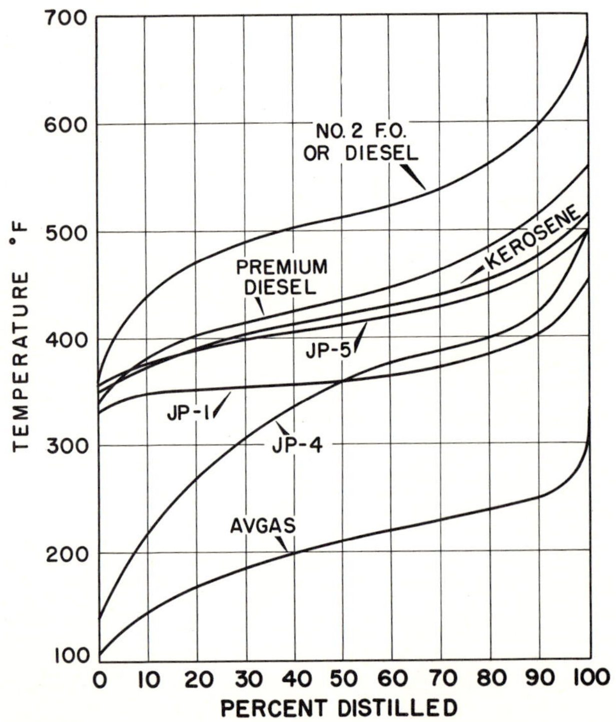

Figure 7·3 ASTM distillation curves for gas-turbine fuels.

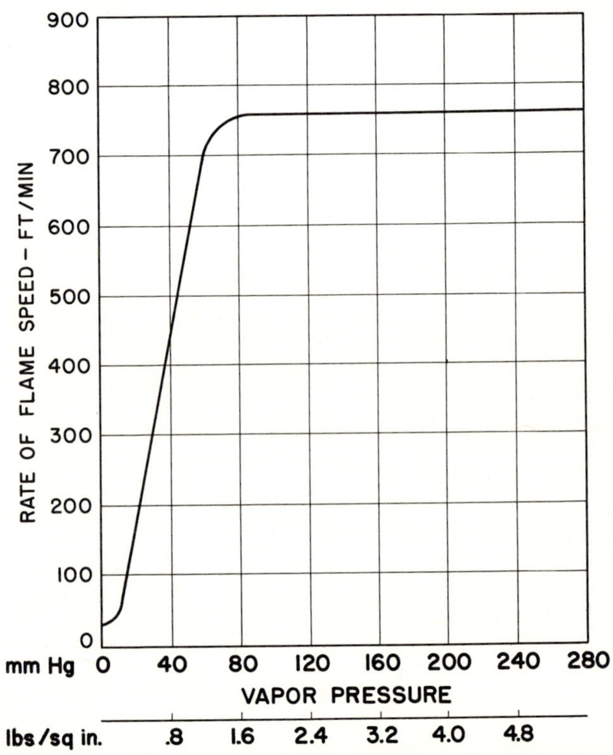

Figure 7·4 Effect of volatility on flame speed.

Kerosene-type fuels will have flame speeds of less than 30 ft per min, whereas that of JP-4 or gasoline will be greater than 760 ft per min.

Conditions resulting in explosive mixtures are of interest, as shown in Fig. 7 · 5. These are old data but are shown specifically for the purpose of qualifying them, as they have been seriously misinterpreted. These data are applicable only to quiescent conditions involving low-energy ignition source, and they therefore fall far short of providing all the needed information. For example, when sufficient ignition energy is present, no lean limit exists. Likewise, with tank breathing, which occurs with changes in altitude, we can only assume that almost any condition occurs in tanks at some time.

Considering not only information available from specific aviation investigations but also the background of experience in utilizing all petroleum products, we can conclude that use of low-volatility, high-flash-point fuels makes possible less hazardous operation than is possible with high-vapor-pressure fuels. This statement is not intended to imply that, to realize fully the advantages of the high-flash fuels, some special hardware may not be desirable. For example, further study of lightning strikes may indicate the need for tank inerting (induction of an inert gas into the fuel tank) or the use of explosion-suppression devices.

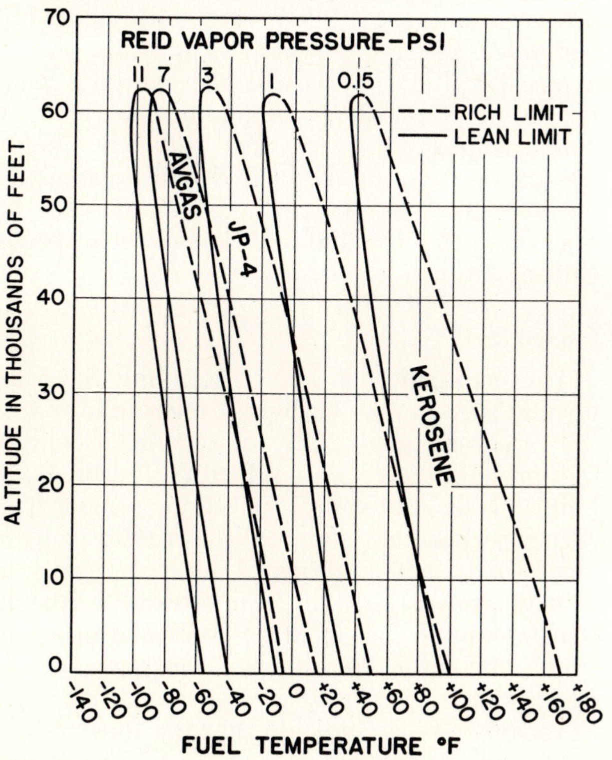

Figure 7 · 5 Ignition limits of fuels with low-energy ignition.

Characteristics at Low Temperature. Liquids flow readily at working viscosities of less than 1,000 centistokes. The **centistoke** is a unit of viscosity. **Viscosity** is the resistance to flow which a fluid may possess.

As shown in Fig. 7 · 6, the fuels under consideration have viscosities of less than 30 centistokes, just above their pour points. This means that their viscosities are quite low until the time they reach their freezing points. For this reason, low-temperature viscosity does not constitute a problem, assuming that the fuel is heated prior to reaching the combustors. A possible exception to this statement is the cold starting condition where cold, high-viscosity fuel is delivered to the fuel nozzles.

Unfortunately, the **pour point** of fuels cannot be dismissed so lightly. Pour point is only a rough indication of pumpability limits, because the delicate wax structure formed during cooling may, in some cases, easily be destroyed. Pour point must suffice, however, pending the availability of a better yardstick. In this connection, some work has shown that cloud point, which usually occurs a few degrees above pour point, is a better index of filtration limits. **Cloud point** is the temperature at which small crystals of wax begin to form in the fuel, thus causing the fuel to have a cloudy appearance.

In order to maintain fuel at a temperature which will satisfy the requirements of combustion, it is necessary to insulate and heat the fuel tanks of high-altitude jet aircraft. The degree to

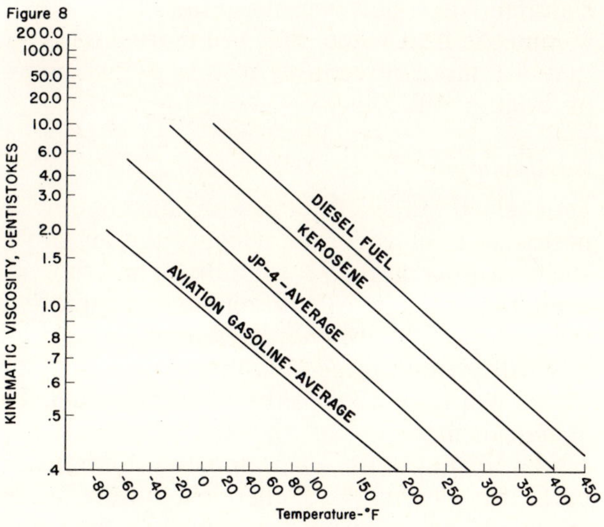

Figure 7 · 6 Variation of viscosity of fuels with temperature changes.

which this must be done depends somewhat upon the type of fuel being used. A fuel with a low freezing point is desirable, since it will not require as much energy for heating.

A discussion of fuel handling would not be complete without reference to the problem of filter icing. Warm fuel can hold in suspension substantial amounts of water. Actually, it may be said that the water is dissolved in the fuel. As fuel cools, this water tends to condense and separate from the fuel. When the temperature of the fuel falls below the freezing point of water, the water may crystallize and clog any filter through which it is passed. This problem has been met by the use of heated fuel filters.

Combustion Characteristics. Even though there is considerable controversy regarding the combustion characteristics of a gas-turbine fuel, certain facts are evident, and upon these we can base a few conclusions.

Broadly speaking, it can be stated that as the molecular weight or carbon/hydrogen ratio increases, clean and efficient combustion of the fuel becomes more difficult. For example, natural gas is easier to burn than residual fuel oil. Likewise, it appears that modification of engines to improve specific output and fuel consumption tends to result in more critical combustion problems.

Some of the combustion characteristics which affect the selection of fuels for gas turbines are blowout limits, reigniting, smoking, combustion efficiency, carbon formation, and effects on engine components. The more volatile fuels approaching the characteristics of gasoline burn cleaner and provide better starting and relight characteristics; however, these fuels have a lower volumetric heat value, they are more costly, and they are more susceptible to loss by "slugging" or boiling off.

Availability

In selecting fuels for a gas-turbine engine, it is necessary to consider the effect of the major fuel specifications on fuel availability. The crude oils available to the petroleum refiner vary greatly in their properties and characteristics. Some crude oils will contain large amounts of gasoline, and others will be suitable only for bunker fuel. The chemical make-up of the hydrocarbons in a crude oil may differ also, and the difference in both the physical and chemical nature of the crude oils will affect the amount of jet fuel which may be obtained therefrom. The refiner may convert some of the hydrocarbons to a different molecular type by processes such as catalytic cracking, or he may separate and concentrate one type of hydrocarbon by processes such as solvent extraction. Such processing inevitably adds to the cost of a fuel. The bulk of jet fuel is prepared by simple distillation.

Kerosene and diesel fuel are the commercial products of most interest as potential components of aviation turbine fuel. We may examine the availability of these two products on the basis of past experience. The petroleum industry in the United States has consistently produced about 6 per cent kerosene from crude-oil runs to the stills over the past 15 years. Crude-oil production is steadily being increased as the petroleum industry expands to meet the demands of our domestic economy. Estimates of production and demand for kerosene indicate a possible surplus of 135,000 barrels per day by 1965. This should be adequate for at least 100,000 barrels per day of aviation turbine fuel. Diesel-fuel utilization has grown in the past years and is expected to double again in volume by 1965. Past production of diesel fuel has been tied closely to demand, and additional quantities could have been produced if required. This situation will probably continue in the future, and the demand is expected to rise to about 7 per cent of the crude-oil run.

Domestic requirements for other distillate fuel oils exceed those of kerosene and diesel fuel combined and are expected to increase also. It is possible that imports will be required to meet the total domestic demand before 1965. It would appear, however, that sufficient quantities of kerosene or diesel fuel will be available to meet the requirements for commercial turbine fuel provided that highly restrictive specifications are not adopted.

Economics

The economic aspects of selecting the ideal turbine fuel cannot be lightly regarded because the price and availability of the fuel is one of the primary determining factors in the feasibility of jet transport operation. When it is realized that one modern jet airliner carries more than 23,000 gallons of fuel for a range of somewhat over 7,000 miles, it is readily apparent that jet-engine fuel must be produced at a fantastic rate to keep our jet transport fleet in operation.

Various reports indicate that the fuel cost for operating jet airliners is from 15 to 25 per cent of total operating cost, depending upon the type of aircraft and the load factor. This is somewhat less than early estimates, and it is a tribute to the engineers and manufacturers who are responsible for the quality of modern gas-turbine powerplants. Pure jet engines have been improved to the point where they are more than twice as efficient as early models. This increase in efficiency has been greatly enhanced in recent years by the advent of the turbofan engine, which is as much as 30 per cent more efficient than the pure jet engine.

Prior to the time of commercial jet aircraft, the petroleum industry was always able to meet the great changes which come about periodically as the result of new types of powerplants. During World War II it suddenly became necessary for the industry to produce vast quantities of high-grade aviation fuel to feed the swarms of fighter and bomber aircraft which were sent against the enemy. This challenge was met with a comparatively minor disruption of normal gasoline supplies. Since the war, the petroleum industry has been required to make another major shift in the percentage of gasoline, kerosene, and "in-between" fuels. This has been accomplished with no apparent inconvenience to the public and without the same increase in cost which has been applied to almost every other manufactured product because of inflation.

Fuel handling

Handling of gas-turbine fuels does not present any more problems than does the handling of aviation gasoline for reciprocating engines. In fact, because of the higher flash point of turbine fuels, they are generally safer to handle.

Even under the most extreme cold conditions, turbine fuel should be stored so that it does not fall to a temperature below its pour point. This requirement will cause no problem with fuels such as JP-4, since the pour point of these fuels is lower than would be encountered at any airport.

Fuel-storage facilities should be designed to provide for adequate settling time for fuel, in order to remove suspended solids and water. If warm fuel is pumped into storage tanks, considerable water may be in solution with the fuel. After the fuel cools, this water will separate from the fuel. The venting or "breathing" of fuel storage tanks also causes water to accumulate. When fuel is pumped out of a tank the fuel is often replaced by warm air which may have a substantial water content. As the air cools within the tank, the water condenses on the sides of the tank and at the fuel surface. Thus it becomes necessary to pump the water out of the bottom of the tank at regular intervals.

Many modern turbine-powered aircraft are designed for "single-point" refueling. This system provides for a great saving of time, since it is not necessary to change fuel nozzles to several different tank openings. The line from the fuel truck is connected to the single refueling valve and the fuel is pumped under high pressure and volume to all fuel tanks. The tanks are, of course, interconnected with cross-feed lines, and the valves are set so that fuel can be taken into as many fuel compartments as desired.

Precautions against the accumulation of static electricity must be carefully observed in all fuel-handling operations. All units in the operation should be connected by ground wires: truck to aircraft, nozzle to aircraft, truck to ground, and aircraft to ground.

STUDY QUESTIONS

1. What are the general requirements for gas-turbine fuels?
2. In gas-turbine fuels, which is more important—heat content per unit volume, or heat content per unit weight?
3. What operations require a fuel of high volatility?
4. Compare high-flash-point fuels with high-vapor-pressure fuels from the safety standpoint.
5. Explain "pour point" and "cloud point."
6. What is the most widely used fuel for gas-turbine engines?
7. Compare the handling of gas-turbine fuels with that of reciprocating-engine fuels.

CHAPTER 8

Fuel Systems and Controls

Problems of fuel-system design

The rapid development in turbojet and turboprop engines since the advent of the Whittle engine has, with the multiplicity of new requirements, added such functional complexities to the fuel system as to make its engineering one of the major tasks in the design of airplanes equipped with either of these types of powerplants.

Among the problems which have arisen are location of fuel tanks, provision for relief of fuel vapors caused by boiling fuel (resulting from rapid decrease in atmospheric pressure when the aircraft climbs to high altitudes), development of automatic scheduling of fuel from the many fuel tanks, and development of engine fuel-control units to meet the complex requirements of the engine. In order to relieve the pilot of the many details required in engine operation and control, it has been necessary to develop controls which would automatically adjust fuel flow to the needs of the engine under all conditions.

With modern high-pressure-ratio engines, the problem of compressor stall and surge has become critical. This has necessitated additional control for fuel flow so that the engine would not be permitted to enter the area of surge and stall.

In airplanes powered with reciprocating engines, the fuel system with which the air-frame engineering staff is concerned begins with the fuel tanks and ends at the carburetor inlet. The only control required between these points is fuel pressure. The rate of delivery is automatically controlled by the carburetor as a function of air-mass flow in the induction system. Thus, the fuel-system engineer has only the relatively simple problem of delivering fuel to the carburetor at the required pressure (which is low) and in sufficient volume to meet the engine demand. Moreover, he has a rather wide choice of proved and reliable fuel-system accessories to do the job.

On the other hand, the gas-turbine fuel system, while beginning at the fuel tanks, has no carburetor and may extend to and even include the fuel nozzles, since some of the control functions may include fuel-nozzle characteristics. The rate of fuel delivery is no longer a function of air-mass flow alone but must be controlled as dictated by several other conditions, some of which may be required to override others according to their momentary relative importance. Moreover, the required fuel pressure ranges from five to twenty times the pressure needed for reciprocating-engine fuel systems, a fact that adds appreciably to the fuel-pump and accessory design problems.

On modern gas-turbine fuel systems, amazing progress has been made in solving the many problems relating to the fuel systems and fuel controls. Adequate fuel control has proved to be one of the principal factors in the development of high-performance gas turbines; however, the modern hydromechanical fuel-control units have performed wonders in meeting the requirements of these engines.

Types of fuel systems

In general, the gas-turbine engine fuel systems found thus far to be the most satisfactory may be divided into two parts:

1. THE AIR-FRAME SYSTEM. This consists of fuel tanks, with their connecting lines; float-operated fuel transfer control valves; selector and shutoff valves to establish an emergency system; fuel-tank booster pumps; heaters; and flow meters.
2. THE ENGINE SYSTEM. This includes all filters, fuel-control units, engine-driven fuel pumps, flow dividers, pressurizing valves, drain valves, after-burner fuel controls, and fuel nozzles.

The engine fuel system usually includes an emergency system to supply fuel to the engine in case of main-system failure. In some cases the emergency system is a duplicate of the main system, while in others the emergency system is not fully automatic and must be controlled carefully by the pilot. With the nonautomatic type of emergency system, the pilot must accelerate and decelerate slowly or there will be danger of rich blowout, lean blowout, stall, or overheating of the turbine area.

THE AIRPLANE FUEL SYSTEM

Requirements

Airplane fuel systems for turbine-powered aircraft usually vary in design from one aircraft model to another; however, there are certain essential requirements which almost any system must meet.

1. Continuous fuel flow must be provided for each engine. The individual engine draws fuel from its own tank, which is called the "engine service tank." This tank is kept filled automatically from the auxiliary tanks until the latter are exhausted.
2. It is necessary to establish a predetermined order for the emptying of the auxiliary tanks so that an automatic system may be designed to control the operation in the most desirable order. This system is controlled by means of floats or electronic sensing devices which signal for the operation of pumps and valves as necessary. The order of emptying may vary between large and small airplanes, or in airplanes of the same size but of different design, to meet stress or trim requirements, but the sequence in all cases should be fixed and fully automatic.
3. Booster-pump equipment should be in all fuel tanks except those which are externally mounted. Fuel may be forced from externally mounted tanks by air pressure. External tanks used with fighter aircraft are often of the "droppable" type and are used merely to give the aircraft a greater range. As soon as the fuel is consumed from these tanks, the tanks are released in order to reduce aircraft weight and drag.
4. Emergency fuel feed to the engine fuel system should be arranged so that fuel from any auxiliary tank may be supplied under adequate booster-pump pressure directly to the engine fuel system.
5. Provision should be made whereby the normal fuel supply to the engine fuel system at boost pressure will be uninterrupted for a specified time in any operating altitude of the airplane.
6. Automatic air-supply cutoffs on the air-frame side of external or internal droppable fuel-tank disconnect joints are needed to shut off the air pressure when the tank is emptied or dropped.
7. Purging systems should be provided to fill the empty tanks with inert gas so that the danger of fire or explosion is reduced. This precaution is particularly important for combat aircraft.
8. It is well to provide for the automatic establishment of emergency fuel feed to the engine in the event of failure of the engine-service-tank booster pump. This automatic emergency system should come into operation at any time that fuel pressure from the engine service tank fails.

Typical airplane fuel systems

Figures 8 · 1, 8 · 2, and 8 · 3 show basic schematic diagrams for one- and two-engine airplane fuel systems which provide for continuous flow and emergency operation. It will be noted that the emergency system operates from the auxiliary tank and that it is a separate system in itself; that is, it can operate independently of the main system. It is obvious that the selector valve and the auxiliary tank shutoff valve must be set properly before the emergency system can oper-

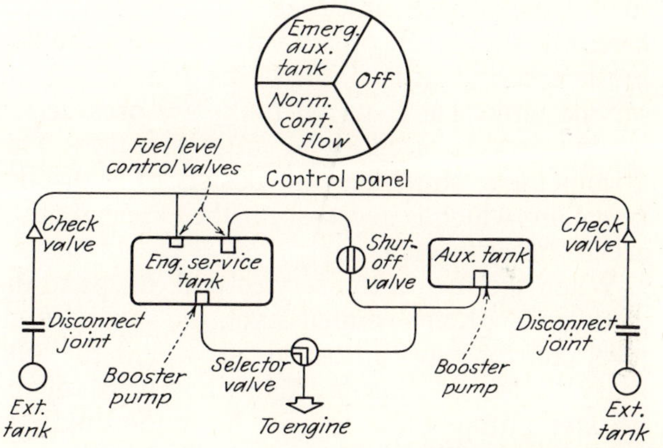

Figure 8 · 1 *Single-engine airplane fuel system.*

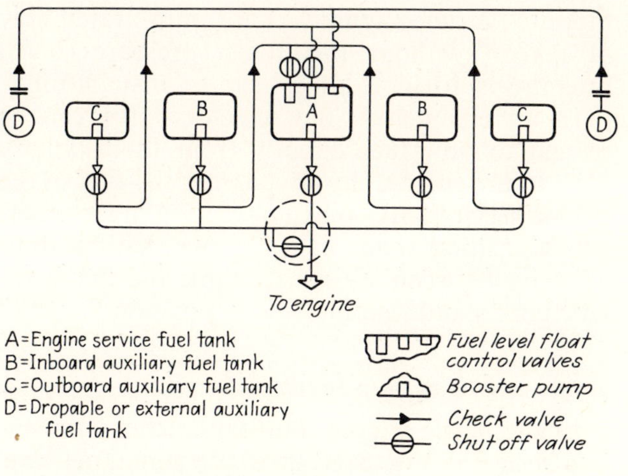

Figure 8 · 2 *Multitank fuel system.*

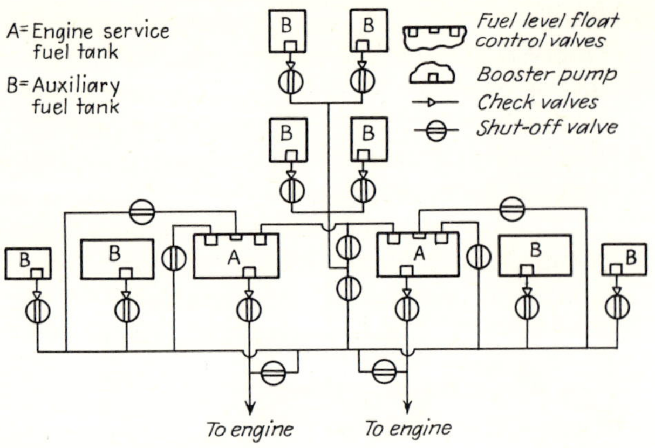

Figure 8 · 3 Two-engine airplane fuel system.

ate. These may be arranged so that they are automatically operated by the emergency control system.

Single-engine system

The fuel flow to the engine in Fig. 8 · 1 originates from the engine service tank. When the fuel in the service tank reaches a predetermined level, the fuel-level control valve permits the fuel in the external jettisonable tanks to flow into the service tank. The external tanks are pressurized by compressor bleed air to cause fuel flow. This arrangement simplifies the design of a disconnect joint which is necessary if the external tanks are to be jettisoned.

When the fuel in the service tank drops to the next lower predetermined level, the second fuel-level control valve automatically opens the auxiliary fuel-tank shutoff valve and turns on the booster pump. Flow is now from the auxiliary tank to the service tank until the fuel in the auxiliary tank is exhausted. The booster pump in the service tank is continuously pumping fuel into the engine, since the main fuel pump mounted on the engine will not pump rated flow at takeoff without this boost pressure (boost pressure is 10 to 15 psig in most installations).

In an emergency, the selector valve can be turned to flow fuel directly from the auxiliary tank to the engine, thus bypassing the fuel in the engine service tank. When operating in the emergency position, it becomes obvious that the shutoff valve between the service tank and the auxiliary tank should be in the OFF position.

Single-engine system (multitank)

In the single-engine multitank schematic system shown in Fig. 8 · 2, there are three fuel-level float control valves set for different predetermined fuel levels in the engine service fuel tank. During normal operation, fuel flow to the engine is obtained first from the engine service fuel tank *A*. The next fuel to be consumed is the fuel in the droppable or external fuel tanks *D*. When the second predetermined fuel level is reached, the fuel level float control valve opens the shutoff valve and turns on the booster pump of the outboard auxiliary tanks *C*.

This operation is repeated when the fuel arrives at the third predetermined level and then fuel flows from the inboard auxiliary fuel tanks *B*. If the booster pump in the service tank should fail, or if anything else should cause the fuel flow to stop, the shutoff valve in the dashed circle could be turned on and fuel flow to the engine would be directly from the auxiliary tanks. During this type of operation, the shutoff valves at the entrance to the top of the service tank would be shut off.

Two-engine system

In Fig. 8 · 3, a typical two-engine airplane fuel system is illustrated schematically. It will be noted in this system that, by using a multitude of shutoff valves, the fuel can be made to flow wherever desired. By utilizing fuel-level float control valves as described in the preceding sections, the fuel flow is normally from the auxiliary tanks *B* to the engine service fuel tanks, and thence to the engines. For emergency service, by manipulating the various shutoff valves, fuel can be made to flow directly from the auxiliary tanks to either engine.

Fuel systems for four-engine aircraft operate in a manner similar to that for a two-engine system; however, it is necessary to use a greater number of valves and pumps. It must be emphasized again that, in any complex system of interconnecting tanks, the arrangement must be such that fuel may be withdrawn from any tank as desired.

On some aircraft, a system such as that shown in Fig. 8 · 3 is equipped with all electrically operated valves and controls, and the electric circuit is such that the system operates in a fully automatic manner, even to the emergency functions.

Military aircraft fuel system

Modern jet-fighter aircraft which operate at supersonic speeds (Mach 2 plus) utilize much more complex fuel systems than those employed in earlier types of aircraft. One such system is shown in Fig. 8 · 4. This system is installed in a Convair F-106 Delta Dart interceptor. This airplane is equipped with a Pratt & Whitney J75

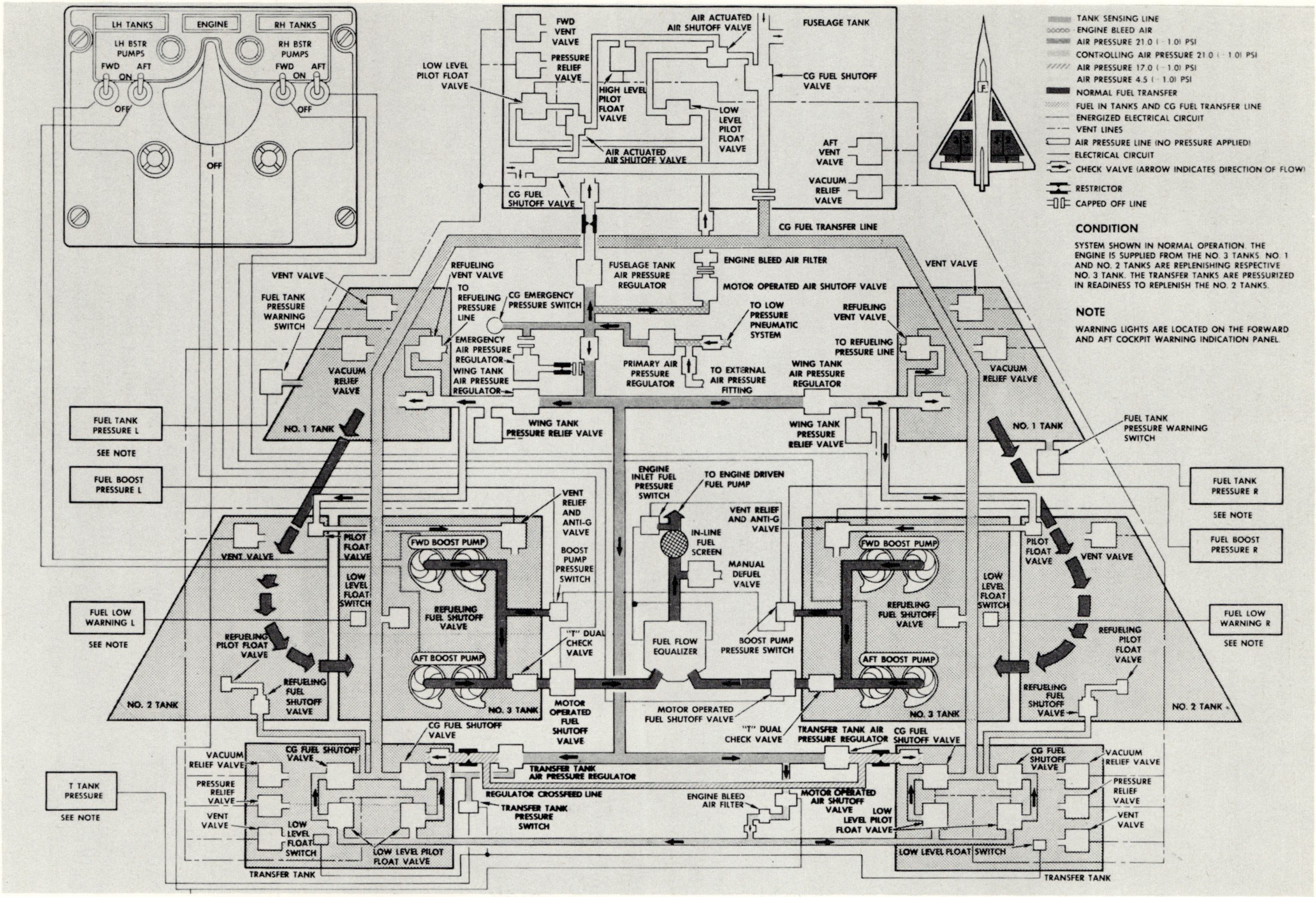

Figure **8 · 4** *Fuel system for Convair F-106 Delta Dart.*

(JT4) engine which can produce more than 26,000 lb of thrust with the afterburner operating and water injection. Without the afterburner operating, this engine consumes more than 11,000 lb of fuel per hr, hence the fuel system must have tremendous capacity. During operation with the afterburner the fuel consumption is at the rate of more than 55,000 lb per hr or nearly 140 gpm.

The fuel for the F-106 airplane is stored in six main wing tanks, three in each wing; two transfer tanks; and one fuselage tank. In the system shown in Fig. 8 · 4, the fuselage tank is not in use. The main tanks are pressurized by engine bleed air to 4.5 $\pm$ 1 psi in order to provide positive pressure for the booster pumps. This is particularly important for high-altitude operation where atmospheric pressure may be lower than 1 psi. The transfer tanks are pressurized to about 17 psi because they are not equipped with booster pumps. The pressurization of the tanks is controlled by means of pressure regulators as shown in the diagram.

During operation of the system, fuel is drawn from the No. 3 tanks by the fuel booster pumps and delivered to the engine pumps through the in-line fuel screen and the fuel flow equalizer. The equalizer regulates and equalizes the flow of fuel from the left and right wing fuel systems. This results in keeping the fuel weight distribution equal between the wings. Fuel enters the equalizer from each wing system, at which time the flow rate of one supply system is made to equal the flow rate of the other system. This is accomplished by the aid of internal valve positioning and control working pressure supplied by the engine fuel pump. The fuel then passes out a single outlet on the forward side of the equalizer and is routed to the engine fuel inlet. The control working pressure for the equalizer is obtained by tapping fuel pressure from the engine stage of the engine fuel pump. Provisions are made within the equalizer so that in case of failure of the control pressure source the equalizer will permit passage of fuel but the unit equalizing action will be inoperative.

As fuel from the No. 3 tanks is used, additional fuel flows from the No. 1 and No. 2 tanks to the No. 3 tanks. This fuel flow is such that the CG (center of gravity) of the airplane is not seriously disturbed. When the fuel level in the No. 2 tanks drops to a predetermined level, the fuel from the transfer tanks is permitted to flow into the No. 2 tanks.

The fuel system is provided with facilities for rapid refueling from a single point. During refueling the pressure is relieved from the tanks through automatic vent valves. When a tank reaches full capacity, the fuel flow is cut off automatically by means of refueling shutoff valves.

The fuel system shown in Fig. 8 · 4 is not the same for all models of the F-106 airplane; however, this system will provide understanding of the requirements and general arrangement of fuel systems for modern jet-fighter aircraft.

Multiengine transport fuel systems

The problems of fuel stowage in the turbojet or turboprop airliner are not so acute as in the case of the fighter-type airplane. The airliner will normally carry all fuel in the wings in order to utilize the fuselage space for passengers and baggage or freight. It is also a requirement that fuel stowage be as far from the passenger compartments as possible.

A typical fuel system for a large propjet (turboprop) airliner is shown in Fig. 8 · 5. The system illustrated shows the installation in one wing only, since the installation for the other wing is similar. The single-point refueling inlet and control panel will normally be located in only one of the wing nacelles.

In the illustration, two large wing tanks are shown built integral with the wing structure. The capacity of these two tanks is 2,640 gallons, thus making the total fuel capacity for the airplane 5,280 gallons.

Fuel flow to the engines is provided by booster pumps located at the rear inboard portion of the tanks. It is pumped through a check valve, a fuel-tank shutoff valve, and a firewall shutoff valve. These valves are manually operated. In the engine nacelle, the fuel passes through a heater, a strainer, and a flowmeter before continuing to the engine-driven fuel pump.

Cross-feed lines and valves are provided so that any engine can draw fuel from any fuel tank and so that fuel may be transferred from one tank to another. The cross-feed valves are electrically operated. In order to transfer fuel from No. 1 tank to No. 2 tank, the cross-feed valves for both tanks are opened. The No. 1 fuel-booster pump is turned on, and the No. 2 booster pump is turned off.

Single-point fueling is accomplished through the **fueling adapter valve** located at the rear of the right inboard nacelle. Fuel from an external tank is directed under pressure through the fueling adapter valve to fuel lines leading to all tanks. At the rear of each tank is a pressure-operated fueling valve and a check valve. When fueling pressure is applied, the pressure-operated

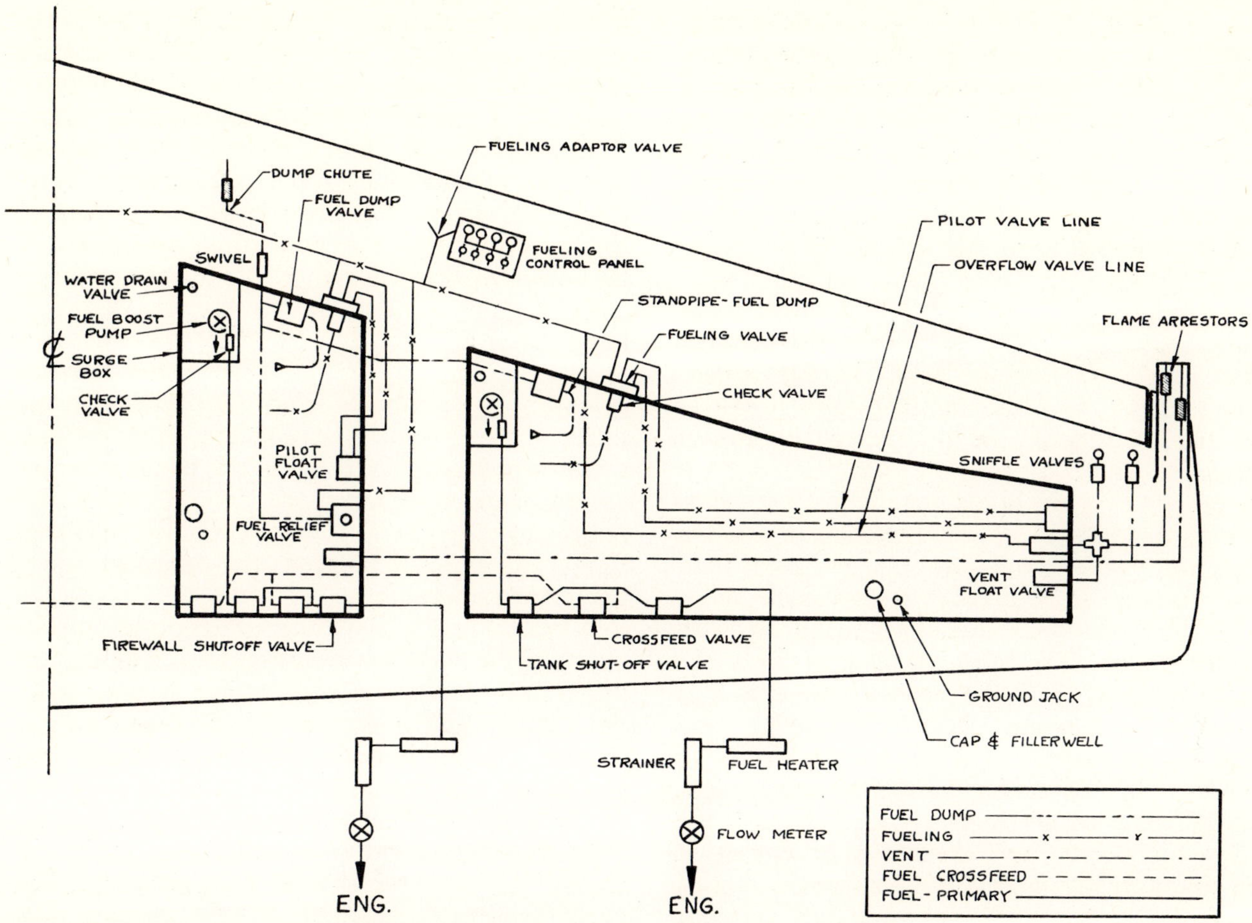

Figure 8 · 5 Fuel system for a turboprop airliner.

valve opens and allows fuel to flow into the tank. The check valve permits fuel to flow **into** the tank but prevents flow outward.

The fueling system is so designed that fuel flow to any tank will be shut off automatically when the tank is full. This is accomplished through the operation of the **fueling pilot float valve** which causes the fueling valve to close. The system is provided with a **fueling relief valve** to prevent damage from excessive pressures.

The **fueling control panel** is located near the fueling adapter valve at the rear of the right inboard nacelle. A drawing of this panel is shown in Fig. 8 · 6. The control panel contains fuel quantity indicators, indicator lights to show condition of overflow valves and pilot valves, tank shutoff valve switches, and the **fueling-system power switch.** With this panel are the following instructions for the operation of the system:

1. To fuel airplane, actuate fueling-system power switch to ON and apply fuel pressure at nozzle with tank shutoff valve switch in FUELING position. Make pilot-valve check-out when tanks are approximately one-quarter and three-quarters full. Fuel-flow shutoff is automatic at full level when pilot valves function properly.
2. To shut off fueling flow at tank levels below full, actuate tank shutoff-valve switch to MANUAL SHUTOFF, return to MANIFOLD EVACUATION for evacuating fueling manifold.
3. For pilot-valve check-out, actuate tank shutoff valve to MANUAL SHUTOFF and wait 10 sec. If pilot valves are functioning, indicator light should be illuminated.
4. To defuel airplane, actuate tank-shutoff valve switch to DEFUEL. After defueling operation, tank-shutoff valve switch must be returned to the normal position (FUELING and MANIFOLD EVACUATION).
5. Fueling-system power switch must be returned to OFF position at all times when fueling operations are not being conducted.

Fueling can also be accomplished through overwing filler openings on the top of each tank. These are provided so that fueling operations may be performed at airports where single-point

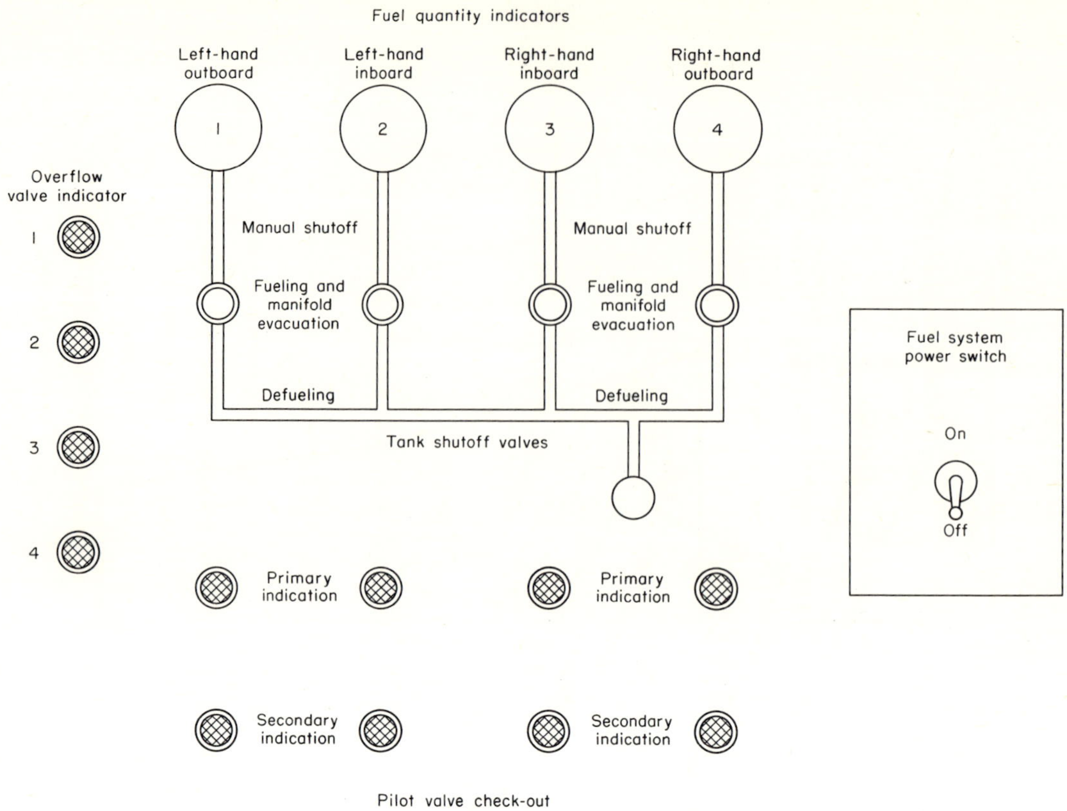

Figure 8 · 6 Fueling control panel for turboprop airliner.

fueling equipment is not available. Ground jacks are provided near each overwing fuel cap so that the fuel nozzle may be grounded to the airplane.

The fuel tanks in this system are equipped with vents to prevent excessive pressures from developing in the tanks. The vent lines are connected to the tanks through vent float valves. The vent discharge passes through flame arresters at the wing tips.

Provision for the dumping of fuel is made through manually operated dump valves which are connected to a common retractable dump chute. The tank outlet to the dump valve is through a standpipe located in each tank. The standpipe prevents complete dumping of all fuel and resulting engine failure from fuel starvation.

Water drains are located in the lowest portion of each tank which is the rear inboard corner. These water drains should be opened daily and during preflight inspection.

The fuel system illustrated in Fig. 8 · 5 is probably typical of systems for both turbojet and turboprop airliners, although the system for each type of airplane will have some differences. For exact details of any particular system, the manufacturer's manual should be consulted.

FUEL SYSTEM FOR JET AIRLINER

The fuel system for a modern jet airliner is shown in the diagram of Fig. 8 · 7. This diagram shows the arrangement of the fuel system in the left wing of a Boeing 720 airliner. The system in the right wing is the same in general arrangement.

This system consists of fuel tanks and component systems which ensure that the engines are supplied with fuel in the proper amount at all times. The fuel is distributed in four main tanks, two reserve tanks, and a center wing tank. Fuel tanks and all system components are suitable for use with any acceptable fuel which conforms to the engine manufacturer's specifications. The most commonly used fuel is jet-type kerosene.

Fuel tanks

All the fuel tanks for the 720 airliner are located in the interspar area of the wing structure between the front and rear spars, as shown in Fig. 8 · 8. The wing skin and ribs form the walls of the tanks, and intermediate ribs serve as baffle plates. The four main tanks and two reserve

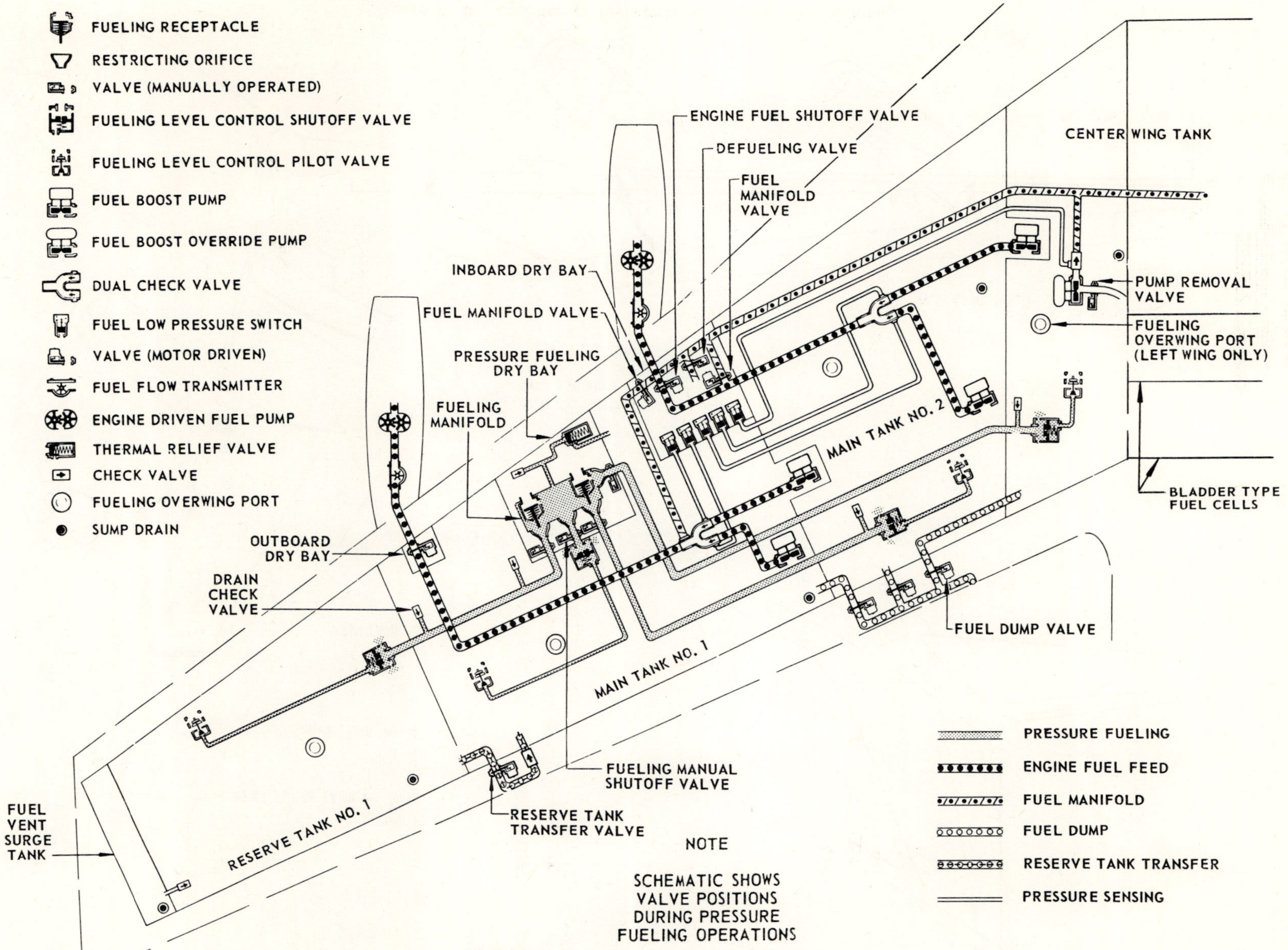

Figure 8 · 7 Fuel system for Boeing 720 airliner.

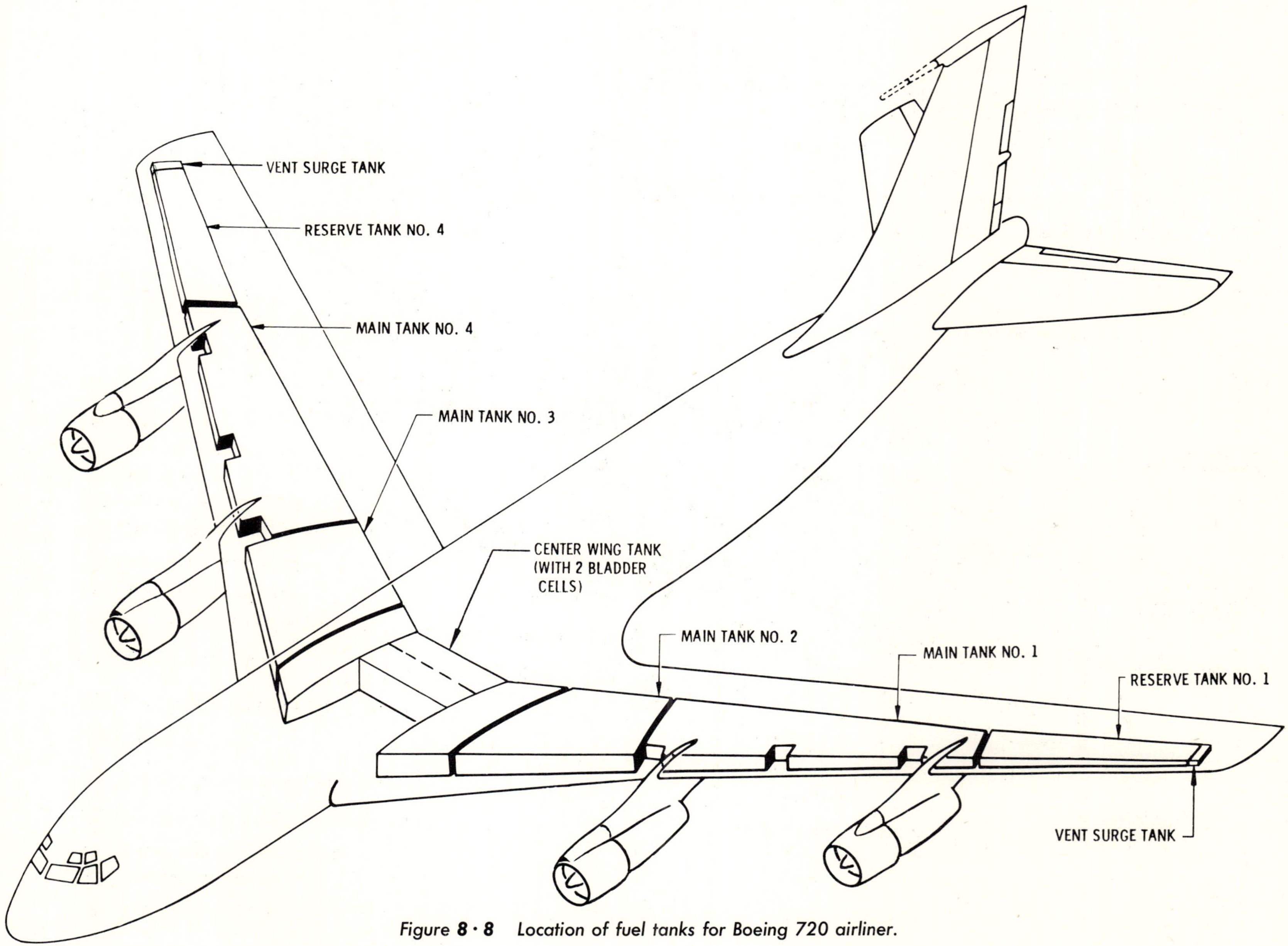

*Figure **8 · 8** Location of fuel tanks for Boeing 720 airliner.*

tanks are completely integral with the wing structure. The center wing tank consists of two integral sections at the wing roots and a center wing cavity containing bladder-type fuel cells. All fuel tanks are sealed fluid tight by means of sealing compound and sealed fasteners.

Vent systems

The fuel vent system provides positive venting to atmosphere of all fuel tanks, fuel cells, and cavities, thereby preventing excessive internal or external pressures across the tank walls during all flight maneuvers. Venting of the four main tanks, two reserve tanks, and integral sections of the center wing tank is accomplished with sealed spanwise upper-wing-skin stiffener ducts interconnected with appropriate drains and vent tubing which connect to vent ports located in critical areas of the tanks. This interconnected system is vented into a surge tank at the wing tip and overboard through a vent scoop. The center wing cavity is vented overboard through a separate vent and drain system. The cavity vent system maintains atmospheric pressure on the outside of the bladder-type fuel cells. The fuel cells are vented into the integral sections of the center wing tank.

Each fuel tank is provided with two vents located to assure that one vent will be open at all times. The vents are equipped with vent valves which close when the fuel level rises to a point where fuel might flow out of the vent. The vent system feeds into a plenum chamber at the end of the wing. This chamber or surge tank is pressurized by means of a ram-air scoop, thus assuring positive pressure in the fuel tanks.

Engine fuel-feed system

The engine fuel-feed system consists of fuel lines, pumps, and valves which distribute the fuel to the engines. It consists of four tank-to-engine fuel-feed systems which are interconnected by a fuel manifold such that fuel may be delivered from any main tank or the center wing tank to any or all engines. The two reserve tanks store fuel and supply the fuel by gravity flow to main tanks No. 1 and No. 4 through electrically operated transfer valves. The fuel-feed line from each main tank is pressurized by two booster pumps which are controlled by separate switches and independent circuits so that engine operation will not be affected by power failure to any single booster pump. The center-wing-tank booster pumps, known as the fuel-boost override pumps, will override the main-tank booster pumps to supply fuel through the fuel manifold to the engines. The distribution of fuel to the engines is controlled by electric-motor-driven slide valves in the fuel lines. The valves are classified into three groups: the **engine-fuel shutoff valves** that shut off fuel to the engines; the **fuel manifold valves** that control manifold distribution; and the **reserve-tank transfer valves** that transfer fuel from the reserve tanks to main tanks No. 1 and No. 4. They are controlled by manually-operated switches located on the flight engineer's lower panel in the control cabin.

Pressure-fueling system

Pressure fueling is accomplished from a station located on the lower surface of each wing between the inboard and outboard engine nacelles. Each pressure-fueling station consists of two fueling-nozzle ground connectors, two fueling receptacles, and a fueling manifold. Four **fueling manual shutoff valves** at each fueling station permit the fuel tanks to be serviced individually. Each station services the main and reserve tanks on its respective side of the airplane. The center wing tank is serviced from either wing station. The maximum fuel delivery pressure is 50 psi. A restricting orifice plate at each fueling manual shutoff valve limits the flow rate of the respective tank. The approximate flow rates are as follows: main tanks No. 1 and No. 4, 234 gpm; main tanks No. 2 and No. 3, 227 gpm; and reserve tanks No. 1 and No. 4, 65 gpm. The center wing tank rate is 175 gpm per side.

Each tank has a **fueling level control pilot valve** which actuates a **fueling level control shutoff valve** to automatically shut off the fuel flow when the maximum tank capacity has been reached. A fueling preset system permits the tanks to be fueled to a predetermined level without constant observation of the fuel quantity indicators. The fueling preset system controls are located on the flight engineer's panel. Preset switches and potentiometers are actuated in conjunction with fuel quantity indicators. When it is desired to preset the fueling, the preset switch is placed in the FUEL position. The fueling preset potentiometer is then adjusted until the fuel gage shows the amount of fuel which is to be added to the tank. As fuel is pumped into the tank, the gauge needle moves toward the zero mark. When the needle reaches 0, the fueling preset indicator light comes on and shows that the desired quantity of fuel has been added.

Fuel-dump system

The fuel-dump system consists of lines, valves, dump chutes, and chute operating mechanisms and provides the means for dumping fuel in flight. Each wing contains an extendable dump chute which must be fully extended before fuel can be dumped. Each chute is fed fuel by dump lines from the main and center wing tanks on the respective side of the airplane. The reserve tank fuel is transferred to and dumped with the main outboard tank fuel. The controls are located on the flight engineer's auxiliary panel. When all dump valves are open, the system will dump all fuel in excess of a predetermined amount.

Indicating systems

The fuel-quantity indicating system incorporates electric capacitance tank probes mounted internally in each fuel tank. The probes have a compensator for fuel density variations, thus assuring an accurate indication under all conditions of density. The indicators are calibrated to show the weight of fuel remaining in each tank and are located on the flight engineer's lower panel. The fuel-flow indicating system provides a visual indication at the pilot's station of the fuel consumption rate for each individual engine. A flow transmitter is located in each engine nacelle strut.

The fuel-temperature indicating system provides a means for checking the temperature of fuel in main tank No. 1 and in the fuel line at each engine. The system consists of an indicator and a five-position selector switch on the flight engineer's lower panel. A temperature bulb is located in main tank No. 1 and in the outlet side of the fuel filter at each engine.

Engine fuel de-icing system

Each engine has a manually controlled engine fuel de-icing system which removes any ice formed in the fuel before it reaches the fuel-control unit. The principal components of each de-icing system are a fuel filter, an air control valve, and a fuel de-icing heater. The control switches and icing indicator lights are located on the flight engineer's lower panel.

The fuel filter is fitted with a pressure warning switch which detects filter icing and then turns on the fuel icing indicator light on the flight engineer's lower panel. The air control valve regulates the flow of high-pressure discharge air to the heater through which all the fuel passes. A fuel-temperature bulb is located on the outlet side of the filter to monitor the fuel-out temperature.

Miscellaneous components of the fuel system

The fuel system includes a variety of units and devices called miscellaneous components. These include fuel filler cap assemblies, baffle check valves, defueling valves, fuel sump drain cocks, and drip sticks. All fuel tanks have an exterior filler cap for individual tank servicing, or the tanks can be individually or simultaneously filled at the pressure fueling stations. The main tanks have baffle check valves installed in the bottom of the inboard ribs to allow fuel to flow inboard toward the booster pumps during normal flight attitude and to prevent fuel from flowing away from the pump area during turbulence or airplane maneuvers.

Pressure defueling of the main tanks or center wing tank is accomplished through a defueling valve located in the inboard **dry bay** of each wing. The dry bay is an area inside the wing which is separate from the fuel tank areas and is the location of certain valves and other components which must be accessible outside the fuel tanks. The fuel booster pumps in the main tanks or the override pumps in the center wing tank deliver fuel to the defueling valves through their respective fuel manifold valves. The defueling rate is approximately 50 gpm for each tank. Alternatively, fuel may be drawn from the tanks through the defueling valve by means of the truck defueling pump. The reserve tanks are defueled through the reserve-tank transfer valve to the adjacent main tanks. Residual fuel in each tank may be drained through fuel sump drain cocks located in the bottom of each tank.

In addition to the quantity indicating system, provisions are made to determine the quantity of fuel in each tank, with the airplane in the taxi attitude, by calibrated **drip sticks** located in the wing lower surface. These sticks consist of scaled tubes which normally extend from the bottom of the wing upward into the fuel tanks. When it is desired to determine the amount of fuel in a tank, the stick is turned to disengage a bayonet lock, and then the stick is withdrawn downward from the wing slowly. As soon as fuel appears at the drip hole, a reading is taken from the scale on the stick to determine the depth of the fuel in the tank. The upper end of the stick is open, and when it is withdrawn to the level of the fuel, fuel flows into the stick and out through the hole at the bottom.

Fuel lines

Where possible, fuel lines for the Boeing 720 airplane are contained within the tank boundaries. They are fabricated mainly from standard aluminum alloy aircraft tubing and are connected to the various fuel system components with standard NAS or MS unions and couplings. Access to all tank equipment and fuel lines is provided externally by access panels in the lower wing surface and internally by openings in the tank ribs which are large enough to allow maintenance technicians to move through the ribs for inspection and maintenance.

Fuel-tank sealing

In order to provide fuel-tight integral tanks in the wings of the aircraft, the tank seams and fasteners are carefully sealed during manufacture. Sealing is accomplished by close metal-to-metal fit of all parts and fasteners and by the application of sealant compound to the seams as required. Repairs to damaged fuel tank sealant must be made with approved materials, using approved methods and tools. The sealant materials used are synthetic rubber base and are essentially nontoxic. When properly cured, the sealant is resistant to heat and extreme cold as well as to fuel. Complete instructions for repairing leaks in the fuel tanks are given in the manufacturer's maintenance manual for the airplane.

GAS-TURBINE FUEL CONTROLS

Need for fuel control

The control of the reciprocating engine is accomplished by varying the airflow to the engine. When the throttle is advanced or retarded, the airflow is correspondingly increased or decreased. The carburetor is designed to meter the correct amount of fuel automatically in proportion to the amount of airflow. In the case of the gas-turbine engine, the amount of air consumed is so great that a mechanism suitable for controlling the airflow would necessarily be heavy and cumbersome. For this reason, it is much more convenient and economical to control fuel flow. This, of course, makes the jet-engine fuel system entirely different from that of the reciprocating engine. The familiar reciprocating-engine carburetor has been eliminated, and a new assortment of controls and devices has been developed.

Elements affecting fuel control

The operation of a gas-turbine engine requires that a number of variable conditions be taken into consideration to provide for safe, efficient operation. Among these conditions are engine rpm, acceleration, tailpipe temperature, compressor inlet temperature, compressor inlet pressure, compressor discharge pressure (burner pressure), and throttle or power control setting. All these conditions affect or are affected by fuel flow; hence the design of a fuel-control unit or system must include devices responsive to these conditions.

The controlling of a gas-turbine engine is a complex process, and as is usually the case, there are several different types of systems that may be developed to do the job. The basic function of each, however, is the same: to provide the correct fuel flow for the most effective and safe engine operation.

Development of fuel control

It would be possible to control a gas-turbine engine manually with a simple throttling valve and obtain fair results. In fact, this has been done satisfactorily for ground-testing operations, and with some modification, it is done in emergencies in flight. It is necessary that instruments be provided to show tailpipe temperature, rpm, and burner pressure. The operator can watch the operating conditions carefully and adjust the fuel flow to meet the required conditions. Manual operation is not satisfactory for the varying flight conditions encountered with modern, high-speed jet aircraft. Every change in altitude, airspeed, temperature, and aircraft position usually requires an adjustment in fuel flow. For this reason, it is beyond the capabilities of a pilot to make the necessary accurate adjustments as rapidly as they may be required; hence automatic or semiautomatic controls are necessary.

Early fuel-control systems utilized separate devices to make the necessary fuel-flow correction to control rpm, pressures, and temperatures; however, later systems have incorporated all the principal controlling devices in one automatic fuel-control unit.

The barometric, or constant-displacement, system

Figure 8 · 9 is a schematic diagram of the early barometric system used with a number of different turbojet engines. It is a semiautomatic system, generally described as a "constant-displacement system."

The booster pump for the system supplies fuel at 20 psi, through the low-pressure filter, directly to the fixed-displacement, gear-type main fuel pump. This pump discharges fuel at a pressure up to 600 psi, depending on flow requirements. A barometric unit and an **overspeed governor** are connected in parallel with the main pump. The barometric compensates for ambient pressure and ram pressure, and the governor prevents overspeeding of the rotor section. The high-pressure fuel flows through the high-pressure filter to the piston-type control valve which is actuated by the throttle. From the control valve, fuel flows through the drip valve to the fuel manifold

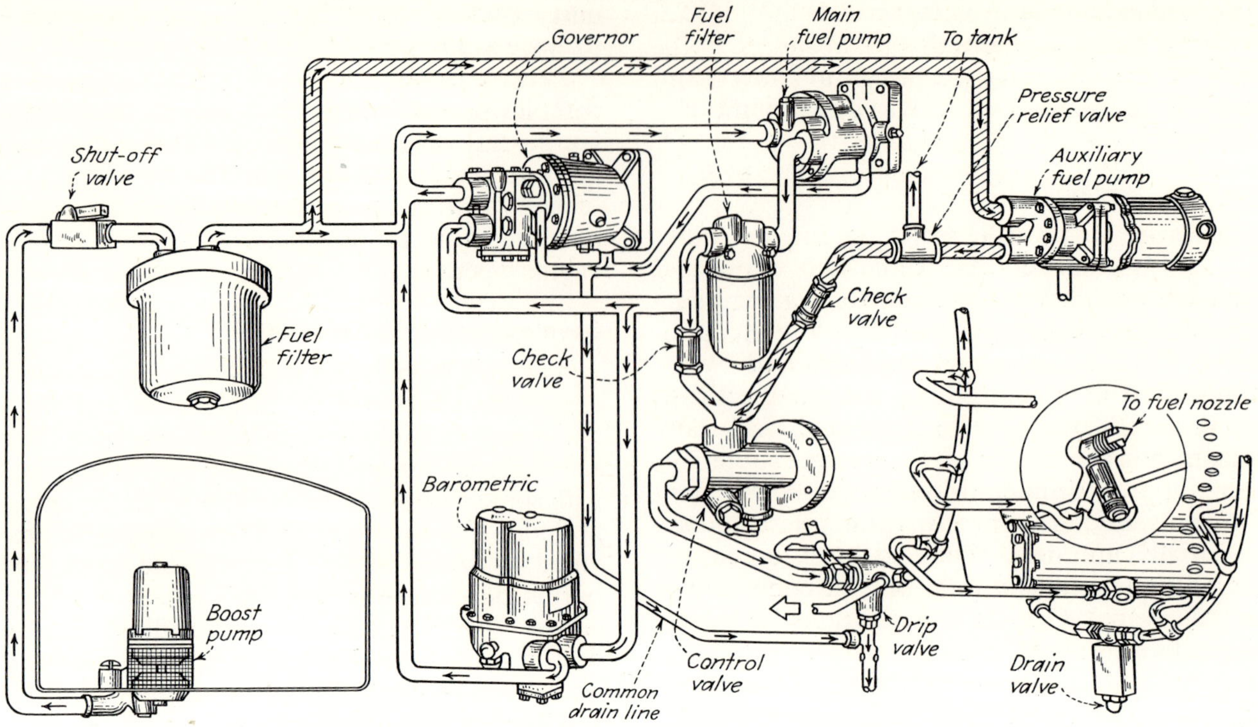

Figure 8 • 9 Fuel system with barometric control and governor.

and thence to the nozzles in the combustion chambers. A high-pressure system is utilized to obtain atomization of the fuel in the combustion chambers.

It must be remembered that the booster pump shown in the fuel tank is a part of the airplane fuel system. This pump is in operation at all times in order to pressure-load the main fuel pump.

A **fuel filter,** capable of removing foreign particles on the order of 10 to 25 microns (millionths of a centimeter), must be installed at the entrance to the engine fuel system to protect the engine-driven fuel pump and the various control devices. An additional filter of suitable design should be installed immediately upstream to the fuel nozzles. Modern fuel-control units are designed to "eat dirt" to a certain extent; however, it is desirable to prevent the entrance of foreign matter to the degree possible with an effective filter.

The **main fuel pump** shown in the illustration of Fig. 8 • 9 is mounted on the accessory section and is engine-driven. Reduction gearing in the accessory section reduces the pump speed to its rated rpm. It is a fixed-displacement pump, employing gears for the pumping elements.

Fixed-displacement pumps have the disadvantage of requiring a relief valve to care for the excess fuel pumped during periods of low demand by the engine. Since the relief valve, in effect, causes recirculation of the fuel (often at high pressure), considerable heat is evolved that may cause the temperature of the fuel at the pump inlet to rise to a point where fuel-pump performance suffers by reason of vapor lock or cavitation. At extreme altitude, this condition is relieved to some extent by incorporating an altitude element in the original relief-valve mechanism that will cause the relieving pressure to be lowered as altitude is gained.

In practice, this appears to be a correct procedure, since the lower fuel rates prevailing at high altitude do not require the high pressure needed by the much higher fuel rates at low altitudes. Moreover, the lower pressure makes it easier for all controls to function at low fuel rates. Regardless of the improved operation gained by the introduction of altitude depression of the fuel pressure, sufficient recirculation temperature rise may still occur to cause malfunctioning of the fuel pump with fuels having vapor pressures appreciably higher than that of kerosene. Cooling of the recirculated fuel is a logical recourse, though both weight and drag problems are introduced. There is also the matter of space availability, which may be difficult to provide. Of the fixed-displacement pumps, the gear type is the simplest and most popular but it has a short life, particularly when pumping the higher vapor-pressure fuels.

The **barometric unit** is designed to control the fuel-pump outlet pressure as a function of ambient pressure and ram pressure. This is accomplished by bypassing fuel from the pressure side of the pump back to its inlet. The remainder of the fuel passes through the control valve to the nozzles.

Figure 8 · 10 shows how the barometric unit functions. The control valve is loaded by the control-valve spring within the control piston. The metering characteristics of the control valve are obtained by the angle or shape of the V notches cut in its circumference. When the pressure below the valve rises, as a result of throttling the engine, the valve rises against the spring force, thus enlarging the effective opening adjacent to the V notches and allowing more fuel to pass back to the pump inlet. This portion is no more than a conventional relief valve. The compensation of fuel pressure, as a function of ram air and pressure altitude, is accomplished through changing the spring force by means of the control piston. The piston is moved by fuel pressure controlled through the pilot valve and the various linkages.

During a descent or an increase in airspeed, it is necessary to increase fuel flow to maintain rpm. Compensation for ram pressure, which gives both altitude and airspeed compensation, is accomplished through the differential-bellows system. The upper bellows is evacuated and the lower bellows is vented to ram air at the compressor inlet. The ram-air pressure elongates the lower bellows against the force of the adjustable bellows spring. This increases the force on the control-valve spring through the pilot valve and control piston. The result is that less fuel is relieved back to the pump inlet, and thus more fuel flows to the nozzles. This action takes place either for a descent or for an increase in airspeed.

The lever mechanism that moves the pilot valve has a fulcrum point on the movable control piston. This provides a follow-up effect to bring the control piston gradually to its new position.

The barometric unit can be called an automatic altitude-compensating relief valve. Although mounted on the accessory section of the engine, it is a nondriven accessory.

Figure 8 · 11 illustrates a typical overspeed governor. This particular governor is designed to limit fuel flow in case of overspeed conditions. It does not function at normal or lower speeds.

In operation the overspeed governor is mounted on an accessory pad and is driven by a splined shaft. Centrifugal force causes the flyweights to move outward, thereby raising a spindle against the tension of a governor spring. This motion is transmitted through the adjusting lever and the linkage bars, and moves the pilot valve to the right.

Overspeed of the engine causes the pilot valve to move so far to the right that small portholes are uncovered. High-pressure fuel flows through the portholes to the piston and causes it to move to the left. The motion of the piston is transmitted through linkage to the governor valve which rotates to the open position, thus permitting fuel to be returned to the suction side of the fuel pump.

The pilot-valve position **during normal speeds** is such that fuel is introduced on the spring side of the piston, thus keeping the governor valve closed.

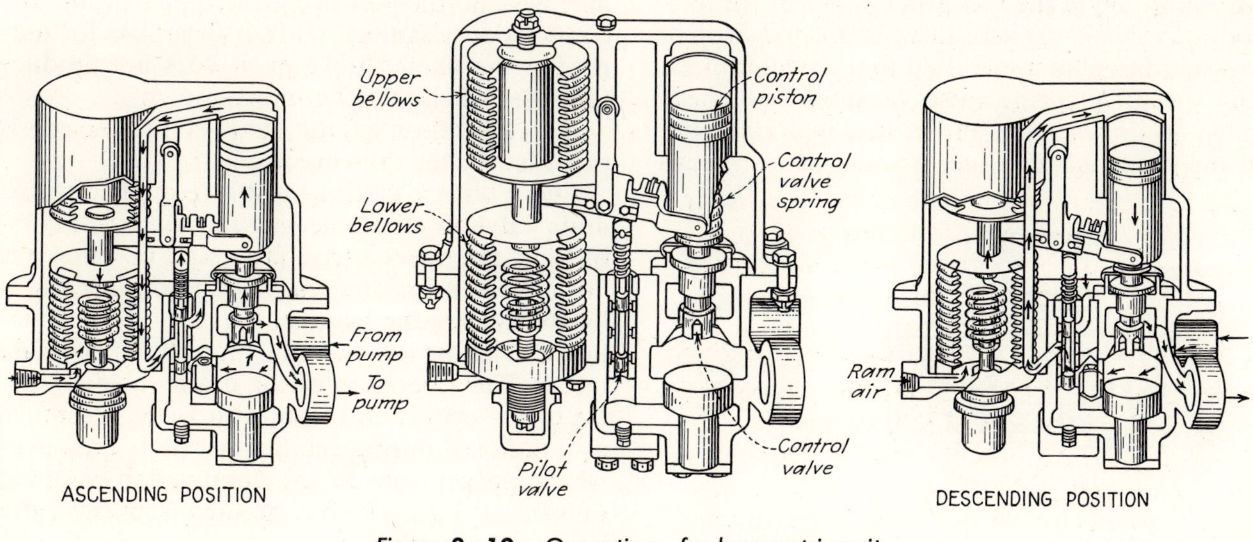

Figure 8 · 10 Operation of a barometric unit.

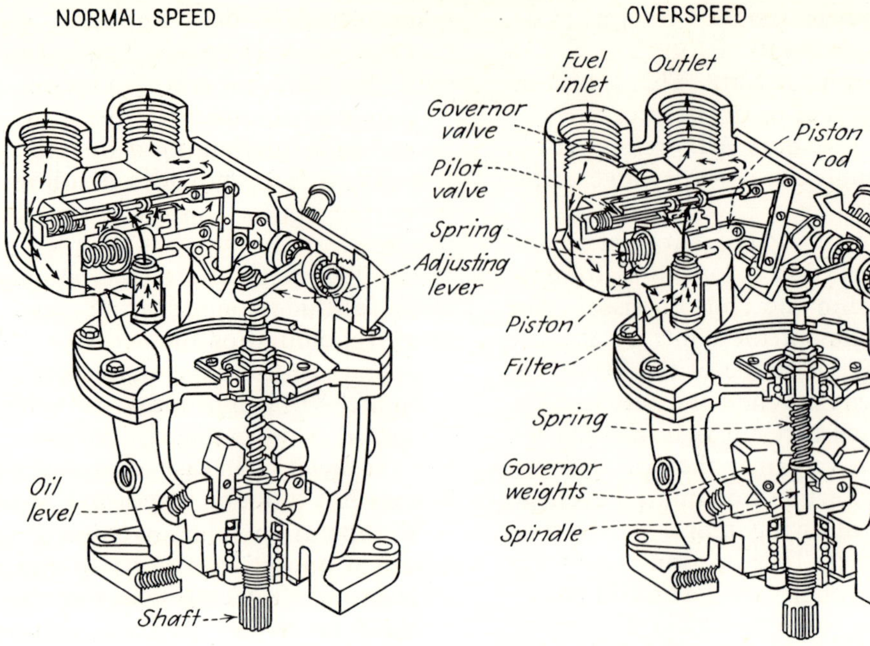

Figure 8 · 11 Overspeed governor.

The **control valve** in the barometric system is operated by the throttle. The function of the control valve is to adjust pressure to the fuel manifold and thus control the amount of fuel flowing into the combustion chambers. Provision is made for setting idle speed by means of an adjusting screw that regulates the metering of fuel through an idling passage.

The **drip valve** is located in the lowermost portion of the fuel manifold. Its function is to give a clean cutoff of fuel flow in the combustion chamber when the control valve is placed in the CLOSED position, thereby preventing fuel dribble into the combustion chamber with consequent carbonization of the fuel nozzles. When the fuel-control valve is CLOSED, there is a rapid drop of pressure in the fuel manifold to a predetermined low value. A spring raises a small ball-check valve off its seat, and all the fuel is drained out of the manifold. The fuel is dumped overboard instead of dribbling into the combustion chambers; hence a clean cutoff of fuel results.

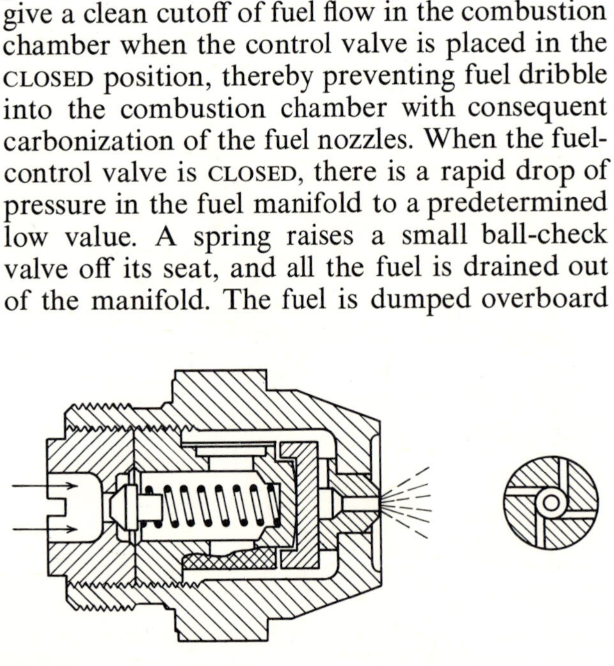

Figure 8 · 12 Simplex fuel nozzle.

The **fuel nozzle** for the system illustrated in Fig. 8 · 9 is the simplex type with a fixed orifice. The simplex nozzle was one of the first forms used in a turbojet fuel system for aircraft, and it still may be found in some systems. It is generally considered to be rather inadequate because a single orifice cannot produce a satisfactory spray pattern under the wide variations of fuel pressure and flow encountered in a turbojet engine.

At low pressures, the arc of spray from the orifice is approximately 70 to 80°, which is desirable for starting. At higher pressures, the arc narrows and the flow is extended farther into the combustion chamber. This is desirable for narrowing the flame; however, it does not produce the most efficient fuel combustion.

An illustration of the simplex fuel nozzle is provided by the drawing in Fig. 8 · 12.

Every turbine engine must be equipped with a **drain valve** to eliminate raw fuel from the combustion chambers after a false start or when fuel has collected for any reason. The drain valve is located under the lowermost combustion chamber, and all the other combustion chambers drain into it. The operation of the drain valve is similar to the operation of the drip valve, inasmuch as it is closed during engine operation and opens when the pressure in the combustion chamber falls below 2 psi. The fuel drained from this valve is dumped overboard.

The emergency system

The emergency system utilized with the barometric fuel system is simply an auxiliary pump which bypasses the overspeed governor and the barometric unit. If any of the units of the main system fail, the emergency pump is turned on. It is then the responsibility of the pilot to control manually the fuel flow to the engine.

It is apparent that the emergency system should be used only for returning the airplane to base after failure of the main system. The pilot would be unable to cope with the many adjustments required during combat operations.

Emergency systems for more advanced fuel systems require less effort on the part of the pilot because of automatic features. These will be discussed later in this chapter.

VARIABLE-DISPLACEMENT FUEL SYSTEM

General description

Variable-displacement fuel systems have been used with several different gas-turbine engines, two of these being the J47 and the Rolls-Royce Dart. Such a system is shown in Fig. 8 · 13.

During operation, fuel flows from the boost pump in the fuel tank, through the filter, and to the main fuel pump. The variable-displacement pump delivers a fuel pressure to the engine proportional to a control oil pressure provided by the regulator. Since the system employs oil as a working fluid, the system may also be called a hydraulic fuel-control system. The regulator varies a control-oil pressure to hold a desired engine speed and to prevent overtemperature as indicated by the action of a thermocouple installed in the tailpipe.

Mounted on the fuel pump is a double check valve to provide fuel to the emergency system. From the fuel pump, flow is to the stopcock which provides for positive shutoff in the closed position. From the stopcock the fuel flow is to the flow divider, to the double drip valve, and finally to the fuel nozzles. The flow divider is used to proportion fuel into the low-pressure and the high-pressure lines leading to the small-slot and large-slot sections of the fuel nozzles. The double drip valve drains the fuel manifold after a false start and when shutting down the engine.

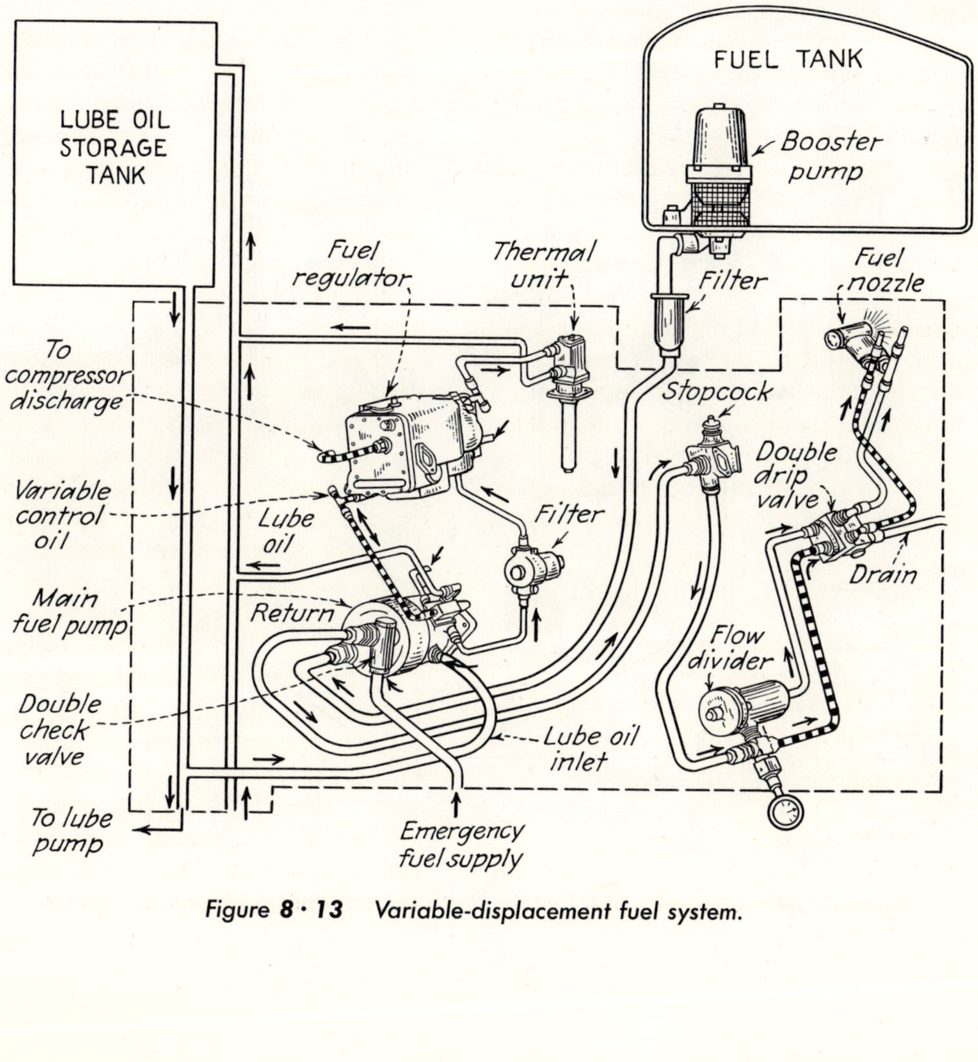

Figure 8 · 13 Variable-displacement fuel system.

Description of units

Variable-displacement Pump. This multipiston pump utilizes a cam plate or wobbler to give a reciprocating motion to the pistons. The angle of the wobbler determines the length of the piston stroke; hence it also determines the output of the pump. This angle is established by the movement of a small servo piston in the pump. The servo piston moves according to the oil pressure received from the fuel regulator.

The Fuel Regulator. Figure 8 · 14 is a schematic diagram illustrating the operation of the fuel regulator as it controls the output of the variable-displacement pump.

The fuel-discharge pressure from the pump is sensed at all times by the **restored hydraulic relay** which regulates fuel pressure in constant ratio to the control-oil-outlet pressure from the main hydraulic relay in the regulator. This main hydraulic relay may be actuated in any one of three ways:

1. THROTTLE CONTROL. Note that a movement of the throttle in the direction of the arrow diverts oil from the pilot valve of the main hydraulic relay into the regulator case. This reduces the variable-control-oil pressure, and the engine slows down because less fuel flows through the burner nozzles.
2. THERMAL CONTROL. When overheated, the thermal unit will drop the pressure acting on its power piston and through the linkages will, in turn, reduce the pressure in the main hydraulic relay.
3. GOVERNOR CONTROL. Through the governor power piston, the governor can override the setting of the throttle. The governor has a topping action and thus provides constant speed control.

Compensation for altitude and airspeed is provided by the motion of a rod joining the two bellows, one being an aneroid and the other being vented to compressor-discharge pressure. The bellows motion is hydraulically relayed through a servo piston and nutcracker linkage to modify the restoring effect of the main spring. This changes the range of control-oil pressure and hence fuel pressure. In this way, the governor droop is scheduled out, so that the rpm is maintained nearly constant for a given throttle setting for varying airspeeds and altitudes.

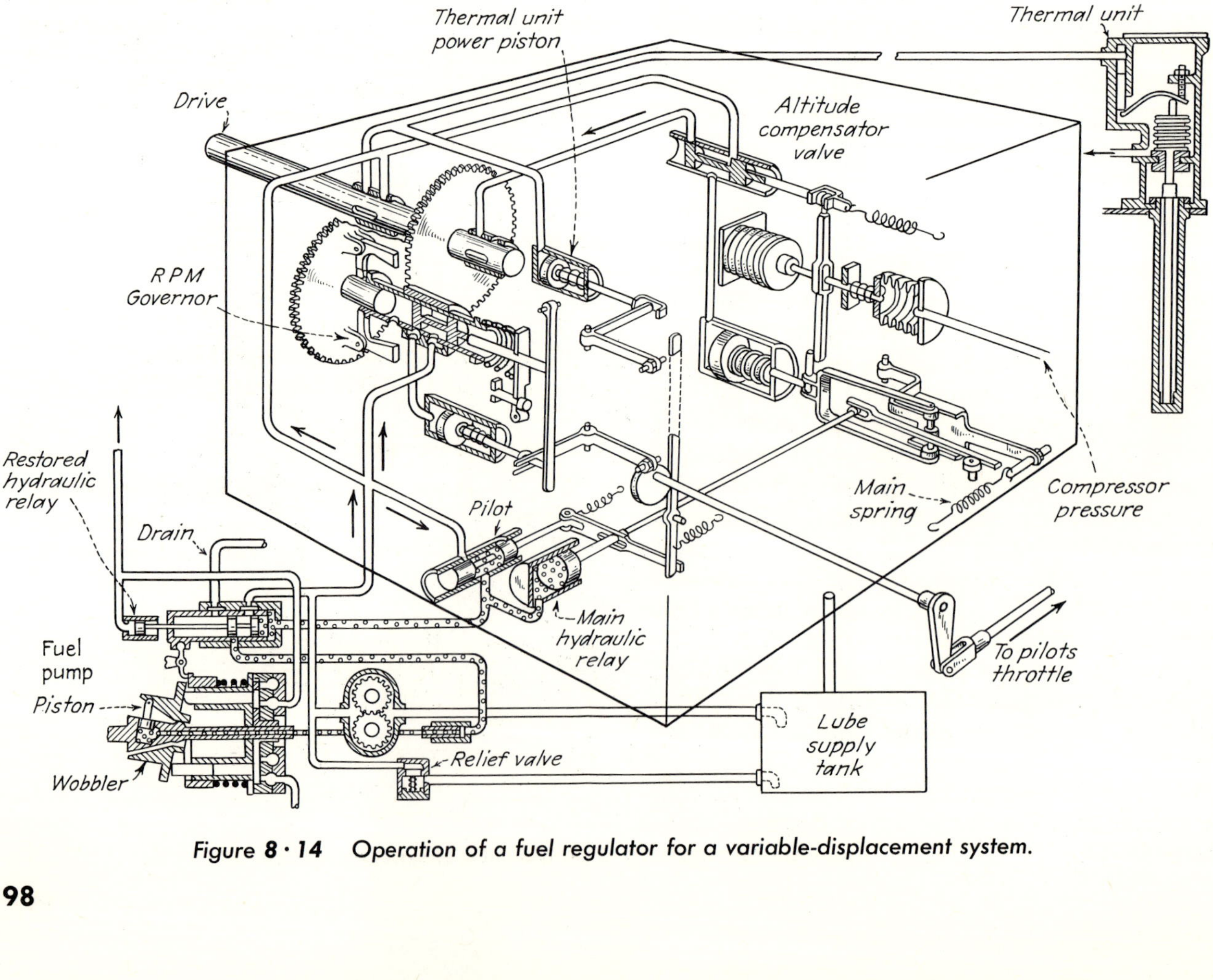

Figure 8 · 14 *Operation of a fuel regulator for a variable-displacement system.*

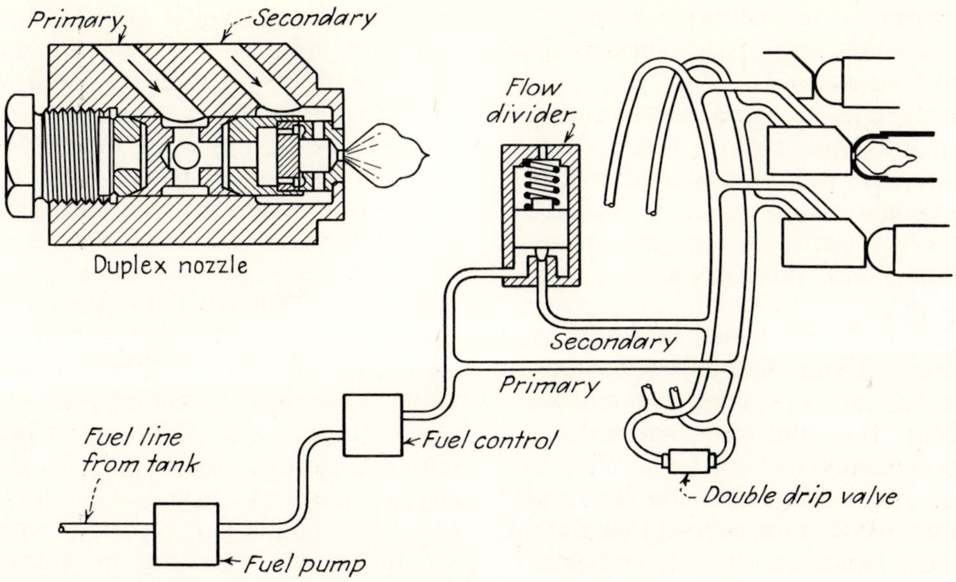

Figure 8 · 15 *Duplex nozzle system with flow divider.*

The Flow Divider and Duplex Fuel Nozzles. Since the flow divider and the duplex fuel nozzle work hand in hand, the description of these units will be combined. Figure 8 · 15 is a schematic illustration which shows the interconnections between the units.

As the flow range of the simplex nozzle was found inadequate, the duplex nozzle was developed for production, using an additional fuel manifold and a flow divider. Duplex systems of this sort are used on a number of American and British engines. The advantages of the duplex nozzle are that (1) it permits application of enough energy for good atomization in the simplest manner, (2) uniformity can be controlled by fixed orifices, and (3) the system employs a minimum of moving parts.

Upon initial starting of the engine, fuel pressure is low and fuel flow will be only through the primary manifold (also called the "small-slot line"). This small-slot pressure flow sprays through the center of the nozzle and produces a rather wide spray pattern. The wide spray is desirable for starting, since atomized fuel is sprayed into the spark-plug area.

As the engine accelerates, the fuel pressure builds up. When the fuel pressure reaches a predetermined value, a passage is opened in the flow divider, and this permits fuel flow through the secondary or large-slot manifold. Thus, both manifolds are in operation above a certain pressure.

With fuel flowing through both manifolds into the fuel nozzle, the arc of spray decreases considerably. This, of course, is desirable for high-speed operation, since the heat is concentrated in the center of the combustion chamber. Flow is controlled for adjustment purposes by lapping the plate in the nozzle.

Duplex Nozzle for Single Manifold. The single-line duplex nozzle with built-in flow divider is shown in the drawing of Fig. 8 · 16. Because this nozzle can be used with a single manifold, it is particularly attractive for use where a large number of nozzles are required.

At low pressures, this nozzle utilizes only the primary orifice; however, as pressure increases, a poppet valve is unseated and fuel is then permitted to flow through both passages.

HYDROMECHANICAL FUEL-CONTROL UNITS

General requirements

As the name implies, the hydromechanical fuel-control unit utilizes hydraulic and mechanical forces and devices to regulate fuel flow to the

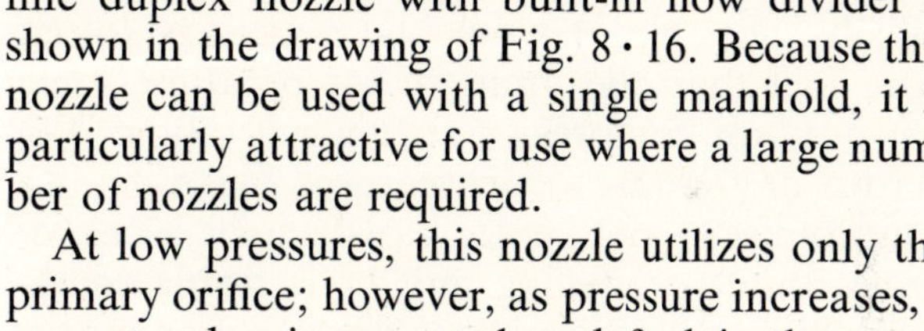

Figure 8 · 16 *Single-line duplex fuel nozzle.*

engine. The principal elements requiring adjustment of fuel flow are speed, acceleration, pressures, and temperatures.

The transient states of acceleration and deceleration demand close control of fuel during the interval of change. It is desirable that the changes be accomplished as quickly as possible; however, there are limits of temperature and fuel-air mixture which must be met to avoid damage to the engine or "blowout."

Fuel-flow limits are illustrated in Fig. 8 · 17. It is necessary to limit maximum flow to prevent damage to the turbine because of high temperature, to prevent "rich blowout," and to avoid stall. Thus the **maximum fuel** line for acceleration is established. The **steady-state** flow line indicates the amount of fuel flow needed to maintain a given rpm. The **minimum** flow line comes into play during deceleration. The fuel-control unit must be designed so that it is impossible for the fuel flow to drop below the minimum flow line until the engine is shut down. If fuel flow is reduced below the minimum flow limits, it will be likely to cause a "lean die-out."

The curves shown in Fig. 8 · 17 are meant for illustration purposes and should not be construed as actual fuel-flow charts.

To achieve the most rapid acceleration theoretically possible, it would be necessary to use more fuel than the limits of temperature, "rich blowout," or compressor stall would allow. The control of the rate of increase of fuel flow is a compromise which gives satisfactory acceleration without producing any undesirable conditions. Similarly, the rate of flow change during deceleration is a balance which prevents lean die-out while producing acceptable operating characteristics.

Simplified hydromechanical fuel system

Figure 8 · 18 is a simplified schematic diagram of a hydromechanical control unit in a fuel system. This illustration is only for the purpose of explaining how various elements of engine operation are utilized to regulate fuel flow.

During operation of the system, fuel flow starts with the boost pump in the fuel tank, passes through an external boost pump (this pump is usually not employed), flows through a filter, and enters the main fuel pump. The main fuel pump pressurizes the fuel and sends it on through the high-pressure filter and into the fuel control unit.

Fuel from the main pump passes through the **throttling valve** in the fuel-control unit. This valve maintains a constant pressure differential across the control. While outlet pressure varies with operating conditions, the difference between control outlet and inlet pressures remains the same. This differential is the metering head for fuel-flow regulation. In many controls, a bypass valve is used to maintain the metering head across the main metering valve.

The engine operating rpm, hence the thrust, is controlled by the **governor valve** illustrated in Fig. 8 · 19. The position of this valve is determined by the balance between two forces: (1) the **governor spring force,** which is determined primarily by the position of the **power lever;** and (2) the governor flyweight force, which is determined by the control rpm.

During acceleration, the governor valve is at first fully open, since the spring is compressed by the forward movement of the power lever and there is little counteracting flyweight force. Then, as the engine speed increases, the flyweights tend to close the valve until equilibrium is reached at the desired rpm.

During deceleration, the power lever motion releases the spring force and allows the valve to move in a closed direction until the minimum flow stop is reached. As the rpm decreases, the flyweight force diminishes and finally permits the spring to open the valve again to an equilibrium position.

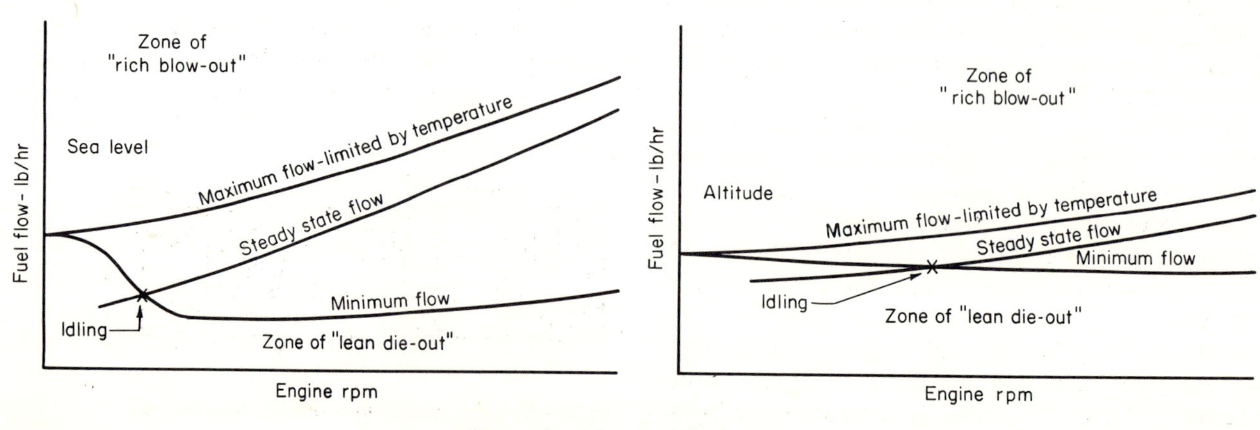

Figure 8 · 17 Fuel flow limits.

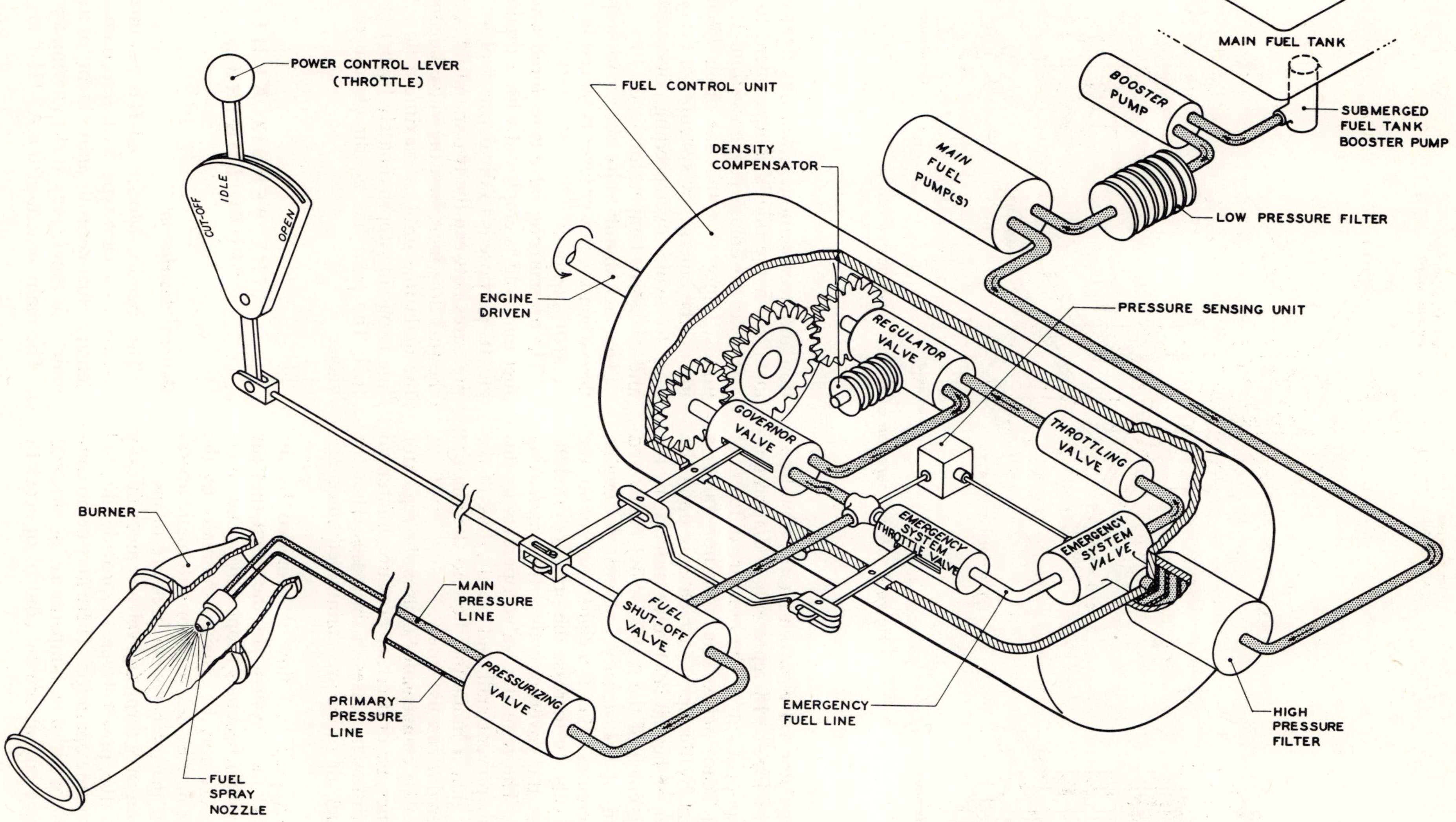

Figure 8 · 18 *Simplified drawing of a hydromechanical fuel-control unit.*

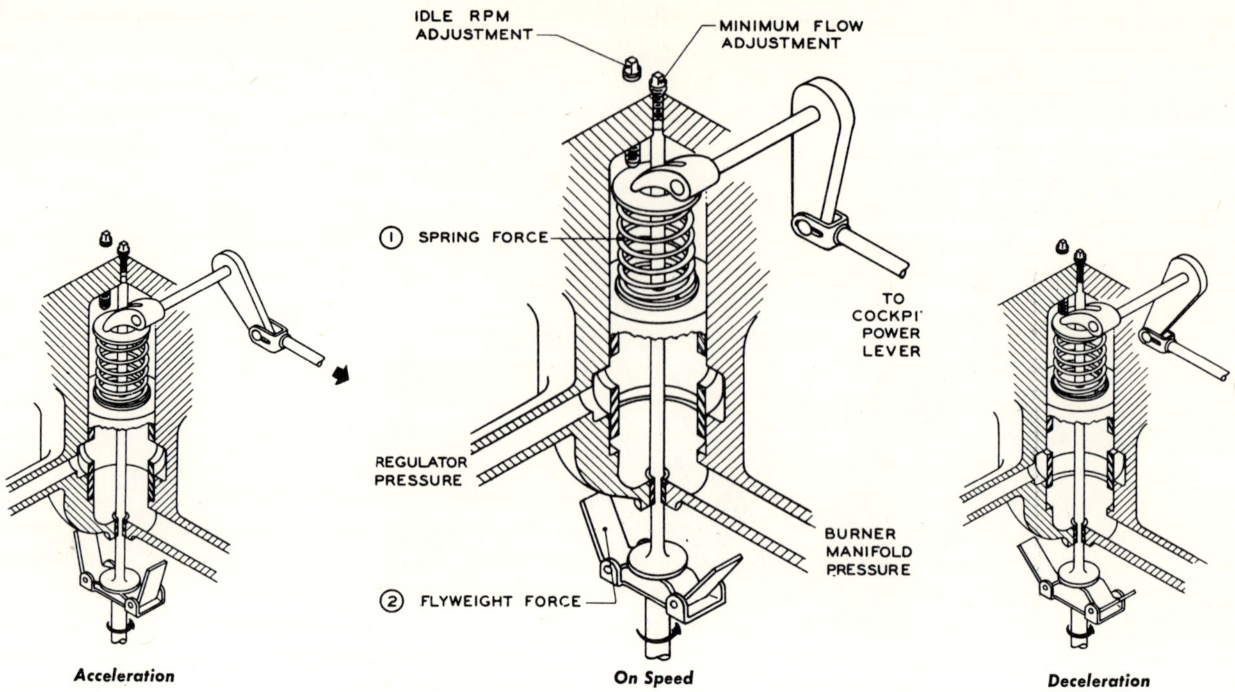

Figure 8 · 19 Operation of an rpm governor valve.

The **regulator valve** shown in the simplified drawing of Fig. 8 · 18 represents a diaphragm-operated poppet valve. The pressure differential (metering head) across the governor valve is established by the spring which acts upon the diaphragm of the regulator valve, and also by the modifying effect of the regulator flyweights and the action of the density aneroid. As engine speed increases, the flyweights act upon the regulator valve to increase the pressure differential across the governor valve, thus increasing fuel flow. The regulator valve operates in this manner to prevent stall and overheating of the engine in that it limits fuel flow to the amount that the engine can use safely.

The **density compensator** causes the regulator valve to increase or decrease pressure differential as mass airflow increases or decreases. Thus the proper fuel/air ratios are maintained automatically.

Uninterrupted fuel flow is provided by the **emergency fuel system** in the event that the normal system becomes inoperative. The emergency system may be selected manually by the pilot, or it may be designed for automatic operation when the main-system pressure drops.

An **emergency throttling valve** is mechanically linked to the power lever to provide the full range of engine operation. There may be no compensation for altitude, temperature, or airspeed, nor is there any automatic control of acceleration and deceleration. A warning light is usually installed in the pilot's compartment to indicate a transfer to emergency operation. It is of the utmost importance that the pilot, when operating on the emergency system, use every precaution to avoid overspeeding, overacceleration, and excessive temperatures.

The **fuel shutoff valve** serves to stop all fuel flow when the power lever is placed in the cutoff or STOP position.

The **pressurizing valve** is required when a dual fuel manifold is employed with a dual fuel nozzle. Its purpose is to divide the fuel with correct pressures between the two manifold sections. As the fuel flow increases, the valve begins to open the main line until at maximum flow the main line is passing approximately 90 per cent of the fuel. The pilot has no direct control over this valve.

THE BENDIX AJ-H1 FUEL CONTROL

General description

The Bendix Model AJ-H1 gas-turbine fuel control is a cam-type hydromechanical speed-density fuel-control unit incorporating both main and emergency fuel metering systems.

The main system of the AJ-H1 fuel control is

designed to deliver to the engine the correct fuel flow under any condition within the operating range of the engine. The fuel flow is scheduled to (1) enable the engine to bypass the compressor stall zone during acceleration, (2) maintain engine temperatures within desired limits during acceleration, (3) prevent lean die-out during deceleration, and (4) maintain any fixed engine speed within the operating limits of the engine, regardless of altitude.

The emergency system of the fuel control, which is a means of bypassing the main system, allows direct manual control over fuel flow if there is a failure in any part of the main system.

The main fuel system of the AJ-H1 gas-turbine fuel control meters fuel in response to signals of **engine speed, burner pressure, ambient air pressure,** and **compressor inlet temperature (CIT).** The position of the pilot's control lever varies the load on the governor spring, thus varying the fuel schedule.

During emergency operation the rate of fuel flow is selected manually by the position of the pilot's control lever; however, the **metering head** across the **emergency throttling valve** is controlled by action of the **emergency altitude bellows** and **needle** in conjunction with the **emergency bypass valve.**

Principles of operation

The air supply to the combustion chambers of a turbojet engine varies with engine speed, aircraft speed, air temperature, and air density. These variations are sensed by the gas-turbine fuel control, and fuel is metered to the engine so as to maintain the correct engine speed, temperatures, and pressures for any combination of varying signals.

The following discussion of the principles of operation of the Bendix AJ-H1 gas-turbine fuel control will be coordinated with the schematic diagram of Fig. 8 · 20. Operation will be explained by tracing fuel flow through the unit and indicating forces that affect the flow.

Fuel under engine main-fuel-pump discharge pressure P_1 is supplied to the control inlet port, where it enters the core of a screen-type main fuel filter. If the screen should become clogged, increased pressure at the fuel inlet will cause the filter to move from its seat against spring pressure and bypass unfiltered fuel to the control, thus assuring an uninterrupted fuel supply to the engine.

As fuel passes through the filter element, a pressure drop occurs; therefore, unmetered fuel pressure after the filter is identified as **P_2 pressure.**

As fuel enters the control, pressure is regulated by the **main bypass valve,** which maintains a constant pressure drop between unmetered pressure P_2 and metered fuel pressure P_3. This is accomplished by bypassing fuel in excess of control demand back to the interstage of the main fuel pump.

The bypass valve is of double-ported design and is actuated primarily by a spring-loaded diaphragm exposed to the $P_2 - P_3$ pressure differential. Double-ported valve design minimizes fuel-flow effect on valve action. The $P_2 - P_3$ head across the bypass valve diaphragm equals the force exerted by the bypass valve spring. The bypass valve configuration includes a series of bimetallic temperature disks located in such manner that disk reaction from fuel temperature changes will influence the effective length of the bypass valve spring. At high fuel temperatures the bimetallic disks become more curved, owing to expansion of the convex side of the disks, thereby increasing the overall length of the disk stack and compressing the bypass valve spring. At low fuel temperatures the bimetallic disks flatten out and decrease the overall length of the disk stack, decreasing load on the bypass valve spring.

An increase in the temperature of a given fuel results in a decrease in the specific gravity of that fuel. Therefore, to deliver the same fuel flow by weight (pounds per hour) that was delivered at a lower temperature, the volume of fuel delivered must be increased. In like manner, a decrease in the temperature of a given fuel will increase its specific gravity, and a lower volume of fuel must then be delivered by the control to hold the weight flow (pounds per hour) within limits. The bypass valve, therefore, must supply more fuel to the metering valve at higher fuel temperatures. This is accomplished by increasing the length of the bimetallic disk stack and the force of the spring, causing the bypass valve to move toward a more closed position, thus bypassing less fuel. Low temperature fuel will shorten the disk stack and reduce the spring force. Under this condition the bypass valve will move toward a more open position, bypassing more fuel.

The $P_2 - P_3$ pressure drop across the **metering valve** is established by bypass spring force. Fuel flow from $P_2 - P_3$ through the metering-valve variable orifice is established by positioning the valve in its sleeve. This position is determined by a combination of variables affecting the valve: **pilot's control lever angle, engine speed, compressor inlet temperature, ambient air pressure,** and **burner pressure.**

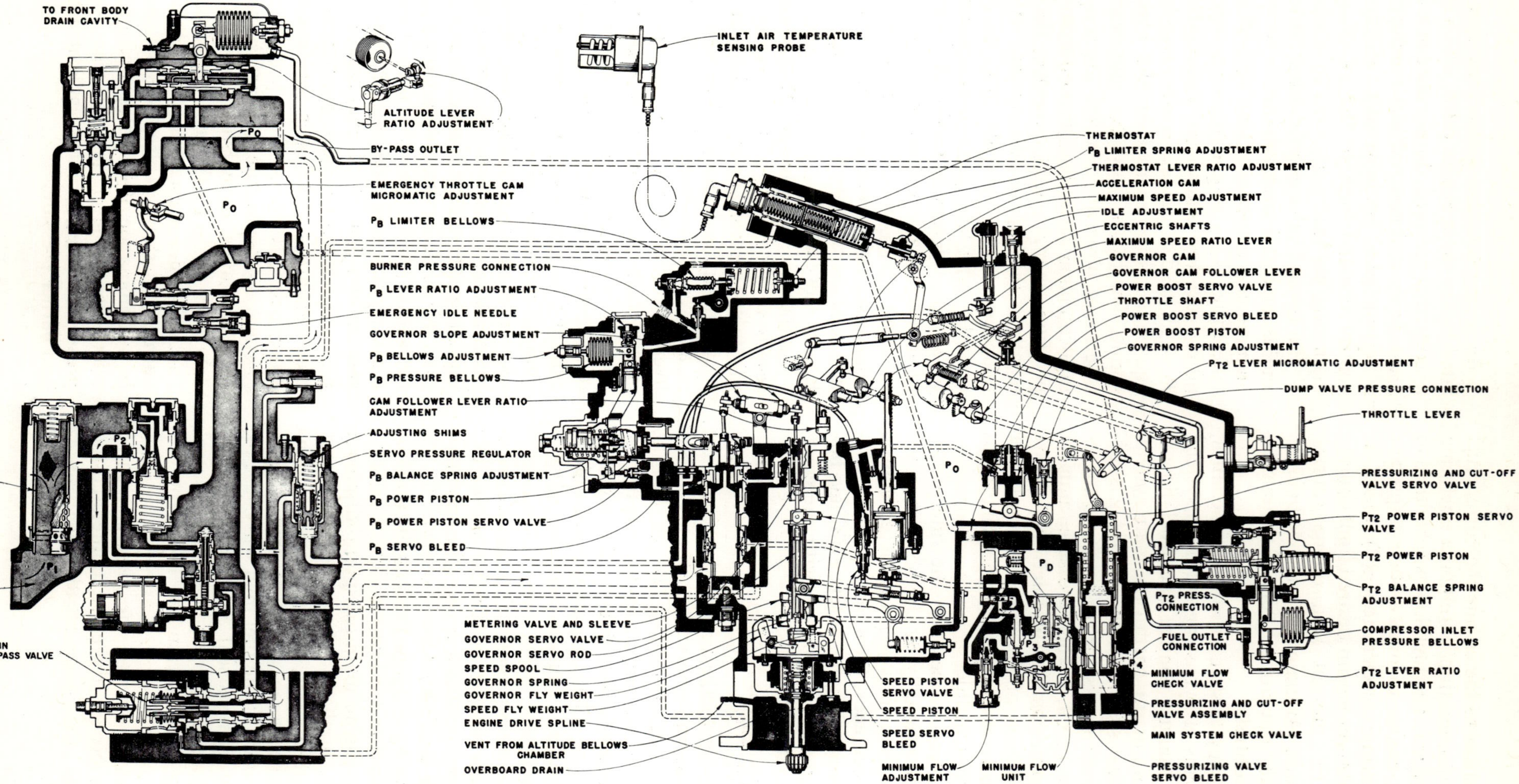

Figure **8 · 20** *Schematic diagram of the Bendix AJ-H1 fuel control.*

The **servo pressure regulator valve** is furnished with fuel at P_2 pressure. The function of this valve is to maintain a constant $P_R - P_0$ pressure level throughout the servo supply system under all operating conditions. Servo supply is designated as P_R on the schematic diagram. All servo systems, except the metering valve and speed system, operate on P_x pressure which is an intermediate regulated pressure between P_R and P_0 pressures. The metering valve and speed system use a combination of P_R and P_x pressures in their operation.

The **metering valve** is positioned axially ("A" dimension) as a result of throttle position, engine speed, and acceleration. All these factors are integrated, and the resultant forces determine the axial position of the metering valve and the corresponding size of its metering orifice at any given instant.

The **control drive shaft** rotates a weight table on which are mounted two pairs of centrifugal weights. The governor weights contact the governor tip. As the speed of rotation increases, these weights apply increasing force to the governor spring.

The pilot's power requirements are transmitted to the control manually, in terms of degrees of throttle angle, through a sector gear attached to the throttle shaft. The throttle shaft in turn rotates the governor cam and through the **governor cam follower lever** actuates the **power-boost piston** to impose a power loading on the governor spring.

The power-boost system is composed mainly of a piston assembly with a hollow stem which incorporates a servo valve at the end. The governor cam follower lever contacts the valve through the **maximum speed ratio adjustment lever.** The power-boost piston and **servo valve assembly** contacts a lever and roller assembly which contacts the governor spring through the governor spring retainer.

Servo pressure P_x is applied to the back side of the piston and flows through the hollow stem to the half-ball valve where bleed off of P_x pressure is controlled by the governor cam follower position. Body pressure P_0 plus governor spring force is applied to the other side of the power-boost piston.

Moving the governor cam follower in a direction to increase the spring force (increased control lever angle) will tend to close the half-ball slightly, decreasing the servo pressure P_x being bled out from the back side of the power-boost piston, thereby allowing a servo pressure buildup on the back side of the power-boost piston and increasing the pressure differential across the piston.

Increased servo pressure on the back side of the power-boost piston will assist movement of the control lever, governor cam, and cam follower in repositioning the linkage to increase governor spring load and assist in holding the new position. The opposite effect is created when the control lever is retarded.

Loading the governor spring (pilot's control lever advance) moves the governor rod away from the governor servo valve allowing P_x pressure release on the top side of the metering valve through the governor servo half-ball valve. This results in $P_x - P_R$ pressure unbalance and allows P_R pressure to move the metering valve toward an open position. Metering-valve-opening movement will continue until the metering valve, through the governor slope adjusting lever, has moved the governor servo piston down to contact either the acceleration cam control stem or the governor rod to establish a $P_x - P_R$ pressure balance in the metering-valve sleeve by closing the governor servo valve to its floating position.

The **acceleration cam** is positioned rotationally by speed signals, while axial positioning of the **acceleration cam follower** reflects changes in compressor inlet temperature (CIT). Rotational position of the acceleration cam is a result of speed-piston movement, which will be dealt with in later paragraphs.

The acceleration cam remains stationary in its axial plane. A sliding acceleration cam follower lever mounted on a shaft which is parallel to the cam axis transmits compressor inlet temperature signals to the metering valve during engine acceleration. The acceleration cam follower lever is positioned axially on the acceleration cam and transmits an integrated signal of speed and compressor inlet temperature to the metering valve through the governor servo valve.

Centrifugal governor weights, driven from the engine drive spline, transmit weight-force output as a function of engine speed to the tip of the governor rod. Increased engine speed increases centrifugal weight force to a point where weight-force output overcomes the throttle-selected governor spring load. When governor weight force tending to close the governor servo valve equals governor spring force tending to open the valve, stabilized metering-valve axial dimension results. This provides an all-speed governing fuel flow for the engine.

The servo system which controls the metering-valve axial dimension is actuated by the **governor rod** or **acceleration servo valve contactor.** As governor spring load is increased by an advance in control lever angle, a state of unbalance is created between the governor flyweights and governor spring. The governor rod controls a servo valve which regulates bleed-off of servo pressure P_x on the back side of the metering valve. As governor spring load is increased by control lever advance, the governor valve rod is moved away from the servo valve causing P_x pressure bleed off on the back side of the metering valve. Higher $P_R - P_x$ differential pressures open the valve in a linear direction. The opposite effect is created when the control lever is retarded.

The **ambient air pressure sensing system** notes changes in ambient air pressure, that is, air pressure in the engine compartment, and transmits signals of these changes through a system of levers to the governor cam follower. The governor-setting cam is contoured to bias fuel flow properly in relation to ambient air pressure. The changes in cam rise result in an increase or decrease to the governor spring which controls the governor rod position through the power boost and which will affect fuel flow by controlling the metering valve through the governor servo system.

The **thermostat bellows and probe assembly** responds to changes in compressor inlet air temperature and repositions the acceleration cam follower on the acceleration cam in a linear plane. This cam is contoured to provide proper maximum acceleration fuel flow for any given ambient air temperature.

The acceleration cam follower is repositioned on the acceleration cam through thermostatic bellows action. A series of levers is used to move the follower across the cam in an axial direction. The station of the cam follower on the acceleration cam actuates a plunger and lever which control the servo valve in the metering-valve servo system during engine acceleration.

The position of the metering-valve servo valve, which controls pressure on the back side of the metering valve, either increases or decreases fuel flow during engine acceleration in relation to ambient air temperature by repositioning the metering valve in a linear direction through the metering valve servo system.

The compressor inlet air temperature (CIT) signal is transmitted to the thermostat bellows assembly through a liquid-filled probe and tube. The thermostat assembly in this installation incorporates a compensating bellows which compensates for fuel temperature changes by "nulling out" (reducing to zero) any fuel temperature change signal that is reflected by the motor bellows of the thermostat assembly.

The **speed system,** sensing changes in engine speed by means of speed flyweights, repositions the acceleration-setting cam in a rotational direction to control the maximum open position of the metering valve in relation to engine speed. The acceleration cam is so designed that rotating movement will effect cam follower movement to provide acceleration fuel flow at predetermined compressor inlet air temperature. In this manner the acceleration-limiting action prevents a greater flow of fuel than the engine is ready to utilize. As engine speed increases, the speed system permits the metering valve to open further and thereby increase fuel flow.

The **speed weights,** driven from the engine drive spline, operate against an actuating spool assembly attached to a lever which actuates another lever incorporating a ratio adjustment device. The ratio adjustment lever contacts the **speed-piston spring** and a servo half-ball valve. Spring force, P_R pressure, and P_0 pressure is applied to the open side of the **speed piston.** The opposite side of the piston is attached to a gear rack which engages the gear on the acceleration cam. Servo pressure P_x is applied to the rack side of the piston, and P_x bleed-off is controlled by the half-ball valve. Therefore, the spring force plus P_R and body pressure P_0 tend to move the piston in one direction (decreased speed), while weight force controlling P_x pressure in the servo system tends to move the piston in the opposite direction. At any time that the forces are in balance, the piston will be held in a stabilized position.

An increase in engine speed will increase speed flyweight force which tends to close the half-ball valve restricting P_x bleed-off on the rack side of the piston, thus increasing P_x pressure. The piston then moves in an increased speed direction simultaneously increasing spring force until P_x pressure force is equal to P_0 plus spring force, and the forces thus balanced hold the piston stabilized in its sleeve.

The speed piston and rack, operating through the acceleration cam, perform their function by controlling P_x servo pressure on the top side of the metering valve. A spring-loaded stem and lever, actuated through the cam follower, control the position of the governor servo valve during acceleration. The acceleration cam rise dictates higher fuel flow by allowing P_x pressure bleed-off at the governor servo valve, while the stem spring dictates lower fuel flow when the acceleration cam follower is on the lower areas of the acceleration cam.

The metering valve is positioned **rotationally** ("B" dimension) by the effect of engine burner chamber pressure applied to the control as an input variable. Burner pressure is applied to a chamber containing an evacuated P_B bellows. Metering-valve positioning is accomplished by the action of the bellows through a lever, servo, and power-piston arrangement.

One end of the P_B bellows is anchored to the casting of the control unit, and the other end is connected to a lever which operates a servo valve. Changes in burner pressure cause the bellows to transmit more or less force to the servo valves, moving the servo valve in a closing or opening direction from its normal position and resulting in movement of the **burner pressure actuator piston** by varying P_x pressure. The piston movement is transmitted by a rack gear to the metering-valve sector gear attached to the metering-valve assembly. Rotation of the metering valve varies the opening of the metering orifice independent of axial movement of the valve.

A decrease in burner pressure, such as would be caused by an increase in altitude, will cause the evacuated bellows to release pressure on the servo valve. This release of pressure is transmitted by the lever attached to the movable end of the bellows and causes the servo valve to move toward its OPEN position.

An increase in burner pressure results in a force increase exerted by the evacuated bellows and in movement of the servo valve toward its CLOSED position, thus permitting P_x pressure to increase. The piston then moves until, at a predetermined position, the lever is actuated by the piston spring to return the servo valve to its initial position. This equalizes the force of the spring and P_x pressure, at which point the metering valve stabilizes in its new rotational position to provide a corresponding increase in fuel flow.

When burner pressure reaches a predetermined value, the differential pressure between burner pressure and engine nacelle pressure at the outside of the fuel control overcomes the restraining force of the **burner pressure limiter spring** through the limiter bellows and opens the valve. This arrangement is a safety device designed to prevent excessive differential pressures from occurring across the engine case under ram conditions. When the burner pressure limiter valve is forced open, air escapes through the valve, causing a drop in the pressure inside the limiter bellows. As a result the bellows contracts and allows the limiter spring to move the limiter valve toward a closed position. Reduction of P_B pressure by the limiter results in decreasing the "B" dimension opening. This reduces engine speed with a resultant reduction in burner pressure until a predetermined safe value for the engine is reached. Lower burner pressure allows the burner pressure limiter valve to close, and normal operation is resumed. Any time that burner pressure increases beyond the safe operating limits of the engine, the burner pressure limiter overrides the signal applied to the P_B bellows and reduces fuel flow and burner pressure to the predetermined safe maximum value.

A **minimum flow** regulating device is installed in the control in parallel with the metering valve. The bleed is always open to provide altitude idle and deceleration flow when the main metering valve is nearly closed.

A diaphragm and lever control a servo valve to maintain a constant ($P_3 - P_D$) pressure differential across the minimum flow bleed while the engine is operating in the minimum flow range. This type of minimum flow system is a distinct advantage over a system which incorporates only an adjustable bleed. By keeping the minimum flow circuit inoperative at higher fuel flows, the linearity of the flow curve can be maintained.

The servo operated **cutoff valve,** mechanically linked to the pilot's control lever, cuts off fuel flow to the engine by controlling P_D pressure which is used to open and hold the pressurizing and cutoff valve against pressurizing valve spring pressure. When the servo valve is moved to allow release of P_D pressure in the pressurizing and cutoff valve, the pressurizing valve spring closes the valve to interrupt fuel flow to the fuel manifold.

Fuel-control operation

During operation of an engine equipped with an AJ-H1 fuel-control unit, the desired engine speed is selected by the position of the pilot's control lever. For best acceleration, the fuel control is required to deliver the maximum amount of fuel the engine can burn without exceeding permissible turbine temperatures or encountering stall conditions.

For any changes in engine speed caused by the pilot's control lever movement, a corresponding change takes place in various operating parts of the control. The curve of Fig. 8·21 shows fuel flow at sea level for the entire operating range.

During a ground start the engine is rotated to

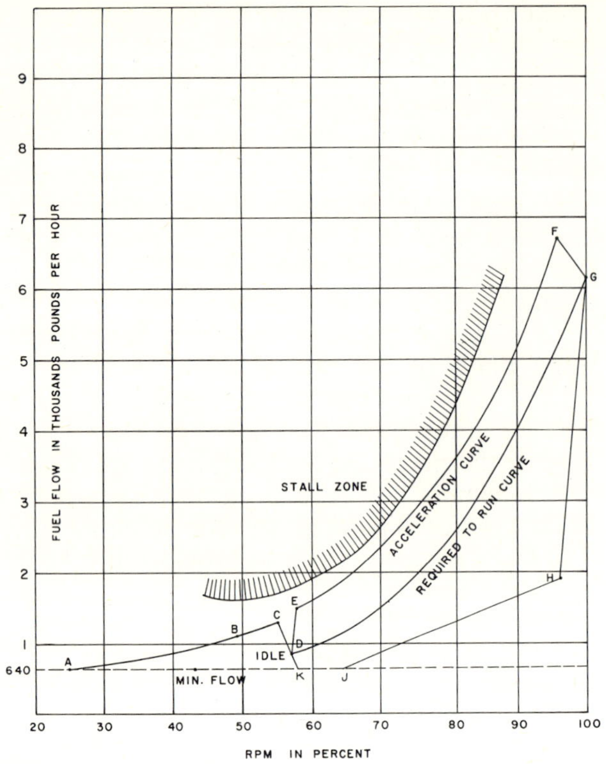

Figure 8·21 Fuel flow curve for the Bendix AJ-H1 control.

a speed of approximately 12 to 16 per cent of maximum rpm, at which point the pilot's control lever is moved into the IDLE position. The fuel flow is now at point A, and because of increasing burner pressure the metering valve rotates toward an OPEN position and meters fuel along line ABC. Up to this point the governor rod has been moved away from the metering-valve servo piston and valve, allowing bleed-off on the back side of the metering valve until the metering valve movement through the walking beam moves the governor servo piston assembly and servo valve into contact with the acceleration cam follower lever stem. As engine speed increases, governor weight force moves the governor rod toward the servo piston and the servo piston toward the governor rod through the interconnecting walking beam. At this point increasing governor weight force moves the governor rod against the servo valve to restrict P_x bleed-off on the back side of the metering valve. Increased P_x pressure moves the metering valve toward its CLOSED position until $P_R - P_x$ pressures are balanced at IDLE. This is point D on the curve.

During acceleration from idle to maximum rpm the pilot moves the control lever to the maximum OPEN position, increasing the load on the governor spring. This moves the rod away from the governor servo valve and creates a state of unbalance between P_R and P_x in the metering valve because of P_x pressure bleed-off through the open governor servo valve. The metering valve moves axially toward its OPEN position because of high P_R pressure and lower P_x pressure. The $P_2 - P_3$ differential momentarily decreases, causing the main bypass valve to decrease fuel flow to P_0. Main fuel flow increases along line DE. At point E fuel flow is restricted by the limit of metering-valve travel set by the acceleration cam. Increasing P_B now operates to open the metering valve rotationally, and the acceleration cam controls the axial opening of the valve-acceleration fuel flow along line EF, skirting the compressor stall zone shown in Fig. 8·21, while providing maximum permissible fuel flow for acceleration.

As engine speed increases during acceleration, governing weight force increases until at point F it begins to close the governor servo valve and restrict fuel flow along line FG. Higher P_x pressure moves the metering valve toward its CLOSED position and the governor servo piston away from the governor rod through the governor slope adjusting lever. When $P_x - P_R$ pressures become equalized, fuel flow is stabilized at 100 per cent rpm at point G on the **required-to-run** (steady-state) curve.

During deceleration from maximum rpm to idle, the pilot moves the control lever to the IDLE position, thereby decreasing the load on the governor spring. Governor weight force moves the metering valve toward its fully closed axial position through the governor servo system. Fuel flows through the metering slots, which are open in a rotational direction, being controlled by burner pressure. As engine speed decreases, the value of burner pressure decreases until at point H the metering valve begins to close rotationally along line HJ. When the valve is nearly closed, the minimum flow bleed circuit controls fuel flow. At point K governor spring force balances governor weight force and opens the metering valve, through the servo system, along line KD, and at point D the idle fuel flow is reestablished.

During locked-throttle climb, the governor system positions the metering valve axially in its sleeve, and the P_B system positions the valve radially. As altitude increases, burner pressure is reduced owing to decreasing air density. Decreasing burner pressure P_B is sensed by the P_B bellows and applied to the metering valve as rotational movement to decrease fuel flow proportionate to decreased air density. At the same time ambient air pressure P_A is reduced, because of decreased air density, to reposition the governor cam follower on the governor-setting cam and bias governor spring load position. The governor servo system repositions the metering valve axially to reduce fuel flow with a subsequent change in rpm. Decreasing P_A results in a change in maximum engine speed at altitude.

Emergency operation

During emergency operation of the fuel control, all normal control effects on fuel flow are eliminated and the pilot assumes manual control of fuel flow. The only automatic compensation during emergency operation is altitude compensation up to an altitude of approximately 30,000 ft. Beyond that altitude the pilot must retard the control lever to maintain a constant rpm. The fuel control may be transferred from main to emergency system only by manual operation of the emergency motor actuator switch in the pilot's cockpit. No provision is made for automatic transfer.

When the emergency motor actuator switch is placed in the EMERGENCY position, the pilot in the control is moved by the motor actuator to interrupt P_2 pressure to the **shuttle-valve piston** allowing the spring to reposition the shuttle valve to the emergency position. Shuttle-valve movement in opening the emergency circuit interrupts the flow of P_2 pressure to the main circuit. With no differential pressure $P_2 - P_3$ across the main bypass valve diaphragm, bypass spring force closes the main bypass valve. Interruption of P_2 pressure in the main system and subsequent interruption of P_3 pressure allows the spring to close the main system check valve in the metered fuel circuit of the main system.

Emergency fuel pressure is regulated by the emergency bypass valve, which maintains the required pressure drop across the emergency throttle valve. This is accomplished by bypassing fuel in excess of control demand back to the inlet side of the main fuel pump.

With the pilot valve in EMERGENCY position, a flow is established from P_2 past the emergency altitude needle to P_E and through a bleed in the emergency bypass valve cover to emergency metered fuel (P_3 emergency). Regardless of altitude, the pressure differential $P_2 - P_E$ across the emergency altitude needle remains at a relatively constant value to which the emergency bypass valve spring is set. Flow through the emergency altitude needle orifice will vary as a function of needle position, which varies the pressure drop across the emergency altitude bleed. This pressure drop across the bleed controls $P_2 - P_3$ across the emergency throttle valve by controlling the emergency bypass valve position.

The **emergency altitude needle** is actuated by a lever and shaft, which are attached to the emergency altitude bellows. The emergency altitude bellows senses compressor inlet pressure to position the altitude needle in its sleeve. Decreased compressor inlet pressure, such as occurs at altitude, causes the bellows to reposition the emergency altitude needle in its orifice, decreasing $P_2 - P_E$ flow across the emergency altitude needle. Decreased P_E flow to the fixed orifice of the emergency altitude bleed results in an increased $P_2 - P_E$ differential across the emergency bypass valve diaphragm, which moves the bypass valve further open. This increases the amount of fuel being bypassed back to the pump, thereby decreasing fuel flow to the emergency throttle valve.

Unmetered fuel P_2 is directed from the shuttle valve into the emergency system through the emergency bypass valve to the emergency throttle valve. The emergency throttle needle is mechanically linked to the pilot's control lever. Movement of the emergency throttle needle in its orifice controls the flow of fuel through the valve, thus providing manual control of fuel flow to the engine. An adjustable bleed is installed in parallel with the emergency throttle valve to provide for emergency idle flow adjustments.

THE HAMILTON STANDARD JFC25 FUEL CONTROL

General description

The Hamilton Standard JFC25 fuel-control unit is designed for use with Pratt & Whitney turbojet or turbofan engines. They operate in accordance with principles already described to provide the engine with the correct fuel flow under all conditions of operation.

Fuel is metered according to a predetermined flow schedule which varies as a function of the pilot's throttle lever (power lever) position, burner pressure, and engine rpm. The control is designed for minimum size and weight consistent with maximum reliability and ruggedness. The JFC25 unit is a hydromechanical type, that is, it is made up of hydraulic and mechanical components. The operating forces required within the control are delivered by servos which operate on approximately a 2:1 pressure ratio.

The JFC25 fuel-control unit, shown schematically in the drawing of Fig. 8·22, includes the following major subassemblies: **minimum pressure** and **shutoff valve, pressure-regulating valve, thrust reverser, speed-sensing governor, throttle valve, burner pressure sensor assembly, ratio regulator assembly, main housing, auxiliary housing,** and **main cover assembly.**

The minimum pressure and shutoff valve, shown in the upper right portion of the drawing, consists essentially of a plunger-type valve, spring-loaded to the CLOSED position. The valve is opened by throttle valve downstream pressure overcoming a spring force plus pump interstage pressure during normal operation. It is closed by porting high-pressure fuel to assist the spring during shutoff. The valve will not open until control inlet pressure rises to the minimum value required to ensure that sufficient pressure is available for operating the servos and valves of the fuel control at low flow conditions.

The pressure-regulating valve system consists of a sensor and a pressure-regulating valve assembly. The pressure-regulating valve assembly includes a housing, a valve sleeve, a spring, a reset piston, a cylinder, and two filtered orifices. The valve is exposed on one side to control inlet fuel pressure and on the opposite side to spring force and throttle valve downstream pressure. The function of the pressure-regulating valve is to maintain a constant pressure drop across the throttle valve so that fuel flow through the throttle valve will be directly proportional to the orifice area of the throttle valve. The bypass port of the valve is connected to the engine-driven fuel pump second-stage inlet (interstage) to permit bypassing of fuel in excess of that needed to maintain the required pressure differential. The sensor assembly, located in the auxiliary housing, compares throttle valve downstream pressure with control inlet pressure and positions the pressure-regulating valve reset piston accordingly.

The thrust-reverser group consists primarily of a cam, a pilot valve and housing, a detent lock, a screw adjustment, and the power lever support housing. The thrust reverser supplies a pneumatic signal to the engine thrust direction control apparatus to actuate this system for forward or reverse thrust according to power lever position. The power-lever-controlled cam is "locked" in the IDLE REVERSE position by the detent mechanism. An increased force on the power lever is then required to move it in the increase reverse thrust direction or in the idle forward direction, thus indicating that a reverse thrust condition exists.

The speed-sensing governor, shown at the bottom right of the drawing, includes a sleeve assembly with two filter screens, a valve, a flyweight assembly, and a head support plate. The drive shaft of the governor is splined to mate with an accessory drive on the engine. Therefore, the speed-sensing governor is driven at a speed relative to engine rpm and, by means of the centrifugal action of the flyweights, positions a pilot valve. When engine speed changes, the flyweight force varies, and the pilot valve is positioned to port the speed servo piston chamber to high-pressure or low-pressure fuel, resulting in movement of the speed servo piston. The motion of this piston repositions the pilot valve until the speed-sensing system returns to equilibrium. The position of the servo piston is therefore indicative of actual engine speed. This piston incorporates a rack which meshes with a gear segment of the **acceleration-limiting cam.** The piston position is also utilized to control selected engine speed.

The throttle valve, similar to main metering valves in other controls, consists of a valve rod and lower cylinder assembly, a feedback spring, and a bracket and lever assembly. The throttle valve meters the fuel flowing to the engine in accordance with engine requirements. Spring force plus high-pressure fuel tends to close the valve, while the action of throttle servo pressure on the large area of the servo piston can exert sufficient force to move the throttle valve toward the INCREASE FUEL FLOW position. The throttle valve, which moves within a sharp-edged orifice, schedules a definite fuel flow because a constant pressure drop is maintained across the valve by the pressure-regulating valve. A minimum fuel flow adjustment is provided to permit maintaining flow at a low limit adequate for engine operation at this level.

The burner pressure sensor assembly consists of a pair of matched bellows, one evacuated and one open to burner pressure, a water drain tube, a linking lever, an adjustment, and the sensor housing. The evacuated bellows provides a standard reference for the motor bellows so that the force exerted by the bellows assembly is proportional to absolute burner pressure.

The throttle valve pressure ratio regulator consists of a large piston, a small piston, and a housing. The regulator is a hydraulic amplifier of infinite gain which operates over the entire range of the throttle valve fixed-orifice flapper-valve system. The action of the pressure ratio regulator improves throttle valve metering accuracy by decreasing the sensitivity of the control to changes in pump discharge pressure.

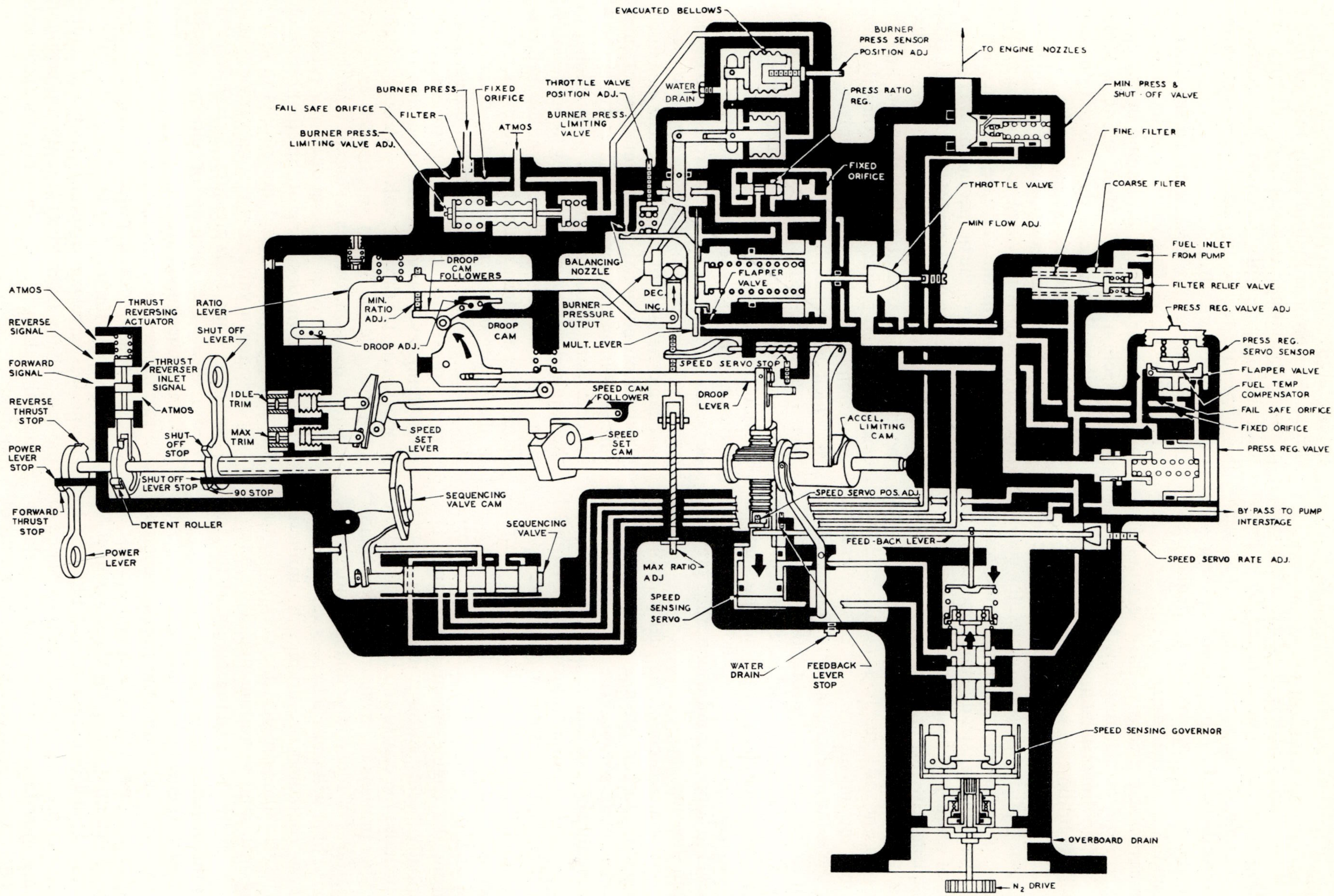

Figure **8 · 22** *Schematic diagram of the Hamilton Standard JFC25 hydromechanical fuel-control unit.*

Operation

The JFC25 fuel control is designed to schedule the quantity of fuel flow required by a high-thrust jet engine to deliver the designed amount of thrust as dictated by the position of the power lever in the cockpit and by the particular operating conditions of the engine.

As mentioned previously, the operation of a jet engine requires that fuel flow be maintained within certain limits which vary according to operating conditions. The variables sensed by the JFC25 fuel control are **burner pressure** and **engine rpm.** Utilizing these variables, the control is capable of accurately maintaining the desired engine rpm during steady-state operation by the use of a **droop system** in conjunction with a speed-sensing governor. During acceleration and starting, the control senses burner pressure and engine rpm and, as a result, schedules fuel flow to permit the maximum rate of acceleration within the allowable engine temperature limits without compressor surge or "rich blow-out." During deceleration the fuel control schedules fuel flow as a function of burner pressure to ensure the maintenance of fuel flow at the minimum level to support combustion, thus preventing "lean die-out."

The operating system of the fuel control may be considered to consist of a **metering section** and a **computing section.** The metering section selects the rate of fuel flow to be supplied to the engine burners in order to provide the required engine thrust output, subject to engine operating limitations as sensed by the control computing section.

High-pressure fuel is supplied to the control inlet from the engine-driven fuel pump. This fuel initially passes through the filter system, which includes the coarse and the fine filter. The coarse filter protects the metering section against large particles of fuel contaminants. Clogging of the coarse filter will cause a bypass valve to open and allow continued operation with unfiltered fuel. The fine filter protects the computing section against solid contaminants. This filter is fuel cleaned by the washing action of the fuel flow to the metering section.

The fuel next encounters the pressure-regulating valve. This valve and the pressure-regulating sensor are designed to maintain a constant pressure differential across the throttle valve. All high-pressure fuel in excess of that required to maintain this pressure differential is bypassed to the pump interstage by the pressure-regulating valve. The valve is servo-controlled by means of the actual pressure drop across the throttle valve orifice. Any error is hydraulically relayed to the pressure-regulating valve spring, altering the force balance on this valve so that sufficient high-pressure fuel is bypassed to maintain the desired pressure drop.

The high-pressure fuel then passes through the throttle valve. This valve consists of a contoured plunger which is positioned by the computing section of the control within a knife-edged orifice. By virtue of the constant pressure drop maintained across the valve, the fuel flow is proportional to the position of the plunger. An adjustable stop is provided to limit the motion of this plunger in the decrease fuel direction to permit selection of the proper minimum fuel flow.

The computing system of the control positions the throttle valve to control fuel flow during steady-state operation, acceleration, and deceleration. The positioning of the throttle valve is accomplished through a multiplying system by which the signal for acceleration, deceleration, or a steady-state speed control is multiplied by a signal proportional to burner pressure to provide the required fuel flow.

Burner pressure is sensed by means of a metallic bellows which is externally exposed to burner pressure. The resulting force is opposed by an evacuated bellows of equal size. The net force, which is proportional to absolute burner pressure, is transmitted through a lever system to a set of rollers whose position is proportional to the required ratio of metered flow to burner pressure. These rollers ride between the bellows-actuated lever and a multiplying lever as shown in the upper center portion of the diagram in Fig. 8 · 22. A force proportional to burner pressure is transmitted through the rollers to the multiplying lever, and a spring transmits a force proportional to throttle valve opening to the multiplying lever. When the throttle is in the required position, there is a state of equilibrium. Any change in roller position or burner-pressure signals causes an upsetting of the equilibrium of the multiplying lever, and this results in opening or closing a variable-metering orifice which is supplied with high-pressure fuel through a fixed orifice. The opening or closing of the metering orifice modulates fuel pressure between the two orifices, and this pressure is used to control the position of a piston attached to the throttle valve plunger. The motion of this piston compresses or relaxes a spring which will return the multiplying lever to its equilibrium position when the throttle valve reaches the required position.

Deceleration control is accomplished by placing an adjustable minimum-ratio stop on the roller-positioning linkage. This provides a linear relationship between fuel flow and burner pressure conforming to engine operating characteristics.

Acceleration control is accomplished by placing a maximum-ratio stop on the roller position. This stop is positioned by the acceleration-limiting cam, which is a three-dimensional cam, rotated by a signal proportional to engine speed. The acceleration cam is contoured to define a schedule of the fuel flow to burner-pressure ratio vs. engine speed for each value of burner pressure. This combination will avoid the overtemperature and surge limits of the engine without compromising engine acceleration time.

The speed-sensing governor controls movement of the speed servo piston through a pilot valve. The governor is of the flyweight type and is engine-driven. When speed changes, the flyweight force changes proportionately, and the pilot valve is positioned to meter either low- or high-pressure fuel to the speed servo piston. The motion of this piston repositions the pilot valve until the speed-sensing system returns to equilibrium. The position of the servo is indicative of actual engine speed. The piston incorporates a rack which meshes with a gear segment on the acceleration three-dimensional cam to provide the speed signal for acceleration limiting. The piston position is also utilized to hold selected engine speed within safe limits by repositioning the droop cam.

The final component of the fuel control to act on metered fuel is the minimum pressure and shutoff valve. One of the primary functions of this valve is to quickly shut off the flow of metered fuel to the engine when the shutoff lever is at a given angle from the OFF position. The **sequencing valve** ports high-pressure fuel to the spring side of the shutoff valve forcing it down against a seat, thus shutting off the flow of metered fuel to the engine. When the shutoff lever is moved into the ON position, the high-pressure signal is replaced by pump interstage pressure, allowing metered fuel pressure to overcome the spring force and open the valve. This valve will provide a minimum operating pressure within the fuel control, ensuring that adequate pressure is available for operation of the servos and valves at low fuel flow conditions.

The sequencing valve, in addition to providing the shutoff signal to the minimum pressure and shutoff valve, provides the pressure signal to the manifold drain valve. The manifold drain ("dump") valve signal is normally set 5 to 10° below the nominal shutoff angle. As the shutoff lever enters the dump-valve-actuation-angle range in being moved to the shutdown position, high-pressure fuel is directed to the spring side of the shutoff valve. Fuel pressure in conjunction with spring pressure provides immediate shutdown of fuel to the engine. The pilot valve is positioned by a shutoff-lever-operated cam so that the signals are generated at the desired lever positions. The pilot valve also provides a windmill bypass feature which bleeds throttle valve downstream pressure to provide fuel flow through the throttle valve when the shutoff valve is closed. This allows the pressure-regulating valve to continue to operate normally, thereby preventing damage to fuel pumps from excessive fuel pressure during engine windmilling.

Some models of the JFC25 fuel-control unit are equipped with a special cam and switch to initiate water injection for takeoff. The water-injection cam is mounted on the power-lever shaft in a position to actuate the water-injection switch when the power lever is placed in the takeoff power position. Water pressure is fed back to the control where it applies force to the ratio lever in such a direction that fuel flow is increased while water pressure is on. This, of course, provides a substantial increase in engine thrust without danger of overheating any part of the engine.

STUDY QUESTIONS

1. Compare a gas-turbine fuel system with the fuel system for a reciprocating engine.
2. What are the requirements for the airplane fuel system with a gas-turbine installation?
3. Why has it become necessary to pressurize fuel tanks as altitude is gained?
4. What factors affect the fuel flow of a gas-turbine engine?
5. What is the function of the barometric control in the constant-displacement fuel system?
6. What unit of a fuel-control system prevents over-speeding of the rotor?
7. Why is it necessary to utilize a drip valve in the fuel system?
8. What precautions must be observed in the operation of the emergency system?
9. What is the advantage of a dual fuel nozzle?
10. What undesirable conditions may develop during acceleration of a gas-turbine engine?
11. What does the hydromechanical fuel-control unit do to prevent these conditions?

12. What are the principal elements found in a modern hydromechanical fuel-control unit?
13. How is constant head pressure maintained across the metering valve?
14. What is the function of the acceleration-scheduling cam?
15. How does the function of the governor flyweights differ from the function of the speed flyweights?
16. What is the significance of the "required-to-run" curve on the flow chart?
17. What prevents "lean die-out" when the power lever is retarded?
18. Compare the **metering** section of a fuel-control unit with the **computing** section.
19. How is acceleration control accomplished in the JFC25 fuel control?
20. Explain the action of the **sequence valve.**

CHAPTER 9

Gas-turbine Starting and Ignition Systems

General requirements

The starting of a turbojet, or any other gas-turbine engine, requires that the engine be rotated at a speed which will provide sufficient air for fuel combustion and that the engine be accelerated to the point where the power developed by the turbine is adequate for engine self-rotation. In general, the low speeds of starting provide a low compressor discharge pressure; hence the engine efficiency is low. Starter systems must be designed so that they will bring the engine up to self-sustaining speed before starting torque is cut off.

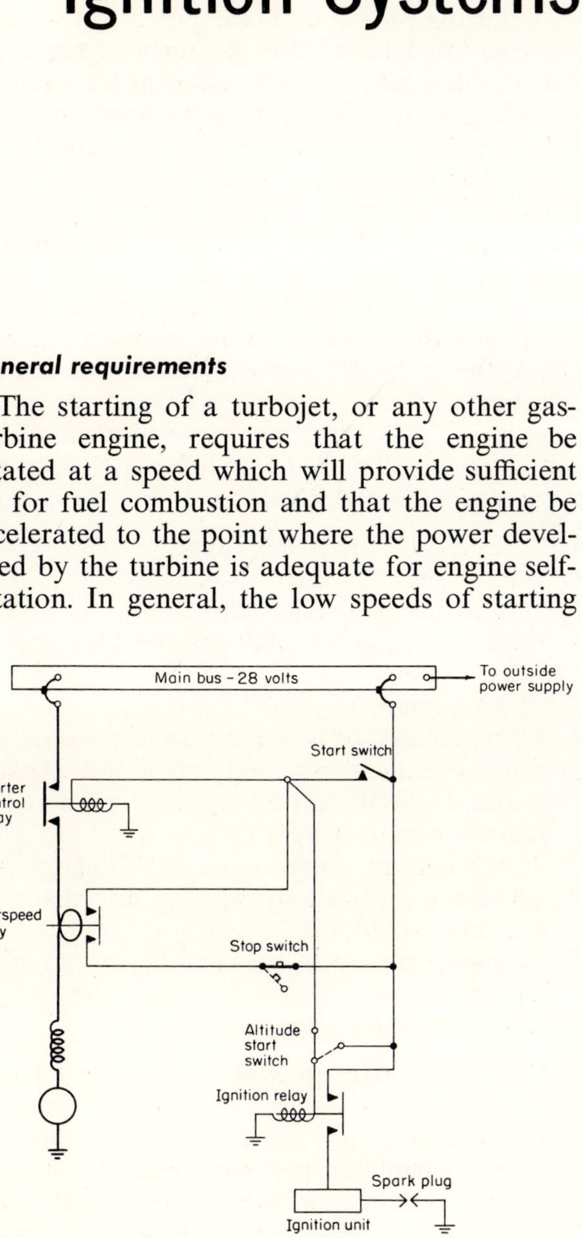

Figure 9 · 1 Electric starting system.

When electric starting systems are used, the starter motor continues to provide torque to the engine even after combustion has started. The systems are so designed that the power circuit to the starter remains energized until engine speed reaches a predetermined safe level. At this time a holding (overspeed) relay opens and de-energizes the main starter relay.

Other types of starters, such as the air-turbine starter and the fuel-air combustion starter, are designed so that cutoff occurs only when engine speed is adequate for self-sustained operation.

ELECTRIC STARTING SYSTEMS

Description and operation

Electric starting systems are of two general types. One type utilizes an electric motor for starting only, and the second type employs a starter which becomes a generator when the engine is operating at normal speeds. The two systems may operate with an identical electric system, but when the system employs the motor for starting only, it is necessary to provide a clutch which will automatically disengage the starter when engine speed is sufficient. When a starter-generator is used, the unit remains permanently engaged to the engine shaft through accessory drive gears.

Figure 9 · 1 is a typical starting circuit for an electric starting system. Usually an electric starting system requires certain basic elements. Among these are the **external power supply,** the **starting switch,** a **stop switch,** a **starter control relay,** an **overspeed relay,** an **ignition system,** and an **altitude start switch.**

Since an electric starter motor may draw from 1,000 to 2,000 amp when first energized, it is necessary that a high-capacity power supply be provided externally. This power supply may be either an engine-driven generator or batteries with sufficient capacity. If batteries are used, it normally requires five 6-volt units connected in series. Each unit should have a capacity of approximately 300 amp-hr. Engine-driven starter power units should be capable of supplying 28 to 30 volts and 2,000 amp. At initial acceleration, the voltage should not drop below 18 volts at the starter terminals.

The starter switch may be a simple toggle switch, spring-loaded for momentary contact. A "button" switch is also satisfactory. The starter switch serves to provide initial power to the coil of the starter control relay. When the starter switch is closed, the starter control relay closes and power is supplied to the starter motor. Current to the starter motor passes through the holding coil of the overspeed relay. The relay closes and provides a complete circuit to the starter control relay, bypassing the starter switch. The starter switch may be released, but the starter control relay will continue to be energized because of the circuit through the overspeed relay.

After combustion commences (light-off) the starter continues to supply torque to the engine until engine speed reaches a predetermined level where engine power is sufficient to maintain acceleration. At this time, the starter motor is generating sufficient back voltage so that current through the holding coil of the overspeed relay is reduced to a point where the holding coil will not keep the relay closed. The relay opens because of spring tension on the contact assembly, and thus the starter control relay is deenergized. Power to the starter is cut off, and the entire system becomes inoperative.

The starter control relay must be a heavy-duty unit capable of carrying a load of 2,000 amp across the main power contacts. Also, the main power leads must be heavy cable which will carry a load of 2,000 amp for the starting period without overheating. Usually the starting period is less than 30 sec.

A stop switch is provided in the starting circuit so that the starting operation can be interrupted if necessary. This switch is connected in series with the circuit through the overspeed relay which supplies power to the starter control relay. Opening the stop switch deenergizes the starter control relay and interrupts the starting operation.

Ignition power is supplied to the ignition unit or units through the ignition relay. This relay is energized at the time that the starter switch is closed; thus ignition is furnished to the spark (igniter) plugs from the beginning of the starter operation. This assures that ignition power is present before fuel flow starts, thus helping to eliminate the danger of a "hot" start.

An altitude start switch is provided in the system so that the engine may be started in flight without energizing the starter motor. When the altitude start switch is placed in the START position, ignition is furnished to the spark plugs. The engine rotation is provided by windmilling, utilizing the force of ram-air pressure.

Electric starting systems are usually designed so that the airplane battery is disconnected from the power bus during starting operations. This is necessary so that power cannot be drawn from the airplane's batteries. These batteries have a relatively low capacity and would be damaged easily by the extreme overload of starting.

Starting procedures

A starting procedure is the process of supplying rotational force to a gas-turbine engine so that it will attain a speed sufficient for starting, providing ignition to start combustion, and then injecting fuel into the combustion chambers. Many starting systems today provide for automatic fuel control, so starting is a relatively simple process. With a system of this type, the operator need only turn on electric power, press the starter button, and move the power control lever into the START or IDLE position. In any event, the manufacturer's instructions must be followed.

A "manual" start is somewhat more involved, but it is not difficult. The following steps are typical of a manual start:

1. Connect external power supply.
2. Turn on master switch.
3. Turn on fuel booster switches.
4. Press starter switch and observe "per cent of power" gauge.
5. Turn on ignition.
6. When per cent of power is 8 to 10 per cent, advance throttle slowly and watch fuel pressure gauge. Light-off will be indicated by a rise in tailpipe temperature.
7. Watch tailpipe temperature (TPT) gauge and advance throttle slowly, making sure that TPT does not exceed limit.
8. At approximately 16 per cent of power, release starter switch.
9. Advance throttle slowly to IDLE position.

The chief danger present with a manual start is overheating the turbine section of the engine. This is called a "hot start." If combustion does not start when fuel pressure rises, the starting procedure should be terminated at once. If excess fuel is accumulated in the combustion chambers when light-off occurs, severe overheating may result.

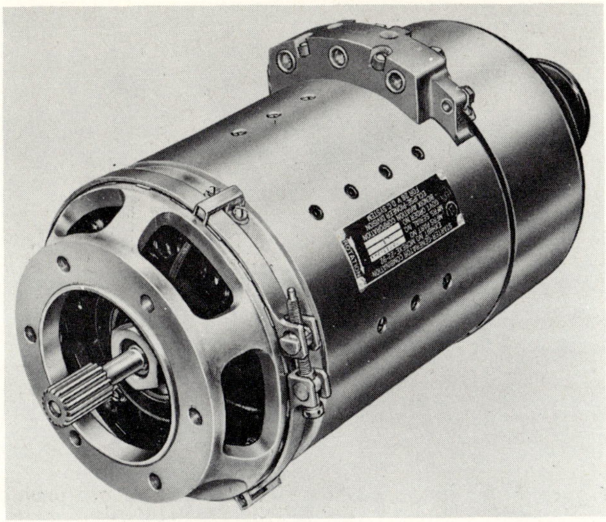

Figure 9 · 2 Starter-generator.

The starter-generator

Figure 9 · 2 illustrates a typical starter-generator. The unit shown is an Eclipse-Pioneer Type 30E13 starter-generator which will provide a torque of 115 lb-ft at an amperage of 1,200.

The starter-generator is constructed with a heavy series winding in addition to the standard generator windings. The series winding is connected so that power from the starter circuit flows through the winding and creates a very strong field to provide a high starting torque. The starting winding is not utilized when the unit is operating as a generator.

Figure 9 · 3 is a circuit typical of the type employed with starter-generators. An examination of the circuit will show that the generator function is independent of the starter function. As soon as engine speed increases sufficiently, the starter section is deenergized and the generator portion begins to operate. As a generator, the unit is capable of producing 30 volts at 500 amp.

AIR-TURBINE STARTERS

General description

Air-turbine starters were designed to reduce the weight of airborne equipment, inasmuch as an air-turbine starter may weigh but one-fourth as much as an electric starter having the same power. A still more striking example may be provided by comparing the Eclipse-Pioneer Type 36E33 air-turbine starter with the 30E13 electric starter-generator. The 36E33 air-turbine starter produces 355 lb-ft of torque and weighs but 22 lb. The 30E13 starter-generator produces 115 lb-ft of torque and weighs 85 lb.

An air-turbine starter consists basically of a turbine section, a high-ratio gear reduction unit, and an automatic engaging and disengaging mechanism. Air enters the annular nozzle chamber and is passed with high velocity through the turbine. The turbine rotates at exceedingly high speeds, often exceeding 45,000 rpm.

The starter is engaged to the engine through a jaw and friction clutch mechanism. It remains

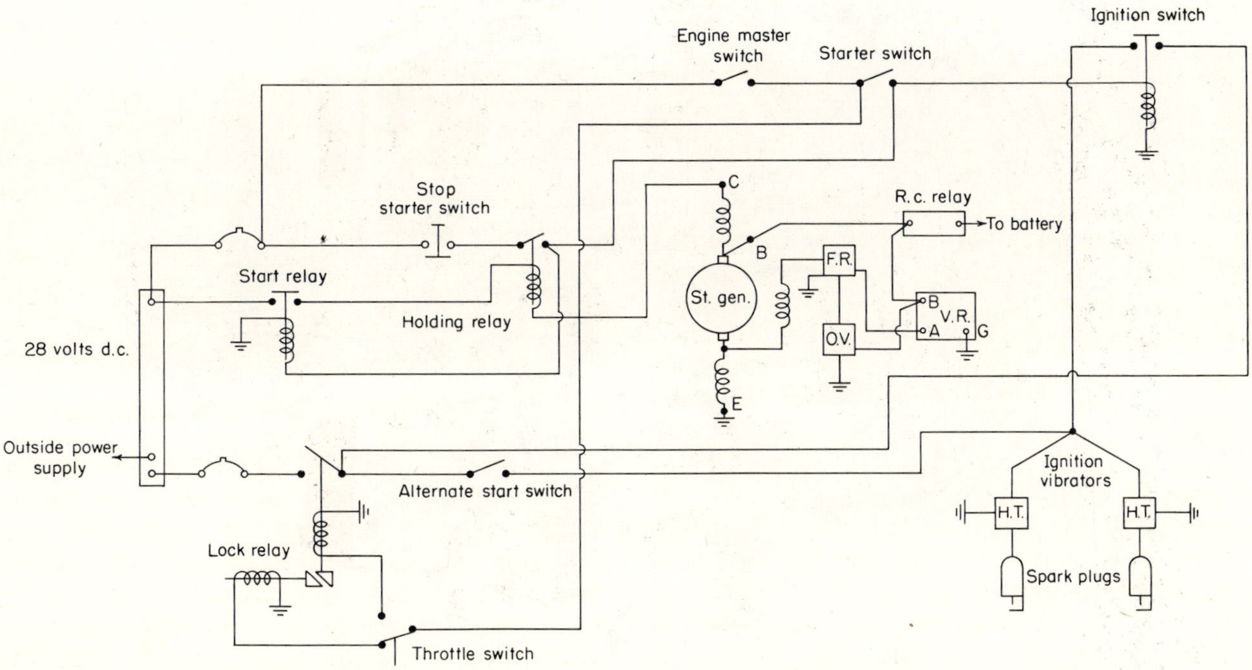

Figure 9 · 3 Starting circuit for starter-generator system.

Figure 9·4 AiResearch air-turbine starter.

Figure 9·5 Bendix-Eclipse Model 36E62 air-turbine starter.

engaged until the engine attains the required cut-off speed.

Figure 9·4 is a photograph of an air-turbine starter manufactured by the AiResearch Manufacturing Division of the Garrett Corporation. This is a lightweight but powerful unit capable of starting large gas-turbine engines in a very few seconds. Another air-turbine starter is illustrated in Fig. 9·5. This unit is the Bendix Type 36E62, manufactured by the Utica Division of Bendix Corporation, Utica, N.Y. It is also a lightweight but powerful starter, capable of cranking large engines in a few seconds.

The Hamilton Standard air-turbine starter

Figure 9·6 is an illustration of the Hamilton Standard PS-500-1 pneumatic starter, designed

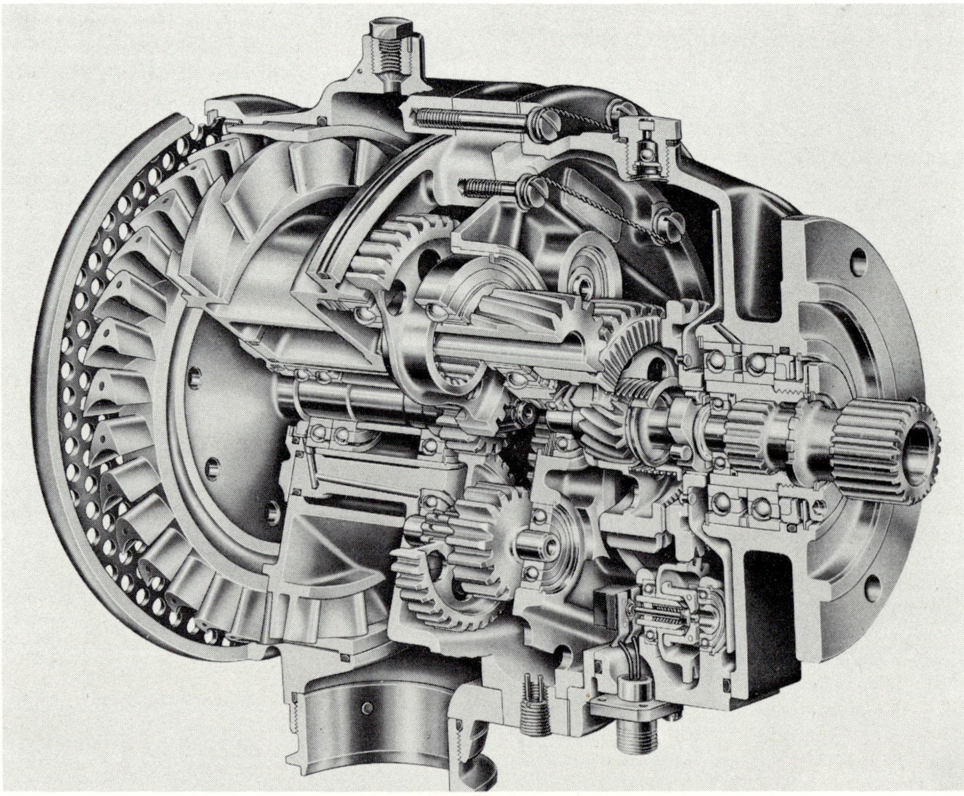

Figure 9·6 Hamilton Standard pneumatic starter.

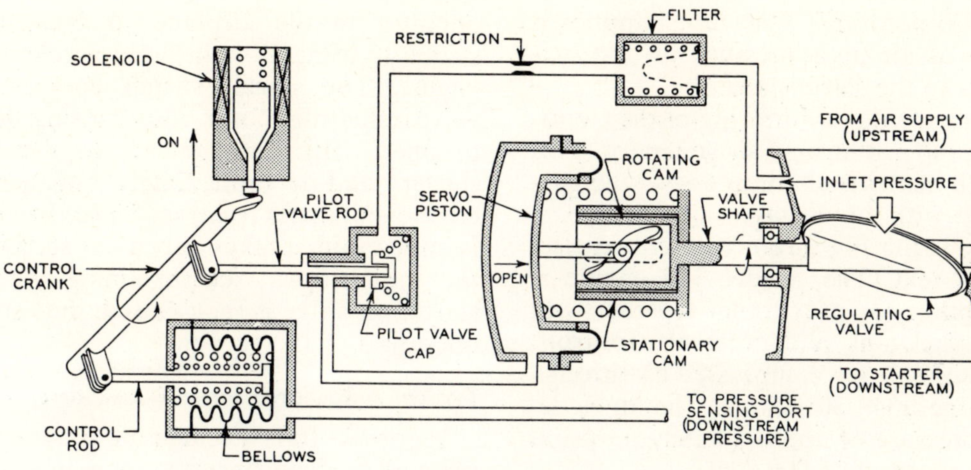

Figure 9 · 7 Pressure-regulating valve and control assembly.

for starting large turbojet engines. This starter consists of an axial-flow turbine, a reduction-gear train of approximately 13.6:1 from the turbine to the output shaft, a clutch assembly, and two centrifugal cutoff switches. The regulated-pressure (approximately 45 psia) and high-temperature (700°F maximum) air is passed through the turbine, accelerating it from zero to approximately 43,000 rpm at the starter cutout speed. A two-stage spur gear reduction is provided to give the desired speed-torque relationship at the output pad. A jaw-type clutch is used to permit starter disengagement when the supply air is shut off. The centrifugal cutoff switches break the circuit to the solenoid of the pressure-regulating valve.

The starter installation includes two major assemblies: the **combination pressure-regulating and shutoff valve,** and the pneumatic starter. The starter mounts on the engine drive pad so that its output drive shaft will mate with the engine starter drive. The combination valve mounts on the starter inlet or at any point in the inlet line.

The pressure-regulating- and shutoff-valve assembly consists of two assemblies: the **pressure-regulating valve,** and the **pressure-regulating-valve control.** These are illustrated schematically in Fig. 9 · 7. The regulating-valve assembly consists of the **valve housing,** the **regulating valve,** the **servo piston,** and the **valve shaft assembly.** The piston moves under air pressure exerted against a diaphragm and in so doing causes rotation of a cam, which in turn changes the position of the regulating valve to direct inlet air into the starter. The slope of the cam track in the rotating cam is designed to provide small initial travel and high initial torque, when the starter is actuated, to aid in breaking any ice formation resulting from the freezing of condensate in the air supply. Also this slope aids in providing more stable valve action by increasing the opening time of the valve. The control assembly is mounted on the regulating-valve housing and consists of the **solenoid,** the **control housing,** the **bellows,** the **control crank,** and the **pilot valve.** The position of the pilot valve is controlled by the action of the solenoid and the bellows, which is vented to downstream air. Air directed by the pilot valve operates the servo piston, which in turn operates the regulating valve.

Air-supply units

The air supply for air-turbine starters usually consists of a small gas-turbine engine mounted on a cart or "dolly" so that it may be moved around as necessary. Figure 9 · 8 illustrates one of these small gas-turbine engines, manufactured by the AiResearch Manufacturing Division of

Figure 9 · 8 AiResearch gas-turbine air generator.

the Garrett Corporation. This unit supplies a large volume of air at a pressure of approximately 45 psia to the starter inlet.

Figure 9 · 9 is a cutaway drawing of the Continental Model 140 gas-turbine air generator. This unit is a small centrifugal-compressor-type gas-turbine engine with a two-stage axial turbine. A similar engine is the J69 turbojet described elsewhere in this text. The Model 140 engine is mounted within a compact trailer together with an 80-gallon fuel tank and necessary controls. The compressor delivers compressed air through a 3½-in. flexible duct, 30 ft long. The air is delivered at a pressure of approximately 53 psi at a compressor speed of 35,000 rpm.

Operation of the air-turbine starting system

The operation of air-turbine starters is comparatively simple. As far as the operator of the aircraft is concerned, there is very little difference between starting with an electric starter and starting with an air-turbine starter, with the exception that the air-turbine starter effects a much quicker start.

The ground power unit consists of an air supply as previously described. This air supply is connected to the air inlet on the airplane with a quick-disconnect union. With the air-supply unit in operation and connected to the air inlet of the airplane, it is merely necessary for the operator in the airplane to press the starter switch to bring about a rapid acceleration of the engine. The starter switch energizes the air-valve-actuating motor, thus causing the air valve to open. This supplies air to the air-turbine starter, and it immediately engages with the engine and starts rotation. Since the acceleration is quite rapid, it takes only a few seconds to bring the engine up to self-sustaining speed and the starter air valve is then closed, thus stopping the operation of the starter.

Typical AiResearch air-turbine-starter system

Figure 9 · 10 is a schematic drawing of a complete air-turbine-starter system as installed in an airplane. This system utilizes components manufactured by the AiResearch Manufacturing Division of the Garrett Corporation. These components are the **air-turbine starter,** the **air-pressure rate-of-change control,** and the **starter control and shutoff valve.**

Initial actuation of the starter control and shutoff valve is through a normally open start switch which is held closed for approximately 3 sec to initiate a starting cycle. From then on, sequence operation of the valve and starter is automatic. A normally closed stop switch serves as a means of stopping the starter in an emergency.

When the starter switch is closed, the solenoid

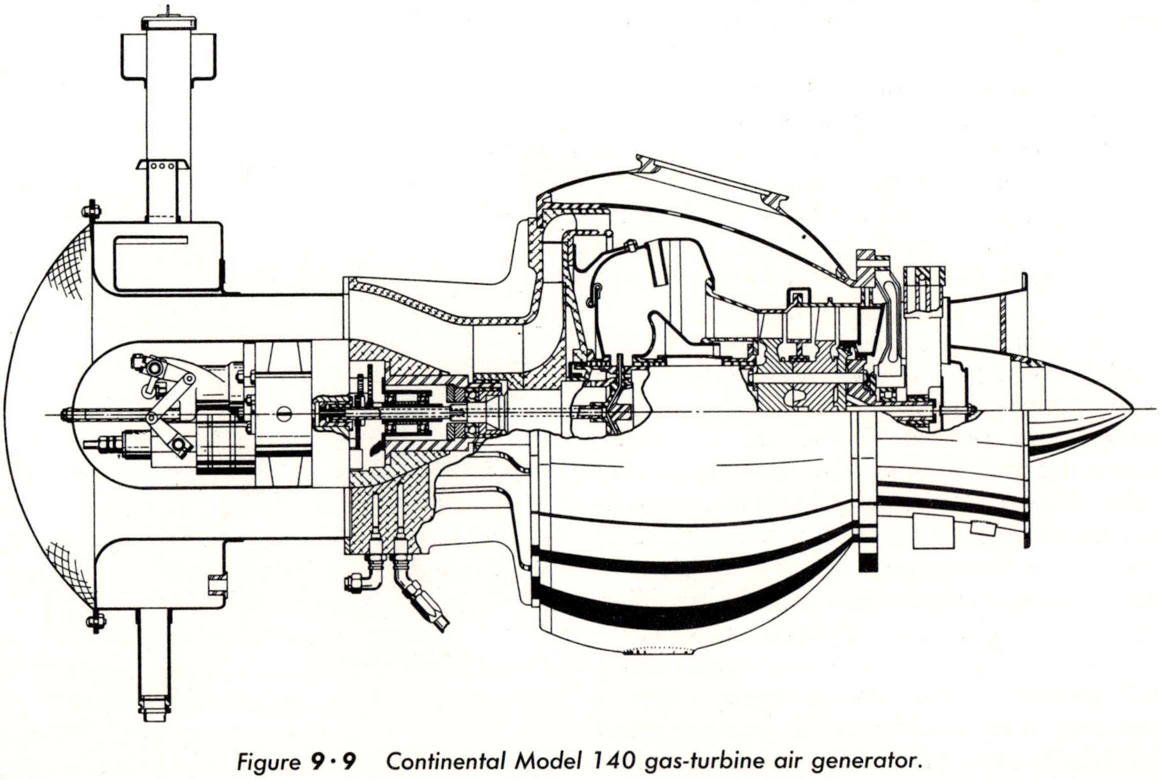

Figure 9 · 9 Continental Model 140 gas-turbine air generator.

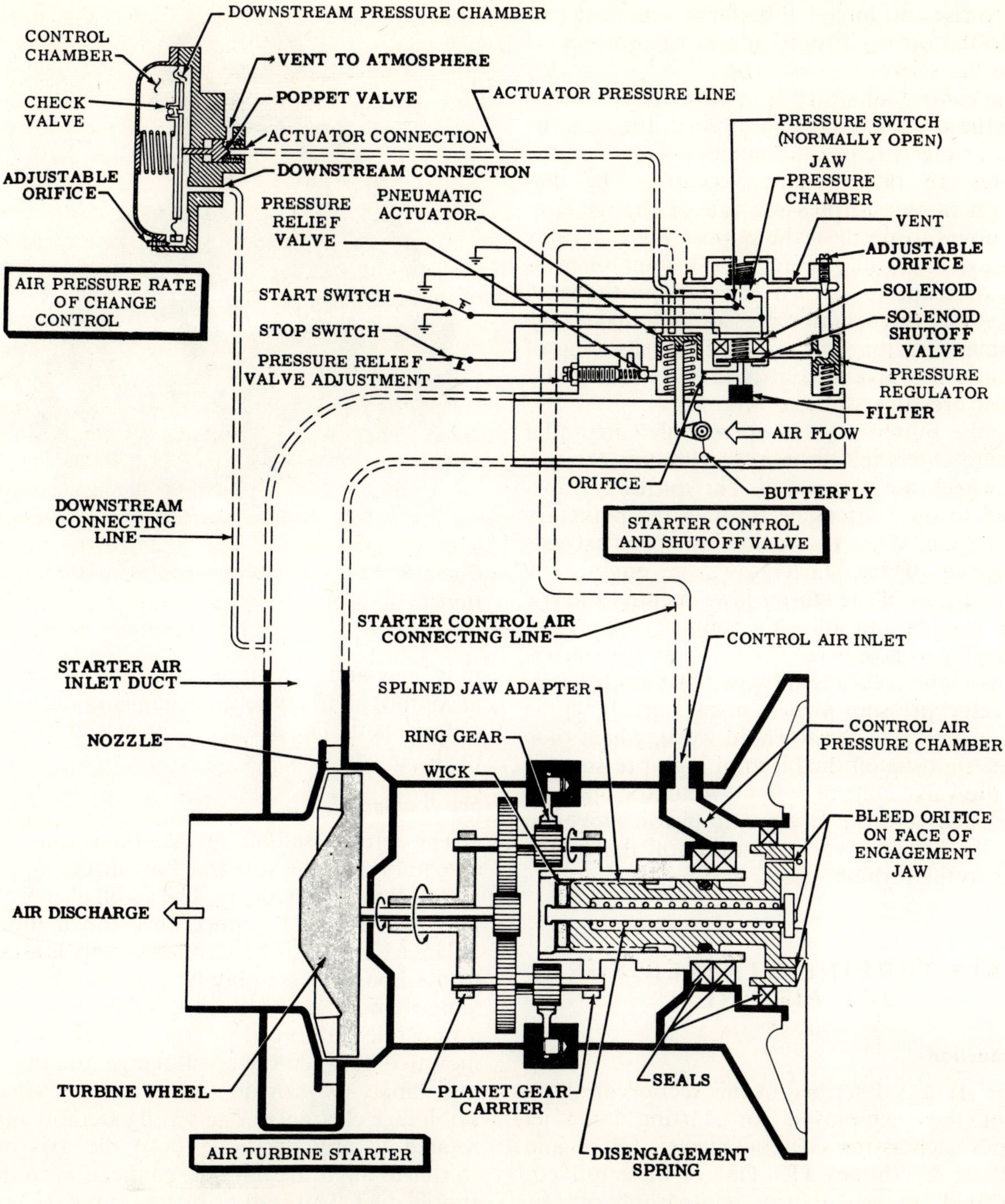

Figure 9 · 10 AiResearch air-turbine-starter system.

valve within the starter control and shutoff valve opens and admits high-pressure air from upstream of the valve into the control system. The high-pressure air is supplied from the external gas-turbine air compressor. From the solenoid valve, the control air passes through a piston-type pressure-regulating valve which limits the control air pressure to the required value. From the pressure-regulating valve, the control air passes through an adjustable orifice, then through a pressure switch, and is admitted to the butterfly-actuating piston. By means of the starter-control connecting line, control air is also admitted to the sealed section of the starter behind the engagement faces of the jaw. The pressure obtained is sufficient to force the jaw outward against the mating jaw in the drive pad on the engine. When the starter engagement jaw mates solidly with the mating jaw, the bleed orifices are blocked, causing the control air pressure to rise and force the butterfly actuating piston to the open position, admitting compressed air to the starter.

The control air which is applied to the butterfly actuating piston is also applied, through the actuator pressure line, to the poppet valve in the air-pressure rate-of-change control. The diaphragm in the air-pressure rate-of-change control moves to position the poppet valve to open or close, permitting more or less actuator pressure to escape to ambient air through the vent. This varies the actuator pressure applied to the pneumatic piston and results in positioning of the butterfly valve to control the rate of increase of inlet pressure to the turbine wheel.

As the butterfly valve opens, inlet air under pressure enters the turbine, accelerating the turbine wheel to a high speed. The torque is transmitted to the starter jaw through the planetary gear system. When the engine-jaw speed exceeds the speed of the starter jaw, the engine jaw moves ahead of the starter jaw, opening the jaw bleed orifices and allowing control air pressure to bleed to atmosphere. The spring in the starter-jaw assembly retracts the jaw. The starter-control-valve pressure switch opens, breaking the electric circuit to the solenoid valve, which then closes, shutting off the bleed of high-pressure air into the valve control system. Actuator pressure is reduced, and the pneumatic piston moves to close the butterfly valve, shutting off the supply of air to the turbine wheel.

AIR-TURBINE STARTER FOR JET AIRLINER

Introduction

The starters described in this section are typical of those employed for starting large jet engines such as the General Electric CJ805 and the Pratt & Whitney JT3. One of these models, illustrated in the cutaway photograph of Fig. 9 · 11, is the ATS-100 for the Pratt & Whitney JT3 engine. This starter is manufactured by the AiResearch Manufacturing Division of the Garrett Corporation.

A similar model of the ATS-100 air-turbine starter, designed for use in the Convair 880 airplane with the General Electric CJ805 engine, is illustrated in the drawing of Fig. 9 · 12. This is a lightweight turbine-type air motor and consists essentially of a scroll assembly which houses the turbine wheel and flow-directing nozzle, a housing assembly, a heat barrier, oil seals, a bearing carrier, the reduction-gear system, an overrunning jaw-and-pawl-type engaging mechanism, an output shaft switch mechanism, and a spline output shaft. The details of the starter are shown in the illustration.

Figure 9 · 11 AiResearch Model ATS-100 air-turbine starter.

Scroll assembly

The scroll assembly houses the turbine-wheel assembly and the integral flow-directing nozzle as mentioned previously. The scroll also provides inlet and exhaust connections and a threaded balance line port. The scroll assembly is attached to the housing assembly by a pair of half-moon flanges and eight bolts. A number of shims and a phenolic reinforced fiberglass heat barrier are mounted between the scroll flange and the housing flange to provide the necessary wheel-to-scroll face clearance. The scroll assembly may be rotated to any position about the axis of the starter to align the inlet air connection to the air supply duct. This is done by loosening the attaching bolts, rotating the scroll to the desired position, and tightening the bolts to the required torque.

Turbine-wheel assembly

The turbine-wheel assembly consists of an integral wheel and shaft with an exducer pinned on one end of the shaft against the front face of the turbine wheel. A slinger, spacer, spur gear, and nut are mounted on the shaft end of the turbine-wheel assembly and make up the entire rotating assembly. A rollpin through the nut and shaft provides for a secure installation.

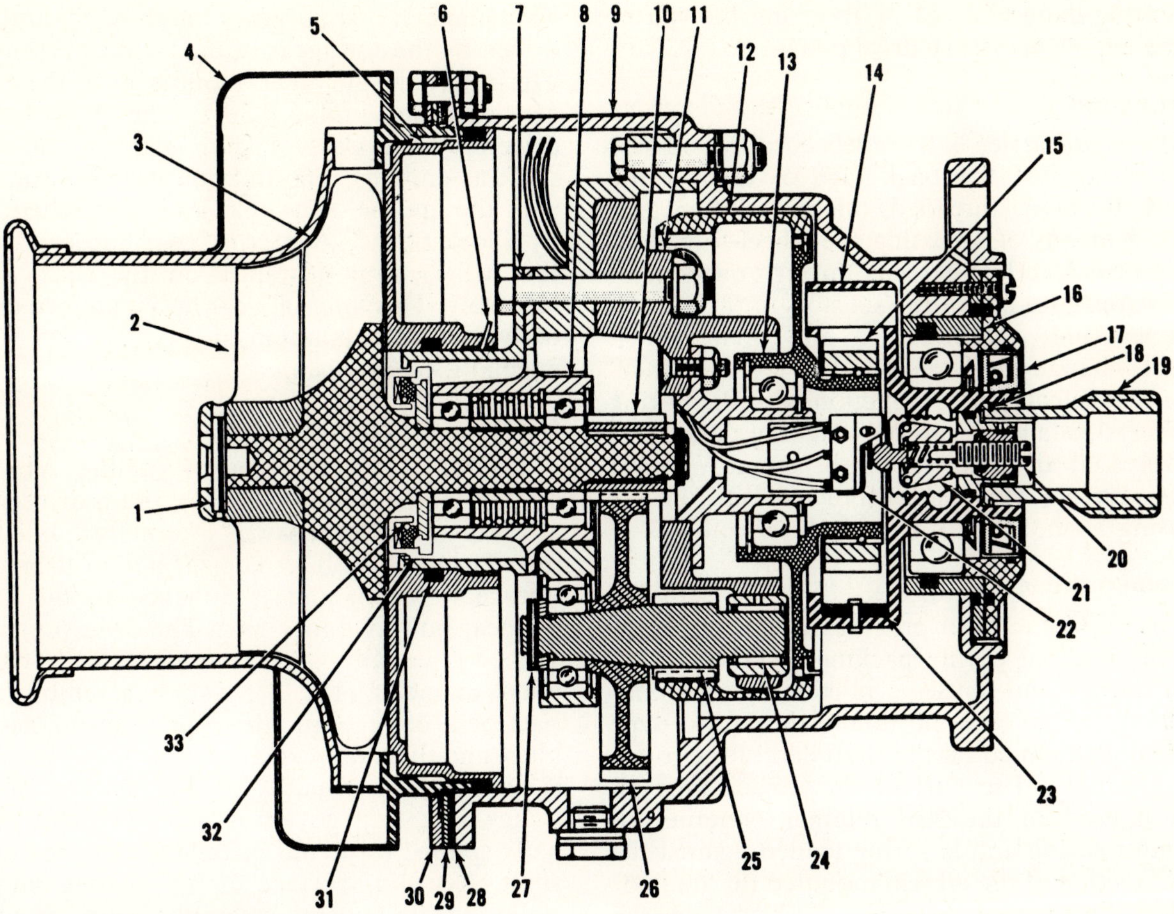

1. Rollpin
2. Exducer
3. Turbine wheel
4. Scroll
5. Heat barrier
6. Rollpin
7. Oil-seal retainer
8. Bearing carrier
9. Starter housing assembly
10. Gear
11. Gear carrier assembly
12. Gear
13. Gear hub
14. Drive shaft assembly
15. Jaw
16. Retainer
17. Seal
18. Lock ring
19. Output shaft
20. Governor adjustment screw
21. Flyweight
22. Output shaft switch assembly
23. Spring
24. Bearing
25. Gearshaft assembly
26. Gear
27. Rollpin
28. Shim
29. Barrier
30. Flange
31. Packing
32. Gasket
33. Seal

Figure 9 · 12 Cutaway drawing of the Model ATS-100 starter.

Housing assembly

The housing assembly, which attaches to the scroll assembly, provides a sealing surface for the O-ring packing installed on the outer diameter of the heat barrier. The housing assembly encases the heat barrier, oil seals, bearing carrier, reduction-gear system, overrunning jaw-and-pawl-type engaging mechanism, output shaft switch mechanism, and output shaft. It also provides oil drain and fill ports and an electrical receptacle. A name and modification record plate are installed on the housing. Elongated keyhole-type holes are provided in the housing mounting flange for aid in attaching the starter to the engine accessory drive pad.

Heat barrier

The heat barrier is threaded onto the oil-seal retainer against the back face of the turbine wheel. By being threaded, this installation provides a means of adjusting the wheel-to-barrier clearance. A rollpin extends through one side of the barrier into one of the six slots in the oil-seal retainer and prevents any movement which might change the wheel-to-barrier clearance. The heat barrier, in conjunction with the O-ring packings installed on it, prevents the passage of compressed air from the scroll assembly to the housing assembly or the entrance of oil from the housing assembly into the scroll assembly.

Oil-seal retainer

The oil-seal retainer provides a sealing surface for the internal O-ring packing installed in the heat barrier and houses a portion of a carbon-type oil seal. Another portion of the carbon-type oil seal is mounted on the shaft end of the rotating assembly. The part of the carbon-type oil seal housed in the seal retainer contains an O-ring packing and is spring-loaded against the other section of the oil seal mounted on the shaft end of the rotating assembly. This provides the sealing effect. Note the numbers 6, 7, 31, and 33 in the drawing of Fig. 9 · 12.

Bearing carrier

The bearing carrier provides for installation of the two ball bearings in which the rotating assembly is mounted. The bearing carrier is bolted to the gear-carrier assembly of the reduction-gear system.

Reduction-gear system

The reduction-gear system consists of a gear-carrier assembly, three spur gear shaft assemblies, an internal gear hub, and an internal spur gear. The gear-carrier assembly is bolted to the inside of the housing assembly and provides for the installation of ball bearings and needle bearings which support the three gear-shaft assemblies. The internal gear hub mounts on a ball bearing which is provided in the gear-carrier assembly. Each of the three spur gear-shaft assemblies consists of an integral spur gear and tapered shaft, a planetary spur gear, a bearing, and a nut. The planetary spur gear has a tapered bore and is friction-mounted on the tapered shaft. It is held in position by the bearing and nut. The nut is secured to the shaft by a rollpin. The planetary spur gears mesh with and are driven by the spur gear of the rotating assembly. The internal spur gear mounts over the three gear-shaft assemblies and meshes with and is driven by the integral spur gears of the gear-shaft assemblies. The internal gear hub meshes with the internal spur gear and is attached to it by a lock ring. The internal gear hub rotates on the ball bearing installed on the gear-carrier assembly. This internal gear hub is splined externally for the installation of the drive jaw of the engaging mechanism.

Engaging mechanism

The engaging mechanism of the ATS-100 starter consists of a drive jaw and a drive-shaft assembly. The drive jaw is internally splined for installation of the internal gear hub of the reduction-gear system and is attached to the internal gear hub by a lock ring. The drive jaw has a series of equally spaced ratchet teeth on its outer diameter. The drive-shaft assembly consists of a drive shaft, three pawl-spring assemblies, and three drive-shaft pawls. The drive shaft is threaded internally for installation of the switch-actuating governor assembly and internally splined to permit installation of the output shaft. A sealing surface for the sealing ring installed in the housing assembly is also provided by the drive shaft. Each pawl-spring assembly consists of a number of varied length leaf-type springs riveted to the inside diameter of the drive shaft. The drive-shaft pawls are installed in keyhole-shaped slots in the drive shaft and are forced into engagement with the ratchet teeth of the drive jaw by the action of the pawl-spring assemblies. The drive-shaft assembly rotates in a ball bearing installed in the housing assembly. An oil baffle is secured to the drive-shaft assembly against the outer face of the bearing by a lock ring. Shim washers are installed between the lock ring and bearing and provide the proper preload on the oil baffle.

Output shaft switch mechanism

The output shaft switch mechanism consists of a snap-action switch assembly and a switch-actuating governor assembly. The switch assembly is installed in the inner diameter of the hub of the gear-carrier assembly. It consists of a normally closed snap-action switch and a flapper to actuate the switch. The switch-actuating governor is threaded into and rotates with the drive shaft. It consists essentially of two spring-loaded flyweights and an actuating pin. Provision for regulating the calibrated cutout speed is made by providing an adjustment screw in the governor body. The adjustment screw is accessible through the center of the output shaft.

Output shaft

The output shaft is splined on both ends. It meshes on one end with the drive-shaft assembly and on the other end with the engine accessory drive pad. The output shaft is secured to the drive-shaft assembly by means of a retaining ring. An internal slot in the output shaft engages a pin in the governor body and prevents any change in the governor calibration which might be caused by the governor's turning in the drive shaft.

Starter operation

During a typical starting operation the air-turbine starter converts the energy of compressed air into rotational mechanical force. This force is delivered through the starter output shaft to an aircraft gas-turbine engine. Compressed air is admitted through the starter scroll inlet connection and is forced into contact with the turbine wheel through the integral nozzle in the scroll assembly. The energy of the compressed air is expended in causing the turbine wheel to rotate at high speed. The air is exhausted at the scroll exhaust connection. The reduction-gear system reduces the high rotational speed, thereby increasing the torque at the starter output shaft. The engaging mechanism provides a means of coupling the reduction-gear system to the output shaft for engine starting. When the engine starts, the output shaft and drive shaft rotate with the engine, and the pawls of the engaging mechanism begin to ratchet. At a predetermined speed the cutout switch opens the electric circuit, which automatically cuts off the compressed air supply. With the air supply cut off and the pawls ratcheting, the turbine wheel and reduction-gear system cease their high-speed rotation. The schematic diagram for the starter system is shown in Fig. 9 · 13.

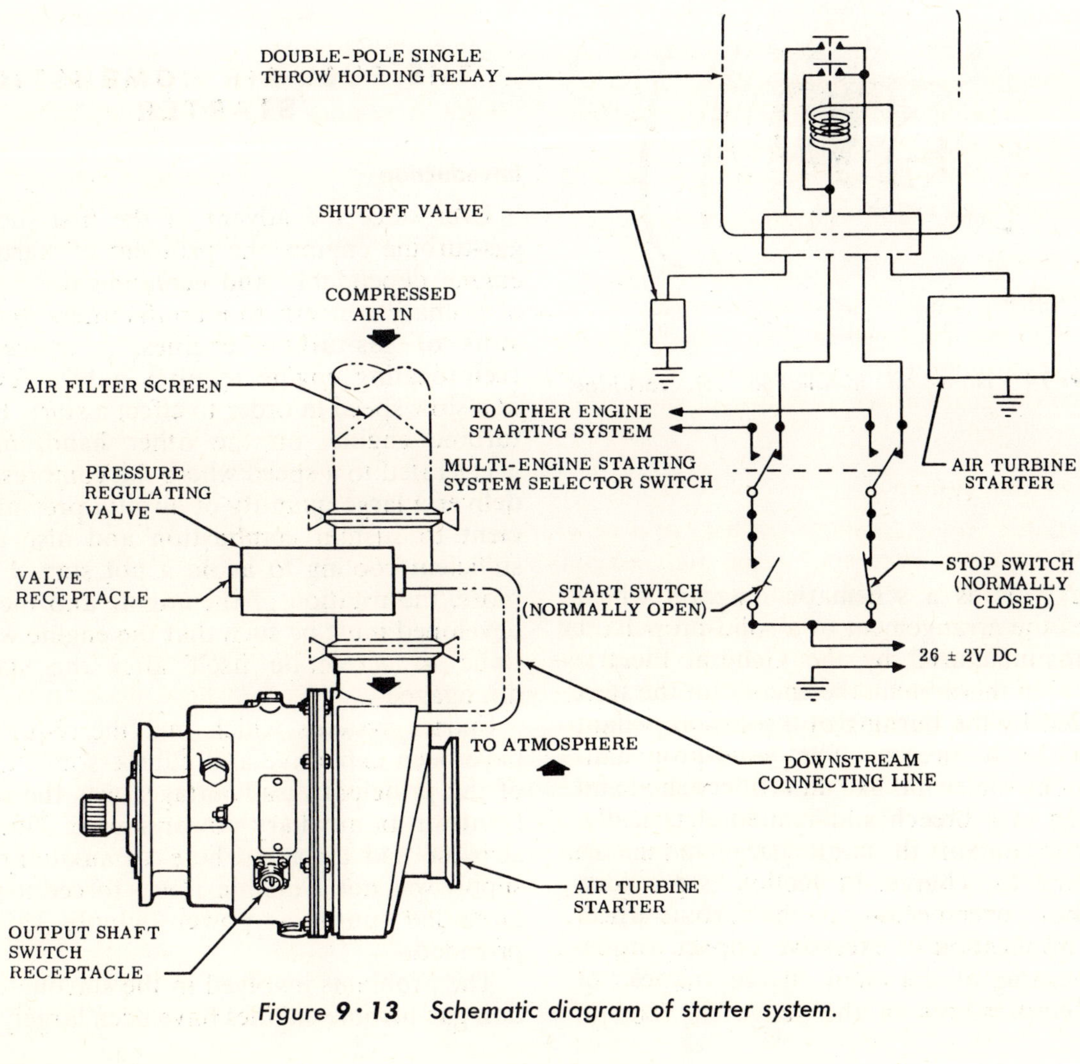

Figure 9 · 13 Schematic diagram of starter system.

THE SOLID-PROPELLANT STARTER

Description

The solid-propellant gas-turbine starter is very similar to the air-turbine starter in operation; however, there are two important differences. The solid-propellant turbine must be constructed to withstand high temperatures, and instead of being driven by an external air supply, it is driven by an integral solid-propellant combustion which generates an adequate supply of hot gases under pressure. Figure 9 · 14 is a schematic drawing showing how the solid-propellant starter may be installed on the nose pad of an engine.

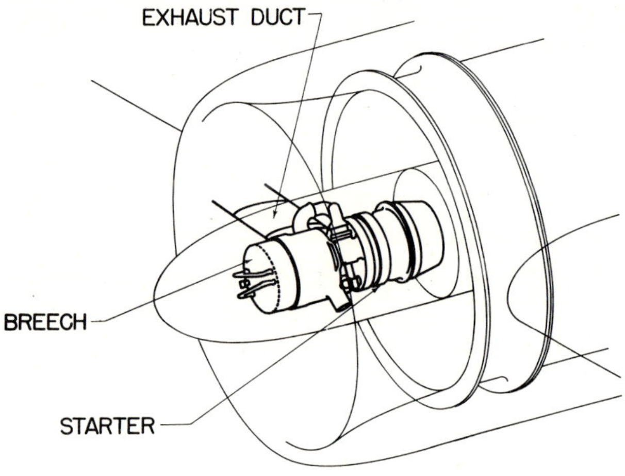

Figure 9 · 14 Nose-pad installation for cartridge starter.

Operation

Figure 9 · 15 is a schematic diagram which illustrates the arrangement of a solid-propellant starter manufactured by the General Electric Company. In this system, the energy for the start is provided by the burning of a solid-propellant cartridge in a breech. The solid-propellant charge looks somewhat like an artillery shell and is inserted in a breech and ignited electrically. The breech contains the main charge and means for igniting the charge. Protection is provided against both overspeeding of the turbine wheel and the application of excessive impact torque. Rapid burning of the main charge creates hot gas at temperatures in the range of 1800 to 2000°F, and pressures in the range of 500 to 1,500 psi.

The burning time of the cartridge can be adapted to match the engine requirements. The starter is so designed that it will automatically declutch from the engine as soon as the cartridge combustion is completed. During the time that the starter is operating, it applies an almost constant torque to the engine. This torque is sufficient to rotate the engine to idling speed within a very few seconds.

The starter is equipped with safety features such that the cartridge cannot be accidentally fired before the breech cap is threaded into place. The breech is equipped with an overpressure valve which prevents the gas pressure in the chamber (breech) from becoming high enough to burst the chamber.

The advantages of the solid-propellant starter are twofold: the starter is light and compact, and it is completely self-contained within the aircraft. Thus the aircraft is independent of an external power source.

THE FUEL-AIR COMBUSTION STARTER

Introduction

Ever since the advent of the first successful gas-turbine engine, the problem of starting the engine dependably and economically has been of primary concern to manufacturers and operators of gas-turbine engines. A conventional reciprocating engine requires only a few turns at a slow speed in order to effect a start. The gas-turbine engine, on the other hand, must be accelerated to a speed where the compressor will deliver a large quantity of air at a pressure sufficient to sustain combustion and also provide sufficient cooling to avoid a hot start. Furthermore, the rotation of the engine and the power developed must be such that the engine will continue to accelerate itself after the starter is disengaged.

Starter systems which met the requirements have been expensive and cumbersome, and one of the principal disadvantages was the requirement for an auxiliary power source. Thus, if an airplane had to land where an auxiliary power supply was not available, it was forced to remain grounded until a power supply could be provided.

The problems involved in the starting of aviation gas-turbine engines have been largely solved

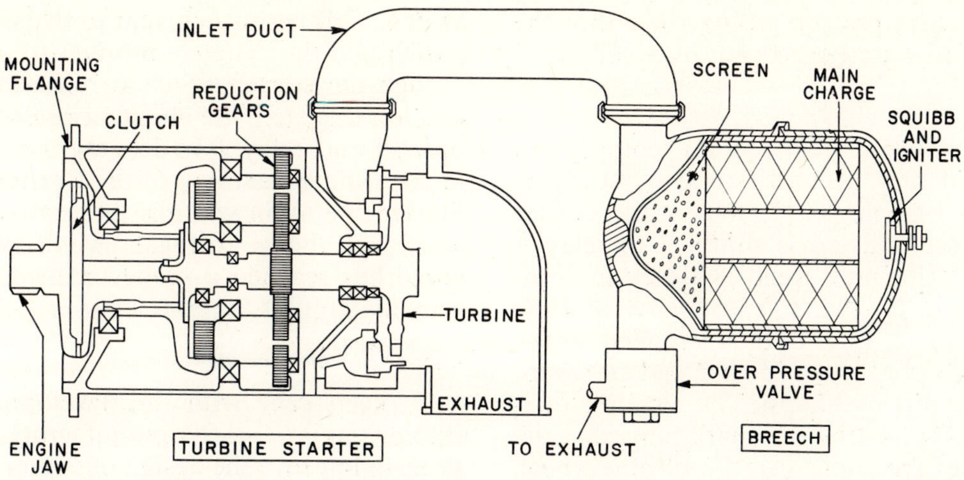

Figure 9·15 General Electric cartridge starter system.

through the development of ground-based air supply units, high-pressure air supplies aboard the aircraft, and fuel-air combustion starters. The fuel-air combustion starter consists basically of a fuel-control system, ignition system, combustion chamber, turbine-driven section, high-ratio reduction-gear unit, clutch system, and fuel and air storage units. Thus the starter is completely contained within the airplane.

Description

A cutaway illustration of the Bendix Model 36E23 fuel-air starter is shown in Fig. 9·16. All the principal components of the starter are clearly visible. The starter is essentially a small gas-turbine engine which delivers power for starting through a high-ratio reduction-gear assembly and clutch arrangement. The compressed air required for combustion is stored in a Fiberglas bottle which is light in weight, shatterproof, and noncorrosive.

The 36E23 starter will accelerate a typical gas-turbine engine to approximately 2,000 rpm in about 3.5 sec. The torque delivered to the engine is approximately 800 lb-ft.

Fuel for the starter is stored under pressure in a piston-type accumulator which is mounted as an integral part of the starter assembly. The fuel is pressurized to 350 psi by air from the high-pressure air supply. The air is compressed by an

Figure 9·16 Bendix model 36E23 fuel-air combustion starter.

engine-driven compressor arranged so that the starting system is recharged automatically.

Operation

The operation of the fuel-air combustion starter system may be understood by studying the diagram of Fig. 9·17. When the starter switch is closed, current is supplied to relay *A* and to the ignition unit. Relay *A* closes and provides current to the fuel solenoid valve and the air solenoid valve. Fuel is thus supplied to the fuel nozzle at a pressure of 350 psi, and air is supplied at a pressure of 300 psi. The resulting fuel-air mixture is ignited by the spark plug, and the resultant gases are applied to the turbine wheel. The starter jaw engages the engine and acceleration begins. The pressure of the air entering the combustion area actuates the chamber pressure switch. This permits current to bypass the starter switch and the system continues to operate even though the starter switch is released. When the fuel is exhausted, the chamber pressure switch is opened and relay *B* is deenergized. This results in immediate cessation of the starting operation. Starter operation will also be terminated by the action of the centrifugal switch when engine speed has reached a predetermined level of approximately 2,000 rpm.

Recent developments

To keep pace with the development of gas-turbine engines, starter manufacturers must continue to improve the design of starters. The Utica Division of the Bendix Corporation has followed this practice and supplies starters for all types of gas-turbine engines. The Model 36E26 starter

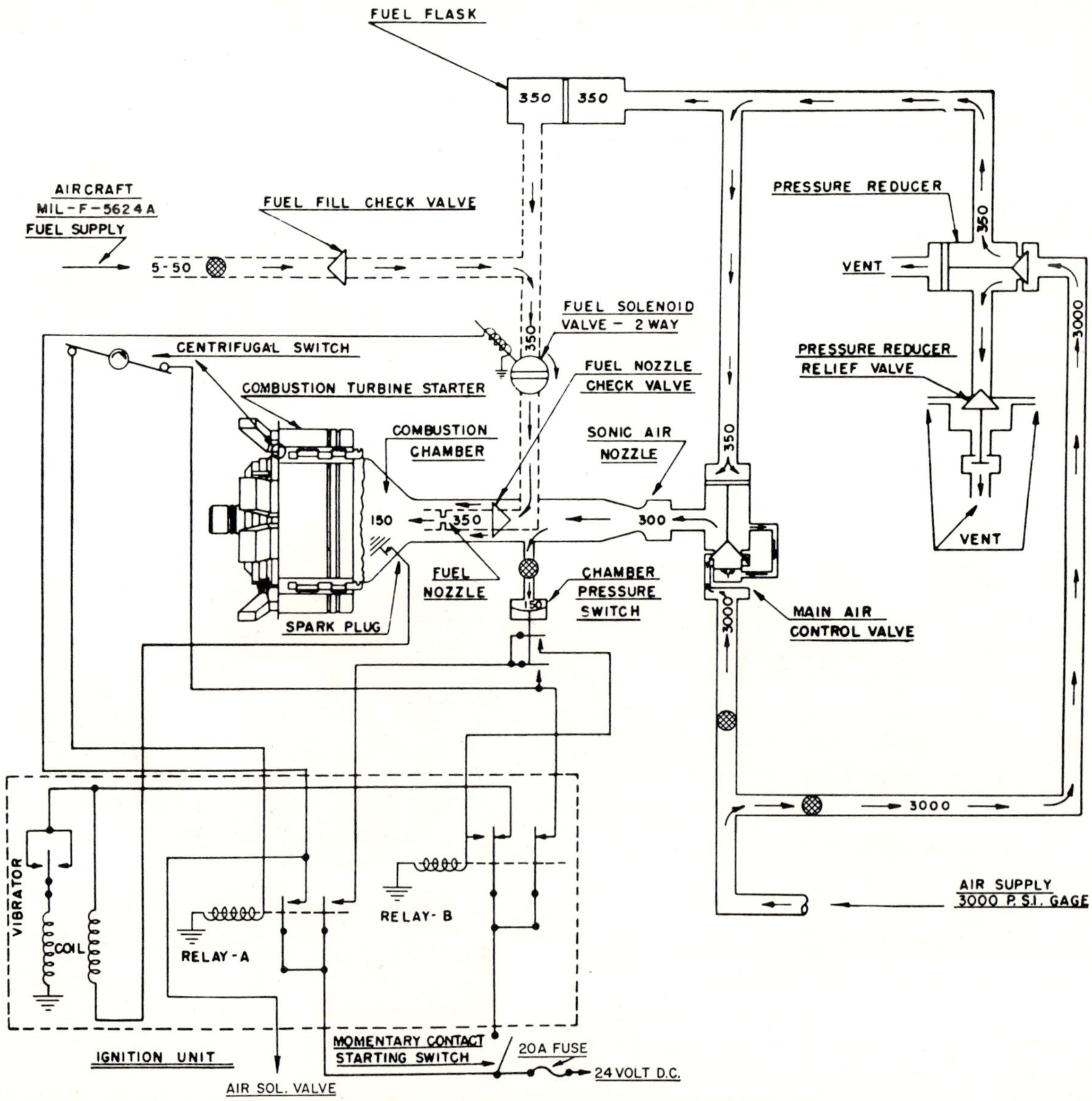

Figure 9·17 Schematic diagram of fuel-air combustion starter.

Figure 9·18 Bendix Model 36E26 fuel-air combustion starter.

shown in Fig. 9·18 is designed to produce a very high torque for starting large gas-turbine engines. This starter operates in a manner similar to that of the 36E23 previously described and will deliver 1,375 lb-ft of torque to accelerate a large engine in 3.5 sec.

Split-compressor engines, such as the Pratt & Whitney JT3, require a longer acceleration period than single-compressor engines because the low-pressure section of the compressor must be accelerated by the airflow produced by the high-pressure section. For engines of this type the Model 36E44 starter has been developed. This starter is designed for a rated torque of 420–435 lb-ft and will accelerate the engine to 2,450 rpm in 13.8 seconds.

The Model 36E44 starter requires a Bendix-Utica 38E15 air valve to regulate the high-pressure air from the air storage bottle to a constant 300 psi for use in the starter. Fuel is taken from the aircraft fuel tanks and stored in the fuel accumulator for each start. A cutaway photograph of the starter is shown in Fig. 9·19.

IGNITION SYSTEMS

Introduction

Ignition systems for gas-turbine engines serve the same purpose as those for reciprocating engines; however, they are required to operate for only the few seconds during which the engine is started. The combustion in a gas-turbine engine is self-sustaining; hence the ignition system is deenergized as soon as the engine is started. In general, the ignition of the fuel-air mixture in a gas-turbine engine is somewhat more difficult to accomplish than in a reciprocating engine because of the turbulence and velocities encountered in the combustion chamber of the gas-turbine engine. Also, gas-turbine fuels are less volatile, as a rule, than the high-performance-number fuels used in reciprocating engines; hence vaporization is not so rapid.

To provide for the starting conditions in a gas-turbine engine, the ignition system is designed to

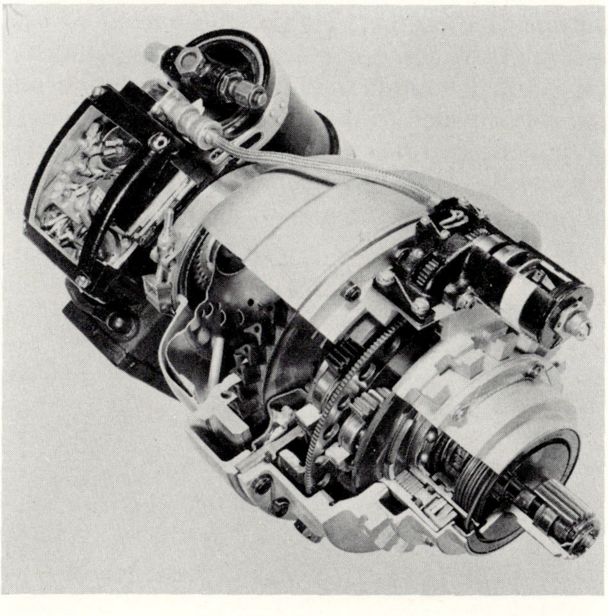

Figure 9·19 Cutaway view of Bendix 36E44 fuel-air starter.

produce a high-voltage, high-energy spark across a wide spark gap. Various types of ignition systems have been used with gas-turbine engines, some utilizing alternating current as a power source and some utilizing direct current.

A-c ignition systems

One of the simplest ignition systems for a gas-turbine engine utilizes 115-volt 400-cycle alternating current as a power supply and, by means of high-ratio transformers, steps the voltage up to approximately 20,000 volts. A system of this type is shown in Fig. 9·20. When the power supply for starting is turned on, an inverter is energized. This inverter supplies the 115-volt 400-cycle alternating current to the ignition relay. When the starter switch is closed, the ignition relay is energized by direct current from the d-c power supply. This closes the a-c circuit to the ignition transformers, and these transformers deliver high-voltage (20,000-volt) alternating current to the igniter plugs.

Another a-c ignition system which has been used with some engines is called the "opposite-polarity system." The circuit for this system is shown in Fig. 9·21.

The igniter plugs for an opposite-polarity system are insulated single-electrode plugs. Two plugs are mounted opposite each other in the combustion chamber. The plugs extend into the chamber, so that the electrodes of the two plugs form a gap at the center of the combustion chamber. The ignition transformers are provided with two secondary windings to each unit, with the direction of winding such that the output of one secondary winding is opposite in polarity to the output of the other. Hence, the output of one secondary winding is connected to one ignition plug, and the output of the other secondary winding is connected to the opposite plug. The effect is that the two secondary windings are in series with each other and their combined voltage is impressed across the gap formed by the igniter plugs.

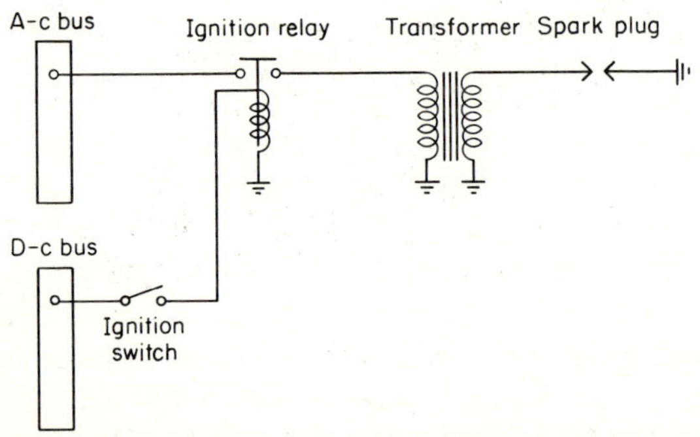

Figure 9·20 Simple a-c ignition system.

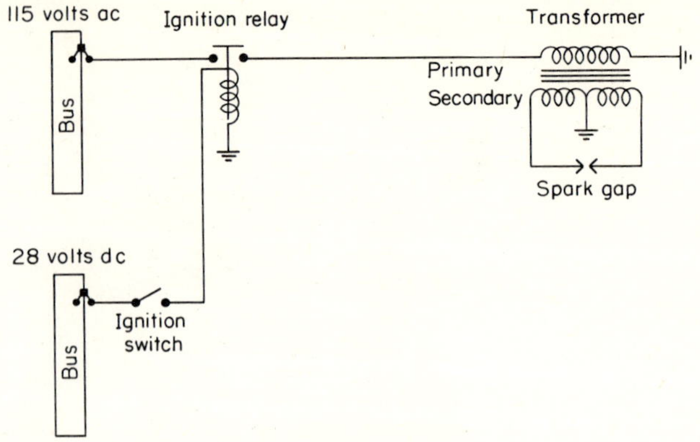

Figure 9·21 A-c opposite-polarity ignition circuit.

Simple d-c ignition system

A simple d-c ignition system consists of a vibrator, a transformer, and ignition plugs, together with necessary wiring, shielding, and attachments. A system of this type is shown in Fig. 9·22. The vibrator serves to change the direct current of the power supply to a pulsating current so that the voltage may be stepped up through transformers. The output of the vibrator is directed through the primary winding of the transformer. This causes a high voltage to be induced in the secondary winding. The secondary voltage is applied to the spark plugs as shown.

Opposite-polarity d-c system

The opposite-polarity d-c system is similar in function to the opposite-polarity a-c system; however, the d-c system employs a vibrator to develop a pulsating current which may be stepped up by the transformer. A schematic diagram of the system is shown in Fig. 9·23.

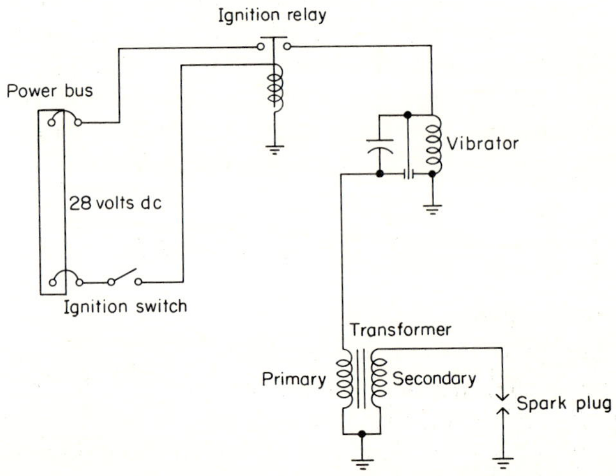

Figure 9·22 Simple d-c ignition circuit using a vibrator.

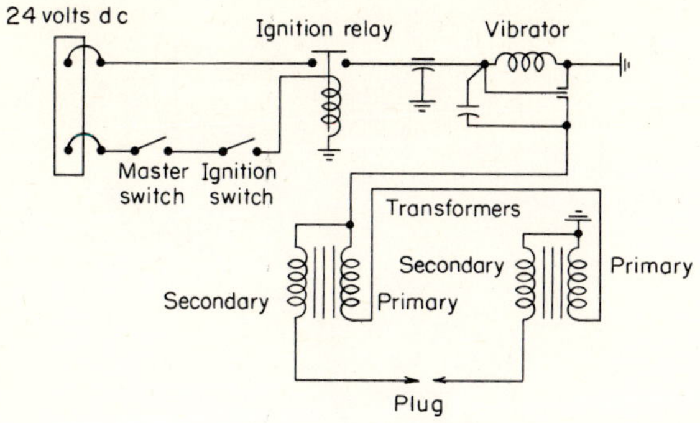

Figure 9·23 D-c opposite-polarity ignition circuit with a vibrator.

High-energy ignition systems

Many aviation gas-turbine engines today utilize the high-energy, capacitor-type ignition system which is capable of starting engines with fouled spark (igniter) plugs and which makes starting at high altitudes much easier.

Figure 9·24 is a schematic diagram of the ignition system for one spark plug. Normally two such systems will be used for each engine, since two of the combustion chambers are provided with spark plugs.

The system shown will operate on 16 to 30 volts, direct current, which is fed to the ignition system through a filter to suppress radio interference. In the filter, the induction coils in series tend to reduce alternating current, but they permit direct current. The capacitors in parallel permit alternating current through ground, but they prevent direct current. Thus, the alternating current is bypassed to ground while the direct current is permitted through the breaker points to the autotransformer T_1. Since the points are opened and closed by the action of the cam M, the current through the transformer is a pulsating direct current.

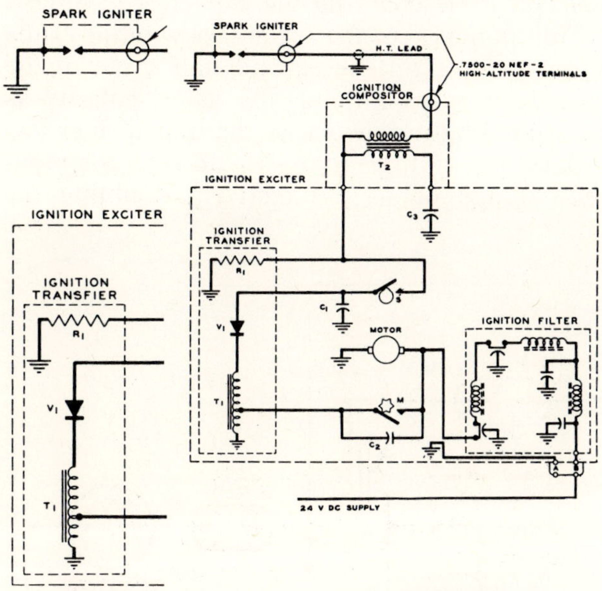

V_1	Rectifier	M	Multilobe cam
T_2	Trigger transformer	C_3	0.25-mfd capacitor
T_1	Stepper transformer	C_2	1.0-mfd capacitor
S	Single-lobe cam	C_1	2.0-mfd capacitor
R_1	2000-ohm resistor		

Figure 9·24 Cam-operated high-energy ignition system.

By self-induction, the transformer T_1 increases the supply voltage of 24 volts to approximately 2,000 volts. This increased voltage drives the current through the selenium rectifier V_1 to charge the capacitor C_1. The rectifier is used to prevent a reverse current which would normally occur with a transformer. If a reverse flow were permitted, it would not be possible to bring the capacitor to a full charge.

Periodically (several times per second) the cam S closes and allows a surge of direct current from the capacitor C_1 through the primary winding of transformer T_2 to the capacitor C_3. This surge of current induces a high voltage in the secondary of transformer T_2, thus breaking down the air gap at the spark-plug electrodes. This ionizes the air gap and produces a low-resistance path for the lower-voltage and higher amperage current which follows. The resistance R_1 and the capacitor C_3 provide an oscillatory current through the primary of T_2 after the first surge of current from C_1. This oscillatory current induces the voltage in the secondary for the high-energy spark at the spark plug.

Another type of high-energy ignition circuit is shown in Fig. 9·25. This system utilizes a conventional vibrator-type spark coil (T_1) to produce the first increase in voltage. The current in the primary circuit of T_1 must pass through the breaker points. When this occurs, the magnetism produced by the primary coil opens the breaker points, thus interrupting the flow. The points close again by spring action, and the action is repeated continuously as long as the power is supplied to the coil. The output of T_1 is an alternating current, but the action of the gas-filled rectifier tubes allows the flow in only one direction. Actual current is always from the ground connections, through the units in the circuit, to the power supply.

The surges of direct current charge capacitor C_4 until sufficient voltage is built up to break down the gap G_2. When this occurs, it causes the gap G_1 to ionize and provide a low-resistance path for C_4 to discharge through the primary of T_2. The capacitor C_6 and the resistor R_2 form

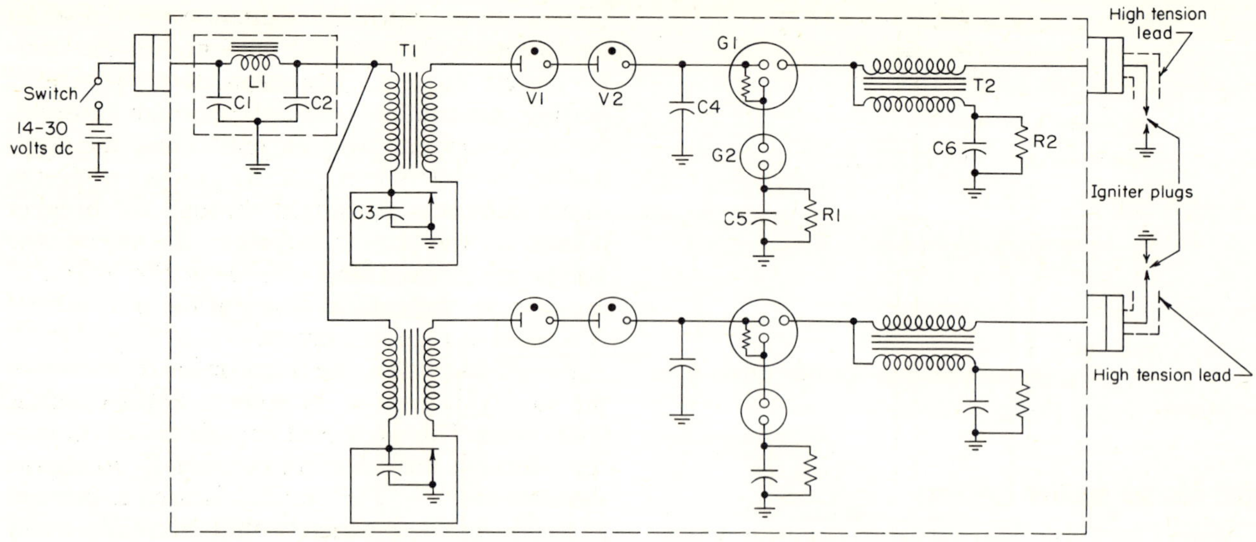

Figure 9·25 High-energy ignition system with gas-filled rectifier tubes.

an oscillatory circuit with the primary of the transformer, thus causing a high-frequency alternating current in the primary. The primary current induces a high-energy voltage in the secondary winding, which is delivered to the spark plugs.

The purpose of a high-energy ignition system is to deliver a comparatively high-amperage spark to the spark plugs, thus producing sufficient heat to overcome most obstacles to fuel ignition.

Typical high-energy ignition system

The ignition system for the CJ805-23 turbofan engine is typical of high-energy types for current gas-turbine engines. A schematic diagram of this system is shown in Fig. 9·26.

This system is supplied with 110-volt 400-cycle electric power which first passes through a radio-frequency filter L_1 and C_1. From the filter, current passes through the primary of the voltage-doubler transformer T_1. The voltage is multiplied more than 30 times by the transformer, and the resulting voltage is doubled through the action of the tubes $V_{1\text{-}1}$ and $V_{1\text{-}2}$. The output voltage of the transformer secondary winding is about 3,500 volts. Tube $V_{1\text{-}2}$ conducts during the first half cycle charging the capacitor to almost the full potential of the secondary winding. Tube $V_{1\text{-}1}$ conducts on the second half cycle of the secondary output. Being the same polarity as the second half cycle across the transformer secondary, the voltage across C_2 adds to the rectified voltage of $V_{1\text{-}1}$, effectively doubling the

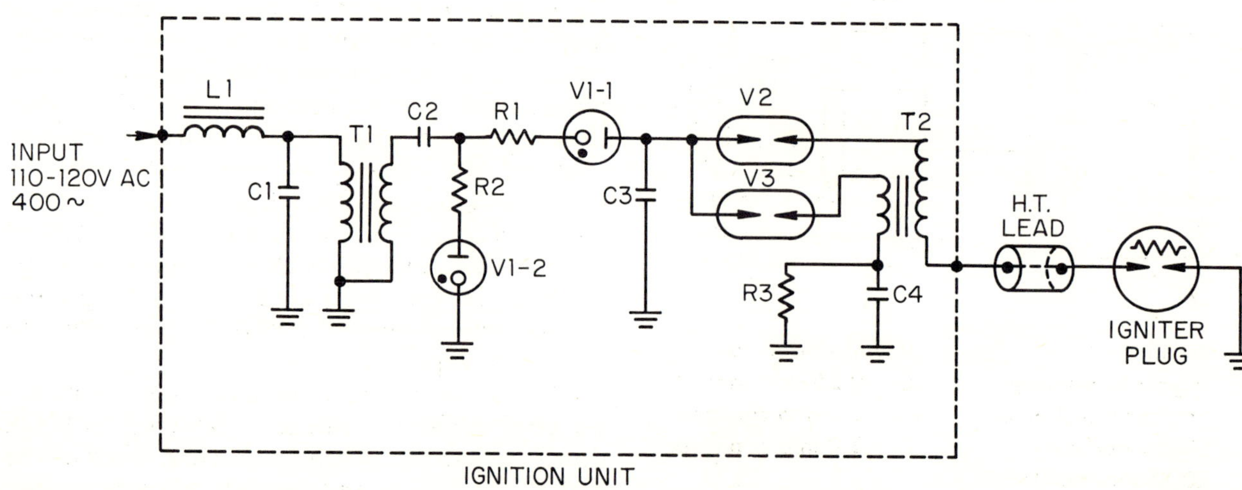

Figure 9·26 High-energy ignition system for CJ805-23 turbofan.

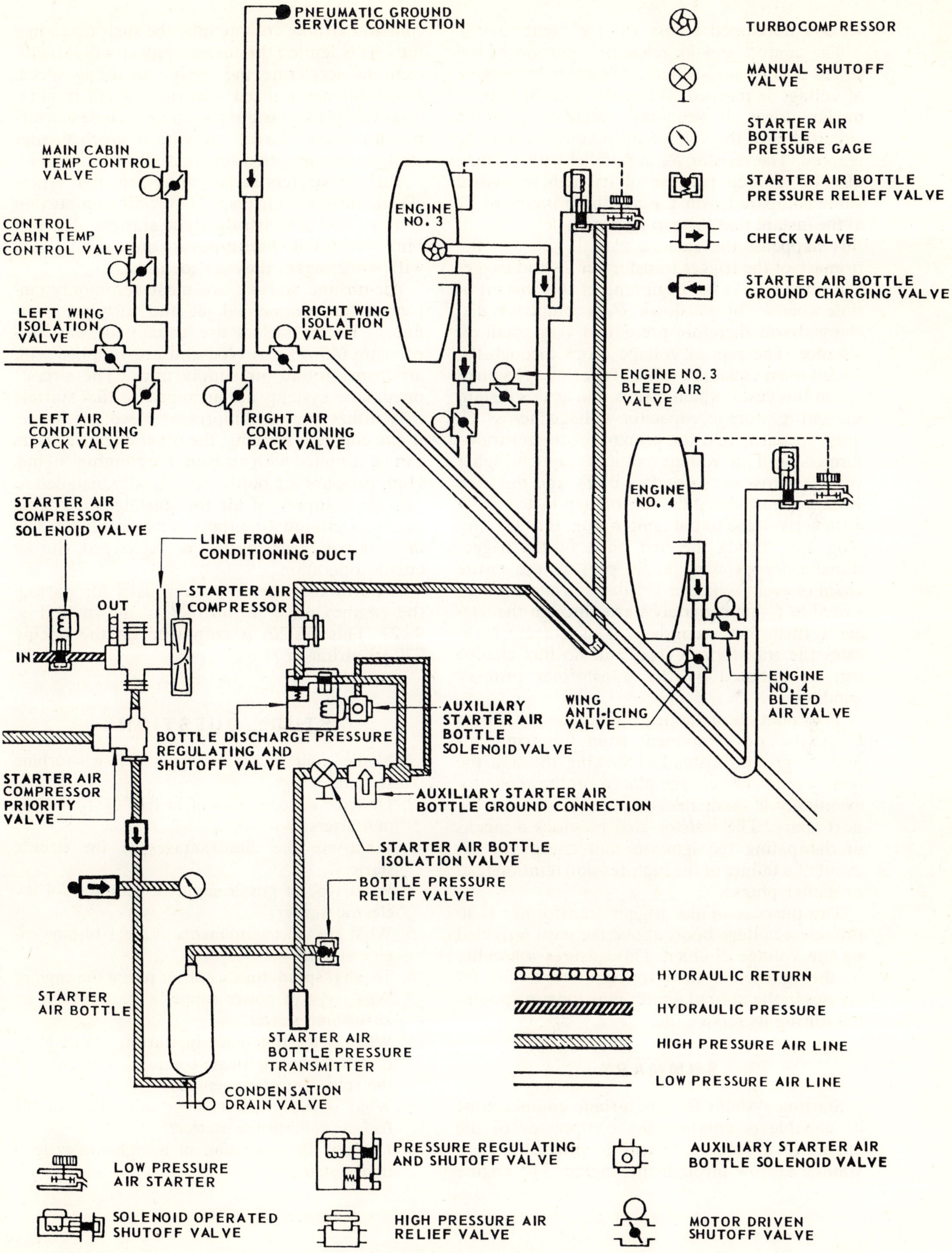

Figure 9·27 Starter air system for Boeing 720 airliner.

voltage developed across storage capacitor C_3.

The control gap V_3 releases a portion of the stored energy in C_3 when a predetermined value of voltage is reached. When the control gap V_3 breaks down, it becomes effectively a short circuit; hence the voltage drop across it may be ignored. The resistor R_3 in parallel with C_4 may also be ignored because of its high resistance value compared with C_4 and the primary of T_2 at the instant that the gap V_3 breaks down. When this happens, the current path is through the primary of the trigger transformer T_2 and trigger capacitor C_4. At the beginning of this instant of time when V_3 breaks down, C_4 is completely discharged and therefore presents a very small resistance. The sum of voltage drops through the circuit must equal the value of the voltage source (C_3 in this case), which means that at this instant the entire storage capacitor voltage across C_3 must appear across the primary of trigger transformer T_2. This voltage induces a much higher voltage across the secondary of T_2, and this high voltage breaks down the control gap V_2 to supply a trigger voltage to the igniter plug. If the igniter plug fires during the first pulse of the trigger transformer primary as is proper, the entire chain of events changes because the current then stored in C_3 immediately starts through the trigger transformer secondary. This current saturates the transformer core, and no flux change can be initiated by the transformer primary winding.

The function of the trigger capacitor C_4 is to block the a-c component from following this path to ground instead of flowing through the igniter plug. The charge placed on the capacitor eventually is dissipated before the onset of the next spark. The resistor also provides a means of dissipating the ignition unit energy in the event of a failure of the high-tension ignition lead or igniter plugs.

The purpose of the trigger transformer is to provide a voltage boost above the level provided by the voltage doubler. This ensures reliability in the ignition system to compensate for increases in the normal igniter plug ionizing potential during its service life.

SUMMARY

Starting systems for gas-turbine engines must be capable of rotating the compressor of the engine at a speed sufficient to provide a high volume airflow through the engine. The engine rpm and airflow volume must be such that when the fuel is ignited the turbine power will be sufficient to accelerate the engine to idling speed. Early engines utilized electric starting motors; however, present engines employ turbine starters because these starters provide a much greater starting torque with a substantially lower weight.

Turbine starters are of three principal types: (1) air-turbine starters, (2) fuel-air combustion starters, and (3) cartridge-type starters. The turbine units for all three types are quite similar, and with some makes they are identical.

Air-turbine starters are most commonly employed for commercial jet and turboprop airliners. These starters are usually of the low-pressure (45–50 psi) type and are supplied with air from ground air supply units. The aircraft pneumatic systems are arranged so that starters may utilize engine compressor bleed air; hence, if one engine is running, the other engine starters can be supplied with air from the running engine. High-pressure air bottles are usually installed to provide a supply of air for starting one engine in case a ground air supply is not available. The high-pressure air bottle is recharged during engine operation.

A portion of the air system used for starting the engines on a jet airliner is shown in Fig. 9·27. This system is employed on the Boeing 720 jet airliner.

STUDY QUESTIONS

1. Name four different types of gas-turbine starters.
2. Describe the operation of each of the four types of starters.
3. What are the disadvantages of the electric starters?
4. What type of power supply is required for the electric starter?
5. What are the requirements of a gas-turbine engine starter?
6. To what speed must a starter rotate the engine?
7. What type of power supply is required for an air-turbine starter?
8. What is the difference in starting acceleration time between the single-compressor engine and the split-compressor engine?
9. What is the greatest single advantage of the fuel-air combustion starter?
10. Describe the operation of a high-energy ignition system.

CHAPTER 10

Lubricants and Lubricating Systems

AVIATION GAS-TURBINE LUBRICANTS

The problem of lubricating oils for gas-turbine engines was relatively simple during the first few years of gas-turbine experience, because lightweight petroleum-base oils were suitable for gas turbines as well as other types of engines. Most of the early engines utilized lubricating oil conforming to MIL-O-6081A, Grade 1010. Engines requiring an extremely light oil were operated on MIL-O-3519, Grade 1005. These were conventional petroleum oils of high quality and light weight which met the requirements of all the older engines. Today, there are very few gas-turbine engines in use which operate satisfactorily on the petroleum-base oils.

Because of the continuous struggle for greater thrust and power, gas-turbine engines have been designed to operate at higher temperatures and pressure ratios. Since some of the temperatures encountered are considerably above the flash point of the petroleum oils, it is obvious that a high-temperature lubricant of some kind had to be developed. In this chapter we shall devote most of the discussion to these high-temperature lubricants.

Temperature requirements

Typical temperatures which may be encountered in the operation of a gas-turbine engine are shown in Fig. 10 · 1. These will vary to some extent with different engines and under different conditions; however, they do point up the problems which exist.

As in all types of powerplants, bearings and gears must be cooled and lubricated by the circulated oil. The unique part of the problem of lubricating a jet engine is that many of these parts are located next to burners and turbine wheels where temperatures can be as high as 1800°F, while external temperatures may be as low as −65°.

It is obvious, then, that the conditions under which the lubricant must work are quite different from those experienced in the reciprocating engine or in ground equipment. The temperature at which the lubricant must satisfactorily operate for both bearings and gear lubrication range from −65°F and below to something in excess of 450°F. Also, bearing and gear speeds are high and vary considerably with engine design. Bearing loads are generally light and gear loads vary, being the lightest on the turbojet engine accessory gears and the heaviest on the turboprop nose gears. Table 10 · 1 shows typical operating conditions. With this background of operating conditions, it should be possible to define in general terms what qualities are required in the lubricant to be used in these engines.

Quality requirements

It is desirable to have one lubricant which will satisfy all the gas-turbine engines, both turbojet and turboprop. We must therefore satisfy the load-carrying requirements of the turboprop engines, which are the critical group in this respect. It has been found in most applications that the load-carrying ability of a Grade 1100 aviation engine oil which has 100 SSU viscosity

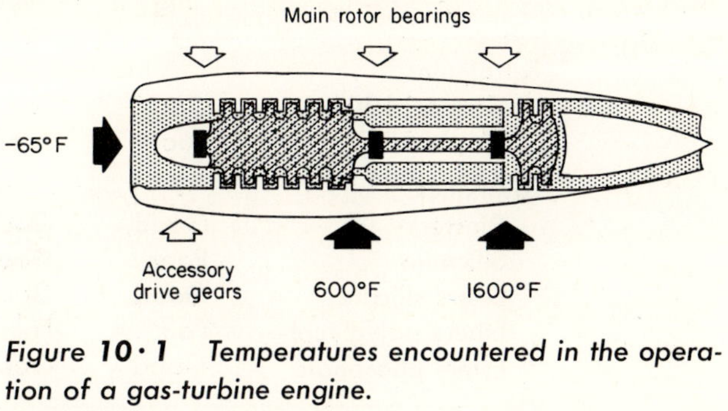

Figure 10 · 1 Temperatures encountered in the operation of a gas-turbine engine.

Table 10 · 1 Operating conditions

Part	Temperature range, °F	Speed, rpm	Loads
Bearings:			
Plain	−65–300	6,000–12,000	Light
Antifriction	−65–450	6,000–20,000	Light
Gears:	−65–400	1,000–20,000	Medium to heavy

at 210°F is satisfactory for this application. We should therefore have a product which is equivalent, if possible, to Grade 1100 oil. If only turbojet accessory gears are to be considered, 45 to 50 per cent of the load-carrying capacity of Grade 1100 oil should suffice.

Because of the altitude at which the aircraft are to operate and the resultant low atmospheric pressure around them, coupled with the high temperatures of the oil, we desire to have a low-volatility product to prevent losses by evaporation. In addition, the oil should not foam and should be essentially nondestructive to elastomers (rubber and synthetic rubbers) used as seals in the lubricating system. All these qualities are necessary if oil consumption is to be held to a minimum.

Many metals are involved in the construction of a satisfactory aircraft gas-turbine engine, and all must be considered in developing a lubricant because of possible corrosive effects of the lubricant. Also, with high-speed antifriction bearings we cannot tolerate excessive carbon or varnish deposits on the bearings. We must therefore have a compound which is thermally stable, oxidation-stable, and noncorrosive at the temperatures of operation.

The effect of ambient temperature on the restarting of an engine at high altitudes must be considered. Also, the problem of starting engines on the ground in arctic regions is of prime importance, since our military forces must maintain defensive interceptor aircraft in these areas. For these reasons, we must have lubricants with a satisfactory low viscosity and a low pour point. Minimum viscosities for satisfactory starting have been cited by various authorities and range from 5,000 to 20,000 centistokes, since there is a considerable variation in the pumps and piping in the lubrication systems on various engines and aircraft. It is desirable that we provide for the worst condition so far as the starting problem is concerned, and this requires a viscosity of less than 5,000 centistokes at the lowest operating temperature.

Choice of lubricant

In Table 10 · 2 it will be noted that Grade 1100 engine oil and Grade 1010 engine oil have been rated for load carrying, viscosity-temperature relationship, volatility, high-temperature oxidation stability, and cost. The ratings are relative, and since we have already stated that Grade 1100 oil would be satisfactory on load-carrying ability, we may now see how these other materials in the classification of synthetics rate against this product. This list of synthetic materials is by no means complete because there are hundreds of compounds and thousands of combinations possible under each of the classifications. We would select from this list of synthetic materials the diesters, silicate esters, and phosphate esters for load-carrying qualities.

Table 10 · 2 Comparison of various petroleum and synthetic lubricants

	Load carrying	Volatility	High temp oxidation stability	Viscosity temp	Cost per gallon
Petroleum:					
Grade 1100	Good	Fair	Poor	Fair	$.75
Grade 1010	Poor	Poor	Poor	Fair	$.55
Synthetics:					
Diesters	Good	Good	Good	Good	$5–10
Silicones	Poor	Good	Excellent	Excellent	$30–40
Esters silicate	Good	Good	Poor	Excellent	$10–12
Ethers polyglycol	Fair	Fair	Fair	Good	$2–6
Esters phosphate	Good	Good	Good	Fair	$4–12

In considering the consumption to be expected with these products, an examination of volatility ratings disqualifies the polyglycol ethers.

The high-temperature oxidation-stability ratings give us an indication of the deposit-forming tendencies of the fluids considered. We would therefore exclude the silicate esters and the polyglycol ethers. Low-temperature starting is rated here by viscosity-temperature relationship. Of these synthetic classes, the phosphate esters would be excluded.

The most important task is to get something to operate satisfactorily regardless of initial cost. If the oil does its job properly, its life and volatility will be such that the utility of a given charge of oil is considerably greater than experienced with the petroleum products such as Grade 1010, and relative maintenance costs can probably be made equivalent or even less with the synthetic. This has been shown to be the case in several flight-test programs.

In general, then, we can say that the diesters are the best class of products to investigate, since they are rated "good" on all four qualities that we are considering. Since hundreds and even thousands of compounds are of possible use, the screening program in the laboratory involves many tests of many compounds. In the case of the development of a satisfactory MIL-L-7808 product, this laboratory work required several years of intensive research both in the United States and in Great Britain.

MIL-L-7808 oil compared with quality requirements

Considering the quality requirements we have set forth for this lubricant, how well have we done in the development? Load-carrying ability is now measured by high-speed gear rigs both in this country and in Great Britain. The test unit specified in the military specifications is the Ryder Gear and Lubricant Tester. This is a four-square rig which operates at 10,000 rpm, with the oil heated to 165°F. Gears are run under increasing loads until 22½ per cent of the gear-tooth area is scuffed, and the gear load is reported in pounds per inch of tooth width. Grade 1100 oil gives an average value of 3,200 lb per inch, and Grade 1010 engine oil gives an average value of 180 lb per inch. The MIL-L-7808 product which is presently approved gives an average rating of 2,000 lb per inch of tooth width. It should be pointed out that a companion product which is used in Europe, and to a limited degree in this country, does give the equivalent of 100 per cent of the load-carrying value of Grade 1100 oil but is of higher viscosity, since the British require operation only to −40°F whereas we require operation to −65°F.

The next consideration is the volatility of the product. This can best be illustrated by the plot of vapor pressure of a typical Grade 1010 oil and the MIL-L-7808B as shown in Fig. 10·2. At a temperature of 300°F, which is not unreasonable for bulk oil temperature in later models of turbojets, the synthetic product has a vapor pressure of 0.1 mm Hg. The Grade 1010 oil has a value of 25 mm Hg, or over 200 times that of the synthetic oil. Expressing it in another way, if the bulk oil temperature is 450°F, the boiling altitudes for Grade 1010 and the synthetic oil are about 47,000 ft and 78,000 ft, respectively.

Still other important considerations for a lubricant are its deposit-forming tendency and its corrosiveness. Deposit-forming tendency is measured by a unit developed by Pratt & Whitney Aircraft which is known as the "panel-coking test." In this test the oil is splashed against a heated aluminum plate by a metal shaft with wires attached in the form of bristles. The plate is heated to 600°F by an electrical strip heater, and the weight of coke formed on the panel is measured at the end of 8 hr. The requirement of the specification is that there be less than 100 mg of coke. The presently approved MIL-L-7808A oil averages about 60 mg.

The low-temperature properties required depend to some extent upon the individual lubricating systems. Grade 1010 oil has been generally satisfactory from the standpoint of low-

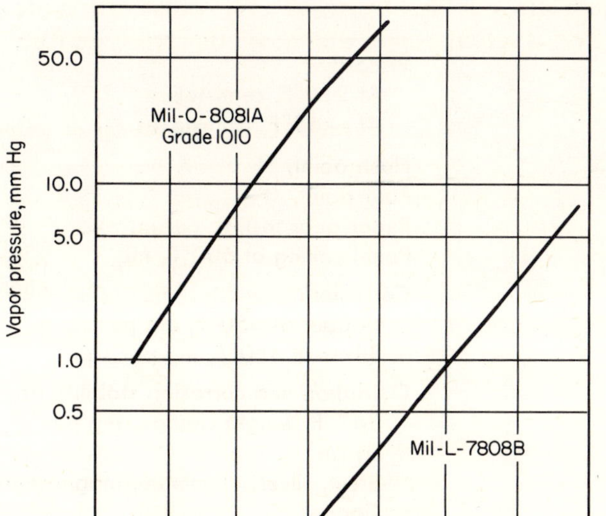

Figure 10·2 Comparison of vapor pressures.

temperature starting characteristics. If 5,000 centistokes is taken as the minimum viscosity for proper low-temperature operation, the synthetic oil shows nearly a 10°F advantage over the mineral oil Grade 1010. If 20,000 centistokes is the limiting viscosity, an even greater temperature difference exists in favor of the synthetic oil, since the two viscosity-temperature lines diverge as viscosity increases from 400 centistokes. The pour point for both these products is in the vicinity of −80 to −90°F.

Status of military specifications

A new amendment to Specification MIL-L-7808B includes a lead-corrosion test at 375°F and a qualification test of 100 hr of satisfactory operation in a J57 engine. Table 10 · 3 compares the typical inspection data of the oil with the major requirements of the specification.

MIL-C-8188A(ASG) specifies a corrosion-preventive version of MIL-L-7808. The only difference between the two grades of oil specified, so far as requirements are concerned, is that Grade B is inferior to Grade A regarding rubber swell. Comparison of test results is shown in Table 10 · 4. Grade B was established to allow the use of a more available base oil derived from petroleum in the less critical engines. Grade A is made from the same base materials as the MIL-L-7808 oil. Since rubber swell is a function of temperature as well as oil composition, Grade A has been designated as the high-temperature oil and Grade B as the low-temperature oil. Humidity-cabinet-life tests are also required as an indication of the corrosion-preventive characteristics of both these oils. An amendment to this specification requires a lead-corrosion test at 325°F and a 20-hr engine test in a J65 engine for qualification.

Materials for use with synthetic lubricants

One of the biggest problems in the application of synthetic lubricants has been to determine what materials can best be used with them. The interaction of the oil and metals was the primary point of investigation, since it was believed that such materials as plastics, elastomers, and paints could be compounded to incorporate the necessary resistance.

Of the metals which the oil might contact, cadmium plate has shown the least resistance. In many cases this lack of resistance is believed to be due to the inadequacies of the cadmium-plating job itself rather than to the corrosive action of the oil, since similar difficulties have been encountered with mineral oils. It is suggested that cadmium plate be deleted from all applications where contact with synthetic oils is probable. Of the more common metals, copper is probably the next least resistant to attack by

Table 10 · 3 Comparison of synthetic oil with specification requirements

	MIL-L-7808B	
	Requirements	*Inspections*
Viscosity:		
At 210°F, centistokes	3.0 (min)	3.6
At −65°F, centistokes	13,000 (max)	12,600
Flash point, °F	350 (min)	430
Pour point, °F	−75 (max)	−90
Ryder gear test, lb per in.	1,700 (min)	2,000
Panel coking at 600°F, mg	100 (max)	59
Corrosion:		
Copper at 450°F, mg per sq in.	3.0 (max)	1.6
Silver at 450°F, mg per sq in.	3.0 (max)	0.7
Oxidation and corrosion stability at 347°F; weight change, mg per sq cm:		
Steel, silver, aluminum, magnesium	±0.2 (max)	Complies
Copper	±0.4 (max)	Complies
Total acid No. increase	2.0 (max)	0.8
Viscosity change at 100°F, per cent	−5 to +15	+2.0

Table 10 · 4 Comparison of synthetic corrosion-preventive lubricant with specification requirements

	MIL-C-8188A (ASG) requirements	Inspections	
		Grade A	Grade B
Viscosity:			
At 210°F, centistokes	3.0 (min)	3.8	3.3
At −65°F, centistokes	18,000 (max)	16,000	12,500
Flash point, °F	350 (min)	430	410
Pour point, °F	−75 (max)	−75	−80
Type H rubber swell, per cent:			
Grade A	12–35	21	37
Grade B	12–40	. . .	
Ryder gear test, lb per in.	1,700 (min)	2,100	1,850
Panel coking at 600°F, mg	125 (max)	56	21
Corrosion:			
Copper at 450°F, mg per sq in.	3.0 (max)	1.1	2.5
Silver at 450°F, mg per sq in.	3.0 (max)	3.0	2.7
Oxidation and corrosion stability at 347°F, weight change, mg per sq cm:			
Steel, silver, aluminum, magnesium	0.2 (max)	Complies	Complies
Copper	0.4 (max)	Complies	Complies
Total acid No. increase	3.0 (max)	1.6	1.6
Viscosity change at 100°F, per cent	−5 to +25	5.7	5.7
Humidity cabinet life, hr	144 (min)	168	168

oxidation products formed from the synthetic oils. This is indicated by the specification requirements which give a ±0.4 mg per sq cm allowable limit in the oxidation-corrosion test weight loss for copper, whereas aluminum, magnesium, steel, and silver plate can show only ±0.2 mg per sq cm in the same test. It is suggested that all copper or copper-containing alloys be deleted as far as possible because of this corrosive effect as well as the catalytic effect of copper on oil oxidation. In anti-friction-bearing applications, plating of the bronze retainers with silver or lead has accomplished this end.

With respect to plastics, one general rule can be expressed: the thermosetting-type plastics are more resistant to softening action than are the thermoplastic. The problem of proper materials for use as oil seals where vibration is encountered has been a very bothersome one. In fact, until this problem is solved, the development of both lubricants and engines of higher output is restricted. It may be solved either by the exclusion of elastomers entirely or by the compounding of elastomers usable at temperatures equal to the highest oil temperatures which are possible with present engines. This problem has been presented to all the major compounders of elastomers for the aircraft industry, and research has led to products which are greatly improved over those available at the end of World War II.

The solvent effect of the synthetic oils has been very evident on painted surfaces, which generally become stripped of all signs of paint in a relatively short time following oil contact. Several paint manufacturers have studied this problem and have developed paints which appear to be very resistant to the oils.

Toxicity

The last item on the list of materials, which should probably not be classed as a material at all, is people. There are many highly publicized rumors about the toxicity exhibited by synthetic lubricants. With regard to external contact, there is no more problem with the synthetic oils than with light mineral oils or solvents. Some drying of the skin has been noted where personnel are in fairly constant contact with the oil, and this has been attributed to a skin-defatting process, which can be counteracted quite satisfactorily by common personal-hygiene methods and the application of skin creams.

The problem of inhalation of hot oil fumes and products of high-temperature breakdown has become of importance because most of the aircraft utilize compressor-bleed air for cabin pressurization and heating. This is the aspect of the toxicity problem which has been most highly publicized. Investigations by authorities in the field have shown that the effect of the fumes from MIL-L-7808 oil on animals is similar to that obtained with mineral oils under the same conditions. Investigation has also led us to believe that the level of toxicity in aircraft under the worst conditions is such that no permanent damage could possibly be done to any person subjected to these fumes. In fact, the problem has been classed as one of nuisance rather than toxicity by many in the field. It is recommended that, where possible, the air-frame industry utilize a two-stage type of operation for cabin pressurization, taking outside air into the cabin and using compressor-bleed air as the power source only.

Experience with synthetic lubricants

Excellent results have been obtained with MIL-L-7808-type lubricant, and we have accumulated millions of flying hours with engines using the lubricant. Almost all turbojet engines from the J40 through the J85 specify the use of a synthetic lubricant; hence it is obvious that this type of lubricant is a necessity for any high-performance type of turbine engine. Undoubtedly, improvements will be made which will permit higher operating temperatures, and these in turn will lead to still greater power output for the engines of the future.

Typical synthetic lubricants

As previously stated, almost all modern gas-turbine engines require the use of a lubricating oil conforming to MIL-L-7808 or the equivalent. A number of companies manufacture oils to this specification or other oils which are superior in performance. For engines which may operate at temperatures above those normally encountered in present-day jet engines, a high-temperature synthetic lubricant has been developed. This lubricant is manufactured to conform to MIL-L-9236 and may be used in place of MIL-L-7808. The MIL-L-9236 oil is specified for the General Electric J93 engine which powers the RB-70 supersonic bomber.

Among the synthetic lubricants used at present for turbojet and fanjet engines are Sinclair S-15 turbo oil, conforming to MIL-L-7808E, and Esso oils TJ-15 and WS-5251. Esso WS-5251 is a high-temperature oil which can withstand 600°F for a short time and 350 to 400°F continuously.

LUBRICATING SYSTEMS

Lubricating systems for modern gas-turbine engines are relatively simple in design and operation, but their function is of vital importance. The principal purposes of the lubricating system are (1) to reduce friction and (2) to cool the areas with which the lubricant comes in contact.

The main units of the typical system are the reservoir or tank, the pressure pump, one or more filters, flow dividers, scavenging pumps, and the oil cooler. The operation of these units is very similar to units of a reciprocating-engine oil system.

It is the purpose of this section to discuss the various units of lubricating systems and then to see how these units are used in typical systems.

Oil tanks

Oil tanks for aviation gas-turbine engines may be separate from the engine, attached to the engine, or a structural part of the engine. They are usually constructed of lightweight welded sheet metal and are shaped to fit the locations in which they are to be placed. If an oil tank is separate from the engine, the design would be the responsibility of the air-frame manufacturer.

In the systems for the Pratt & Whitney J57 and the Wright J65 engines, the oil tank is designed to fit the contour of the compressor case. Thus, the oil tank is designed by the engine manufacturer so that it will present as few problems as possible in engine installation.

The oil tank for the Rolls-Royce Dart engine is actually a part of the engine structure; that is, it forms a part of the air-intake casing. This is explained in Chap. 17.

Oil pumps

Oil pumps for turbine engines are usually of the positive-displacement gear or vane type, with a relief valve to prevent excessive pressure. A modified gear-type pump is called the "gerotor pump." The principle of this pump is shown in Fig. 10·3. Two kidney-shaped slots are cut in the metal surface against which the gear and

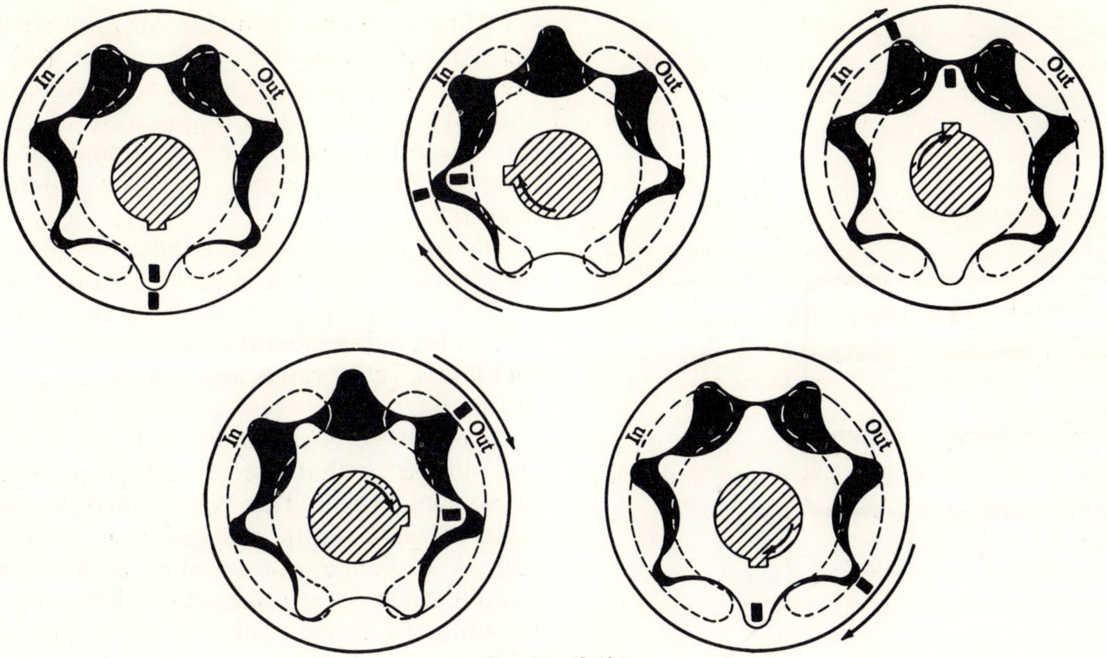

Figure 10 · 3 Operation of a gerotor pump.

rotor turn. During rotation of the pump elements, the space on the "in" side of the pump increases in volume, thus drawing oil into the space. On the "out" side, the volume of the space is decreased and the oil is forced to leave the pump through the "out" line.

Multiple oil pumps are constructed with two or more gear elements. Construction of this type is to provide pressure for different sections of the system.

Scavenge pumps are usually constructed in much the same manner as the pressure pump. They are used to draw the oil from the sumps at the bearings, accessory housings, and other drainage points and return the oil through the oil cooler to the tank.

Filters

Oil filters or strainers are constructed in many designs. The main objective of a filter is to remove all foreign particles from the lubricant without creating excessive back pressure against the pressure or scavenge pump. Filters are usually provided with bypass valves to permit the flow of oil in case the filter becomes clogged.

Filters may consist of fine metal screens, serrated disks, plain disks, screen disks, and many other configurations. One of the most popular engine systems employs a filter which consists of a stack of screens in disk form separated by inlet and outlet spacers. The disks are mounted on a perforated tube which extends through the center of the stack. The filter may be disassembled easily for cleaning.

Oil coolers

Oil coolers for aviation gas-turbine engines may either be simple oil radiators with ram-air cooling or be of the type which utilizes fuel as the cooling medium. The latter type of unit is most commonly used with modern turbine engines.

The fuel-oil cooling unit is a heat exchanger which transfers the heat in the oil to the fuel flowing to the fuel nozzles. To some extent, this process salvages some of the heat energy and returns it to the combustion chamber of the engine. Figure 10 · 4 is a schematic diagram of a typical oil cooler using fuel as the cooling medium. This unit is equipped with a temperature-control valve which allows cold oil to bypass, thus providing for more rapid warm-up. This valve also protects the cooler from damage due to excessive pressure which may be developed when the oil is cold.

Typical lubricating systems

Figure 10 · 5 illustrates a lubricating system for a General Electric J47 turbojet engine. Lu-

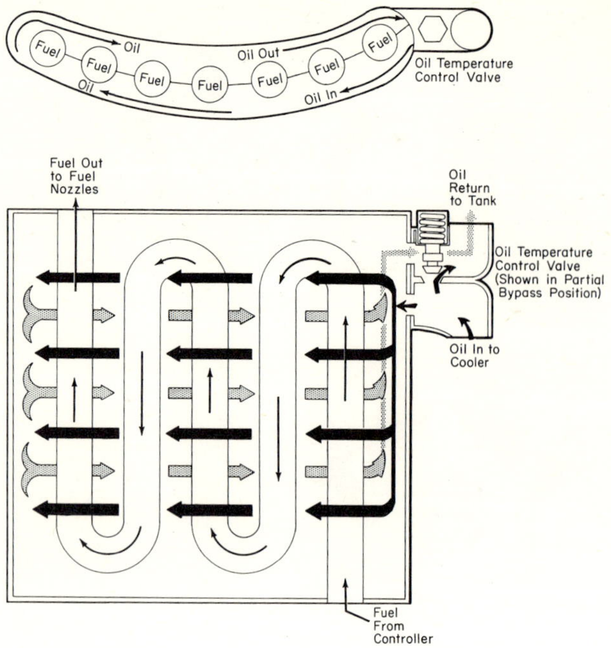

Figure **10·4** *Schematic diagram of a fuel-coolant oil cooler.*

bricating oil flows from the oil tank to a two-element pressure section of the pump. One of the pressure elements supplies oil to the compressor and turbine bearings at a pressure of 10 to 25 psi. The other pressure element supplies all the accessory bearings and the front compressor bearing.

The oil from the various bearings drains into sumps, from which it is picked up by the scavenge elements and returned to the tank. The system is equipped with a fuel coolant oil cooler, fuel filters, relief valve, and other units as shown in the diagram.

Figure 10·6 is a schematic diagram of a system similar to that used in a high-performance twin-spool engine. This system incorporates one main pressure pump, marked *P* in the diagram, and six scavenge units marked *S*. The pressure pump directs oil at a pressure differential of approximately 45 psi. The bearing voids and the oil tank are joined by breather lines, and the entire system is pressurized to maintain approximately sea-level pressure at all altitudes. This

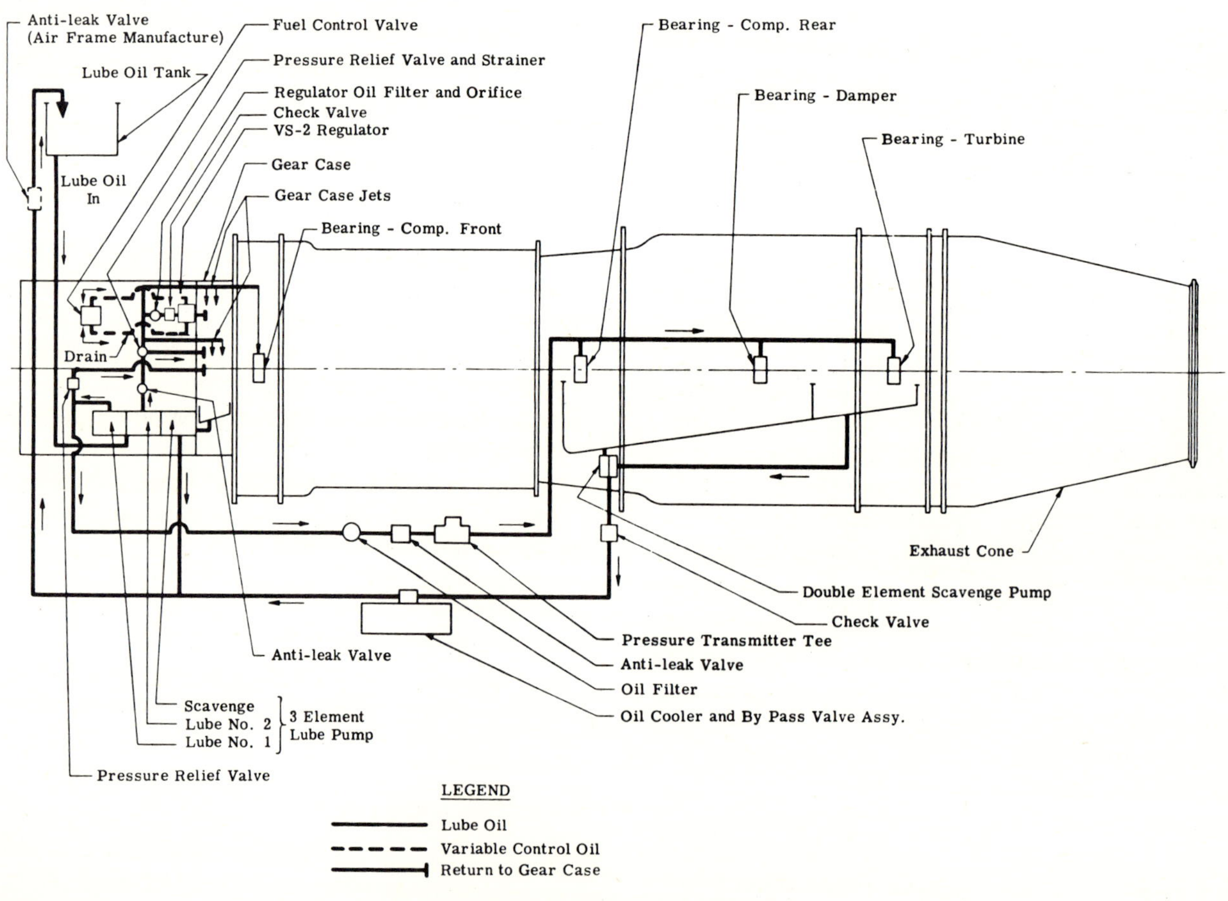

Figure **10·5** *Schematic diagram of a J47 lubricating system.*

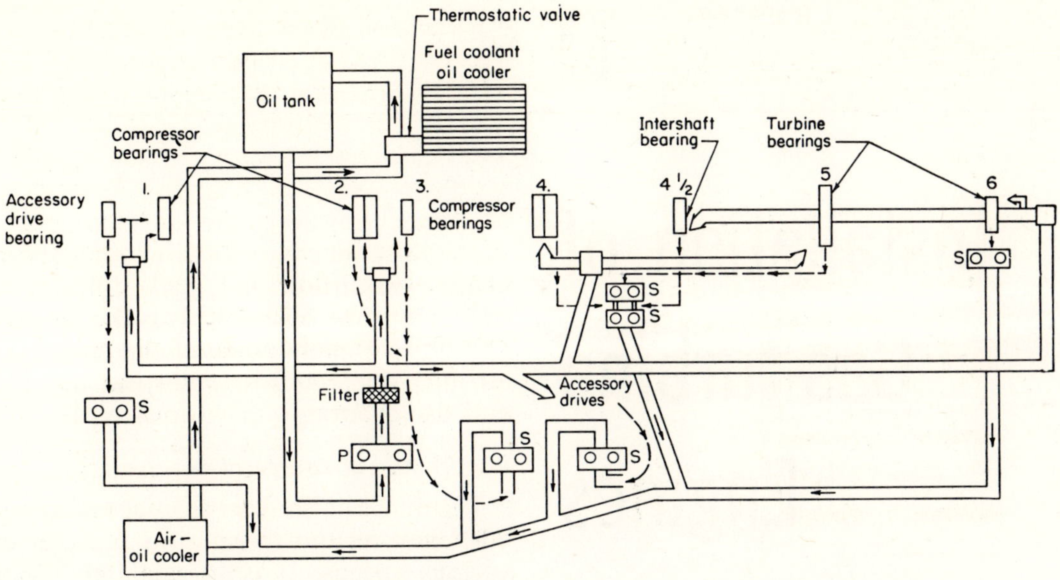

Figure 10·6 Schematic diagram of oil system for a JT3 turbojet engine.

is accomplished by means of a breather pressurizing valve which is open at sea level but begins to close as altitude is increased. The pressure air bleeds through the labyrinth seals from the compressor pressure and is distributed throughout the system.

Scavenge oil from all the bearings in the engine drains into sumps from which it is picked up and returned through the oil coolers to the tank. In the diagram, the scavenge oil is indicated by dotted lines and arrows. The system is equipped with both an air-coolant oil cooler and a fuel-coolant oil cooler.

The main bearings in the engine are numbered from 1 to 6. Numbers 1 and 2 are the front-compressor rotor bearings; Nos. 3 and 4 are the rear-compressor rotor bearings; No. 4½ is the intershaft bearing; No. 5 is the front-turbine bearing; and No. 6 is the rear-turbine bearing.

It should be noted that all lubricating systems are similar in function but that there are differences in all systems. The operator of gas-turbine engines must understand the system with which he is working and must be sure that the system is serviced according to the manufacturer's instructions.

STUDY QUESTIONS

1. Why has it become necessary to develop synthetic lubricants?
2. To what temperatures may gas-turbine lubricants be subjected?
3. What are the quality requirements of a gas-turbine lubricant?
4. What is the military specification for synthetic turbine oil?
5. Why must volatility be low for a gas-turbine lubricant?
6. What is the importance of a low pour point?
7. What is the effect of synthetic lubricants on rubber and paint?
8. Compare synthetic lubricants with mineral oils in toxicity.
9. Name a standard synthetic lubricant meeting Spec. MIL-L-7808.
10. What are the principal units in a typical lubricating system?

CHAPTER 11

Installation of Gas-turbine Engines

INTRODUCTION

The installation and mounting of gas-turbine engines in aircraft involve a number of factors, all of which affect the performance or reliability of the engine. Among these factors are the design of the air-inlet system, exhaust system, cooling arrangement, and mounting structure. These elements are largely the responsibility of the airframe manufacturer; however, the engine manufacturer is also involved because he must supply the engine data which enable the airframe manufacturer to design a suitable installation. Exhaust systems, including thrust reversers, noise suppressors, and conventional nozzles, are the concern of both the engine manufacturer and the airframe manufacturer. The design, however, must be such that it adapts itself to the needs of the aircraft in which the engine is installed.

AIR INLETS

When it is realized that a modern turbojet engine may devour as much as 350 to 550 tons of air per hour, the importance of efficient air-inlet-duct design immediately becomes apparent. Because of structural considerations and the placement of cockpits and equipment, the efficiency of the air-inlet duct has suffered at times; however, the development of high-pressure-ratio engines, with their sensitivity to changes in airflow, has made efficient inlet-duct design a prime objective. It is the purpose of this section to explore some of the characteristics of air-inlet ducts and to point out the advantages and disadvantages of various types.

Importance of structural design

Figure 11 · 1 shows the effect of air-inlet-duct efficiency on the thrust of a turbojet engine at various speeds. It is immediately evident that inefficient duct design takes a heavy toll, especially as airspeed increases. It will be noted that, for a given duct efficiency, the losses at 600 mph are about three times as great as those at 300 mph. In addition to duct losses, serious performance penalties may also result from the external effects of the inlets, either because of added form drag or because of prematurely induced compressibility effects. These conditions also become more pronounced as speed increases.

It is indicated that the aerodynamic requirements for the ideal air-induction system are (1) that it delivers air to the engine with no reduction in available energy, (2) that it does not increase the drag of the body on which it is placed, (3) that it does not compromise the high-speed characteristics of the airplane, and (4) that it provides adequate air induction in all reasonable flight attitudes. To these requirements might be added, in the case of axial-flow installations, the necessity of maintaining a fairly uniform pressure distribution over the compressor face.

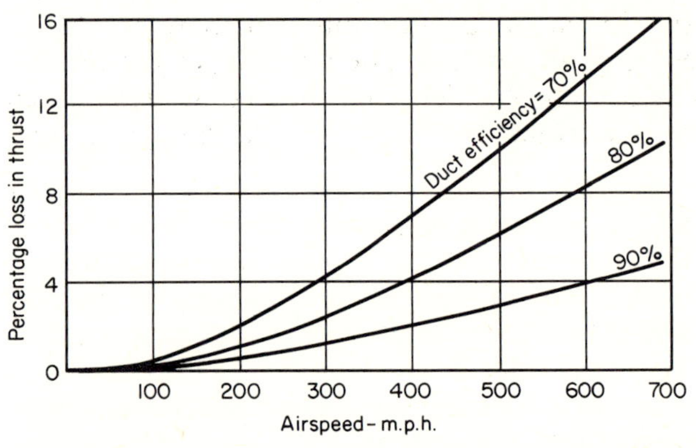

Figure 11 · 1 Effect of duct efficiency on sea-level maximum thrust.

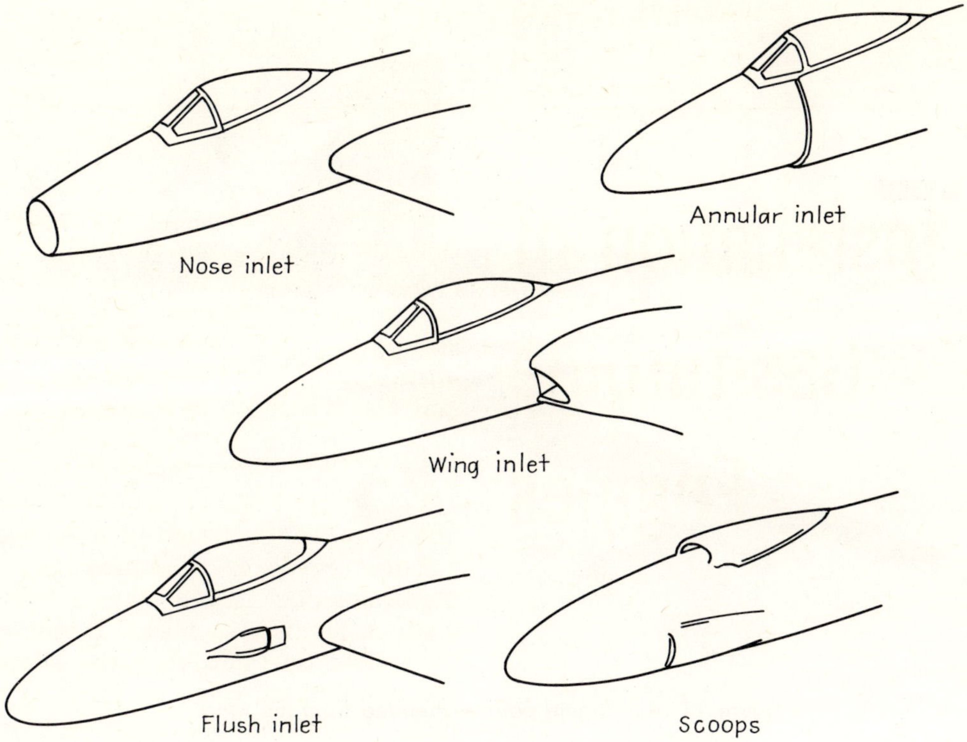

Figure 11 · 2 Various types of air inlets.

Unfortunately, these criteria must be weighed against structural and design considerations in arriving at the details of any given installation. For this reason, we have almost as many variations in induction systems as we have airplanes, each one of which meets the requirements for an ideal system to a greater or lesser degree, depending upon the extent of compromise necessary on overcoming particular design problems.

Types of air inlets

In the past a wide variety of air inlets have been employed or considered for jet engines installed in aircraft. Five types which were considered possible for subsonic fighter aircraft are shown in the drawings of Fig. 11 · 2. Of the five types shown, the nose inlet and wing root inlet have been found to be most suitable.

For subsonic aircraft such as the F-100 the nose inlet was found to be quite satisfactory. The inlet adds no aerodynamic drag to the aircraft, and it permits a reasonably direct flow of air to the compressor of the engine. The disadvantages of this type of duct lie in the length of the duct and the difficulty in providing adequate strength. The nose-inlet duct also offers problems when it comes to locating the cockpit and various items of required equipment.

Modern fighter aircraft such as the F-102, F-104, F-105, and F-106 utilize wing root air inlets. These inlets are especially designed to meet the requirements of supersonic flight and the effects of shock waves which form at the inlets. The inlets are set out slightly with an open gap between the inlet and the fuselage to eliminate the effects of the boundary layer. A general idea of the arrangement of such air inlets is shown in the diagram of Fig. 11 · 3.

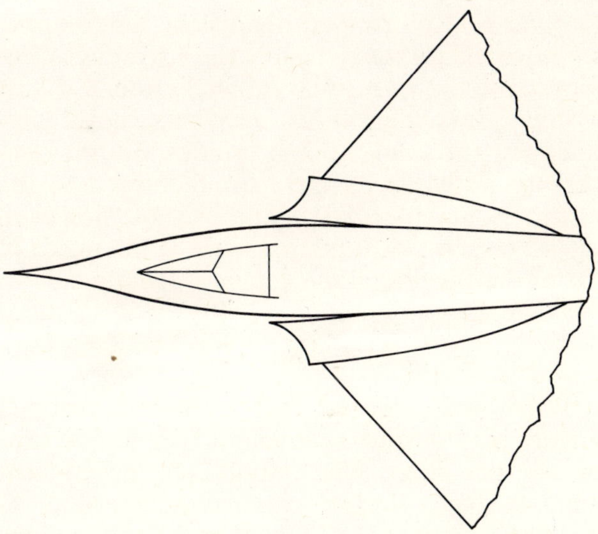

Figure 11 · 3 Drawing of a supersonic air inlet at wing root.

Figure 11 · 4 Engine pods suspended from the wing.

The simplest and most efficient air inlet is provided by the pod engine mounting used for modern jet airliners. These pods, whether suspended from the wing as shown in Fig. 11 · 4 or attached to the side of the fuselage as in Fig. 11 · 5, provide a no-drag airflow directly into the compressor of the engine. As the engine pods are mounted on struts to allow an air gap between the pod and the wing or fuselage, there is a minimum of aerodynamic interference with the smooth flow of air over the structure adjacent to the engine pod.

Figure 11 · 5 Engine pods at side of fuselage.

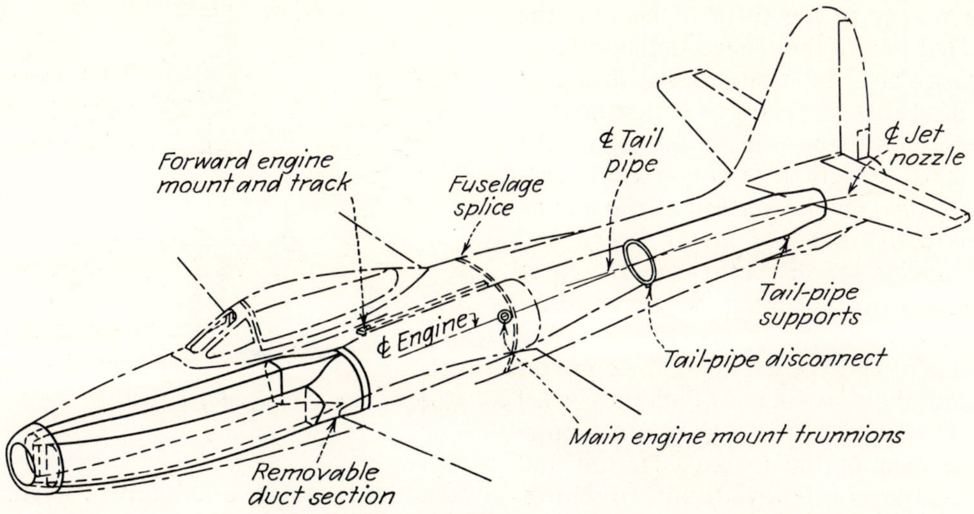

Figure 11·6 Turbojet engine installation.

COOLING AND EXHAUST SYSTEMS

Typical installations

A typical powerplant installation in an early fighter-type airplane is shown in Fig. 11·6. In this installation, a circular nose inlet branches into two ducts just inside the nose. The ducts join just ahead of the engine to form an annular section that attaches to the compressor inlet. The engine extends from the compressor inlet back to the quick-disconnect clamp that fastens the engine exhaust cone to the tailpipe. The tailpipe terminates in a jet nozzle located immediately forward of the aft end of the fuselage, as shown in Fig. 11·7, to form the primary of a cooling-air ejector.

Cooling air enters the airplane through the air-inlet louvers and flows rearward between a stainless-steel shroud and the hot surfaces of the powerplant. No other insulation is used for the engine, exhaust cone, tailpipe, or shroud. This method of protecting the fuselage structure from the high exhaust-gas temperatures was chosen mainly for two reasons: (1) a comparison of shroud weights with the weight of insulating blankets indicated that a shroud would be lighter; and (2) an ejector was used rather than a ram scoop in order to assure sufficient airflow for cooling and burner compartment ventilation during ground operation.

Investigations have shown that it is not a good practice to cool the exhaust system too much. The hot gases produced by the engine contain

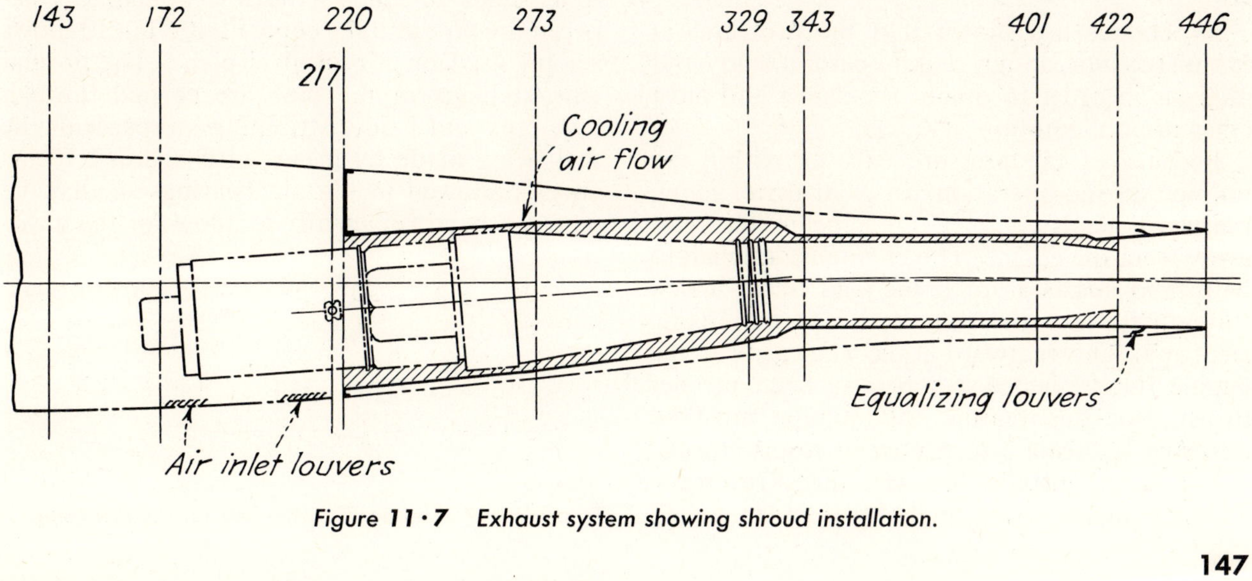

Figure 11·7 Exhaust system showing shroud installation.

considerable energy in the form of heat. If the gases are cooled before they leave the jet nozzle, they lose energy and the engine loses thrust. It is therefore desirable to retain as much heat as possible in the jet stream without damaging the engine or aircraft structure. It is common practice to insulate the exhaust section of the engine with blankets constructed of metal screen and foil. These blankets retain the heat in the gases and prevent it from escaping to the aircraft structure.

Aircraft structures exposed to high temperature are commonly constructed of stainless steel or titanium. It is thus possible to retain a large degree of the heat in the exhaust stream and still prevent damage to the aircraft structure.

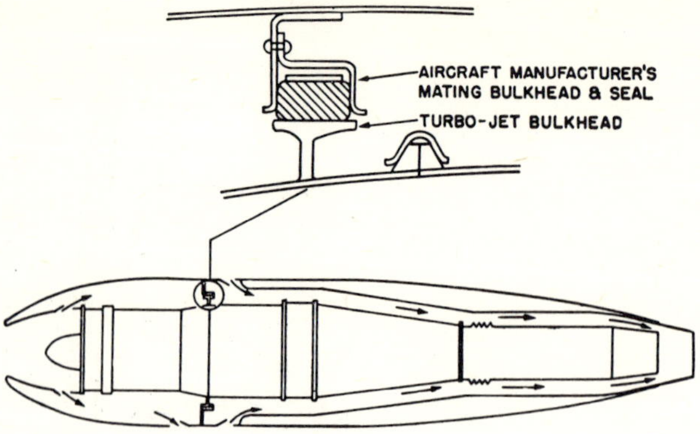

Figure 11 · 8 Firewall installation.

Firewall or bulkhead

As systems were developed and put into operation, the requirement for an external bulkhead to separate the compressor and turbine sections was established. This bulkhead is commonly called a **firewall.** Its purpose is to prevent fuel from entering the region around the combustion, turbine, and exhaust sections in the event that fuel leakage occurs in the compressor and accessory sections of the engine as a result of battle damage or fuel-line failure.

The firewall bulkhead on the engine and the mating bulkhead provided by the aircraft manufacturer are shown schematically in Fig. 11 · 8. The addition of the bulkhead across the nacelle at this point requires that the cooling air-inlet louvers for the aft section of the nacelle be located on the rear side of the bulkhead. Louvers are also generally provided ahead of the bulkhead, through which is discharged the ventilating air from the compressor section.

Problems of design

Experience has shown that the problems of exhaust-system design require careful study and analysis in order to obtain the safest and most efficient configuration.

Because of the pressure existing within the tailpipe during operation, an axial force, acting rearward, tends to force the pipe to the rear, away from the engine. The engine is designed to withstand this axial force, the load being carried through the exhaust cone and into the engine structure. The rearward force acts against the engine thrust; hence it subtracts from the net thrust. For this reason, the tailpipe must be designed to create a minimum of reverse thrust.

For engine installations with thrust reversers, a more complex arrangement is required for both structural design and control system arrangement. The thrust reverser is mounted in such a manner that the reverse loads are carried through the engine to the engine mounts and thence to the wing or other aircraft structure to which the engine is mounted.

The exhaust-system design problems are comparatively simple for those installations where the jet nozzle is fastened directly to the end of the exhaust cone, as shown schematically in Fig. 11 · 9, or where only a short length of pipe is used between the engine and jet nozzle, because for these cases it is not necessary to support the exhaust system from the airplane structure.

Tailpipe mounting

Another problem becomes evident when the tailpipe mounting provisions are considered. The pipe may be too long and heavy to be cantilevered from the engine exhaust cone and then must be supported from the airplane structure. This would be fairly simple except for fuselage deflections that occur because of control surface and landing loads. It is therefore necessary in some cases to make a flexible attachment between the pipe and the cone. Figure 11 · 10 shows a joint used on a typical airplane. The normal tension loads in the pipe are carried through four pins and load-distributing straps, with the seal being made by means of a bellows. Clearance is allowed in the pin bearings so that the pipe can bend sufficiently to allow for the maxi-

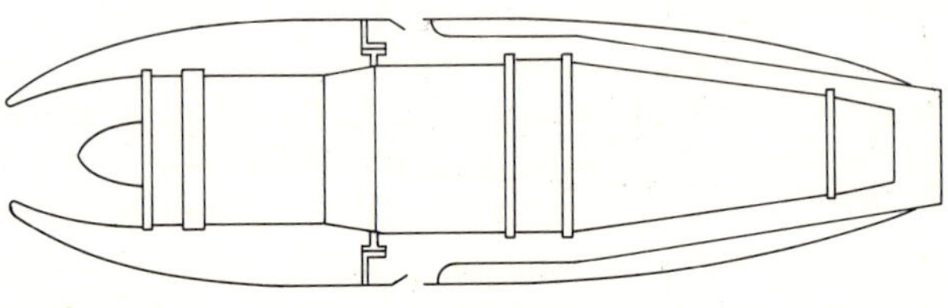

Figure 11 · 9 Jet nozzle attached to exhaust cone.

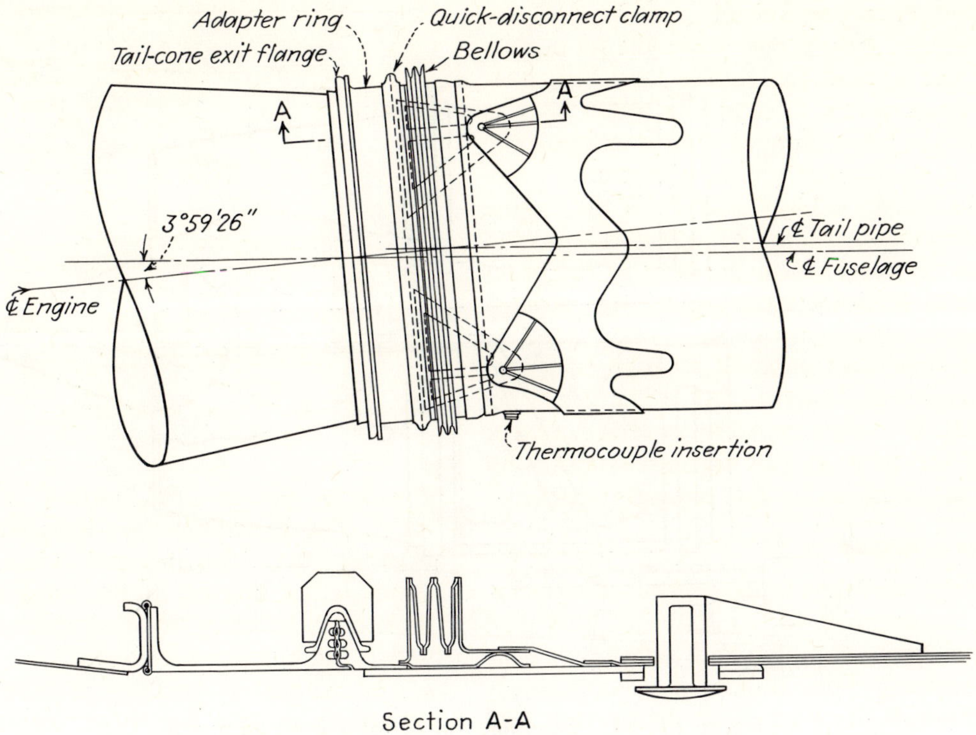

Figure 11 · 10 Flexible tailpipe joint.

mum fuselage deflections. The flexible joint also permits expansion and contraction without distortion.

JET NOZZLES

Relation to fuel consumption

The jet nozzle raises a considerable number of interesting design problems. During the initial phase of design, the engine manufacturers pointed out several benefits that could be gained by using a variable-area jet nozzle. The most attractive of these advantages was the possibility of reducing the engine specific fuel consumption at high-altitude cruise. Initial estimates showed that as much as 15 per cent improvement could be realized at 35,000 ft altitude.

Variable-area nozzle

Figure 11 · 11 shows an experimental variable-area nozzle. An actuator was used to move the plug through a mechanical linkage installed in a supporting strut. After the first run, all the internal rollers, which allowed the plug to move on a supporting tube, seized, and plug motion could be obtained only after the rollers were filed flat. The really disturbing point of these tests, however, came when the data were found to show that specific-fuel-consumption improvements were considerably less than originally estimated. As a result, further development of this variable-area jet nozzle was discontinued. Further investigations have shown that when thrust augmentation devices such as afterburners are used, the variable-area jet nozzle becomes a necessary feature. A number of successful variable-area nozzles have been developed, the two most commonly used being the eyelid type and the iris type. These will be discussed more fully in Chap. 12, Thrust Augmentation.

Effect of nozzle diameter

One of the most baffling problems associated with jet-nozzle design arose because of the extreme sensitivity of the performance of all gas-turbine engines to minute changes in jet-nozzle area. On a typical engine, for example, a change of ⅛ in. in nozzle diameter results in a change in takeoff thrust of approximately 135 lb.

This problem was recognized at Edwards Air Force Base shortly after the first flights of the Bell P-59A. A simple thrust stand was erected to check the performance of the I-16 engines. It was found that the engines were not delivering the same thrust after correction for barometer and ambient-air-temperature differences as General Electric had measured at the factory. Approximately 50 to 100 lb of thrust was appar-

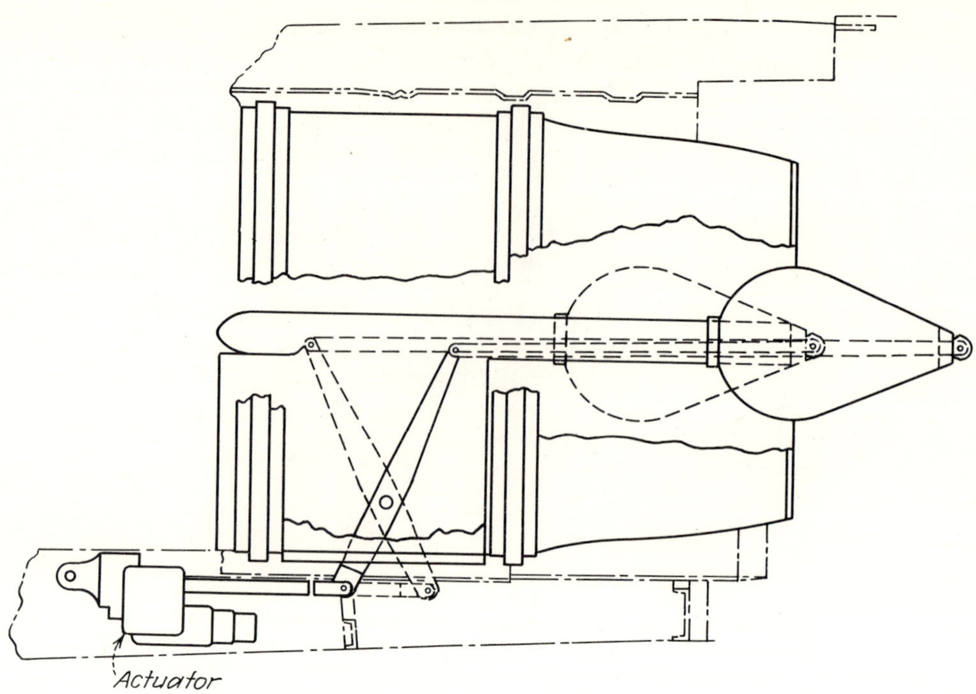

Figure 11·11 Experimental variable-area jet nozzle.

ently being lost between the factory and the test site. After checking the test stand thoroughly, the next step was to check the jet-nozzle areas, which were known to be critical even in those early days. The nozzles used with about five engines were taken off and the areas measured. The results were within a fraction of a square inch of the areas recommended by GE.

Although it should have become evident that the tailpipe temperatures needed to be compared between the factory tests and those at Edwards Air Force Base, this was not done, and about 6 months elapsed before the missing thrust was located. The factory tests, as it finally turned out, were conducted with a nozzle whose shape or contour was slightly different from the jet nozzle used on the Bell P-59A airplane installation. This caused a variation in airflow sufficient to affect the thrust output. Although the physical jet areas were identical, the factory tests were being run with an effective jet area less than that used on the engine in the field test. As a result of this and similar episodes, the correct jet-nozzle area for a turbojet engine is now determined more by observing the tailpipe temperature than by checking the specified physical area. Because of the variations in manufacturing tolerances, the jet-nozzle area is stamped on the data plates of many engines. To obtain the rated thrust of any particular engine, the jet-nozzle area must be the same as that shown on the data plate.

Tailpipe temperature effect

Some early tailpipes terminated in a short conical nozzle section with scribe marks on the surface. The net-nozzle area was adjusted by cutting strips from the conical section. Each scribe mark represented a 1-sq-in. increase in jet-nozzle area. This method was quite satisfactory but was rather tedious.

It was, moreover, a one-way proposition. As more time was spent on an engine, the tailpipe

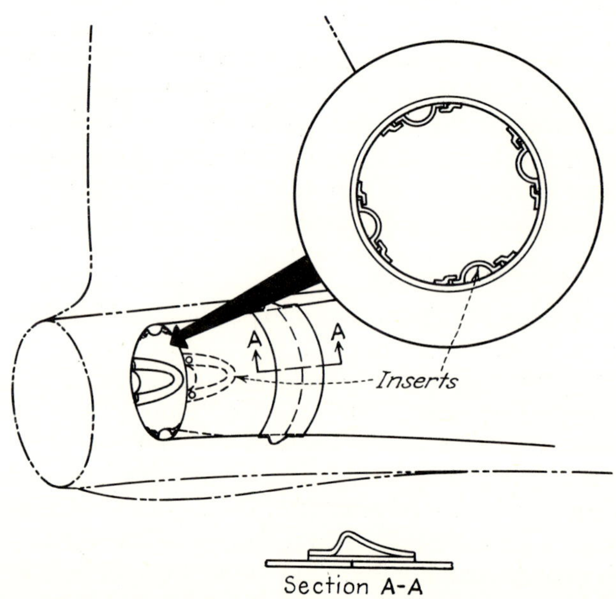

Figure 11·12 Removable jet-nozzle inserts.

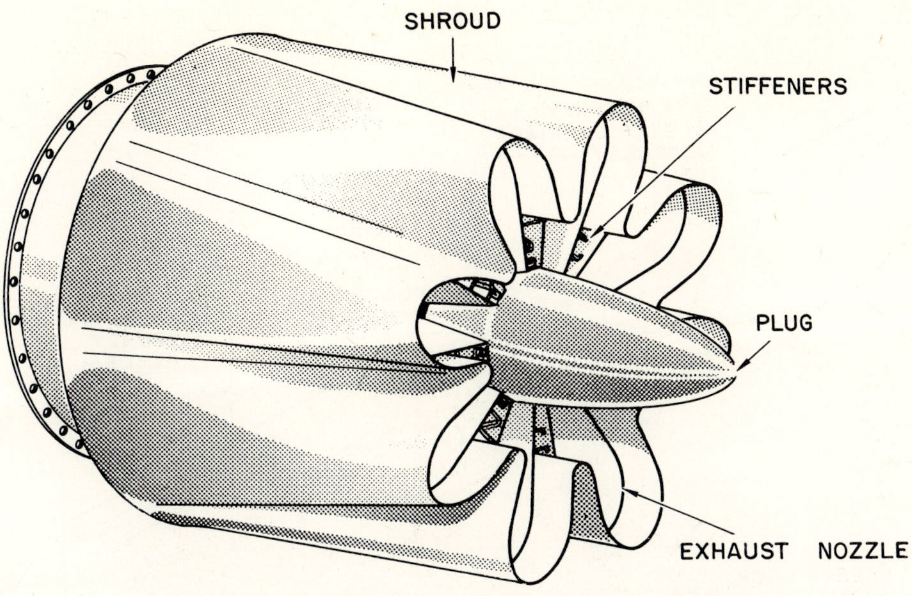

Figure 11 · 13 Sound suppressor. (General Electric Co.)

temperature went up and the nozzle had to be trimmed. Whenever a new engine was needed for an airplane, it invariably required a smaller jet nozzle, and a mad scramble for a new tailpipe ensued. It is obvious that this method would not be suitable for production orders.

Use of inserts

A production method developed to vary the jet-nozzle area by the use of inserts is shown in Fig. 11 · 12. These inserts, often called mice, fit inside a joggled retainer and are held in place by two stainless-steel screws. A set of inserts, designed for the engine, is provided with each airplane where this method is applicable. The inserts have different curvatures, thus making it possible for the technician to adjust the effective nozzle area for maximum thrust. The technician runs the engine at full power, checks the tailpipe temperatures, and then makes adjustments as necessary.

Noise suppressors

Because of the high noise level which accompanies the operation of jet engines, especially on takeoff, the industry has developed noise suppressors to take the place of the conventional jet nozzle.

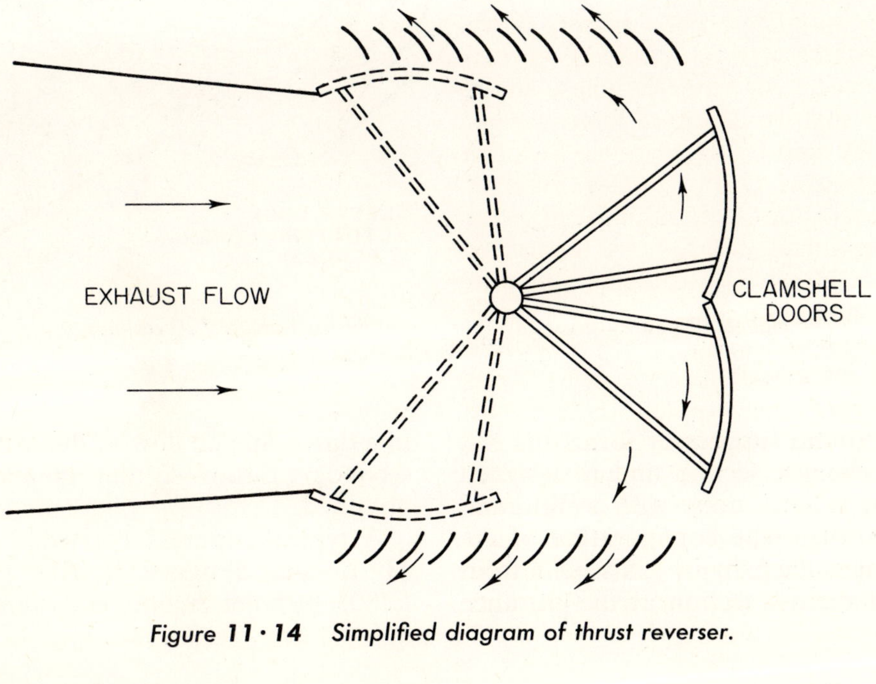

Figure 11 · 14 Simplified diagram of thrust reverser.

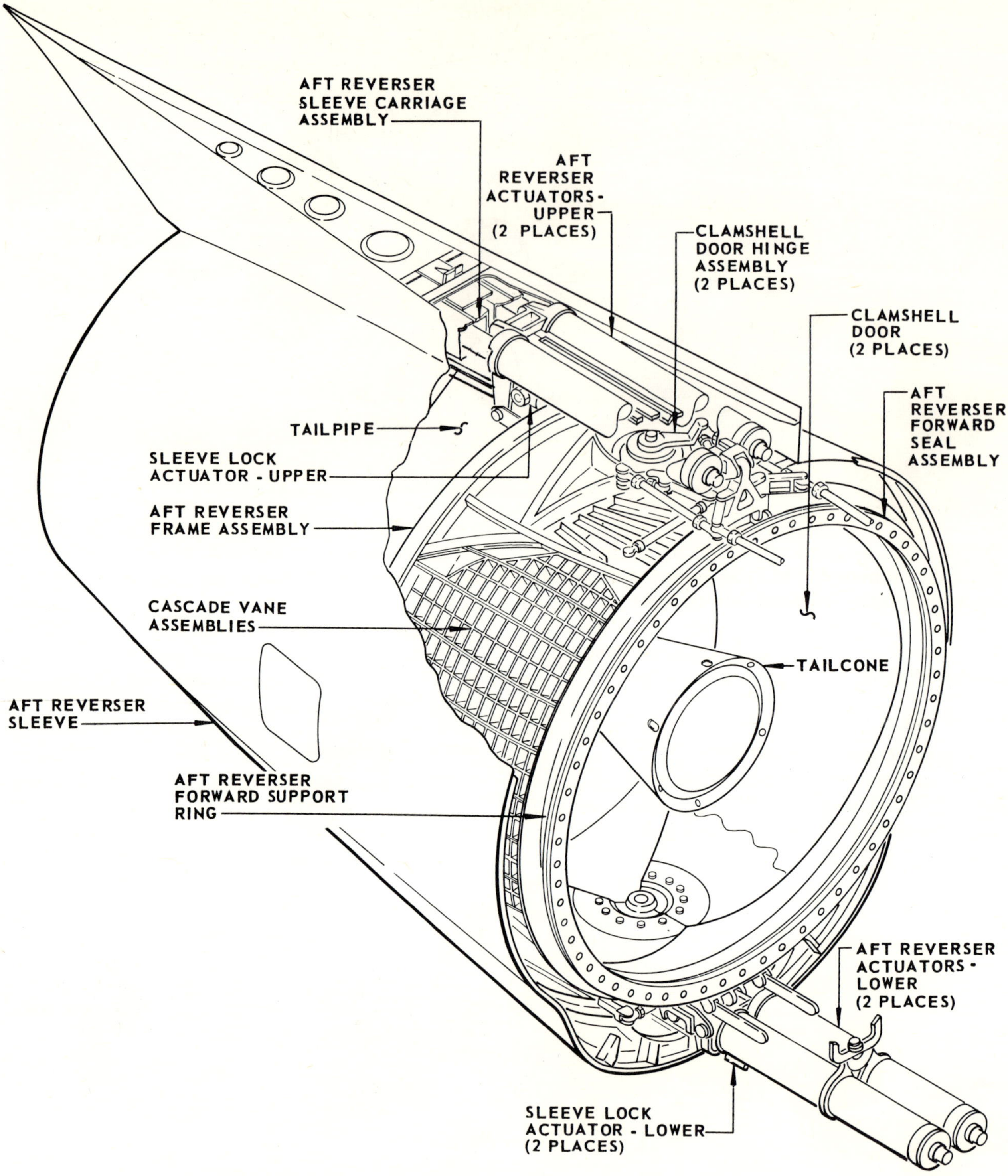

Figure **11 · 15** *Thrust reverser assembly for Boeing 720 airliner.*

The noise or sound suppressor forms the exit portion of the exhaust section and is designed to reduce engine exhaust noise with a minimum loss of engine thrust. The configuration of the suppressor changes the exhaust gas stream from a flow of circular cross section at the entrance to a daisy-shaped flow at the exit and introduces secondary airflow so that the maximum thrust is obtained.

A typical suppressor assembly, shown in Fig. 11 · 13 and designed for the General Electric CJ805 turbojet engine, is a stainless-steel fabri-

Figure **11·16** *Separation of F-100 fuselage to provide for engine installation. (United States Air Force photograph.)*

Figure **11·17** *Air Force crew working on a J57 engine in F-100 airplane. (United States Air Force photograph.)*

cation and consists of an exhaust nozzle and a shroud. The nozzle is formed to the desired shape and welded. A streamlined plug is bolted to the inner ends of the daisy petals, and stiffeners are welded into position between each petal. The shroud is then bolted to the stiffeners, completing the fixed construction assembly. The forward end of the shroud mates with the aft end of the airframe nacelle, and the flange at the forward end of the exhaust nozzle is bolted to the aft flange on the thrust reverser.

The noise created by jet engine exhaust is due, primarily, to the violent mixing of the high-velocity gases with the surrounding air. The daisy configuration of the exhaust nozzle divides the exhaust gas stream into eight smaller lobe-shaped streams. Because the eight smaller streams have a larger mixing area, the mixing of the exhaust stream with the surrounding air is less violent, and thus a reduction in noise level is achieved. The suppressor also changes the frequency of the sound to a higher level which is more quickly attenuated by the atmosphere.

The suppressor illustrated in Fig. 11 · 13 continues the pod contour and serves as the external surface of an ejector nozzle. This type of nozzle requires a source of secondary air. At cruise conditions, aft nacelle cooling air is routed between the shroud and the exhaust nozzle, thus increasing the mass flow of the exhaust gas stream. At operating points other than cruise, additional secondary air may be required. If so, the additional air must be supplied through inlets in the airframe nacelle. With the ejector-type suppressor installed, engine thrust losses will not exceed 1.5 per cent of the engine guaranteed gross thrust. Provisions for exit area trimming and exhaust nozzle total pressure instrumentation are included on the suppressor.

The sound suppressor described in this section is only one of many types which have been designed. The basic principle, however, is similar to the others because all types break up the main stream of the exhaust into smaller portions.

Thrust reversers

Because of the relatively high landing speeds of jet aircraft and because there is no propeller to provide drag or reverse thrust upon landing, it is necessary to create a reverse thrust through the action of the jet exhaust stream. For this purpose jet thrust reversers have been developed.

A jet thrust reverser is simply a device for blocking the jet exhaust stream which normally flows to the rear of the aircraft and causing it to be directed forward. A thrust reverser consists of clamshell vanes which are caused to rotate to a position where they block the jet exhaust stream and turn it outward through vanes which direct it forward. This action is explained in the simplified drawing of Fig. 11 · 14.

The aft thrust reverser assembly for a Boeing 720 jet airliner is shown in the drawing of Fig. 11 · 15. This drawing shows the actuating mechanism and the clamshell doors inside the reverser sleeve. When the pilot moves his controls to the REVERSE THRUST position, the aft reverser sleeve moves to the rear and exposes the cascade deflector vanes. At the same time the clamshell doors rotate to the position to close off the main exhaust stream and direct the gases through the vanes. The aft reverser operates simultaneously with the forward fan reverser, and both are operated by means of pneumatic actuating cylinders controlled through the pilot's thrust lever.

TYPICAL ENGINE INSTALLATIONS

The actual procedure for installing gas-turbine engines in aircraft must in all cases follow the instructions provided by the manufacturer in the Aircraft Maintenance and Overhaul Manual. As noted in the preceding portion of this chapter, there are many different types of installations, depending upon the purpose for which the aircraft is designed and the type of engine installed.

Military aircraft engine installations

An example of how a turbojet engine is installed in a military fighter aircraft is shown in Fig. 11 · 16. As shown in the photograph, the airplane fuselage can be separated and the engine removed or installed quickly and easily. The photograph illustrates an F-100 fighter plane with a J57 engine. The engine is equipped

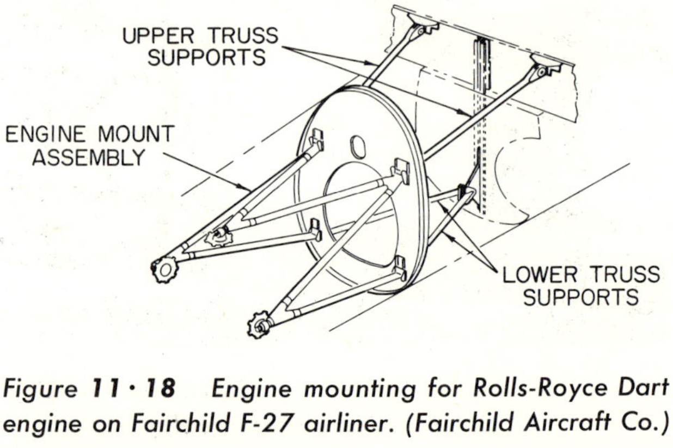

Figure 11 · 18 Engine mounting for Rolls-Royce Dart engine on Fairchild F-27 airliner. (Fairchild Aircraft Co.)

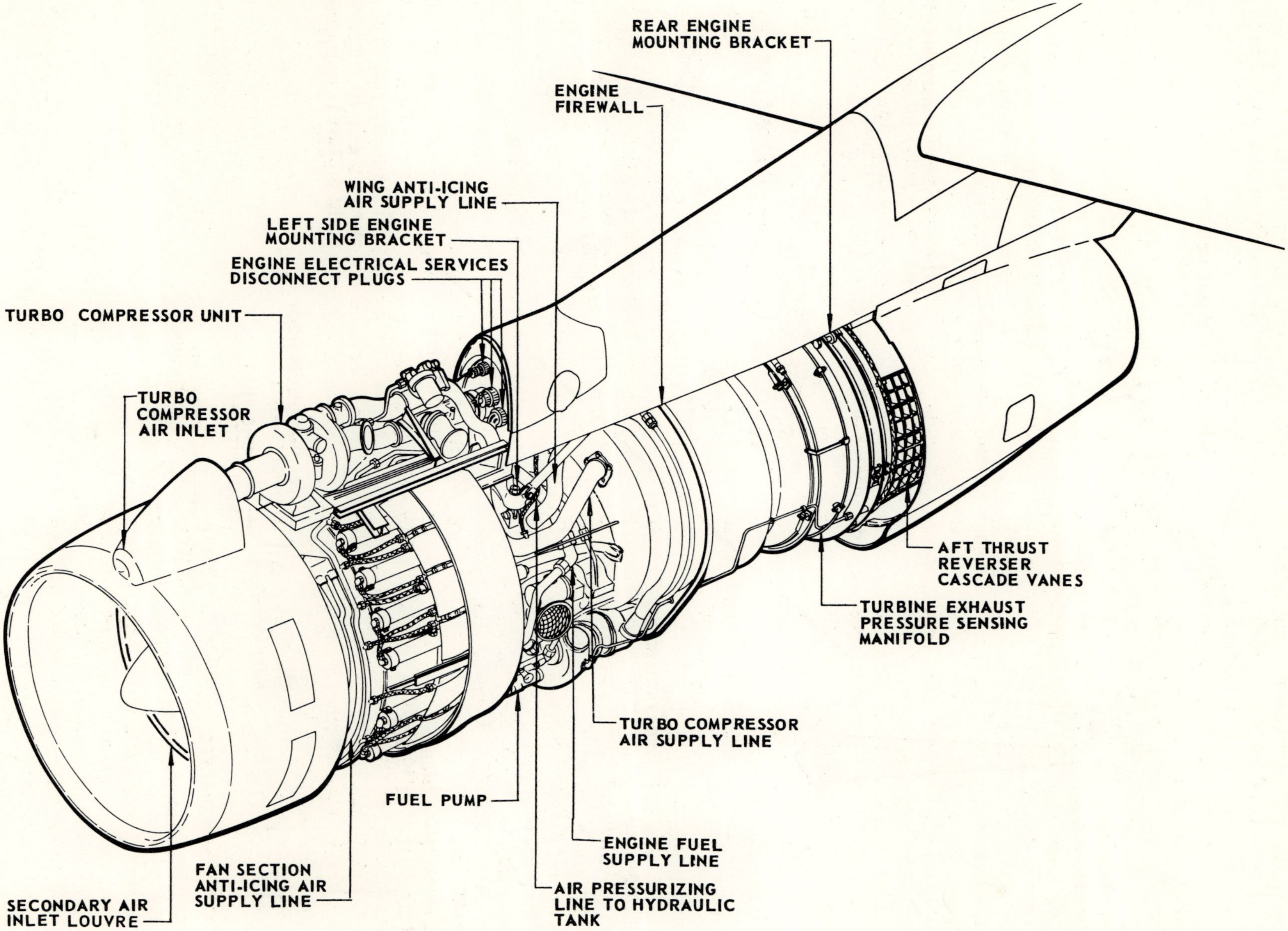

Figure **11 · 19** *Installation of JT3D engine on Boeing 720 airliner.*

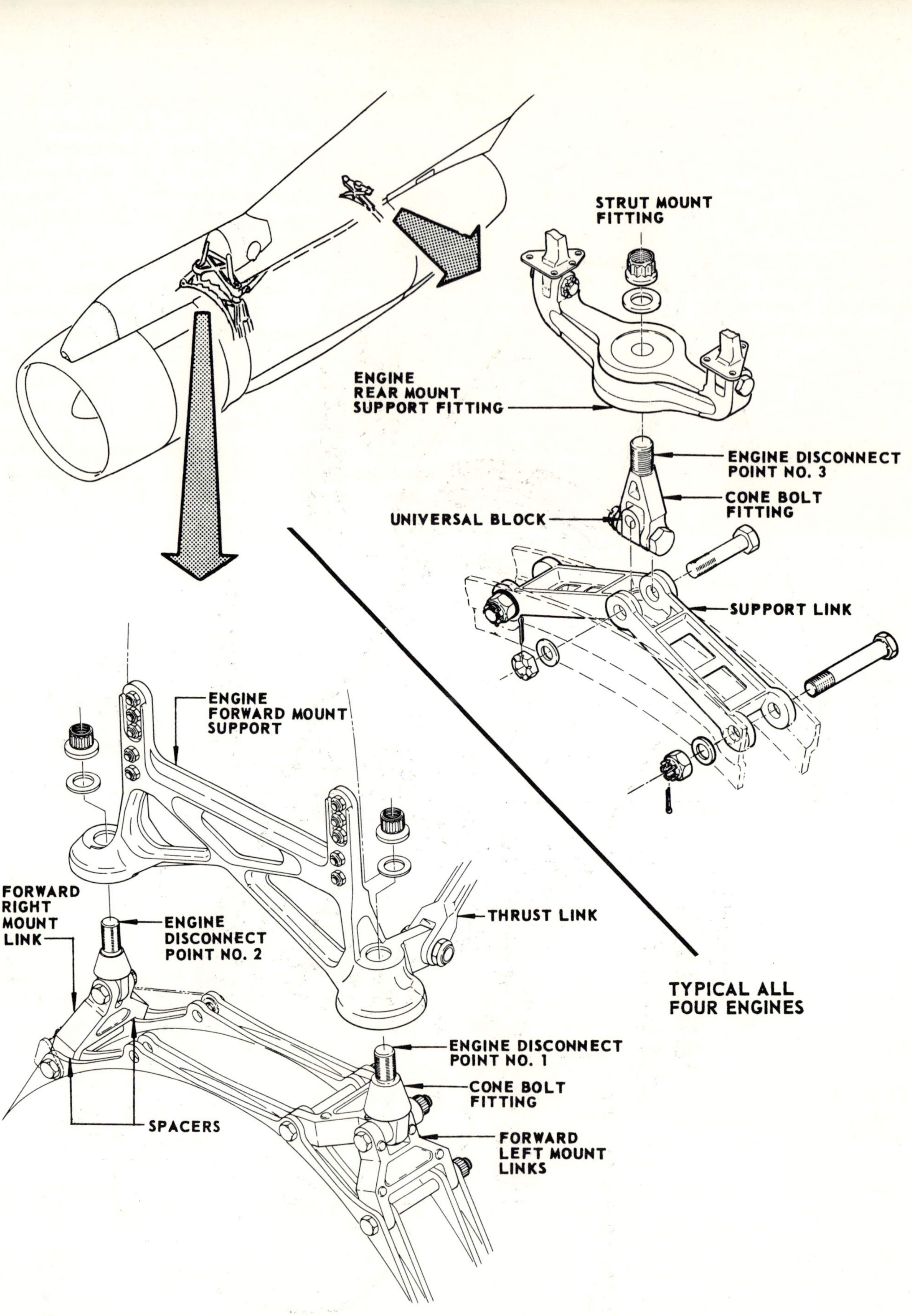

Figure **11 · 20** *Mount fittings for JT3D engine on 720 airliner.*

with an afterburner, which is a requirement for the majority of combat aircraft.

The design of military jet engine installations is such that an engine can be replaced in a very short time. With some aircraft, an expert crew can remove and replace an engine in less than fifteen minutes. This is made possible by the fact that the airplane fuselage can be separated and that all system connections (fuel, pneumatic, electrical, mechanical, etc.) are of the quick-disconnect type. The engine change equipment, such as engine dollies, hoists, and tail dollies, is designed to make the operation quick and easy.

Figure 11 · 17 illustrates the ease with which a crew can perform maintenance on an engine when the rear section of the fuselage is removed. The crew in the picture is working on the afterburner section of a J57 engine.

Mounting for a turboprop engine

An example of the type of engine mounting employed for a turboprop engine is shown in Fig. 11 · 18. These drawings show the tubular steel engine mount together with enlarged views of the attachment fittings for the Rolls-Royce Dart engine installed on a Fairchild "Friendship" F-27 airliner. It will be observed that the mount is similar to a typical mount for a reciprocating engine and consists of six steel tubes welded to seven fittings to form a W-shaped structure.

Three fittings, welded at the three forward points, are attached to pedestal-type mounting brackets installed on the engine intermediate compressor casing. Each engine mount fitting is secured to a mounting bracket by a longitudinal hollow tension bolt that passes through a seating washer and the mounting bracket and is then threaded into the tapped engine mount fitting. The hollow tension bolt is locked in place by a concentric left-hand threaded locking bolt that is also threaded into the engine mount fitting. A lock ring or tabwasher, installed under the head of the locking bolt, engages the hollow-tension bolt and maintains bolt torque.

The four aft fittings support the powerplant on the forward face of the firewall. Each fitting is attached to a firewall mounting bracket with transverse AN bolts in shear. The engine mount tube structure is utilized to support the engine control linkage, fire detector cable, electrical harnesses, fuel heater, and various hose and heatshield assemblies.

Pod mounting for a turbofan engine

The installation of a Pratt & Whitney JT3D turbofan engine on a Boeing 720 airliner is illustrated in Fig. 11 · 19. This illustration shows the left side of a typical engine installation and reveals a number of the details associated with such an installation.

The engine is secured to the strut at three points. The forward mounting position has an arrangement of links bolted to the forward flange of the diffuser case and terminating in two cone bolts which are rigidly secured into the strut fitting. The aft engine mount is comprised of two links bolted to the top of a double flange around the exhaust section. These terminate in a single cone bolt secured into a strut-mounted bracket arranged to compensate for changes in engine length due to thermal expansion. The mount fittings and their locations are shown in the drawings of Fig. 11 · 20.

STUDY QUESTIONS

1. Approximately what weight of air may be consumed by a modern jet engine in 1 hr?
2. Discuss the effect of air-inlet-duct efficiency on thrust.
3. List the aerodynamic requirements for the ideal air-induction system.
4. What are the advantages of a nose inlet?
5. What is the simplest and most efficient type of inlet?
6. Why is it not good practice to "overcool" the turbine and exhaust section of a jet engine?
7. Give two functions of a blanket wrapped around the exhaust section of a gas-turbine engine.
8. What is the function of the bulkhead between the compressor section and turbine section in the engine nacelle?
9. Discuss the effect of nozzle area.
10. How does a noise suppressor accomplish its purpose?
11. Explain the operation of a thrust reverser.
12. Describe the engine mount for the Dart engine in an F-27 airliner.
13. How does the Fairchild F-27 engine installation compare with the installation of a reciprocating engine?
14. Why is a universal (flexible) fitting used for the rear mounting of the JT3D engine in the wing pod?

CHAPTER 12

Thrust Augmentation

Methods of increasing the thrust of turbojet engines for short periods of time have become of increasing importance in improving the effectiveness of this type of aircraft powerplant. Thrust-augmentation methods have found their principal application in improving the takeoff and climb characteristics of jet-propelled aircraft, as well as improving the combat and high-speed performance of many military aircraft. In addition to the use of rocket assist, two principal methods of thrust augmentation are in common use: afterburning (reheating), and water-alcohol injection.

There is another method for increasing the thrust of a turbojet engine; however, it may be considered a part of the engine design. This method is the bypassing of a portion of the compressed air around the combustion chamber and returning it to the airstream, thus increasing the mass flow. A similar principle is used in the aft-fan engine where the extra air is taken in through an annular inlet just forward of the turbine section. In the CJ805 engine a separate and independent turbine wheel is equipped with extension blades which extend into the outer duct. These blades increase the mass airflow, thus increasing the thrust.

THE AFTERBURNER

Effects of increased temperature

The volume of air passing through a turbojet engine is four to five times as much as is required for the combustion of fuel. The excess air is necessary to provide cooling, since the heat of combustion may exceed 3700°F. The additional air forms a blanket around the burning area and then mixes with the burned gases. This results in a stream of hot gases, 1400 to 1500°F, flowing through the turbine into the tail cone. In this stream is a large quantity of unburned air suitable for the combustion of additional fuel. When a large quantity of fuel is sprayed into the hot exhaust stream, the resulting combustion increases the temperature to as much as 2500°F. The addition of the fuel increases the mass of the exhaust gases, and the addition of heat increases the velocity. Since thrust is proportional to mass times change in velocity, the thrust of the engine is increased as much as 50 per cent.

The exhaust stream does not always have sufficient temperature to ignite the fuel which is injected into the afterburner; hence it is necessary to provide a source of ignition in these cases. One method used is called the "hot streak." With this type of ignition, a special fuel nozzle is placed at a point nearer the turbine than the location of the fuel spray bars. Fuel from this nozzle is ignited by the hot gases, and the flame carries to the area where the main fuel injection takes place. With other ignition systems, a spark plug is placed in the afterburner to ignite the afterburner fuel.

Development

Afterburning is not a new idea. Like jet propulsion, it threads back through history, and no individual or organization can be considered the sole pioneer in either achievement. Afterburners, turbojets, gas-turbine engines, steam turbines, and internal-combustion engines are the brain children of dozens of scientists and engineers who have labored during the last 300 years. Each has contributed knowledge which makes our present powerplants possible.

In spite of the many claims and counterclaims, it can be observed that a number of organizations in the United States performed significant work leading to modern afterburner designs. These include NACA, Ryan, Solar, General Electric, Westinghouse, Wright Aeronautical, Pratt & Whitney, Allison, and others.

Undoubtedly, the first afterburner in America

was built by NACA in 1944. This research was conducted in the NACA Lewis Flight Propulsion Laboratory and began with the delivery of the nation's first turbojet engine, General Electric's I-A.

Ryan Aeronautical Company began afterburning research in the following year, under contract with the Navy's Bureau of Aeronautics. Ryan produced the first commercially built afterburner in this country at that time and demonstrated it on Apr. 25, 1946. The engine used was the GE I-16, which was also the jet powerplant for the composite-engined Ryan Fireball fighter, the Navy's first plane to have a jet engine.

Afterburning is such an old idea that it is impossible to know who thought of it first. Among engineers, afterburning is more accurately known as "reheating." In England, the afterburner is called a "reheat engine." Reheating is defined as "appreciably improving the efficiency of an engine by reheating the working fluid after it has partially flowed through the engine." Actually, afterburning is a well-known thermodynamic technique, capitalized upon by many inventors, and extending back almost to the very first heat engine devised.

Not long after James Watt produced the first practical steam engine in 1764, reheating was incorporated into its design. The steam was reheated between expansions in the cylinders in order to extract more mechanical energy from it. The same strategy was followed after Sir Charles Parsons introduced his commercially successful steam turbine in 1884. Reheating has been universally used as a means for increasing steam-turbine powerplant efficiency by adding heat to the steam after it has partially passed through the engine.

Sir Frank Whittle, of England, described the same principle in the "dual thermal cycle" engine which he patented in 1936. This device employed a diesel engine and compressor to supply air and combustion products to a turbine which drove the main compressor. The turbine exhaust was used to obtain jet propulsion. Whittle proposed "adding heat to the turbine

Figure 12·1 NACA altitude wind tunnel test for afterburner.

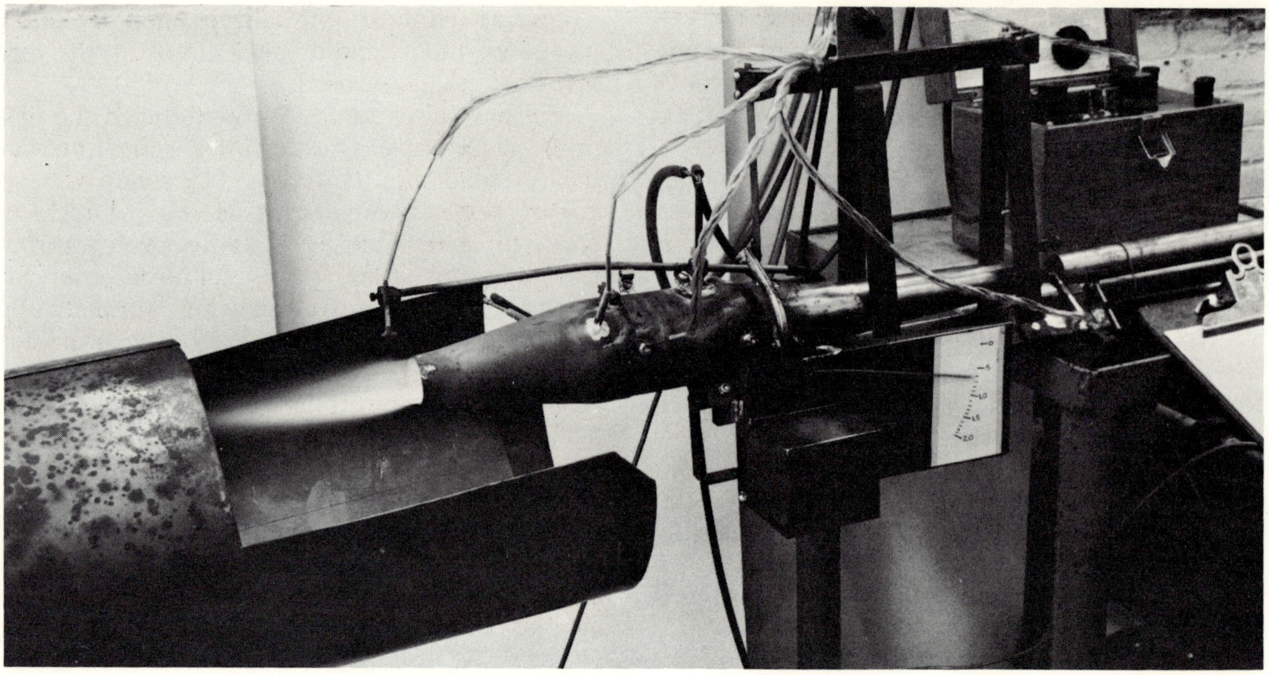

Figure 12·2 Testing a ramjet combustion chamber.

exhaust before final expansion in order to provide further thrust for carrying temporary overloads."

On Aug. 27, 1940, an all-metal monoplane designed by the Italian engineer S. Campini, of the Caproni Company, made a 10-min flight from Forlanini airport in Milan. This was the second airplane to fly successfully with jet propulsion, the first having flown the year before in Germany. Campini used a piston engine, probably a radial Isotta Fraschini, to power a compressor which supplied air to the combustion chambers of his jet engine. In his plans for later versions of this aircraft and engine, he envisaged the advantages of burning fuel in the tailpipe to attain increased thrust.

In 1940, Whittle produced a successful turbojet engine, the W1X, described in the first chapter of this text. In his accomplishment, Whittle drew heavily upon the achievements of his predecessors. These included the Englishman John Barber, who patented the first gas turbine in 1791; Sir Charles Parsons, developer of the axial-flow compressor; America's Dr. Sanford Moss; Swiss Dr. Adolph Meyer, who produced the first modern gas turbine in 1930; and many others.

In 1944, NACA began work on the first United States afterburner, using General Electric's first turbojet engine, the I-A, as a powerplant. In NACA's first model, the basic requirements of efficient diffusion from turbine outlet to burner inlet, proper fuel preparation, and flame-holding regions were recognized. Although combustion was successfully initiated, internal losses were so high and combustion

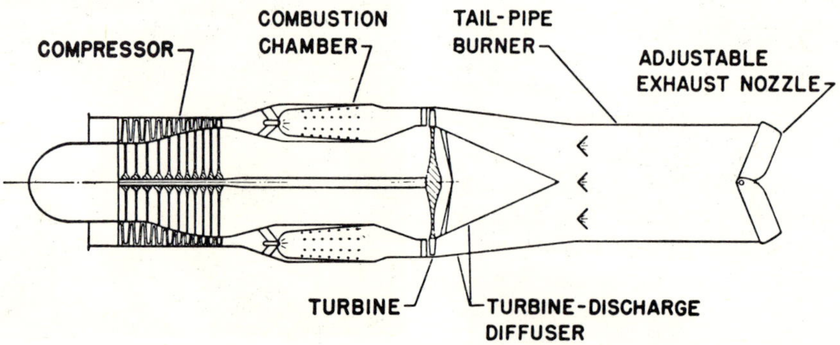

Figure 12·3 Schematic diagram of a turbojet engine with an afterburner.

efficiency so low that the maximum thrust realized was only 100 lb.

In 1945, a blower rig was established in which afterburner inlet conditions were simulated by forcing air through a preheater, inlet diffuser, and afterburner by a blower system. Numerous designs were tested in this rig, leading to the configuration which was tested with the GE TG-180 (J35) engine in 1946. In this test, a thrust of 40 per cent was obtained.

The first altitude wind-tunnel tests of afterburners were made in NACA's laboratory in 1946. The arrangement for these tests is shown in Fig. 12 • 1. Here the afterburner is operated in connection with a TG-180 engine. A number of configurations were investigated, and successful operation was accomplished at 30,000 ft altitude with increases of thrust of over 100 per cent at high simulated Mach numbers.

During 1944, the Ryan Aeronautical Company launched a company-sponsored confidential research project to develop a successful ramjet combustion chamber for use with helicopter rotor blades. In this work, Ryan engineers constructed experimental combustion chambers and tested them by burning fuel within them in a high-velocity stream of air. The chambers were fired on a test stand, as shown in Fig. 12 • 2. The developed thrust was measured by instruments connected to the engine in the stand.

Essentially, a ramjet is a compressorless jet engine which involves the same problems of igniting fuel in a high-velocity airflow, maintaining combustion, and developing thrust that are found in the afterburner. Likewise, an afterburner may be considered a ramjet engine attached to the exhaust end of a turbojet engine. The turbojet engine "rams" the air into the ramjet, or afterburner. Sometimes, the combination of turbojet engine and afterburner is called a "turbo-ram-jet." The operation of ramjet engines is explained in Chap. 5.

Description

A schematic diagram of a turbojet engine equipped with an afterburner is shown in Fig. 12 • 3. Since the afterburner inlet-air velocities must be sufficiently low to avoid excessive pressure losses and to ensure satisfactory combustion, a diffuser is provided between the turbine discharge and the afterburner inlet. The **diffuser** is an expansion area which causes the velocity of the gases to decrease. The diffusion is normally accomplished by placing a cone immediately to the rear of the turbine. An early type diffuser cone is shown in Fig. 12 • 4. An exhaust

Figure 12 • 4 Early type diffuser cone.

cone of more recent type is shown in Fig. 12 • 5.

Fuel nozzles are located at the burner inlet plane or, in some installations, at various locations in the diffuser. Figure 12 • 6 illustrates an early type fuel-injection grid designed by the Ryan Aeronautical Company. Modern afterburners are equipped with fuel spray bars or perforated fuel rings as shown in Fig. 12 • 7. The

Figure 12 • 5 Typical exhaust cone.

Figure **12·6** *Early type Ryan fuel grid.*

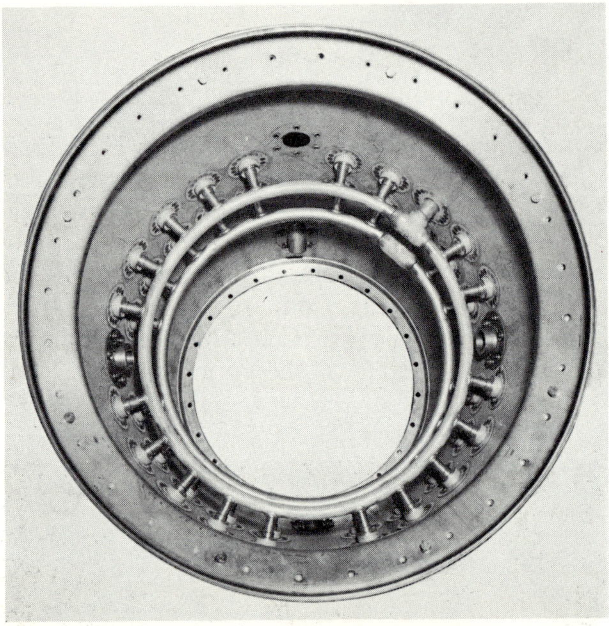

Figure **12·7** *Fuel rings in an afterburner.*

Figure **12·8** *Flame holders attached to diffuser cone.*

function of the fuel-injection system, of course, is to deliver a spray of fuel into the afterburner, uniformly and in large quantities. The quantity of fuel may be twice as great as that consumed by the engine without the afterburner.

Even though the airflow velocity through the afterburner is decreased by the diffuser, it is still necessary to provide flame holders to prevent the flame from being blown out the exhaust of the engine. The flame holders provide the stagnation regions and the turbulence necessary for combustion of the fuel to the extent required to allow complete combustion within the afterburner. Figure 12 · 8 illustrates the flame-holder rings attached to the diffuser cone for a particular afterburner. Note that the cross section of the flame-holder ring would form a V.

An essential requirement for every afterburner is a variable-area nozzle. The jet nozzle of a turbojet engine is very critical in cross-section area. An area slightly too large will result in a substantial loss in thrust, because the velocity of the gases at the exhaust nozzle will be lower than with a smaller-area nozzle. The nozzle must not be too small, or overheating will result and too great a back pressure may be developed. Therefore we see that there is an optimum nozzle area for particular conditions. It is obvious that large quantities of fuel burned in the afterburner will require a larger-than-normal exit area for the burning gases. To provide for the requirements of nozzle area, the variable-area nozzle was developed. Figure 12 · 9 illustrates an early type of "eyelid" nozzle used on one of the first Ryan afterburners. This type of nozzle variation was quite satisfactory, as indicated in the illustration of a later-model General Electric afterburner shown in Fig. 12 · 10.

In a typical afterburner of current design, the fuel nozzles are a group of spray bars arranged radially, like the spokes of a wheel, near the rear of the diffuser section. A short distance to the rear of the afterburner diffuser cone are the flame holders. These are three concentric rings held in place by two steel bars mounted crosswise in the afterburner duct. The rings are nearly U-shaped in cross section, with the open part of the U facing the rear. As has been previously explained, these rings serve to provide the stagnation region where the combustion of fuel may take place. The rear portion of this afterburner is constructed with a double wall to provide for cooling. The engine exhaust gases flow between the walls and cool the afterburner section. The cooling gases are at a temperature of about 1000°F, whereas the afterburner gases are at a temperature of more than 2500°F.

Afterburner operation

The principal purposes of the afterburner are to provide additional thrust for takeoff and climb and to allow for brief periods of high-speed performance which would otherwise be impossible. Since fuel consumption is high during afterburner operation, the afterburner is employed only when necessary to accomplish the desired performance.

An afterburner control system may include a special fuel pump, a fuel control, an igniter control, and an exhaust-nozzle actuator control. The **fuel pump** may either be a separate pump or be included in the main fuel pump. In either case, its function is to supply the additional fuel required for afterburner operation.

The **afterburner fuel control** is designed to provide the correct fuel pressure or metering orifice for all conditions of afterburner operation. It may be designed to vary the pressure differential across a fixed orifice, or it may vary the size of an orifice where the pressure differential remains constant. The principal controlling factor for afterburner operation is the burner pressure. This pressure is fed to a chamber containing a bellows to which is linked the metering valve. As burner pressure increases, the metering valve is opened to provide additional fuel flow.

The **igniter control** serves to provide "hot-streak" ignition to "light off" the afterburner fuel. The control is arranged so that it is activated at the time that the afterburner is turned on. At this time it injects a stream of fuel into the engine, where it will be ignited by the engine combustion. This stream of fuel provides a flame which is carried to the area where fuel injection is taking place, thus providing ignition for the afterburner fuel.

The variable-area exhaust nozzle is also activated when the afterburner is turned on. The nozzle is opened immediately and automatically by the nozzle actuator or actuators whenever the afterburner is turned on.

Figure 12·9 Early type "eyelid" nozzle.

WATER INJECTION

Types of water injection

Basically, there are two types of water injection for gas-turbine engines. Figure 12·11 illustrates a type in which the water is injected at the compressor inlet. The water should be as pure as possible to prevent the buildup of mineral deposits on the compressor blades. Pure alcohol may be added to prevent freezing and to provide more heating to compensate to some extent for the heat lost through the evaporation of the water-alcohol mixture. The heat is supplied when the alcohol burns.

Compressor-inlet injection

In the compressor-inlet-injection method, the evaporation of the injected liquid extracts heat from the inlet air, and thus it is theoretically possible to cool the inlet air to the saturation temperature of the fluid mixture before it enters the compressor. When liquid in excess of that required to saturate the inlet air is injected at the compressor inlet, further cooling is obtained by additional evaporation during the mechanical

Figure 12·10 Improved type eyelid nozzle.

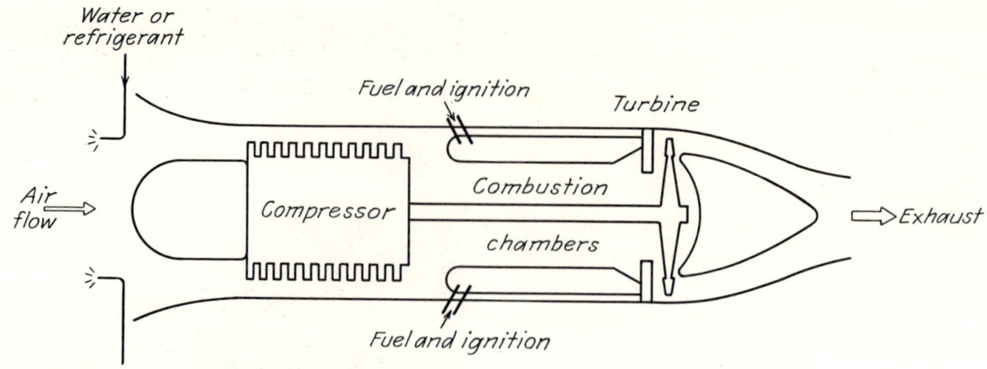

Figure 12 · 11 Water injection at compressor inlet.

compression process. The temperature of the fluid is, therefore, reduced.

An increase in both the compressor-pressure ratio and airflow is obtained. This is reflected throughout the engine as an increase in total mass flow and exhaust-nozzle pressure ratio, both of which increase the thrust produced.

The injection of water at the compressor inlets is divided into two parts: (1) the case where sufficient water is injected to saturate the compressor-inlet air; and (2) the case where sufficient water is injected at the compressor inlet to saturate the air at some point during the mechanical compression process.

In some centrifugal-type engines, water injection is accomplished in the compressor section by feeding water and alcohol into the center of the compressor rotor, which injects the liquid by centrifugal force into the airstream. This method of injection may produce an increase of thrust of as much as 25 per cent.

The second most commonly employed method for water injection is the type in which water is injected into the diffuser case of the engine. This method is employed together with the compressor-inlet method for the Pratt & Whitney JT3 engine.

Water-injection system for the JT3 engine

Since thrust increase is particularly desirable at takeoff when the engine is required to produce its greatest output of power, the water-injection system for the JT3 engine operates only at high engine power. Water is injected into both the compressor inlet and the diffuser case only at temperatures above 40°F. Below this temperature water is injected only into the diffuser case because of possible ice formation at the engine inlet.

The water-injection system supplies water from tanks located in the keel-beam area of the main wheel wells. This water is pumped through necessary piping to the manifolds from which the engines are supplied. A schematic diagram of the system on a Boeing 707 airliner is shown in Fig. 12 · 12. For the JT3 engine, only pure water is used, and no alcohol or other additives are permissible.

Two electrically driven pumps submerged in the water tank deliver water under pressure to the manifolds which carry the water to the engine-driven pump on each engine. Water-injection shutoff valves, upstream of the pumps at each engine, prevent water from being discharged into the engines until the engine power levers are advanced to maximum power settings. The inlet-water-injection shutoff valves are controlled by a separate switch on the copilot's instrument panel, and the tank pumps are controlled from an adjacent switch on the same panel. The system operates automatically in that if it is "armed" it will initiate water injection at the proper power setting and will shut itself off when the water supply is exhausted.

To operate the water-injection system for the JT3 engine, the water-injection pump control switch on the copilot's instrument panel is placed in the ON position while the engines are at idle. If the ambient air temperature is above 40°F, the inlet-water-injection shutoff valve switch is moved to the OPEN position. The system is now armed; that is, the water-injection pump relays are energized, and the electrically driven water-injection pumps start. Water pressure then actuates the water-injection pressure switches, holding the relays energized and furnishing power to the normal pressure indicating lights on the copilot's instrument panel.

Advancing the thrust lever to maximum power then actuates a switch in each fuel-control unit, thus opening the engine water-injection shutoff valves. With the valves open, water from the electrically driven water-injection pumps will flow into the engine-driven

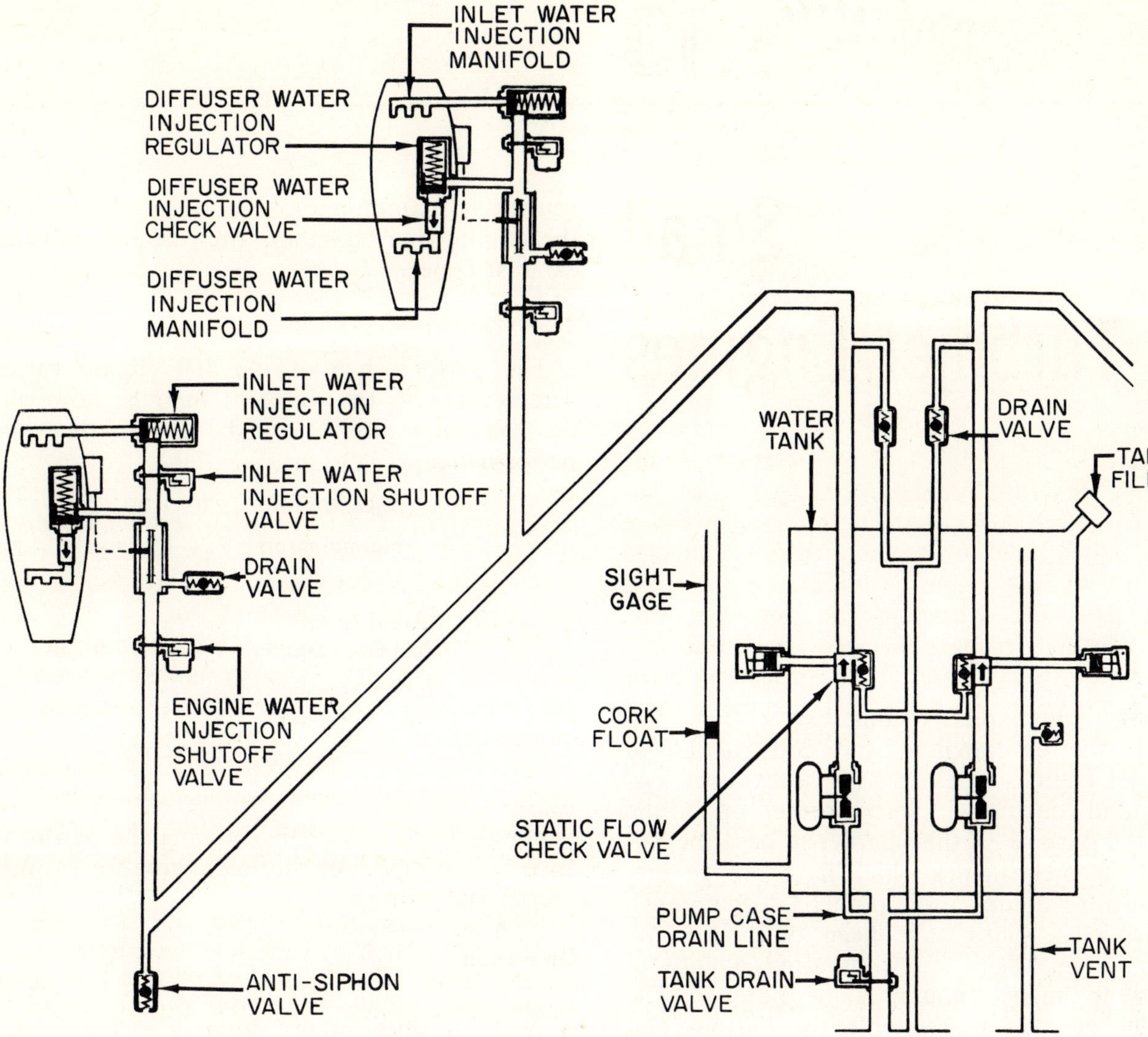

Figure 12·12 Diagram of water-injection system for Boeing 707 airliner.

pumps and be discharged to the engine manifolds through the water-injection regulators. The regulators maintain a constant flow of water by holding a fixed pressure differential across an orifice. When the water supply is exhausted, the water pressure drops, thus allowing the pressure switches to relax. This deenergizes the water-injection pump relays and indicator lights. The water-injection manifold drain valves then open and allow all residual water to drain from the system.

STUDY QUESTIONS

1. Name two commonly used methods for producing thrust augmentation.
2. Explain afterburning, or reheat, operation.
3. How is it possible to burn additional fuel in the exhaust stream?
4. By what organization was the first afterburner operated in the United States?
5. Compare an afterburner with a ramjet.
6. Why is a diffuser necessary with an afterburner?
7. Explain the function of a flame holder.
8. Approximately how much thrust is gained by the use of an afterburner?
9. Why is it necessary to use a variable-exit nozzle with an afterburner?
10. Why does the injection of water, or water and alcohol, into the inlet of a gas-turbine engine increase the thrust?
11. Give a brief description of the water injection system for a Pratt & Whitney JT3 engine.
12. Why must the water for a water-injection system be as pure as possible?

CHAPTER 13

Small Turbojet Engines

In preceding chapters of this text, we have discussed the development and construction of turbojet engines together with accessories, systems, and controls necessary for their operation. It is the purpose of this chapter to describe two small turbojet engines which have proved successful for military and commercial operations.

Turbojet engines have been built in many sizes, ranging from a few hundred pounds of thrust to many thousands of pounds. The smaller engines are suitable for liaison- and training-type aircraft, while the large engines are installed in high-performance fighter aircraft, bombers, transports, tankers, and commercial airliners.

THE CONTINENTAL J69 TURBOJET ENGINE

General description

The J69 engine is a small turbojet type suitable for use in trainer aircraft, missiles, and drone-target aircraft. It consists of a single-stage centrifugal compressor, a straight (through-flow) annular combustor, and a single-stage axial-flow turbine. The engine is manufactured in several different models for use in different types of aircraft.

Figure 13 · 1 is a cutaway view of the J69-T-9 model. The basic engine configuration for each model is quite similar. In the drawing can be seen the air inlet; the single-sided, single-stage centrifugal compressor; the diffusing section; the combustion chamber; the turbine; and the exhaust cone.

Specifications

The performance of the J69 engine varies somewhat with the different models; however, the following table gives the general level of performance:

Takeoff thrust	920–1,000 lb
Normal thrust	725–795 lb
Dry weight	317–345 lb
Takeoff specific fuel consumption	1.13–1.29 lb/lbt/hr
Normal specific fuel consumption	1.115–1.27 lb/lbt/hr
Specific weight	0.32–0.37 lb/lbt
Airflow	17.5–20.1 lb/sec

Figure 13 · 2 is a chart showing the performance of one model of the J69 engine for various speeds and altitudes.

Operation

The operation of the J69 engine may be understood by a study of the drawing in Fig. 13 · 3. Ram air enters the engine through the **air inlet** (1) and continues rearward to the **compressor** (2). Here it is flung outward by the action of the compressor through the radial **diffuser** (3). At this time it makes a 90° turn into the **axial diffuser** (4). The purpose of the diffusers is to change the velocity of the air to pressure. This pressure forces the air into the combustion chamber, where it takes three different paths. Primary air flows rearward between the **outer combustion chamber casing** and the **outer combustor shell** (7), thence inward through the hollow vanes of the **turbine inlet nozzle** (8) and forward inside the **primary air admission shell** (6). Here it flows through perforations and mixes with the fuel which is being ejected from the **rotating fuel injector** (5). Primary air also flows into the combustor area through openings in the forward end of the outer combustor shell.

Secondary air for cooling flows through slots and perforations in the outer combustor shell to mix with the burning gases and be ejected

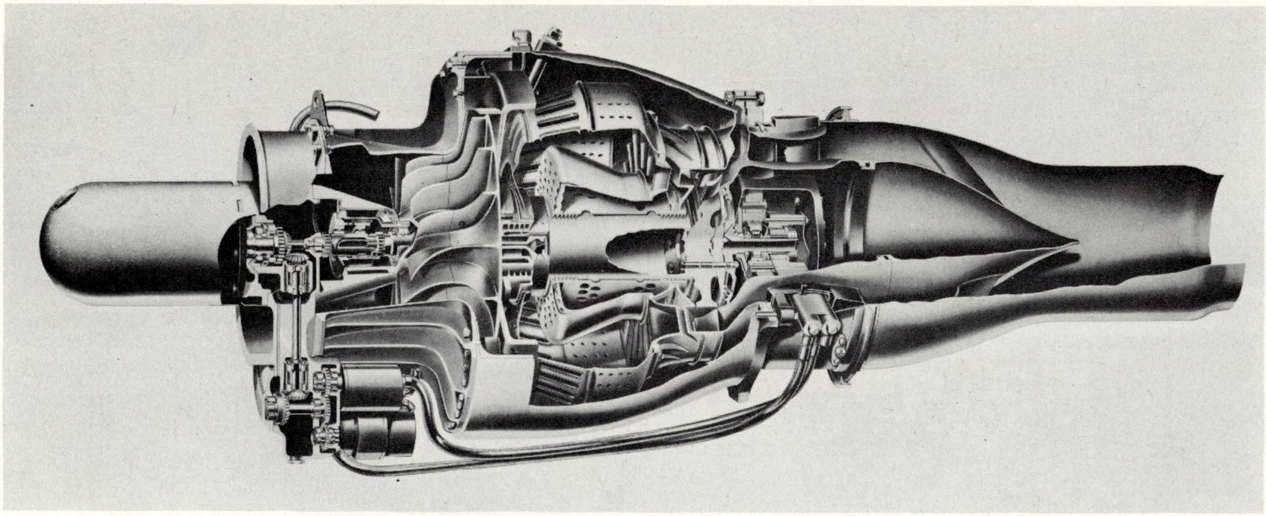

Figure 13 · 1 The Continental J69 engine.

through the turbine inlet nozzle. The turbine inlet nozzle directs the hot gases to impinge on the turbine blades. The turbine extracts sufficient power from the gases to operate the compressor and accessories, and the balance of the gas energy produces jet thrust as the gas is ejected through the jet nozzle.

Controlled fuel flow is introduced into the compressor-turbine shaft assembly near the **control island** (12). It flows rearward through the shaft to the rotating fuel injector, where it is thrown outward in a radial spray to mix with primary air.

For starting purposes, a fuel primer system is incorporated in the engine. This assures positive, low-temperature light-offs and good starting

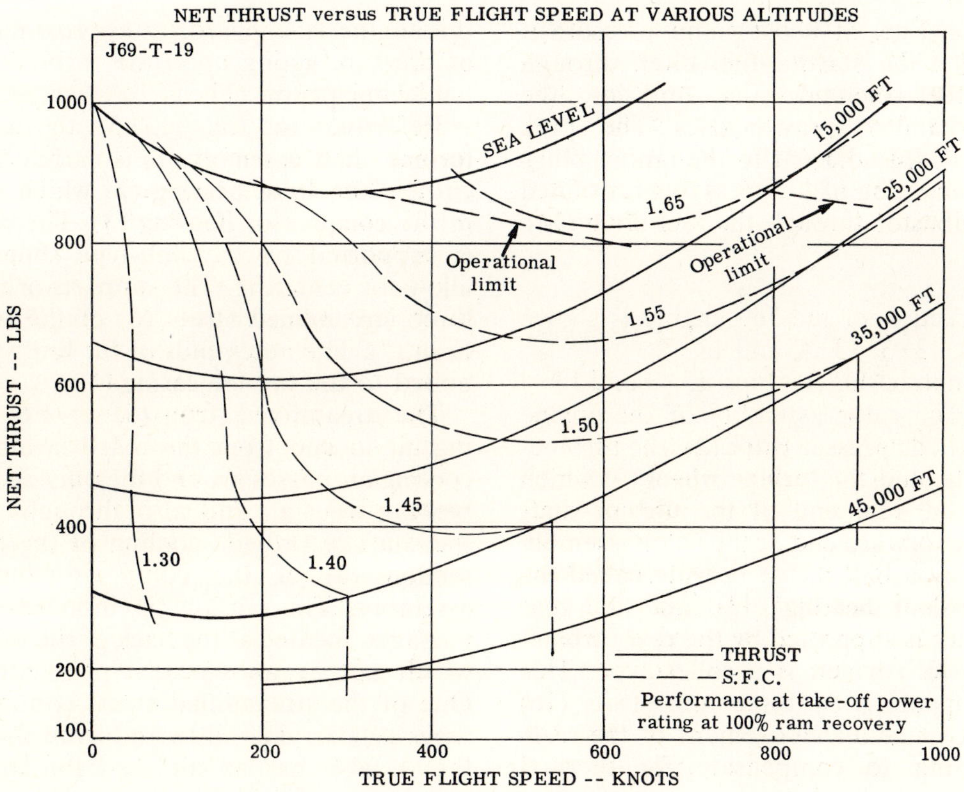

Figure 13 · 2 Performance curves for a J69 engine.

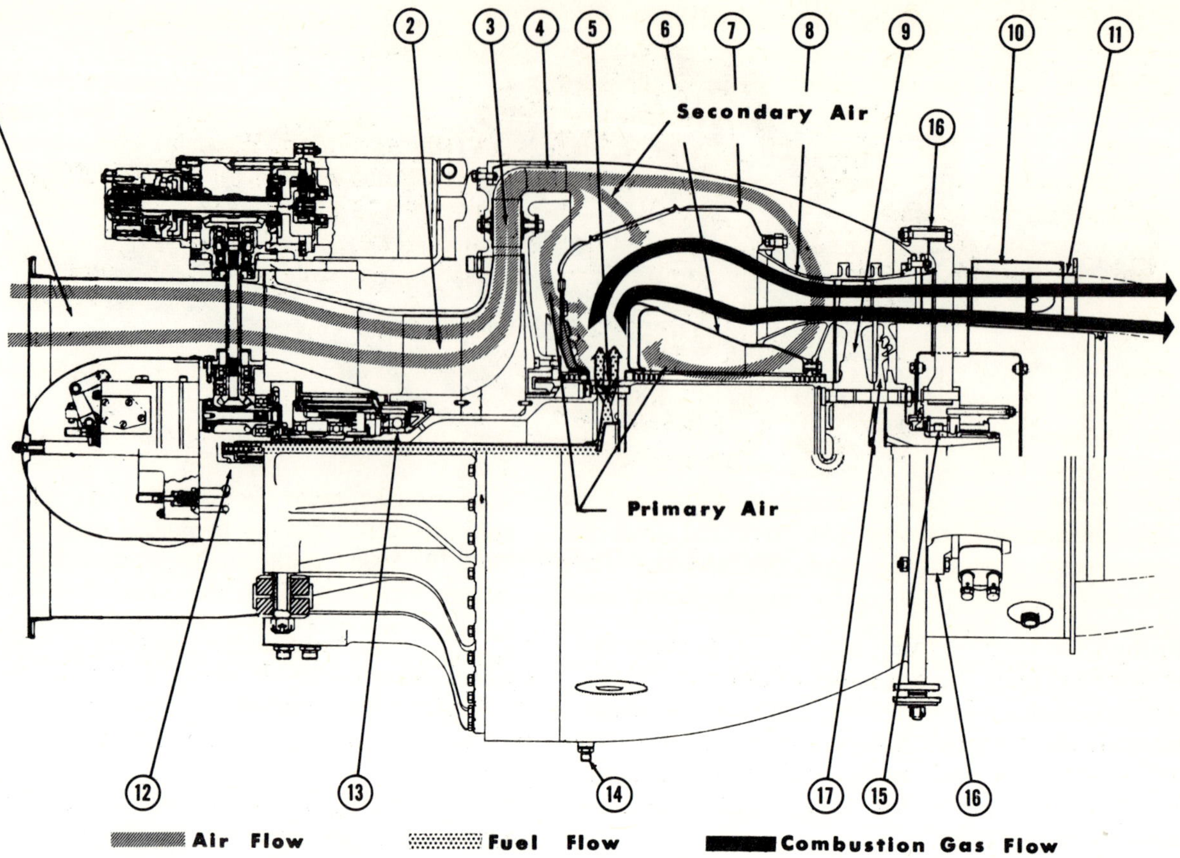

*Figure **13·3** Fuel-air flow diagram for the J69 engine.*

characteristics. Fuel at boost-pump pressure is passed through the starting-fuel filter, through the starting-fuel solenoid valve, and into the combustion-chamber spray nozzles. The spray nozzles are located adjacent to the igniter plugs in the same mounting pad. Excess fuel is drained from the combustor through the **fuel drain** (14).

Construction

The construction of the J69 engine is shown in Figs. 13·1, 13·3, 13·4, and 13·5.

The index numbering for Figs. 13·3 and 13·4 is identical. The rotor assembly of the engine consists of the **compressor rotor** (2), the turbine-shaft assembly, and the **turbine wheel** (9) which is bolted to the rear end of the turbine-shaft assembly. The forward end of the rotor assembly is supported by a ball thrust bearing called the **front turbine-shaft bearing** (13), and the rear end of the rotor is supported by the **rear turbine-shaft bearing** (15) which is a roller type. This bearing is supported by three link struts (16) which permit tangential movement of the rear-bearing housing to compensate for thermal expansion forces.

The forward thrust of the turbine shaft within the engine is opposed by approximately 2 atm of pressure acting upon the turbine side of the balancing piston (17).

Referring to Fig. 13·5, the **compressor-turbine shaft assembly** (1) is carried at the inlet end by the **front bearing** (2) which is mounted in the **compressor housing** (3). The **rear bearing** is supported by the **link-type supports** (5) to allow for centering. The supports are encased in three **streamlined struts** (6) of the **exhaust diffuser** (7). The outer ends of the link supports are bolted to the **turbine casing** (12).

The streamlined struts (6) serve to carry cooling air to and from the rear-bearing area. The cooling air passes around the outside of the rear-bearing housing and also through passages in the shaft to provide cooling of the hollow rear section (8) of the compressor-turbine shaft assembly. This airflow is induced by tubular passages, located at the back of the turbine rotor, which act as an ejector-type centrifugal fan. One of the streamlined struts contains oil passages (9) carrying oil to and from the rear bearing. It also has an air vent for balancing air pressures at the rear bearing.

The forward end of the compressor-turbine

shaft assembly is hollow, thus providing a passage through which the main fuel supply for the engine is fed. The fuel flows through the fuel tube to the **fuel distributor** (10) at the **combustion chamber** (11). The centrifugal effect of the rotating distributor sprays the fuel into the combustion chamber.

The **turbine housing** (12) is the structural center of the engine. The annular combustion-chamber structure is installed inside the turbine housing over the center portion of the compressor-turbine shaft assembly. The hot gases from the combustion chamber are directed into the **turbine inlet nozzle** (13), through the **turbine rotor** (14), and into the exhaust diffuser. The front of the turbine housing is enclosed by the **compressor housing** (3), installed over the **compressor rotor** (15) and the **inducer** (16). The compressor housing carries the front bearing and the **gear train** (17) which drives the various accessories at the intake end. The starter-generator and its drive are also mounted on the compressor housing.

The **control island** (18) carries the **oil pump** (19), the **governor** (20), the **fuel pump** (21), and the **fuel-flow altitude compensator** (22). The fuel and oil pumps are gear-type units. The governor is a plunger-type valve actuated by external flyweights. The complete island is enclosed in a streamlined cover.

Fuel system

A schematic diagram of the fuel system for the J69 engine is shown in Fig. 13 · 6. Initial fuel pressure is provided by the boost pump, which draws the fuel from the tank and pumps it through the **main fuel-cutoff valve** to the **main fuel filter.** The filter is provided with a bypass so that fuel may continue to flow if the filter becomes clogged. From the filter, fuel is directed to the main fuel pump and thence to the **metering valve.** The position of the metering valve is

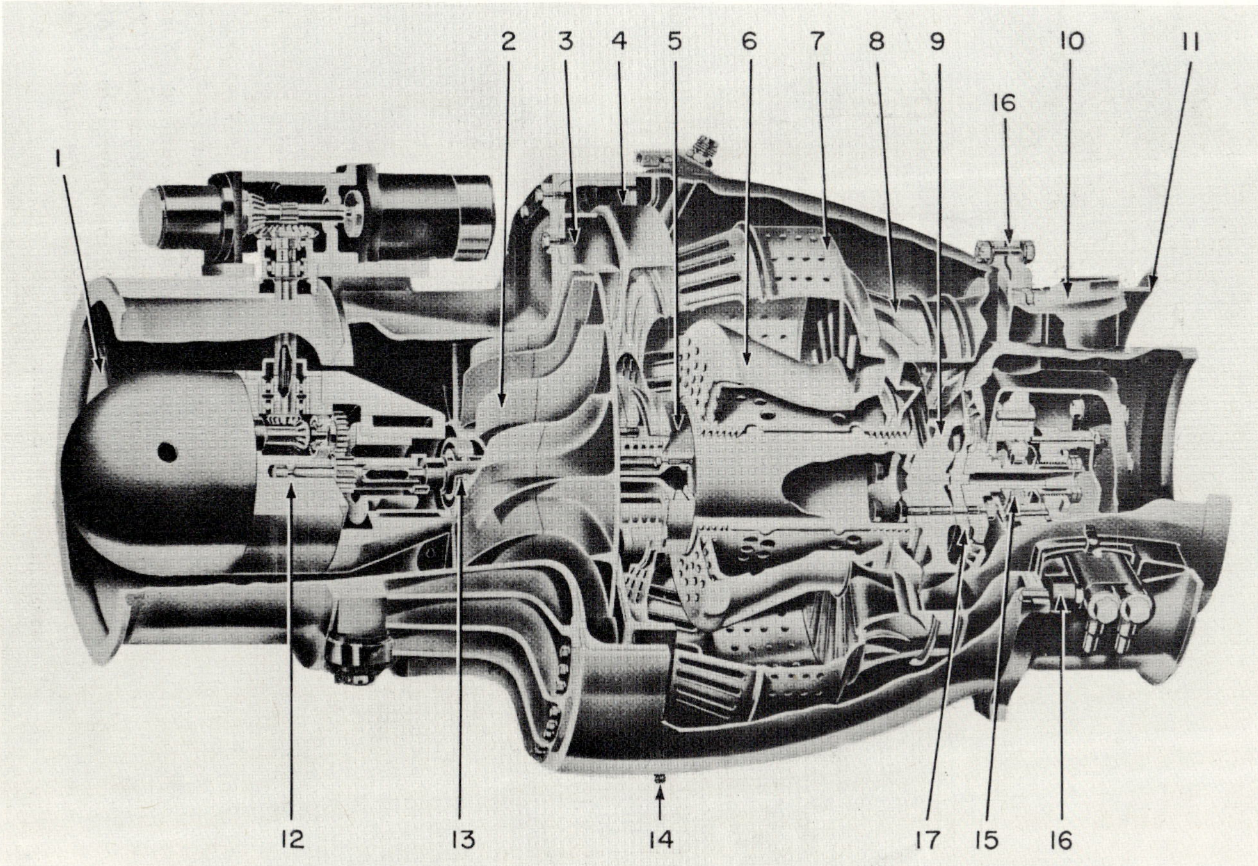

Figure **13 · 4** *Principal parts of a J69 engine: (1) air inlet; (2) compressor rotor; (3) radial diffuser; (4) axial diffuser; (5) fuel injector; (6) primary air admission shell; (7) outer combustion shell; (8) turbine inlet nozzle; (9) turbine wheel; (10) streamlined strut; (11) exhaust diffuser; (12) control island; (13) front turbine shaft bearing; (14) fuel drain; (15) rear turbine shaft bearing; (16) link struts; (17) balancing piston.*

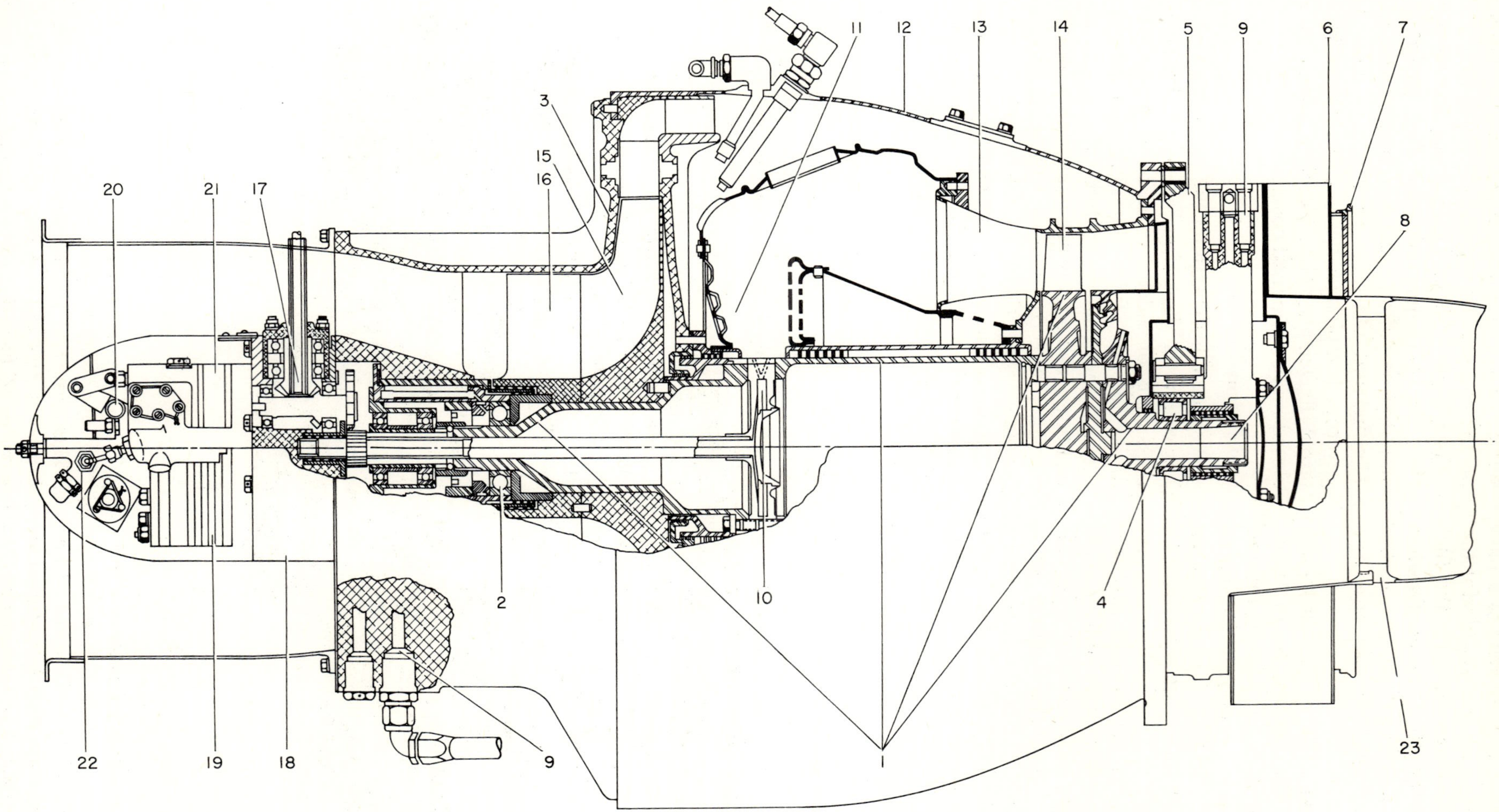

Figure **13·5** *Construction details of the J69 engine: (1) compressor-turbine shaft assembly; (2) front ball bearing; (3) compressor housing; (4) rear roller bearing; (5) bearing support; (6) streamlined strut; (7) exhaust diffuser; (8) rear shaft; (9) oil passages; (10) fuel distributor; (11) combustion chamber; (12) turbine housing; (13) turbine inlet nozzle; (14) turbine rotor; (15) compressor rotor; (16) inducer rotor; (17) gear train; (18) accessory base; (19) oil pump; (20) governor; (21) fuel pump; (22) altitude compensator; (23) inner tail cone.*

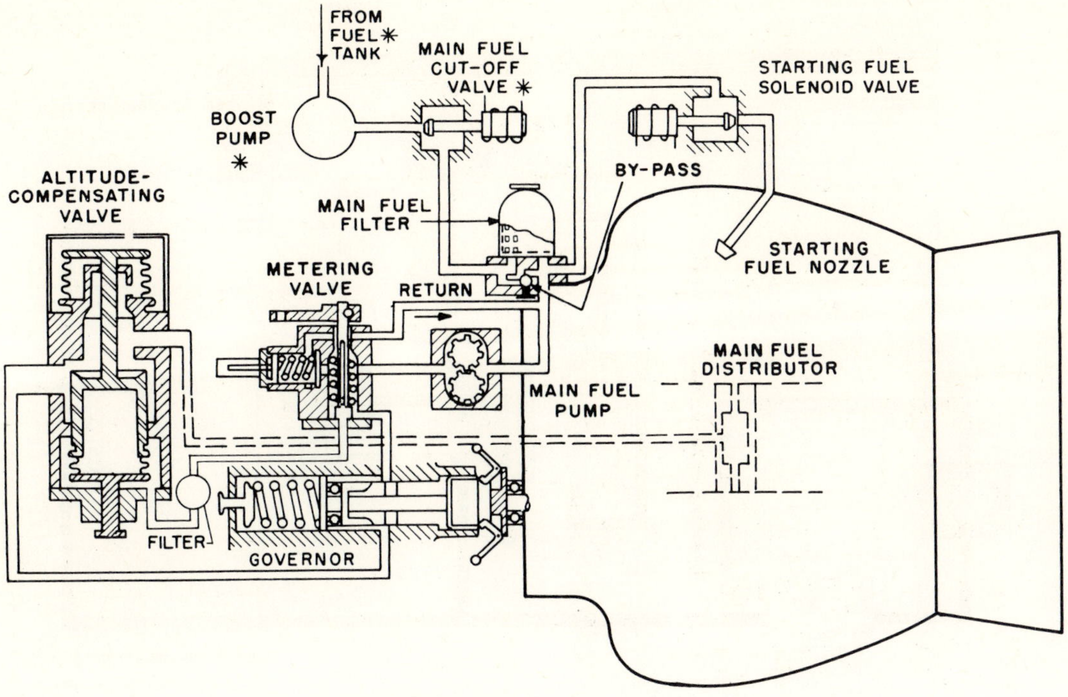

Figure 13·6 Fuel system.

determined by the throttle setting, through the fuel valve control shaft. The governor valve spring tension is also established through the same linkage from the throttle. Thus the metering valve and the governor setting are both determined by the throttle setting.

The **governor** acts to limit engine speeds above 17,000 rpm. The governor spring balances the effect of the governor flyweights in accordance with throttle setting. As the throttle is advanced, the governor spring tension is increased, thus permitting the engine to attain a higher rpm before the flyweights act to reduce fuel flow. Fuel from the governor flows through the **altitude-compensating valve** before being delivered to the fuel tube in the turbine shaft. The purpose of the altitude-compensating valve is to reduce fuel flow as altitude increases and as barometric pressure changes.

Regulated fuel enters the fuel tube and flows rearward to the fuel distributor, where it is injected into the combustion chamber by centrifugal force and pump pressure.

As was mentioned previously, starting fuel is supplied to the starting fuel nozzles in order to make sure of a good start. Pressure for this fuel flow is provided by the fuel boost pump.

Lubrication system

The lubrication system for the J69 engine is shown schematically in Fig. 13·7. Oil from the engine oil tank enters the engine at the connection beside the oil filter to feed the main oil pump. The pump feeds oil under pressure back to the oil filter, which contains both a pressure-regulating valve and a valve to bypass the filter element if it should become clogged.

From the oil filter, oil is directed through an external hose to the rear bearing, then back by another external hose to one scavenge section of the oil pump. Another external hose feeds oil to the top of the starter-generator drive case and gear train. The scavenge pump for this area is built into the gear train. The third path from the main oil filter leads oil through passages in the compressor housing to the front ball bearings and gearing. Two scavenge pumps serve this area, one for normal and the other for inverted flight. All scavenge pumps exhaust into the main oil return line to the tank. Air vent lines connect the front bearing section and the starter drive case to the top of the oil tank.

The lubricating oil used by the J69 engine conforms to Specification MIL-O-6081A, Grade 1010. This oil is applied to the system by the oil pump at a pressure of approximately 60 psi.

Electrical system

The electrical system for the J69 engine is shown schematically in Fig. 13·8. The principal purposes of the electrical system are starting the

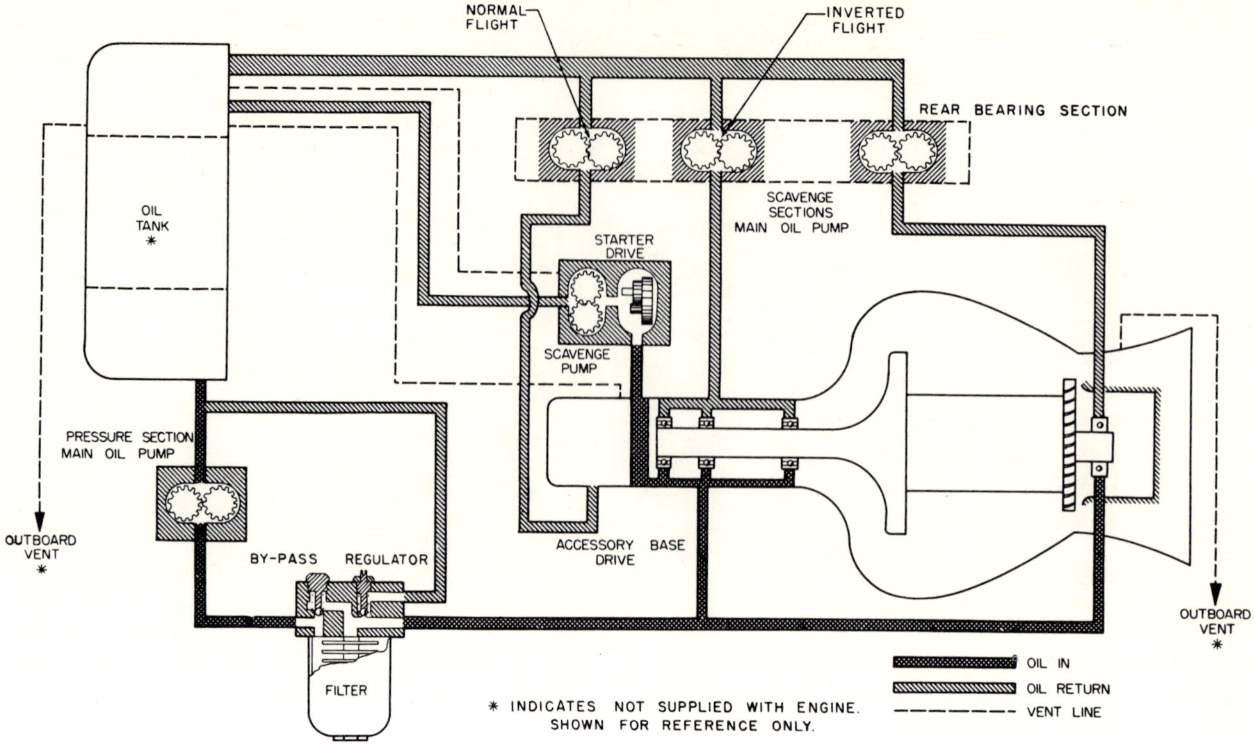

Figure 13·7 Lubrication system.

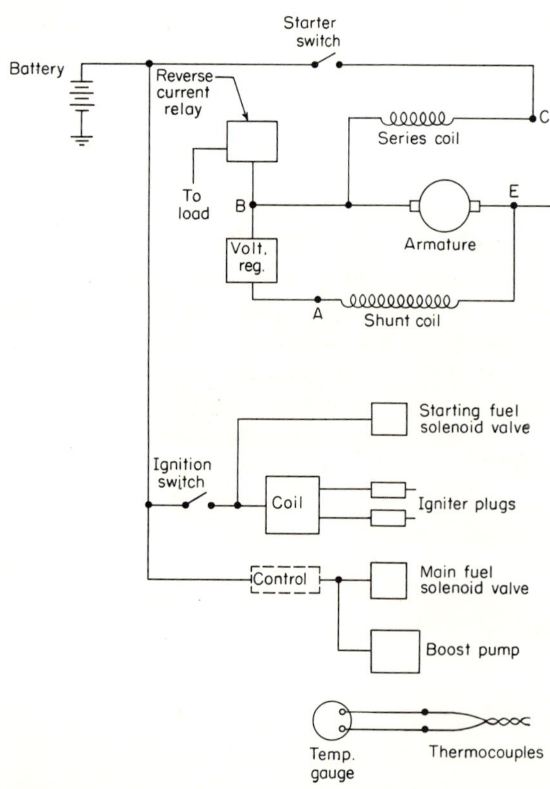

Figure 13·8 Electrical system.

engine, ignition, power generation, and fuel control and boost.

Electric power generation is accomplished by a standard generator arrangement including a voltage regulator, a reverse-current relay, and the generator portion of the starter-generator. This is typical of d-c power systems.

The generator is made a starter by passing the starting current through a series field to provide a strong starting torque. This is accomplished when the starter switch is closed.

Ignition is provided through a high-energy ignition coil to two igniter plugs located on opposite sides of the combustion chamber. The starting fuel solenoid valve is energized when the ignition switch is closed.

The main fuel solenoid valve and the fuel booster pump are continuously energized when the engine is in operation.

A separate electric circuit carries current from the thermocouples to the temperature gauge. The thermocouples are installed in the exhaust diffuser area.

To start the J69 engine, the following steps are performed:

1. Open the fuel shutoff valve.
2. Turn on the starter switch.

3. After 10 sec (approximately 2,000 rpm), turn the ignition switch on.
4. Advance the fuel-control lever (throttle) to IDLE position.

The maximum allowable exhaust temperature during acceleration is 1300°F for 3 sec. Exhaust temperature must in no case exceed 1350°F.

After the engine is started and has accelerated to approximately 5,000 rpm, the starter and ignition switches are turned off.

THE GENERAL ELECTRIC CJ610 TURBOJET ENGINE

Description and specifications

The General Electric CJ610 engine is the commercial version of the military J85 turbojet. This engine is equipped with an 8-stage axial-flow compressor, an annular combustor with 12 fuel nozzles, and a 2-stage turbine. The engine produces approximately 8 lb of thrust for each pound of engine weight, making it one of the most efficient engines of its type with respect to weight. The dimensions, weight, and performance of the engine are as follows:

Weight	381 lb
Length	40.5 in.
Diameter	17.7 in.
Thrust	2850 lb
Sfc	0.99 lb/lbt/hr
Thrust/weight ratio	7.5 lbt/lb

The basic configuration of the CJ610 (J85) turbojet engine is shown in Fig. 13·9. This cutaway illustration shows the 8-stage axial compressor, the combustion section, and the 2-stage turbine. The engine is equipped with variable-inlet guide vanes and controlled interstage bleed. All engine casings are steel.

Front-frame section

The front-frame section of the CJ610 engine serves the same purpose as that of other turbojet engines. It houses the inlet guide vanes and the No. 1 roller bearing and serves as the principal forward structure.

The principal component of the front-frame section is the front frame itself, which is basically a fabrication of permanently joined sheet metal parts. It consists of an inner and outer shell joined by 15 hollow struts, the ends of which are brazed to the two shells. Structural reinforcement is provided by doublers which are brazed to both ends of the struts and to the inner and outer shells. A drawing of the front-frame section (Fig. 13·10) shows the movable hinged portion of the inlet guide vanes by which the vanes are made variable in angle with respect to the axis of the engine.

The No. 1 bearing housing is brazed to the aft end of the inner shell and reinforced at the forward end by three stiffening ribs which are brazed to the bearing housing and the inner shell. A sump cover is bolted to the forward face of the bearing housing preventing oil leakage

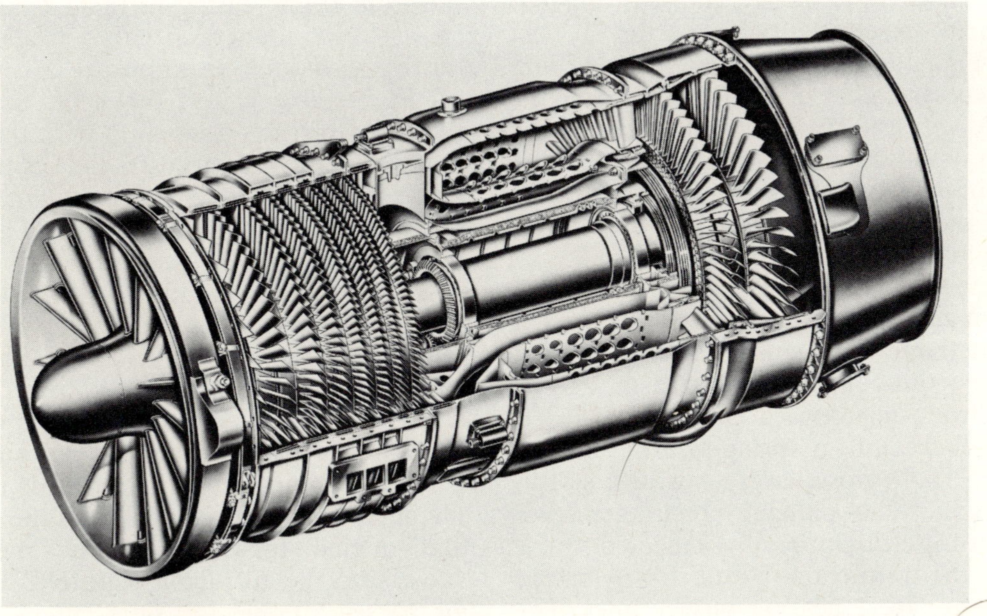

Figure 13·9 General Electric CJ610 turbojet engine. (General Electric Co.)

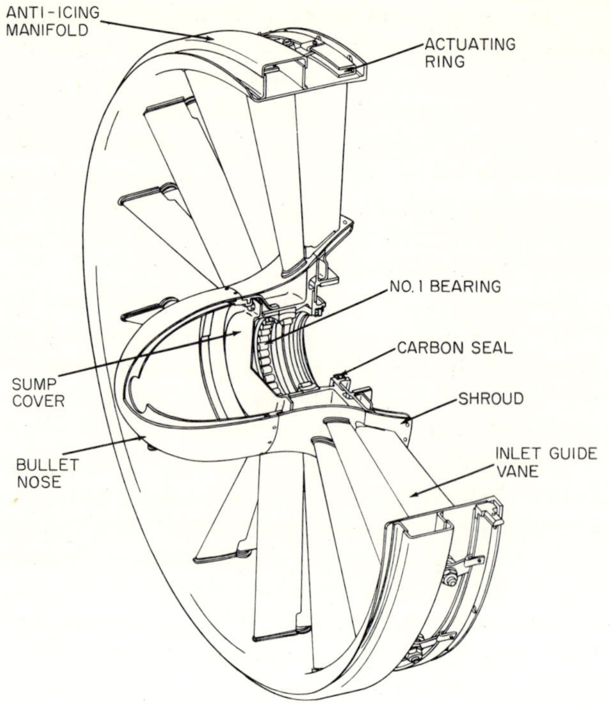

Figure ***13 · 10*** *Front-frame section. (General Electric Co.)*

from the bearing area and also serving as a retainer for the outer ring, outer race, and rollers of the No. 1 bearing. The bearing is a spherical, self-aligning roller bearing, designed to absorb any misalignment of the forward compressor shaft. Rotation of the outer race in the bearing housing is prevented by a pin located in the housing which engages with a slot in the outer ring of the bearing.

The inlet-guide-vane shroud and seal ring and the No. 1 carbon seal are bolted to the aft face of the bearing housing. The shroud and seal ring is a one-piece assembly providing shrouding for the inboard ends of the inlet guide vanes and also incorporating the stationary part of the forward compressor air seal. It is properly aligned by means of a pin which is located on the aft face of the bearing housing.

A double-skinned dome or bulletnose is mounted on the forward end of the inner shell of the front frame by means of an interlocking quick connect-disconnect mechanism. It is assembled from the forward end by rotating it circumferentially until a spring-loaded plunger, built into the bulletnose, engages with a slot in the inner shell. The plunger prevents further rotation until it is intentionally deflected.

Three of the front-frame struts are oversized and house the No. 1 bearing service lines. The 10 o'clock strut houses the bearing oil supply line; the 2 o'clock strut houses a sump vent tube; and the 6 o'clock strut houses an oil scavenge tube. All these service tubes are brazed to the bearing housing and terminate in external fittings which are brazed to the outer casing. Lubrication for the bearing is supplied through an oil nozzle bolted to the inside of the bearing housing between the bearing and the seal. The spherical surfaces of the No. 1 bearing are lubricated through two holes 180° apart in the outer ring and grooves on the inside of the outer ring leading away from these holes. Oil reaches the holes in the outer ring through a small, shallow groove which is machined circumferentially around the inside of the bearing housing in the same plane as the holes.

Fifteen variable-inlet guide vanes are positioned directly downstream of, but not touching, the front-frame struts. Miplon bushings are fitted to the inner stems of the vanes which are supported in slots machined into the aft face of the bearing housing and inner shell, and they are held in place by the shroud and seal ring. The outer stems of the vanes extend through bosses which are brazed to the outer shell of the front frame. Each vane stem has a square cross section which mates with a square hole in one end of an actuator lever. The levers are fastened to the vane stems by spring washers, castellated nuts, and cotter pins. The other ends of the levers are attached by means of pins and clips to a two-piece actuator ring. Circumferential movement of the actuator ring causes a synchronized opening or closing of the inlet guide vanes.

A series of axial scribe lines on the outside of the front-frame casing at the 1 o'clock position is used in conjunction with a single scribe line on the actuator ring to ensure a correct setting of the variable vanes. When the scribe line on the actuator ring is lined up with the scribe line marked 0 on the front-frame casing, the inlet guide vanes are at a full open or zero degree position. The inlet guide vanes will actuate from this position to a 35° closed position.

An anti-icing manifold is brazed externally and circumferentially around the front-frame outer casing over the ends of the 15 hollow struts. Anti-icing airflow is controlled by an electrically operated valve which is mounted on the aft side of the manifold at the 10 o'clock position. Compressor discharge air is bled from the customer bleed manifold on the main frame and ducted forward, through the anti-icing valve, into the anti-icing manifold. Some of this air passes aft through slots cut in each of the IGV supporting bosses, through the hollowed-out vane stems, and into the hollow IGVs. Inside the IGVs the air flows radially inward, around baffles, then outward through holes in the outboard ends of the vanes and into the airstream. The remainder of the air passes through the struts into the inner hub, thus anti-icing the struts. A portion of this air is directed forward through a hole at the forward end of the inner skin of the dome, then aft between the two skins and out through radial holes at the aft end of the outer skin. The rest of the air in the inner hub passes aft through internal passages in the bearing housing and into the double-skinned IGV shroud portion of the shroud and seal ring. This air anti-ices the shroud ring and then passes out into the airstream through radial holes at the aft end of the shroud.

The complete front frame, the IGVs, and the shroud and seal ring are coated with corrosion-resistant epoxy paint.

Compressor section

The compressor section of the CJ610 engine is shown in the drawing of Fig. 13 · 11. It consists of two major units, the eight-stage **compressor-rotor assembly** and the **compressor-stator assembly.**

The compressor-rotor assembly consists of a forward shaft, eight stages of disks and blades, seven spacers located between the stages, a main drive shaft, several seals, and mounting hardware. The forward shaft, first-stage disk, and first-stage spacer are machined from one piece. A bearing journal on the shaft supports the No. 1 bearing inner race and carbon seal runner, which are held on the shaft by a snap ring. The stationary part of the No. 1 carbon seal and the rollers and outer race of the No. 1 bearing are mounted in the front frame. The serrated rotating half of the compressor forward labyrinth air seal is riveted to the forward shaft.

The first- and second-stage rotor blades are coated with corrosion-resistant epoxy paint. At assembly the first-stage rotor blades are moment-weighed, and the second- through fourth-stage blades are pan-weighed. The blades are then mapped and assembled to their respective disks with equal moment or pan weights opposite each other. This procedure is to assure correct balance of the completed assembly.

The first-stage blades are attached to the first-stage disk by pins and snap rings. The remaining stages of blades are dovetailed to the disks and are held in the rim of the disks by the

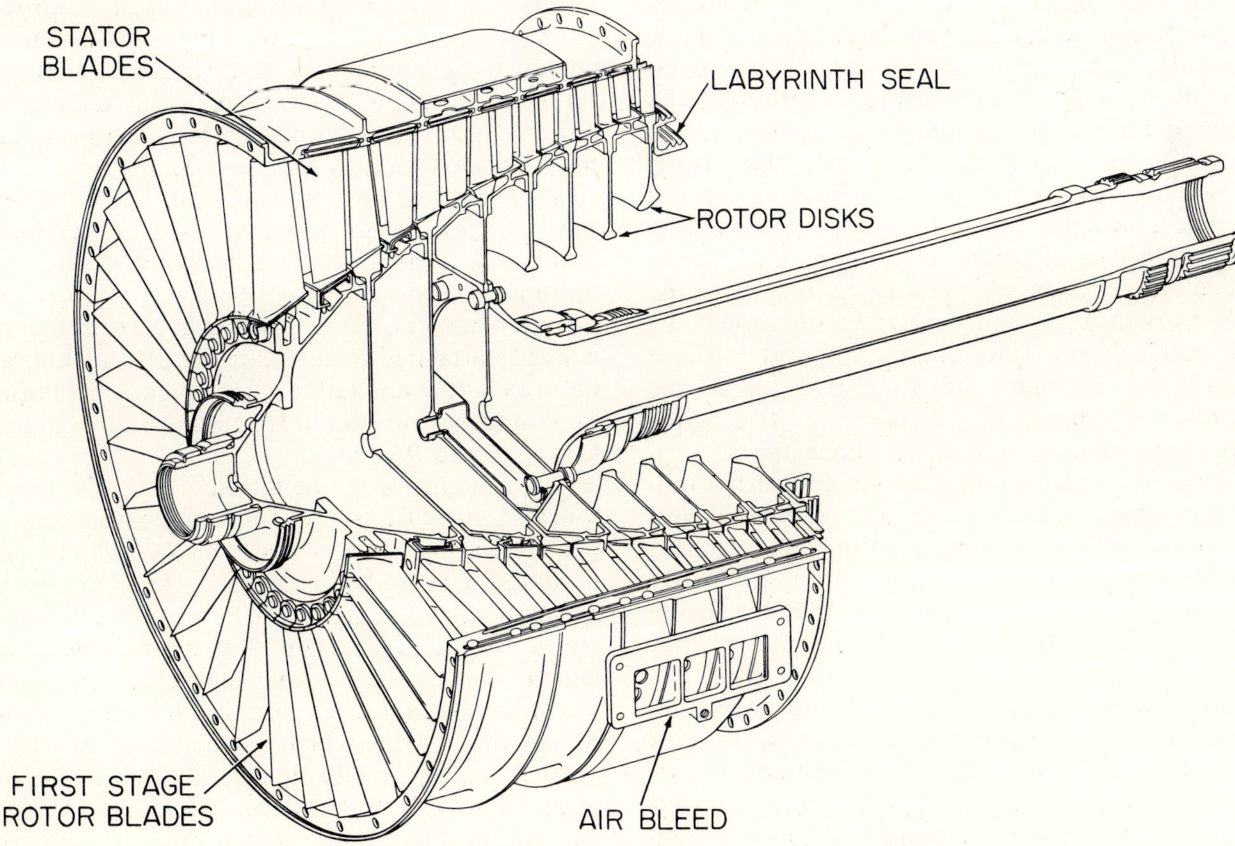

Figure 13 · 11 Compressor section with drive shaft. (General Electric Co.)

spacers. Each blade and disk assembly is balanced individually, and the rotor is balanced periodically during assembly. Unbalance is corrected by interchanging blades and by assembling balance washers. Damaged first-stage blades may be replaced with new blades of equal moment weight without unstacking the rotor or rebalancing. The rotating half of the eighth-stage dual labyrinth seal is bolted to the aft face of the eighth-stage disk.

The compressor drive shaft is bolted to the aft face of the fourth-stage disk and supports the No. 2 bearing inner race and carbon seal runner, which are held on the shaft by a tabwasher and a locknut. Two external splines on the aft end of the shaft mate with internal splines on the power takeoff driver bevel gear and the turbine forward shaft. Labyrinth teeth, which are machined into the first- and second-stage spacers, mate with the shroud rings on the first- and second-stage stator vanes and form a seal preventing recirculation of air inside the compressor. Mounting hardware consists of locknuts and close-tolerance bolts and studs.

The compressor-stator assembly is located between the front frame and the mainframe. The casing itself is split and flanged along the horizontal center line so that it is possible to remove the upper or lower half for rotor inspection. Close-tolerance horizontal flange bolts provide proper alignment of the casing halves. Reinforcement of the casing is provided by several stiffening ribs which are machined externally and circumferentially into the casing. There are three sets of air-bleed holes in the casing skin between four of the stiffening ribs, radially outboard of stator-vane stages 3, 4, and 5. Sheetmetal strips are welded to the stiffening ribs in such a way as to provide a separate bleed air manifold for stator stages 3, 4, and 5. These manifolds terminate in four pads near the horizontal split line. Each pad has three openings, one for each of the three stages.

Individual stator-vane airfoils are brazed into outer support bands which are segmented into 12 equal sections per stage. Seven stages of these stator-vane segments slide circumferentially into tracks which are machined into the inside of the casing. The first- and second-stage vanes are brazed to inner bands or shrouds which are also segmented. There is a hole between each pair of stator vanes in the outer support bands for stages 3, 4, and 5, and the holes make it possible to bleed air from the inside of the compressor into the manifold. Rotation of the vane segments is prevented by two locking keys which fit in slots machined in the stator casing tracks at the 3 o'clock position on the upper half and at the 9 o'clock position on the lower half. The locking keys mate with notches cut in the outer bands of two vane segment stops in each of stages 1 through 7. The vane segments are assembled in the compressor stator halves with the vane segment stops at the 3 and 9 o'clock positions. Assembly of the two casing halves holds the locking keys in place.

At assembly, shroud rings are riveted to the inner shrouds of stages 1 and 2. These rings constitute the static halves of two interstage labyrinth seals and are split at the horizontal center line to permit removal of either casing half.

Variable-geometry components

The variable-geometry components consist of two fuel-driven actuators, two bellcrank links, synchronizing cables, variable-inlet guide vanes and actuator rings, two compressor bleed valves, and main fuel control. The purpose of these components is to adjust airflow and pressure through the compressor, thus preventing surge and stall and providing for the most effective operation of the compressor. The variable-geometry components are shown in Fig. 13 · 12. The two actuators are piston-operated and are mounted to brackets which are, in turn, mounted on the compressor casing. High-pressure fuel, which is scheduled by the variable-geometry servo in the main fuel control, is introduced to either side of a piston inside the actuator, causing a shaft which is attached to the piston to either extend or retract. This shaft is connected, through linkage, to one arm on a master bellcrank link. There are two bellcrank links, one mounted and pivoting about a clevis pin on each actuator bracket. Each link has five arms, one of which is attached to the actuator piston shaft as described above. Another arm on each bellcrank in connected to the inlet-guide-vane actuator ring at 3 and 9 o'clock.

The amount of air being bled from the third, fourth, and fifth stages of the compressor is controlled by two bleed valves which are mounted at 3 and 9 o'clock on the compressor casing. Each valve has six openings, two for each stage. The valve is operated by a single pushrod which moves two cam plates in opposite directions inside the valve. Each cam plate moves three gates which slide across the valve ports opening or closing the bleed valve. The pushrod travels forward or aft in a direction parallel to the engine center line. A rod end which threads onto each pushrod is connected

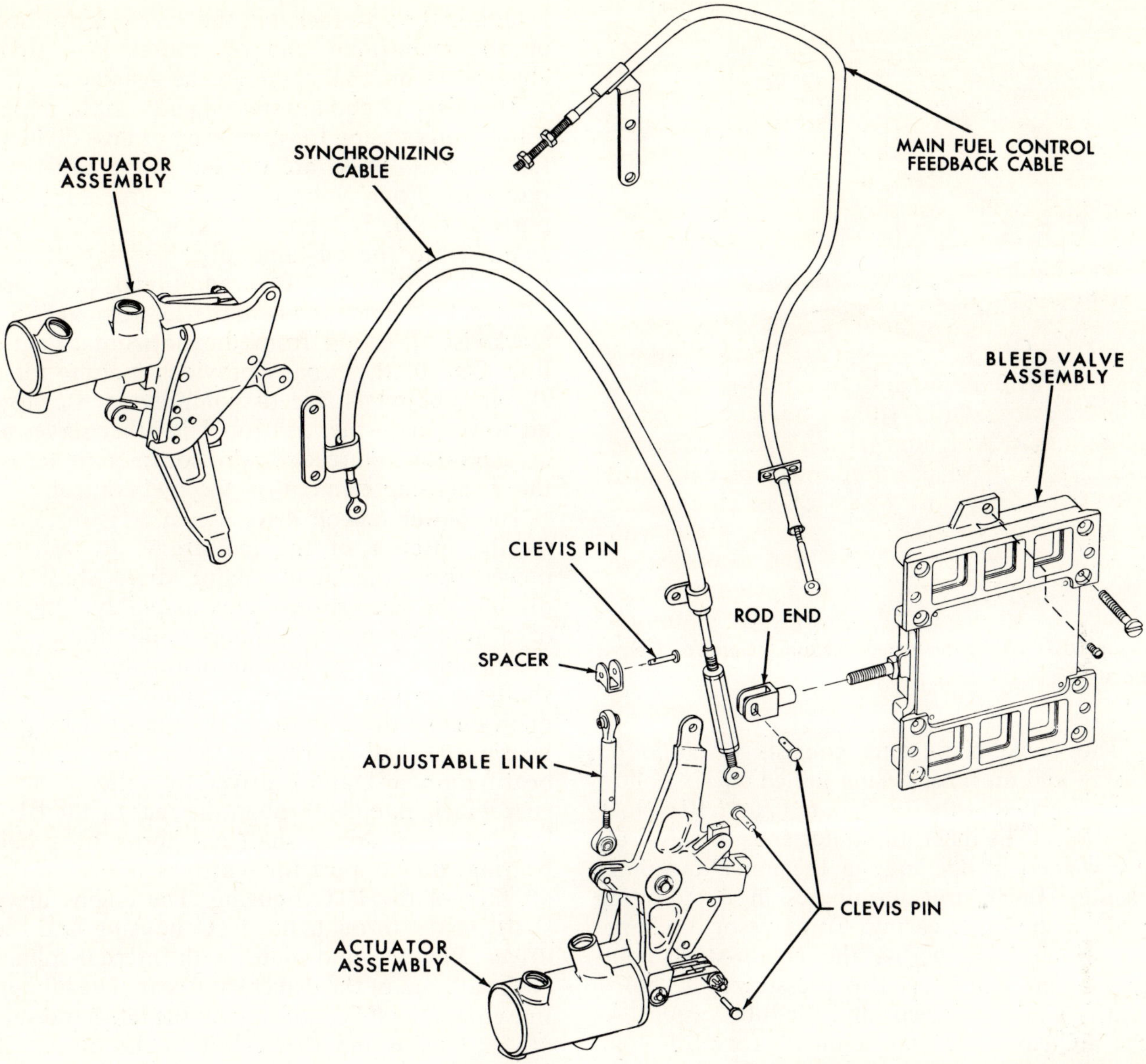

Figure **13 · 12** *Variable geometry components. (General Electric Co.)*

by means of a clevis pin to a third arm on the bellcrank link.

A two-piece (upper and lower) synchronizing cable is pinned to another arm on the bellcrank links. This cable is housed in a two-piece conduit which is attached through brackets to the compressor casing. The purpose of the synchronizing cable is to ensure simultaneous movement of the two bellcrank links. The fifth arm on the left-hand bellcrank is not used. On the right-hand side of the engine the fifth arm on the link is attached to a teleflex feedback cable which operates a lever on the main fuel control.

When the inlet guide vanes are at a full open or zero degrees position, the bleed valves should be fully closed. It is necessary at assembly, therefore, to adjust the variable-geometry components to ensure proper synchronization of the system.

Mainframe section

The **mainframe section,** shown in Fig. 13 · 13, serves as the main structural component of the engine. It consists of a mainframe casing which houses the power takeoff drive assembly and provides a mount for the twelve-flow divider-fuel nozzles, the fuel manifold, two eighth-stage leakage air poppet valves, and the transfer gearbox assembly. The eighth-stage compressor stator vanes and exit guide vanes are mounted on 12 segments which slide into a track machined into the forward end of the mainframe.

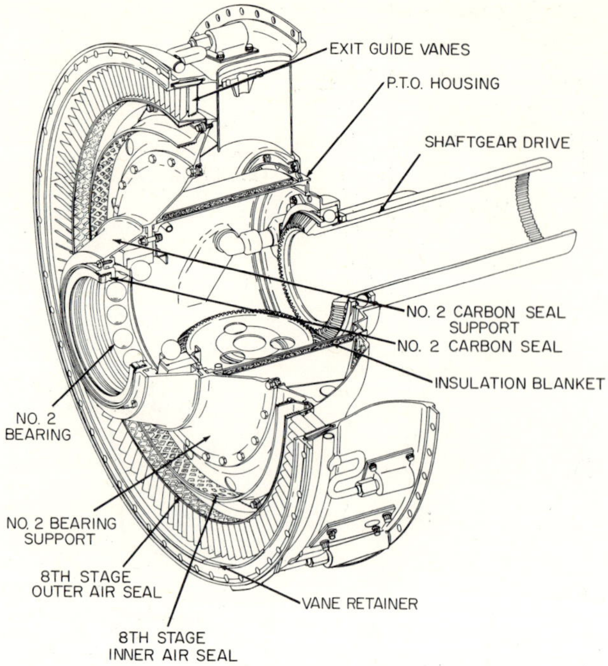

Figure 13·13 Mainframe section. (General Electric Co.)

The basic mainframe consists of an outer casing and an inner casing joined by six hollow struts located at the 12, 2, 4, 6, 8, and 10 o'clock positions. The inner and outer ends of the struts are welded in openings in the inner and outer casings. Each strut terminates in a mounting pad on the outer casing. Compressor discharge air which leaks across the eighth-stage compressor air seal is used to pressurize the No. 1 and No. 2 carbon seals and the lube system. As the air leaves the seal, it moves forward along the No. 2 bearing housing and seal support and pressurizes the No. 2 carbon seal. Seal-leakage air also passes through holes, located between the compressor drive-shaft mounting bolts, into the rotor. The air moves forward through the rotor and out through holes in the forward compressor shaft and pressurizes the No. 1 carbon seal. The lube system itself is pressurized by this same air which leaks across the carbon seals. Poppet valves, mounted on the 2 and 10 o'clock mainframe strut pads, control the pressure of the seal leakage air. The 12 o'clock strut is not used during engine operation and is therefore blanked off. The 8 o'clock strut and pad provide mounting and passage for the lube-in line and the power takeoff (PTO) oil scavenge line. The 4 o'clock strut houses the No. 3 bearing scavenge line and the main sump vent line. The 6 o'clock strut houses the radial drive shaft and its shield. The transfer gearbox is mounted on a bracket at the 6 o'clock position on the mainframe, and the radial drive shaft engages an internal spline in the gearbox.

There are 12 equally spaced pads on the mainframe outer casing for mounting of flow divider-fuel nozzles. These are located at the half-hour positions. Four compressor-discharge-air bleed ports at the 1, 5, 7, and 11 o'clock positions are available to the customer for bleeding air for cabin conditioning. Two additional pads are located 180° apart in positions 23° removed in a clockwise direction from the horizontal center line. One of these pads provides mounts for a P_3 air tube which bleeds compressor discharge air to the main fuel control. The other serves as an aspirator to direct compressor inlet air across the T_2 sensing element on the fuel control.

The **power takeoff drive assembly** is installed on the interior of the mainframe to transfer power from the main engine drive shaft for driving the accessories. This assembly consists of a housing which supports and houses all components. At the bottom of this housing is a shaftgear whose axis is perpendicular to the main engine drive shaft. It is supported by two bearings, a ball bearing outboard and a roller bearing inboard, and is driven by a driver shaftgear which extends through the rear of the PTO housing. The driver shaftgear turns in a ball bearing, the support for which is bolted to the aft face of the PTO housing. The engine drive shaft passes through the PTO housing and the driver shaftgear and mates with internal splines at the aft end of the driver shaftgear. The lube-in line enters the PTO housing on the left-hand side of the engine and carries oil to the oil-nozzle assemblies, which spray the No. 2 bearing, the No. 3 bearing, the bearing which supports the driver shaftgear, and the point of contact between the two shaftgears. Another connection on the left side provides passage for the PTO oil scavenge line. Connections on the right side of the PTO housing provide for the passage of the No. 3 bearing oil scavenge and the PTO vent line. An insulation blanket covers the exterior of the PTO assembly to shield it from excessive heat. The PTO housing is secured at the forward end by eight bolts which pass through the No. 2 bearing support to the carbon seal support. An alignment pin properly positions the PTO housing on the No. 2 bearing support.

The bearing support for the No. 2 or main engine thrust bearing is aligned and bolted to the forward inner mainframe flange and supports the outer race of the No. 2 bearing, the PTO housing, and the No. 2 carbon seal support. The bolts for the bearing outer race retainer and the PTO housing forward flange pass through the bearing support and engage in channel nuts in the carbon seal support. The No. 2 carbon seal, which rides on a seal runner on the compressor drive shaft, is mounted on the forward end of the seal support. An oil nozzle is located between the seal and the bearing. The bearing itself is a ball bearing with a split inner race.

Engine-accessories section

The **transfer gearbox** is mounted on studs at the 6 o'clock position on the mainframe and contains a bevel shaftgear which is driven by the radial drive shaft at axis A. This shaft has an integral bevel gear which engages with another bevel gear mounted on the driven shaft at axis B. Axis B is parallel to the engine center line and perpendicular to axis A. The driven shaft at axis B is splined internally at both the forward and aft end. The aft pad on the transfer gearbox is available to the customer for mounting an aircraft accessory. It rotates at 7,811 rpm in a clockwise direction facing the pad. A transfer shaft is splined to the forward end of the axis B shaft and drives the **accessory drive gearbox,** as shown in Figs. 13 · 14 and 13 · 15.

The accessory drive gearbox is mounted on a bracket below the engine front frame. It has a total of six pads, three forward and three aft, all

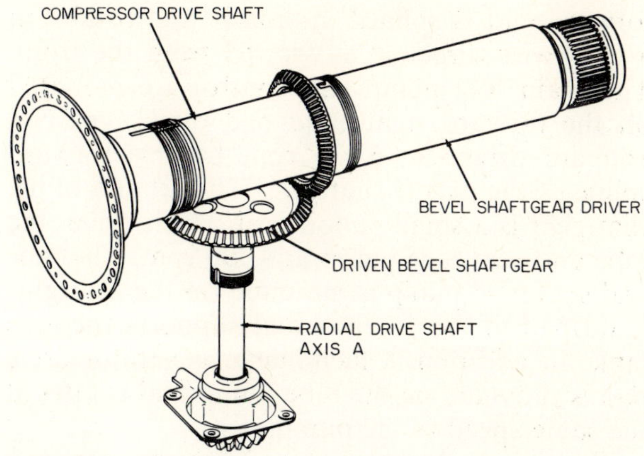

*Figure **13 · 14** Power extraction system. (General Electric Co.)*

of which are serviced by four shaftgears identified by location at axes B, C, D, and F. The axis B shaftgear is located at the center of the gearbox and is driven in a clockwise direction at 7,811 rpm by the transfer shaft at the aft center pad. An overspeed governor is mounted on the forward center pad and driven at the same speed by the axis B shaftgear.

The axis D shaftgear is driven by the axis B shaftgear at a maximum of 6,075 rpm in a clockwise direction. Both ends of this shaftgear are available for driving engine accessories. The

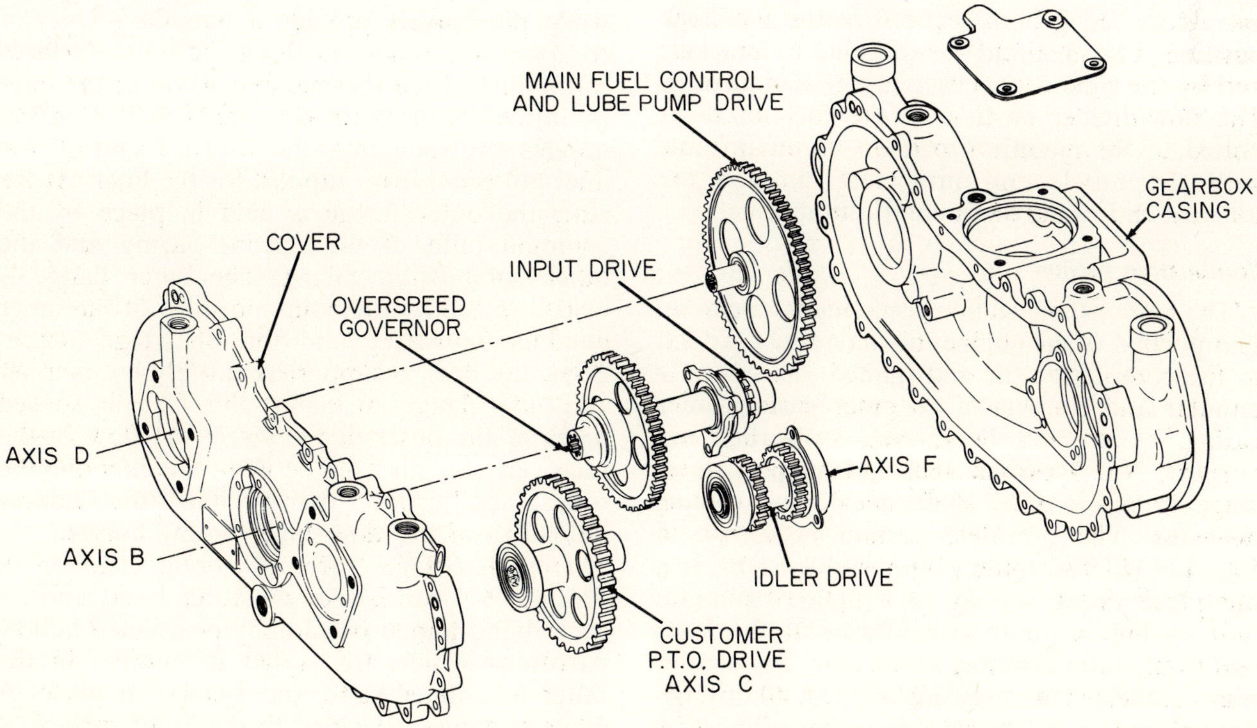

*Figure **13 · 15** Gearbox assembly.*

forward end is splined internally and rotates in a clockwise direction as viewed from the front. The main fuel pump and control are mounted on the forward right-hand pad of the gearbox and are driven at 6,075 rpm by this forward spline of the axis D shaftgear. The aft end of the shaftgear is a small pinion gear which drives the lube and scavenge pump at 4,190 rpm. The lube and scavenge pump is mounted on the aft right-hand pad of the gearbox and supports the lube tank. In addition, a tachometer-generator drive pad is provided on the lube pump and rotates at the same speed as the pump.

The axis F shaftgear is an idler gear which is driven at 15,947 rpm in a counterclockwise direction by the axis B shaftgear. This idler gear is located to the left of the axis B gear (aft looking forward) and drives the axis C shaftgear through a gear machined at the aft end of the idler gear shaft.

The axis C shaftgear is located at the extreme left of the gearbox and is driven, as described above, by the idler gear at axis F. It rotates at a maximum of 7,088 rpm in a counterclockwise direction as viewed from the rear and is available for customer usage at the forward and aft left-hand pads on the gearbox.

All shaftgear speeds in the foregoing description are at 100 per cent of engine speed or 16,500 rpm.

A two-piece fuel manifold is mounted around the mainframe with a fuel pressurizing and drain valve connecting both sections at the 6 o'clock position. The manifold is supported by brackets and by the rigid connections to the fuel nozzles. The flow-divider portion of the fuel nozzle is bolted to the mounting pad on the mainframe with the nozzle end projecting through the forward end of the main combustion liner.

Combustion section

The combustion-section components make up that portion of the engine in which fuel is added to the compressed air and ignited. This area is annular and consists of an **outer casing, inner casing, combustion liner, No. 3 bearing** and **support, No. 3 carbon seal** and **support, first-stage turbine nozzle, shaft shield,** and **cooling deflector.** The complete section is shown in Fig. 13 · 16. The outer combustion casing is a one-piece casing serving as a major structural unit by bolting to the mainframe at the front and to the turbine casing at the rear. Two igniter plugs are seated in the casing and extend through the top and lower left sides into the combustion liner. The annular inner combustion casing is

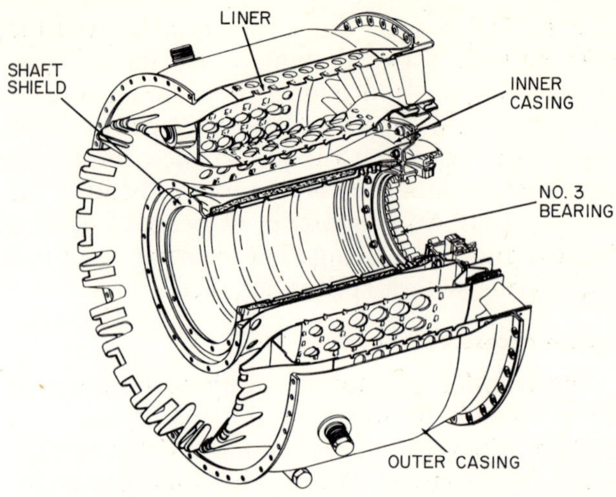

Figure 13 · 16 Combustion section.

also a one-piece fabrication. At the front, it bolts to the inner casing of the mainframe, and at the rear, to the No. 3 bearing support. Holes in the rear flange form passages for air entering the balance piston chamber.

The combustion liner consists of a cowl section, a dome section, an outer shell, an inner shell, and the outer and inner flanges. All of these components are made of high-temperature alloy sheet and are welded and riveted to form a one-piece fabrication. Air from the compressor section enters the liner through perforations designated as thimble holes and louvers. The thimble holes direct air into the burning area while the louvers provide a boundary layer of comparatively cool air along the inner surfaces of the liner. Free thermal expansion of the liner is allowed by the method of support. Twelve fuel nozzles protrude into the forward end of the liner but provide no support for the liner. At the rear, the outer flange is held in place by the common joint of the turbine casing and the outer combustion casing. The inner flange is held in place by the common joint of the inner combustion casing and the first-stage nozzle. Thus, the liner is supported entirely by its inner and outer flanges at the aft end. Equally spaced holes in the outer flange direct the flow of the relatively cool air from between the liner and the outer combustion casing into the hollow partitions of the **first-stage turbine nozzle.**

The first-stage turbine nozzle, shown in Fig. 13 · 17, consists of an outer band and an inner band joined by radially positioned hollow partitions which are seated in cutouts in the outer and inner band and welded in place. A support flange is welded to the inner surface of the inner band. The outer band has a flange with

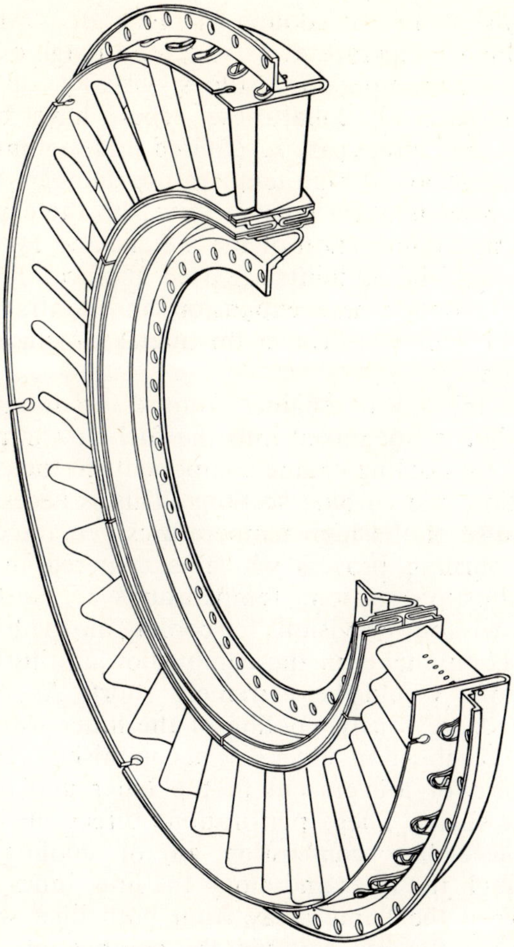

Figure 13 · 17 First-stage turbine nozzle.

bolt holes spaced to receive bolts that fasten the outer combustion casing, the nozzle, and the turbine stator casing together. A cooling deflector is mounted at assembly between the liner aft inner flange and the inner flange of the first-stage turbine nozzle. The deflector extends radially outward and then aft, overlapping the inner band of the nozzle. Holes in the liner and the deflector provide passages for secondary combustion air to cool the nozzle inner band. The inner support flange is bolted between the turbine stationary seal on the aft side and the No. 3 bearing support on the forward side. Slots at the rear of the inner band direct cooling air, which flows down through the partitions, against the shanks of the first-stage turbine blades. The No. 3 bearing support retains the outer race and rollers of the No. 3 bearing and supports the oil-nozzle assembly which directs lube oil to the No. 3 bearing. The No. 3 or rear seal support retains the No. 3 carbon seal and the inner turbine air seal. The No. 3 bearing support, the forward face of the first-stage turbine wheel, and the inner and outer turbine labyrinth seals form the **balance piston chamber.**

The forward flange of the shaft shield is bolted to the aft face of the PTO housing, and the aft flange bolts to the No. 3 bearing support. Its functions are to protect the compressor drive shaft and oil lines from heat generated in the combustion chamber and to connect the No. 2 bearing area to the No. 3 bearing area, forming one common sump for the two bearings.

Turbine section

The turbine section, shown in Fig. 13 · 18, consists of a **turbine-stator assembly** and a **turbine-rotor assembly.** The primary function of the turbine is to extract energy from the heated air to drive the compressor.

The turbine-stator assembly consists of several components, mounted on an annular outer casing that is split and flanged along its horizontal center line. The second-stage turbine nozzle and shrouds for both the first- and second-stage turbine wheels are mounted on tracks on the inside of the turbine casing. The split second-stage turbine nozzle consists of an outer support band and an outer band joined to an inner band by welded partitions. When the turbine stator casing is assembled to the engine, the slotted flange of the inner band retains and supports the stationary half of the turbine interstage seal. Rotation of the nozzle is prevented by two locating pins which lock the nozzle to the casing.

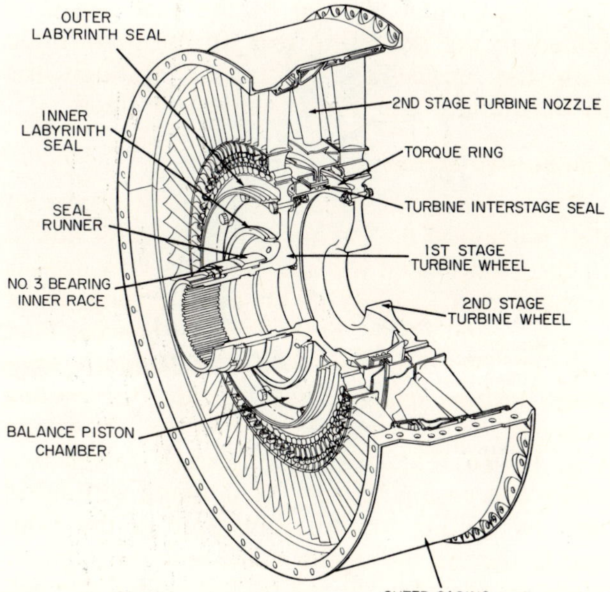

Figure 13 · 18 Turbine section.

The major components of the turbine rotor assembly are two **turbine-wheel assemblies,** a **turbine-interstage-seal assembly,** and a **torque-ring assembly.** These are shown in Fig. 13 · 19. The first-stage turbine wheel is integral with an internally splined shaft which mounts the entire turbine-rotor assembly on the compressor drive shaft. Turbine blades are inserted into dovetails in the circumference of the wheel and are held in place by locking strips. Baffles are inserted between turbine blade shanks to prevent cross flow of gases and to dampen vibration. The rotating half of the inner forward turbine air seal is riveted to the forward face of the first-stage turbine wheel, while the rotating half of the outer seal is bolted to the wheel by the same bolts that fasten the torque ring to the wheel. A locknut and tabwasher retain the No. 3 rubbing seal runner and the inner race of the No. 3 roller bearing. Separating the two turbine wheels is the torque ring, which has an integral baffle with holes that allow the passage of cooling air. The labyrinth portion of the turbine interstage seal is also machined into the torque ring.

Basic engine cooling

Because of the intense heat developed within a gas-turbine engine, by both compression and combustion, it is necessary to make elaborate

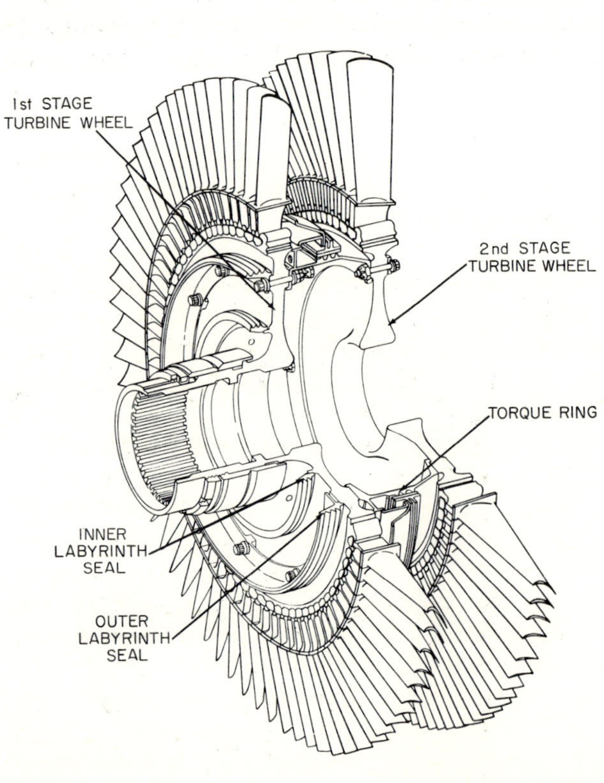

Figure 13 · 19 Turbine rotor assembly. (General Electric Co.)

arrangements for cooling in those areas where the high temperatures exist. Even though excellent high-temperature alloys, usually of the nickel-base or cobalt-base types, have been developed, great care is required in designing an engine to avoid high-temperature damage. Further consideration must be given to the expansion and contraction of the metals, and for this problem, sliding joints are used in place of rigid attachments where expansion and contraction would otherwise cause the metal to crack or buckle.

In the CJ610 engine, compressor discharge air that is not mixed into the fuel-air charge is used for cooling engine components in the combustion and turbine sections. This is necessary because of the high temperatures generated by the burning process which is centered in the combustion section. Temperatures are so high that without provisions for cooling air the life of the components in the combustion and turbine sections would be extremely short. As mentioned previously, cooling of the inner walls of the combustion liner is accomplished by a pattern of louvers cut in the inner and outer liner shells. These perforations direct the flow of **secondary combustion air** or cooling air through the liner and along the liner inner skin to keep the flame away from both liner walls. The air flowing between the combustion liner and the outer combustion casing cools the outer wall of the liner and then passes inward through a series of holes in the aft outer flange of the liner (refer to Fig. 13 · 16). A portion of this air then flows outward through radial holes in the first-stage turbine nozzle outer band and aft to cool the turbine casing and first-stage turbine shroud, and there it reenters the main gas stream. The remainder of the air flows through the hollow partitions in the first stage-turbine nozzle into the inner band of the nozzle and is expelled aftward, through slots in the inner band, against the shanks of the first-stage turbine buckets. Some of this air, as it passes through the partitions, is expelled through small holes drilled in the concave surface and cools the trailing edges of the partitions.

Secondary combustion air which flows between and cools the combustion liner and the inner combustion casing is directed into the balance-piston chamber. This chamber is located just forward of the first-stage turbine wheel. The air enters the chamber through a series of holes spaced between the bolt holes in the rear flange of the inner combustion casing. It pressurizes the chamber and acts on the forward face of the first-stage turbine wheel, utilizing the wheel as a piston. This action offsets the forward thrust moment of the compressor rotor, reducing the thrust loading on the No. 2 bearing. Balance-piston air leaves the chamber through twelve holes in the first-stage turbine wheel. These holes are located between the first-stage turbine-wheel torque-ring bolt holes. Six of the holes bleed air back toward the second-stage turbine wheel cooling the aft side of the first-stage wheel, the inner diameter of the torque ring, and the forward face of the second-stage wheel. This air passes through the inner diameter of the second-stage wheel, cooling the aft side of this wheel, and reenters the main gas stream. The other six holes in the first-stage turbine wheel direct air to a chamber formed by the aft side of the first-stage wheel and the baffle ring which is integral with the torque ring. A series of axial holes in the baffle ring allow the air to continue aft through the turbine interstage seal and onto the second-stage bucket shanks and turbine wheel outer rim. Some of the air that passes through the baffle ring moves outward past the baffle on the stationary half of the interstage seal and cools the aft side of the first-stage wheel outer rim and the bucket shanks.

Air also leaves the balance-piston chamber through the inner forward turbine labyrinth seal into a chamber at the root of the turbine forward shaft. At this point it pressurizes the No. 3 carbon seal and passes aft through holes in the first-stage turbine wheel. It combines with air between the two turbine wheels, aids in cooling this area, and then passes through the bore of the second-stage wheel and out into the main gas stream.

Lubrication system

The CJ610 turbojet engine has a pressurized closed-circuit lubrication system designed to furnish oil to parts requiring lubrication during engine operation. After oil has been supplied to these parts, it drains to the sump, from which it is recovered and recirculated throughout the system. All system components are engine-furnished and engine-mounted. Relatively few external oil lines are used as most oil lines are cored or drilled passages. The main components of the lubrication system are located beneath the front-frame section. They include a lubrication and scavenge-pump assembly, mounted on the aft right-hand pad of the gearbox, an oil reservoir mounted on the lubricating and scavenge pump, an oil cooler mounted on and against the right-hand side of the oil reservoir, and an oil filter mounted in the lubrication and scavenge pump. A schematic diagram of the lubrication system is shown in Fig. 13 · 20.

The **oil reservoir** is a sheet-steel fabrication and is mounted on the aft flange of the lubricating and scavenge pump. Included within the oil reservoir is an air chamber which is isolated from the remainder of the oil reservoir, a dwell chamber, and a system of vent tubes. The oil-reservoir filling port is located on the aft face, and the oil level is indicated by a dip stick graduated in pints to be added. A vent relief-valve boss, a remote-fill vent boss, and a combination check-valve boss are located on the reservoir. The lubricating and scavenge pump mounting flange, oil cooler mounting flange, and a check-valve boss are located on the forward face. The drain boss is located in the bottom of the reservoir. A remote-filling boss is located on the aft side of the oil reservoir. Beneath the reservoir is a deep well which forms a recess for the tachometer-generator unit, which is driven from the rear pad of the lubricating and scavenge-pump assembly.

The reservoir capacity is 4.0 quarts and may be serviced from the left-hand side of the engine by an optional remote-fill tube which has a filler cap incorporating the dip stick. Oil that overflows during filling is collected in a scupper and drained overboard through a drain port.

The **lubricating and scavenge-pump assembly** is a positive-displacement pump consisting of six guided-vane-type pump elements mounted along a common drive shaft and contained within a common housing. The No. 1 and No. 2 pump elements are 3-gpm scavenge elements that service the main engine sump. The No. 3, No. 4, and No. 5 pump elements are 1-gpm elements that service the gearbox and No. 1 bearing sump. Element No. 6 is a 2.5-gpm pressure element which supplies oil under pressure to all parts of the engine. The operating elements of the pump extend aft into the oil reservoir from the pump flange on which the oil reservoir is mounted. The oil passes between the pump and the gearbox through one of six oil ports on the pump mounting flange which mate with six ports on the lubricating pump mounting pad on the gearbox. All scavenge oil from the pump is discharged into the oil-reservoir dwell chamber. Oil from the reservoir enters the pressure element through a pendulum-type swivel pickup tube that extends from the aft right-hand side of the pump body. It is delivered by the pressure element directly to the oil cooler, through the oil filter, and into the gearbox where

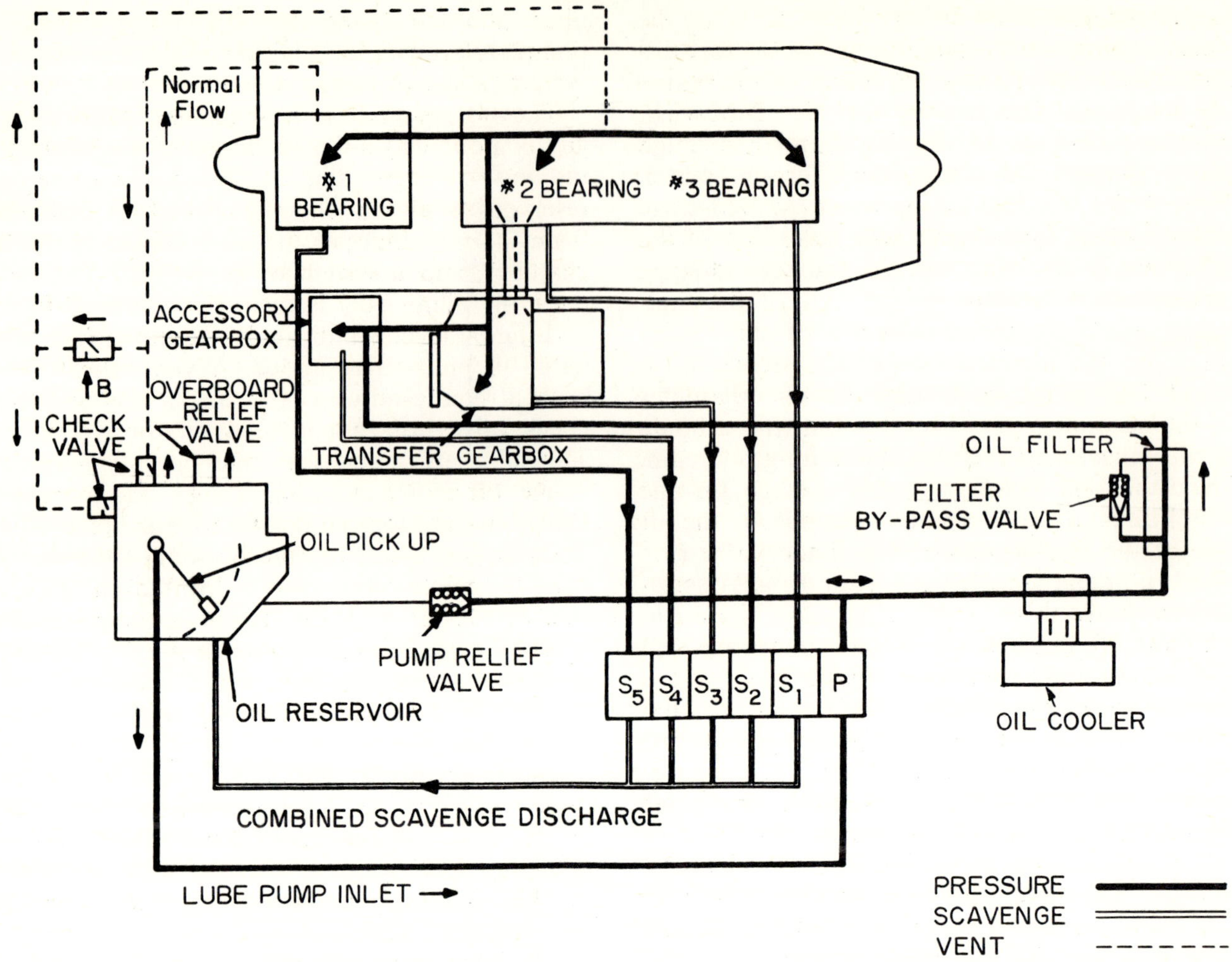

Figure 13·20 Schematic lubrication system. (General Electric Co.)

it is distributed throughout the system. A pressure relief valve mounted in parallel with the oil-discharge passage is included to prevent damage that might result from excessive oil pressure due to cold starting or restriction of normal oil flow. The relief valve is factory adjusted to open at a pressure differential of 90 ± 5 psi across the valve and will return oil directly to the reservoir if this pressure differential is reached.

The **sump vent system** is designed to maintain a controlled positive pressure in the front bearing sump, main sump, gear case, and oil reservoir. The purpose of maintaining a positive pressure above ambient is to make the lubrication system insensitive to altitude. Controlled carbon-seal leakage is the source of pressurizing air. This air is normally recirculated except for excessive air which is vented out of the system through the reservoir relief valve.

The **oil-filter assembly** is mounted in the pump housing. It is a full-flow in-line-type filter with a corrosion-resistant screen element of corrugated steel. The screen filters out oil contaminants over 40 microns in size and is removable for cleaning. A filter bypass valve is included in the core of the filter element. If the pressure difference between oil entering the filter and oil leaving the filter exceeds 21 to 23 psi, the valve opens to permit a direct flow of oil through the unit without filtration.

The **oil cooler** is a typical liquid-to-liquid-type heat exchanger mounted on the oil-reservoir mounting flange and against the right-hand side of the oil reservoir. It consists of numerous longitudinal passages arranged in a honeycomb pattern. Both fuel and oil flow simultaneously through adjoining passages, and an exchange of heat occurs between hot engine oil and cool fuel. Oil enters and leaves the cooler through ports located in the housing for the pressure bypass valve. When the pressure across the valve exceeds 20 ± 4 psi owing to clogging, the valve will open and the oil will bypass the heat-exchanging elements.

During operation of the lubrication system, oil is taken from the reservoir through the pendulum-type pickup by the pressure section of the pump. This is shown as section *P* in the diagram of Fig. 13 · 20. From the pump, oil flows through the cooler, oil filter, and service lines to the bearings and gears requiring lubrication. The oil then drains into the sumps and is returned to the reservoir through the scavenge sections of the pump.

Electrical system

The electrical system for the CJ610 engine includes provisions for ignition, a tachometer generator, an electrically operated anti-icing valve, a fuel flowmeter, and an oil pressure transducer. The actual configuration of the electrical system may vary to some extent, depending upon the aircraft manufacturer's requirements.

The **ignition system** is of the high-energy type, similar to that described in Chap. 9 and illustrated in the diagram of Fig. 9 · 26. The construction of the igniter plug for the ignition system is shown in Fig. 13 · 21.

The **tachometer-generator** unit is engine-mounted in a recessed well on the aft side of the oil reservoir and is driven by the lubrication pump. This unit produces an electric signal proportional to engine speed. The tachometer generator is mechanically integral with the alternator, but electrically it is a separate unit. It is a permanent magnet rotor-type having two poles and three field windings. A three-phase output is produced, the frequency of which is one cycle per revolution of the generator rotor.

The **anti-icing air valve** is electrically operated by means of a cockpit-mounted switch. The valve is mounted on the engine front frame at approximately the 10 o'clock position. Actuation of the valve is provided by a d-c energized solenoid. With electric power to the solenoid, the valve is OFF. The valve is ON when the solenoid is electrically deenergized, thus ensuring a flow of anti-icing air in the event of an electric-power failure. Electric power to the valve is supplied by an aircraft 24-volt d-c system.

The **flowmeter** provides the pilot with an indication of the weight of fuel being consumed by the engine. Fuel passing through the mass flowmeter produces an electric signal proportional to weight flow in pounds per hour. The flowmeter is customer-furnished equipment and is engine-mounted on a flowmeter bracket below the main fuel pump. The flowmeter consists of a constant-speed motor, impeller, turbine, restraining spring, and transmitter. Functionally, the flowmeter converts rate of flow to angular momentum, to torque, to displacement, to electric signal. The electric power required by the flowmeter is 115-volt 400-cycle alternating current supplied by a customer-furnished generator.

The **oil-pressure transducer** converts lubricating oil pressure to an electric signal. The transducer which may be mounted on the engine, utilizes the Autosyn principle to provide a remote indication of oil pressure. In an Autosyn system the transducer rotor movement is duplicated by the movement of the rotor in the indicator.

Exhaust gas temperature (EGT) is sensed by eight chromel-alumel thermocouples connected in parallel. This produces an output which is the arithmetical average of the temperatures sensed by the eight thermocouples.

Main fuel system

The main fuel system for the CJ610 turbojet engine is similar in many respects to systems for other turbojet engines. The system includes valves, filters, fuel-control unit (General Elec-

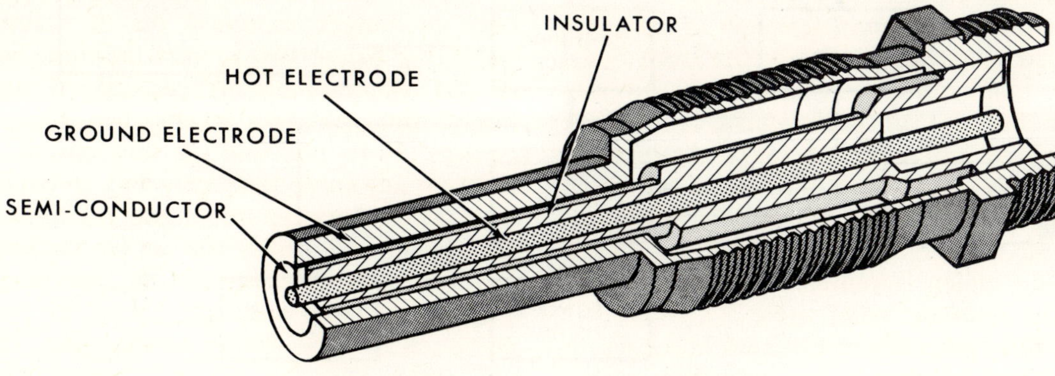

Figure 13 · 21 Construction of igniter plug.

tric MFC-2), overspeed governor, fuel pumps, fuel/oil cooler, pressurizing and drain valve, and fuel nozzles incorporating flow dividers. The fuel-control unit is the heart of the system; it compensates for variations in temperatures and pressures and regulates fuel flow to maintain speed limits throughout the range of engine operation.

The **main fuel-control pump** supplies high-pressure fuel to the main fuel control, which is mounted on the pump. Its secondary purpose is to port high-pressure fuel to the overspeed governor.

The main fuel pump is mounted on and is driven from the forward right-hand pad of the accessory drive gearbox. The pump assembly consists of a booster pump, fuel screen, main pump, and pump relief valve. The booster pump is a centrifugal-impeller type located at the fuel-inlet port of the pump housing. It ensures proper fuel flow through the fuel screen to prevent cavitation of the main pump. The fuel screen is designed with a bypass feature such that fuel flow will continue even though the screen may become clogged. The main pump element is a positive-displacement gear-type pump which supplies high-pressure fuel to the fuel-control unit. Bypass fuel from the control unit is fed back to the inlet of the pump. The pump relief valve, located in the pump housing between the main pump outlet and inlet, allows excess fuel to recirculate to the pump inlet when discharge pressure exceeds 750 psig.

The **main fuel-control unit** for the CJ610 turbojet engine is similar in operation to the hydromechanical fuel controls described in Chap. 8. This type of unit generally consists of a metering section and a computing section. The metering section selects the rate of fuel flow to the engine fuel nozzles in accordance with information received from the computing section. The engine-operating parameters are integrated in the computing section to provide proper selection of fuel flow in every engine operating regime without exceeding safe engine-operating limits.

High-pressure fuel is supplied to the fuel-control inlet by the engine-driven pump assembly described previously. The main fuel supply is tapped to supply fuel for operation of the compressor-inlet-temperature servo, variable-geometry servo, governor servo, and main metering-valve servo. Servo fuel is passed through a 10-micron filter which is equipped

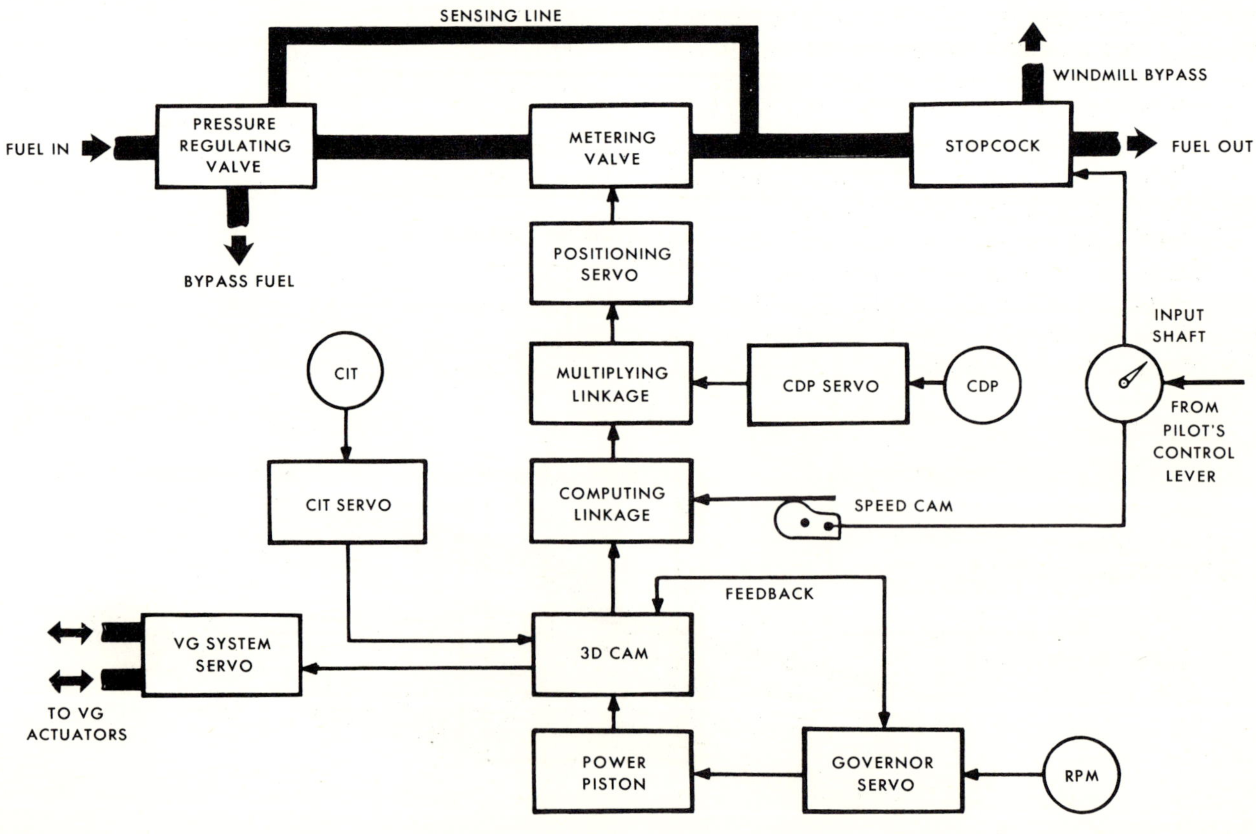

Figure 13 · 22 Block diagram showing fuel-control operation.

with a relief valve so that fuel pressure will be supplied to the servos in case of a clogged filter.

The main fuel flow continues to the **pressure-regulating valve,** at which point some of the fuel is bypassed back to the fuel pump inlet. The remainder of the main fuel flow continues through the **metering valve** and **stopcock** to the engine fuel nozzles.

The operating principles of the fuel-control unit may be understood in part by referring to Fig. 13·22. This is a block diagram to show the relationships among the various components of the unit. The pressure-regulating valve maintains a constant pressure differential across the metering valve, hence fuel flow will depend upon the area of the orifice in the metering valve. This area depends upon the information supplied by the computing section which acts through the **positioning servo** (main metering-valve servo) to adjust the metering valve.

Since the computing section positions the metering valve piston, it selects the fuel flow for each condition of engine operation. Selected steady-state conditions are maintained in conjunction with limiting and biasing functions such as **top speed, acceleration limit, CIT bias,** and **minimum fuel flow.** Engine-operating parameters such as **input shaft position, engine rpm, compressor discharge pressure** (CDP), and **compressor inlet temperature** (CIT) are integrated, and a single output signal representing fuel flow W_f is transmitted to the metering valve piston. The link between the computing section and the metering section is the positioning servo. The action of the computing section with respect to the metering section is indicated in the diagram of Fig. 13·23.

The **engine-overspeed governor** system consists of a hydromechanical self-contained governing device which mounts on the front center pad of the engine gearbox. The governor senses engine speed through a mechanical connection to the gearbox. Under normal operating conditions all engine fuel passes through the governor unhindered; however, if an engine overspeed occurs, the governor is set to cut in at 103.5 ± 0.5 per cent speed and bypass the excess amount of fuel to the main pump inlet.

During engine operation the **fuel-pressurizing valve** performs two functions: (1) it maintains a minimum fuel pressure for the operation of the main fuel-control servo and (2) it provides proper fuel pressure to the nozzles throughout the range of engine operation. The drain valve

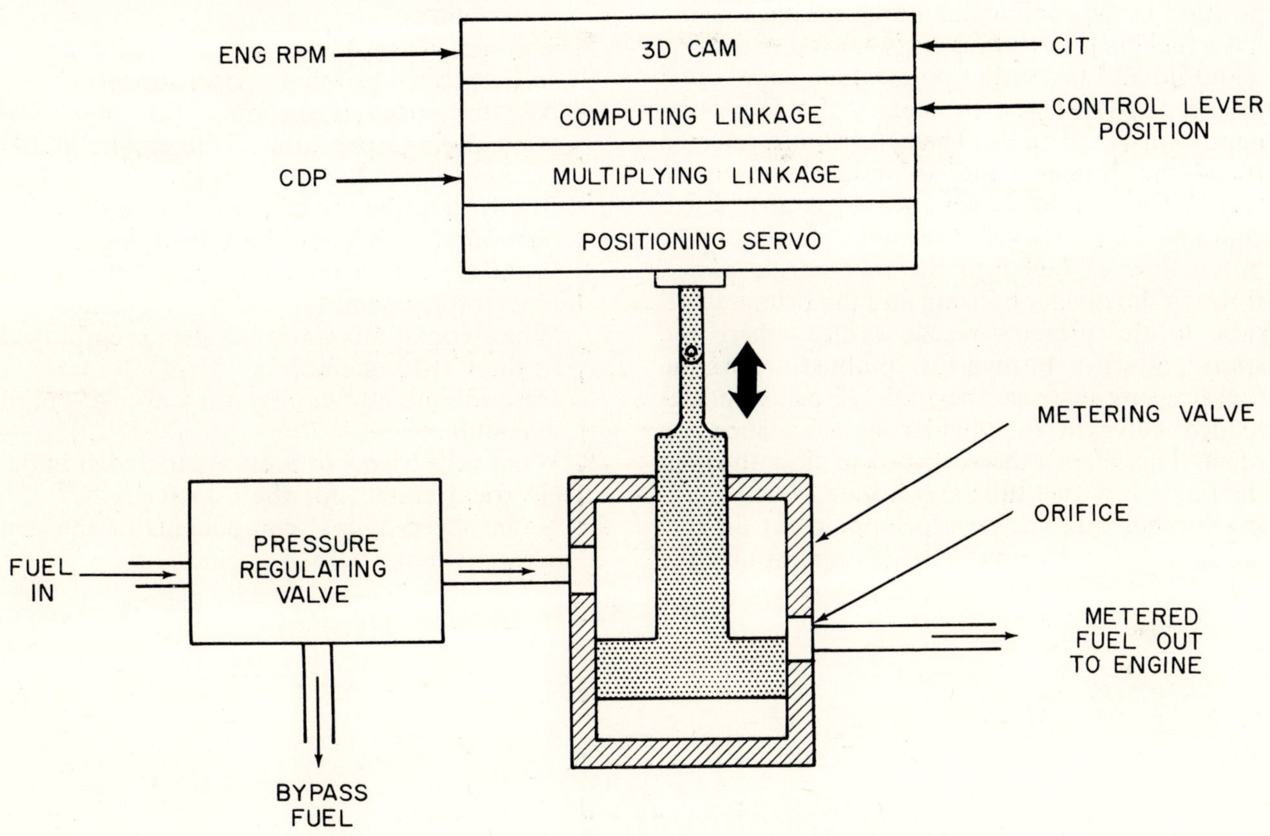

Figure 13·23 Control of metering valve by computing section.

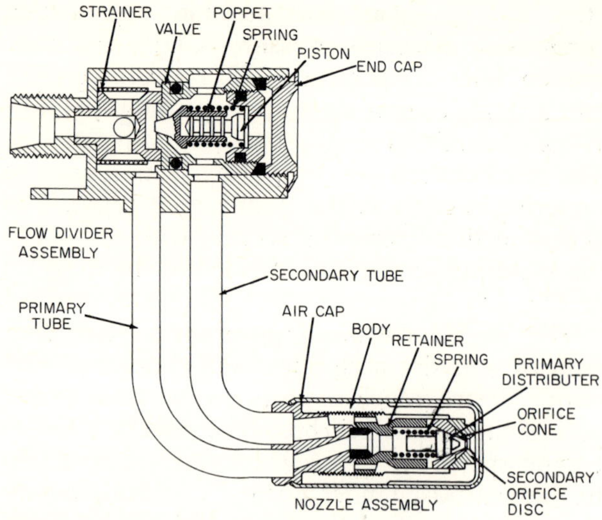

Figure 13 · 24 Flow divider and nozzle.

drains the fuel manifolds during engine shutdown.

The 12 **fuel nozzles** provide the correct fuel spray pattern to the combustion section for the entire operating range of the engine. Each fuel nozzle is divided into two major sections, the **flow divider** and the **nozzle,** shown in Fig. 13 · 24. The flow-divider housings are attached to pads on the mainframe which retain the nozzles in position in the combustion section. This housing has a fuel inlet port which is connected to the fuel manifold and two outlet ports which supply fuel to the nozzle section by means of two permanently attached tubes. These tubes are referred to as the primary and secondary fuel-nozzle tubes. The nozzle section, which is air-cooled, contains the primary and secondary orifices. The initial flow of fuel into the fuel nozzle passes through the divider housing and the primary fuel tube to the primary nozzle orifice where the spray pattern is formed for combustion. As the fuel pressure increases beyond 122 psi, a spring-loaded valve in the divider housing starts to open. This allows the fuel also to flow through the secondary fuel tube and nozzle orifice forming another spray pattern to supplement the fuel requirements. During engine operation the primary fuel flow remains constant and the secondary fuel flow increases to satisfy engine operating requirements.

STUDY QUESTIONS

1. What is the maximum takeoff thrust of the J69 engine?
2. What type of compressor is utilized by the J69 engine?
3. Describe the unusual features of the fuel system for the J69 engine.
4. What is the military model number for the General Electric CJ610 engine?
5. What is the weight of the CJ610 engine and how much thrust does it produce?
6. Give a brief description of the front frame section for the CJ610 engine.
7. Describe the No. 1 bearing.
8. Through what angle will the inlet guide vanes rotate?
9. Explain how the compressor rotor blades are attached to the rotor disks.
10. How are stator-vane airfoils secured inside the stator case?
11. Explain the function of the variable-geometry components.
12. Describe the mainframe section and explain its functions.
13. Through what device do the engine accessories receive power for operation?
14. At what speed is the main fuel pump rotated during operation of the engine at 100 per cent speed?
15. Briefly describe the construction of the combustion chamber for the CJ610 engine.
16. What are the major components of the turbine rotor assembly?
17. What type of lubrication system is employed in the CJ610 engine?
18. Describe the lubricating and scavenge-pump assembly.
19. What subsystems or units are included in the electrical system for the CJ610 engine?
20. Name the principal components of the fuel system for the CJ610 engine.

CHAPTER 14

The General Electric J47 Turbojet Engine

Since the beginning of aircraft gas-turbine development in the United States, General Electric has been a leading research and manufacturing company for the development and production of aircraft gas turbines. Turbine design and development was a familiar activity with General Electric long before the advent of the aviation gas turbine. For many years the company was involved in the utilization of turbines for driving electric generators. During World War II, General Electric supplied the Air Force with thousands of turbine-driven superchargers which enabled our bombers to operate at altitudes high enough to avoid any effective antiaircraft fire.

In the fall of 1941, the Whittle W1X engine was brought to this country and given to General Electric for development. This development reached a climax in October of 1942 when the first American jet-propelled airplane, the Bell XP-59A, flew for the first time at Muroc, Calif. The engines which powered this plane were designated I-A and produced 1,200 lb of thrust at 16,500 rpm.

The I-16 was a later development of the I-A and had a thrust rating of 1,600 lb. This engine utilized a centrifugal compressor and reverse-flow combustion system, and it went through a series of refinements until the I-40 evolved. The I-40 engine had a centrifugal compressor but had a through-flow type of combustion system and developed 4,200 lb of thrust at 11,500 rpm. The I-16 engine has a USAF designation of J31 and the I-40 is the J33. The J33 was turned over to the Allison Division of General Motors for further development and production, and it served the Air Force well in the Lockheed Shooting Star fighter and the T33 jet trainer. The J33 engine has also been used in other fighters and missiles.

From the centrifugal-type engine, General Electric turned to the development of an axial-flow engine of 4,000-lb-thrust at 7,600 rpm. This engine was designated the J35 and was the predecessor of a successful line of axial-flow engines, the J47 series, each of which has a higher thrust rating than the preceding one. Following the J47 series are the J73, J79, J85, and J93, all of which have many improvements.

The J35 engine was also turned over to the Allison Division of General Motors for further development and production. The J35 found extensive use in the Northrop Scorpion night interceptor and in some models of the Republic F-84 Thunderjet.

In this text we shall describe in detail the J47 and the J79 (CJ805), since these offer features which are typical of the most successful turbojet engines.

THE J47 TURBOJET ENGINE

The GE J47 series of turbojet engines produced a remarkable record in the history of aircraft powerplants. These engines have experienced more combat time than all other turbojets combined. They were used to power the famous North American F-86 Sabrejet airplanes used so effectively against the Russian MIG-15 in the Korean War, and they also powered the Boeing B-47 Bomber which was the backbone of the Strategic Air Command intercontinental defense until the advent of the B-52 and later bombers.

During 1955 and into 1956, the J47 engines were logging an average of approximately 11,000 hr per day flight time. This is the equivalent of 458 engines operating 24 hr per day every day.

Because of the success of the J47 engines and because they are commonly used for instruction, we have included this description of their design, construction, and systems.

Table 14 · 1 Sea-level static performance

Thrust	Rpm, %	Rpm (actual)	Jet thrust, lb	Spec. fuel consumption, lb per hr per lb	Temperature, °C	Temperature, °F
Maximum	100	7,950	7,500	2.30	705	1300
Maximum, screen in place	100	7,950	7,160	2.33	705	1300
Military	100	7,950	5,425	1.15	705	1300
Normal (continuous)	100	7,950	4,990	1.13	655	1210
90 per cent normal	98.0	7,790	4,490	1.12	655	1210
75 per cent normal	92.2	7,330	3,730	1.11	655	1210
Idle	37.8	3,000	400 (max)			

Basic engine description

The Model J47-GE-17 engine is an improved J47 engine equipped with a **reheat exhaust system;** that is, it incorporates an **afterburner** for thrust augmentation. The basic engine has a 12-stage axial-flow compressor directly coupled to a single-stage, impulse-type turbine wheel. The combustion section is of the "can" type, with eight through-flow combustion chambers arranged around the midsection of the engine. Dual fuel nozzles are installed in the midsection of the engine, and they extend into the forward end of each combustion chamber.

The reheat section, or afterburner, is equipped with two eyelids which are actuated to provide a variable-area exhaust nozzle.

Performance

The performance is presented for standard NASA atmospheric conditions, with no allowance being made for inlet-duct losses, loading of accessory drives, or any compressor air bleed other than that required for the engine itself (see Table 14 · 1).

Figure 14 · 1 is a cutaway view of the J47-GE-17 engine showing the arrangement of the parts from the front (accessory section) to the exhaust nozzle. Figure 14 · 2 is an exploded drawing showing the main sections of the engine. We shall describe the engine beginning with the most forward components and proceeding completely through the engine to the exhaust nozzle.

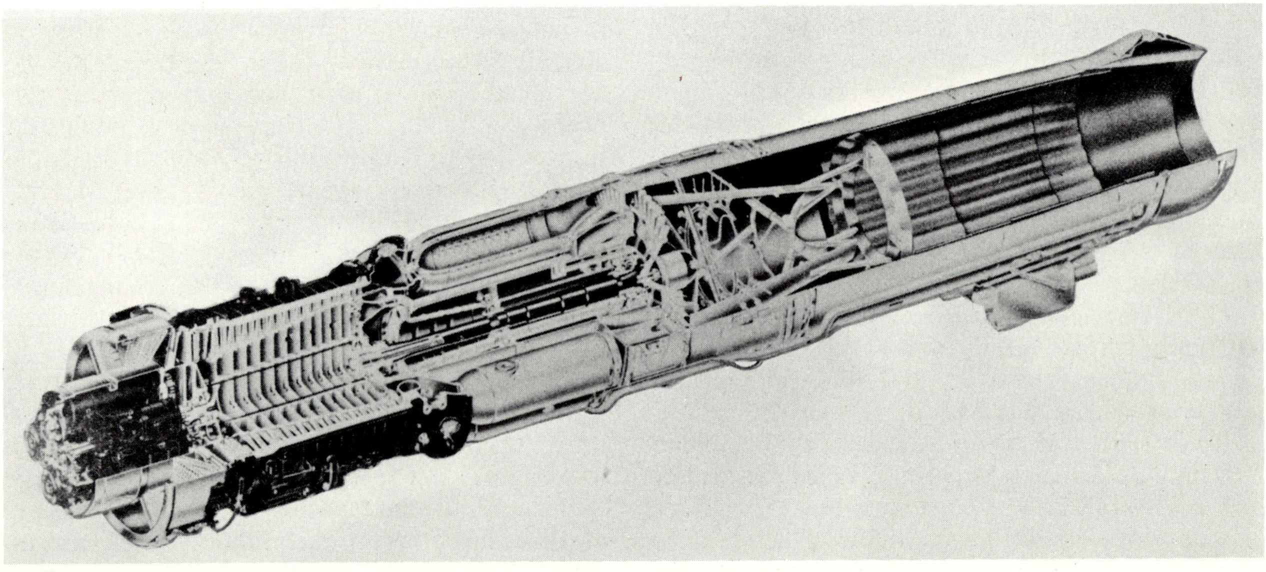

Figure **14 · 1** *Cutaway view of the J47-GE-17 turbojet engine.*

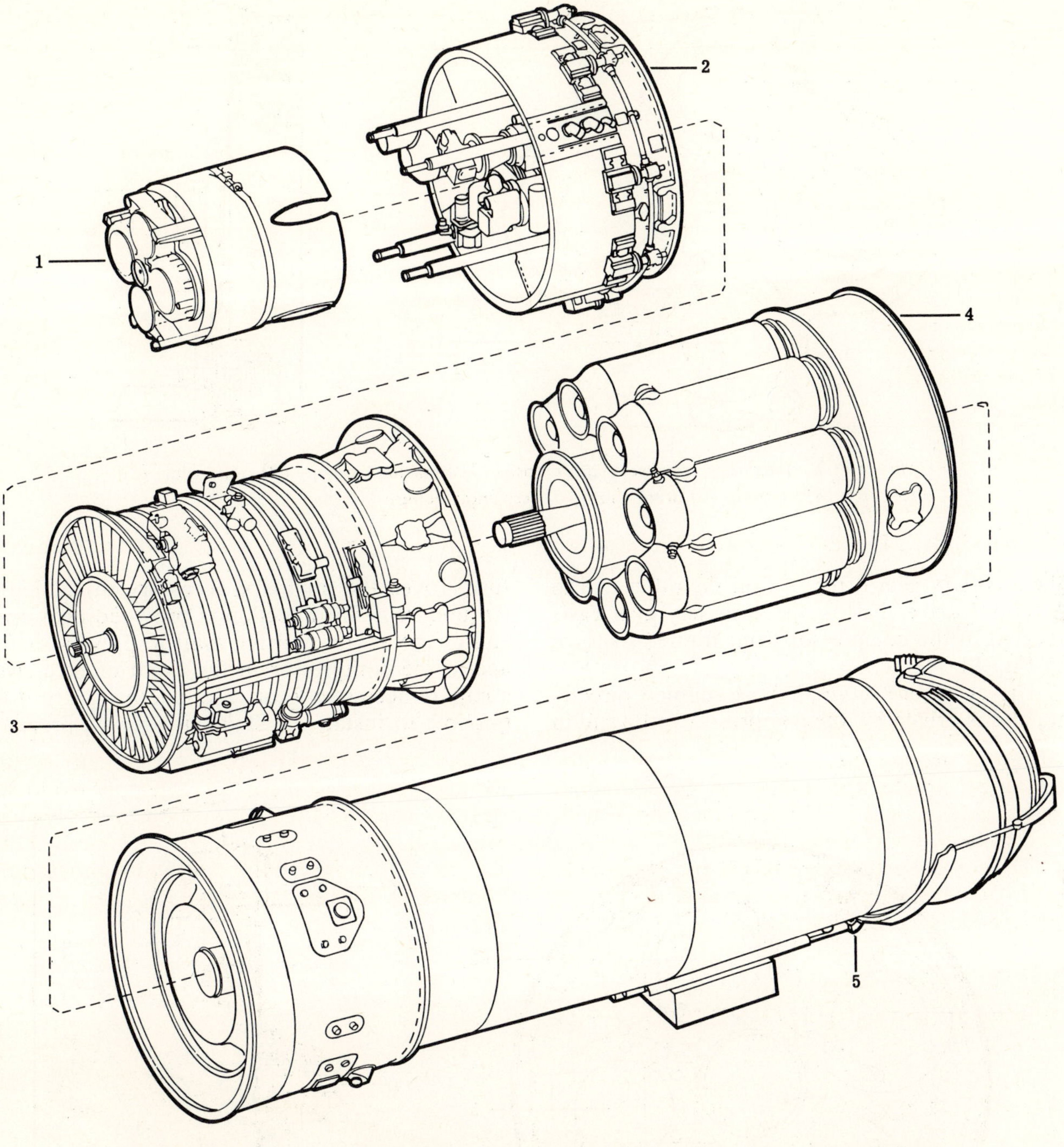

Figure ***14·2*** *Major sections of J47-GE-17 engine: (1) auxiliary drive section; (2) front frame accessory and air inlet section; (3) compressor section; (4) turbine section; (5) exhaust section.*

Auxiliary accessory gear case

The auxiliary accessory gear case is the foremost part of the engine and is mounted to the accessory gear case by four tubular supporting legs. It consists of a magnesium-alloy housing and cover plate which encase the drive gears for four power-extraction pads as shown in Fig. 14·3. The purpose of the auxiliary accessory gear case is to provide power-extraction pads to drive accessories necessary for aircraft operation.

A typical installation will have an alternator mounted on the top vertical pad, a generator mounted on the bottom vertical pad, and hydraulic pumps mounted on the right and left horizontal pads. The internal gears are driven by a quill shaft which is splined on one end to

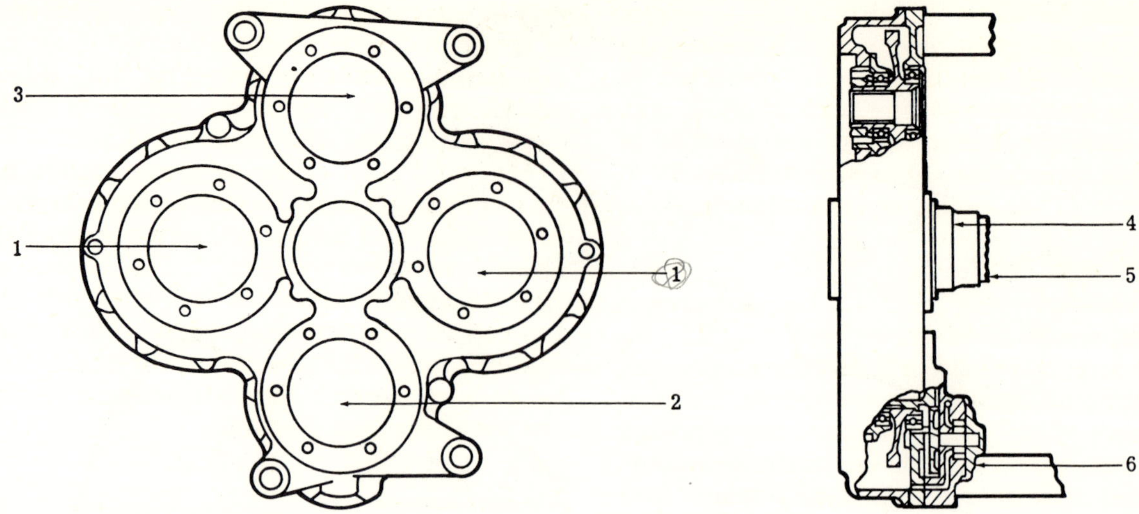

Figure **14·3** *Auxiliary accessory gear case: (1) hydraulic pump; (2) generator; (3) alternator; (4) housing; (5) drive shaft; (6) scavenge pump.*

the auxiliary gear drive and on the other end to a central drive shaft in the **accessory gear case,** located in the compressor front frame as shown in Fig. 14·4.

The accessory drive shaft is coupled directly to, and is driven by, the compressor rotor and in turn drives the auxiliary accessory drive shaft. An undercut shear section is provided near one end of the auxiliary drive shaft to prevent damage to the compressor in the event of excessive auxiliary accessory gear case loading. Since it is possible to install the shaft either way, precau-

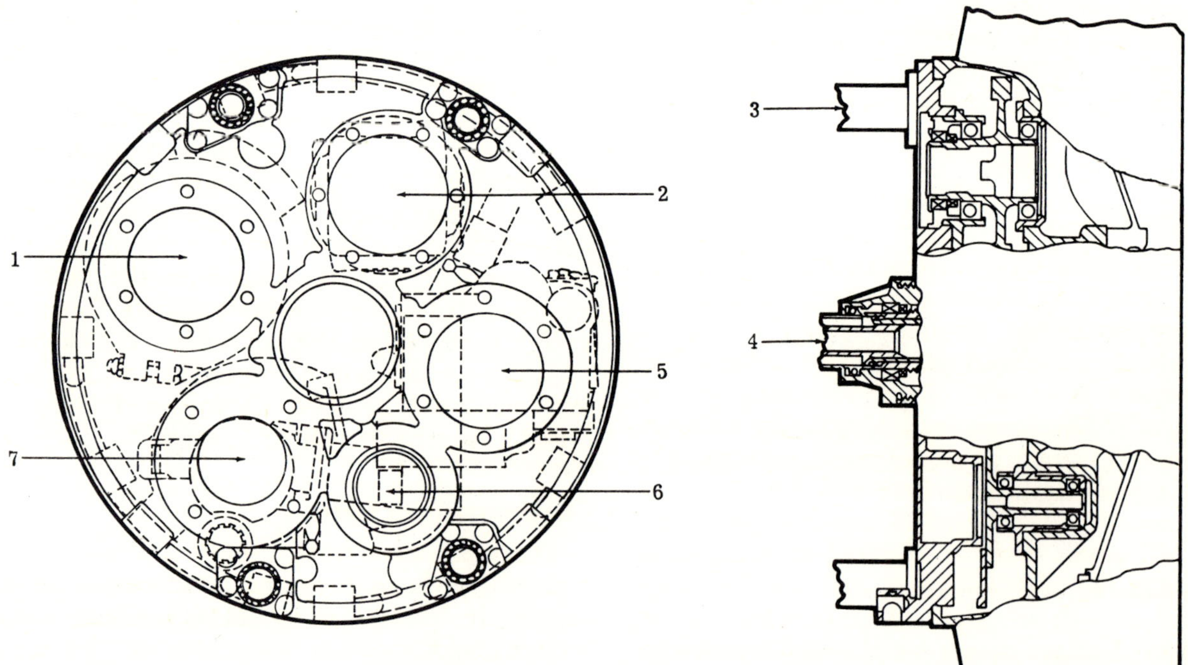

Figure **14·4** *Accessory gear case: (1) starter generator; (2) amplidyne with control tachometer; (3) auxiliary accessory support mounting; (4) auxiliary accessory drive shaft; (5) tandem pump with control valve; (6) lube pump and indicating tachometer; (7) booster pump.*

tion should be taken on installation to ensure that the shear section of the shaft is nearest the compressor. The rotating shaft is protected by a housing fitted with O-ring seals at either end.

The auxiliary accessories are not installed at the engine factory but are to be installed prior to installation of the engine in an airplane.

Accessory gear case

The accessory gear case is a recessed area in the compressor front frame and contains the gearing necessary to drive the engine accessories. The engine accessories are located on the gear-case cover plate which encloses the gear-case area. These accessories are the starter-generator, overspeed governor, amplidyne, main fuel-control valve, and main fuel pump.

Since the accessory gear case is an integral part of the compressor front frame, the gear case cannot be removed intact, as a unit. The gear-case cover, the accessories, and the gears may be removed. The cover plate is bolted to the compressor front frame, and the two parts are line-bored in one operation to ensure proper alignment of the bearings and gears. They are then matched pieces, and no other cover plate may be installed on the same compressor front frame without danger of misalignment.

The engine accessories rotate in a direction opposite to the compressor rotation. A magnetic plug is installed in the bottom of the gear case, and it is accessible from the front of the engine. It is removed on inspections to check for the presence of steel particles, which may indicate an impending failure.

Air-guide section

The air-guide section is located in front of the compressor front frame around the accessories. The purpose of the air-guide section is to duct air in through the compressor front frame to the compressor. It consists of an inner shell and an outer shell. The inner shell completely encloses the space occupied by the engine accessories between the auxiliary gear case and the accessory gear case. Two sections of cowling form this inner air-guide shell. A channel or scupper is formed in the bottom of the shell to collect any oil, fuel, or hydraulic fluid which may leak in this area. Any leakage passes out through a port into the bottom compressor front-frame strut, through the hollow portion of the strut, into the overboard drain system through a casting bolted to the bottom of the compressor front frame. This drain is commonly called the **front-frame drain.** The compressor front-frame drain line may be attached to the largest port on either side of the drain casting. The smaller port in the casting accommodates an anti-icing line, through which anti-icing air flows up to a drilled passageway in the front of the strut.

The outer shell of the air-guide section consists of four quadrants, bolted to the compressor front frame and held together by cover plates. These cover plates are called "island covers" and serve as junction points for fuel, lube, anti-icing, and certain electrical connections. The connecting lines are brought through the inlet air stream by passing through four streamlined fairings, one for each island position.

Air-inlet screens

There are eight individual screen sections located in the air-guide section of the engine. The purpose of these screens is to prevent damage to the engine from foreign matter which may be picked up when the airplane is on the ground or in low-level flight. The screens are retractable, and the retraction mechanism is driven by a single electric motor located near the top of the engine on the right-hand side of the compressor. The motor drives a series of flexible cables which extend from the motor in both directions. The flexible shafts drive worm and jackscrew units located at each screen position.

A clutch is incorporated in each jackscrew unit to prevent damage in case any of the screens should become jammed.

The inlet screens should be retracted during icing conditions. The retraction mechanism is energized by means of a manual three-position switch in the cockpit. When the switch is in the ENGINE EXTEND SCREEN position, the screens will be extended; with the switch in the RETRACT position, the screens will be retracted. When the switch is placed in the ACCESSORY COVER position, the screens will be retracted and the anti-icing valve will be open, allowing airflow to the bullet nose and the island fairings.

Compressor front frame

The compressor front frame is a single casting and bolts directly to the compressor stator casing. It consists of an inner and an outer section held together concentrically by eight streamlined hollow struts, to form an annular passage for the inlet air.

Support for the accessory drive gears, the drive shaft housing, No. 1 bearing housing, and allied parts is provided by the inner (center) portion of the frame.

The compressor front frame is the main supporting structure for the front end of the compressor rotor, since it holds the compressor-rotor front bearing. It comprises four external engine trunnion mounts: two on the horizontal center line and two on the vertical. It also houses the inlet guide vanes, the balance chamber, and the No. 1 bearing air and oil seals. The eight hollow streamlined struts, in addition to supporting the inner and outer sections, are utilized in the following manner: The top or No. 8 strut vents directly into the gear case; this vent is not used and should be capped at all times. The No. 1 strut, together with struts 2, 5, and 7, vent the area between the No. 1 bearing and oil seals. The hollow portions of struts 3 and 6 channel into the balance chamber; these are no longer used on the J47-GE-17 engine, and their diamond-shaped ports should be kept capped at all times to prevent loss of air pressure from the balance chamber. No. 4 strut, in addition to accommodating the compressor front frame drain, has a drilled passageway in the back of the strut which is fitted with a drain line, which also vents and drains the area between the No. 1 bearing air and oil seals. The Nos. 2, 4, 6, and 8 struts also contain anti-icing passageways.

Inlet-guide-vane assembly

The inlet-guide-vane assembly is required to direct inlet airflow into the first stage of the compressor at the proper angle so that the correct angle of attack will be established for the first row of compressor blades. It also meters the air into the compressor by the restriction presented to the airflow by the angular setting of the vanes.

The inlet-guide-vane assembly is a sheet-metal-fabricated part consisting of 37 hollow blades mounted between two concentric rings. It is bolted to the inner section of the compressor front frame. The hollow guide vanes are manifolded together on their outer ends, and twelfth-stage-compressor air is supplied to the manifold. The primary purpose of the air fed through the hollow vanes is to provide air pressure to the balance chamber. This air pressure counteracts a part of the forward thrust acting on the compressor rotor, thus relieving the thrust bearing of part of the load.

The warm air flowing through the inlet guide vanes also serves to prevent the formation of ice on the vanes.

Anti-icing system

Every aircraft or engine designed for all-weather flying must have provision for the prevention of ice formation or for the removal of ice when it forms. Turbojet engines utilized the compressor air for this purpose, since the air is heated by compression. The anti-icing system of the J47 engine is illustrated in Fig. 14·5.

Warm air is supplied continuously to the inlet guide vanes and the struts in the compressor front frame which are not covered by island fairings. The air to these uncovered struts is taken from the balance piston lines and routed through drilled passageways in the front of the struts. Air is discharged out the front of the strut under a cap, where it leaks around the strut and back through the compressor.

Warm air is supplied to the island fairings and bullet nose from the No. 3 extraction port, on the cabin-pressurization manifold, located in the compressor rear frame. A valve, located on the left compressor stator casing horizontal split line near the compressor rear frame, controls this airflow as dictated by the same three-position switch in the cockpit that controls the air-inlet screens. Air, as controlled by the valve, is piped externally forward, through the island covers into a partitioned area in the front of each of the island fairings, from which it is exhausted back

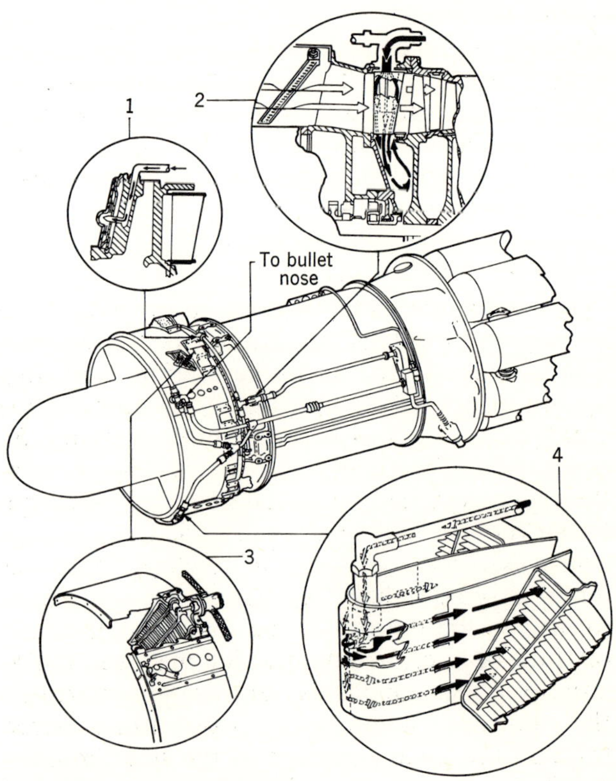

Figure 14·5 Anti-icing system: (1) front frame strut anti-icing; (2) inlet-guide-vane anti-icing; (3) retractable screen; (4) island fairing anti-icing.

through the compressor. Air is also piped from the valve through No. 4 island into the bullet nose. The air supplied to the anti-icing system from either the cabin-pressurization manifold or the twelfth stage of compression will be (at its source) at a temperature of 485 to 500°F and under a pressure of approximately 5 atm at 100 per cent engine speed. When the valve is open, the pressure downstream of the valve should be approximately 25 psig. When the valve is closed, the pressure downstream of the valve should be zero to ensure that there is no loss of air mass through this system when anti-icing air is not needed.

The balance chamber

As mentioned previously, the balance chamber serves to take a part of the axial load from the No. 2 compressor bearing. It is a space between the compressor front frame and the forward wheel of the compressor rotor and is surrounded by the inner ring of the inlet-guide-vane assembly.

In drawing air into the engine, the compressor receives a forward thrust, but the No. 2 ball thrust bearing serves to transmit this thrust to the compressor rear frame, near the midsection of the engine. The balance chamber is supplied with twelfth-stage-compressor air which is fed through the inlet guide vanes. This air exerts a rearward pressure on the first wheel of the compressor, thus taking a part of the thrust load from the No. 2 bearing. Air escapes from the balance chamber through the No. 1 wheel outer seal and through the No. 1 bearing air seal, located around the compressor shaft.

No. 1 bearing air and oil seals

Because of the air pressures that exist within a gas-turbine engine during operation, provision must be made to control air leakage and to direct the compressed air in proper passages and areas where it will serve the functions required. If air pressure leaks into some areas, it can cause loss of lubrication and otherwise disturb the smooth functioning of the engine.

The No. 1 bearing air and oil seals are labyrinth-type seals machined into a plate which bolts to the rear of the inner section of the compressor front frame. This arrangement is shown in Fig. 14·6. The mating rotating surfaces, which complete the seals, are formed by a hub. This hub, along with the No. 1 bearing inner race, is retained on the forward end of the compressor shaft by a locking nut. The purposes of this air and oil seal arrangement are to prevent

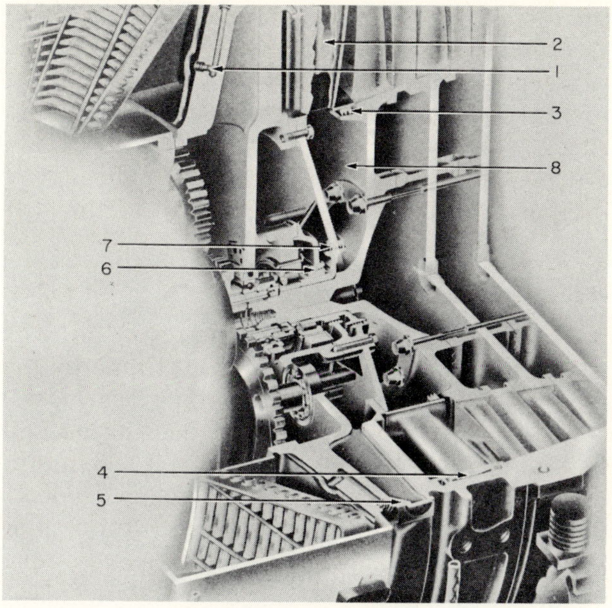

Figure 14·6 No. 1 bearing area: (1) front frame strut; (2) hollow IGV; (3) balance-chamber air seal; (4) No. 1 wheel outer air seal; (5) No. 1 bearing air- and oil-seal vent; (6) No. 1 bearing oil seal; (7) No. 1 bearing air seal; (8) balance chamber.

loss of air from the balance chamber, and to prevent loss of oil from the No. 1 bearing area. Air from the balance chamber which leaks through the No. 1 bearing oil seal is vented overboard in five places. The action of the scavenge element of the main lubricating and scavenge pump creates a pressure in the gear case which is slightly negative in relation to atmospheric pressure. Since the pressure between the air and oil seals will be greater than the pressure in the gear case, the air between the air and oil seals will blow through the oil seal into the gear case and prevent oil from leaking out in the opposite direction.

Air leakage into the gear case from any other source will destroy the pressure differential across the oil seal. Once this differential pressure has been destroyed, the oil will work its way through the No. 1 bearing oil seal and be blown out the No. 1 bearing air and oil seal vents by the air escaping from the balance chamber through the No. 1 bearing air seal. In extreme cases of excessive wear, the air and oil seals will allow air from the balance chamber to enter and pressurize the gear case, resulting in oil leakage around the spindles of the accessories mounted on the gear case cover plate. However, if oil is found to be leaking from the No. 1 bearing air and oil seal vents, it is wise not to assume that the air and oil seals are worn until it has been ascertained that no source of air leakage into the gear case, other than through the oil seal, is present. The most logical and easiest place to check would be the square cover plate on the top of the compressor front frame which caps the passage through the No. 8 strut to the gear case.

No. 1 bearing

The forward end of the compressor rotor is supported in the compressor front frame by the No. 1 bearing. This is a roller-type bearing located just forward of the No. 1 bearing air and oil seal assembly, as shown in Fig. 14 • 6, and it is designed to take only radial loads. The outer race and rollers are retained in the compressor front frame, and the inner race is retained on the compressor shaft.

In order to remove the compressor front frame and attached parts from the compressor section, all that is necessary is to remove the bolts on the compressor forward flange and the interconnecting lines. The No. 1 bearing inner-race retaining nut need not be removed prior to removing the compressor front frame.

Compressor section

The compressor consists of a rotor with 12 rows of compressor blades and a stator having 13 rows of vanes. The stator is the external portion, and it is bolted to the compressor front frame and the compressor rear frame. The rotor is the internal portion and is mounted on bearings located in the compressor front frame and the compressor rear frames. These are the No. 1 and the No. 2 bearings, respectively.

Two types of compressor rotors are used in J47 engines. The earlier type is a shaft-type rotor in which the 12 wheels are shrunk onto a single shaft. For the purposes of this discussion, we shall describe only the later design, called the **curvic-coupling** type.

The curvic-coupling rotor differs from the shaft type in that there is no centrally located through-shaft. The wheels are held together by 12 through-bolts which extend completely through the compressor rotor. The contacting, driving surface between the wheels is a circular row of serrations or gearlike teeth machined on a flange on each side of the individual wheels. The No. 1 and No. 12 wheels have these gearlike teeth or serrations on one side only. On the other side of these wheels are stub shafts that serve as the shaft for the rotor assembly.

The connection between the turbine wheel and the curvic-coupling compressor rotor is accomplished in the same manner as is used with the shaft-type rotor except that, since there is no through-shaft, a bolt is inserted through the twelfth-stage wheel, on assembly of the rotor, to accommodate the installation of the compressor nut. The turbine shaft splines into the large hub machined on the twelfth-stage wheel. The twelfth-stage wheel, then, is directly driven by the turbine shaft, and the other wheels are driven by virtue of their connection to the twelfth-stage wheel through the series of circular gearlike teeth. Gearing the wheels together in this manner eliminates the possibility of slippage of the wheels, and the subsequent unbalance of the rotor. Some additional advantages of the curvic-coupling rotor are that assembly, disassembly, and maintenance are much easier.

A careful examination of the illustrations of Figs. 14 • 1, 14 • 6, and 14 • 7 will help to provide a clear understanding of the construction of the compressor rotor. Figure 14 • 8 shows how the rotor appears with the blades and spacers removed. A cutaway view of the assembled compressor section together with the accessory sections is shown in Fig. 14 • 9.

To prevent warpage of the rotor wheels from the normal heat of compression and to prevent pressure build-up between the rotor wheels, a cooling airflow is provided through the last seven wheels of the compressor rotor. Air enters the area between the fifth and sixth wheels through four ¼-in. holes in the spacer ring between these two wheels. This air is allowed to flow aft through the remaining wheels by four 1-in. holes in each of the last seven wheels. The cooling air together with the twelfth-stage seal leakage discharges into the area behind the twelfth-stage wheel in the compressor rear frame and may be vented overboard through a screen and three designed vents on the aft side of the compressor-rear-frame firewall. The area behind the compressor rotor should be vented in at least two of the places provided, and if ducting is used it should be designed so as to ensure that not more than 5 lb above atmospheric pressure be imposed in this area. More than 5 lb back pressure may critically reduce the compressor cooling airflow. Air leakage through these vents will be approximately 1.4 per cent of the entire engine airflow. These cooling-air features are the same in the shaft and the curvic-coupling-type compressor rotors. While there are only three vents designed as such, the trunnion mounts on the compressor rear frame, when not used as mounts, may be used as vents in lieu of the three

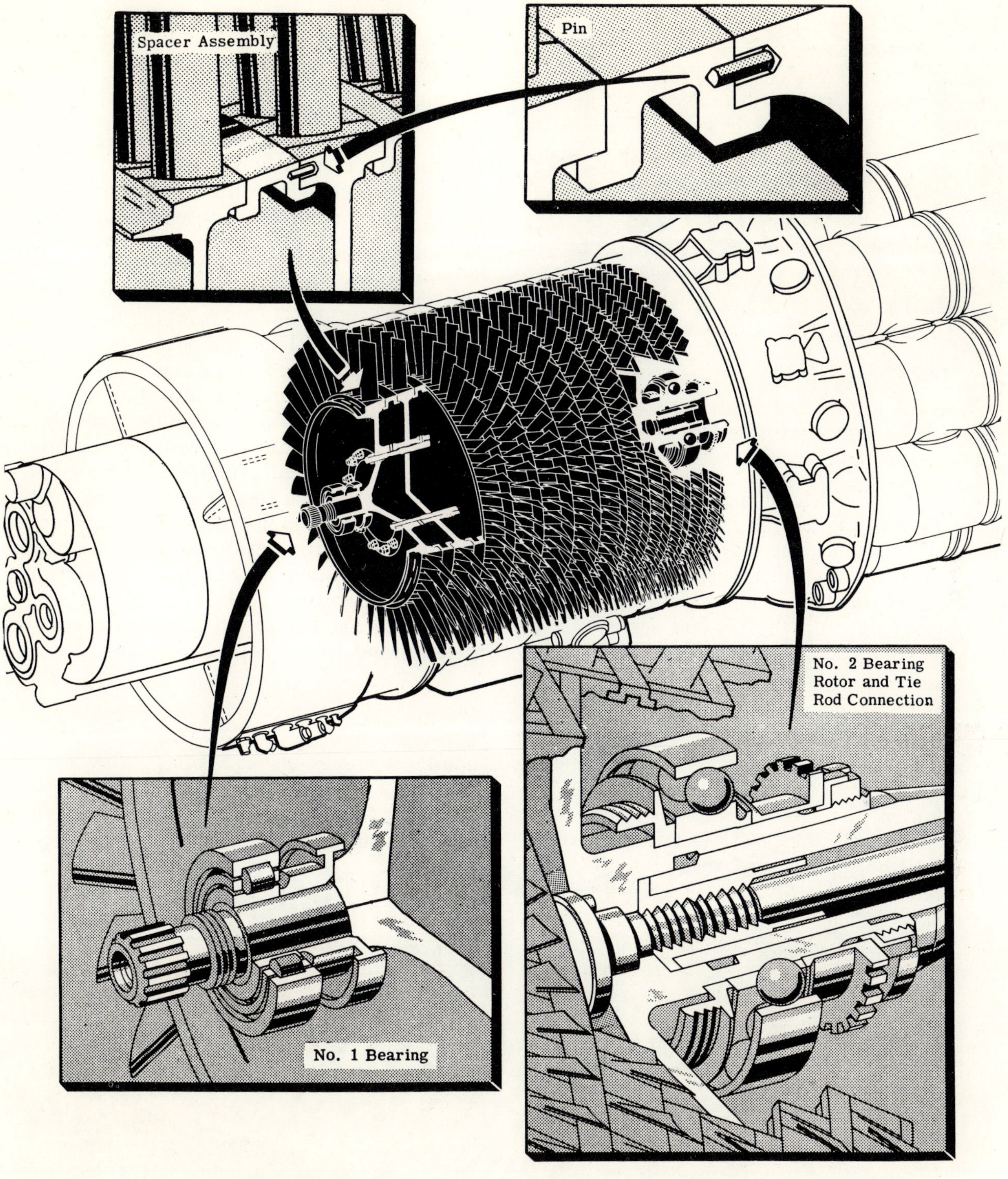

Figure 14·7 Compressor section with curvic-coupling rotor.

designed vents on the compressor-rear-frame firewall.

In the shaft-type compressor rotor a compressor nut threads directly onto the compressor shaft inside the large hub of the twelfth-stage wheel. In the curvic-coupling rotor, the compressor nut threads to a short bolt which extends through and is doweled to the twelfth-stage wheel. This is shown in Fig. 14·10 and also in Fig. 14·7. In either case, a tie bolt threads into the other end of the compressor nut and bottoms directly against the compressor shaft or the short bolt, depending upon which is used. The tie bolt extends entirely through the hollow turbine

Figure 14·8 View of curvic-coupling rotor and wheel.

shaft to the rear face of the turbine wheel. The installation of the tie bolt nut on the aft end of the tie bolt, against the rear face of the turbine wheel, retains the turbine-wheel shaft in the hub of the twelfth-stage compressor wheel. The compressor nut and the tie bolt serve only to hold the compressor and turbine assemblies together. The torque load is taken by the spline connection between the compressor hub and the turbine-wheel shaft.

Twelfth-stage air seal

The twelfth-stage air seal is a series of three labyrinth-type seals bolted to the forward face of the compressor rear frame, as shown in Fig. 14·11. The mating surfaces which complete the seals are flanges machined on the rear of the outer edge of the twelfth-stage rotor wheel. The purpose of the twelfth-stage air seal is to reduce the amount of air leakage from the twelfth stage of compression. This air leaks between the rotating twelfth-stage wheel and the stationary compressor rear frame and into the area behind the compressor rotor. From this area twelfth-stage seal leakage, together with the compressor cooling air, is vented overboard. Since a loss of compressed air is a loss of energy, the amount of air leakage through the twelfth-stage air seal directly affects the performance of the engine.

The twelfth-stage air seal may be referred to as the "compression seal." It is the largest seal in the engine and is subjected to the heat of compression and to the highest pressure in the en-

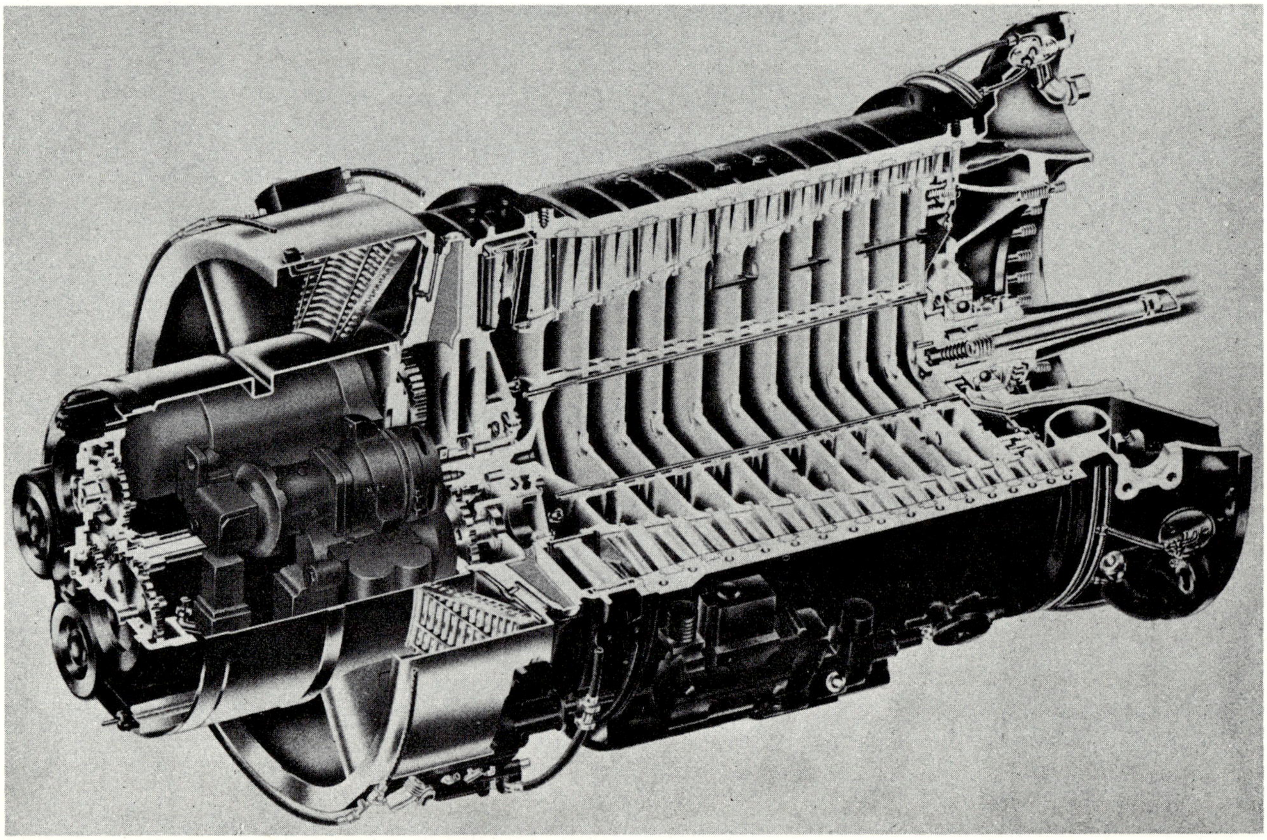

Figure 14·9 Quarter-cutaway view of forward half of engine.

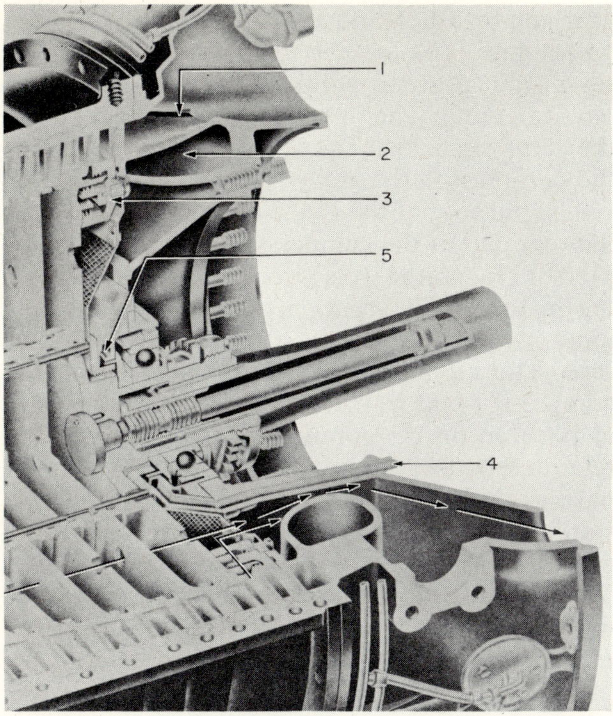

Figure 14·10 No. 2 bearing area: (1) supply port to cabin-pressurization manifold from diffuser passage; (2) cabin-pressurization manifold; (3) twelfth-stage air seal; (4) No. 2 bearing air- and oil-seal vent; (5) chamber between No. 2 bearing air and oil seals.

gine. These conditions lead to the possibility of considerable air loss at this seal.

After initial installation, there may be some signs of rubbing until the seal has become properly seated by engine operation. A slight twelfth-stage seal rub may also be heard on engine shutdown. The extent of seal wear may be judged by placing some sticky substance around the compressor rear frame vents and noting the amount of aluminum dust deposited thereon. No signs of twelfth-stage air-seal wear should be apparent after the first few hours of engine operation.

Labyrinth-type seals are used throughout the engine where the expansion and contraction brought about by large temperature changes make the application of a more positive seal impractical.

Compressor-stator case

The compressor-stator casing is located around the compressor rotor between the compressor front frame and the compressor rear frame. The installation is shown in Fig. 14·9.

The compressor-stator casing consists of a split cylindrical aluminum casing into which 13 rings of blades are dovetailed. In addition to providing a support for the stationary compressor blades and encasing the compressor rotor, it also supports and aligns the compressor front and rear frames and is a convenient mount for many of the engine accessories.

The stator blades are dovetailed into semicircular rings which are slid into grooves machined in the inside of the casing halves. To ensure rigid support and proper alignment of the compressor front and rear frames and to prevent air leaks, all mating surfaces are machined fits. The stator casing halves are matched pieces and are used together as a unit. The top or bottom half may be removed for inspection of the compressor rotor and blades, but the casing halves should not be replaced individually. The end bolt holes and the sixth hole from the end, on both sides, on the horizontal split line are slightly smaller than the other holes and are used as dowel holes to give perfect alignment between the casing halves.

Compressor rear frame

The compressor rear frame is a single casting located between and supporting the compressor-stator casings at its forward face, and the turbine frame and combustion chambers at its rear face. The No. 2 bearing is located in and supported by the center section of the compressor rear frame. The diffuser passages and the cabin-pressuriza-

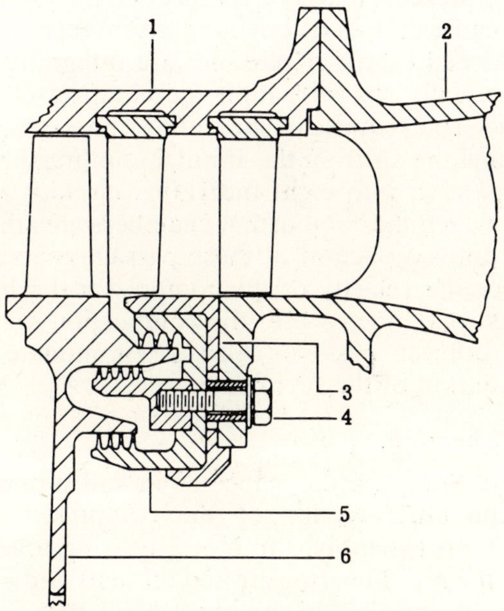

Figure 14·11 Twelfth-stage air seal: (1) compressor stator; (2) compressor rear frame; (3) steel retainer ring; (4) mounting bolts; (5) aluminum seal teeth; (6) twelfth-stage rotor.

tion manifold are cast integrally within the compressor rear frame. Equally spaced within the shell of the frame are eight diffuser passages through which air from the compressor discharge flows to the combustion chambers. Eight duplex fuel nozzles are mounted on pads and extend directly into the diffuser passages, with their tips extending into the inner combustion-chamber liners.

The compressor rear frame also contains four engine mounting pads, two on the horizontal center line and two on the vertical. These mounting pads are not always used, however.

The cabin-pressurization manifold is a hollow doughnut-shaped passageway cast within the compressor rear frame and fitted with four air-extraction ports equally spaced around the frame. A portion of the cabin-pressurization manifold may be seen in the illustration of Fig. 14 · 10.

The purpose of the cabin-pressurization manifold is to provide for extraction of compressor discharge air for use with aircraft equipment such as cabin pressurization and aircraft de-icing systems. Air is also taken from the cabin pressurization manifold to run the reheat (afterburner) fuel pump, for anti-icing of the engine air-inlet area, and to cool the rear face of the turbine wheel.

Air enters the cabin-pressurization manifold through slots in the inner side of each of the diffuser passages. Air extraction from the manifold represents a loss of compressed air; hence it also causes a loss of engine performance.

The eight diffuser passages are integrally cast and equally spaced within the compressor rear frame. The purpose of the diffuser passages is to channel the air from the annular compressor discharge area into eight individual circular areas onto which the combustion chambers are affixed. The diffusing action of these passages serves to reduce the velocity of the compressor discharge air as it is directed into the combustion system. The diffuser passages also accommodate the installation of the fuel nozzle.

No. 2 bearing

The No. 2 bearing is located in and supported by the inner section of the compressor rear frame. In assembly, the No. 2 bearing together with the No. 2 bearing air and oil seals and allied parts are installed in the compressor rear frame. The bearing installation is illustrated in Fig. 14 · 10. The rear hub of the compressor rotor is drawn in through the bearing and allied parts by a pulling tool. A rear scavenge pump drive gear, a tabwasher, and a locking nut are installed on the hub. The locking nut holds the compressor hub in position within the inner race of the No. 2 bearing. The nut must be removed before the compressor rear frame can be removed from the compressor rotor.

The purpose of the No. 2 bearing is to support the rear end of the compressor rotor in the compressor rear frame. It is also the only thrust bearing in the engine. Since it is a ball bearing, its inner race cannot move in relation to the outer race. The outer race is held in position in the compressor rear frame, and the inner race is held in position on the compressor rotor hub by the retaining or locking nut. Therefore, the rotating parts are held in a fixed position in relation to the stationary parts by the No. 2 bearing. The forward thrust loading of the engine shaft, imposed by the tendency of the compressor to move forward as it draws air into the engine, is transmitted by the No. 2 bearing to the stationary parts. The high thrust loading imposed on this bearing is evidenced by the fact that approximately three times more oil is supplied to this bearing than to any other bearing in the engine. Because of the extent of the expansion and contraction within the engine, it is not practicable to use more than one thrust bearing.

The thrust bearing is located in the center of the engine to allow expansion in both directions, thus decreasing the axial clearances necessary between rotating and stationary parts.

No. 2 bearing air and oil seals

As explained in the case of the No. 1 bearing, it is necessary to seal the spaces surrounding the bearings in order to prevent air leakage and blowout of oil in the bearings. The No. 2 bearing air and oil seals are located in the compressor rear frame around the twelfth-stage wheel hub. These may be seen in Fig. 14 · 10. They are initially installed in the compressor rear frame and when the compressor is pulled into the frame, they will lie between the No. 2 bearing and the rear of the compressor rotor. The labyrinth air seal is machined in an inner flange on a plate which bolts to the seal housing and encloses the vented area. The seal housing in turn bolts to the compressor rear frame. The labyrinth-type oil seal is machined in a flange on the No. 2 bearing oil deflector. The oil deflector is held between a shoulder on the twelfth-stage wheel hub and the compressor-rear-frame scavenge-pump drive gear by the lock nut. The oil seal extends between a flange on the seal housing and the flange of the air seal.

The mating surface for the oil seal is the inner flange of the seal housing. Since the air seal and the seal housing are bolted to the compressor rear frame, they are stationary parts. The oil seal being held to the compressor is a rotating part and rotates between the air seal and the seal housing. The teeth of the air seal point inward toward the compressor hub. The teeth on the rotating oil seal point outward toward the inner flange of the seal housing.

The purpose of the air seal is to isolate the oil seal from the positive pressure created behind the compressor twelfth-stage wheel. This pressure is created by the flow of twelfth-stage leakage and compressor cooling air into this area. The purpose of the oil seal is to prevent the leakage of oil from the No. 2 bearing area forward toward the compressor. The purpose of the seal housing is to serve as a mounting for the air seal and to form an annular vent area and passageway so that the area between the air and oil seals may be vented overboard. Air which blows through the air seal escapes through two passages cast in the compressor rear frame. These passages exit through the flange on the turbine frame which serves to bolt the turbine frame to the compressor rear frame. These two ports are located in the turbine frame flange 180° apart at the 3 and 9 o'clock positions. Most of the air that leaks through the air seal, and into the area between the air seal and the oil seal, goes overboard through these vents. Some of this air leakage, however, is drawn through the oil seal and into the turbine frame by the vacuum created by the turbine frame vent. Air being drawn through the oil seal prevents oil leakage in the opposite direction into the vented area and then out the vents.

Compressor rear frame oil shield assembly

A sheet-metal-fabricated housing, called the "compressor rear frame oil shield assembly," is installed in the inner portion of the rear frame. Its purpose is to retain the oil in the area around the No. 2 bearing and allied parts and to act as the sump which facilitates the scavenging of the oil by the rear scavenge pump.

The rear scavenge pump mounts on the rear of the compressor rear frame, and the pump drive shaft extends through the compressor rear frame casting into the oil shield assembly. The pump sump is incorporated in the oil shield. The rear scavenge pump gear train is encased by the oil shield and is driven by a gear keyed to the compressor hub.

Turbine frame

The turbine frame is a cylindrical fabricated structure flanged and bolted to the rear of the inner section of the compressor rear frame. Figure 14 · 12 is an illustration of the unit in a vertical position. The turbine frame extends rearward around the center line of the engine and encases and supports the No. 3 and No. 4 bearings which in turn support the turbine shaft and wheel. The turbine frame supports the combustion chamber inner liners, transition liners, and turbine nozzle diaphragm.

Turbine casing

The turbine casing bolts to the turbine frame and encases the turbine wheel, nozzle diaphragm, and transition liners. The main engine mounting trunnions are incorporated on the turbine casing. The turbine casing then, in supporting the turbine frame, supports the engine itself.

No. 3 damper bearing

The No. 3 damper bearing is a straight roller-type bearing surrounding the turbine shaft within the turbine frame. It is located axially almost midway between the No. 2 bearing and

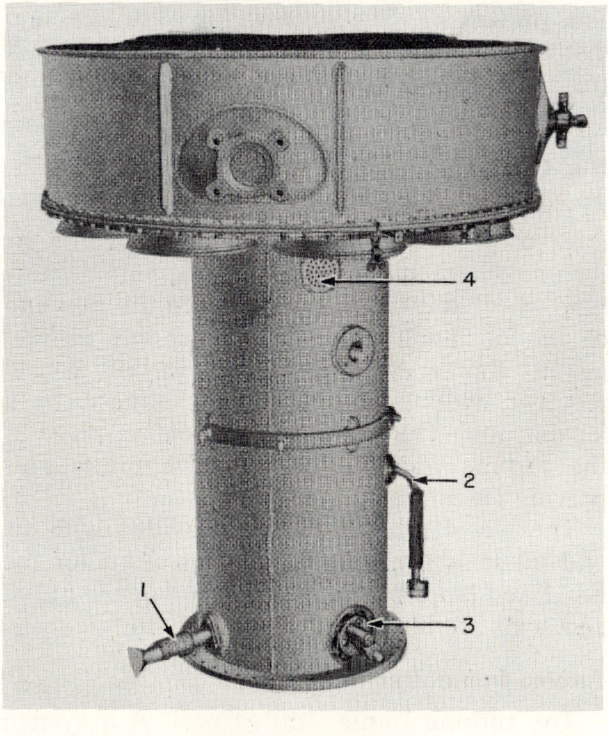

Figure 14 · 12 Turbine frame: (1) turbine-wheel cooling-air bellows; (2) No. 4 bearing-oil scavenge line; (3) oil-flow divider mounting port; (4) No. 4 bearing air- and oil-seal vent (one of four).

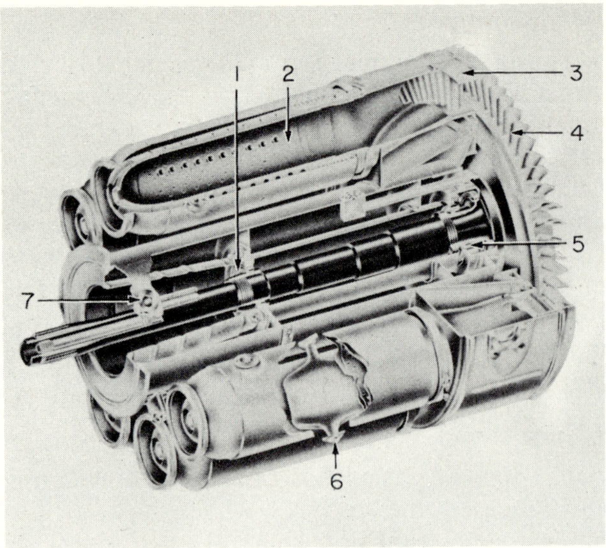

Figure 14 · 13 Turbine and combustion section: (1) No. 3 bearing; (2) burner; (3) nozzle diaphragm; (4) turbine wheel; (5) No. 4 bearing; (6) turbine frame vent; (7) eighth-stage cooling air.

the No. 4 bearing. The purpose of the No. 3 bearing is to dampen any whipping or bending force imposed upon the turbine shaft.

The bearing is mounted in such a manner that it will give a small amount under severe loads, thus preventing the concentration of stresses and distributing the loads throughout the length of the shaft. Figure 14 · 13 shows the mounting of the No. 3 bearing within the turbine frame.

No. 4 bearing

The No. 4 bearing is a straight roller bearing located on the turbine shaft just forward of the turbine wheel. Its purpose is to support the turbine wheel and the turbine shaft in the rear end of the turbine frame. The inner race is held against the forward face of the turbine wheel, together with the No. 4 bearing spacer, by a locking nut. The outer race is held in place in the turbine frame by a retaining plate. This bearing takes only radial loads.

The No. 4 bearing area is provided with air and oil seals similar to those described for the No. 1 and No. 2 bearings. The No. 4 bearing area and seals are illustrated in Fig. 14 · 14.

Turbine frame vent

The turbine frame vent is part of a system whereby air is drawn from the area surrounded by the cylindrical section of the turbine frame. This area is enclosed on the forward end by the No. 2 bearing oil seal and on the rear by the No. 4 bearing oil seal. Air is drawn from the turbine frame through a port located at approximately the 7 o'clock position on the cylindrical section of the turbine frame.

The purpose of the turbine frame vent is to scavenge oil vapors from the turbine frame and to control oil leakage through the No. 2 and the No. 4 oil seals. Oil vapors must be scavenged from this area to prevent the possibility of explosions.

The turbine-frame interior is kept at a negative pressure of between 3 and 9 in. H_2O by means of an ejector located at the end of the vent line. This device applies a small amount of suction to the line.

Combustion system

The combustion system consists of eight direct-flow combustion chambers ("cans") and inner liners equally spaced around the engine between the compressor rear frame and the turbine area. The chambers are held in place by quick-disconnect clamps. Figures 14 · 13 and 14 · 15 illustrate the arrangement of the combustion chambers in the engine.

Since combustion chambers 3 and 7 are the only ones containing ignition electrodes, interconnecting cross-fire tubes are provided to propagate the flame throughout the combustion system when the engine is started. The combustion is then continuous, and ignition is discontinued after the engine reaches a self-sustaining speed. The inner liners are positioned in the chambers by the cross-fire connections. They are supported on their forward ends by the fuel

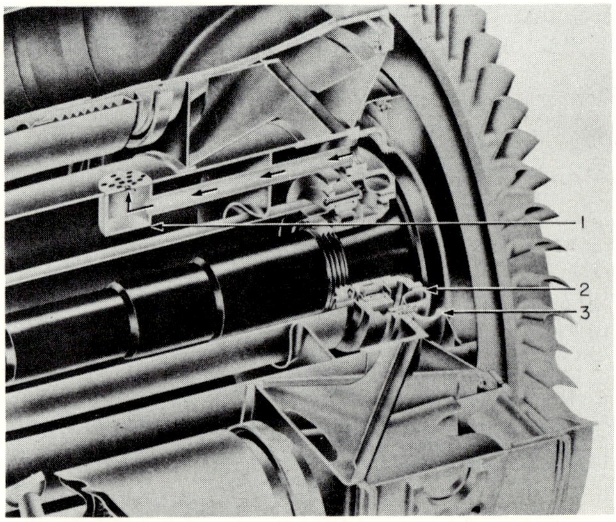

Figure 14 · 14 No. 4 bearing and seals: (1) bearing air- and oil-seal vent; (2) oil seal; (3) air seal.

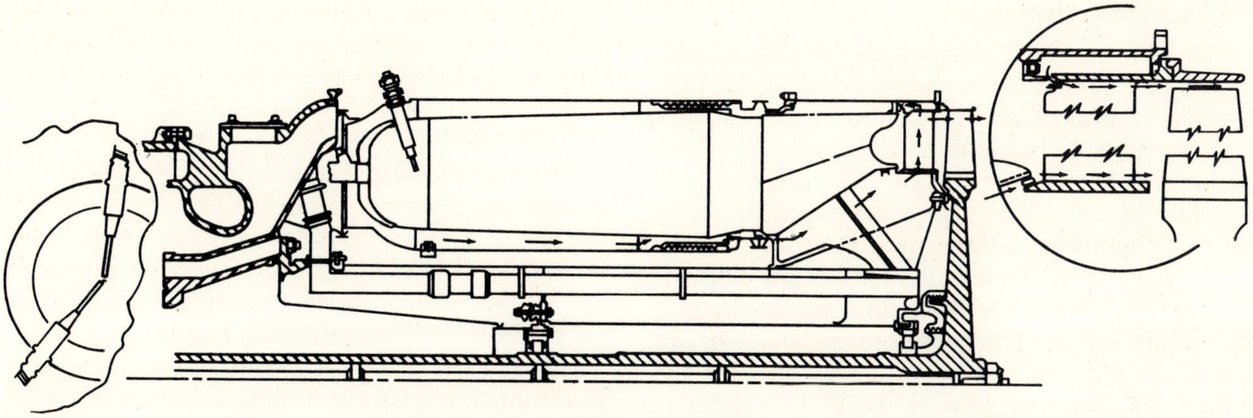

Figure 14·15 Turbine and combustion section.

nozzles and on their aft ends by their fit into the transition liners. The inner liners are not rigidly fixed at either end and can expand in either direction from the cross-fire tube connection. Expansion in the outer chambers is allowed for by a built-in bellows arrangement.

Air flows from the compressor into the combustion chambers between the inner liner and the outer combustion chamber. Primary air passes through a series of large circular holes into the inner liner and mixes with the fuel for combustion. Secondary air passes through louvers provided in the inner liner and forms a cooling blanket of air within the inner liner. All combustion takes place within the inner liner. Air in the inner liner is expanded by the heat of combustion and expelled through the transition liners and nozzle diaphragm onto the turbine wheel.

Some of the air in the combustion chamber does not enter the inner liner but passes around the inner liner and into the turbine frame. The amount of air that enters the holes in the inner liner and the amount of air that passes around the inner liner and into the turbine frame are determined by a metering device fixed to the aft end of the inner liner. This device will be either a metering ring or a metering flange depending upon the type of inner liner used. The air which does not enter the inner liner, but passes between the inner liner and chamber, flows into the turbine frame, where it serves to cool the transition liners and the hollow nozzle diaphragm blades.

The newer elliptical combustion system is so named because in it is incorporated an elliptical-shaped cross-fire connection. The connection between adjacent combustion chambers consists of elliptical-shaped adapters which are welded to a flange and extended through the chambers and onto the inner liners. The flanges of the adjacent adapters fit together flush and are held together between flanges on the combustion chambers by Marmon clamps. These elliptical-shaped cross-fire passages allow for a smooth airflow around them and between the inner liner and the chamber. This increases the cooling effect of the louvers in the inner liners by decreasing the turbulence behind the cross-fire connection.

The elliptical combustion system also incorporates the thimble-type inner liner. The thimble-type inner liner is one in which the large holes in the liner (which admit air into the liner for combustion) are indented slightly into the liner. This projects the air farther into the center of the liner, producing a more combustible fuel-air mixture. The tendency of the engine to smoke at low altitudes is decreased by the use of the thimble-type inner liners.

Transition liners

The eight transition liners are sheet-metal-fabricated passageways located at the aft end of the combustion chambers and encased and supported by the turbine casing. The purpose of the transition liners is to duct the heated air from each of the eight circular-shaped combustion chambers and distribute it onto the annular-shaped nozzle diaphragm. Each transition liner is supported at its forward end by a port in the turbine frame. The combustion-chamber inner liners also project through these ports and into the transition liners. The aft end of the transition liner is supported on its inner surface by a flange on the nozzle diaphragm and on its outer edge by pins which project into a channel welded inside the turbine casing. The transition liners are cooled by air which passes around the metering device at the aft end of the combustion-chamber inner liners.

Turbine nozzle diaphragm

The turbine nozzle diaphragm consists of hollow blades welded between two concentric rings. A small section of the turbine nozzle diaphragm is shown in Fig. 14 · 13. The inner ring is bolted to the rear face of the turbine frame. The outer ring fits into a recess in the turbine casing. The entire assembly is encased by the turbine casing.

The purpose of the nozzle diaphragm is to accelerate and direct hot gases onto the buckets of the turbine wheel. The partitions or blades of the nozzle diaphragm are so constructed and installed that the exit area between the blades is much smaller than the inlet area. The velocity of the air must increase through the diaphragm in order to exit at the same rate of flow as it entered. The velocity increase through the nozzle diaphragm increases the force applied to the turbine wheel, since the turbine wheel is an impulse type and works principally on the velocity drop across it. Since this reaction of the velocity increase takes place in the nozzle diaphragm, the nozzle diaphragm is called the "reaction type." The velocity increase also results in a temperature drop to the turbine wheel which prolongs the life of the turbine buckets and allows higher operating combustion-system temperatures.

The exit area of each nozzle diaphragm is measured at the factory, and the area in square inches is stamped on the rear of the inner ring. The areas of nozzle diaphragms to be used in the J47-GE-17 engine range from 116 to 120 sq in. In replacing the nozzle diaphragm, the exit area should be duplicated within specified limits. If the nozzle diaphragm is replaced with one having an exit area smaller than the specified limits, the airflow through the engine may be restricted to the point that compressor discharge pressure will be increased beyond normal operating limits. This, of course, may result in the tendency of the compressor to stall, especially during accelerations. Smaller exit areas will also result in a slower rate of acceleration, since the compressor will have to work against an increased back pressure, making it harder for the engine to come up to speed. Increasing the nozzle diaphragm area will result in faster acceleration, less tendency to stall, and an increase in specific fuel consumption.

The hollow nozzle diaphragm vanes are cooled internally by airflow from around the outside of the transition liners. This air collects inside the inner ring of the nozzle diaphragm and flows outboard through the hollow nozzle vanes. The air then flows forward across the outer ring, down between the outer ring and the turbine frame, and back past the diaphragm vanes, over the tips of the turbine buckets, and out rearward with the exhaust gases.

Turbine wheel

The turbine wheel and shaft are of one-piece construction, with a strengthening rim welded to the wheel. A portion of the construction is shown in Fig. 14 · 14. It is located in and supported by the turbine frame, with the wheel rotating just aft of the nozzle diaphragm. The purpose of the turbine wheel is to convert the force of escaping gases into rotary motion to drive the compressor rotor. The turbine shaft splines into the hollow hub of the twelfth-stage compressor wheel and is in this way supported at the forward end. The No. 4 bearing supports the turbine shaft and wheel in the turbine frame at its aft end and the No. 3 damper bearing supports the shaft approximately at the center.

The turbine wheel is principally an impulse wheel and works on a velocity drop through the buckets. The inlet and exit areas between the buckets are approximately the same near the roots of the buckets. This allows for the continuance of the velocity as increased by the nozzle diaphragm and contributes to bucket-root cooling. The reaction principle is utilized to some extent at the tips of the turbine buckets by incorporating a twist in the buckets which makes the exit area smaller than the inlet area. This difference in inlet and exit areas imparts an increased velocity to the exhausting gases and adds turning force to the wheel.

Turbine-wheel cooling

The air used to cool the forward face of the turbine wheel is extracted through the top half of the compressor-stator casing at approximately the 10 o'clock position on the eighth stage of compression. Air is routed externally to the rear through a short pipe and down through the compressor rear frame. Thence it goes into the turbine frame by way of a short expansion bellows. Internal piping conveys the air through the turbine frame and onto the face of the turbine wheel through a single port located just outboard of the air-and-oil-seal assembly. The rate of airflow and the efficiency of the cooling of the turbine wheel are determined by the clearance between the turbine wheel and the two circular flanges mounted to the rear of the turbine frame. The air passes out over the face of the wheel, under the flanges, and aft between the roots of the turbine buckets. This airflow also serves to cool the roots of the buckets.

Cooling air to the rear face of the turbine

wheel is extracted from the cabin-pressurization manifold at approximately the 1 o'clock position on the compressor rear frame. Air is piped externally, over the combustion system and down through the No. 1 strut in the exhaust cone. In the inner exhaust cone the air flows forward and around the hub of the rear of the turbine wheel. Here the air is metered by flanges as it passes outboard over the rear face of the turbine wheel and out with the exhaust gases.

Turbine buckets

The turbine buckets are forged from a heat-resistant alloy and are dovetailed into the rim of the turbine wheel. Each of the 96 buckets is prevented from sliding out of the dovetail by a pin which is inserted from the rear side of the wheel into the bottom of the bucket root. The purpose of the buckets is to provide the contoured surfaces necessary to convert the energy of the exhaust gases into rotary motion of the wheel.

Since it is impossible to manufacture a large number of buckets all having the same exact weight and the same weight distribution, the relative effect of any particular bucket on the balance of the wheel is indicated by a number stamped on the bucket root. This number is called the **moment weight.** Two buckets may weigh exactly the same, but the weight distribution throughout the buckets might vary. If the difference is great enough, they will not carry the same moment-weight number. The actual weight difference between a bucket with a moment weight 1 and a bucket with a moment weight 60 will be approximately 30 g. The higher the number, the heavier the bucket. This system of moment weights allows any bucket to be replaced without rebalancing the wheel if the bucket is replaced with one having the same moment weight.

Reheat section

The reheat (afterburner) section of the engine is located just aft of the turbine section and consists of an outer cone, inner cone, flame holder, pilot dome, two rows of spray bars, and the necessary connecting lines and attaching parts. Figure 14 · 16 is a cutaway illustration of the reheat section.

The outer portion of the reheat section, or the outer exhaust cone, is bolted to the rear flange of the turbine casing. The inner portion of the reheat section, or the inner exhaust cone, is supported by the outer exhaust cone with four supporting tubes or struts. These supporting struts are fixed to the inner cone and are fastened on their outer ends to the outer cone by mounting supports. Short pipes are welded to these removable mounting supports, and they extend into the outer ends of the inner-cone supporting tubes. The rear section of the inner exhaust cone is removable and contains the pilot dome and also supports the flame holder. The flame holder is welded to four supporting struts. These struts

Figure 14 · 16 Reheat and exhaust cone.

are parallel to the center line of the engine and are welded to the rear portion of the inner exhaust cone. The inner cone and flame holder extend downstream into the exhaust nozzle. The forward section of the inner exhaust cone contains the two internal fuel manifolds which supply fuel to the two rows of spray bars. The spray bars are fitted around the outside of the inner exhaust cone and project into the exhaust stream.

At the point where fuel is sprayed into the airstream by the spray bars, there is available the necessary air to support combustion because the main combustion system uses the oxygen in only about one-fourth of the total airflow. The velocity of the gases exceeds the burning rate of the fuel-air mixture so that the burning cannot take place at this point. The pilot dome and the flame holder are located downstream of the spray bars. After the fuel has been vaporized in the exhaust stream and reduced in velocity, the pilot dome and flame holder cause turbulence so that combustion and proper flame propagation will occur.

Variable-area exhaust nozzle

The variable-area exhaust nozzle bolts to the reheat section of the engine at the extreme rear end as shown in Fig. 14 · 16. The orifice at the aft end of the nozzle is equipped with eyelids or clamshells to provide a variable exit area. The eyelids are pivoted at the top and bottom of the nozzle and pivot on needle bearings. They are actuated by an electric motor through a jackscrew arrangement.

The exhaust nozzle is equipped with a ceramic-coated inner liner. This liner is of corrugated sheet-metal construction and is riveted to the inside of the exhaust nozzle. The purpose of the exhaust nozzle inner liner is to prolong the life of the nozzle by protecting it from the high temperatures caused by the operation of the afterburner. The use of the ceramic-coated inner liner allows an increased reheat fuel flow schedule without warping and buckling the exhaust nozzle.

The rear section of the exhaust nozzle, to which the eyelids are attached, consists of an inner and outer shell. During reheat operation air is supplied to the area between these shells from the cabin-pressurization manifold. This cooling airflow is necessary to prevent warping of the exhaust nozzle and eyelids.

The main fuel system

There are three fuel systems on the J47-GE-17 engine. The main and emergency systems supply fuel for the engine, and the reheat fuel system furnishes fuel for the reheat section when it is in operation. The three systems have a common fuel supply, but each has a separate control.

The main system supplies fuel for all normal operations. Some of the components of the system are mounted internally on the engine, and some are mounted externally. A drawing of the fuel-system components is shown in Fig. 14 · 17.

The engine-driven booster pump, the main fuel pump assembly, the differential pressure switch and the main fuel-control valve are mounted internally on the forward face of the accessory gear case. A stopcock, oil cooler, flow divider, drip valve assembly, and fuel manifolds are mounted externally on the engine. Fuel nozzles are mounted in the compressor rear frame and protrude into the combustion chambers.

In normal engine operation, filtered fuel from the fuel tank mounted booster pumps enters the engine through the No. 3 island. The fuel passes through the engine-driven booster pump to the main fuel-pump assembly. A differential pressure switch senses internal pressure in the main fuel pump. Fuel from the main pump flows directly into the main fuel-control valve. This valve is controlled by the electronic control system and regulates the amount of fuel which is supplied to the combustion section of the engine. Excess fuel is returned to the inlet of the engine-driven booster pump as bypass fuel. The bypassed fuel passes through the emergency fuel regulator before it reaches the engine-driven booster pump. The controlled fuel leaving the fuel-control valve may flow in two directions. Normally the overspeed governor is closed, and the fuel must flow to the stopcock. Fuel passes through the manually operated stopcock to the oil cooler. From the oil cooler, it flows into the flow divider, where it is separated into large- and small-slot fuel flow to be discharged from the large and the small slots of the fuel nozzles. In the fuel nozzle, the fuel from the large- and small-slot manifold is combined and sprayed into the combustion chambers.

Engine-driven booster pump

The engine-driven booster pump, mounted on the accessory gear case, ensures a constant supply of fuel to the main fuel pump and also eliminates vapors from the fuel.

Fuel vapor forms in a ring on the periphery of the impeller, and this vapor is drawn off into the integral vapor pump, where it is pressurized back into liquid fuel. It is then internally ported to the discharge side of the booster pump.

The booster pump, by eliminating fuel vapors,

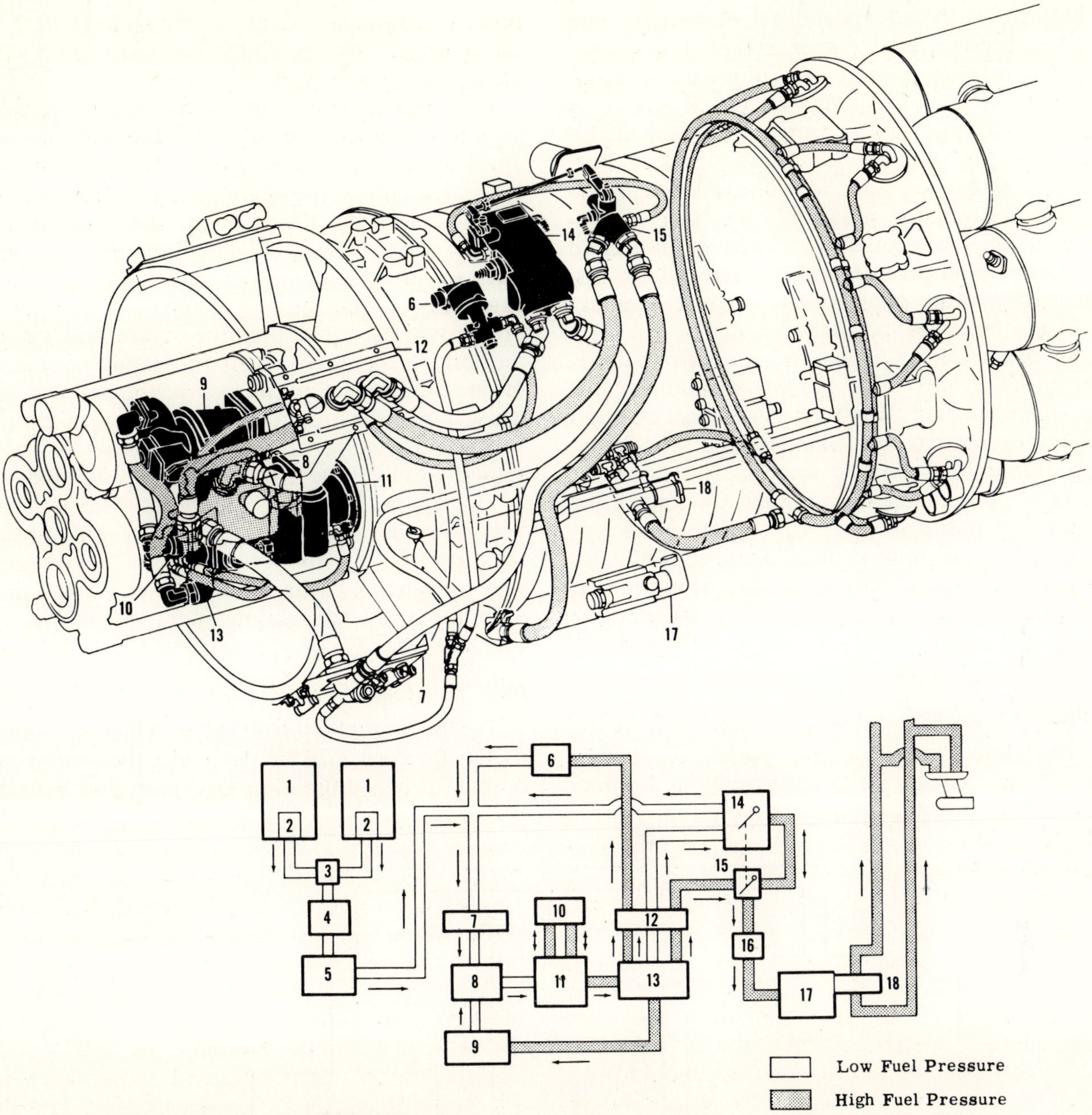

Figure **14 · 17** *Main fuel system: (1) fuel cell; (2) booster pump; (3) North American flow equalizer; (4) fuel-shutoff valve; (5) inlet filter; (6) shutoff valve; (7) island 3; (8) booster pump; (9) overspeed governor; (10) pressure differential switch; (11) main fuel pump; (12) island 4; (13) main fuel-control valve; (14) emergency regulator; (15) stopcock; (16) flow meter; (17) oil cooler; (18) flow divider.*

prevents vapor lock which might otherwise cause failure of the main pump. The booster pump is designed to deliver 40 gallons of fuel per minute under 15 psi at 100 per cent engine speed. This amount of fuel is considerably in excess of the maximum fuel required by the engine.

Main fuel-pump assembly

The main fuel-pump assembly is mounted on the accessory-gear-case cover. Its purpose is to provide more than enough fuel to operate the engine on either the main fuel system or the emergency fuel system, and to protect the system against excessive fuel pressures.

The main fuel pump is a positive-displacement, dual-element, gear-type pump, and its drive shaft incorporates a double shear section which will allow either element to operate without interference because of seizure of the other element. Fuel from the engine-driven booster pump enters the inlet port of the main fuel pump and is pressurized by gear action. Pressurized fuel from each element is individually manifolded into its own filter and check valves. A pressure-sensing tap is provided in each manifold for the pressure differential switch. Fuel flows through the filter and check valve into a common discharge manifold. An internal passage from the common discharge manifold houses a high-pressure relief valve which discharges to the pump inlet when the pump pressure becomes too high.

In the event of main-system control failure, the entire output of the pump is available to the emergency system. The pump is so designed that in the event that one element fails on takeoff, the output of the other element is sufficient to maintain full engine thrust. Each element of the main fuel pump is designed to deliver 17 gallons of fuel per minute under 600 psi at 100 per cent engine speed. Under certain climatic and operating conditions the output of one element alone may not be sufficient to operate the engine at full power in flight.

Differential pressure switch

The differential pressure switch is located within the accessory-gear-case section. Its purpose is to indicate a failure of either element of the main fuel pump, and it is considered a part of the emergency warning system.

A bellows within the differential pressure switch is balanced through external and internal application of fuel pressures. The pressure of each element of the main fuel pump is sensed through a pressure tap on the discharge manifold of each element. These pressures are applied to the bellows, one internally and the other externally. When the two elements of the pump are providing equal pressure, there will be no motion of the bellows. In the event of failure of one pump element, motion of the bellows will cause an electric contact to complete a circuit to the cockpit warning light. The switch is designed to close the circuit when approximately 25 to 40 psi pressure differential exists. The bellows will withstand a maximum differential pressure of 1,000 psi.

A schematic diagram of the pump and differential switch arrangement is shown in Fig. 14 · 18.

Main fuel-control valve

The main fuel-control valve, which is bolted to the forward face of the main fuel-pump assembly, is a combination throttling and bypass-

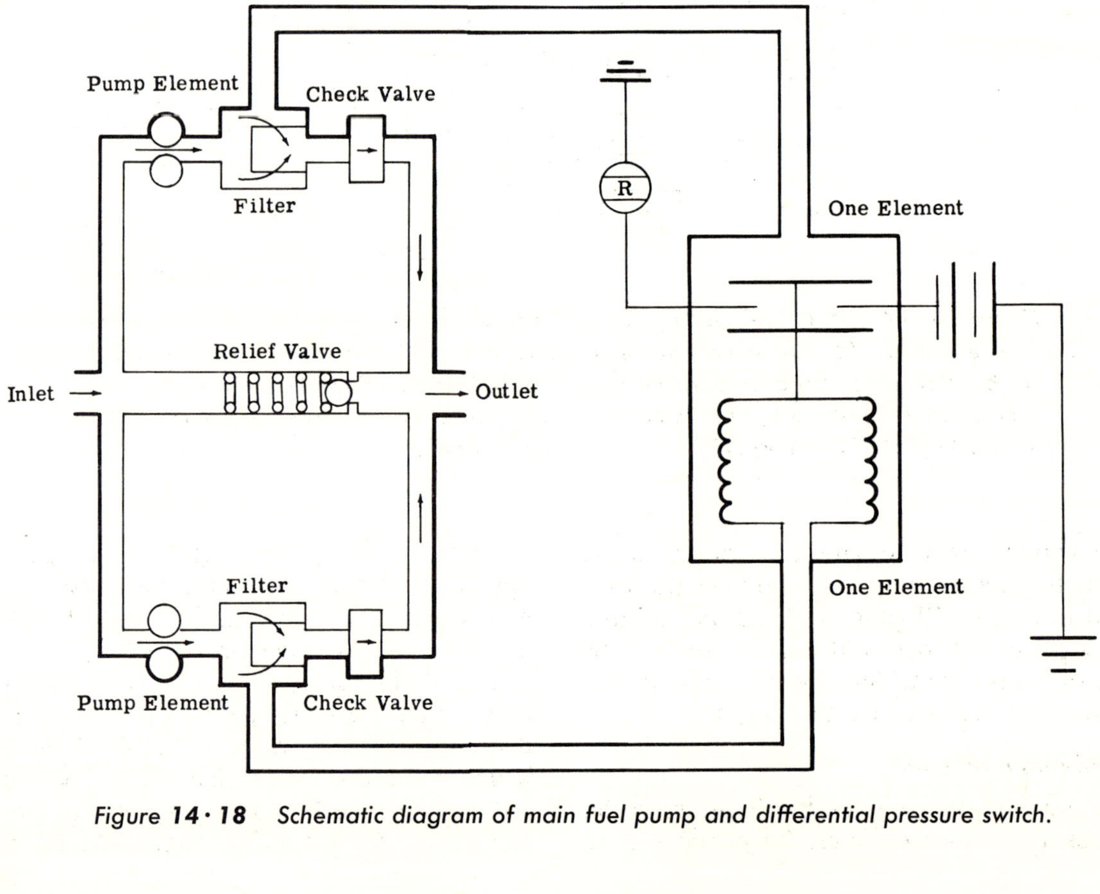

Figure 14 · 18 Schematic diagram of main fuel pump and differential pressure switch.

ing type of valve. The position of the throttle valve is a function of signals received from the integrated electronic control; that is, the electronic control provides the signals which determine the fuel-control valve opening or orifice size.

The main fuel-control valve is designed to maintain a relatively constant fuel flow to the engine, and the value of this flow depends upon the throttle valve setting. The valve will attempt to maintain this flow regardless of discharge pressure up to a preset limit of approximately 600 psi.

The throttling valve is positioned by a throttle valve cam which is driven by a two-phase a-c electric motor through a reduction gear box. The integrated electronic control controls the power and direction of rotation of this motor. Under full amplifier power, the motor will move the throttle valve from the full-open to the full-closed position in approximately ¾ sec. The fuel-control valve is provided with a minimum flow stop for the throttle valve so that a minimum of fuel will continue to flow, regardless of signals to the control valve.

Stopcock

A stopcock is provided in the fuel system to permit positive fuel shutoff, and also to provide metering during emergency starts. The valve is manually controlled and when in the closed position will completely shut off the fuel. At positions from 0 to 14° of shaft position, the stopcock will meter fuel. Above 14° of shaft rotation, the stopcock permits fuel flow for 100 per cent engine speed. The stopcock is placed in the fuel line on the engine side of the main fuel-control valve.

Flow divider

In order to provide proper amounts of flow for the large and small slots of the fuel nozzles, a flow divider is incorporated in the fuel system just before the fuel enters the fuel manifolds. Within the flow divider is a valve which at low fuel pressures is closed by a heavy spring. This permits fuel to flow only to the small-slot manifold when the fuel pressure is below 60 psi. As fuel pressure rises above 60 psi, the valve will permit fuel to flow in increasing amounts to the large-slot manifold.

Drip valves

It is essential that the fuel manifolds be drained when the engine is stopped; otherwise fuel would drip into the combustion chambers and create a fire or explosion hazard. For this reason, both fuel manifolds are equipped with drip valves.

The drip valves are held open by a light spring pressure. When fuel pressure increases above a predetermined value, approximately 3 to 6 psi, the valves close and prevent the fuel from draining. Both drip valves are controlled by small-slot-manifold pressure.

Fuel nozzles

A duplex fuel nozzle is installed so that it directs fuel into the forward end of each of the inner combustion chambers. In each nozzle are passages for fuel from the large- and small-slot manifolds. The small-slot manifold provides fuel for a correct spray pattern at low fuel pressures, and the large-slot manifold provides the additional fuel flow necessary for normal operation. During all speeds above starting, fuel is supplied through both the large- and the small-slot manifolds. Fuel-nozzle operation is shown in Fig. 14 · 19.

Small-slot fuel is directed through two tiny jets into the spin chamber, where the fuel is rotated and sprayed out the orifice into the combustion chambers. The large-slot fuel flows into the same spin chamber through larger-diameter holes and provides a larger volume of flow for normal operation.

The fuel-nozzle tip is provided with an air shroud which directs a stream of air across the tip of the nozzle to prevent carbon formation and to aid in the elimination of smoking.

Emergency fuel system

An emergency fuel system is provided so that the engine may be supplied with fuel in case the main system fails. The emergency system consists of three principal parts: the **overspeed governor,** the **shutoff valve,** and the **emergency fuel regulator.**

The emergency fuel regulator is positioned by mechanical linkage from the pilot's lever and is independent of the electronic fuel-control system. It is also independent of the aircraft electrical system and will operate if the electrical system fails.

Although the emergency system is independent of the electronic system, it will operate in conjunction with it. In this case, the emergency system will control main fuel flow to the engine, but the integrated electronic controls will regulate exhaust-gas temperature and reheat-fuel flow.

When the engine is operating on the main sys-

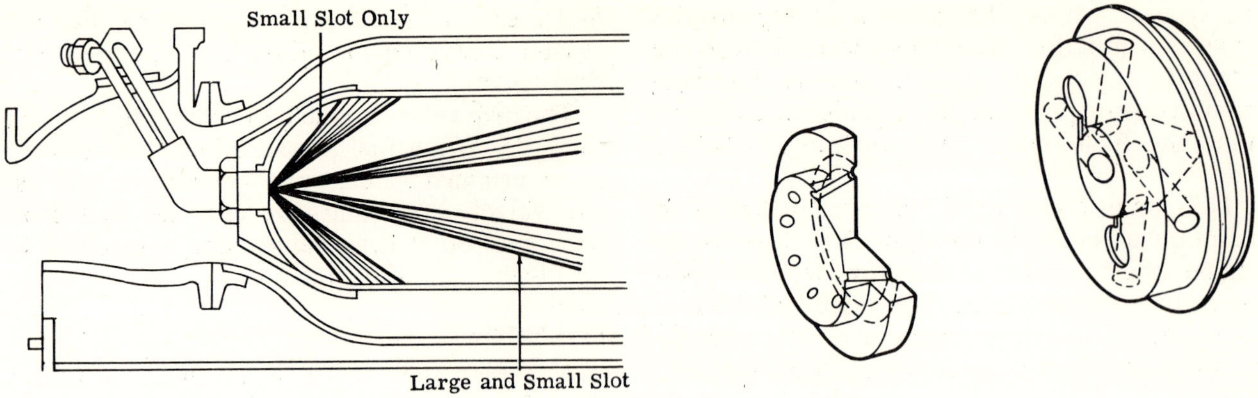

Figure 14 · 19 Fuel-nozzle operation: spray pattern (left) and plates in fuel nozzle (right).

tem, the emergency fuel system will not interfere with the engine operation. There is an amber emergency warning light on the instrument panel which will be lit when the emergency fuel regulator is scheduling fuel. This light will serve as a warning if the emergency system is operating. Under these conditions, all throttle motions must be made with caution.

Overspeed governor

Figure 14 · 20 illustrates the mechanical overspeed governor. Its purpose is to bypass fuel if engine speed exceeds 104 per cent. In simple terms, the governor flyweights operate a pilot valve which controls servo fuel pressure to a power piston. This piston, in turn, controls a bypass valve. Operation begins at about 100.4 per cent engine speed. As engine speed increases, the bypass valve will be opened farther, and at 106 per cent engine speed, the bypass valve is completely open.

Shutoff valve

The purpose of the shutoff valve is to deactivate the main fuel control valve by releasing pressure behind the main bypass valve. Normally, the shutoff valve is held closed by an energized solenoid. When the solenoid is deenergized, the shutoff valve opens and allows the operating pressure behind the bypass valve to be returned to the booster pump inlet. This allows the bypass valve to open, and the main fuel valve can no longer operate. At this time, the emergency fuel regulator controls the flow of fuel to the engine by regulating the amount that is bypassed.

Emergency fuel regulator

The purpose of the emergency fuel regulator is to schedule fuel to the engine in case of main system failure. Figure 14 · 21 illustrates the operation of the fuel system when the main fuel system is operating, and Fig. 14 · 22 illustrates the operation of the emergency system.

The emergency fuel regulator is an altitude-compensating pressure regulator. It regulates the pressure in the fuel system by controlling the amount of fuel that is bypassed by the main fuel system.

In the emergency fuel regulator is a bypass valve which is positioned by the emergency fuel regulator. This bypass valve may either restrict the flow of fuel in the bypass line, increasing the pressure, or allow full flow of fuel in the bypass line and lower the pressure.

A solenoid in the emergency fuel regulator is used to render the emergency fuel regulator inoperative. When the solenoid is energized, the emergency fuel regulator cannot operate. When the solenoid is deenergized, the emergency fuel regulator may operate and schedule fuel pressure. The bypass valve in the bypass fuel line is actuated by a power piston which uses fuel pressure from the stopcock inlet to force the valve closed. This fuel pressure is metered to the power piston by a diaphragm-operated pilot valve. The solenoid operates a ball-type valve holding stopcock pressure out when the solenoid is energized. When the solenoid is deenergized, it allows stopcock pressure to enter the diaphragm to the pilot valve assembly and also to actuate the power piston.

When the engine throttle is advanced, a cam attached to the input shaft is rotated, causing a set of rollers to depress the lower lever. This action depresses the diaphragm, opens the pilot valve, and allows stopcock pressure to operate the power piston. This forces the bypass valve closed and increases fuel pressure and engine speed. The resultant fuel-pressure increase,

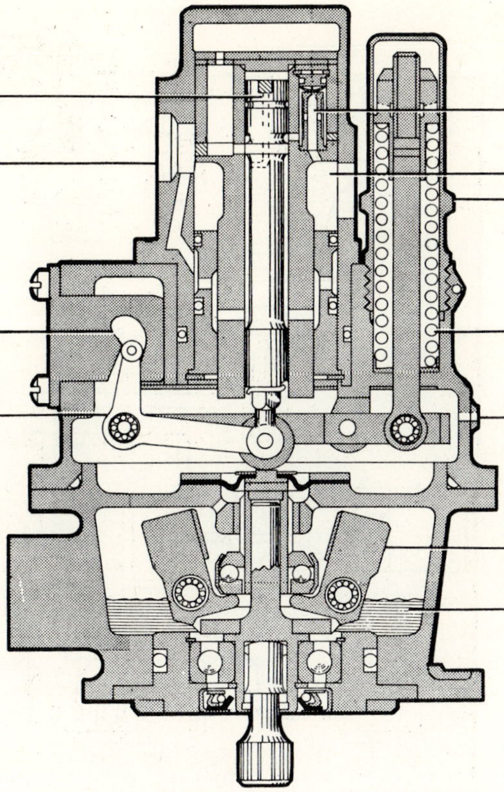

Figure 14·20 Overspeed governor.

sensed at the stopcock, raises the diaphragm-operated pilot valve to a stabilized position. As the engine throttle is retarded, the rollers move back, permitting the lower lever and diaphragm to rise. This closes the pilot valve. Pressure behind the power piston is relieved through an orifice, and the fuel pressure forces the valve open, thus decreasing fuel pressure and engine speed.

The upper lever is pivoted in the middle with an evacuated bellows on one end and a bellows vented to compressor inlet pressure on the other end. This upper lever applies pressure on the roller according to varying compressor inlet pressures which result from changes in altitude and ram. These changes in compressor inlet pressure will cause the fuel pressure to be changed as altitude and ram are changed. Fuel pressure will be increased if altitude is decreased or if ram pressure is increased.

Top-speed adjustment of the emergency fuel regulator is accomplished by turning a ratchet on the top of the emergency fuel regulator. Turning this ratchet raises or lowers the position of the evacuated bellows which will change the effect of pressure of the upper lever on the rollers. Turning the ratchet counterclockwise will decrease engine speed and vice versa. One click of the ratchet will change the fuel pressure approximately 3 psi (1½ psi on later models). This will result in change of approximately 1 per cent (½ per cent on later models) engine speed at 80°F compressor inlet temperature.

It should be remembered that both the stopcock and the emergency fuel regulator are put into operation when controlling solenoids are deenergized. Hence, whenever electric power fails, fuel control is automatically transferred to the emergency system.

System selector switch

Control of the fuel systems is provided by the emergency-fuel-system selector switch. The switch permits a normal fuel-system operation, an emergency fuel-system operation, or a standby position in which both the main and the emergency fuel systems are operating. When the switch is in the NORMAL position, the engine is operating on the main fuel system. When it is in the ON position, or in the TAKEOFF position below 72° of the thrust selector, the engine is operating on the emergency fuel system. In the TAKEOFF position above 72° of the thrust selector, the engine is being operated by the main system but the emergency fuel regulator is in a standby condition. In the TAKEOFF position and the ON position, all throttle motions should be

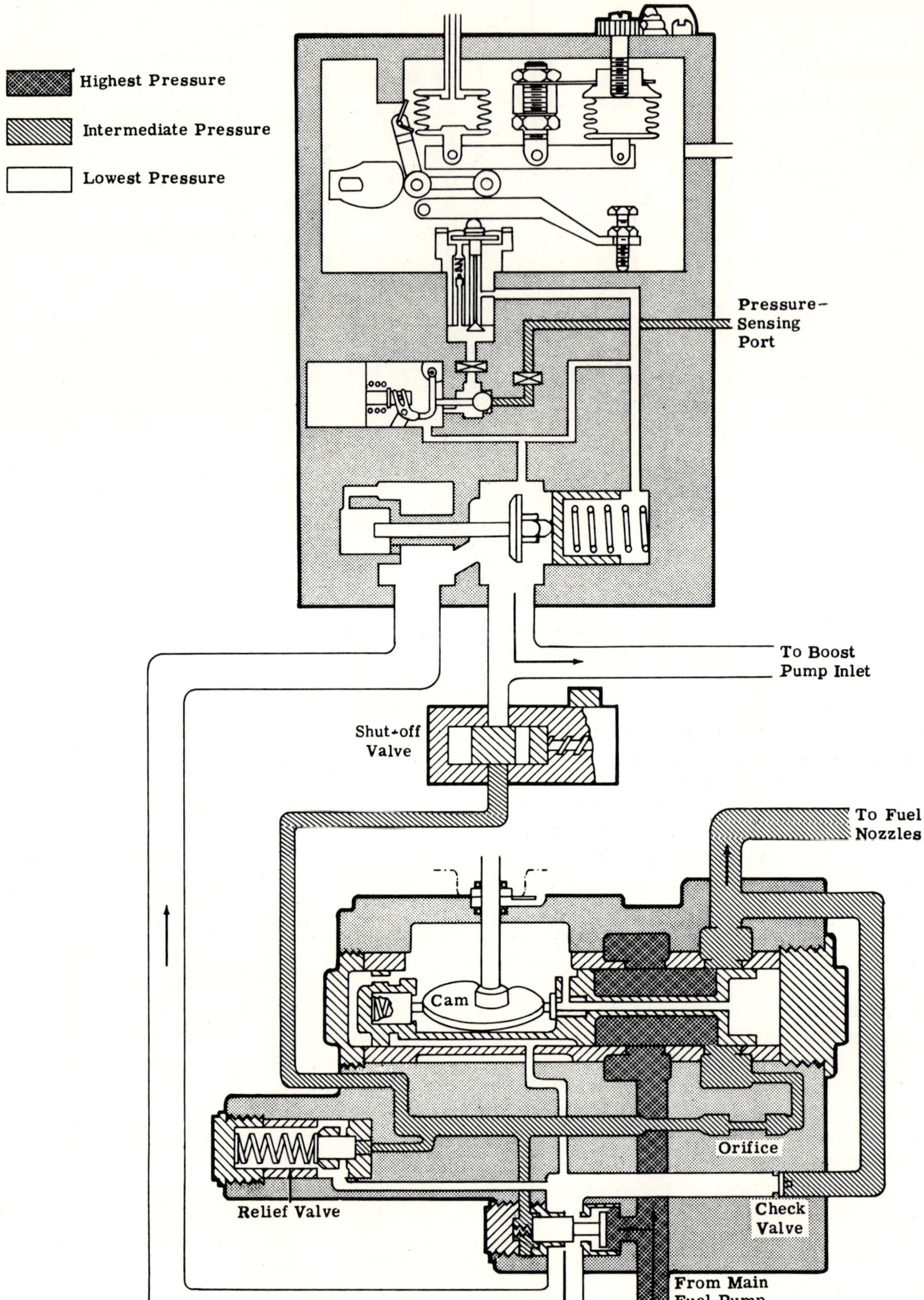

*Figure **14·21** Main-system operation.*

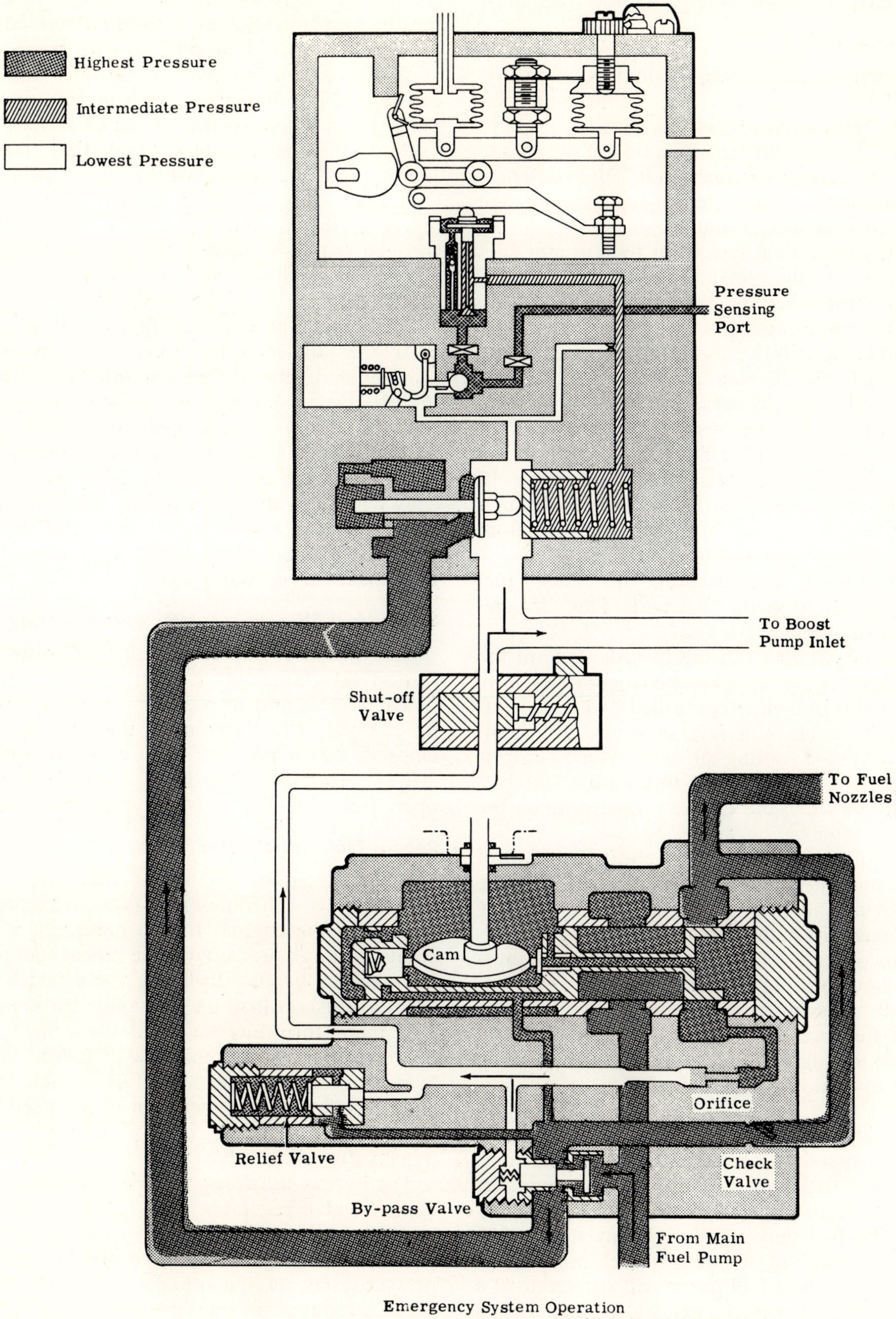

Figure **14·22** *Emergency-system operation.*

made slowly because of the fuel-scheduling characteristics of the emergency fuel regulator.

Reheat fuel system

The reheat fuel system is supplied with fuel from the same tanks that supply the main system. Its purpose, of course, is to furnish properly controlled fuel to the reheat section of the engine when this section is in operation. Three different configurations of the reheat-fuel system are shown in Figs. 14 · 23 and 14 · 24.

During operation, fuel from the airplane fuel tanks flows to the North American flow equalizer and from thence to the reheat fuel pump. From the pump, fuel passes through the reheat filter and shutoff-valve assembly and then to the reheat fuel-control valve. The reheat fuel-control valve is governed by an electronic control. The fuel burned in the reheat section passes from the control valve through a flow divider to primary and secondary manifolds. Each manifold has 19 spray bars, and each bar has 10 orifices through which fuel is sprayed into the reheat section. Excess fuel is bypassed to the main system.

In the later configuration, the reheat fuel pump is located in the fuel tank. This arrangement is shown in Fig. 14 · 24.

Compressed air from the twelfth stage of the compressor is used to drive the reheat fuel pump. This compressed air is controlled by an air-shutoff valve driven by a reversible electric motor. There is also a cooling-air valve which controls the flow of compressed air to cool the variable-area nozzle when the engine is operating on the reheat system.

Controlled fuel from the reheat fuel-control valve flows into the reheat fuel flow divider. The flow divider apportions the fuel to the primary and secondary manifolds. Primary fuel alone is used to operate the reheat section at low pressures, but both primary and secondary fuel operate the system at high fuel pressures. Low reheat fuel pressures are present during initial light-off, at high-altitude operation, and during reheat operation with low fuel flow.

Reheat fuel pump

In the first two reheat-system configurations, the reheat fuel pump is aircraft-mounted. The pump is a centrifugal type, driven by an air turbine. The fuel tank booster pumps supply fuel under pressure to the reheat fuel pump. Unloaded, the reheat fuel pump will accelerate at a rate of about 50,000 rpm per sec. During normal operation the speed is approximately 30,000 rpm.

In the third configuration, the reheat fuel pump is partially submerged in the fuel tank and has the same purpose as the pump described previously; however, it supplies its own boost pressure. The pump has two air turbines and two centrifugal fuel impellers. Each turbine and impeller turns independently of the other. One turbine and impeller are termed the "booster pump," and the other turbine and impeller are called the "high-pressure pump." Compressed air from the air-shutoff valve is directed first through the high-pressure turbine, then through the boost turbine, and finally it is vented overboard.

When the tank-mounted reheat fuel pump is used, the tank-mounted booster pumps supply only the main engine fuel systems. As a result of the change in piping, the booster pumps are rewired to run at one speed only. When the booster pumps supply both the main engine fuel system and the reheat system, they are electrically wired to run at two speeds. The low speed is for the main engine fuel system, and the high speed is for fuel supply to the main engine and the reheat fuel system.

Reheat fuel filter and shutoff-valve assembly

The purpose of the reheat fuel filter and shutoff-valve assembly is to filter the fuel to the reheat system and to provide a positive shutoff for the reheat fuel system. If the filter becomes clogged, and a pressure drop of approximately 20 psi exists across the filter, a relief valve will open and allow the fuel to bypass the filter.

Reheat fuel-control valve

As indicated by the name, the reheat fuel-control valve schedules fuel flow to the reheat section of the engine. It also maintains a constant discharge pressure on the reheat fuel pump by use of a bypass throttling valve which provides a constant load on the reheat turbo pump. The main throttling valve in the reheat fuel-control valve controls the flow of fuel to the reheat section of the engine. Fuel which is not necessary for engine operation is bypassed back to the engine-mounted fuel booster pump by the bypass throttling valve.

The bypass throttling valve in the reheat fuel-control valve regulates the pressure across the main throttling valve. A 50-lb spring plus outlet pressure acts on one side of a piston, and inlet pressure acts on the other side of the piston. Thus it requires an inlet pressure of 50 lb more than outlet pressure to hold the bypass throttling valve in equilibrium. This, of course, results in a

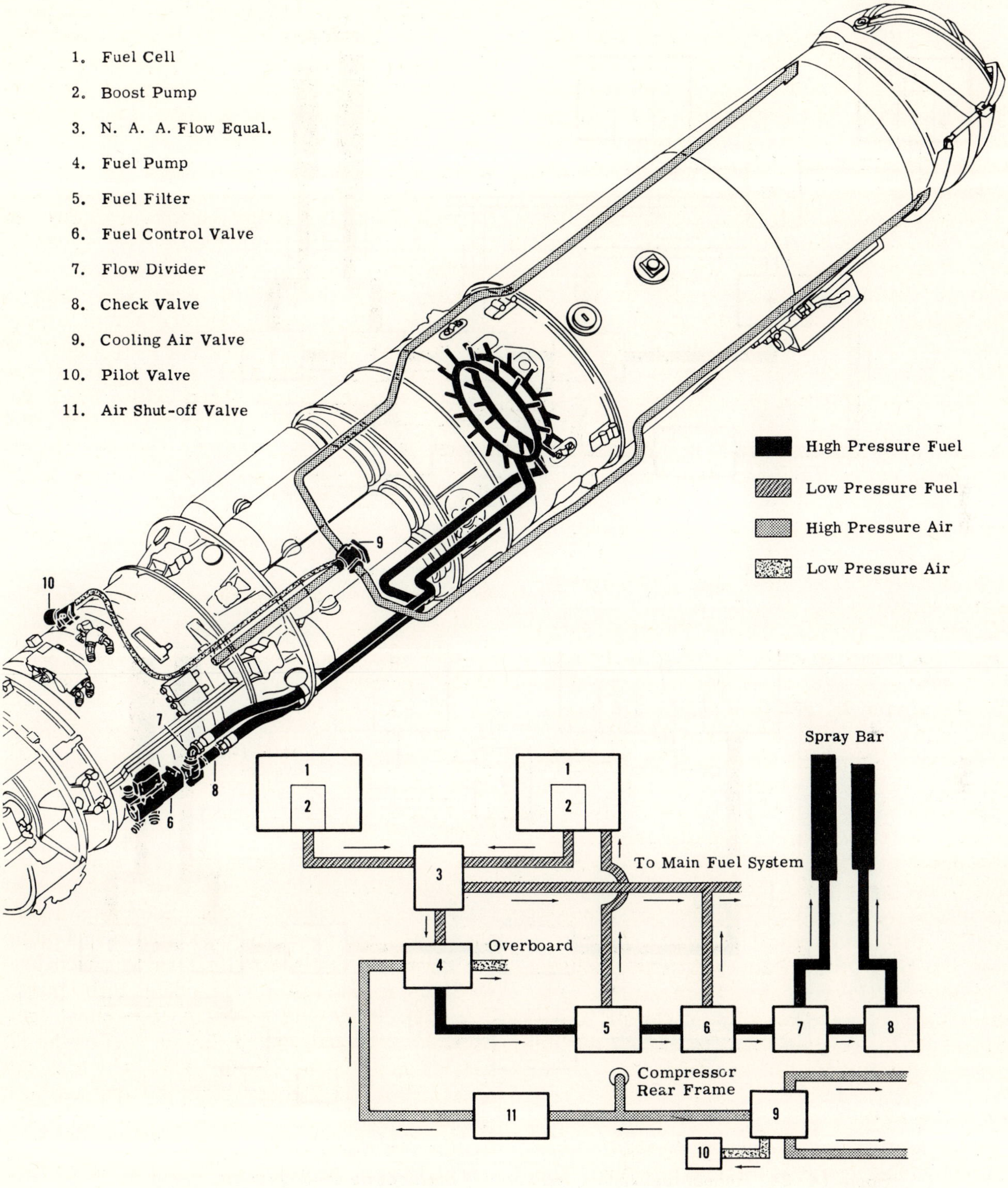

Figure 14·23 Reheat fuel system and variable-area nozzle cooling system, configuration 1.

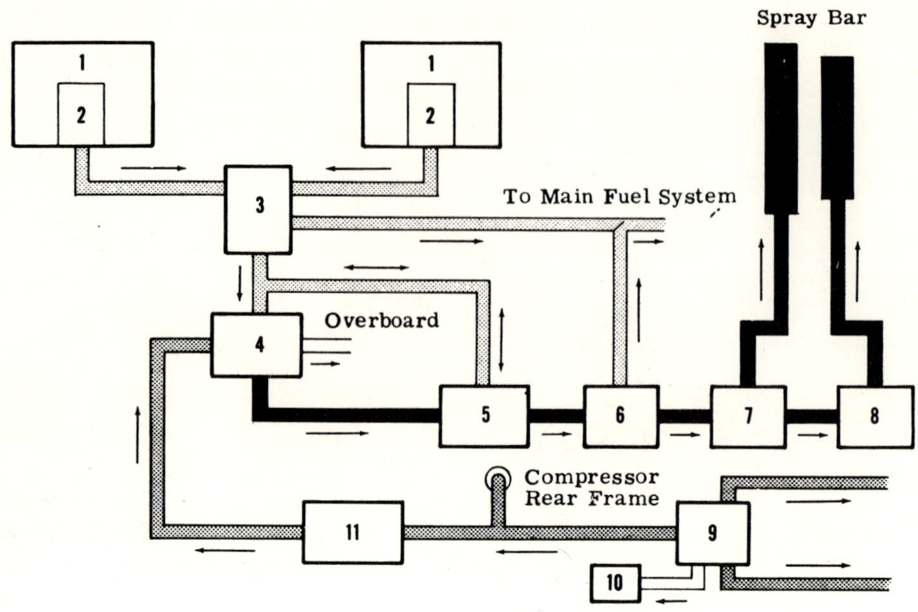

Configuration 2

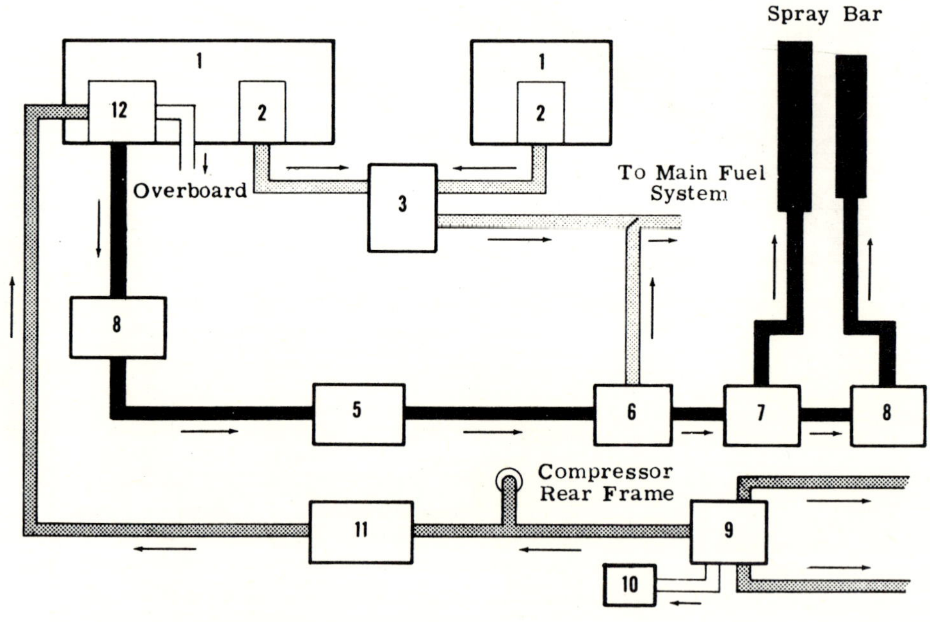

Configuration 3

Figure **14·24** *Reheat fuel system and variable-area nozzle cooling system, configurations 2 and 3: (1) fuel cell; (2) booster pump; (3) North American flow equalizer; (4) fuel pump; (5) fuel filter; (6) fuel-control valve; (7) flow divider; (8) check valve; (9) cooling-air valve; (10) pilot valve; (11) shutoff valve; (12) reheat fuel pump.*

differential pressure of 50 lb across the main throttling valve. A study of Fig. 14 · 25 will aid in gaining an understanding of the reheat fuel-control valve operation.

Reheat flow divider

The purpose of the reheat flow divider is to apportion reheat fuel flow to the primary and the secondary manifolds. When the fuel pressure is low, only the primary manifold receives fuel. The flow divider is mounted on the reheat fuel-control valve.

Variable-area nozzle system

The purpose of the variable-area nozzle system is to provide a means of regulating the exhaust nozzle area so that optimum starting, thrust, specific fuel consumption, and accelera-

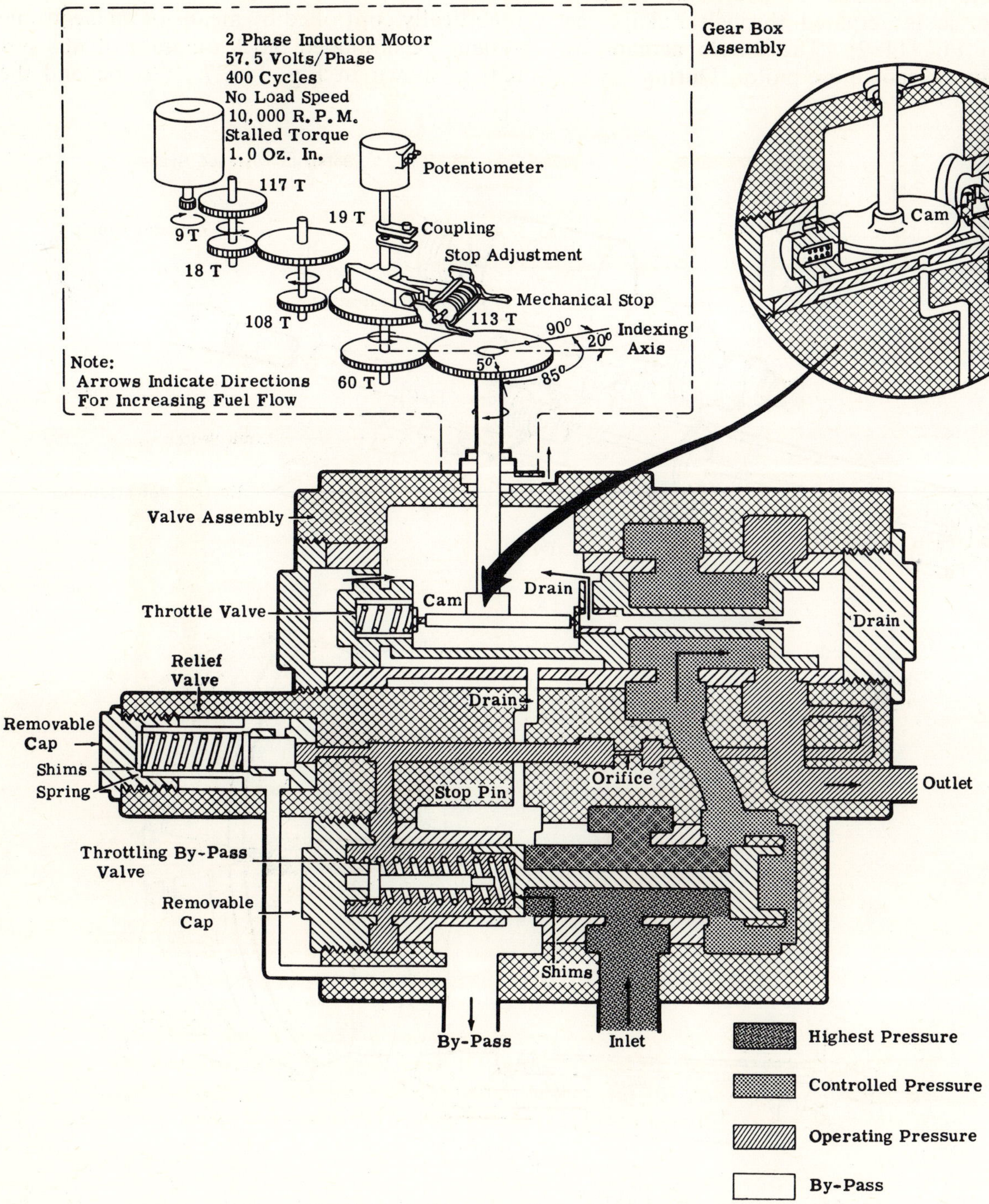

Figure 14 · 25 Schematic diagram of reheat fuel-control valve.

tion characteristics can be obtained. At military and reheat power settings, the nozzle controls the exhaust-gas temperature by varying the exhaust nozzle area to maintain a preset temperature. During such operation, the variable-area nozzle will decrease the area of the exhaust to increase exhaust temperatures and increase the area of the exhaust jet to decrease exhaust temperatures. During rapid accelerations the variable-area nozzle system will remain fixed and enable the engine to accelerate rapidly.

The nozzle is actuated through linkage, as shown in Fig. 14·26. The **nozzle actuator** is driven by a reversible d-c motor. During "automatic" operation, the power to this d-c motor is supplied by an amplidyne generator located in the engine-driven accessory section. The amplidyne output is controlled by an input signal from the electronic control system. During manual operation of the nozzle, the d-c motor is driven by a 24-volt d-c supply.

Integrated electronic control system

As has been previously mentioned, the main fuel-control valve and the reheat system are automatically controlled by means of an electronic system. A simplified block diagram of this system is shown in Fig. 14·27. It is beyond the

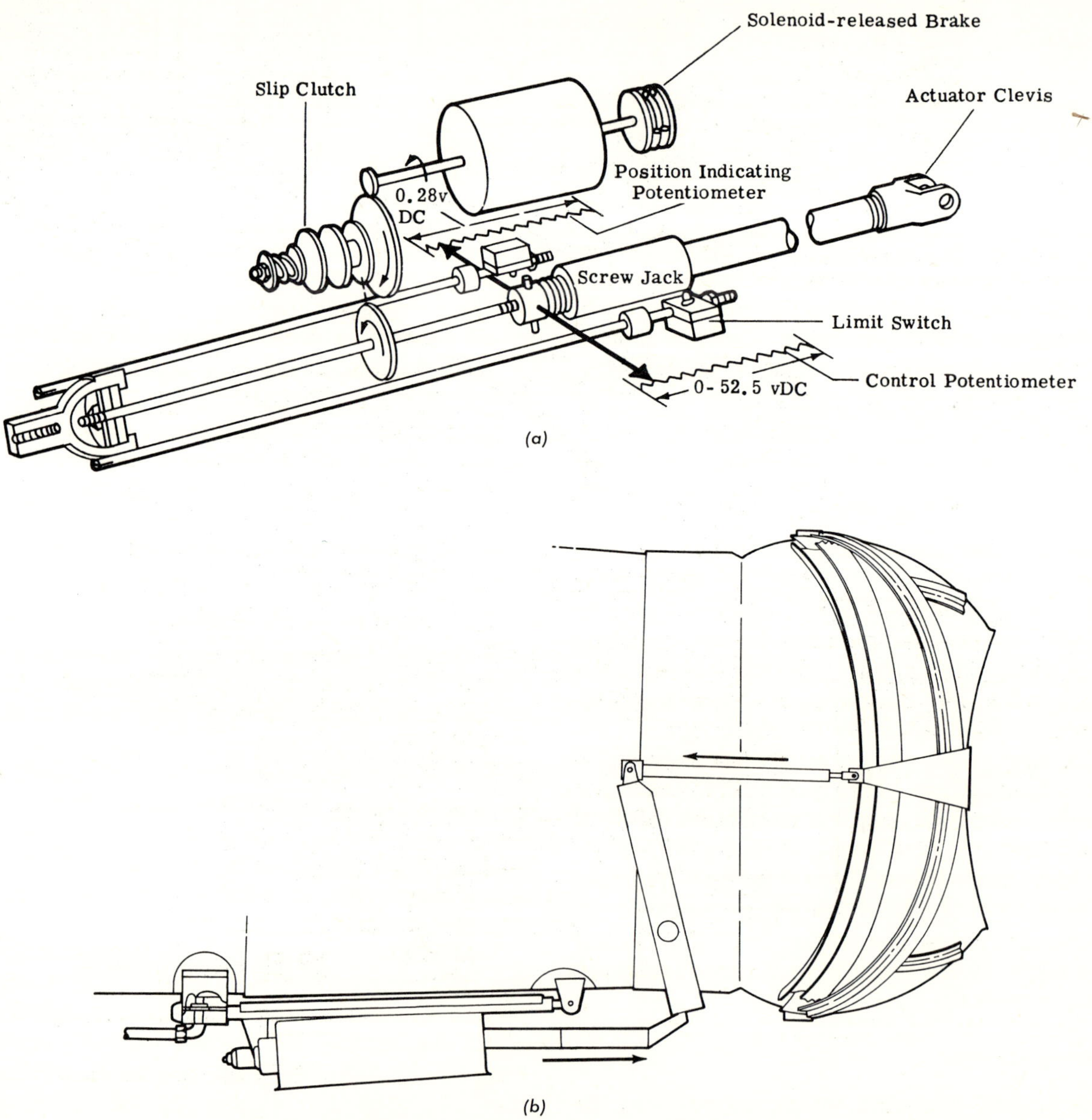

(a)

(b)

Figure ***14·26*** *(a) Linkage and (b) schematic of variable-area nozzle actuator. Arrows indicate directions for increasing nozzle area.*

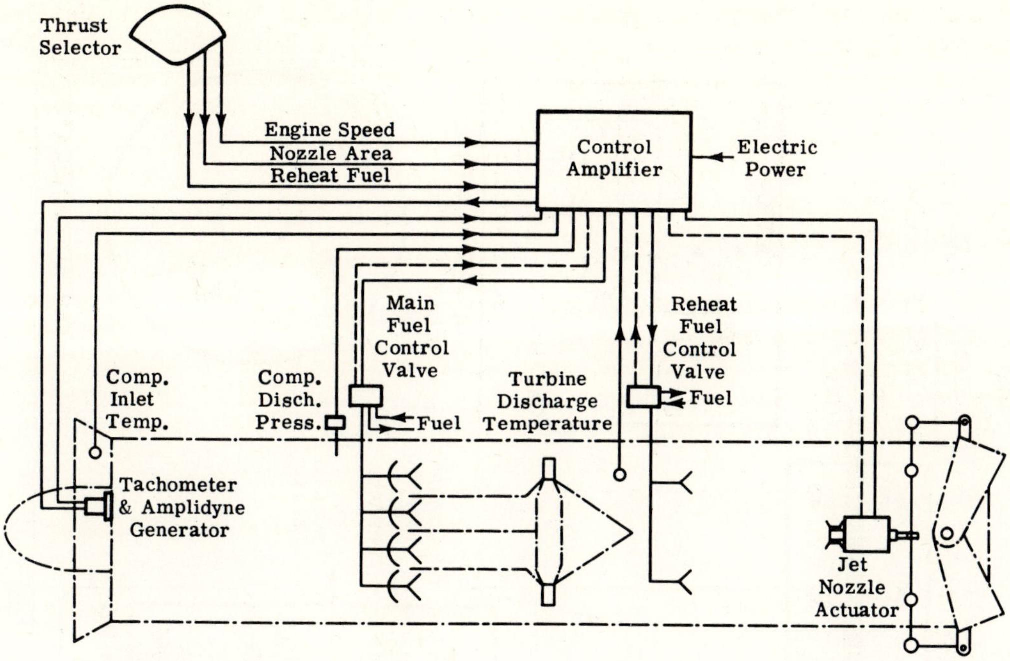

Figure 14 · 27 Simplified block diagram of integrated control on engine.

scope of this text to describe in detail all the electronic elements of the control system; hence we shall discuss only the functions of the system.

As shown in the schematic diagram of Fig. 14 · 27, the electronic fuel-control system consists of a number of different components. The control amplifier shown in the diagram is actually two amplifiers: one for the main system, and one for the reheat system. These amplifiers are designed to integrate the various incoming signals and then send a single signal to the fuel-control valves. The signal sent to the control valve is in the form of a two-phase a-c signal. This two-phase current rotates the actuating motor which moves the throttling valve in the control valve. The direction of the motor rotation will depend upon which phase of the two-phase current leads the other. We may say that if phase 1 leads phase 2, the motor will rotate in a clockwise direction and increase fuel flow; if phase 2 leads phase 1, then the motor will rotate in a counterclockwise direction and decrease fuel flow. It is the function of the amplifier to regulate the phase and strength of the current to the motor in accordance with signals received by the amplifier.

In general, the amplifiers are governed by the following signals:

1. The desired operating thrust, as put into the thrust selector by the throttle position, and transmitted to the amplifiers.
2. The engine speed, sensed by the tachometer.
3. The compressor inlet temperature, sensed by a temperature bulb located in the compressor inlet.
4. The compressor discharge pressure, sensed by the **compressor-discharge-pressure sensor,** which is a dual-bellows aneroid unit that actuates a potentiometer. This sensing unit is located on the engine and receives compressor discharge pressure. It converts the pressure indication to an electrical signal and sends it on to the amplifier.
5. The turbine discharge temperature, sensed by thermocouples located just forward of the engine turbine.

The operation of the electronic control system is such that it causes fuel to be scheduled to the engine in a manner which will prevent damage to the engine, regardless of how the throttle is moved.

The control amplifiers (main and reheat) can be considered as the brain of the electronic control in that they consider all the affecting factors and, through integration of these factors, will signal the proper response to satisfy the desired engine condition.

One of the main operating characteristics of the electronic control is its ability to make completely automatic starts. In making automatic starts, the control will provide for sufficient fuel flow for ignition and then gradually add more fuel to bring the engine up to speed and yet maintain normal and proper exhaust-gas temperatures at all times.

In summary, the electronic control system

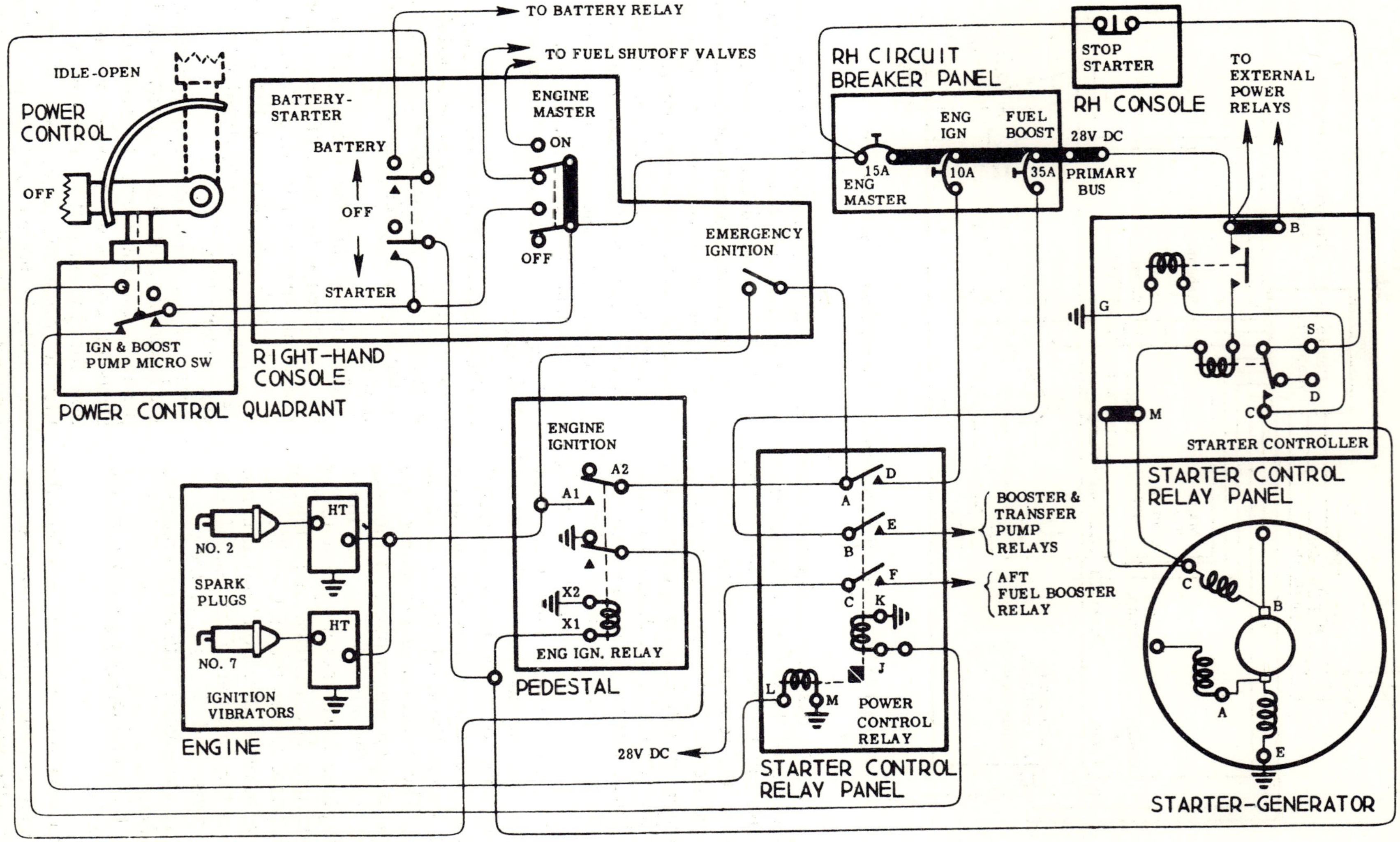

Figure **14·28** *Starting system for J47 engine.*

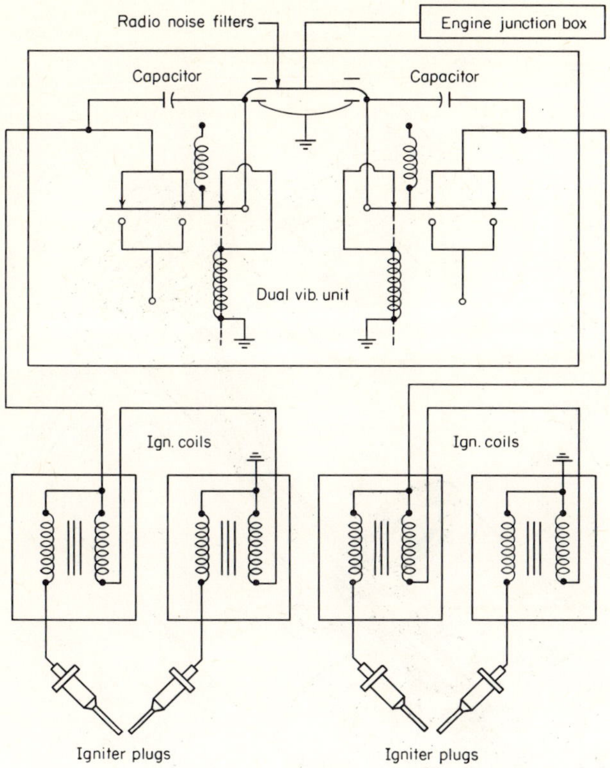

Figure 14·29 Schematic diagram of ignition system.

regulates main fuel flow, reheat fuel flow, and variable-area nozzle position. All these functions are performed automatically in accordance with the desired operation as indicated by the throttle position and according to the conditions existing within the engine.

Starting system

A typical starting system for a J47 engine is shown schematically in Fig. 14·28. This system utilizes a starter-generator which, as the name implies, is used first as a starter and then as a generator when engine speed has reached the proper value.

Starting of the engine is accomplished through the use of a ground power unit with sufficient capacity to meet the high current demands of the starter. Before starting, the ground power unit must be connected to the airplane receptacle, because the starting system is so designed that the airplane batteries cannot be connected to the starter. Furthermore, the airplane batteries do not have sufficient capacity to operate the starter.

When the engine master switch is turned ON, electric power is made available for starting. Also the fuel shutoff valves are opened. The next step in starting is to place the BATTERY-STARTER switch in the STARTER position. This switch need be closed for only a few seconds because of the holding action of the undercurrent relay which is located in the **starter-control relay panel.** Current from the starter switch energizes the **starter-control relay.** This connects the starter to the primary bus, and the starter begins to rotate. Current to the starter passes through the coil of the undercurrent relay and causes it to close the power circuit through the coil of the starter-control relay. Thus the starter-control relay will remain energized and current will continue to the starter. It is then no longer necessary to hold the battery-starter switch in the STARTER position. When the starter is energized, the ignition relay is also energized. Ignition does not start, however, until the engine speed reaches 6 per cent and the power control lever (throttle) is placed in the START-IDLE position. At this time the throttle microswitch is closed and the starter-control relay is energized. This relay furnishes ignition current through the ignition relay to the ignition system. It also energizes the booster-pump relays which furnish power to the fuel booster pumps. At this time the engine "lights off" and begins to develop power.

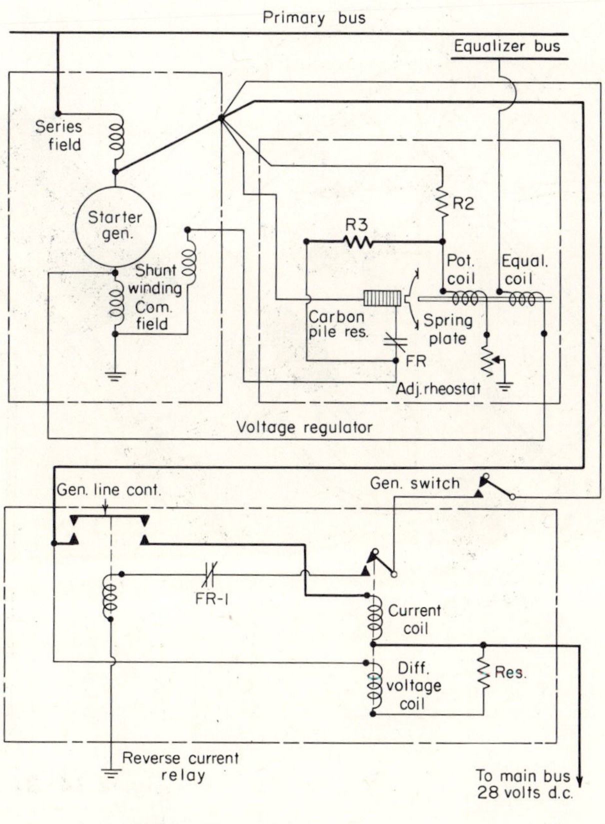

Figure 14·30 Reverse-current relay and voltage-regulator circuits.

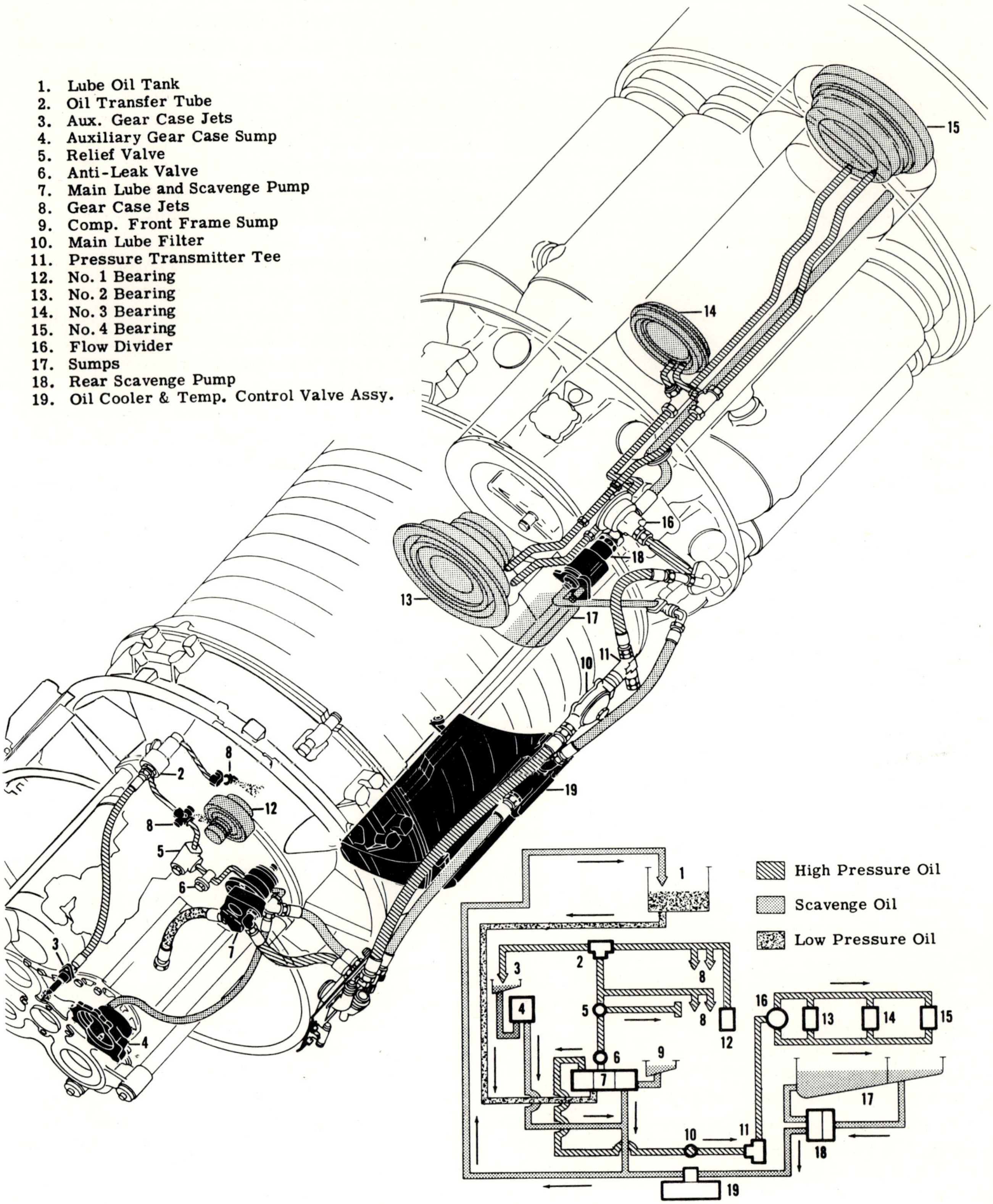

Figure **14·31** *Lubricating system.*

When the engine accelerates to approximately 23 to 25 per cent, the back emf (voltage) of the starter-generator cuts down the current to the starter sufficiently to release the undercurrent relay. The relay opens and disconnects the starter from the power bus.

Ignition circuit

The ignition system used with the J47 engine and shown in Fig. 14 • 29 is known as the "opposite-polarity" type. This means that the two high-tension coils in each system produce a high-tension voltage of opposite polarity. Thus, when one is positive the other is negative, and vice versa. Each individual system consists of shielding capacitors, a vibrator, a double high-tension coil, and two single-electrode igniter plugs.

The current is supplied to the vibrators from the engine junction box when the throttle is placed in the START-IDLE position during starting. The purpose of the vibrators is to change the steady direct current to a pulsating or interrupted direct current. This pulsating direct current into the series-connected primary circuits of the ignition coils produces a rapidly expanding and collapsing magnetic field. This moving field induces a high-tension current in the secondary coils.

The coils are wound in such a manner that the high-tension terminals are of opposite polarity. These opposite-polarity terminals are connected to the two single-electrode igniters. When the igniters are properly installed in the combustion chambers, their electrodes form a gap of about ⅜ to ½ in., across which the high-tension current flows. The spark thus created ignites the fuel.

The two sets of ignition electrodes are installed in the No. 3 and the No. 7 combustion chambers. The electrodes are about 180° apart in the combustion chambers.

Generator system

The generator system used with the J47 engine is a two-generator system, balanced with an equalizer circuit. Each system consists of a generator, a voltage regulator, a differential reverse-current relay, a generator switch, a generator tie-in relay, and the equalizer bus which is common to both systems. One of the generators is the starter-generator used for starting the engine. The circuit for one generator system is shown in Fig. 14 • 30. The generator operates in connection with a field relay which is not shown in the simplified schematic drawing. The contacts FR and FR-1 represent the field-relay contacts.

Lubrication system

The J47-GE-17 engine has a positive-displacement recirculating type of lubricating system. It supplies metered lube oil to the various gears and bearings of the engine for both lubrication and cooling.

The system operates satisfactorily through the complete engine operating range and is limited only by inverted flight and extremely high altitudes. Satisfactory lubrication is obtained at extreme altitude if correct pump inlet pressure is maintained. Inverted flight is limited to about 30 sec because of the large supply pump capacity and the inability of the scavenge pumps to clear the oil sumps while inverted. The oil system, shown in Fig. 14 • 31, comprises a three-element gerotor-type main and scavenge pump, a dual-element rear scavenge pump, a single-element auxiliary-accessory-gear-case scavenge pump, an oil flow divider, an oil cooler, a filter, check valves, and a relief valve.

STUDY QUESTIONS

1. Describe the compressor section of the J47 engine.
2. What type of combustion system is used?
3. Describe the turbine.
4. Where is the accessory section located on the J47 engine?
5. What is the function of the balance chamber?
6. Describe the construction and operation of the air and oil seals.
7. How many bearings are utilized for the rotor of the engine?
8. What method is used to cool the turbine wheel?
9. Describe the variable-area exhaust nozzle.
10. Name the principal units of the fuel system.
11. What type of fuel-control system is used?
12. Describe the starter for the J47 engine.
13. What type of ignition system is used?
14. List the principal components of the lubrication system.

CHAPTER 15

The General Electric CJ805 Aft-fan Engine

BASIC DESCRIPTION AND PERFORMANCE

The General Electric CJ805-23 turbofan engine, shown in Fig. 15 · 1, is essentially a modification of the CJ805-3 engine, shown in the cutaway illustration of Fig. 15 · 2. The basic engine (gas generator) is the same for both engines, and the difference lies in the configuration aft of the turbine section. The CJ805-3 engine incorporates a thrust-reverser and a sound-suppressor mounted at the rear flange of the turbine section, while the -23 engine mounts a free-turning single-stage turbine having fan blades which extend outward into the coaxial duct surrounding the main exhaust duct. The fan blades accelerate a large additional mass of air which results in changing the thrust at takeoff power from 11,650 to 16,100 lb. At the same time, the sea-level takeoff specific fuel consumption is changed from 0.81 to 0.58 lb per lb thrust per hr. The value of the turbofan or aft-fan principle is thus clearly demonstrated.

An important feature of the CJ805 engine is the variable-inlet guide vanes coupled with six stages of variable stator blades in the compressor section. This design enables the engine to maintain a high degree of efficiency throughout a broad range of operating conditions without encountering a stall condition.

The engine has an exceptionally good power/weight ratio because of its load-carrying outer skin, conical-type construction of support members, and sheet-steel construction.

The performance and specifications of the CJ805-23 engine are given in the following list:

Compressor type	17-stage axial flow, single spool
Compressor ratio	13:1
Combustor	Cannular with 10 combustion liners
Turbine section	3-stage main turbine, 1-stage free-turning fan turbine
Weight	3,800 lb
Thrust, sea-level takeoff	16,100 lb
Power/weight ratio	4.24 lbt/lb
Maximum cruise power, sea level	13,600 lb
Sfc at max. cruise	0.53 lb/lbt/hr
Bypass ratio	1.54:1
Bypass pressure ratio	1.65:1

The CJ805-23 turbofan engine consists of five major sections as shown in Fig. 15 · 1. These are the **compressor section,** the **combustion section,** the **turbine section,** the **aft-fan section,** and the **accessory drive section.** Each of the major sections comprises subsections which will be described in those parts of this chapter which describe and explain the major sections.

DESCRIPTION OF MAJOR SECTIONS

Compressor section

The compressor section of the CJ805-23 engine, shown in Fig. 15 · 3, consists of the **front frame, compressor front stator casing, compressor rear stator casing, compressor rotor,** and **compressor rear frame.** The function of the compressor section is to "pump" as much air as possible to the combustion section where it is expanded by means of the burning fuel and then ejected rearward through the turbine section. To pump a large mass of air through the engine, the air must be compressed. In the compressor

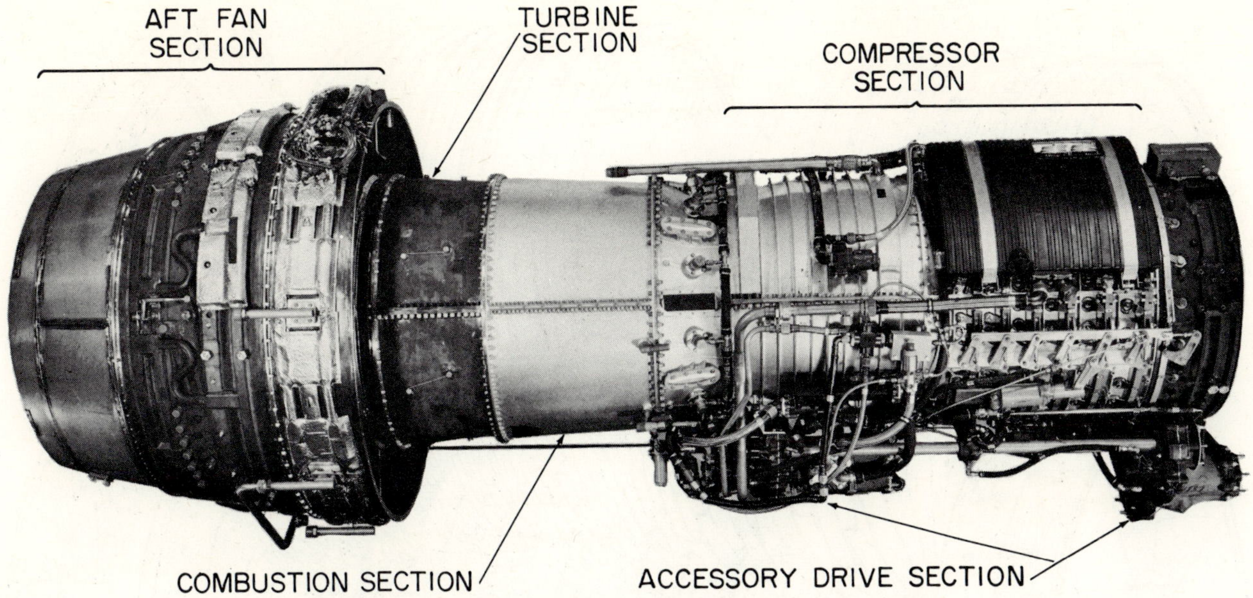

Figure 15 · 1 General Electric CJ805-23 turbofan engine.

of the CJ805 engine the air is compressed at a ratio of 13:1, and the temperature of the air is increased to approximately 850°F at standard sea-level takeoff conditions.

The **compressor front frame,** illustrated in Fig. 15 · 4, is the forward structural member of the engine. Housing the No. 1 bearing which supports the forward end of the compressor rotor, the compressor front frame also serves as the air inlet to the engine. The outer shell of the frame provides trunnion mounts by which the engine may be mounted in the engine

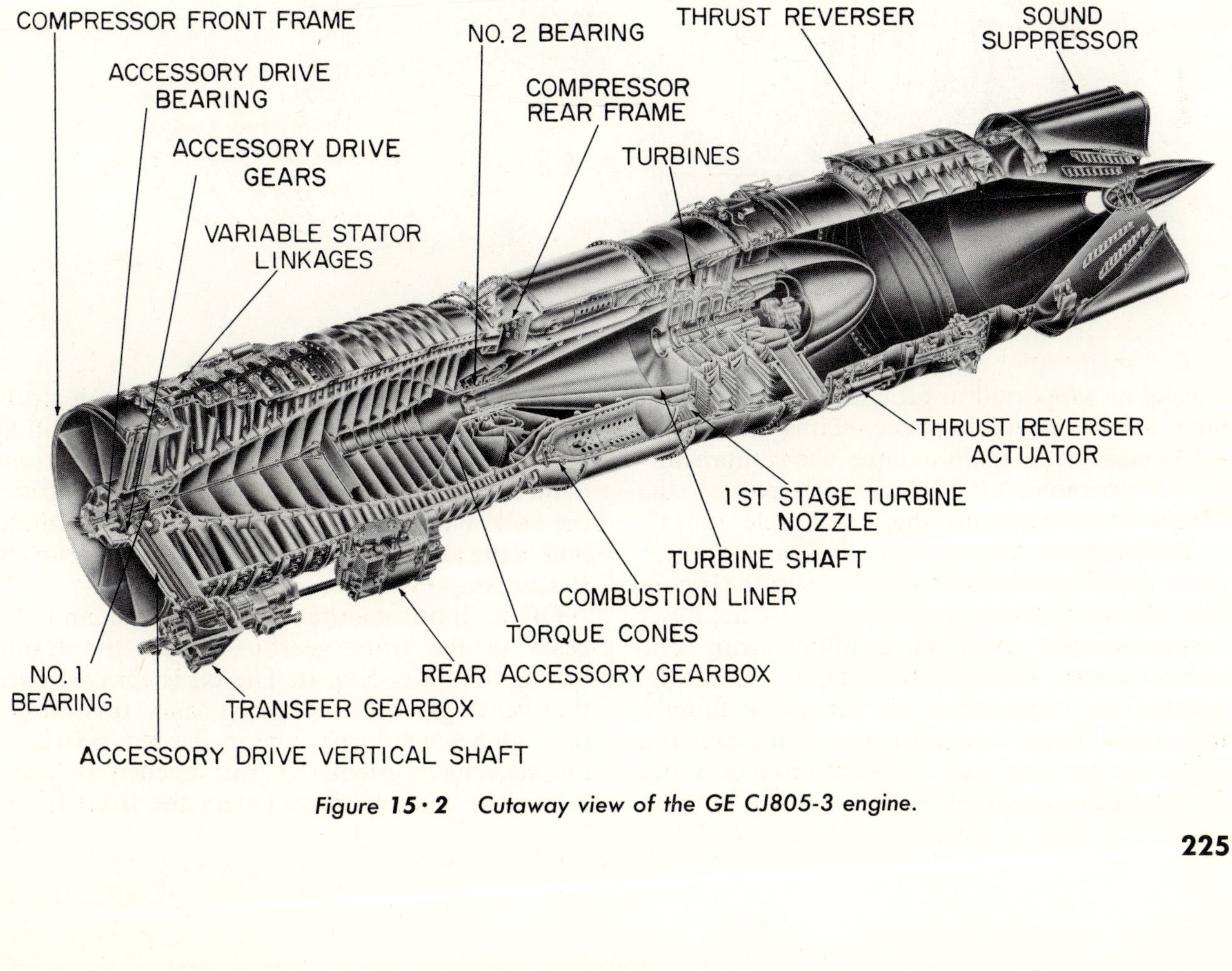

Figure 15 · 2 Cutaway view of the GE CJ805-3 engine.

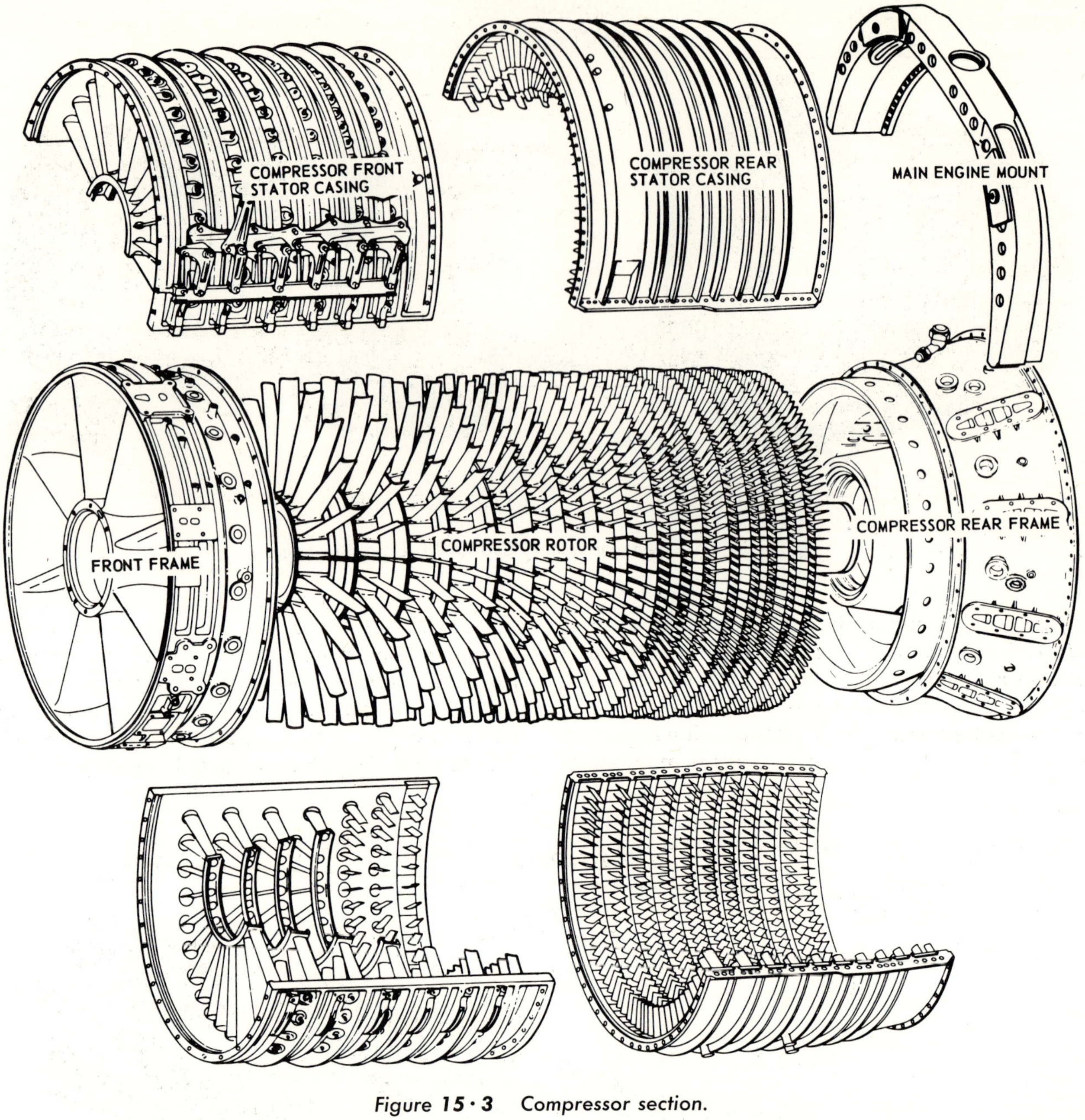

Figure **15 · 3** *Compressor section.*

nacelle or supported in ground handling equipment. The front frame also contains the mounting bosses for the ignition units, the compressor-inlet-temperature (CIT) sensor probe, the transfer gearbox, and the nose cone fairing.

The frame is a magnesium-alloy casting which provides a strong, lightweight structure. It consists of an inner hub connected to an outer casing by eight streamlined, hollow struts. The hub houses the front gearbox, the No. 1 bearing, and the No. 1 bearing oil and air seal. A mounting ring is bolted to the aft face of the hub and supports the inner ends of the inlet guide vanes (IGV) and the bearing carbon rubbing air seal.

Anti-icing air enters the frame through struts 2, 3, 7, and 8. This air then flows into a manifold cast into the rear side of the hub, and from thence it is distributed to the remaining struts, the hollow IGVs, and the nose cone. From these points the air is discharged back into the primary airstream.

Oil to lubricate the No. 1 bearing and the gears in the front gearbox enters the frame through the No. 5 strut. The oil is sprayed over the bearings and the gears and then flows through a second cast tube in the No. 5 strut to the scavenge system. The air needed to pressurize the bearing sump enters the front frame from the compressor rotor stub shaft, leaks across the oil seal, and flows through the scavenge line in the No. 5 strut to the transfer gearbox. This air originates from the seventh stage of the compressor from whence it is fed into the compressor interior.

Power is extracted from the compressor rotor shaft to drive the gears in the front gearbox. This arrangement can be seen in the cutaway illustration of Fig. 15 · 2. This power is transmitted to the transfer gearbox through the radial drive shaft which is housed in the No. 5 strut.

Twenty stainless-steel IGVs, spaced evenly around the hub, are mounted aft of the front-frame struts. These vanes are connected, through half-rings and bellcranks, to the variable stator vanes. The angle of the vanes is variable as a function of compressor inlet temperature and engine speed. Self-aligning bearings, providing for free vane movement under all conditions of operation, are employed in the vane mounts. The IGVs and variable stator vanes are actuated by means of hydraulic cylinders which receive operating pressure from the fuel-control unit.

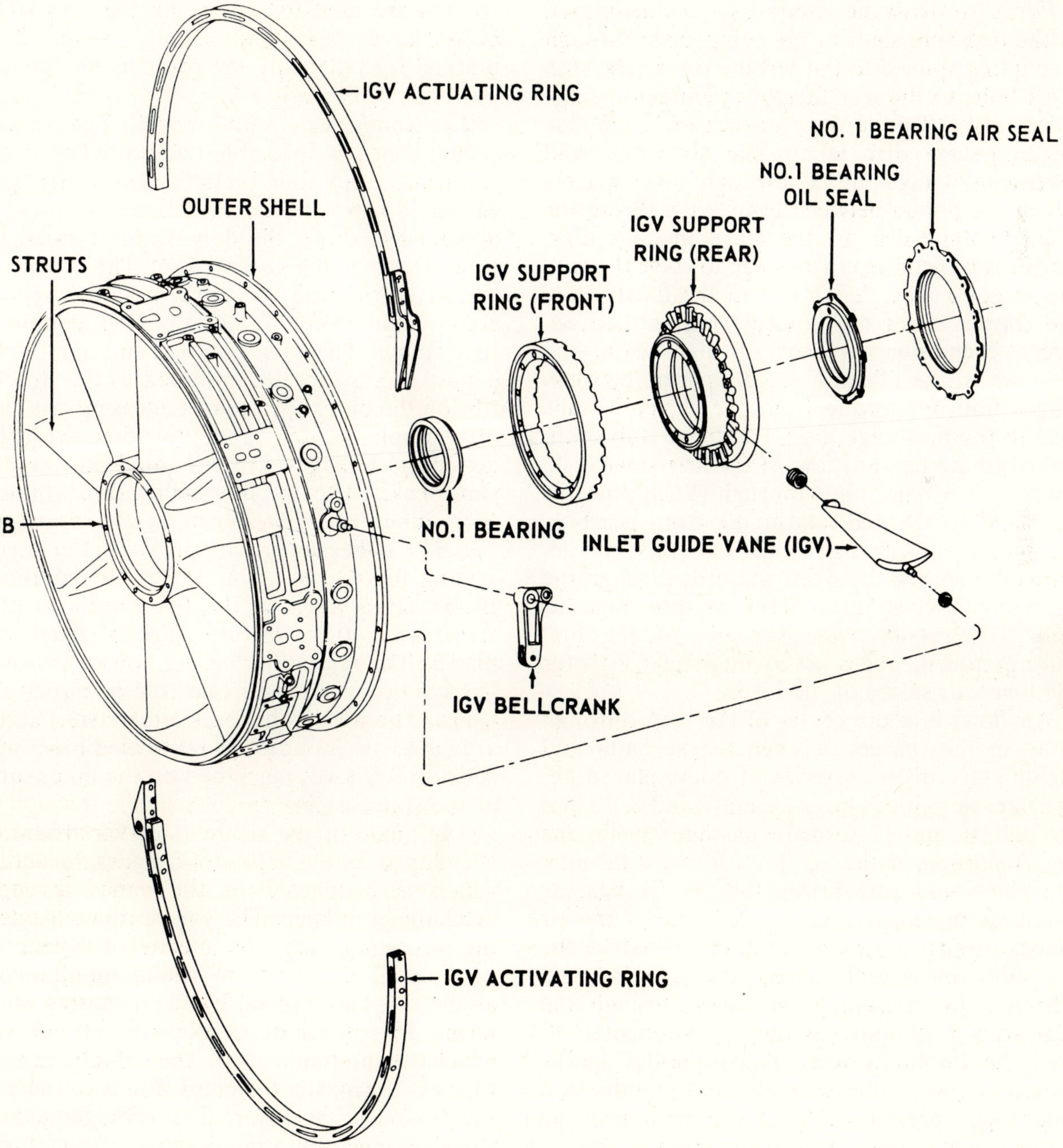

Figure 15 · 4 Compressor front frame.

The **compressor rotor** performs the actual work of compressing the air and delivering it to the combustion section. A cutaway drawing of the rotor is shown in Fig. 15 • 5. The interior of the rotor acts as a passageway for No. 1 bearing sump pressurizing air. In addition, the rotor transmits power from the turbine to drive the accessory gear trains.

The **compressor rotor** is a bolted assembly consisting of disks, blades, spacers, ducts, torque cones, and stub shafts. The rotor blades are machined from stainless-steel forgings, while the remainder of the components are machined low-alloy-steel forgings.

Power to drive the compressor is transmitted to the rear stub shaft of the compressor through a coupling splined to the turbine rotor. The stub shaft bolts to the rear face of the fifteenth-stage rotor disk. Torque is transmitted from the fifteenth-stage disk aft to the sixteenth- and seventeenth-stage disks through the spacers which are bolted between each stage. From the fifteenth-stage disk to the eleventh-stage disk, torque is transmitted from stage to stage through the **torque cones.** As shown in the illustration, the torque cones are conical rings bolted between each adjacent pair of disks from the eleventh to the fifteenth stage. Spacers between stages transmit torque from the eleventh-stage disk to the first-stage disk. The front stub shaft, bolted to the forward face of the first-stage disk, contains a female spline through which power is delivered to the gearsets in the front gearbox. In addition to being torque-transmission devices, the spacers between the fifth through the seventeenth stages also serve as interstage air seals. The stator vane shrouds of the first through fourth stages act as interstage seals for the first four stages of the rotor.

Air flows into the center of the rotor through holes in the spacers between the seventh- and eighth-stage disks. A series of ducts placed between each pair of stages except 7 and 8, 15 and 16, and 16 and 17 forms a passageway for the air. A portion of the air flows forward through the rotor and pressurizes the No. 1 bearing sump, as previously explained. Some of the air bleeds through holes in the ducts to reduce the pressure differential across the spacers and wheels. The remaining air flows through the rear stub shaft and the turbine rotor shaft to cool the turbine wheels. A paddle-like baffle, located between the seventh- and eighth-stage rotor disks, prevents angular acceleration of the air as it enters the center of the rotor.

The compressor blades are mounted on the rotor disks by a single-tang dovetail arrangement. Slots are machined in the rims of the disks to receive the dovetail roots of the blades. Movement of the blades is prevented by the spacers bolted between the pairs of disks at the rims. A ring, mounted on the forward face of the first-stage disk, prevents forward movement of the first-stage blades; and the seventeenth-stage air seal, attached to the rear face of the seventeenth-stage disk, prevents rearward movement of the seventeenth-stage blades.

The bolts used for mounting the stub shafts and the seventeenth-stage air seal and the bolts in the outer surface of the third- and tenth-stage spacers are used for balancing the rotor. These bolts have been individually weighed and marked for balancing the rotor by changing the positions of the bolts.

The **compressor stator** consists of an outer casing which is divided horizontally and circumferentially into four sections. Inside the stator casing 18 rows of stator vanes are mounted and arranged to direct the flow of air from stage to stage through the compressor. The compressor casing carries the structural load between the compressor front frame and the compressor rear frame. The rear gearbox and some of the engine accessories are mounted to the stiffening ribs on the outer surface of the casing. Six stages of variable stator vanes together with their actuating linkages are mounted in the front stator cases. The first four stages are shrouded at their inner ends as shown in the drawing of Fig. 15 • 3. These shrouds provide interstage air sealing for the first four stages. In addition to the variable vanes in the front sections of the stator case, there is one stage of fixed stator blades. The rear stator casing houses 10 rows of fixed stator vanes and one row of outlet guide vanes. The airfoil-shaped stainless-steel stationary vanes are mounted in fabricated bases which slide into grooves machined on the inner surface of the stator casing.

The angle of the six rows of variable stators is changed by the action of a hydraulic actuator which is connected to the vanes through a mechanical linkage. The vanes are connected to the actuating rings by means of levers, thus causing all the vanes to rotate simultaneously as the rings are moved by the actuating mechanism. This mechanism consists of bellcranks which are interconnected. The second-stage bellcrank is the master bellcrank and is connected to the hydraulic actuator. The arrangement of a variable vane mounting is shown in the drawing of Fig. 15 • 6. The outer end of the vane is

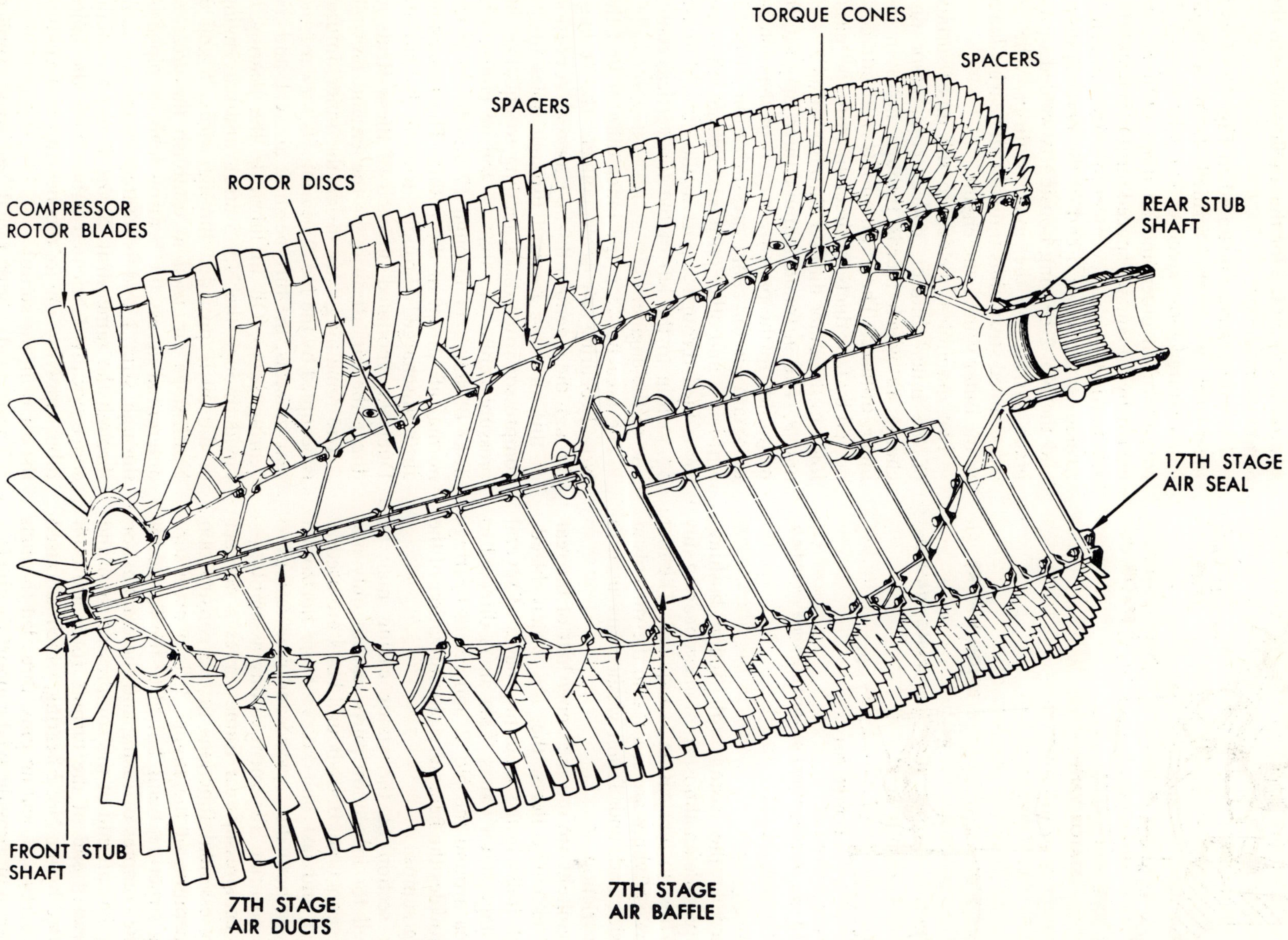

Figure **15·5** *Cutaway drawing of the compressor rotor.*

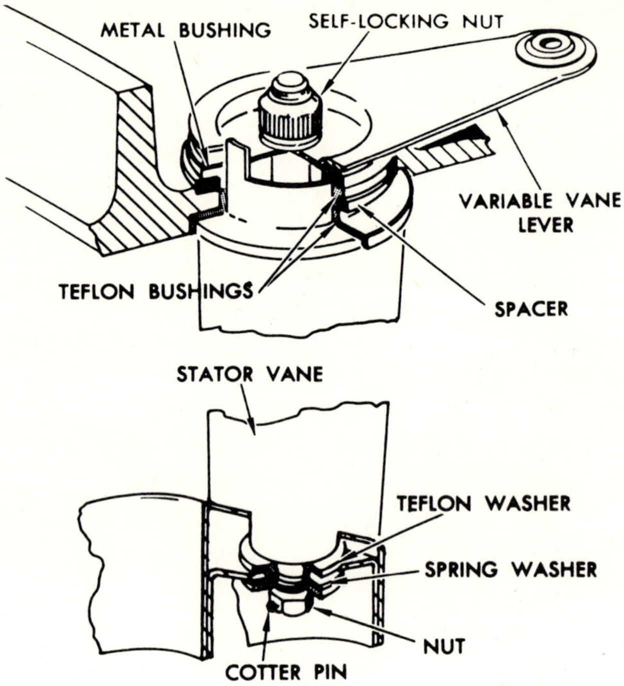

Figure 15 · 6 Arrangement of variable-vane mounting.

mounted in the casing with Teflon bushings which provide an almost frictionless bearing surface. Thus no lubrication is required. The first four rows of vanes are secured to the shrouds at the inside end, and Teflon washers are used to reduce friction.

A manifold around the upper half of the rear stator casing collects the air which bleeds through holes in the mounting bases of several ninth-stage stator vanes. This air is directed to the area around the No. 2 and No. 3 bearings and provides the necessary pressurization of the bearing sumps.

Any section of the stator casing may be removed for inspection purposes but only one section at a time. Body-bound bolts are used in the first and last bolt holes of the horizontal flanges of both the forward and rear casings to ensure accurate alignment during assembly.

The **compressor rear frame** serves a number of very essential functions in the construction and operation of the engine. It is a main supporting structure, being bolted to the flanges of both the compressor casing and the combustion casing. The aft end of the compressor rotor and the forward end of the turbine rotor are both supported by the No. 2 bearing, which is housed in this section. Locating pins, threaded through the outer skin, provide axial support for the forward ends of the combustion liners, while the inner combustion casing is bolted to the aft end of the inner diffuser. Pads for mounting 10 fuel nozzles are provided on the outer shell. In addition to providing support, the rear frame serves as a diffuser passage for the air flowing into the combustion section.

The rear frame, illustrated in Fig. 15 · 7, is a sheet-steel weldment formed from a high-temperature alloy steel called Chromoloy. It consists of an outer and an inner diffuser connected by 10 equally spaced struts. The stationary portion of the seventeenth-stage air seal is riveted to the forward end of the inner diffuser, and the No. 2 bearing housing is welded to the inner surface of the inner diffuser. The 10 streamlined struts thus transmit all bearing loads to the outer skin of the engine.

Two manifolds, welded to the inner surface of the inner diffuser, collect compressor discharge air which bleeds through holes in the inner diffuser. These manifolds are interconnected by four ducts to direct air from the manifolds to the outer diffuser through struts 2, 4, 7, and 9. The airframe manufacturer may attach manifolds to the pads over each of these struts to obtain air for cabin pressurization, airframe anti-icing, and other purposes. Engine anti-icing air is bled through a port in the outer diffuser, 5° 45 min to the left of the top vertical center line of the engine.

The area between the inner diffuser and the No. 2 bearing air seals collects the leakage through the seventeenth-stage air seal. A portion of this air flows forward through holes in the compressor rotor disks, tending to equalize the pressure across the spacers, torque cones, and disks of stages 11 through 17 and to offset the normal aft thrust of the rotors. The remaining air is discharged through struts 3, 8, and 10. Ducts are mounted on each of these struts to direct the air aft through the engine firewall.

The remaining area inside the inner diffuser is occupied by the No. 2 bearing, its forward and rear oil seals, the turbine-shaft air seal, and the bearing sump. Oil to lubricate the bearing is piped through the No. 5 strut, and scavenging of the oil is provided through struts 5 and 6. Ninth-stage air, used to establish the pressure differential across the bearing seals, is piped in through strut 1 and is vented to the oil tank through the same strut. Struts 1 and 5 contain concentrically fabricated tubes with the inner tube supplying oil to the bearing and venting the air from the sump.

Thrust loads, both radial and axial, applied to the ball-type No. 2 bearing are transmitted through the 10 struts to the outer diffuser as

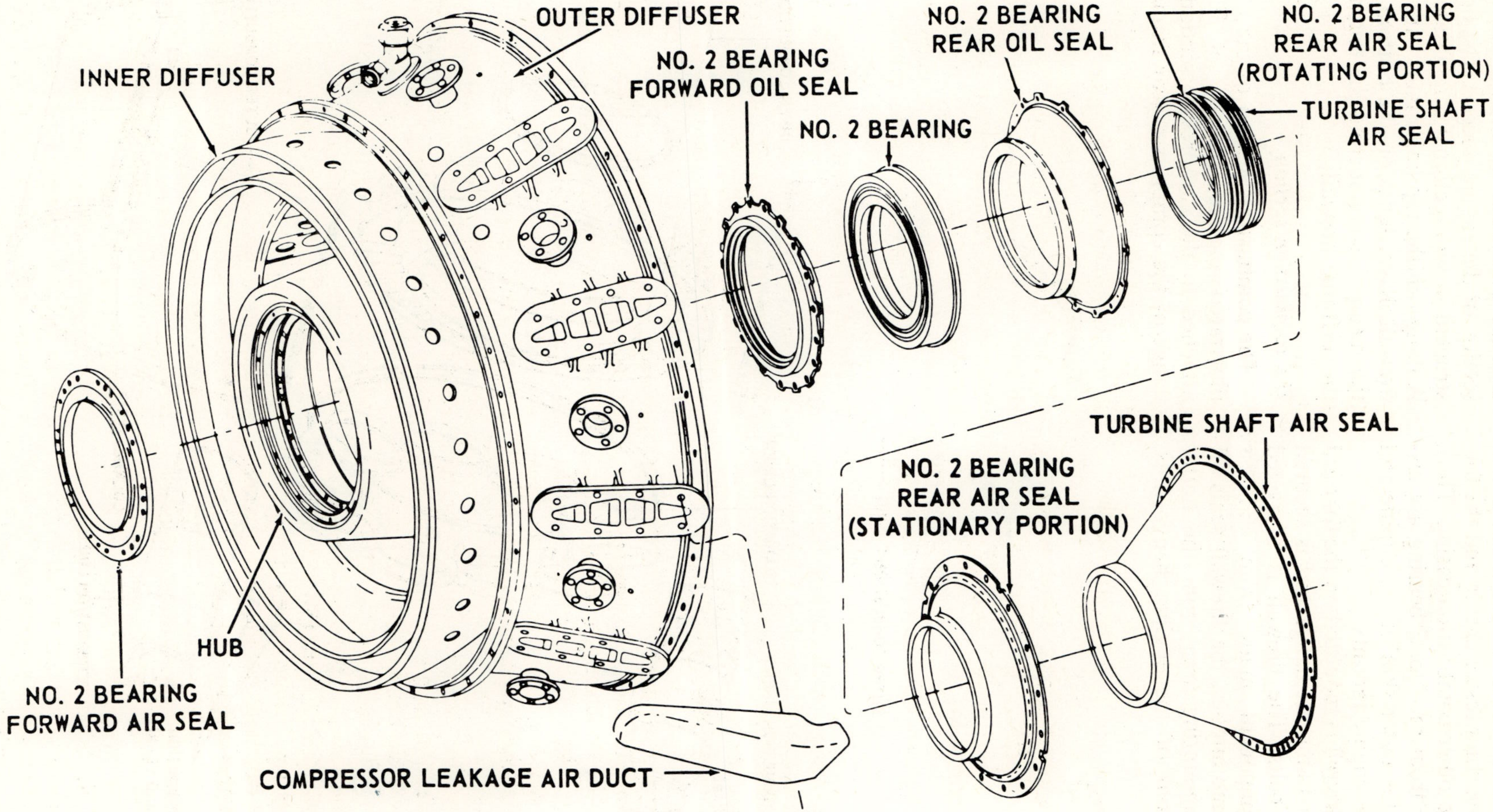

Figure **15·7** *Rear frame.*

previously explained. Flanges at the forward and aft ends of the rear frame bolt to flanges on the aft end of the compressor casing and the forward end of the outer combustion casing. Thus, all thrust loads from the No. 2 bearing are distributed along the outer skin of the engine and terminate at the engine mounts.

The engine **firewall** is mounted on the forward side of the rear-frame aft flange and is secured by the flange bolts. The firewall is a riveted sheet-metal component made from stainless steel. As its name implies, the firewall seals off the accessory mounting area from the hotter operating sections of the engine; hence fires, which are more likely to occur in the hot section, are not permitted to extend into the forward section of the engine nacelle. The firewall is constructed in two sections to facilitate installation. A flange on the top half mates with the airframe firewall. Rivet holes on the bottom half provide the means for connecting the engine firewall to the airframe firewall.

The combustion section

The combustion section of the CJ805 engine, illustrated in the drawing of Fig. 15 • 8, is attached to the compressor section at the rear flange of the compressor rear frame. It consists

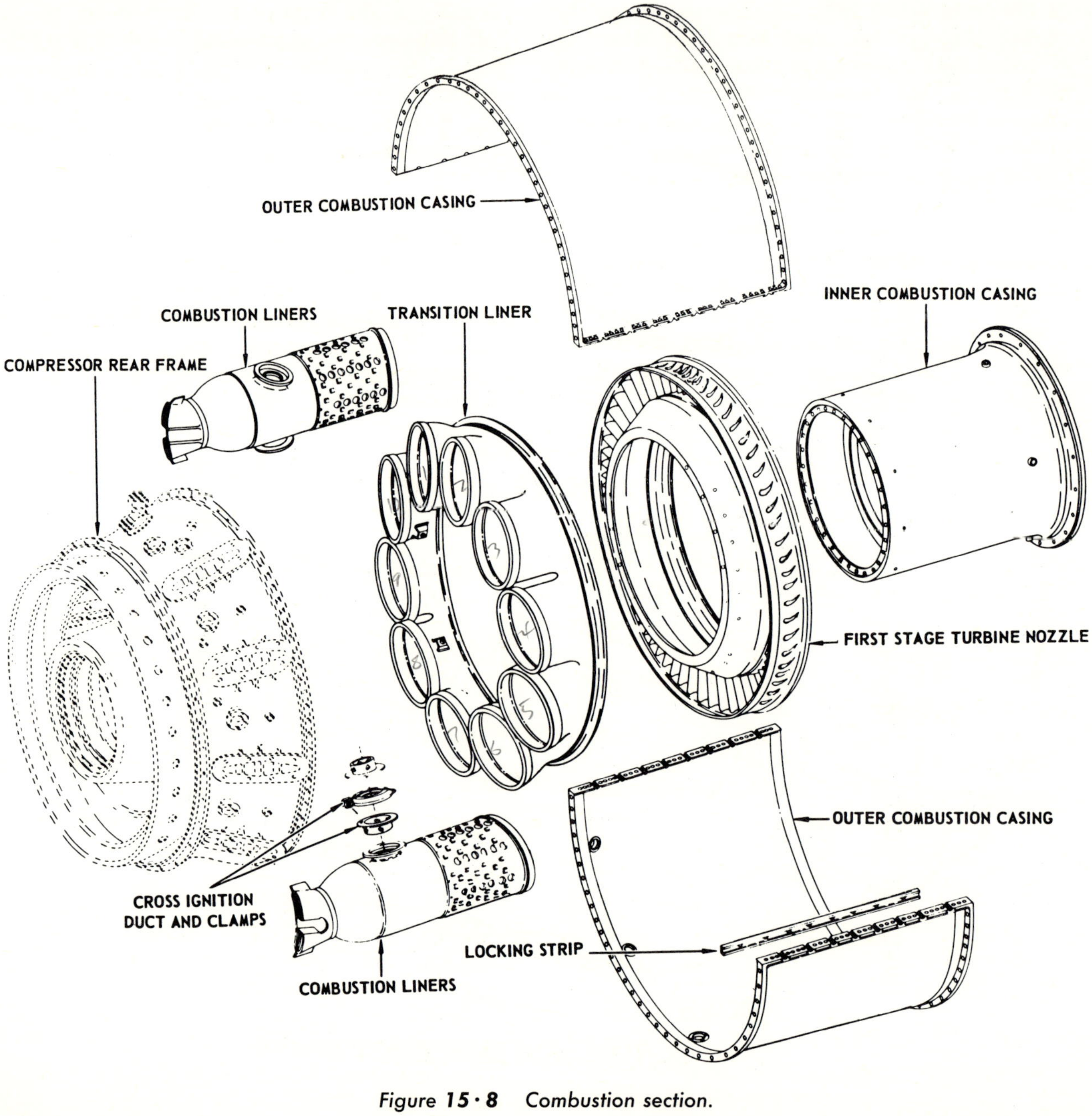

Figure 15 • 8 Combustion section.

of an **inner combustion casing, combustion liners, transition liner, first-stage turbine nozzle,** and **outer combustion casing.** As previously explained, a combustion chamber of this type is called a "cannular" combustion chamber because of the annular passage formed by the inner and outer combustion casings and the placing of individual liners or "cans" in this passage. Air from the compressor flows through the compressor rear frame and into the combustion section. A portion of this air enters the combustion liners where it is mixed with fuel and burned, while the remaining air blankets the inner and outer casings to provide cooling. As the fuel-air mixture in the liners is ignited, it expands, thus substantially increasing the velocity of the gases. The high-velocity gases are then ejected through the transition liner where the 10 individual streams are changed to one annular stream. The first-stage turbine nozzle then directs the gas stream onto the first-stage turbine buckets in the turbine section.

The **inner combustion casing** is a supporting structure for the combustion section and forms the inner wall of the air passage through the combustion section. This casing also acts as a heat shield to protect the turbine shaft from the intense heat of combustion.

The casing is fabricated from Chromoloy sheet stock formed and welded. Three stiffener rings are welded to the inner surface, and a flange is welded to each end. The forward flange bolts to the compressor-rear-frame inner diffuser, while the first-stage turbine nozzle bolts to the rear flange. Five pin bosses are located around the rear portion of the casing. A pin bolt is threaded into the casing, and, when pinned to the transition liner, it provides the radial support for the liner. Air is bled through 10 equally spaced holes in the forward part of the casing to cool the shank portion of the first-stage turbine buckets and to prevent the flow of hot gases around the turbine seals into the area.

Each of the 10 **combustion liners** receives a portion of the air from the compressor, mixes the air with fuel from the fuel discharge nozzle located at the front of the liner, burns the fuel-air mixture, and ejects it to the rear. Each liner is a sheet-steel fabrication, welded and riveted together, consisting of four sections: the dome, the forward outer liner, the inner liner, and the rear outer liner. These sections are shown in Fig. 15 • 9. The dome and inner liner are fabricated from Stellite, which is cobalt-base superalloy designed for extremely high temperatures, while the remaining parts are fabricated from Incoloy, which is also heat-resistant. The inner liner is ceramic-coated to increase its heat-resistant characteristics.

A fuel nozzle is inserted into the self-aligning

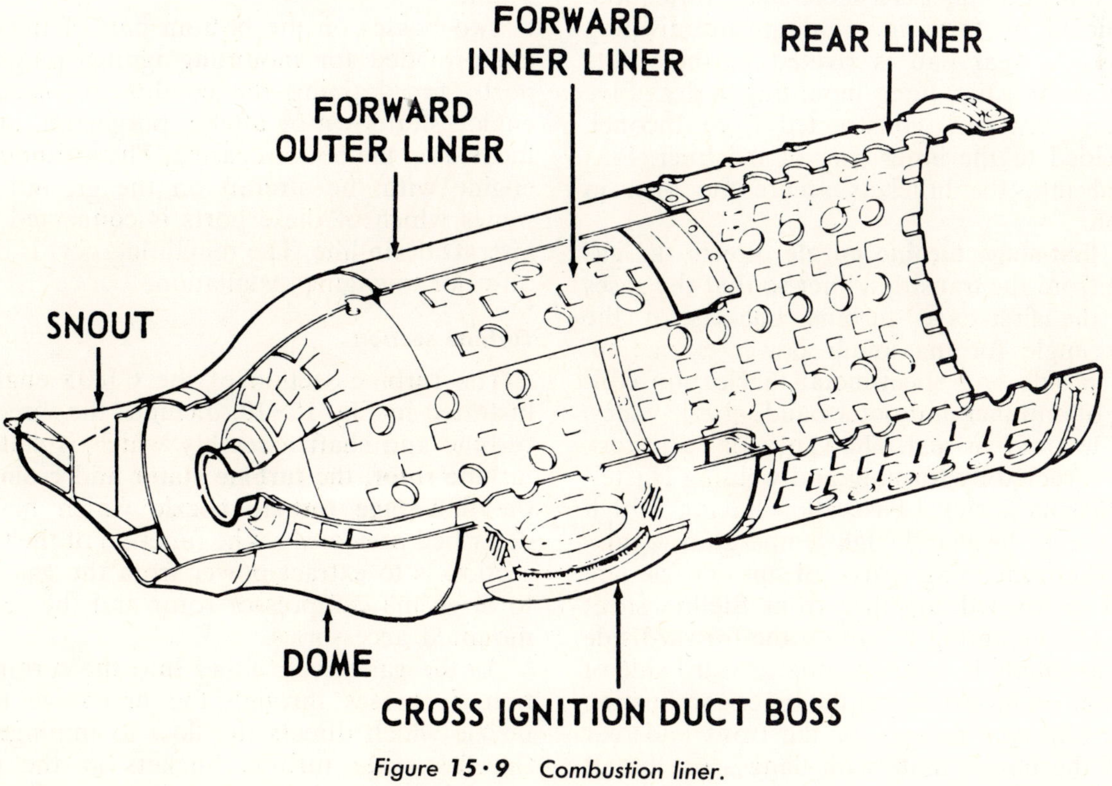

Figure 15 • 9 *Combustion liner.*

eyelet in the forward end of each liner, and the liners are connected by cross-ignition ducts. The fuel nozzles and the cross-ignition ducts, plus the aligning pin which is threaded into the outer diffuser of the compressor rear frame, provide radial and axial support for the liners. The aft end of each liner is slip-fit into holes in the forward face of the transition liner.

The snout at the forward end of the dome directs air into the liners. A portion of this air provides a cooling layer between the inner liner and the forward outer liner. Louvers in the inner liner direct the remaining air into the inner liner. The louvers are positioned and shaped to create the optimum fuel-air mixture. Igniter plugs, located in the No. 4 and No. 5 combustion liners, ignite the fuel-air mixture, and the flame is spread to the remaining liners through the cross-ignition ducts. The air flowing past the snout forms a cooling layer between the liners and the combustion casing. A portion of this air flows through the louvers in the rear outer liner and holds the flame in the center of the liners, thus preventing deterioration of the liners because of heat.

As explained previously, the **transition liner** transforms the 10 streams of hot gases from the combustion liners into a single annular gas flow, thus providing uniform velocity and pressure at the first-stage turbine nozzle. It also supports the aft end of the combustion liners.

The transition liner is a sheet-steel fabrication, constructed by forming, welding, and riveting Incoloy. A wear pad is riveted to the inside of the combustion-liner mounting holes. Five mounting brackets, constructed from Inconel, are welded to the inner ring of the liner. Pins inserted into the brackets secure the liner in position.

The **first-stage turbine nozzle** directs the gas stream from the transition liner so that the gases strike the first-stage turbine buckets at the proper angle for maximum power extraction.

The nozzle is a sheet-metal fabrication consisting of an inner and outer band, nozzle vanes, air baffles, and air seal. The vanes, or partitions, are constructed by forming and welding Hastelloy which is a nickel-base alloy having a high resistance to heat and high-temperature oxidation. The remaining parts of the nozzle are welded or riveted together from Stellite sheet stock. The mounting flange on the forward side of the assembly is bolted to the forward side of the flange on the aft end of the inner combustion casing. Slip-type flanges on the front and rear side of the nozzle mate with flanges on the aft end of the transition liner and the forward end of the first-stage turbine shroud.

Because of the high temperature of the gases passing between the nozzle partitions, cooling of the partitions (vanes) is required. The turbine nozzle is shrouded so that the cooling air flowing aft through the combustion section is directed into the hollow vanes. The seal on the forward side of the nozzle assembly mates with the seal on the transition liner, preventing leakage of the cooling air. The baffles on the lower rear side of the inner band direct the cooling air onto the shank sections of the first-stage turbine buckets and the first-stage turbine wheel.

The **outer combustion casing** forms the load-carrying outer skin between the compressor rear frame and the turbine stator and casing. It also serves as the outer wall of the air passage through the combustion section. The casing is fabricated from Chromoloy sheet stock and is split along the horizontal center line. The horizontal flanges are bolted together and sealed with a locking strip to prevent gas leakage and to reduce hoop stress on the flange bolts. This makes it possible to use smaller bolts and lighter flanges, resulting in a weight reduction. Flanges at the front and rear of the casing bolt to the rear flange of the compressor rear frame and the forward flange of the turbine stator. Either half of the casing may be removed for inspection, but only one half of the casing may be removed at a time.

Two bosses on the bottom half of the casing are provided for mounting igniter plugs. Two ports, for draining the combustion section at engine shutdown or after a purge run, are also located on the bottom casing. The attitude of the engine with the aircraft on the ground determines which of these ports is connected to the aircraft drain line. The remaining port is capped off prior to engine installation.

Turbine section

The turbine section of the CJ805 engine, illustrated in Fig. 15 · 10, includes the three-stage turbine and shaft assembly which is called the **turbine rotor,** the **turbine stator** and **casing,** and the **first-stage turbine nozzle** which has been described previously. The function of the turbine section is to extract power from the gas stream to drive the compressor rotor and the gearbox-mounted accessories.

As the gas stream flows into the turbine section, it passes through the first-stage turbine nozzle which directs the flow to impinge upon the first-stage turbine buckets at the proper

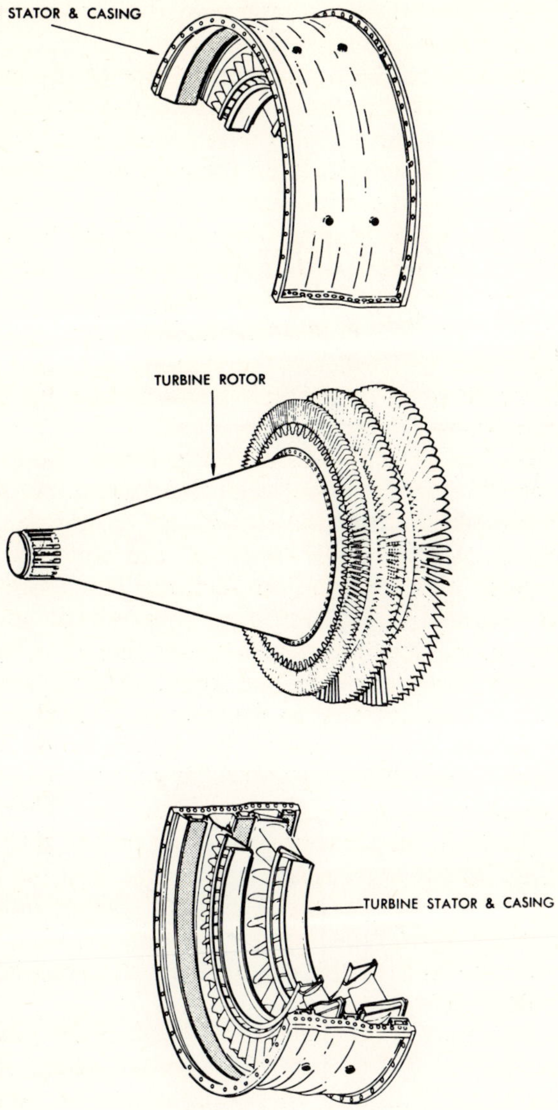

Figure ***15 · 10*** *Turbine section.*

angle. The gas then continues through the second-stage nozzle to the second-stage turbine and on through the third-stage nozzle and turbine. Power is imparted to the turbine rotor as the gases strike each row of turbine buckets. The gas stream then flows into the inner passage of the aft-fan assembly.

The **turbine-rotor assembly,** illustrated in the cutaway drawing of Fig. 15 · 11, consists of a conical hollow forward shaft, three turbine wheels, or disks, with associated buckets, two torque rings through which torque is transmitted from one disk to another, two air baffles, and the rotating portion of the No. 3 bearing forward air seal. The shaft, air baffles, torque rings, first-stage disk, and third-stage disk are made of stainless steel, while the second-stage disk is machined from an iron-base-alloy forging which has high heat-resistant characteristics. The first-stage turbine buckets are machined from cast Udimet, and the second and third-stage buckets are machined from another type of nickel-base-alloy forgings.

The rotor shaft, disks, torque rings, and the stationary portion of the No. 3 bearing forward air seal are bolted together as follows: shaft to forward face of the first-stage disk, torque ring between the first and second-stage disk, torque ring between the second and third-stage disk, and air seal to the aft face of the third-stage disk. The baffles are mounted on support rings attached to the front and rear faces of the second-stage disk near the inner rim of the disk. A bucket assembly, consisting of two buckets, is inserted in each of the serrations in the outer rim of each turbine disk. Locking strips clamp the buckets in position. There are 148 buckets in the first-stage disk, 114 in the second stage, and 84 in the third stage.

The spline on the forward end of the turbine shaft mates with the female spline in the end of the compressor rotor shaft, and the two shafts are held together with a self-locking bolt. Support for the forward end of the turbine shaft and the rear end of the compressor rotor is provided by the No. 2 bearing. The turbine-rotor rear stub shaft, an integral part of the third-stage disk, is supported in the No. 3 bearing. This bearing is mounted in the fan front frame.

Cooling air at seventh-stage pressure flows through the center of the compressor-rotor rear stub shaft and into the interior of the turbine rotor. Here the air cools the fore and aft faces of the first- and second-stage disks and the forward face of the third-stage disk. The air is then routed through the turbine-rotor rear stub shaft and into the fan front frame. Here the air flows outward through all struts except the No. 5 strut and into the hat section on the outer circumference of the frame. The air heats the hat section, decreasing the amount of thermal stress created by the difference in temperature of the two air streams through the frame. This air is vented overboard through a series of slots in the hat section.

The **turbine-stator-and-casing assembly** shown in Fig. 15 · 12, is that section of the engine immediately to the rear of the combustion section. The principal components of the turbine stator and casing are the casing, the turbine shrouds, and the second- and third-stage turbine nozzles.

The casing forms the load-carrying outer skin of the engine between the outer combustion

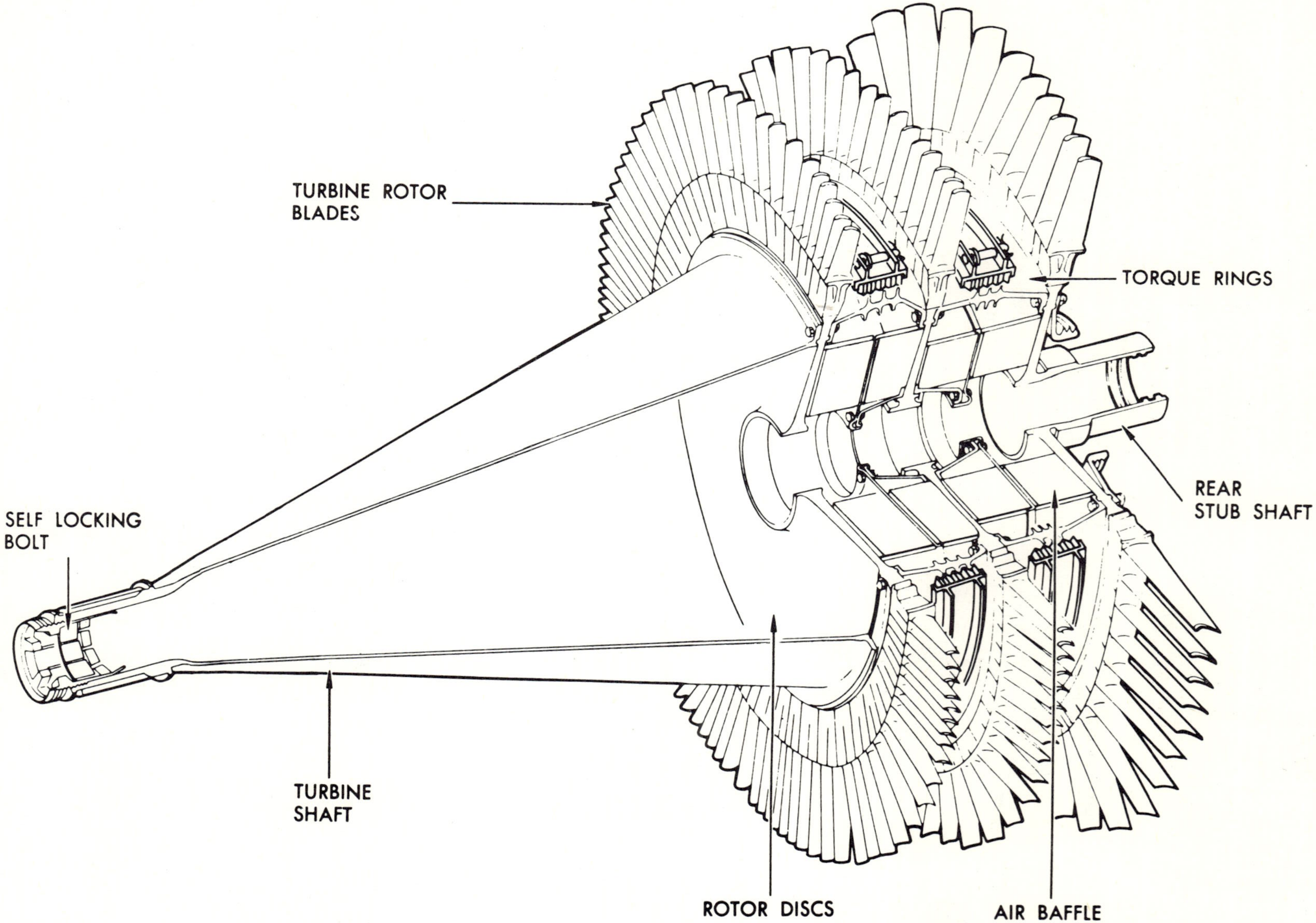

Figure **15·11** *Turbine rotor assembly.*

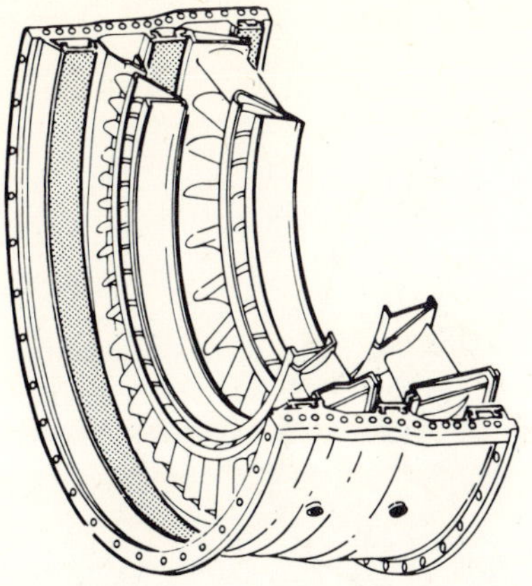

Figure 15 · 12 Bottom half of turbine-stator-and-casing assembly.

casing and the fan front frame, in addition to supporting the turbine shrouds and nozzles. It is constructed of formed and welded sheet stainless steel. The complete assembly is split along the horizontal center line for ease in disassembly, inspection, and repair.

The **shrouds** are made in split rings which surround the turbine blades when the engine is assembled. The inner surface of each shroud is a stainless-steel honeycomb which serves as a seal to prevent the flow of hot gases around the tips of the turbine blades. The honeycomb surface creates a turbulence in the flow of gases, thus effectively reducing the amount of gases passing around the blade tips.

The turbine nozzles for the second and third stages are also constructed as split rings mounted inside the casing. The second-stage nozzle vanes (partitions) are weldments formed from Hastelloy, a high-temperature nickel-base alloy, and the inner and outer bands are fabricated from Stellite sheet stock. Stellite is a cobalt-base high-temperature alloy. The nozzle vanes are hollow to permit the flow of cooling air through the inside. The third-stage nozzle vanes are formed and welded from sheet stainless steel.

An important feature of the turbine casing, from a safety standpoint, is that the casing is designed to provide for turbine-blade containment. This means that if a turbine blade should break free from the turbine disk, the blade will not penetrate the casing and be thrown outward from the engine.

Aft-fan section

The aft-fan section, illustrated in the drawing of Fig. 15 · 13, is a concentric-duct arrangement which houses the fan turbine and serves as the main exhaust section of the engine. It is attached to the turbine-stator-casing rear flange. The aft fan consists of a single-stage free-floating rotor, which is part compressor and part turbine, mounted in bearings which are supported inside the aft-fan section.

Gases flowing aft from the turbine section flow into the inner passage of the fan front frame and through the vanes of the fan-turbine nozzle. The nozzle directs the gases at the correct angle to the turbine blades of the fan rotor. The gases impart energy to the blades to drive the fan rotor and then flow through the inner passage of the fan rear frame and into the exhaust.

Secondary air is ducted into the fan inlet which is the outer air passage of the fan front frame, and then flows to the compressor airfoils (outer sections) of the fan rotor blades. The compressor airfoils compress the air at a ratio of 1.65:1. The air then flows through a row of outlet guide vanes, which straighten the direction of flow, and through the outer passage of the fan rear frame to the exhaust section. The secondary airflow through the fan is 1.54 times that of the primary airflow.

The **fan front frame** forms the air inlet to the fan section. It also serves as a support structure because the No. 3 bearing and the No. 4 bearing are housed in the fan-front-frame hub. The No. 3 bearing supports the aft end of the turbine rotor, and the No. 4 bearing supports the forward end of the fan rotor. A link-pin engine mount is provided on the outer shell of the frame to provide for installation of the engine in the nacelle. Eight thermocouples are mounted in the diffuser and extend into the primary airstream to sample the exhaust gas temperature (EGT).

Concentric flow passages for the primary and secondary air are formed by the outer shell, the inner diffuser, and the hub. The hub and outer shell are connected by eight equally spaced streamlined struts to support the hub in the center of the frame. A small portion of the primary gas flow is routed out through the struts to anti-ice the portion of the struts that extend through the secondary airstream. This is necessary because the air flowing through the secondary duct comes directly from the outer atmosphere and may contain moisture at below freezing temperatures. The struts also serve as passageways for oil service piping to the bearings, for air to pressurize the bearing sumps, and for the venting of turbine cooling air. The stationary portion of the fan front air seal and the fan turbine nozzle complete the fan-front-frame assembly.

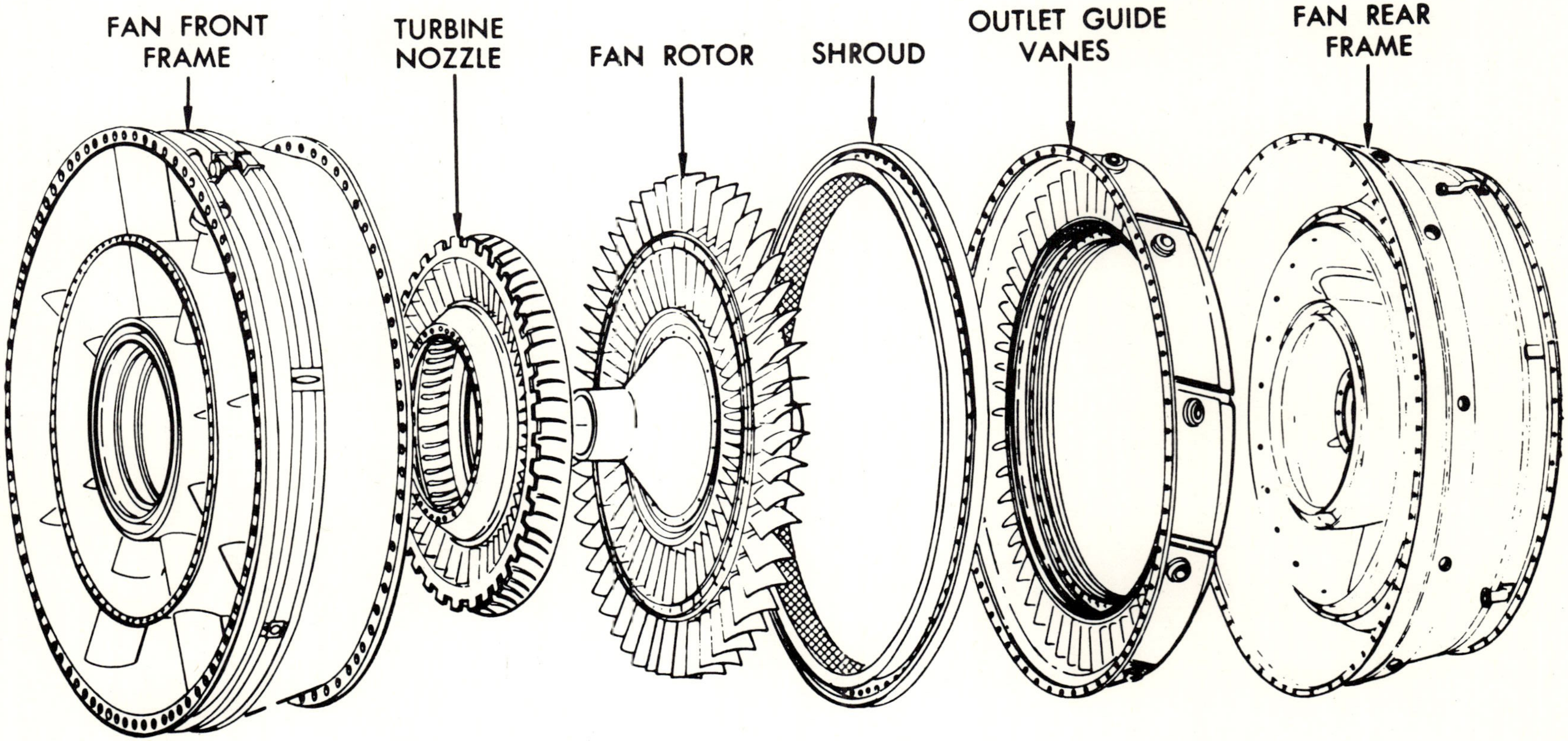

Figure 15·13 Aft-fan section.

The **fan rear frame** serves as a diffuser passage for the two airstreams from the fan rotor. The segmented honeycomb shroud ensures that all the secondary air flows through the fan blades. A row of outlet vanes straightens the airflow from the fan rotor compressor airfoil, while baffles around the inner portion of the rear-frame struts perform this function in the primary gas stream. The frame also serves as the aft-support structure of the engine since the No. 5 bearing, which supports the aft end of the fan rotor, is housed in the rear-frame hub. The inner cone is assembled to the inner flange of the frame and is designed to provide a smooth flow into the exhaust section. The outer shell, inner diffuser, and hub are connected by eight equally spaced streamlined struts and are constructed to form the concentric flow passages. The struts of this section, as in other sections of the engine, serve as passageways for oil and air service lines.

The **fan rotor** consists of two conical shafts and a rotor disk with associated blades. The rotor blades, each consisting of two airfoil sections separated by a dividing platform which seals the blade slot between the inner and outer ducts, are mounted to the rotor disk by inserting the blade bases into dovetail slots in the outer rim of the disk. Projections on the blade mid-platforms interlock the blades, preventing deflection of the blade tips. Serrations on the front and aft sides of the platforms mate with honeycomb segments on the aft side of the turbine nozzle and the forward side of the outlet-guide-vane assembly to form the labyrinth-type airseal separating the two airstreams. The rotor is supported at the forward end by the No. 4 bearing, which is a ball-type designed to absorb axial thrust, and at the rear end by the roller-type No. 5 bearing.

ENGINE AIRFLOW

Seventh-stage airflow

Because of the need for cooling and pressurization within various portions of the engine, air is bled from the main airstream at different locations and ducted to the areas where the air is required. A portion of the major airflow is ducted into the compressor rotor through holes located around the circumference of the seventh-stage spacer. Angular acceleration of the air is prevented by means of a paddle-like device which also distributes the air through ducts in the center of the rotor. The air bleeds through holes in the ducts and tends to reduce the pressure differential across the disks and spacers of the first twelve rotor stages. The air performs the same function in stages 12 through 15; however, the bleed ducts are not required. As previously explained, a small amount of air bleeds through holes in the compressor front stub shaft to pressurize the area between the No. 1 bearing air and oil seals. The flow of air across the air seal prevents the loss of oil; and since a part of the air flows into the gearbox areas, it provides a positive pressure to the scavenge pumps.

The remaining air from the seventh-stage supply flows through the center of the compressor-rotor rear stub shaft and into the interior of the turbine rotor. The function of this air with respect to the turbine and fan front frame was described earlier.

Sump pressurizing air

Air at ninth-stage pressure bleeds through holes in the upper half of the compressor rear casing and into a manifold. From the manifold the air is routed through the No. 1 strut of the compressor rear frame, into the area between the No. 2 bearing air and oil seals, through the No. 5 strut of the fan front frame into the area between the air and oil seals of the No. 3 and No. 4 bearings, and through the No. 3 strut of the fan rear frame to the interseal area of the No. 5 bearing. This air establishes the proper pressure differential across the oil seals to prevent excessive oil consumption.

A portion of the air bleeds into the bearing sumps and provides a positive pressure to the scavenge pumps. Pressure in the sumps is regulated by the tank and sump pressurizing valve. Excess air in the sumps is vented through the No. 1 strut of the compressor rear frame, the No. 2 strut of the fan front frame, and the No. 1 strut of the fan rear frame to the oil tank. From the oil tank, air flows to the **tank and sump pressurizing valve.** This valve controls pressure in the tank and sumps by venting air overboard or by drawing ambient air into the system.

Seventeenth-stage seal leakage air

The seventeenth-stage seal is designed to permit a controlled leakage of compressor discharge air. This leakage flows into the annulus formed by the turbine-rotor forward air seal and the inner diffuser of the compressor rear frame. This air surrounds the No. 2 bearing seals and aids in establishing sump pressure by limiting leakage of ninth-stage air. It also bleeds through holes in the seventeenth- through the twelfth-stage rotor disks, tending to reduce the pressure differential across the spacers, torque cones, and disks of these stages. During engine operation the compressor-turbine rotor generates a force in the aft direction. By porting seventeenth-stage air into the interior of the rotor, the normal aft thrust is offset, thus reducing the load on the No. 2 bearing and contributing to longer bearing life. Excess seventeenth-stage leakage air is ducted outside the casing through struts 3, 8, and 10 of the compressor rear frame.

Seventeenth-stage bleed air

Seventeenth-stage compressor discharge air is bled through holes evenly spaced around the inner diffuser of the compressor rear frame to supply the air needs of the aircraft. This air is collected in two interconnected manifolds inside the diffuser. Figure 15 · 14 illustrates the flow of compressor discharge air into the manifolds and out through the compressor rear frame strut to the bleed port. Air from the manifolds is discharged through struts 2, 4, 7, and 9. The airframe manufacturer can attach a bleed air manifold to any one or all of the ports over these struts. Bleed air may be used for wing anti-icing, windshield rain clearing, cabin pressurization, or other airframe requirements.

A portion of the bleed air is required as the heating agent in the engine-supplied fuel heater. The airframe manufacturer and the engine manufacturer jointly establish the airflow and installation requirements of the fuel heater.

Rear frame heating air

As mentioned in the description of the fan section, a small amount of the basic engine exhaust gases flow into the area between the fan inner diffuser and the fan rotor. It bleeds through holes in the fan rotor and makes possible the design of a lightweight fan rotor wheel and shaft assembly. The remainder of the air flows through holes near the inner ends of the fan rear-frame struts, then outward through the struts and into a manifold around the circumference of the frame. Here the air serves a heating function, decreasing the thermal stresses created by the temperature differences between the two exhaust streams flowing through the frame. The air is vented into the nacelle through a series of slots located between the strut ends on each side of the manifold.

Engine inlet anti-icing

The engine anti-icing system is designed to prevent the accumulation of excessive ice in the compressor inlet. Specifically, the system anti-

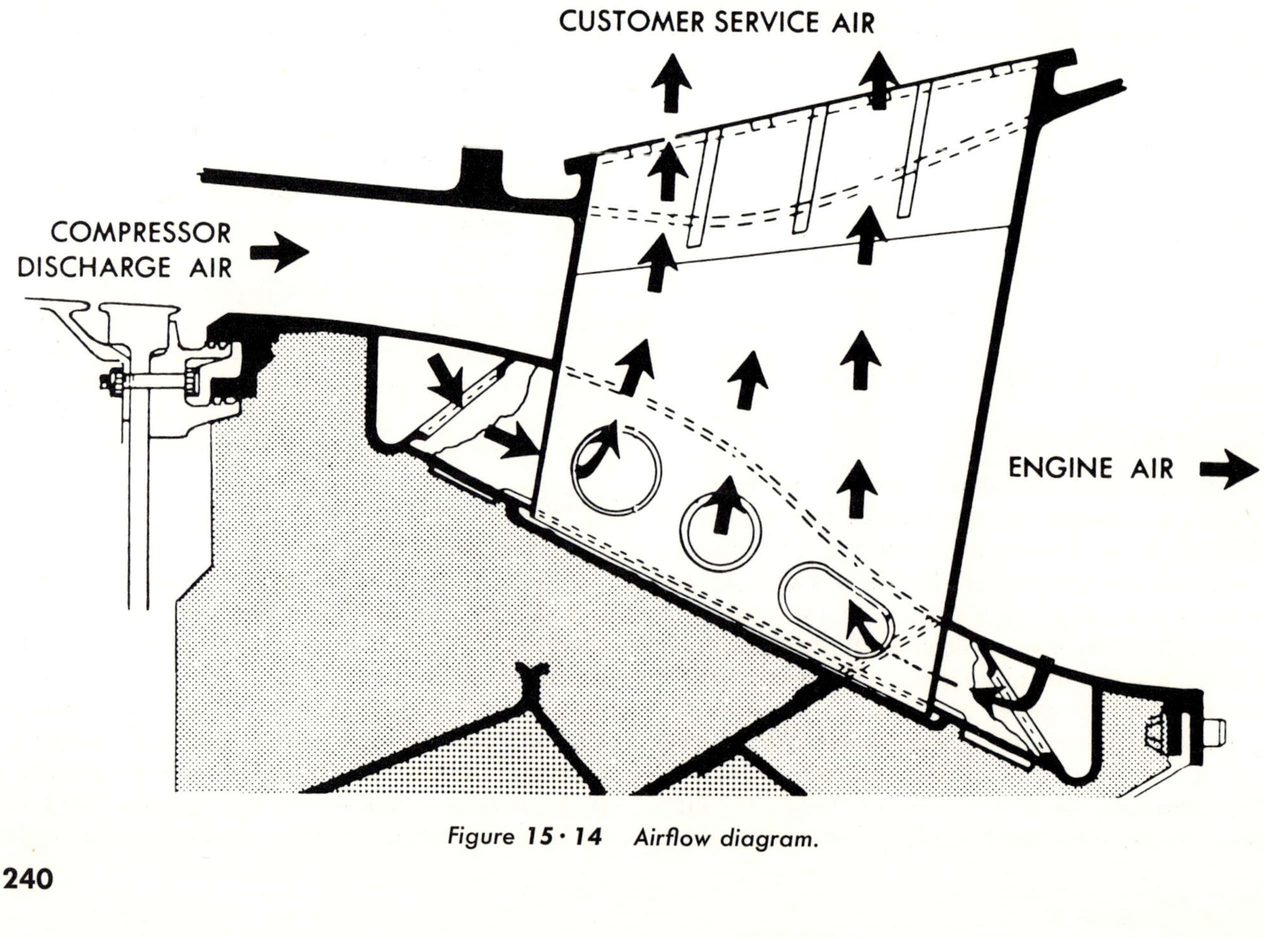

Figure 15 · 14 Airflow diagram.

ices the compressor-front-frame hub and struts, the inlet guide vanes, and the airframe-furnished nose cone fairing.

Seventeenth-stage compressor discharge air is bled through a port in the outer diffuser wall of the compressor rear frame and into a solenoid-operated regulating valve. The valve, which is turned ON and OFF through a cockpit switch, regulates the pressure of the seventeenth-stage air as a function of downstream air pressure. This regulated air performs the anti-icing function.

Air from the regulating valve enters the compressor front frame through the struts numbered 2, 3, 7, and 8. From the struts, the air flows into the manifold formed by the double wall of the front-frame hub and the inlet-guide-vane support. A portion of the air from the manifold heats the remaining four struts of the compressor front frame. This air flows outward through the struts and is vented into the primary airstream through a small hole on each side of the strut near the juncture with the front frame outer ring.

The 20 hollow IGVs are also anti-iced by a portion of the air from the vane support manifold. The air flows outward through the vanes and is discharged into the primary airstream through bleed slots in the trailing edge of each vane.

The remaining air in the manifold is discharged through a port on the face of the inlet gearbox housing at the 12 o'clock position to supply air for the airframe-furnished nose cone fairing.

FUEL AND CONTROL SYSTEM

The CJ805-23 fuel and control system establishes engine speed by metering fuel flow for starting, acceleration, deceleration, and steady-state operation. Since the fan rotor is aerodynamically coupled to the engine rotors, the fuel and control system indirectly controls fan rotor speed. In addition to fuel flow, the system controls the direction and velocity of the air flowing through the compressor by varying the position of the IGVs and the first six stages of stator vanes. Besides being used for combustion, fuel is used as servo fluid within the fuel-control unit, as hydraulic fluid for positioning the variable stator actuators, as a coolant for engine and constant-speed drive oil, and to establish reference pressure in the pressurizing and drain valve.

The complete fuel system consists of a two-element gear pump with an integral centrifugal boost element, a fuel filter, a fuel heater, a bypassing-type fuel-control unit, two fuel-oil coolers, a pressurizing and drain valve, 10 fuel nozzles, and two variable stator actuators. All the tubing used in the system is fire resistant. Figure 15 · 15 is a schematic diagram of the fuel system.

Fuel-flow path

The aircraft boost pump delivers fuel to the inlet of the engine fuel pump where the boost element steps up the pressure from 15 to 45 psi. The fuel flows from the boost element to the two gear-type high-pressure elements which increase the pressure 100 to 850 psig, depending on engine speed and flight conditions. The fuel discharge from the engine pump flows to the fuel heater. The fuel flows through the heater which uses compressor discharge air to maintain the fuel temperature above freezing to eliminate any possibility of ice in the fuel. From the fuel heater the fuel flows by way of the fuel filter to the fuel-control unit.

The **hydromechanical fuel-control unit** meters the flow of fuel to the combustion chambers by varying the area of the metering orifice and, at the same time, maintaining a constant pressure drop across the orifice. The input signals to the fuel-control unit are **power lever angle, engine speed, compressor inlet temperature (CIT),** and **compressor discharge pressure (CDP).** The control integrates these parameters to establish the area of the metering orifice and thus schedule the proper flow of fuel for any operating condition.

The fuel pump always delivers more fuel to the control than is required to sustain the engine operation at any given condition. The control uses some of this excess fuel to operate its servo mechanisms and some to position and cool the variable stator actuators. The remainder (bypass flow) is returned to the inlet side of the high-pressure elements of the pump.

The CDP servo mechanism in the control compares changes in CDP to a constant reference pressure. To maintain this constant pressure, some servo flow may be drained from within the control unit case. This flow is piped externally to the inlet of the fuel pump.

Metered fuel from the fuel control flows through the engine fuel-oil cooler, the constant-speed-drive fuel-oil cooler, and the airframe-furnished fuel-flow transmitter. In the coolers fuel is used as the coolant for the oil, provided that the fuel temperature remains below 127°C (260°F). Each of the coolers has a temperature

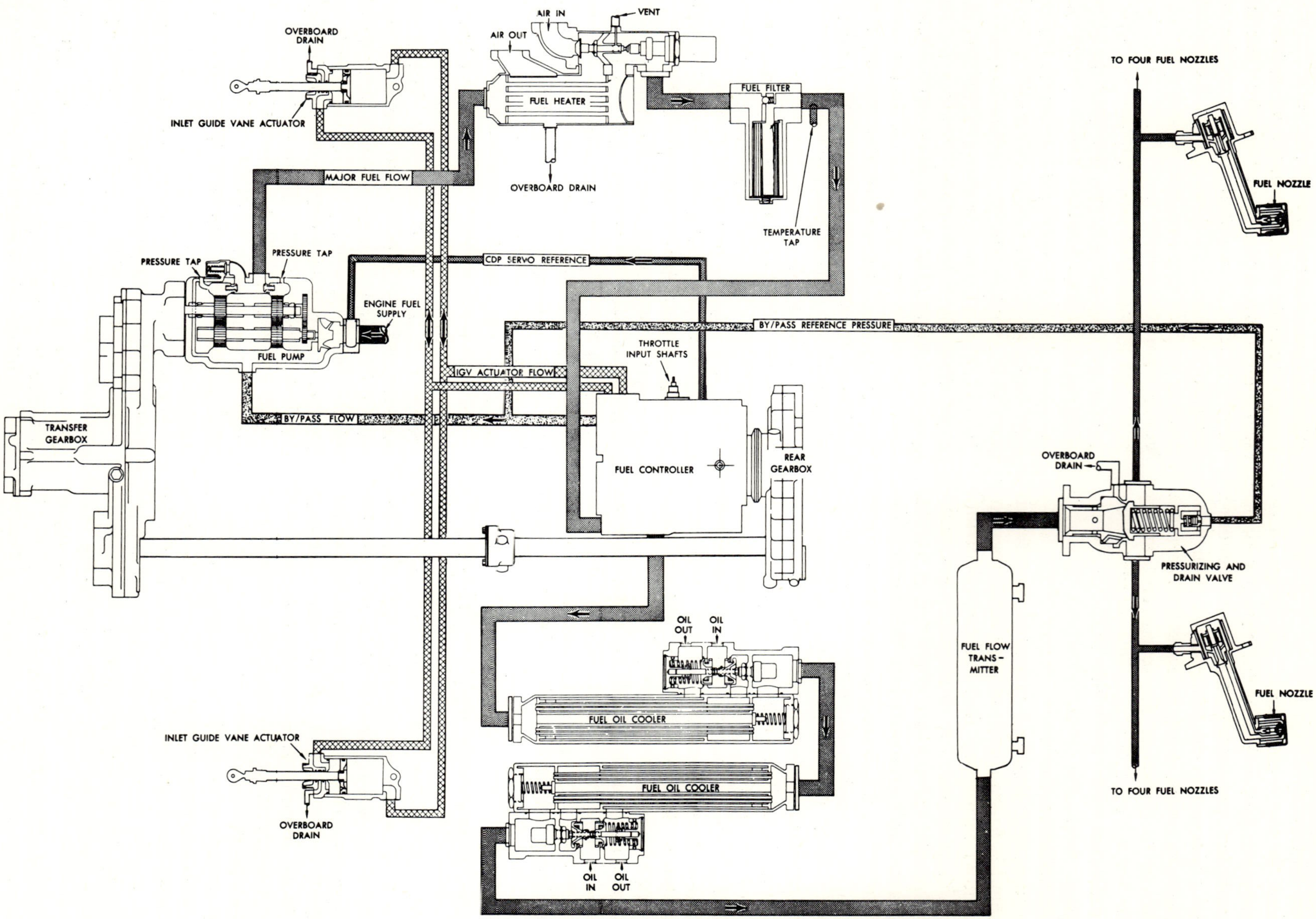

Figure 15·15 Schematic diagram of fuel system.

sensor which may cause the cooler to bypass its oil flow.

From the fuel-flow transmitter, metered fuel is routed to the pressurizing and drain valve. The pressurizing portion of the valve is closed until sufficient pressure is built up to operate the servo assemblies in the control. The pressurizing valve opens when metered fuel pressure is sufficient to overcome the reference pressure and the valve spring force. The reference pressure is directed to the back side of the pressurizing valve from the control bypass flow. At engine shutdown, the drain section of the valve opens, draining the fuel manifold to prevent post-shutdown fires but maintaining the upstream portion of the fuel system primed to permit faster starting. Fuel from the pressurizing and drain valve flows into 10 fuel nozzles which spray atomized fuel into the combustion chambers where it is mixed with air and burned.

Operation of the system

Engine performance points are defined as speed settings or as a percentage of 7,460 rpm, which is referred to as 100 per cent engine speed. **Idle** speed is 4,475 rpm or 60 per cent, **maximum cruise** is 7,236 rpm or 97 per cent, **maximum continuous** is 7,386 rpm or 99 per cent, and **take-off** is 7,684 rpm or 103 per cent. The fuel-control system controls engine thrust by regulating engine speed, and the system itself is controlled primarily by the pilot, who selects the desired engine speed by positioning the cockpit power lever. Other signals, such as CIT, CDP, and engine speed N, provide secondary control by establishing the correct fuel flow for automatic starts, establishing the maximum rate for acceleration, determining the maximum rate of deceleration, and creating the proper sequence of events at engine shutdown. If the control system is operating properly, the engine demonstrates certain normal operating characteristics. The more important characteristics, and the engine parameters that cause them, are reviewed in this section.

The engine is started by rotating it rapidly with an air-turbine starter. As the engine turns, it actuates the engine fuel pump, and pump discharge pressure rises. Fuel flows to the control, and a part of the flow is returned to the pump through the bypass line while operating pressures develop in the fuel control. Advancing the fuel shutoff lever opens the fuel shutoff valve in the discharge section of the control and permits metered fuel to flow to the pressurizing and drain valve.

If the engine has been started previously, the line between the fuel control and the pressurizing and drain valve is already primed. The valve remains closed until metered fuel pressure at the inlet is high enough to overcome the force of reference pressure plus spring force. The valve then opens, and fuel flows to the fuel nozzles. During a start the variable stators are closed, that is, they are rotated toward the closed position, but they do not actually shut off airflow. While starting the engine, the throttle (power lever) is at the IDLE position, and the fuel control attempts to deliver the fuel required for idle operation. However, if this much fuel were delivered to the combustion chambers, a very hot start or no start would occur. The control, then, must compute a maximum fuel flow for each start condition. The parameters of CDP, CIT, and rpm are integrated to establish the correct flow. Since CDP is low under starting conditions, the fuel flow must also be low, that is, approximately 500 to 700 lb per hr. Compressor inlet temperature, which reflects the ambient conditions, determines the specific flow within the above range. On cool or cold days the CIT signal reflects the greater air density and calls for a greater fuel weight through the compressor at a given rpm. With this condition existing, the fuel-control unit meters a higher starting fuel flow than would be the case on a warm day.

During a normal light-off (start of fuel ignition), exhaust gas temperature (EGT) rises sharply to a peak level and then drops slowly as the engine automatically accelerates to idle speed. During a hot start the EGT peak exceeds the maximum peak limit or drops too slowly and is above the maximum temperature specified at 5, 10, or 15 sec after light-off. The likelihood of a hot start increases as the engine continues to rotate without a light-off because both the rate of fuel flow and the total fuel accumulation in the combustion chamber become greater. Hot starts are avoided by monitoring the time that expires after the first fuel-flow indication and aborting the start if no light-off occurs after a maximum specified time or at maximum starting rpm.

During a start the throttle is signaling for idle speed even at minimum throttle position. The control system senses both the rpm due to motoring (rotating with the starter) and the desired rpm which is higher. At light-off a maximum acceleration fuel schedule is generated immediately to correct the off-speed condition. The exact rate of acceleration varies with the value of CDP and rpm at light-off. Compressor inlet temperature is also a factor because it will vary fuel flow according to air temperature as explained previously. During acceleration, engine speed increases, the compressor performs more work, and CDP increases. More air is available in the combustion section, and fuel flow increases. This process continues until the actual speed of the engine equals desired speed.

In a throttle advancement above IDLE, the same sequence of events described above takes place; but at some point in the acceleration, rpm and CIT form a combination which causes the variable stators to begin opening. The stators are full open before the engine reaches top speed, but the full open and full closed points vary with CIT. In some instances of cold temperature the stators may start to open before the engine reaches idle speed.

During a deceleration the throttle is rapidly retarded and the fuel control immediately attempts to decrease fuel flow to satisfy the new throttle setting. If the fuel control succeeded in reducing fuel flow in proportion to the retarded throttle position, the high rpm of the engine would still be producing a high airflow and the fuel-air mixture would be excessively lean. This, of course, would result in a flameout. Since CDP represents the amount of air in the combustion section, the CDP signal is allowed to override the throttle signal. As in acceleration, the decrease in fuel flow during a deceleration is a gradual process. First there is a slight decrease in rpm as a result of retarding the throttle, and this produces a slight decrease in CDP, which is followed by a decrease in fuel flow and a further decrease in rpm. This process continues until the speed indicated by throttle position is reached.

The control system maintains the correct performance of the engine during **steady-state** conditions, that is, when the throttle is set for a continuous rpm as in cruising. If ambient temperature, aircraft speed, or altitude changes, the control automatically adjusts fuel flow for constant rpm.

Advancing the throttle to the maximum stop increases engine rpm to 7,684. This results in an increase in EGT, hence the EGT indicator should be monitored for excessive temperature indications. The higher engine speed increases mass airflow through the engine, and higher EGT increases exhaust gas velocity; hence an increase in thrust results.

When the throttle is retarded below the IDLE position, the thrust-reversing system is brought into operation. The fuel-control unit is signaled for an increase in fuel flow, and the thrust-reverser doors are closed. Engine speed is increased to 7,460 rpm to produce maximum reverse thrust.

With the throttle at the IDLE position, the engine is shut down by retarding the fuel shutoff lever to OFF, closing the fuel shutoff valve in the fuel control. Operating pressures within the fuel control close the variable stators while fuel continues to circulate between the fuel pump and the control as the engine coasts down. Fuel pressure drops in the major fuel line, and spring force in the pressurizing and drain (P & D) valve closes the valve to keep the fuel line primed for the next start. Fuel between the P & D valve and the fuel nozzles drains through the drain port on the P & D valve and also from the combustion section of the engine. In emergencies, the engine can be shut down with the throttle at any position by retarding the fuel shutoff lever to OFF; however, this practice is not recommended for normal operation.

The fan rotor is aerodynamically connected to the engine rotors such that any change in engine speed results in a predetermined percentage change in fan-rotor speed. Fan-rotor speed of 100 per cent is 6,000 rpm. During operation, there is a close correlation between fan speed and engine speed; that is, the fan rotor and engine rotors reach the same performance point in approximately the same time with little or no fan speed overshoot or hunting. Fan-rotor speed lags engine speed percentage-wise at low-speed points. For example, at engine IDLE (60 per cent) fan-rotor speed is approximately 23 per cent. This speed mismatch is overcome above approximately 80 per cent rpm. If an emergency condition were to occur in which the fan rotor attempted to overspeed, retarding the throttle to IDLE or, in an extreme case, shutting down the engine would keep the fan-rotor speed within limits. However, factory testing has demonstrated that it is practically impossible for the fan rotor to exceed the maximum limits.

Fuel pump

The engine fuel pump is illustrated in Fig. 15 · 16. Fuel enters the pump at the centrifugal-type boost element, which is driven at a greater speed than the high-pressure elements. The pressure of the fuel is increased 15 to 45 psi, depending upon engine speed, and is discharged from the boost element to the two gear-type positive-displacement high-pressure elements. Each of these elements discharges fuel through a check valve to a common discharge port. At a

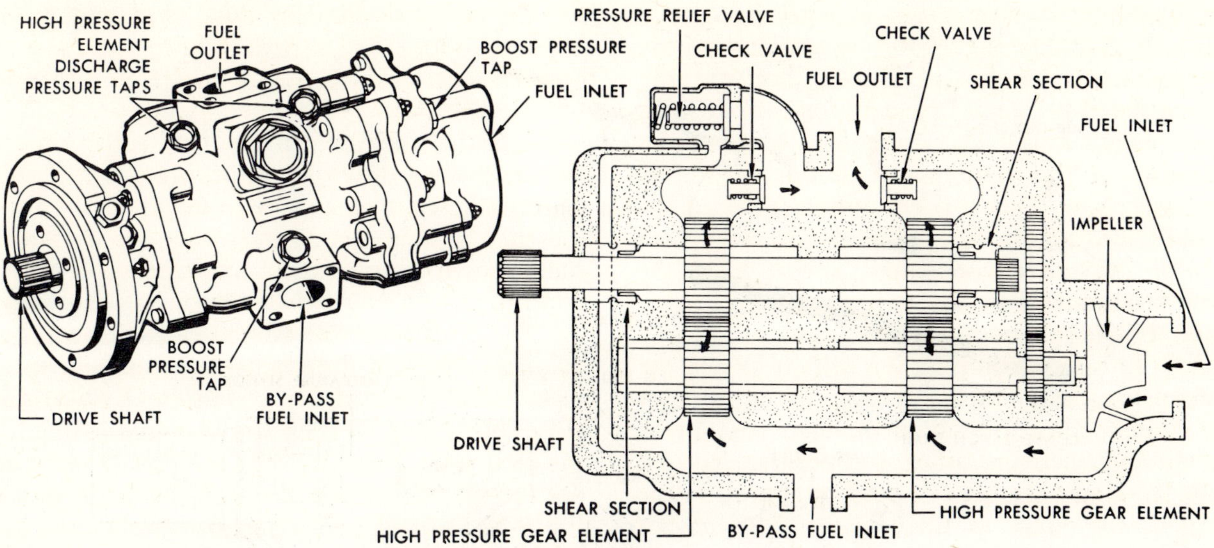

Figure 15 · 16 Engine fuel pump.

discharge pressure of 850 psig, the high-pressure elements deliver approximately 51 gpm.

Shear sections are incorporated in the drive systems of each element; hence, if one element fails, the remaining element remains operative, and the check valves prevent recirculation through the inoperative element. One element can supply sufficient fuel for moderate aircraft speeds. A relief valve in the discharge port of the pump opens at approximately 900 psi and is capable of bypassing the total flow at 960 psi. This permits fuel at pump discharge pressure to be recirculated as a protection against "dead-heading" the pump. The bypass fuel is routed to the inlet side of the two high-pressure elements.

The fuel pump is mounted at 7 o'clock on the rear face of the transfer gearbox and requires a power input of 48 hp maximum. Pump speed is 3,960 rpm at 100 per cent engine speed. Four ports on the pump are provided for attaching pressure-measuring instruments. Two ports, located between the boost and gear elements, may be used to measure pressure at the inlet of the gear elements (engine boost). The two other ports, located downstream of each gear element, may be used to measure discharge pressure of each element. A port on the pump mounting flange provides for drainage of the interstage-shaft seal area.

Fuel heater

The fuel heater, shown in Fig. 15 · 17, is an air-to-liquid heat exchanger. Fuel discharge from the fuel pump is routed into the heater where it flows through the fuel tubes and is warmed by the air flowing around the tubes. The fuel then flows past a temperature sensor and is discharged to the fuel filter.

Seventeenth-stage compressor discharge air is tapped off the compressor rear frame and routed to the air inlet of the heater. The heater incorporates an air shutoff valve which controls the flow of air as a function of heater inlet fuel temperature. Expansion or contraction of the thermostatic element in the fuel temperature sensor opens or closes the air shutoff valve. The valve is closed at fuel inlet temperatures above 7°C (45°F) and full open at fuel inlet temperatures below 0.5°C (35°F). Air flowing through the air shutoff valve is routed into the heater core, and an air baffle ensures that the hot air surrounds the fuel tubes. The air discharged from the heater is exhausted overboard.

Two drain ports are provided on the heater body, one for venting the area between the dual-stage seals on the air-shutoff-valve shaft and the other for removal of any water accumulation in the heater body. Both drains are piped to an airframe overboard manifold.

Fuel filter

The fuel filter, located at the 7:30 o'clock position on the compressor rear stator casing, removes solid contaminants from the fuel. This unit is illustrated in Fig. 15 · 18. It is constructed of sinter-bonded stainless-steel wire cloth and filters out all particles larger than 40 microns.

Fuel enters the filter as shown and surrounds the filter screen. After it passes through the screen into an inner chamber, it flows out the

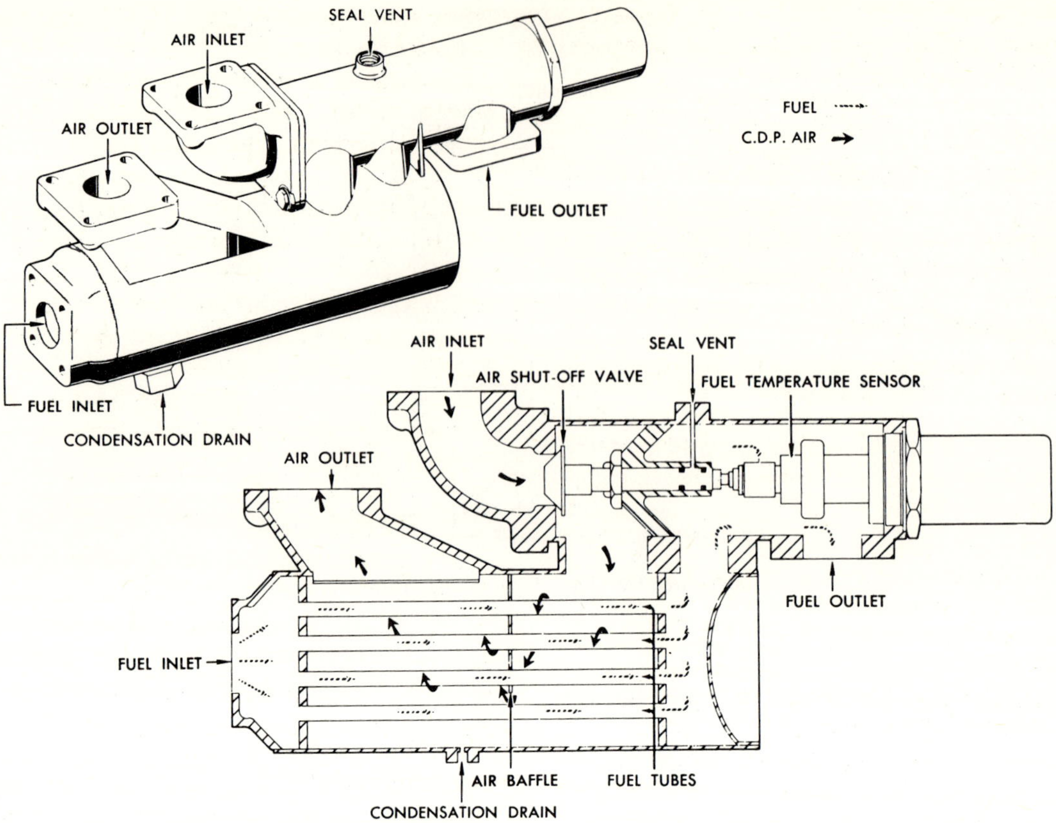

Figure 15·17 Fuel heater.

discharge port. If the filter becomes clogged, the fuel is bypassed through the relief valve, which opens when the pressure differential across the filter is 20 to 24 psi. The pressure drop across a clean filter is 10 psi maximum with fuel at 32 to 38°C (90 to 100°F) and at a flow of 57.5 gpm. If the filter screen becomes clogged, the pressure drop across the filter is 32 psi maximum with full flow through the relief valve.

The filter has a drain plug to facilitate empty-

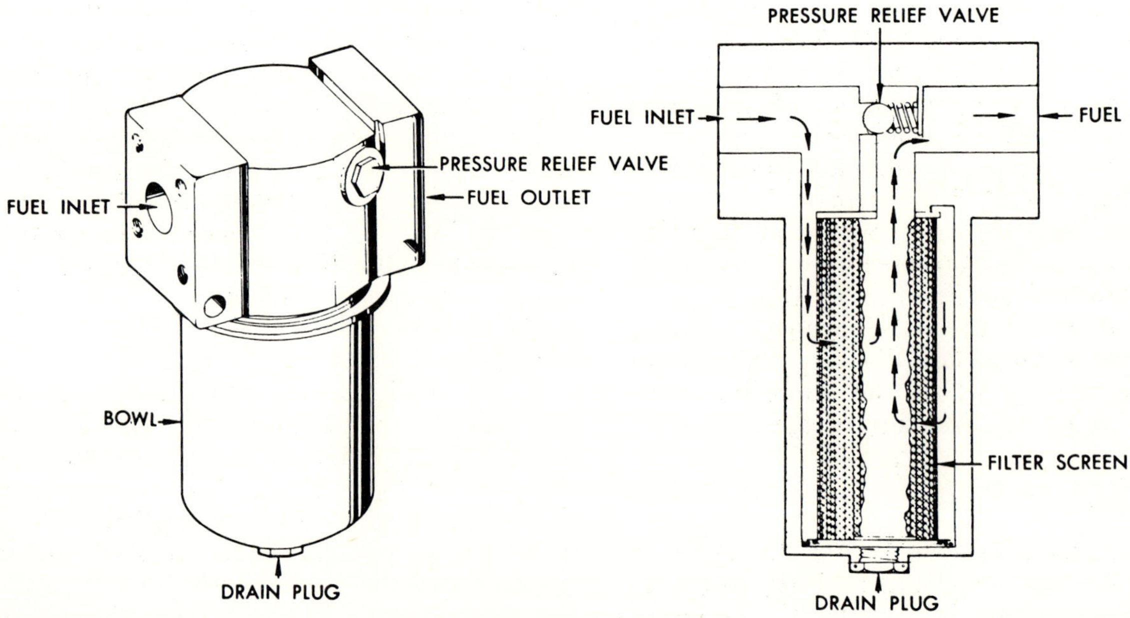

Figure 15·18 Fuel filter.

ing the filter case before removal for cleaning or replacement and to drain water from the unit at specified service periods.

Fuel-control unit

The fuel-control unit, illustrated in the drawing of Fig. 15 · 19, is a hydromechanical fuel-metering device, similar to those described in Chap. 8. It is mounted at the 6 o'clock position on the forward face of the rear gearbox. The functions performed by this unit are typical of controls for any large jet engine. These functions are as follows:

1. Meters fuel to the combustion chambers, thus controlling engine speed during steady-state conditions and allowing rapid yet safe accelerations and decelerations.
2. Schedules variable-stator-vane angle to control airflow into the engine, reducing the possibility of engine stall. The control directs high-pressure fuel to the stator actuators to position the stators as the schedule changes.
3. Prevents fuel surge.
4. Limits turbine inlet temperature to a safe value by establishing fuel-flow limits.
5. Provides a positive fuel shutoff independent of throttle action.

The fuel-control unit used with the CJ805-23 turbofan engine is a Woodward Model 1307 control which responds to throttle position, compressor inlet temperature (CIT), engine rpm, compressor discharge pressure (CDP), and acceleration. It also regulates the position of the variable stator vanes to control airflow in response to engine speed and CIT.

As with previously described hydromechanical fuel controls, this unit maintains a constant pressure differential across a variable orifice which is called the **metering valve.** The primary controlling element for the valve is the throttle; however, the actual position of the valve is also determined by the parameters mentioned in the previous paragraph. The application of these parameters is similar to that of other previously described controls.

Pressurizing and drain valve

The pressurizing and drain valve is mounted on a standoff bracket which is attached to the rear end of the lubricating and scavenge pump. Figure 15 · 20 illustrates both the outside configuration and the interior construction of the unit.

When metered fuel flows to the inlet of the pressurizing and drain valve, the fuel pressure is applied to surfaces *A* and *B* of the drain valve and surface *C* of the pressurizing valve. When inlet pressure rises 80 to 100 psi above reference pressure, which is applied to the rear side of the pressurizing valve, the drain valve snaps up, thus closing the drain port. The pressure also moves

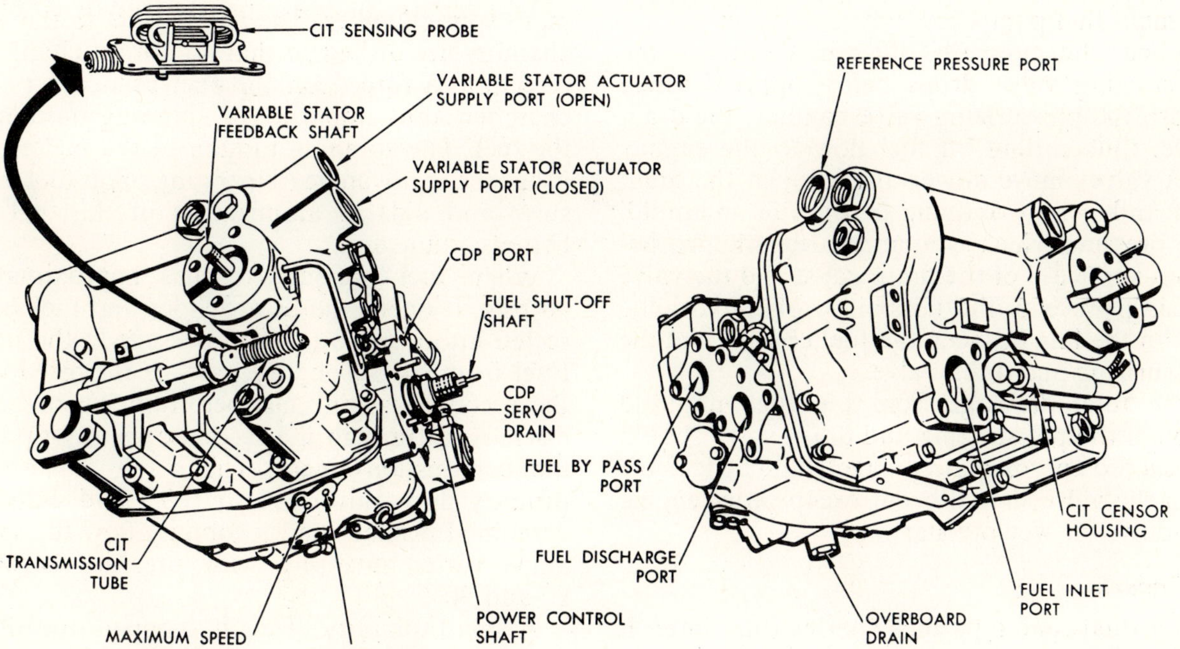

Figure 15 · 19 Fuel-control unit.

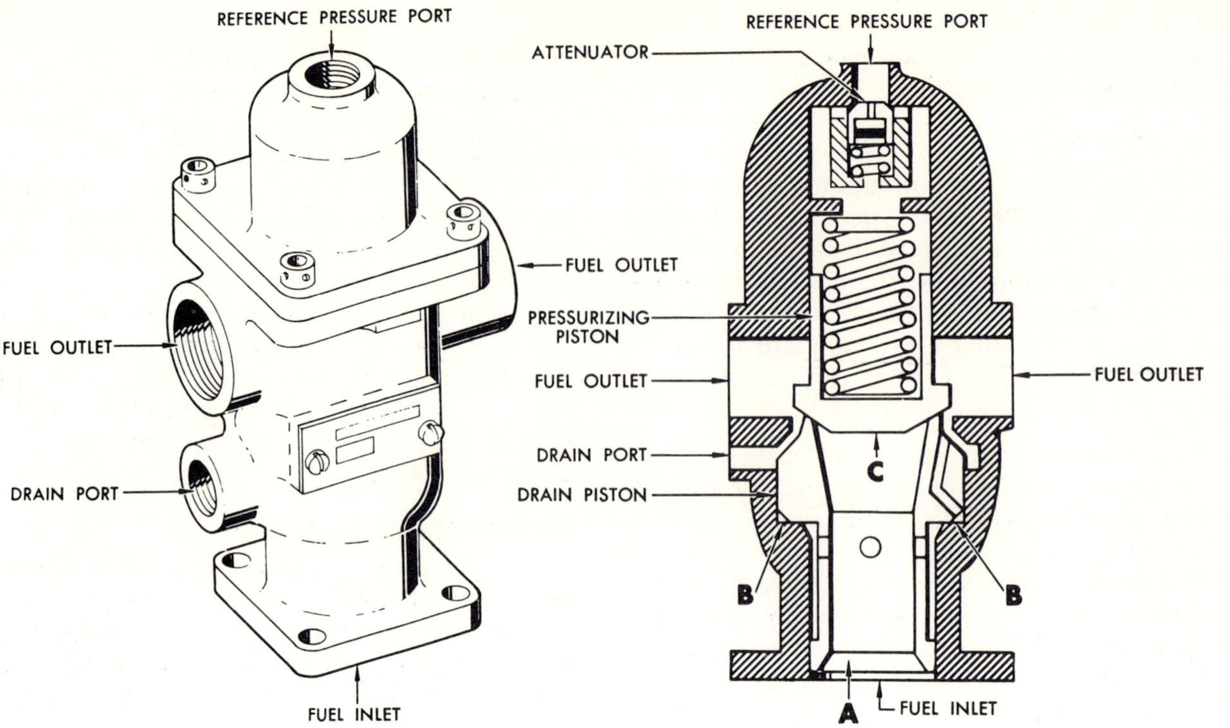

Figure 15·20 Pressurizing and drain valve.

the pressurizing valve up and away from the drain valve, thus allowing fuel flow to the engine. The attenuator in the reference pressure port damps any fluctuations in this pressure, thereby contributing to smoother valve operation. The reference pressure for the valve is taken from the fuel bypass line between the fuel control and the main fuel pump.

When the pressure differential across the pressurizing valve drops below approximately 85 psi, the pressurizing valve contacts the drain valve, thus cutting off fuel flow to the engine. Both valves move downward to open the drain port and permit draining of the fuel manifold. The pressure which is momentarily trapped between surface *B* of the drain valve and the valve housing prevents the possibility of a hydraulic lock in the line between the fuel control and the pressurizing and drain valve.

The total drainage from the fuel manifold through the pressurizing and drain valve should not exceed approximately 450 ml, and no leakage from the valve should occur except after engine shutdown or wet pre-start.

Fuel nozzle

Ten dual-cone-type fuel nozzles (one in each liner) inject an atomized spray of fuel into the combustion liners. A flow divider in each nozzle creates high- and low-pressure supplies which are discharged through separate, concentric spray tips, thus providing the correct spray angle for all rates of fuel flow. A fuel nozzle is illustrated in Fig. 15·21.

Fuel enters the inlet of the nozzle and passes through a 250-mesh double-wrapped screen. A drilled passage in the nozzle stem directs the fuel through a second screen and into the primary spin chamber. The entry ports to the spin chamber are drilled so that an abrupt change in direction is forced on the fuel as it enters the chamber, thus imparting a spinning motion to the fuel. The spinning motion of the fuel establishes the spray angle (depending upon fuel pressure) and aids in atomization of the fuel for better combustion.

When fuel pressure reaches approximately 90 psig, it opens the flow divider and fuel is directed into a second drilled passage in the stem. Fuel from the secondary passage is ported into the secondary spin chamber from whence it is discharged through the secondary spray tip into the combustion liners. The spray angle with primary flow only is 80 to 100°, and with the combined primary and secondary flow the spray angle varies inversely with pressure between 80 and 90°.

A small quantity of air is scooped out of the main airstream by the shroud around the nozzle tip. This air cools the nozzle tip and improves combustion by retarding the accumulation of carbon deposits on the face of the nozzle. The

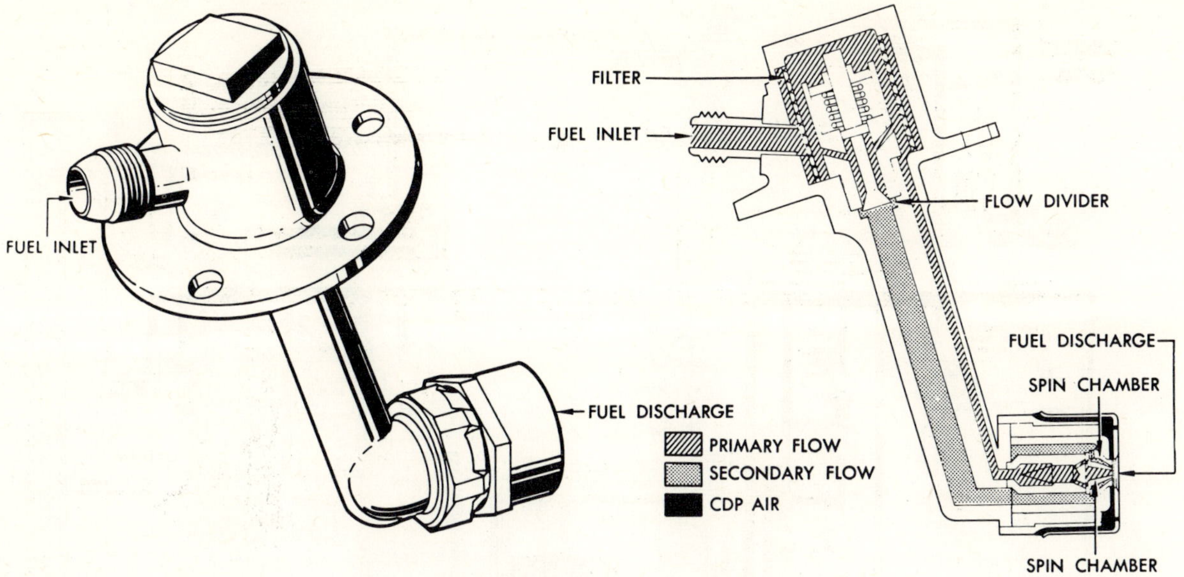

Figure ***15 · 21*** *Fuel nozzle.*

shroud also provides some air for combustion and helps to contain the fire in the center of the liner.

Variable stator actuator

The variable stator actuators are single-ended uncushioned hydraulic cylinders which are driven in either direction by high-pressure fuel. The fuel pressure to the actuators is regulated by the fuel-control unit in accordance with CIT and engine rpm requirements. Pressure applied to the head ends of both actuators forces the pistons to move, extending the actuator rods and closing the variable stator vanes. Pressure applied to the rod ends of the two actuators forces the pistons to reverse direction and open the stator blades.

As shown in the illustration of Fig. 15 · 22, the rod end of the actuator is protected against fuel leakage by dual-stage 0-ring seals. An external port drains any leakage of the first-stage seal. The maximum leakage past the seal and out the drain port is 1 ml per hr with a pressure differential of 1,200 psi across the seal.

A bleed orifice in the actuator piston permits a small amount of fuel to flow through the actuator for cooling purposes. The direction of this flow is reversible, depending upon the direction of the pressure differential across the piston. At a pressure differential of 100 psig across the piston and a fuel temperature of 38°C (100°F), the cooling flow through the orifice is between 800 and 1,200 ml per min.

The actuators are mounted on the engine, one on the upper left side of the compressor forward casing and the other on the lower right side.

LUBRICATION SYSTEM

The lubrication system for the CJ805-23 engine is illustrated in the schematic diagram of Fig. 15 · 23. The system may be divided into subsystems according to function. These are the

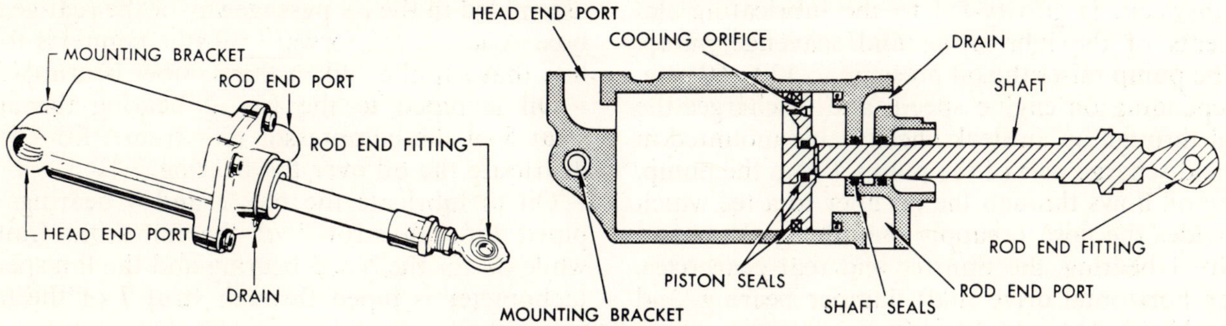

Figure ***15 · 22*** *Variable stator actuator.*

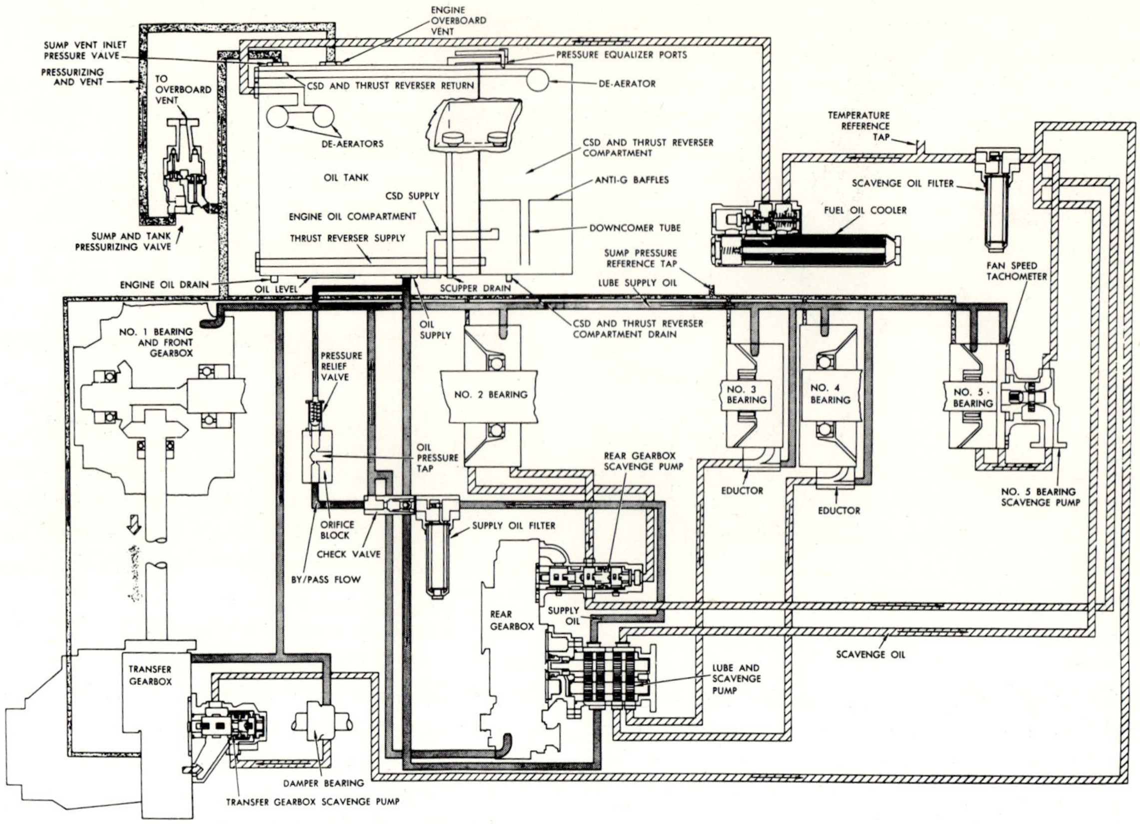

Figure 15·23 Lubrication system.

oil supply subsystem, the oil scavenge subsystem, and the pressurization subsystem.

Oil supply subsystem

The oil supply subsystem delivers oil at the required pressure to the five main bearing areas, the transfer and rear gearboxes, and the damper bearing. The oil is discharged through jet nozzles onto the bearings, gears, and other mating surfaces in these areas.

Oil is stored in a two-compartment engine-mounted oil tank. One compartment supplies oil for the engine thrust-reverser-control system and the aircraft constant-speed drive system, while the other compartment supplies oil to the engine oil supply subsystem. Oil to the engine oil supply subsystem is gravity-fed to the lubricating elements of the lubricating and scavenge pump. The pump raises the oil pressure to 12 to 60 psig, depending on engine speed, and discharges the oil through an antileak check valve mounted in the discharge port of the pump. From the pump, the oil flows through the oil filter to a tee which divides the flow to supply the inlet gearbox and No. 1 bearing, the transfer and rear gearboxes, the horizontal drive shaft damper bearing, and the No. 2, 3, 4, and 5 bearings.

Oil is piped to the inlet gearbox and No. 1 bearing area through a passage in strut 5 of the compressor front frame. Five jets distribute oil in the area. Three jets are on a tube assembly mounted in the inlet gearbox; one jet is a drilled hole in the oil passage of the inlet gearbox casing, and one jet is a drilled hole in the oil passage of the front frame.

Eight jets distribute the oil in the transfer gearbox. Six of the jets are on the oil-tube assembly mounted in the gearbox, and two jets are orifices assembled in the drilled oil passages in the gear case.

Three jets distribute oil in the rear gearbox. Two jets are on the oil-tube assembly mounted in the rear gearbox, and one jet is an orifice assembled in the oil passageway of the rear gearbox. A tee in the forward oil line connects to a line that supplies oil to the damper bearing.

Oil is piped to the No. 2 bearing through strut 5 of the compressor rear frame. Four jets distribute the oil over the bearing.

Oil to lubricate the No. 3 and 4 bearings is piped through strut 7 of the fan front frame while oil for the No. 5 bearing and the fan speed tachometer is piped through strut 7 of the fan rear frame.

During cold weather starting, lubricating pump discharge pressure may go as high as 500 psig; however, the airframe-oil-pressure indicating circuit is not designed to withstand pressures above 200 psig. Therefore, a pressure relief circuit, consisting of a restricting orifice and pressure relief valve, bypasses oil from the downstream side of the oil filter to the inlet of the pump. Enough oil is bypassed through the pressure relief circuit to keep the oil pressure at the airframe indicator from exceeding 200 psig.

The second section of the oil tank (the smaller, forward section) contains the oil used as the hydraulic medium in the thrust-reverser-control system. The auxiliary section of the tank is also used for the oil supply to the constant-speed-drive (CSD) electrical system. All lines and equipment required to connect the CSD, except the CSD oil cooler, are airframe-supplied equipment.

Scavenge subsystem

Lubricating oil, as it is sprayed over the bearings and gears, absorbs heat and picks up any metal particles or contamination from the gears. In order to use this oil again as a lubricant, it must first be cleaned and cooled. The scavenge subsystem performs these functions and returns the oil to the oil tank in condition for reuse in the oil supply subsystem.

The lubricating system is of the "dry sump" design; that is, all sump areas are constantly pumped dry of oil by the scavenge pumps. The capacity of the scavenge subsystem is approximately 2.5 times as great as the amount of the oil distributed in any one area. The balance is made up of sump pressurizing air. The advantages of this type of system in a jet engine are as follows:

1. Reduces oil foaming
2. Reduces gear drag
3. Reduces temperature buildup within gear cases and bearing areas.
4. Contributes to the design of a minimum-volume lubrication system.

During all flight attitudes, scavenge oil from the No. 1 bearing sump and the inlet gearbox drains through strut 5 of the compressor front frame to the transfer gearbox. This oil, plus the oil in the transfer gearbox, is picked up by one element of the transfer-gearbox scavenge pump. The scavenge flow from the damper bearing is piped to the second element of the transfer-gearbox scavenge pump. The discharge from the two elements of the scavenge pump is manifolded and routed through a single line to the scavenge oil filter.

Two elements of the lubricating and scavenge pump scavenge the oil from the No. 3 and No. 4 bearing sumps, one element per sump. The outlet of each sump is passed through an eductor which ensures more positive scavenging. An eductor is a device which creates a suction on the sump outlet by ejecting a portion of the incoming oil through a venturi opening. The common outlet of the two scavenge elements is piped to the scavenge oil filter.

Lubricating oil drains through holes in the bottom of the fan-speed tachometer and accumulates in the No. 5 bearing sump. The No. 5 bearing scavenge pump then pumps the oil from this sump to the outside of the engine through strut 5 of the fan rear frame and to the scavenge oil filter.

All the scavenged oil routed to the scavenge oil filter is discharged through a common port in the filter and flows through the fuel-oil cooler. From the fuel-oil cooler the oil is returned to the engine compartment of the oil tank.

Sump pressurizing subsystem

The sump pressurizing subsystem regulates the pressure of the air in the oil tank, gearboxes, and bearing sumps. These units are pressurized to ensure a positive pressure at the inlet of the lubricating and scavenge pumps.

The sump pressurizing subsystem includes the airspace above the oil level in the oil tank, bearing sumps, gearboxes, and interconnecting lines. The scavenge pumps are also a part of the pressurizing system, at least to the extent that they pump large quantities of air from the sumps to the oil tank and thereby affect air pressures in both areas.

Seventh- and ninth-stage compressor air bleeds across the bearing oil seals and pressurizes the five bearing sumps and three gearboxes. These areas are connected by a manifold and vented to the oil tank through a ¼ psi sump-vent check valve mounted in the oil tank sump-vent port. If sump pressure exceeds tank pressure by ¼ psi, the check valve opens, allowing the excess sump air to enter the oil tank. The valve allows flow into the tank only to prevent oil or tank vapors from backing up into the sump areas.

The oil scavenge pumps are capable of handling considerably more oil than actually exists in the bearing sumps and gearboxes. Because the pumps are a constant-displacement type, they make up for the lack of oil by pumping air from the sumps. Thus, large quantities of air are delivered to the oil tank. Tank pressure is established at 4 to 7 psi above ambient pressure by the sump and tank pressurizing valve. A poppet valve in the sump and tank pressurizing valve maintains the tank pressure at 4 to 7 psi by venting excess tank air to the atmosphere whenever this pressure is exceeded.

At certain engine speeds, the ability of the scavenge pumps to pump air forms a vacuum in the sumps. Since the scavenge pumps can be damaged by operating them in too great a vacuum, a sump vacuum relief valve in the sump and tank pressurizing valve opens at -2 to 3 psi allowing air to enter the sump.

Compressor air ducted to the Nos. 1, 2, 3, 4, and 5 bearing areas comprises the high-pressure side of the pressurization subsystem. Seventh-stage compressor air is bled through openings in the spacer ring between the seventh- and eighth-stage compressor-rotor disks and flows forward through the center of the compressor rotor to the rear side of the No. 1 bearing carbon oil seal. Leakage across this seal pressurizes the No. 1 bearing sump and the gearboxes. A manifold on the compressor rear casing collects ninth-stage air. A line from the manifold directs the air externally along the engine to the compressor rear frame and then to the fan frames. At the compressor rear frame, the pressurizing air is ducted through strut 1 to the No. 2 bearing sump pressurizing air seal. At the fan front frame, the air is ducted through strut 5 to the air seal of the No. 3 and No. 4 bearings and through strut 3 of the fan rear frame to the No. 5 bearing air seal.

Excess air in the sumps and gearboxes is vented to the oil tank. Routing of the air is as follows:

1. Excess air from the gearboxes and No. 1 bearing is vented from the transfer gearbox to the oil tank. The rear gearbox is vented to the transfer gearbox through the horizontal drive shaft.
2. Air from the No. 2 bearing sump is vented through strut 1 of the compressor rear frame.
3. Excess air is vented through strut 2 of the fan front frame from the No. 3 and No. 4 bearing sumps.
4. Air from the No. 5 bearing sump is vented through strut 1 of the fan rear frame.

A series of lines running the length of the engine acts as a manifold to connect all the sump vents and gearboxes together. The manifold is then connected to the oil tank and to the sump and tank pressurizing valve.

ELECTRICAL SYSTEM

The electrical system for the CJ805-23 turbofan engine consists of ignition circuits for engine starting, EGT and rpm signal generators to power the cockpit indicators, and the anti-icing valve circuit. The arrangement of this system as it is installed on the engine is shown in the drawing of Fig. 15 • 24.

Ignition circuit

The ignition circuit provides the spark necessary to ignite the fuel-air mixture in the No. 4 and No. 5 combustion liners. The flame then distributes itself through cross fire tubes to ignite the fuel-air mixture in the remaining liners. Once ignited, the fuel-air mixture continues to burn, hence it is necessary to energize the ignition circuit only during starts.

An ignition system on a reciprocating engine fires in relatively small, confined areas. Also, the fuel-air mixture is spread evenly throughout the combustion areas. In a jet engine, however, the spark generated by the ignition circuit must ignite the gases as they move rapidly toward the exit area. In addition, the fuel-air mixture in which the ignition plug is immersed is composed of layers of varying fuel density. The location of the igniter plug, therefore, must be closely correlated with design of the combustion chambers. If the plug is located in an area in which there is too much or too little fuel, no light-off will occur.

The ignition circuit consists of two high-energy ignition boxes, two igniter plugs, and the high-tension leads connecting the boxes and plugs. A switch in the cockpit controls the electric power to either of the ignition circuits. This switch may be connected to the starter relay or to the cockpit fuel shutoff lever. In either case, the ignition circuit should be deenergized as the engine reaches 3,500 rpm. Deactivating the circuit as soon as combustion is ensured improves the life of the ignition circuit components.

When the contacts of the airframe switches are closed, power is supplied from the 110-volt 400-cycle airframe bus to the capacitor discharge ignition units. In each unit the amplitude and frequency characteristics of the a-c aircraft power are changed into 3,500-volt-peak pulsating direct current. The pulsating direct current is fed into a tank capacitor which discharges approximately two times per second through a voltage-booster circuit. The discharge from the booster leaves the ignition unit and travels

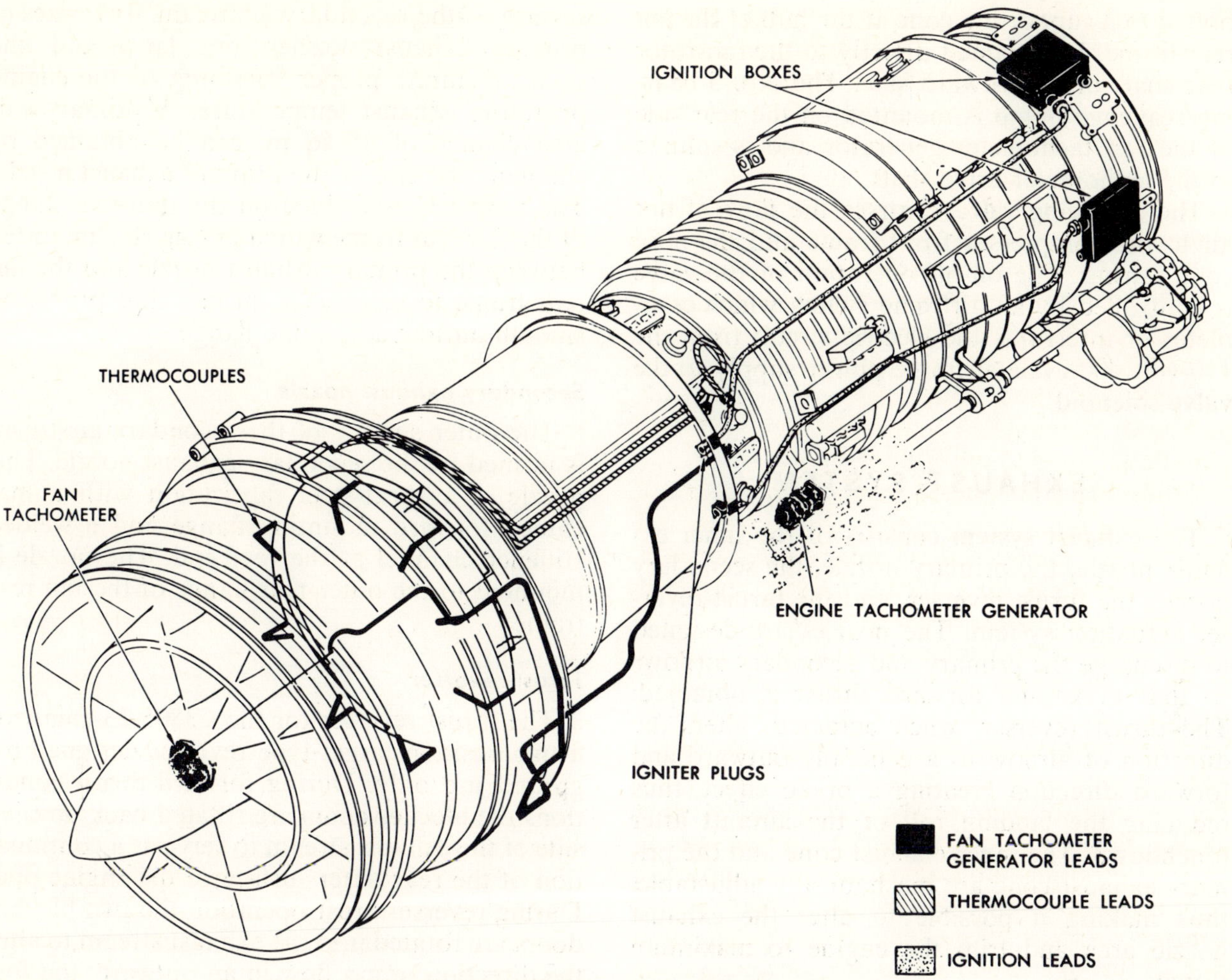

Figure 15·24 Installation of electrical system.

through the high-tension lead to the igniter plug. There it produces an arc of short duration but very high energy. It is this arc which ignites the fuel-air mixture. A complete description of the ignition system is given in Chap. 9.

Indicating and actuating circuits

The **engine-speed** indicating circuit consists of the tachometer generator, associated wiring, and the indicator. The frequency of the tachometer-generator output varies directly with engine speed, and this signal is transmitted through an airframe-furnished electric cable to the cockpit indicator. The indicator translates the frequency signal to rpm.

Exhaust gas temperature (EGT) is indicated in the cockpit to keep the pilot and flight engineer informed regarding the proper operation of the engine. The EGT indicating circuit consists of eight single-loop thermocouples mounted in the inner diffuser of the fan front frame, the necessary wiring, and the indicator in the cockpit. Two thermocouple harnesses, each containing four thermocouples, are mounted in the intermediate skin area between the fan compressor and the fan-turbine-inlet passage areas. The thermocouples extend into the primary gas stream just aft of the engine gas turbine. The signal generated by the thermocouples is transmitted through engine leads to the outer shell of the fan rear frame and forward to the front side of the engine firewall. The cockpit indicator and any additional leads required to complete the circuit are supplied by the aircraft manufacturer.

The **aft-fan-speed** indicating circuit consists of a tachometer generator and the leads connecting it to the cockpit indicator. Being directly driven by the aft fan, the tachometer generator produces an output signal proportional to the rpm of the fan. The signal is translated by the indicator in the cockpit to percentage of maximum fan speed. The tachometer generator is bolted to a supporting cone at the hub of the fan rear frame. It is splined directly to the fan-rotor rear shaft on its forward side. The No. 5 bearing scavenge pump is mounted on the rear side of the fan tachometer generator and is splined to the tachometer drive shaft.

The **anti-icing valve** controls the flow of hot air to the compressor front frame and the nose cone fairing. The on-off operation of the valve is controlled by an airframe switch which completes or interrupts the electric circuit from the 110-volt 400-cycle airframe power supply to the valve solenoid.

EXHAUST SYSTEM

The **exhaust system** consists of the inner exhaust nozzle, the primary nozzle, the secondary nozzle, the thrust reverser, and the thrust-reverser actuation system. The nozzles are designed to discharge the primary and secondary airflows so that maximum forward thrust is obtained. The thrust reverser, when actuated, alters the direction of airflow to a generally outward and forward direction creating a brake effect, thus reducing the landing roll of the aircraft after touchdown. The inner exhaust cone and the primary exhaust cone are mechanically adjustable, thus making it possible to alter the exhaust nozzle area and trim the engine to maximum thrust.

Inner exhaust nozzle

The inner exhaust nozzle forms the inner surface of the primary exhaust stream. It is a cone-shaped component designed to aid in discharging the exhaust gases in a flow of circular cross section. The nozzle is a welded and riveted sheet-steel assembly consisting of an inner cone, a two-piece outer cone, and an adjustment bolt. By turning the adjustment bolt the rear portion of the outer cone can be moved forward or aft. Moving the cone forward increases the primary exhaust exit area. An area change of ± 22.5 sq in. can be obtained by adjusting the outer cone. A slot on the forward outer cone and a key on the rear outer cone prevent radial movement, and a circumferential guide provides proper alignment of the rear outer cone. The adjustment bolt must be lockwired prior to engine operation. The nozzle assembly is bolted to the inner flange of the fan rear frame.

Primary exhaust nozzle

The primary exhaust nozzle forms the outer surface of the primary gas stream and the inner surface of the secondary airstream. Two sizes of primary exhaust nozzles, one large and one small, facilitate proper trimming of the engine operating exhaust temperature. A primary exit area change of 45 sq in. can be obtained by changing the size of the primary exhaust nozzle. The nozzle is assembled on the mid-rear flange of the fan rear frame with a fairing ring mounted between the primary exhaust nozzle and the fan rear frame to serve as a shroud, thus providing smooth airflow across the flange.

Secondary exhaust nozzle

The outer surface of the secondary airstream is formed by the secondary exhaust nozzle. This nozzle is a sheet-steel fabrication with a machined forward mounting flange and a welded stiffener channel at the rear end. The nozzle is mounted to the outer rear flange of the fan rear frame.

Thrust reverser

The thrust reverser for the CJ805-23 engine is a two-position target-type reverser designed for ground use only. During forward-thrust conditions the **blocker doors** are rotated back on each side of the exhaust stream to serve as a continuation of the rear outer surface of the engine pod. During reverse-thrust operation the two blocker doors are rotated into the exhaust stream to alter the direction of gas flow in an outward and forward direction. This action is shown in the drawings of Fig. 15 · 25.

The reverser consists of two contoured blocker doors, the support beams, and the actuating linkages. It is a welded and riveted assembly made of sheet metal and stainless-steel forgings. The reverser support beams are attached to the reinforcing hat sections on the fan frames at the top and bottom positions. The top beam supports the blocker doors, taking all the vertical *g* loads, while both the top and bottom beams transfer reverse-thrust loads to the engine structure.

Movement of the blocker doors from the stowed to the reverse-thrust position is performed by simultaneous translating and rotating action produced by hydraulic actuators which drive motion-multiplying scissors-type linkages. Two actuator and linkage systems are provided, one at the top and one at the bottom of the vertical center line. The two systems are synchronized by a flexible-cable coupling between the actuators. Each actuator and linkage system is mounted to a support beam which incorporates a track for the translating motion of the

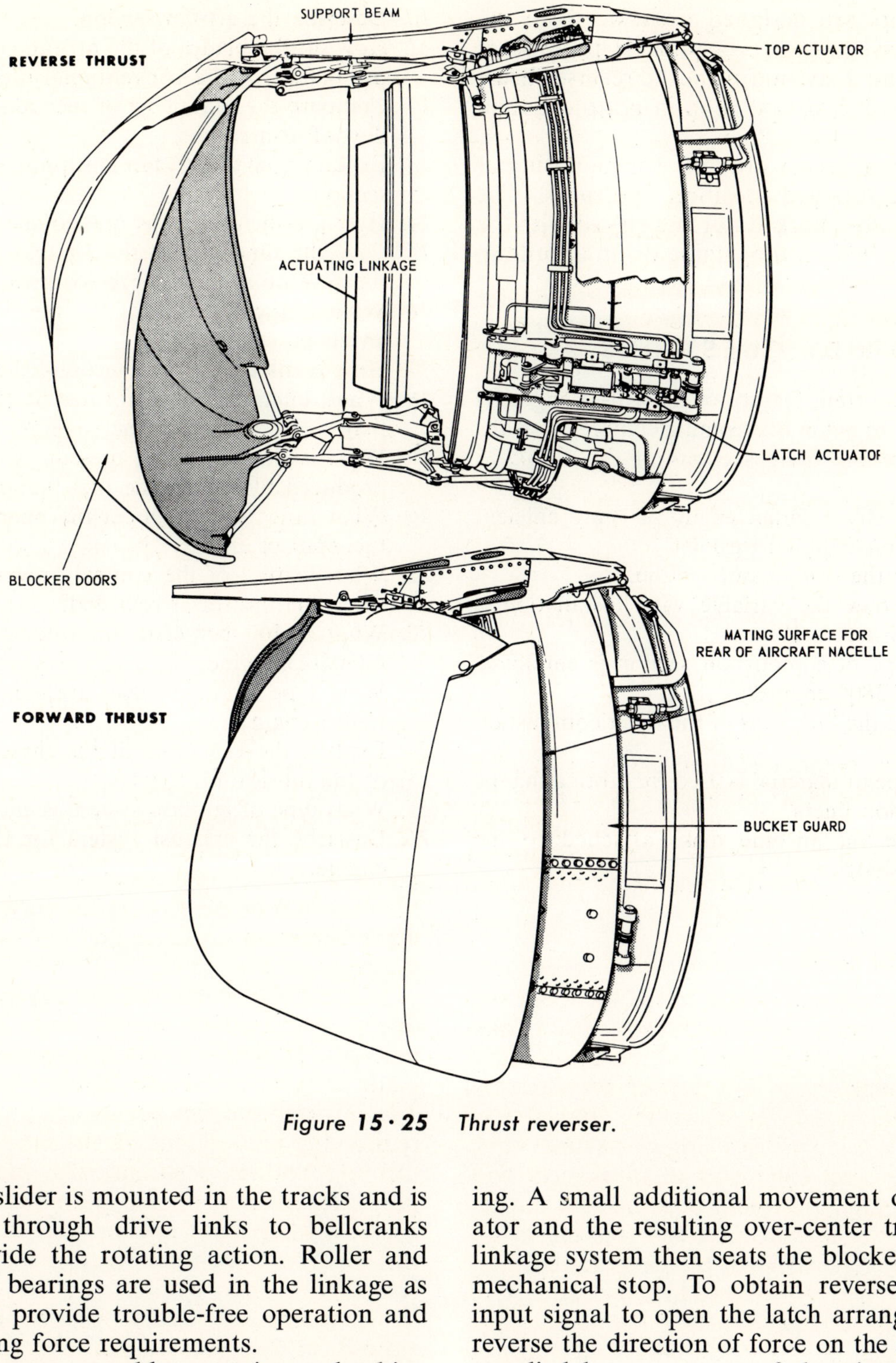

Figure **15 · 25** *Thrust reverser.*

linkage. A slider is mounted in the tracks and is connected through drive links to bellcranks which provide the rotating action. Roller and antifriction bearings are used in the linkage as required to provide trouble-free operation and low-actuating force requirements.

The reverser assembly contains a latching mechanism which assures that the blocker doors will remain in the stowed position even if the hydraulic actuation system fails. The doors are latched by two piston-actuated toggle-type latches, one on each side of the reverser on the horizontal center line. As the doors move from the reverse to the stowed position, a stirrup on the door leading edge contacts the latch and a cam arrangement to provide partial latching. A small additional movement of the actuator and the resulting over-center travel of the linkage system then seats the blockers against a mechanical stop. To obtain reverse thrust, an input signal to open the latch arrangement and reverse the direction of force on the actuators is supplied by movement of the aircraft throttle. Thrust variation in the reverse-thrust position is accomplished by changing engine speed.

The thrust reverser is designed for use at engine power settings from idle to maximum reverse at forward ground speeds from 60 to 200 knots without detrimental reingestion or cross ingestion of exhaust gases. The thrust reverser is designed to operate mechanically in crosswinds up to 35 knots. The thrust-reverser blocker doors are designed to move from the forward thrust to the reverse-thrust position in not more than 2 sec and from the reverse-thrust to the forward-thrust position in not more than 6 sec.

The reverser design provides for mounting of an airframe-supplied position indicator. The assembly can be checked out and rigged with the reverser installed on the engine or on a buildup fixture.

STUDY QUESTIONS

1. What important feature of the CJ805 engine is designed to aid in overcoming compressor stall?
2. What are the basic sections of the CJ805-23 engine?
3. Explain why a fanjet engine is more efficient than a conventional turbojet.
4. Describe the compressor section.
5. Explain how the variable vane installation is accomplished.
6. What type of combustion section is employed for the CJ805 engine?
7. What are the functions of the inner combustion casing?
8. What type of material is used for fabricating the combustion liners?
9. How are the turbine disks attached to the turbine shaft?
10. Describe the aft-fan section.
11. How do the blades of the aft fan compare with the blades of the conventional turbine?
12. Compare the mass flow of secondary air to the flow of primary air.
13. Explain how the aft fan is supported in the fan section.
14. How are the oil sumps pressurized?
15. Give the functions of the air that is bled from the seventh-stage compressor area.
16. What is the function of the bleed air through an air-oil seal?
17. How is inlet anti-icing accomplished?
18. What functions are performed by the fuel other than being burned in the engine?
19. What is the maximum pressure which may be produced by the engine fuel pump?
20. What input signals direct the operation of the fuel-control unit?
21. What feature of the control prevents flameout when the throttle is retarded?
22. What is 100 per cent fan rotor speed in the CJ805-23 engine?
23. What type of fuel nozzles are employed in the CJ805 engine?
24. Explain the functions of the three subsystems of the lubrication system.
25. What type of ignition system is employed?
26. Describe the exhaust system for the CJ805-23 engine.

CHAPTER 16

The Pratt & Whitney JT3D Turbofan Engine

GENERAL DESCRIPTION AND OPERATION

The Pratt and Whitney JT3D turbofan engine is widely used for modern jet transport aircraft. It is an axial-flow design having a multistage split compressor, an eight-can cannular combustion chamber, and a split four-stage reaction-impulse turbine. Figure 16 · 1 is a photograph of a cutaway model used for display purposes. This illustration reveals the principal features of the engine.

Model numbers

Several model numbers have been assigned to JT3D engines; however, the general configuration remains quite similar for all models. The JT3D-1-MC6 and the JT3D-1-MC7 models are those which have been modified from JT3C-6 and JT3C-7 turbojet engines during overhaul by the addition of the fan section at the front of the compressor. The JT3D-1 and JT3D-3 models are new production engines which have the forward fan sections installed at the time of manufacture.

Performance and specifications

The performance and specification figures listed in this section for the JT3D engine are generally applicable to the different models of the engine; however, it must be noted that improvements may be made which will bring about variations. The figures given will nevertheless provide sufficient information for the purposes of this text.

Diameter	53.0 in.
Length (without airframe parts)	136 in.
Weight	4,170 lb
Power/weight ratio	4.31 lbt/lb
Compressor ratio	13:1
Fan pressure ratio	1.66:1
Fan bypass ratio	1.4:1
Fuel consumption (tsfc)	0.53 lb/lbt/hr
Oil consumption	2.9 lb/hr
Takeoff thrust	18,000 lb

Operation

Air to operate the engine and provide thrust enters the engine through the compressor inlet case assembly from the airframe inlet duct which is attached to the front of the inlet case. The inlet case assembly is provided with vane-type multipurpose struts which transmit No. 1 bearing loads to the outer case structure, conduct anti-icing air and lubricating oil to the inner diameter of the engine, and direct the air to the front compressor section.

The compressor section is a split type (twin spool) and consists of two rotor assemblies. Each rotor assembly is driven by an independent turbine, and each rotor is free to rotate at its best speed. As it is necessary to rotate only one of these units during the starting procedure, the selection of the small rear compressor permits the use of a smaller starter than would otherwise be required.

Primary air from the inlet case enters the front compressor which consists of eight rotor stages and seven stator vane stages. The first two rotor stages have extended blades which drive the secondary air through the outer duct. The air path of the front compressor has a constant inside diameter and a decreasing outside diameter. The rotating parts are connected by a drive shaft which passes through the inside of the rear compressor rotor and drive shaft to the three rearmost turbines.

Figure 16·1 Cutaway model of JT3D engine.

Between the front and rear compressors and around the rear compressor rotor and case is the intermediate case. This case has an automatic arrangement for bleeding front compressor air to improve the acceleration characteristics of the engine. The inlet vanes, ninth stage, transmit No. 2 and No. 3 bearing loads to the outer case; conduct, supply, and return oil to the bearings; and direct the compressed air from the front compressor to the rear compressor.

The rear compressor has seven rotor stages, numbered 10 through 16, and six stator vane stages, numbered 10 through 15. It has a constant outside diameter and an increasing inside diameter, and it is driven by the first-stage turbine through an independent shaft concentric with the front compressor drive shaft. The diffuser case is immediately behind the rear compressor.

Compressor exit guide vanes at the front of the diffuser case straighten the airflow and direct the air into the diffuser where it is expanded for entry into the combustion chambers. Struts in the case transmit the No. 4 and No. 5 bearing loads to the outer portion of the case. In addition, these struts conduct bearing oil pressure and suction lines and provide high-pressure air for engine functions such as anti-icing. They also provide a source of clean air for aircraft pressure needs. On engines with water injection, the water injection manifold is mounted on the front flange. The water is dispersed through 16 curled tubes mounted on the periphery of the diffuser case, through mating holes in the case, and injected into the air stream. The outside contours of the compressor and diffuser sections give the engine its "wasp waist" and at the same time provide a convenient location for the accessory section.

The fuel manifold, which consists of eight circular clusters of six fuel nozzles, is located in the diffuser case annulus and injects the fuel into the forward ends of the combustion chambers. From the diffuser the air is channeled into eight portions for burning with the fuel in the eight combustion chambers.

The combustion section consists of eight separate cans arranged annularly, hence the name "can-annular" or "cannular burners." The chambers, as viewed from the rear of the engine, are numbered in a clockwise direction starting with the uppermost chamber as No. 1. These chambers are connected by crossover tubes. The compressed air with fuel injected is ignited by sparkigniters installed in the No. 4 and No. 5 combustion chambers. After "light-up" the flame is perpetuated by the construction of the combustion chamber and the heat generated by previous combustion. The exhaust gases pass through the combustion chamber outlet duct into the turbine-nozzle case.

The turbine-nozzle case houses the front stages of the four-stage turbine and is located aft of the combustion section. The first stage of the turbine drives the rear compressor, and the second, third, and fourth stages drive the front compressor. The turbine-nozzle case also contains the four turbine exhaust nozzles which are made up of a series of stationary vanes designed to direct the exhaust gases through the turbine blades and into the turbine-exhaust case.

To the rear of the turbine-nozzle case is the turbine-exhaust case which houses the fourth-stage turbine disk and blades and through which the exhaust gases are ejected from the engine. The turbine-exhaust case also supports the No. 6 bearing and the No. 6 bearing oil sump.

Main bearing nomenclature

In all overhaul procedures and description of the JT3 main bearings, reference is made by bearing number instead of bearing nomenclature. The numbering of the bearings is shown in Fig. 16 • 2. The following list gives the main bearing nomenclature and its corresponding bearing number.

Bearing nomenclature	Bearing No.
Front compressor front (roller)	1
Front compressor rear (duplex ball)	2
Seal plate (ball)	2½
Rear compressor front (roller)	3
Rear compressor rear (duplex ball)	4
Turbine intershaft (roller)	4½
Turbine front (roller)	5
Turbine rear (roller)	6

ENGINE DESCRIPTION

The detailed description of the JT3D engine may be followed by examination of the cutaway photograph shown in Fig. 16 • 3. This illustration shows the construction of the engine both externally and internally.

Front accessory section

The front accessory section consists of one assembly and related parts, including the front accessory drive support. The case of this assembly is made of magnesium. Sixteen nuts secure the front accessory support to the No. 1 bearing support. The front accessory support has two mount pads on its front face; however, only one is used. This is the mounting and drive for the compressor bleed valve control. The other opening has provisions for mounting and driving a tachometer.

The No. 1 bearing scavenge pump is mounted on the lower rear face of the support. Air and oil for the bleed control are brought out of the No. 1 bearing support and routed through tubing to the control. The front accessory drive gear, which is externally splined, is inserted with an 0 ring seal around it into the front compressor front hub engaging the internal spline in the hub. It is retained by the same nut that holds the No. 1 bearing inner race in position. The front accessory drive gear meshes with and drives the tachometer drive gear, the No. 1 bearing scavenge pump drive gear, and the compressor-bleed-control drive gear.

Number 1 bearing supports and inlet-case assembly

The No. 1 bearing supports and inlet-case assembly are discussed together as the supports are mounted in the inner diameter of the inlet-case assembly, the vanes of which carry the structural load of the No. 1 bearing from the inner shroud to the outer shroud.

The inlet-case assembly consists of 23 or 28 hollow vanes, incorporating foam rubber stiffeners in the center bays, that are welded between hollow double-walled titanium inner and outer shrouds. Each vane extends from the inner wall of the inner shroud to the outer wall of the outer

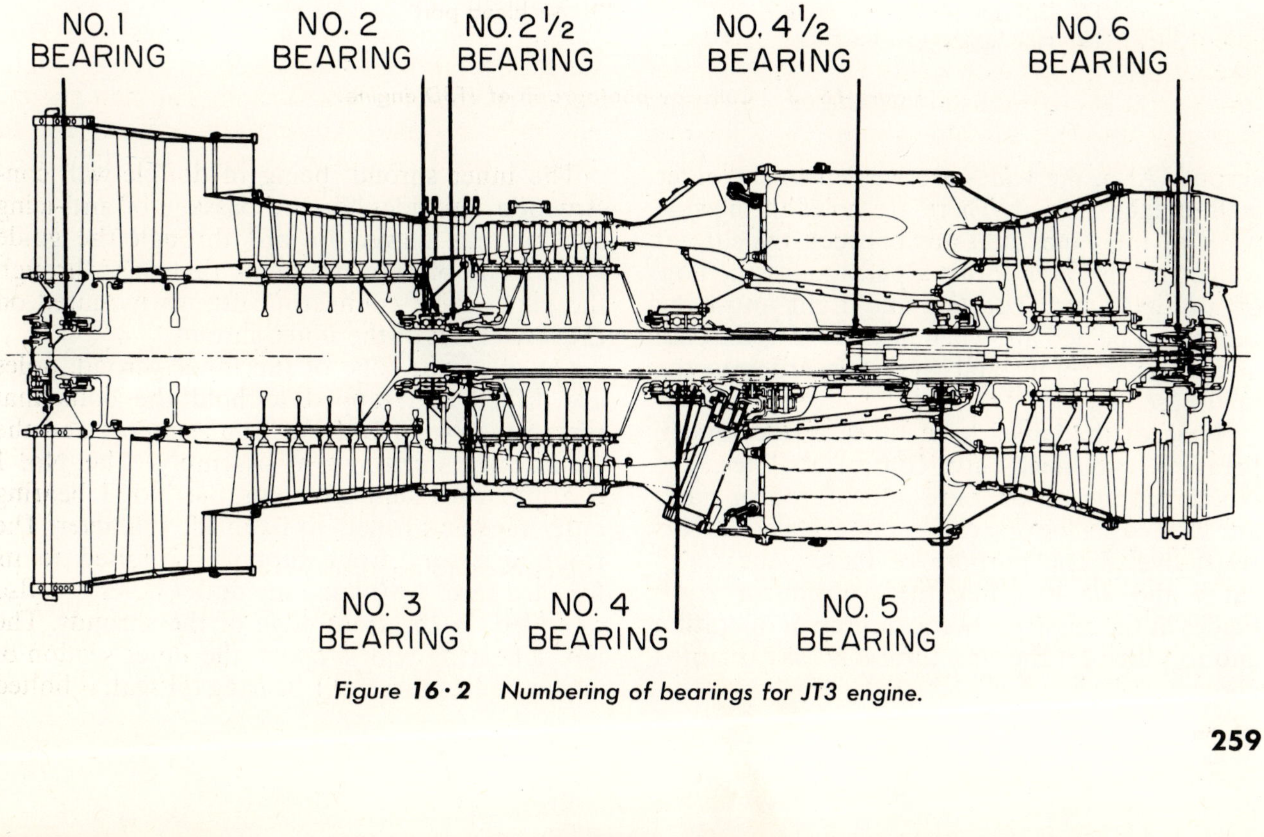

*Figure **16 • 2** Numbering of bearings for JT3 engine.*

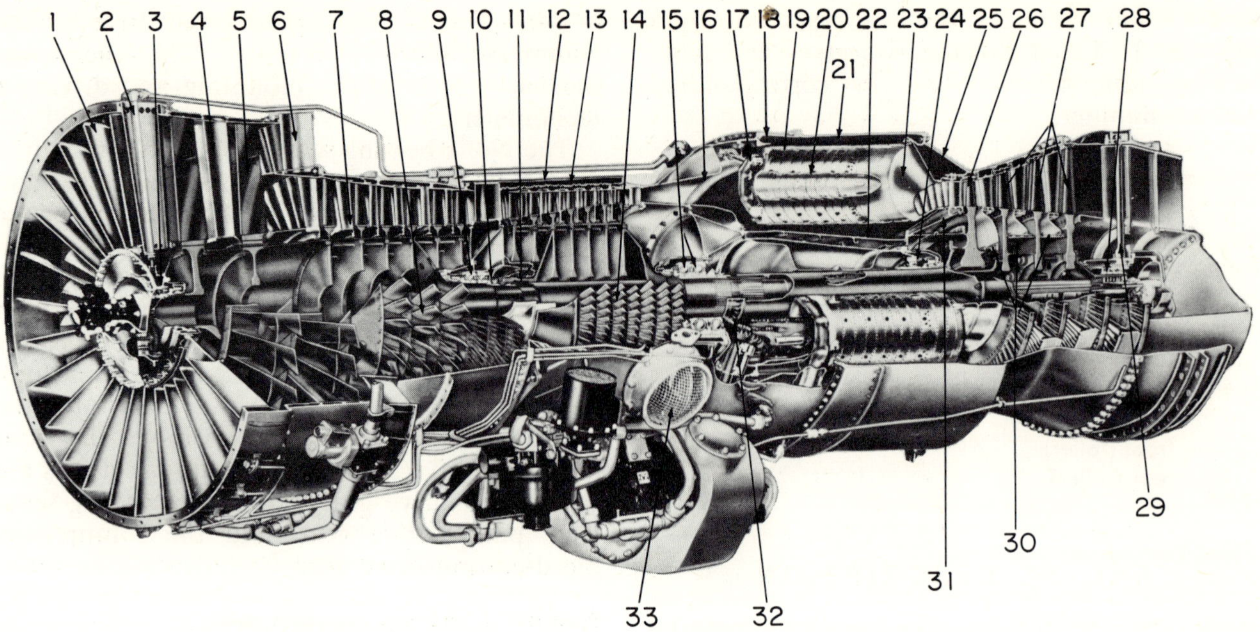

1. Inlet guide vanes
2. Breather tube
3. No. 1 bearing
4. No. 1 stator vane
5. Fan second-stage rotor blade
6. Fan outlet guide vane
7. Compressor stator vane shroud
8. Forward compressor rotor
9. No. 2 bearing
10. No. 2½ bearing
11. No. 3 bearing
12. Compressor intermediate case
13. Rear compressor case
14. Rear compressor rotor
15. No. 4 bearing
16. Diffuser section
17. Fuel manifold
18. Front combustion chamber outer case
19. Combustion chamber
20. Combustion chamber inner liner
21. Rear combustion chamber outer case
22. Combustion chamber inner case
23. Combustion chamber outlet duct
24. No. 5 bearing
25. Turbine nozzle case
26. Rear compressor drive turbine
27. Forward compressor drive turbine
28. No. 6 bearing
29. Support rod
30. Knife-edge seals
31. Turbine inner case
32. Main accessory drive gears
33. Air bleed port

*Figure **16·3** Cutaway photograph of JT3D engine.*

shroud. Also, the vanes are welded to the outer wall of each shroud. There are holes in the side walls of the vanes that are between the shroud walls, and this area forms a passage for the flow of anti-icing air. The left and right anti-icing tubes feed heated air, when desired, into the case and vane assembly and then forward into the inlet case.

Some of the vanes have tubes inside for oil, breather, and bleed-valve-control air lines. The No. 1 bearing oil-pressure and oil-return tubes are located in the 7 o'clock and 5 o'clock struts respectively. The purpose of these vanes is to cause the air to enter the compressor rotor blades at the angle for best compressor operation and to transport the structural load as explained above.

The inner shroud, being of double-wall construction, provides for the passage of anti-icing air that has flowed inward through the guide vanes. The oil and air tubes that pass through the vanes have connecting fittings mounted on the inner wall of the inner shroud.

On the front edge of the inner shroud, holes are drilled and tapped to hold the bolts that secure the No. 1 bearing front support and the compressor inlet-air-seal assembly. The No. 1 bearing front support holds the No. 1 bearing outer race and rollers in its inside diameter. The front accessory front support is bolted to its forward face. Anti-icing air outlet holes are also provided on the front edge of the shrouds. The No. 1 bearing rear support, the inner section of which houses the No. 1 bearing oil seal, is bolted at its outer diameter to the compressor vane inner shroud rear flange.

The outer shroud, being also of double-wall construction, provides for the passage of anti-icing air. The outer wall of the outer shroud forms the support for the outer fittings of the tubes that pass through some of the vanes. A pressure-probe fitting passes through both walls of the shroud to protrude into the airstream between the vanes.

The front rim of the outer shroud is drilled and tapped to receive the bolts which hold the airframe air-inlet duct. The rear rim of the outer shroud is drilled and tapped to receive the bolts that hold the shroud to the front compressor case.

Front compressor

The front compressor, which is housed in the **front compressor front case** and **front compressor rear case,** consists of a rotor having eight rows of blades and a stator assembly made up of seven rows of vane and shroud assemblies located between each pair of rotor blades. There is no second-stage stator numbered as such, and there is no third stage of rotor blades designated as such. The first two rows of blades are considerably larger than the rest and are called "fan" blades.

The airstream is separated into primary and secondary streams by means of a spacer between the second and fourth stages of rotor blades. The primary airstream is directed internally by the third-stage stator while the secondary airstream is exhausted through the exit struts of the fan discharge case. The numbering of the blade stages from front to rear is 1, 2, and 4 through 9, and the numbering of the stator stages is 1 and 3 through 8. The ninth-stage stator vanes are incorporated in the intermediate case and will be discussed later.

The third-stage vane and shroud assembly is of single-piece construction, whereas the fourth through the eighth stages are of the split type. The spacers of the fourth through the eighth stage are integral. The inner shrouds form a seal ring for the two air seals on the outer diameter of each rotor disk spacer.

The compressor is driven by a shaft from the second, third, and fourth turbines. The speed of the front compressor N_1 is roughly two-thirds the speed of the rear or high-speed compressor N_2. The front compressor provides the initial compression to the air that passes through the engine and transmits this air to the rear or high-speed compressor. The stator vanes and rotor blades diminish in size and increase in number from the front to the rear of the compressor. This matches the decreased volume of air as the pressure increases.

The front accessories are driven by a gear attached to the front hub of the front compressor rotor. The rear hub of the front compressor contains the front compressor drive turbine shaft coupling as shown in Fig. 16 • 4. This illustration shows the inner shaft which drives the front compressor from the rear turbine sections. The coaxial outer shaft drives the rear compressor from the first turbine stage. The illustration in the figure represents the JT3C turbojet engine

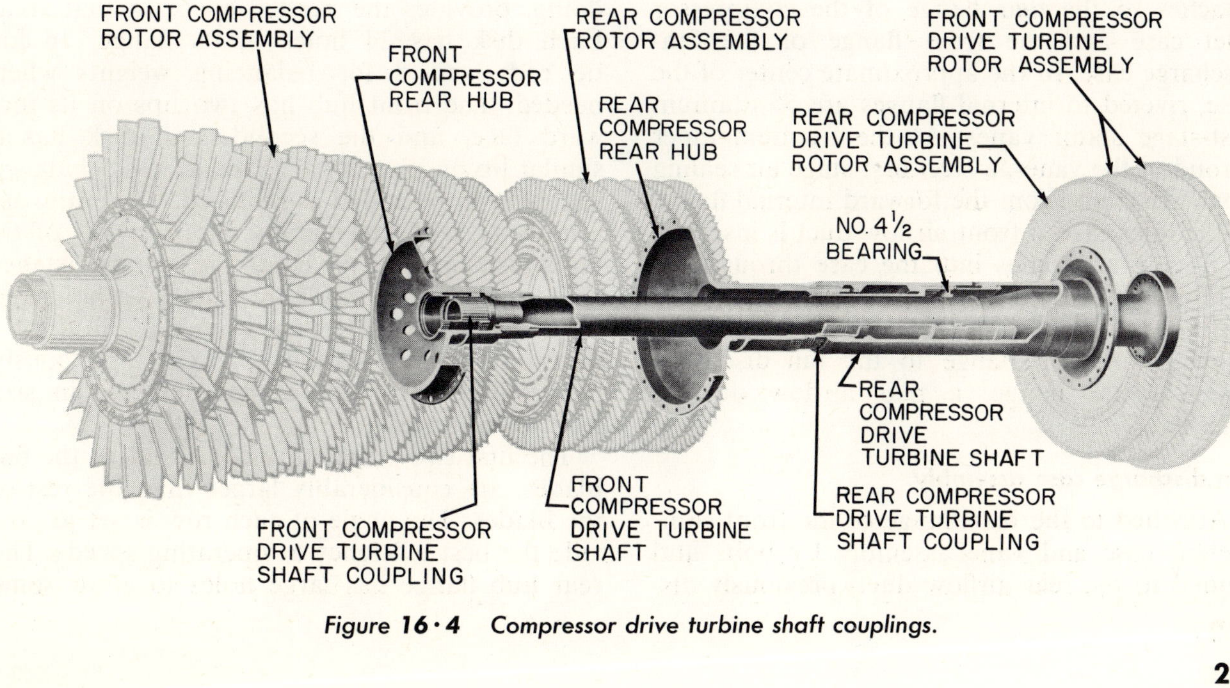

Figure 16 • 4 Compressor drive turbine shaft couplings.

rather than the JT3D turbofan engine; however, the drive shaft arrangement is similar for both engines.

The rear hub of the front compressor rotor is supported by a double ball bearing, No. 2. The two halves of the No. 2 bearing are separated by an oil baffle in which drilled holes direct oil to the forward and aft sections of the bearing. The No. 3 bearing, called the **rear compressor front bearing,** is mounted with inner race and rotors on the end of the rear hub. An oil seal on the rear end seals the bearing compartment from the engine air pressure.

On the front hub the rotor is supported by a roller bearing as shown in Fig. 16 · 3. This is the No. 1 bearing, the liner of which is in the No. 1 bearing front support. An oil seal on the front end seals the bearing compartment from engine air pressure. The bearing support is secured to the inlet-case assembly.

Front compressor cases

There are three front compressor cases: (1) the **front compressor case and vane assembly,** (2) the **fan discharge case assembly,** and (3) the **front compressor rear case assembly.** These cases will be discussed individually in order to avoid confusion. The front compressor cases carry the structural load of the front of the engine. These cases decrease in diameter from front to rear to match the decreasing diameters of the stator rings. The front compressor case and vane assembly and the fan discharge case are part of the **front compressor rotor and case assembly.**

Front compressor case and vane assembly

The front compressor case and vane assembly attaches to the rear flange of the compressor inlet case and the front flange of the fan-discharge case. In the approximate center of the case, riveted to internal flanges are 55 titanium first-stage stator vanes. At the titanium inner shroud of the vanes, a steel first-stage air sealing ring is riveted. From the forward internal flange to the inlet case, a front airflow duct is inserted. Anti-icing air comes into the case through the reinforced openings in the outer wall, circulates, and flows forward into the inlet case. From the rearward internal flange to the fan discharge case forward flange, a rear airflow duct is inserted.

Fan discharge case assembly

Attached to the rear flange of the front compressor case and vane assembly by bolts and pinned to the rear airflow duct, previously discussed, is the **fan discharge case.** This case is constructed of stainless steel and consists of an outer shroud and an inner case with 38 steel ducts between them. The struts are secured in the inner case by a riveted locking plate at the rearward end and wired to the fairing retaining screws at the forward end.

Front compressor rotor and case assembly

The front compressor rotor and case assembly consists of the front compressor rotor, as described below, assembled to include the front compressor case and vane assembly, the airflow ducts, the fan discharge case, and the third-stage vane and shroud assembly.

The front compressor rotor consists of two hubs, seven disks, eight rows of blades, two sets (16 each) of tie rods and associated hardware. Each row of blades is inserted into undercut slots in its disk. Locks are inserted under the blades and bent to secure them in place. The blades do not have a tight fit but are seated by centrifugal force during engine operation. The front hub forms the disk for the first row of blades, but the rear hub is a separate unit that is held fast to the rear face of the seventh-stage disk. The spacers between the disks are internally reinforced with tubes. The tie bolts run through these tubes and through the disks.

The spacers have two knife edges on their outside diameters. These run against the seal platforms on the stator vane inner shroud, which was previously described. The exception to the above is the large spacer between the second- and fourth-stage compressor blades. This spacer serves as a means of joining the fan section to the remainder of the compressor and, in so doing, provides the needed space in that area. Each disk has 24 holes in its flange, 16 for tie rods and 8 for balancing weights when needed. The front hub has two lips on its forward face, and the second-stage disk has a similar lip on its rearward face to which balance weights can be added as required. The entire assembly is held together by the two sets of tie bolts. The smaller set holds the first two stages of blades together, while the longer set holds the third through eighth stages together. Both sections are joined by the spacer, as previously mentioned. The rotor blades decrease in size from front to rear.

The first two stages of blades, that is, the fan blades, are considerably larger than the rest of the blades. The angle of each row is set to provide the best efficiency at operating speeds. The rear hub flange has large holes to allow some ninth-stage air into the compressor rotor. This air serves the dual purpose of providing bearing seal pressurizing and cooling. The front hub and disks of the front compressor are titanium. The rear hubs are made of titanium or steel, depending upon the particular model of the engine. The blades of the first two stages of all models are titanium, however the rest of the blades may be titanium or steel.

Front compressor rear case assembly

Attached to the rear innermost flange of the fan discharge case and to the forward flange of the intermediate case is the front compressor rear case assembly, which is of welded construction. Welded to this case internally are the supports for the vane and shroud assemblies. A center external flange serves as a reinforcement and is also used to support the brackets that carry external items.

Front compressor vane and shroud assemblies

The first-stage stator vanes were discussed under the front compressor case and vane assembly; and as there are no second-stage stator vanes, our description here begins with the third- and fourth-stage stator vanes. These vanes are constructed of aluminum alloy, while the fifth- through eighth-stage vanes are made of stainless steel. The stators are pinned to inner shrouds to which are riveted **inner air seal platforms.** The third-stage stator vanes and shroud are built into a single circular assembly, which is held stationary by a flange inserted between the fan discharge case and front compressor rear flanges. The fourth- through eighth-stage stator vane and shroud assemblies are each made in two halves to facilitate disassembly and assembly. They are prevented from rotating by being pinned to each other and to the third-stage vane and shroud. When these units are assembled, the rings are held in the engine by shoulders on the inside of the front compressor rear case.

The angle at which the vanes are mounted in the shrouds is set to feed air into the following row of rotor blades to give the best compressor efficiency at operating speed. Stator vanes decrease in size from front to rear to match the decreasing size of the rotor blades.

Compressor intermediate section

The compressor intermediate case is illustrated in Fig. 16·5. The forward mounting points are on this case, and the case is attached to the rear flange of the front compressor case. A locating pin is used at top center between the two flanges. The intermediate case surrounds the rear compressor, but it is not considered part of the compressor. It serves to separate the low-pressure front compressor from the high-pressure rear compressor and also serves a structural function of joining the external cases. The intermediate case is of steel construction and has a double wall on the forward end. Guide vanes are welded from the outer wall through the inner wall, extending into a shroud ring. The double mounting lugs are machined from rings, and the case is welded to the rings. The front flange of the case is drilled and tapped for bolts that attach it to the front compressor case, and the rear flange is drilled but not tapped.

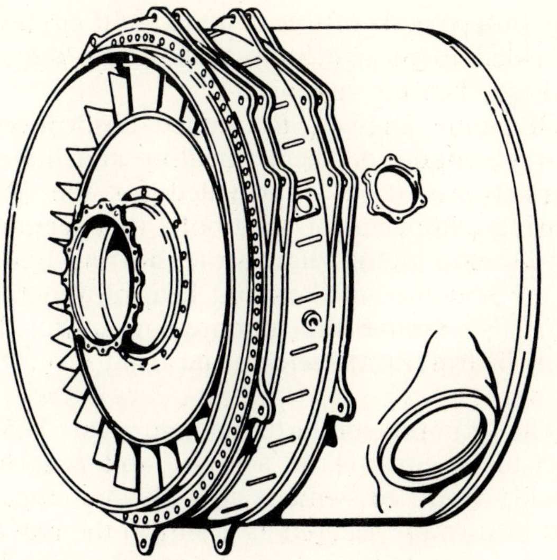

Figure 16·5 Compressor intermediate case.

An oil breather pad is located between the mounting flanges at approximately the 2 o'clock position. An oil fitting is at the 6 o'clock position between the mounting flanges. Toward the rear of the case a hole is provided for air bleeding and the air bleed valve mounting.

The 30 vanes that are on the inside front of the case serve as air inlet guide vanes to the rear compressor. They are hollow steel vanes with openings in their side walls at the outer end. These openings together with the double wall of the case form the breather passage from the upper breather connections on the case to the lower opening. The lower opening connects directly to the main accessory case. Breather and oil tubes for the bearings located in the center of the case are extended through the hollow guide vanes. The inlet vanes are welded to the No. 2 bearing front support whose outer configuration is shaped to form an inner shroud ring for these vanes. At the front (inner) and rear (outer) ends of this support, bolt circles are provided to receive bolts to hold the diagonal and rear bearing supports.

On some engines, to reduce absorption of heat by engine oil, heat-shielding and thermal-blanket insulation is provided for the No. 2 bearing compartment. The bolts that secure the compressor intermediate front bearing housing to the rear face of the front bearing inner support also secure the compressor intermediate front bearing heatshield to the front face of the support.

The compressor intermediate rear bearing heatshield consists of a front and rear heatshield, both of which are secured together and in turn are secured as a unit to the rear face at the inner bolt circle and outer bolt circle of the compressor intermediate bearing rear support. The thermal-blanket insulation is contained within the heatshield.

The rear support holds the No. 3 bearing seal assembly and the rear of the No. 2 bearing housing and is bolted to the rear of the center support and the vane inner shroud. A stepped-edge seal ring is bolted to the support at the outer diameter of the rear support plate. This ring together with its mating knife-edge seal minimizes air leakage out of the front end of the rear compressor. The area inside the ring and behind the rear support plate is exposed to high-pressure air extracted from the twelfth stage of compression. The air pressure outside the seal ring is from the ninth stage.

Rear compressor

The rear, or high-speed, compressor is driven by a hollow shaft from the first-stage turbine. Its function is to compress further the air delivered by the front compressor and then feed this air into the diffuser case and burners.

The rear compressor consists of a stator having six rows of vanes, stages 10 through 15, and a rotor having seven rows of blades, stages 10 through 16. The exit guide vanes, stage 16, are mounted in the diffuser section. In function, these are part of the compressor, but because of their structural location, we will discuss them under the diffuser.

Rear compressor vane and shroud assemblies

There are six vane and shroud assemblies in the rear compressor. The vanes decrease in height from the front to the rear of the compressor. The outside diameter of the air passage formed by these assemblies is constant, and the decreasing size of the vanes is accomplished by increasing the diameter of the inner shroud rings. The vanes are steel and are brazed to the inner shroud and extended through the outer shrouds. On the tenth through fourteenth vane and shroud assemblies, the spacers separating the assemblies are integral, but the fifteenth stage does not provide spacing. However, this is furnished by the seventh stage, located in the diffuser case. Between each shroud, dowel pins are used to lock the series of shrouds together. The rear shroud is pinned to the outer shroud to prevent rotation. An edge of each spacer rests against a case shoulder to center the assembly.

The inside shroud of each row of vanes has a steel ring, with a small step on its inside face, riveted to it. Two knife-edge seals on each rotor spacer ride free of the steps, forming an air seal between the compressor stages.

Rear compressor rotor

The rear compressor rotor, shown in Fig. 16 · 6, has seven rows of blades attached to disks, two hubs, six spacer assemblies, and 16 tie rods with nuts and washers. A knife-edge seal ring is riveted to a lip on the forward face of the tenth-stage disk. This seal ring rides free of the platformed seal, which was referred to in the compressor intermediate case discussion.

The blades are inserted into undercut slots in the disk rims and are held in place by locks inserted under the blades and bent to lock. The blades do not have a tight fit but are seated by centrifugal force during engine operation.

Every disk has 32 holes in its flange, 16 for tie bolts and 16 for balancing weights when needed. The entire assembly is held together by tie bolt heads on one end and a nut and washer on the other end.

The spacers between the disks are internally reinforced with tubes. The tie rods run through the tubes and the spacer internal flanges as well as the disk flanges.

The front hub is attached to the front face of the twelfth-stage disk and the rear hub to the rear face of the sixteenth-stage disk, as shown in the illustration. The hubs are secured to the rotor assembly by the steel tie bolts mentioned above, and a steel tube runs from one hub to the other inside the rotor. The tube is a force fit into the inside diameter of both hubs and permits breathing within the inside. It also keeps twelfth-stage pressure air from the No. 3 and No. 4½ bearings.

The third spacer from the front has holes drilled in it through which twelfth-stage air is bled into the rotor center where it passes through

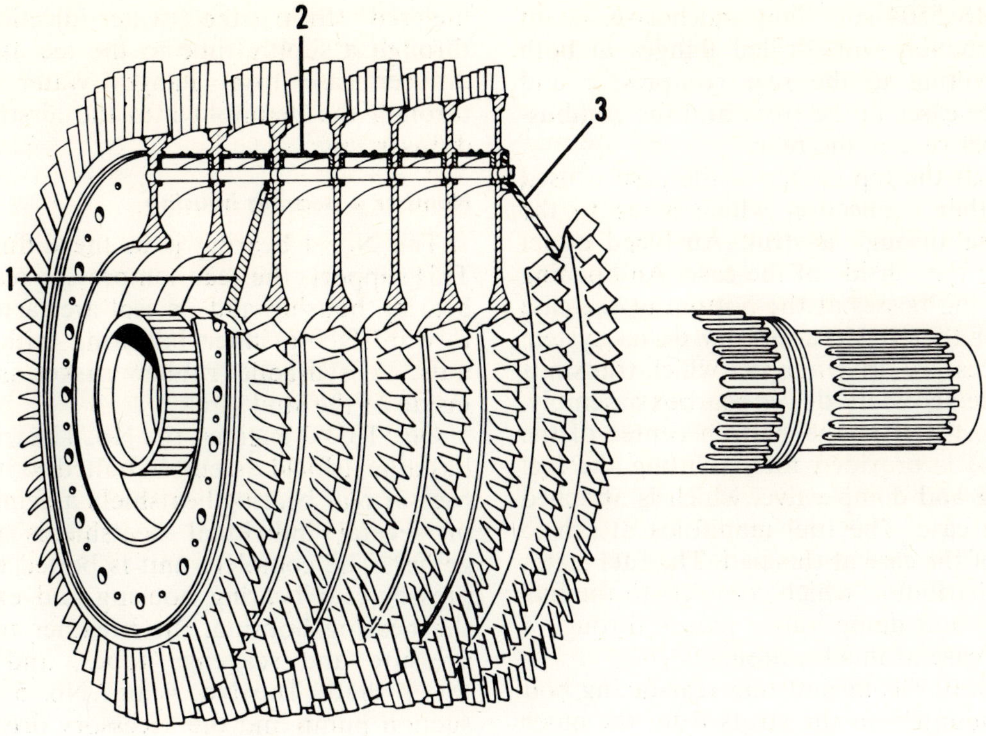

Figure 16·6 Rear compressor rotor.

holes in the front hub. The air is directed forward to the space just behind the No. 3 bearing rear support plate and pressurizes the No. 3 bearing oil seal. As there are no holes in the rear hub disk, the pressure of the air against this disk counteracts part of the rotor forward thrust.

The **rear compressor case** carries no structural bearing loads and is thus made of relatively thin sheet metal. Its purpose is to hold the stator parts of the compressor and act as an air separator. The air pressure increases from front to rear and finally becomes sixteenth-stage pressure. Between this case and the intermediate case which envelopes it, the pressure is ninth-stage air. The case has a flange at its rear end to which the screws that attach the case to the diffuser section are secured. Also, bolt holes are provided to hold the intermediate case to the diffuser section.

Diffuser section

The velocity of the air as it leaves the rear compressor is very high with motion which is both rearward and tangential. Exit guide vanes at the forward end of the diffuser case convert the tangential whirl into pressure energy. After passing through the vanes, the high pressure air has a large rearward velocity. The gradually increasing area of the airflow passages provided by the diffuser case configuration decreases the airflow velocity to a suitable burning speed and increases the pressure.

The diffuser case, shown in Fig. 16·7, is of

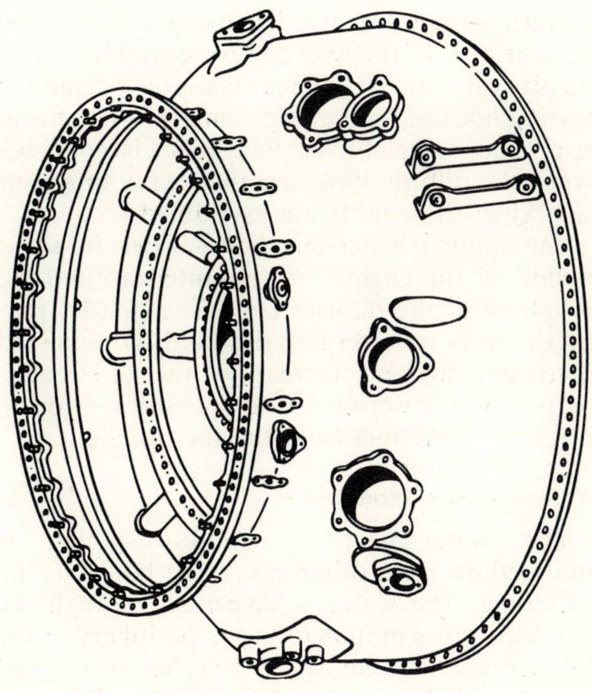
Figure 16·7 Diffuser case.

welded AMS 5504 corrosion and heat-resistant steel construction with drilled flanges at both ends for bolting to the rear compressor and intermediate cases in the front and the combustion chamber case in the rear.

The pad at the top center of the case is used for a breather connection which leads to the bearing area through a strut. Air-bleed outlet pads are on the outside of the case. An opening with mounting bosses at the bottom of the case provides for the angled accessory drives system and the accessory drive adapter which transmits power to the accessory drives gearbox assembly. The rearmost pad at the bottom center of the diffuser case is provided for mounting the fuel pressurizing and dump valve, which is attached outside the case. The fuel manifolds attach to the inside of the case at this pad. The fuel manifold inlet distributor, which connects to the fuel pressurizing and dump valve, passes through a hole in the case at this location.

Two forward-facing and one rear-facing bolt rings are mounted on the struts from the outer case. Sheet metal between these bolt rings and on the inner and outer case add to the strength and form the gradually increasing cross section of the diffuser. A slightly tapered diffuser inner inlet duct is bolted between an inside flange on the inlet-guide-vane inner shroud and the outer forward-facing bolt circle inside the diffuser case. This piece forms a part of the diffuser ducting. The rivet circle that holds the diffuser inner inlet duct to the inlet vane ring extends forward to ride over two knife-edges that are on the rear disk of the rear compressor. This forms the air seal between the rear compressor and the bearing housing area at the center of the diffuser case. The outer shroud flange is clamped between the diffuser case and the rear compressor case when these units are assembled.

The optional water-injection nozzles, for some models of the engine, are mounted around the periphery of the diffuser case. The nozzles permit the entry of water into the highly compressed airstream, thereby increasing the airstream's density and lowering its temperature prior to injection of the fuel from the fuel nozzle.

Water-injection manifold

The water-injection manifold which is mounted on the diffuser case front flange, is the split type. The water is dispersed through the 16 curled tubes mounted on the periphery of the diffuser case. The elbows of the tubes are aligned to the 16 holes in the diffuser case. Water is metered from the water-injection control through a supply tube to the tee fitting on the diffuser case front flange. Water then flows through the manifold into the airstream in the diffuser case.

Number 4 bearing housing

The No. 4 bearing is in the diffuser section. This supports the rear compressor rear hub. The bearing housing and its seal are fastened to the forward-facing inner bolt ring with studs and nuts. A synthetic rubber gasket is used just inside of the stud circle.

On JT3D-3 engines, the No. 4 bearing housing has a heatshield assembly bolted to it. The compressor rear bearing heatshield assembly consists of several individual heatshields welded together. The complete unit is bolted to the compressor rear bearing housing and extends over the bearing housing, the breather tube, the oil pressure and scavenge tubes, and along the tower shaft. The No. 4 and No. 5 bearing oil suction pump and the accessory drive gears are in the lower rear of the diffuser.

Fuel manifold

As viewed from the rear, 16 brackets are welded to the inner structure of the diffuser. These brackets, and brackets that are welded to the inside rear of the diffuser outer case, support the fuel manifold.

The fuel manifold, shown in Fig. 16·8, is split at the inner flange and between the No. 1 and No. 8 clusters. The manifold assembly consists of a primary and a secondary fuel manifold and eight clusters of fuel nozzles. There are six fuel nozzles in each cluster. The fuel manifold incorporates mounting lugs in the inner and outer diameter which are secured to the rear face of the diffuser case. Eight combustion-chamber positioning brackets are mounted on the outer-diameter mounting lug locations.

Combustion section

The combustion section is that part of the engine housing the combustion chambers and located immediately to the rear of the diffuser section. The **combustion-chamber outer case** is secured at the front flange to the rear flange of the diffuser case and at the rear flange to the turbine-nozzle case. Located within the combustion-chamber outer case are eight combustion chambers, the combustion-chamber inner case, the No. 5 bearing support, and the No. 5 bearing housing.

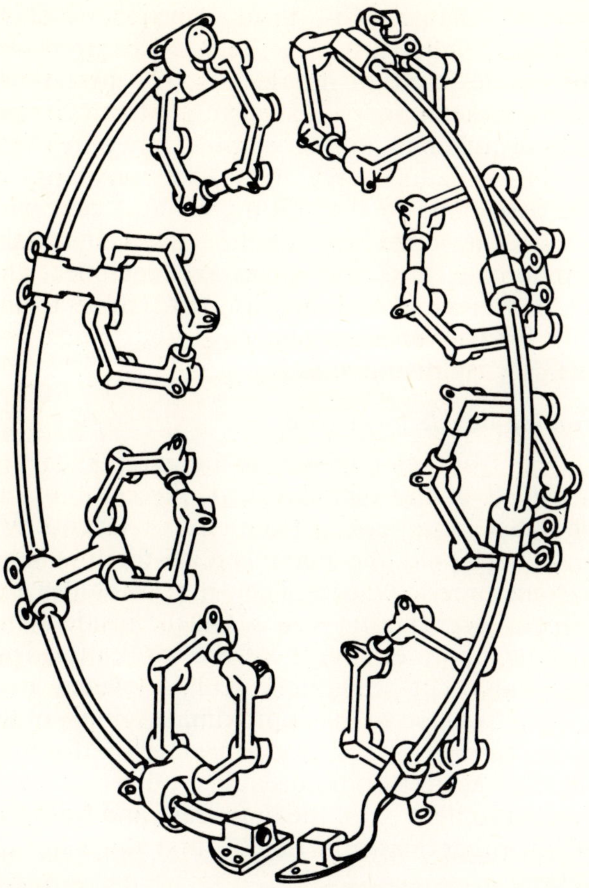

Figure 16 · 8 Fuel-manifold assembly.

Combustion-chamber outer cases

The outer casing for the combustion section is divided into two sections called the **front-combustion-chamber outer case** and the **rear-combustion-chamber outer case.** These casings are constructed of corrosion and heat-resistant steel. The front flange of the front case is bolted to the rear flange of the diffuser case, and the front flange of the rear case is bolted to the rear flange of the front case. The outer rear case rear flange is internal, that is, it extends inward instead of outward. It lies behind the forward external flange of the **turbine-nozzle case** and is bolted to it. Thus, the outer cases, if bolted together, or the rear case by itself can slide rearward over the turbine-nozzle case to give access to the combustion chambers. The outer front case has interlocking lips on its flanges to provide sealing when the flanges are joined.

Combustion chambers

Each combustion chamber for the JT3D engine is a welded assembly of heat-resistant sheet metal, as shown in Fig. 16 · 9. At the head of each chamber are six swirl vanes which surround the fuel nozzles. These vanes provide combustion air and also aid in atomizing the fuel spray from the nozzles. The holes in the walls of the chamber meter the entering air to assure that a cooling blanket of air exists between the flame and the walls of the chamber. An inner liner, open at the front, is provided in the center of the chamber to supply additional air and to maintain uniform burning of the fuel. These liners can be seen in the cutaway illustration of the engine in Fig. 16 · 3.

Near the head end of the combustion chambers, most of the air is used for combustion; however, farther down the chamber the entering air is used to cool the hot gases to a safe temperature for the chambers. The rows of smaller holes admit air that cools the chamber walls directly.

The combustion chambers are radially housed inside the combustion-chamber outer case. As mentioned before, the forward face of each combustion chamber presents six apertures which align the six nozzles of the corresponding fuel-nozzle cluster. The combustion chambers numbered 1, 3, and 7 are interchangeable, and those numbered 2, 6, and 8 are interchangeable. The numbering is clockwise as viewed from the rear and starts with No. 1 at the 1 o'clock position. Chambers numbered 4 and 5 have sparkigniter sleeves and guides, and the No. 4 chamber is provided with an air-pressure-transfer tube. Interconnecting flame tubes join all the chambers to provide cross ignition. The tubes on the odd-numbered chambers are female and have an air shield on the upstream side that overlaps the mating male flame tube. When being removed, the even-numbered chambers must be taken off first.

Each combustion chamber is held in place by a positioning bracket and a clamp. The clamp holds the open end of the chamber to an outlet-

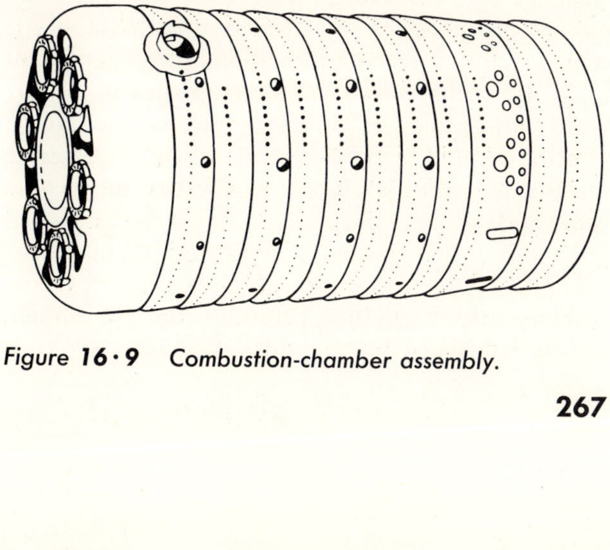

Figure 16 · 9 Combustion-chamber assembly.

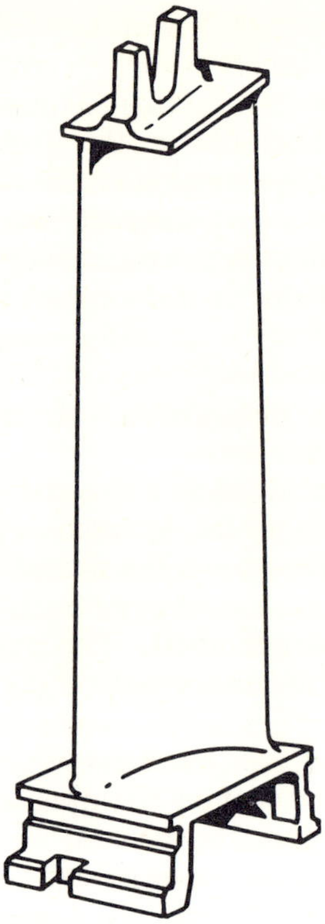

Figure 16 · 10 Stator vane with lugs.

duct assembly which is secured to the turbine case. Each positioning bracket is held in place by two outer-fuel-manifold retaining bolts. The bracket hooks onto a retaining lug on the combustion chamber to prevent axial movement. The fuel nozzle inside the air swirl guide holds the chamber in lateral position.

Combustion-chamber inner case

The combustion-chamber inner case is a sheet-metal fabrication which surrounds the turbine drive shaft. Reinforcements welded to its inner surface help to retain its shape and prevent excessive warpage due to heat. The primary function of the inner case is to form the inner surface of the combustion section, but together with the No. 5 bearing support the case provides an inner passage for the sixteenth-stage air which helps to reduce the effects of radiant heat from the combustion chambers. Holes in its forward flange, in the diffuser, and in the No. 5 bearing support permit the air to flow. A flange on the rear end attaches to the turbine-nozzle inner case.

Turbine section

The turbine section includes the first-stage, or high-speed, turbine which drives the rear compressor; the second-, third-, and fourth-stage turbines which drive the forward compressor; the nozzles; and the nozzle case. The **nozzle case** is constructed of heat-resistant stainless steel and is attached to the combustion outer rear case as explained previously. The rear flange of the case bolts to the turbine exhaust case with a spacer installed between the flanges. Inside the nozzle case, slots and ridges are machined, and stator vanes and seals are inserted. The attaching lugs for the vanes are shown in Fig. 16 · 10, and the case is illustrated in Fig. 16 · 11.

Turbine-nozzle inner case

The front end of the turbine inner case is attached to the rear flange of the combustion-chamber inner case. At the outer rear end, two shoulders hold the inner end of the first-stage turbine nozzle vanes. The front shoulder has pins for securing the vanes. On the inside diameter of the inner case there are two knife-edged air seals. The innermost seal is riveted to a support welded to the approximate middle of the inner case. It rides free of an inner platform on the first-stage turbine disk. The outer air seal is riveted to the rear of the inner case and rides free of an outer platform on the first-stage turbine disk.

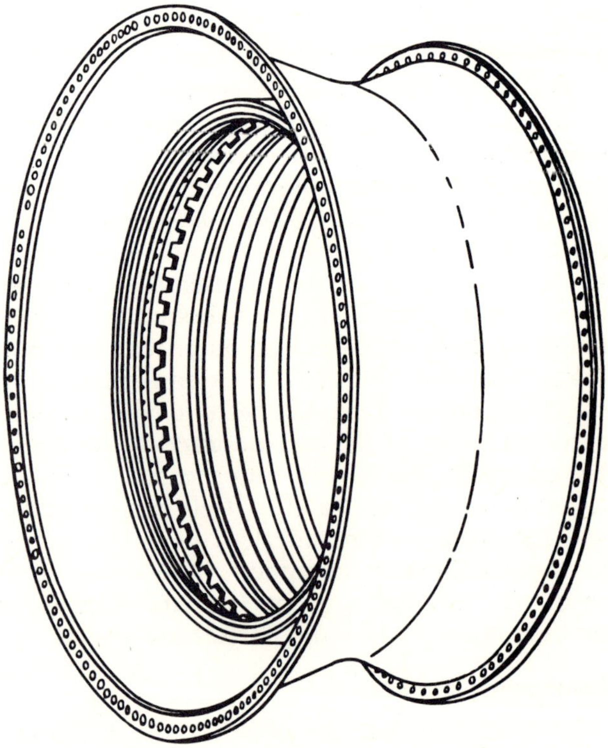

Figure 16 · 11 Turbine-nozzle outer case.

Combustion-chamber outlet duct

The purpose of the combustion-chamber outlet duct, illustrated in Fig. 16 · 12, is to direct the hot gases from the combustion chambers to the first-stage turbine nozzle. This assembly is constructed of welded, heat-resistant sheet metal and is secured in place by the same bolts which attach the combustion-chamber outer case to the turbine-nozzle case. The forward flanges of the duct ports carry the rear edge of the clamps that hold the combustion chambers in place.

Nozzle vanes, shrouds, and seals

The first-stage turbine nozzle vanes are cast from a corrosion and heat-resistant alloy and are coated with a special protective material on all internal and external surfaces. The second-stage vanes are cast nickel-base alloy, the third-stage vanes are cast cobalt-base alloy, and the fourth-stage vanes are cast nickel-base alloy. The first-stage vanes are held in place by pins between the recesses in the outer turbine case and are snapped onto their pins in the turbine-nozzle inner case assembly to seat on the shoulders of the case.

A seal and a spacer are placed in the outer case just to the rear of the first-stage vanes. One knife edge of the first-stage turbine blade shroud runs between the seal's two knife edges.

The second-stage turbine vanes fit into the outer case just behind the first seal spacer. Extensions on the root platform of each vane hold them in place on their forward edges, and a lock ring is used at the outer rear edge. The inner ends of these vanes slip into slots in the shroud ring which has a turbine inner seal riveted to the inside edge. The seal ring carries four knife edges on its inside flange. One of these knife edges runs free against the shoulder on the rear side of the first-stage turbine disk, and the remaining three (rearmost) run free against the flange face of the front compressor drive turbine hub. This arrangement forms a seal between the first- and second-stage disks to force the working gases to pass through the second-stage nozzle vanes. A lock ring, a stepped seal, and a seal spacer are provided in the outer case following the second-stage nozzle vanes. The two knife edges formed by the second-stage turbine blades run free against the stepped seal. The mounting of the third-stage turbine nozzle vanes is the same as that of the second stage. The inner seal ring, however, instead of having knife edges, has a stepped platform. Six knife edges, arranged in pairs, are on the spacer between the second- and third-stage turbine disks. The mounting of the fourth-stage nozzle vanes is identical to that of the third stage. The outer seal is stepped and extends into the turbine-exhaust case. There is no lock ring, but the fourth-stage vanes are held in the rear by the stepped seal, which is held by its fitting between the turbine-nozzle case and the turbine-exhaust-case flanges.

Rear-compressor-drive turbine rotor

The rear-compressor-drive turbine rotor consists of a disk, blades, and a shaft, as shown in Fig. 16 · 13. The name of this assembly is descriptive of its function, that is, driving the rear or high-speed N_2 compressor rotor. The coupling of the turbine and compressor are shown in Fig. 16 · 4. The No. 5 bearing inner race and seal assembly are located forward of the turbine

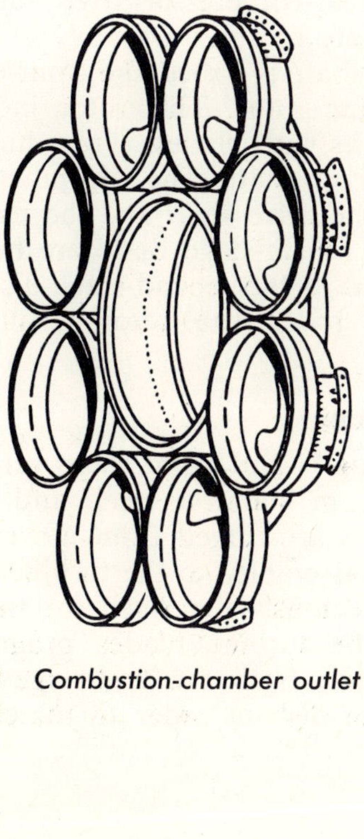

Figure 16 · 12 Combustion-chamber outlet duct.

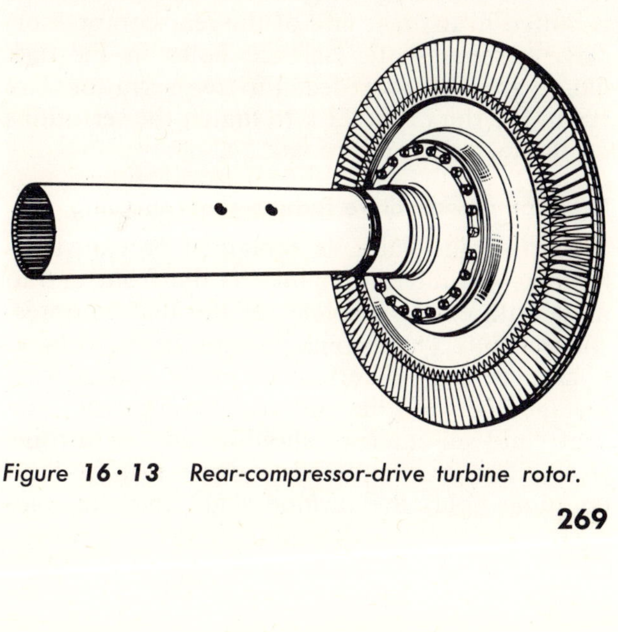

Figure 16 · 13 Rear-compressor-drive turbine rotor.

disk and are secured on the shaft by a nut, tab-washer, and lock ring at the time of assembly. The turbine shaft splines into the rear compressor rotor and is secured to the rotor hub by the **rear-compressor-drive turbine-shaft coupling.** The outer race of the No. 4½ bearing is located inside the turbine shaft.

First-stage turbine blades

The first-stage turbine blades are made of forged-nickel alloy and are attached to the turbine disk by means of "fir-tree" serrations at the blade roots. The blades can only be inserted by a forward and rear movement and are held in place by a rivet at the bottom of the fir tree. The working part of the blade is a curved surface that changes the direction of the swiftly moving gases that pass over it. The reaction of these gases to change in direction of travel produces the force that turns the turbine rotor.

At the outward end of each blade is a wide flat section called a **shroud,** which prevents gas leakage to the blade ends. Each blade has a stepped arrangement by which it locks to adjacent blades, and this arrangement forms a complete shroud ring when the blades are assembled to the disk. This ring gives the rotor greater strength as well as providing a gas seal. Because of the construction of the turbine blades, it is impossible to remove one turbine blade only; at least half of the blades at 90° on either side must be removed. A staggered movement of all these blades must be made before the center-blade fir tree will clear the disk. There are 130 blades in the first-stage turbine rotor.

Turbine disk

The first-stage turbine disk is made of corrosion-resistant, high-temperature nickel alloy. A drilled flange on the front face of the disk is bolted to the rear end of the rear-compressor-drive turbine shaft. Balance holes in the disk flange are also provided. Fir-tree serrations are cut in the rim of the disk to match the serrations on the turbine blade roots.

Rear-compressor-drive turbine-shaft coupling

The turbine shaft, as explained previously, is hollow with internal splines at the front end to mate with external splines of the rear compressor rear hub. The turbine position in the turbine-nozzle case is determined by a spacer ring installed between the end of the compressor hub shaft and an internal shoulder in the turbine shaft. The rear-compressor-drive turbine-shaft coupling holds the turbine shaft and the compressor rear hub together. Outside threads on the forward end of the coupling mate with threads on the inside of the compressor hub shaft, and a shoulder on the outside of the coupling butts against an internal shoulder in the turbine drive shaft. An internal spline at the rear of the coupling is used for wrenching. On later-model JT3D engines, the coupling is locked in with a jam nut having a left-hand thread, whereas in earlier engines the coupling is locked by the No. 4½ bearing outer race and liner. The later-model coupling is shown in Fig. 16 · 14.

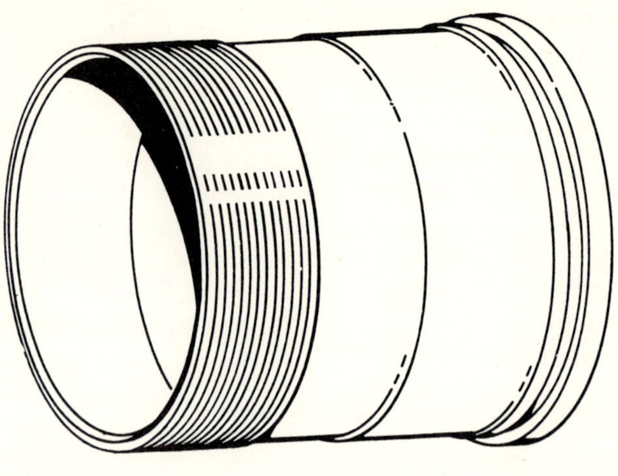

Figure 16 · 14 Rear-compressor-drive turbine-shaft coupling.

Front-compressor-drive turbine rotor

The front compressor N_1 of the JT3D engine is driven by the second, third, and fourth stages of the turbine through a drive shaft rotating inside the rear-compressor-drive turbine shaft. The complete rotating assembly of the turbine disks and shaft is called the front-compressor-drive turbine rotor. This rotor includes three turbine disks with blades, a rear hub, a spacer, and the drive shaft. The second-stage disk is bolted to the drive shaft and the rear hub. The third- and fourth-stage disks are held together and secured to the second-stage disk by means of 10 tie rods. The rotor is illustrated in Fig. 16 · 15.

Turbine blades

The second- and third-stage turbine blades are cast from stainless steel, and the fourth-stage blades are forged from nickel alloy. The blades are assembled to the turbine disks in the manner previously described for the first-stage turbine. The turbine blades progressively increase in size from the first-stage disk to the fourth-stage disk in order to match the blade

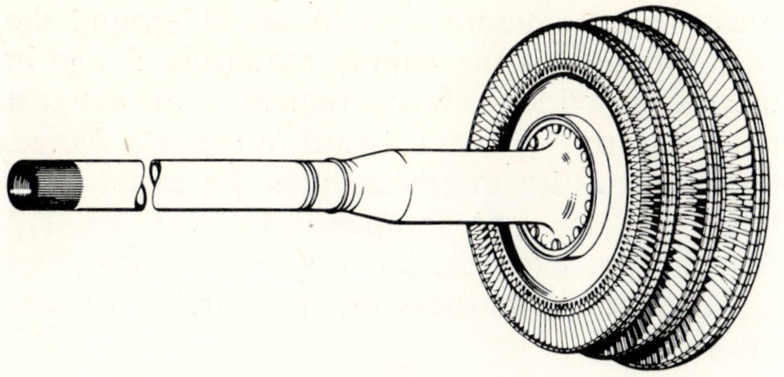

Figure 16 · 15 Front-compressor-drive turbine rotor.

size to the increase in volume of gas as it expands while passing through the turbine. There are 114 blades in the second-stage disk and blade assembly, 108 blades in the third stage, and 80 blades in the fourth stage.

Turbine disks

The second- and third-stage turbine disks are made from corrosion and heat-resistant nickel alloy, and the fourth-stage disk is made from stainless steel. The disks are separated from each other by spacers and are attached to the front-compressor-drive turbine shaft and the turbine rear hub by tie rods, as explained previously. The second-stage disk is constructed with an integral spacer to hold it in the correct position relative to the other disks.

Front-compressor-drive turbine shaft

The front-compressor-drive turbine shaft is made from low-chrome steel alloy. The shaft is comparatively long and thin because it must pass through the first-stage turbine rotor shaft and must reach from the front of the second-stage turbine disk to the rear hub of the front compressor. The front end of the shaft is supported inside the front compressor hub; the middle of the shaft is supported by the No. 4½ bearing inside the rear compressor rotor shaft, and the rear end is supported by the turbine assembly which in turn is supported by a bearing at the turbine rear hub.

The shaft is hollow and provides a passage for ninth-stage cooling air which passes through holes in the rear hub to cool the rear bearing seals, oil pump, and the rear face of the fourth-stage turbine disk. Oil is also piped through the shaft to its middle bearing. The forward end of the shaft is externally splined and internally threaded to provide for coupling to the front compressor rotor. The shaft external splines mate with the internal splines on the front compressor rear hub, and the shaft internal threads mate with the external threads on the front-compressor-drive turbine coupling. A spacer is used between the compressor rear hub shaft and the front edge of the drive shaft to position the shaft correctly.

Front-compressor-drive turbine-shaft coupling

The front-compressor-drive turbine-shaft coupling, shown in Fig. 16 · 16, fits inside the front compressor rotor rear hub where its external shoulder butts against an internal shoulder inside the hub. The coupling lock is externally

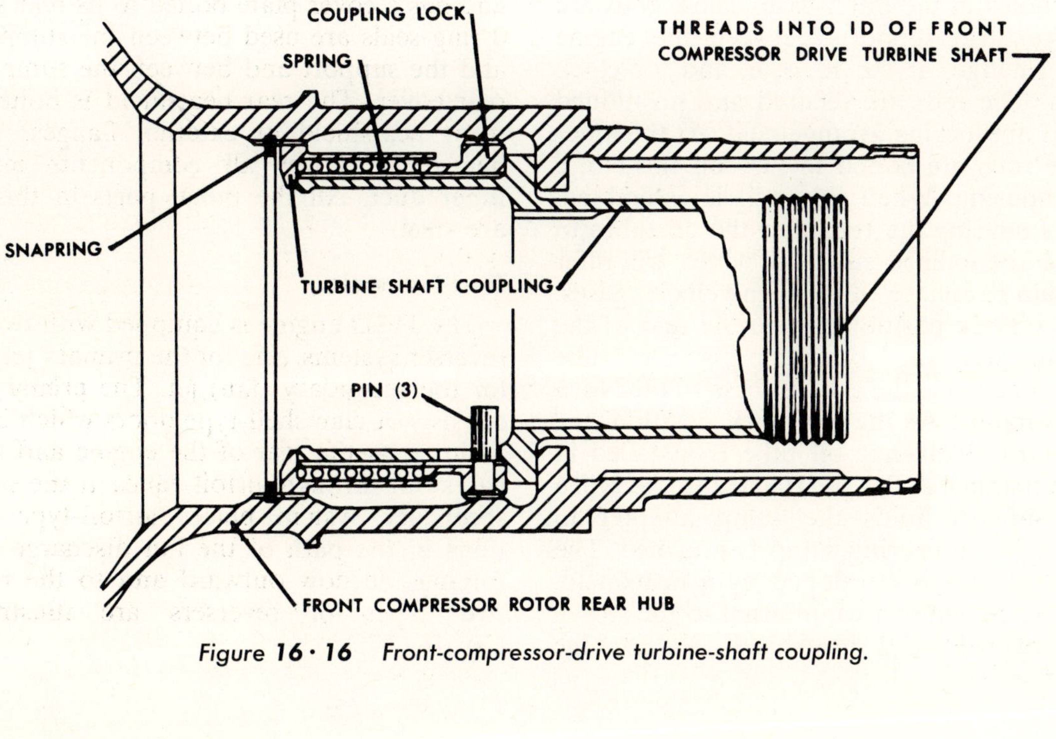

Figure 16 · 16 Front-compressor-drive turbine-shaft coupling.

splined to mate with a short section of splines inside the hub just forward of the shoulder. The coupling lock is held in the engaged position by means of the spring and is disengaged by means of three pins which extend through slots to the inside of the coupling. The special wrench for the coupling engages the pins and pulls them forward to disengage the lock. When this is done, the wrench can turn the coupling by means of internal wrenching splines. This movement unscrews the coupling from the turbine shaft.

Turbine-exhaust case

The turbine-exhaust case is that section of the engine immediately to the rear of the turbines. Its principal functions are to carry the high-velocity gases from the turbine section to the tailpipe and to provide support for the turbine rear bearing (No. 6). The outer case is constructed of sheet-nickel alloy and is bolted to the rear of the turbine case at its front flange. The rear flange is bolted to the tailpipe. Two heavy external flanges near the center of the case are used for engine mounting. The outer case supports all the parts used in the inner case, and a few of the parts are mounted directly on the outer case. The fourth-stage turbine outer seal is located in the front of the outer case with its supporting flange positioned between the rear flange of the turbine nozzle case and the forward flange of the exhaust case.

A turbine-exhaust fairing assembly, consisting of an inner duct and four turbine-exhaust struts, is positioned on the exhaust case. Four No. 6 bearing support rods pass through the struts and through holes in the outer case. These rods are secured in strut supports between the engine mounting flanges at the 12, 3, 6, and 9 o'clock positions. The rods are secured and positioned by a dual nut locking arrangement. At the inner ends, the rods are bolted to lugs on the No. 6 bearing housing. When this unit is assembled, the struts envelop the rods and the oil tubes to provide a streamlined structure which will offer a minimum resistance to the high-velocity gases. At the 3 o'clock position and to the rear of the support rod, the No. 6 bearing pressure oil tube passes from outside the exhaust case to the No. 6 bearing support. At the 6 o'clock position, the No. 6 bearing scavenge oil tube is installed in the same manner. A flange attached to the No. 6 bearing support holds the sump adapter to which the No. 6 bearing sump is attached. The entire sump area is enveloped by a heatshield. The fairing-assembly area, internal to the struts, is also covered by a heatshield. An average pressure rake manifold is mounted around the case forward of the engine mounting flange in order to provide for measurement of the exhaust gas pressure P_{t7}. Just forward of the rear flange, the temperature thermocouples are installed to sense exhaust gas temperature EGT or T_{t7}. Attached to each thermocouple are leads from a thermocouple harness which encircles the case at this point.

Number 6 bearing support and shields

The No. 6 bearing seals buildup and the No. 6 bearing inner race and rollers are mounted on the rear turbine hub. The outer race is held in a housing which also serves as the seal's housing. This housing is attached to and is held by the No. 6 bearing support. Four support rods from the outer case thread into the rear of the support. Attached to the rear of the support is the sump adapter, to which the sump is bolted. The rear heatshield is bolted to the rear-heatshield flange which is retained in the support by the adapter. Bolted to the front of the No. 6 bearing support is the No. 6 bearing air-seal support. A knife edge is riveted to the support's forward extremity to ride on a shoulder which is riveted to the rear hub, thus forming a gas seal.

Number 6 bearing oil sump

The No. 6 bearing oil sump contains an oil scavenge pump which is driven by a pinion gear attached to the rear hub. Inserted through the pinion is the oil nozzle which provides lubrication for the No. 4½ bearing inside the turbine shaft. The sump bolts to the rear of the adapter and has a cover plate bolted to its rear side. The 0 ring seals are used between the sump adapter and the support and between the sump and the rear cover. The rear heatshield is bolted to the front heatshield at external flanges. The two heatshields cover all components inside the inner duct. All the metal parts in this section are steel.

Thrust reverser

The JT3D engine is equipped with two thrust-reverser systems, one for the primary jet and one for the secondary (fan) jet. The **primary** system consists of clamshell-type doors which block the main jet at the rear of the engine and force the gases out through airfoil vanes at the sides. The **secondary** system places airfoil-type obstructions in the path of the fan discharge air, thus forcing the flow outward and to the rear. The two types of reversers are illustrated in Fig. 16 · 17.

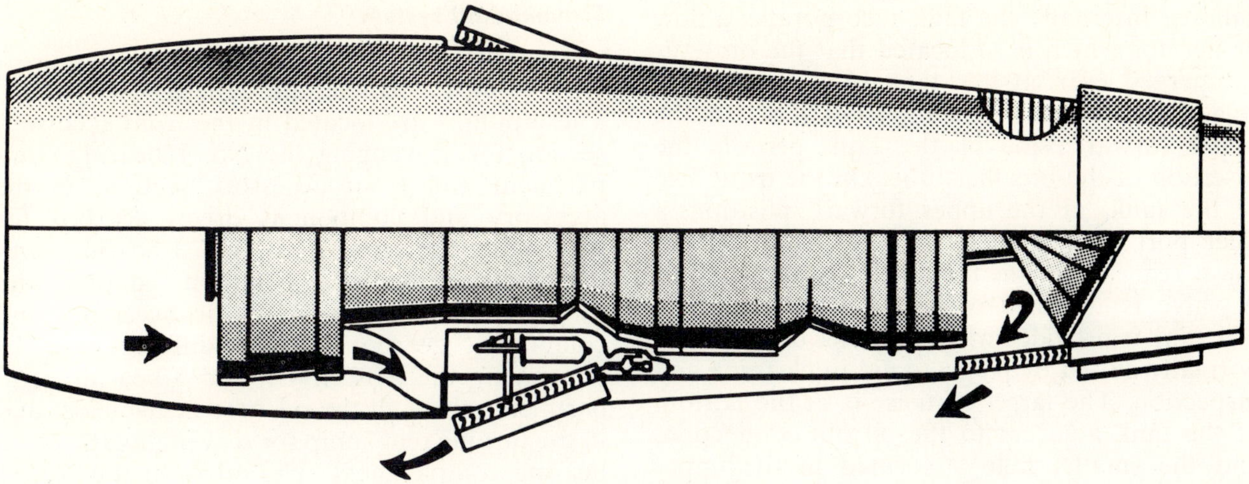

REVERSE THRUST

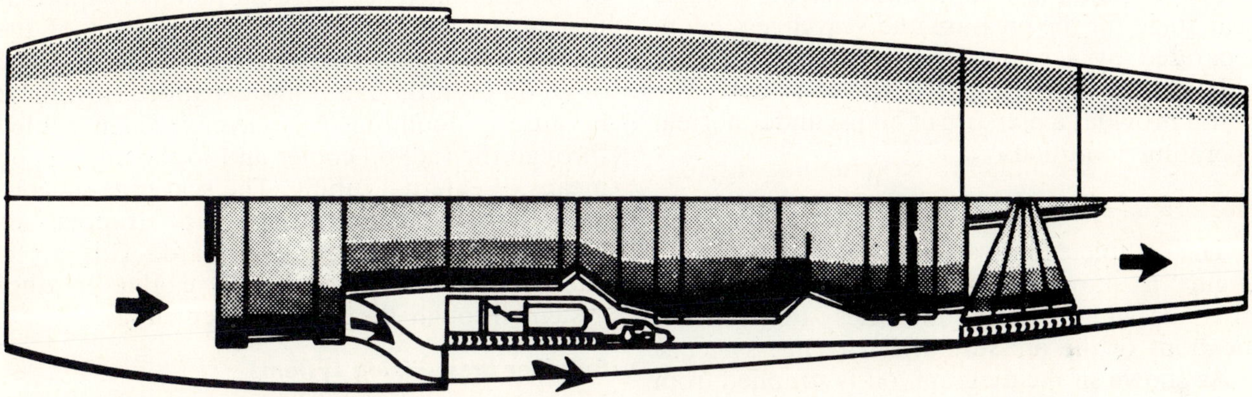

FORWARD THRUST

Figure 16·17 Thrust reversers for the JT3D engine.

LUBRICATION SYSTEM

The lubrication system for the JT3D turbofan engine consists of a pressure system, scavenge system, and pressurization system. It is self-contained and provides high-pressure oil to the parts of the engine requiring lubrication. The scavenge system removes oil from the sumps while the pressurization or breather system maintains the correct pressure for proper operation.

The engine lubricant is PWA-521, a synthetic type conforming to the manufacturer's specification. This lubricant meets and surpasses the requirements of MIL-L-7808. It has a low volatility at low temperatures, a flash point of 430°F, a pour point of −65°F, and weights 7.7 lb per gallon at 60°F. When the oil is hot, it produces toxic fumes which cause irritation to the eyes and nose; hence the technician should avoid breathing the fumes and should not allow the oil to be spilled on the skin.

Oil tank

The engine oil tank is mounted in the upper right quadrant of the intermediate case. It is strapped to the engine at this location by two straps attached to brackets on the front and rear flanges of the case. Strips of resilient material serving as vibration isolators are installed between the tank and the engine and between the tank and the straps. The tank has a capacity of 6.0 gallons with a minimum usable quantity under all operating attitudes of 3.25

gallons. Internally the tank incorporates a flow de-aerator which is so located that the outlet is submerged even at low tank levels to prevent re-aeration of the oil. A connection or boss at the upper forward side of the tank permits the insertion of the breather tube. On the front face of the tank, at the upper forward position, a filler port with a filler cap is located. On the lower rear side of the tank are two connections. A valve is installed on the drain connection. The tank is provided with two holes or openings, normally closed by plugs, to aid in cleaning and inspection. The larger of these is at the bottom of the tank adjacent to the oil out connection, and the smaller hole is located in the upper middle of the tank adjacent to the filler connection.

Oil pump

The oil pump is a duplex unit having a single gear stage for the pressure and scavenge section, separated by a center body, with the pressure unit located forward of the scavenge unit. The pump provides a pressure of 50 psi under normal operating conditions.

Pressure oil system

The complete lubrication system for the JT3D engine is shown schematically in Fig. 16 • 18. This drawing clearly indicates the relative locations of the pressure and scavenge systems.

As shown in the diagram, oil is supplied from the tank to the inlet of the spur gear pressure pump. The pump forces the oil through the oil filter into an adapter and through external tubing to the engine components requiring lubrication and cooling. Proper distribution of the total oil flow among the various components is maintained by metering orifices and clearances. The oil pressure differential is controlled by an oil pressure relief valve.

An oil filter assembly, equipped with a bypass relief valve and located forward of the accessory and components gearbox, assures a clean supply of oil to the lubrication system. The relief valve permits the oil to bypass the filter element in the event that the screens become clogged. The filter assembly is composed of a series of screens in disk form, separated alternately by stamped inlet and outlet spacers and assembled around an inner filter element. The filter assembly is easily removed for disassembly and cleaning. Connections are provided for an oil pressure transmitter and a differential pressure switch which activate with the buildup of contaminants.

Scavenge oil system

The scavenge oil system contains five gear-type pumps located throughout the engine. These pumps are located in the front accessory section for scavenging the No. 1 bearing compartment and front accessory section; in the accessory and component drives gearbox for scavenging the Nos. 2, 2½, and 3 bearing compartments through a common adapter and through external tubing to the accessory and component gearbox; in the diffuser case for scavenging the Nos. 4, 4½, and 5 bearing compartments through the oil tube heatshields; and in the turbine rear sump for scavenging the No. 6 bearing compartment. The oil from the Nos. 1 and 6 bearing compartments flows through external tubing to the gearbox to join the scavenge oil from the Nos. 4, 4½, and 5 bearing areas on its way to the oil tank. The oil handling capacity of the combined scavenge oil pumps is approximately three times the quantity output of the pressure oil pump. This is to provide for removal of all oil even though it has expanded in volume because of foaming. All scavenge oil is routed through the fuel-oil cooler and to the oil tank by means of external tubing. The scavenge oil contains a considerable amount of trapped air which must be vented overboard. This air is handled and dissipated by the engine breather system through the breather centrifuge.

Breather pressurizing system

The scavenge areas of the engine are vented through external tubing and inner passages to a breather chamber formed by the compressor intermediate case annulus and then to a cavity in the accessory and component drives gearbox. The common overboard vent removes air from this cavity through a rotary breather which prevents the oil particles from being carried overboard in the breather airflow and also maintains the correct pressure inside the cavity.

FUEL SYSTEM

The engine fuel-system components have been discussed in other portions of this text. To summarize, however, the sytem includes a hydromechanical fuel-control unit, a two-stage fuel pump, a fuel-air heat exchanger, a fuel-oil heat exchanger, a fuel filter, and necessary plumbing. A simplified diagram showing the fuel de-icing system is shown in Fig. 16 • 19.

As previously explained, the fuel is fed from the control unit to the manifold and nozzles.

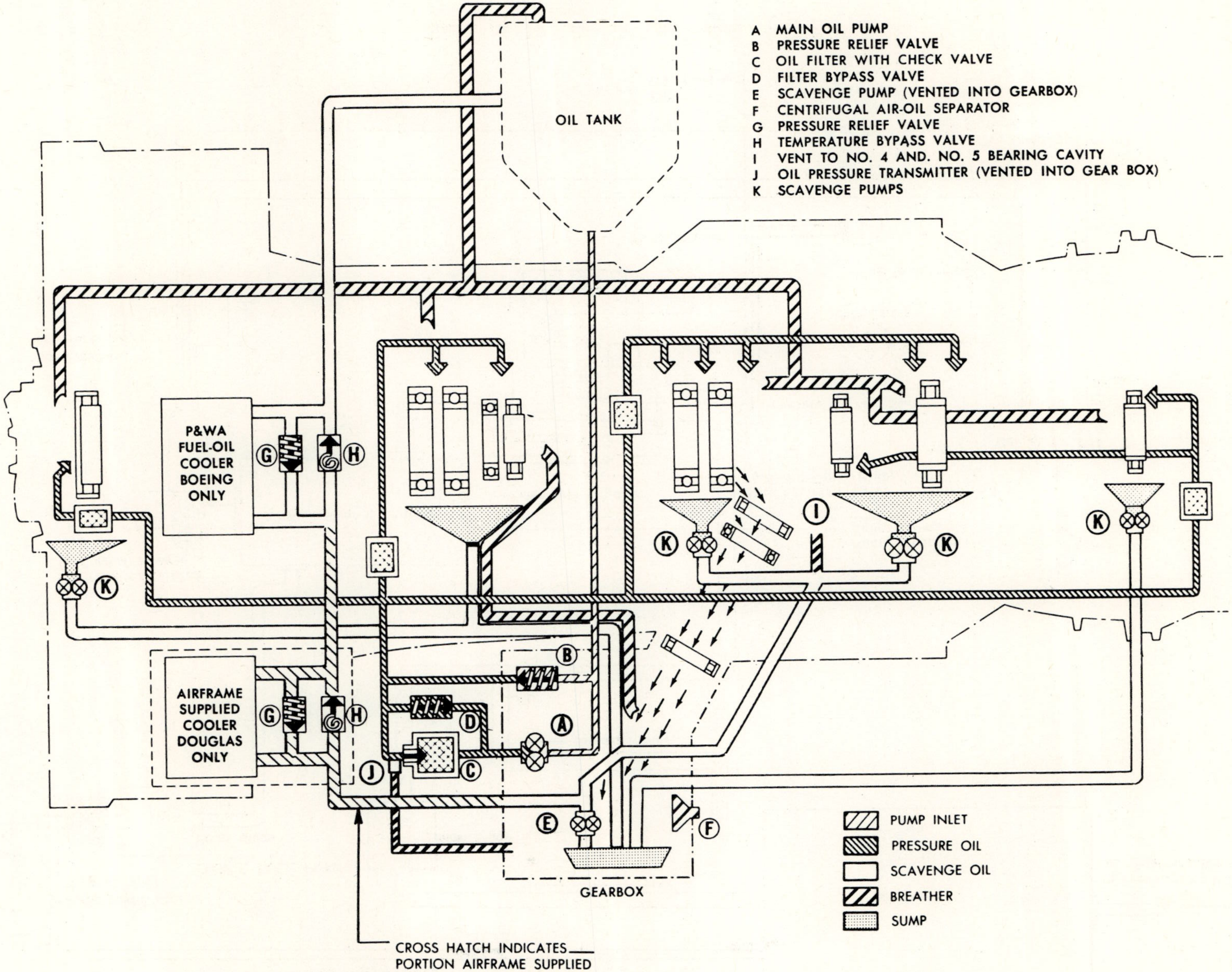

Figure 16·18 *Lubrication system.*

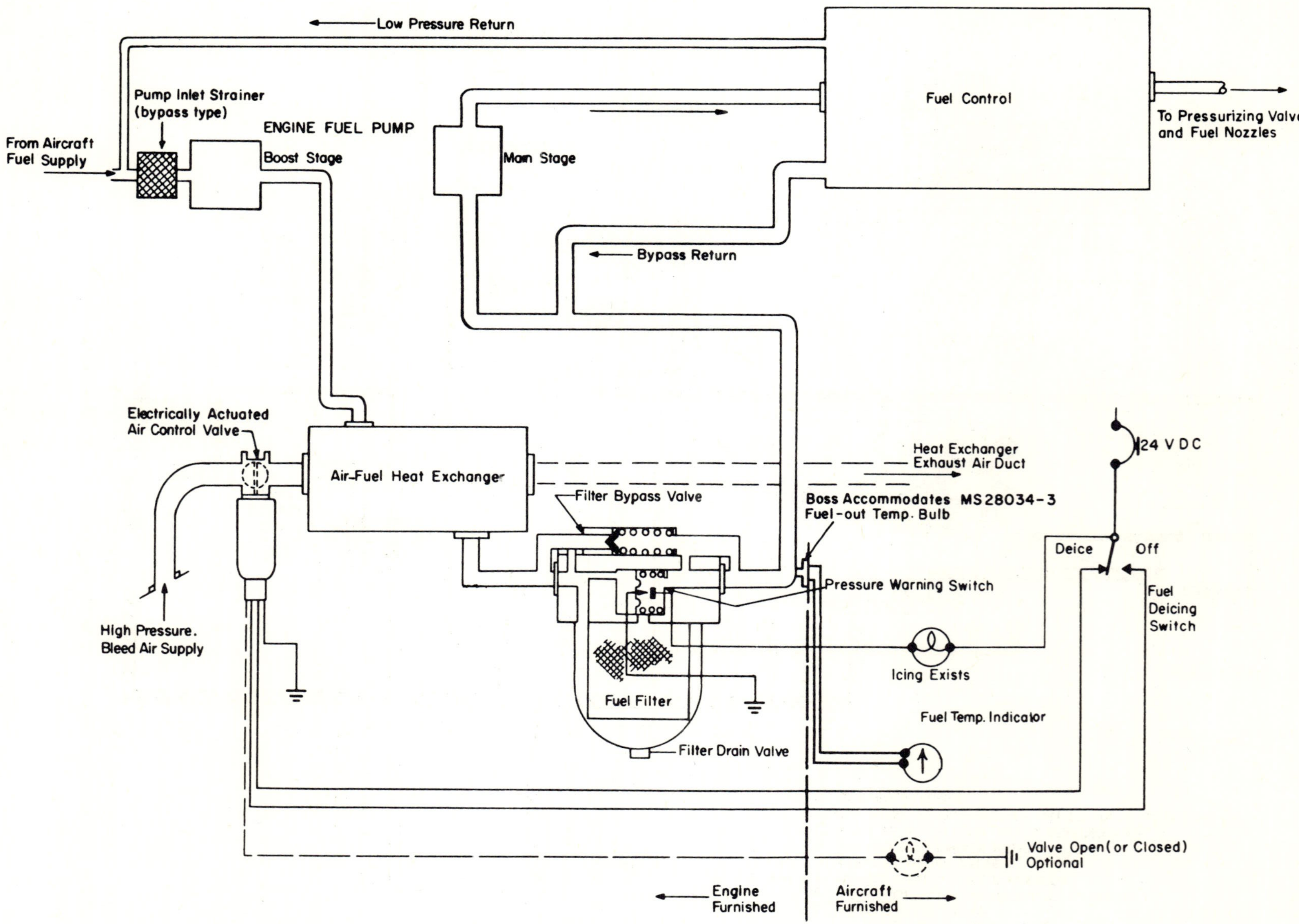

Figure 16·19 Fuel de-icing system.

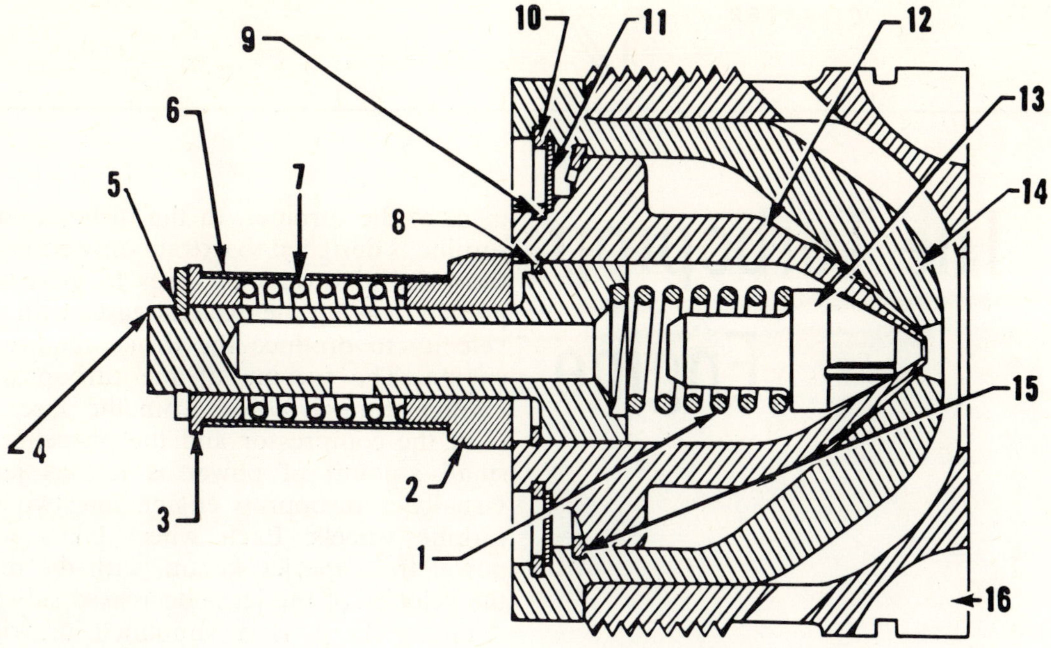

1. SPRING
2. LOWER PRIMARY FERRULE
3. UPPER PRIMARY FERRULE
4. SPRING SEAT
5. PRIMARY STRAINER SNAPRING
6. PRIMARY STRAINER
7. SPRING
8. SPRING SEAT SNAPRING
9. INNER SECONDARY STRAINER SNAPRING
10. OUTER SECONDARY STRAINER SNAPRING
11. SECONDARY STRAINER
12. NOZZLE BODY INSERT
13. PRIMARY PLUG
14. NOZZLE BODY
15. INSERT SNAPRING
16. RETAINING NUT

Figure **16 · 20** *Fuel nozzle construction.*

Each combustion chamber is supplied by a cluster of six nozzles constructed as shown in Fig. 16 · 20. These are dual-type nozzles which produce a swirling spray.

STUDY QUESTIONS

1. Describe the fan section of the JT3D engine.
2. Compare the airflow volume through the fan with that of the primary compressor section.
3. Explain how the turbines and compressor sections are coupled to permit independent rotation of the two compressor rotors.
4. What is the advantage of coupling the starter to the rear compressor only?
5. Give the compression ratio for the twin-spool compressor of the JT3D engine.
6. Describe the combustion section.
7. What is unusual about the design of the combustion chambers and fuel-nozzle arrangement?
8. Describe the bearings associated with the main rotating elements of the engine, and give the bearing numbers.
9. How are the bearings supported at the front and rear of the engine?
10. Describe the compressor intermediate section.
11. Explain how the compressed air is prevented from flowing around the ends of the stator vanes and compressor blades.
12. In which section are the exit guide vanes for the compressor mounted?
13. How are the compressor blades secured to the compressor disks?
14. What provision is made for balancing the compressor disks?
15. What is the function of the rear compressor case?
16. Describe the diffuser case, including the construction material, the effect on airflow, and the items which it supports.
17. Describe the fuel manifold.
18. What unit serves to conduct hot gases from the combustion chambers to the turbine nozzle?
19. Describe the construction of the first-stage nozzle vanes and the material employed for these vanes.
20. What function is performed by the knife-edge seals associated with the turbine section?
21. Discuss the reason for using three turbine stages to drive the front compressor while only one stage is necessary for the rear compressor.
22. How are the turbine blades attached to the turbine disks?
23. Explain how thrust reversal is accomplished in the fan section.

CHAPTER 17

The Turboprop Engine

The combination of a propeller with a gas-turbine engine as a power system for aircraft has been in existence for over 15 years; however, a standard term to describe this combination has not been firmly established. Among the terms being used are **propeller turbine, turbo-propeller, prop-jet, turbo-prop,** and **turboprop.** Of all these terms, **turboprop** appears to be most widely accepted and the most technically correct for a short word to designate the turbine-propeller combination. In this text we shall use the word **turboprop** to designate a gas-turbine engine designed to drive a propeller.

Characteristics of turboprop engines

In general, the turbine section of a turboprop engine is very similar to that of a turbojet engine. The main difference is the design and arrangement of the turbines. In the turbojet engine the turbine is designed to extract only enough power from the high-velocity gases to drive the compressor, leaving the exhaust gases with sufficient velocity to produce the thrust required of the engine. The turbine of the turboprop engine extracts enough power from the gases to drive both the compressor and the propeller. Only a small amount of power is left as jet thrust. Usually a turboprop engine has two or more turbine wheels. Each wheel takes additional power from the jet stream, with the result that the velocity of the jet is decreased substantially.

Figure 17 · 1 is a simplified drawing of a turboprop engine. The two turbine wheels are mounted on the compressor drive shaft; thus the turbine and compressor rotate as a single unit. The rotational speed of the turboprop engine is usually considerably higher than that of a turbojet, being from 11,000 to over 40,000 rpm for different engines. This high rpm requires that the propeller be driven through reduction gears. In the illustration of Fig. 17 · 1 the reduction-gear assembly is located immediately forward of the compressor and is integral with the engine assembly. Some turboprop engines have the propeller reduction-gear case mounted separately and a drive shaft connects the gear case to the turbine power unit.

Turboprop engines having two independently rotating turbines have proved quite successful. One of the turbines drives the compressor, while the other drives the propeller. The independent turbine has some advantages over the direct-

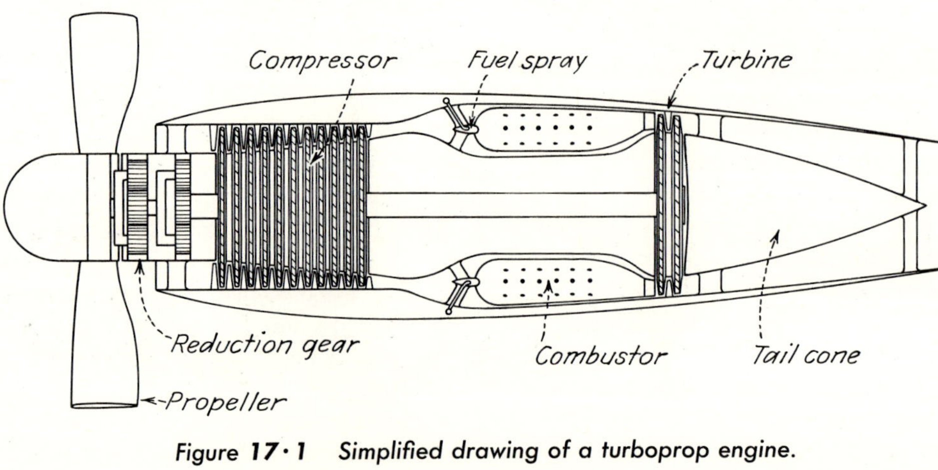

Figure 17 · 1 Simplified drawing of a turboprop engine.

Figure **17·2** *AiResearch Model TPE331 turboprop engine.*

connected type. Both the propeller and the compressor may be operated at rotational speeds that produce the best respective efficiencies. This allows the propeller to operate at high speed during takeoff and climb, thus reducing the tendency of the propeller to stall. Propeller speed can be lowered at altitude, with a subsequent reduction in tip compressibility losses. The independent turbine can be compared with a two-speed reduction gear for reciprocating engine-propeller combinations.

A currently produced turboprop or turboshaft engine for small aircraft or helicopters is shown in Fig. 17·2. This is the AiResearch Model TPE331 (T76), which produces 600 shp and 622 eshp, maximum output, at 41,730 rpm. The maximum continuous output is 500 shp at approximately 39,640 engine rpm or 1,900 propeller rpm. The engine weight is approximately 260 lb, thus producing a power/weight ratio of 2.44 eshp per lb. The sfc is about 0.614 lb per shp per hr.

As shown in the illustration, the engine utilizes a 2-stage centrifugal compressor, a reverse-flow annular combustion chamber, and a 3-stage turbine. The compressor impellers are made of titanium, while the combustion chamber parts, turbines, and nozzle vanes are made of high-temperature stainless steel. The turbine blades and wheels are one-piece castings.

The Rolls-Royce Dart turboprop engine is shown in Fig. 17·3. It utilizes a 2-stage centrifugal compressor, and current models employ a 3-stage axial turbine. Various models of this engine have been used for a decade or more in such airplanes as the Vickers Viscount and more recently in the Fairchild F-27 airliner. This engine will be described more fully in Chap. 18.

Turboprop advantages

Since turboprop engines have reached the stage of development where they are now used extensively as aircraft powerplants, experience has shown that in many cases these engines have far exceeded expectations in performance and economy of operation.

Among the principal advantages of the turboprop engine are the following:

1. SIMPLICITY OR DESIGN AND CONSTRUCTION. The turboprop contains fewer moving parts than a reciprocating engine; hence the number of parts which may fail because of wear is reduced. Furthermore, it is easier to isolate and correct troubles within the engine itself.
2. LOW WEIGHT/POWER RATIO. This characteristic is also termed **specific weight** and is found by dividing the weight of the engine by the equivalent horsepower (ehp). A number of turboprop

Figure **17·3** *Rolls-Royce Dart turboprop engine.*

engines have been developed which have a specific weight of less than 0.50 (½ lb per ehp). Conventional reciprocating engines usually have a specific weight of from 1.00 to 1.75.

3. LOW FUEL CONSUMPTION. Turboprop engines have been developed to attain a specific fuel consumption of less than 0.40. This means that the engine consumes 0.40 lb of fuel per equivalent horsepower per hour. The only reciprocating engine which has bettered this figure is the turbo-compound type. Further development of turboprop engines will undoubtedly lead to still more fuel economy.
4. ECONOMICAL MAINTENANCE. The operational history of turboprop engines shows continued reduction of maintenance cost. Gas-turbine engines are now operating for over 3,000 hr between overhauls with routine inspections and maintenance, and it is expected that continued improvement in design, materials, and operating techniques will lead to major overhaul periods of more than 4,000 hr. The best reciprocating engines do not usually exceed 1,200 hr between major overhauls.
5. LOW-DRAG INSTALLATION. Since turboprop engines are much smaller in diameter than equivalent reciprocating engines, the nacelle containing the turboprop engine can be designed to produce much less parasite drag when the aircraft is in flight. Figure 17 · 4 illustrates the installation of a turboprop engine in a transport airplane.
6. OPERATIONAL FLEXIBILITY. The turboprop engine performs well at sea level and under takeoff conditions, and it also gives good performance at altitudes of over 30,000 ft. It is limited in highspeed operation because the propeller efficiency drops off rapidly when the aircraft speed is more than 500 to 600 mph. At these speeds and above, the pure-jet gas turbine is the most efficient engine.

Power output of turboprop engines

One consideration to bear in mind when evaluating the performance of turboprop engines is the effect of airspeed and altitude on power output and specific fuel consumption.

The power delivered by a turboprop engine increases very rapidly with increasing airspeed because the density and total energy of the air at the compressor become greater. This is because of the ram effect produced as the airplane rushes through the air. Thus, at sea level and at 500 mph, the power delivered by a simple-cycle turboprop engine may be approximately 45 per cent greater than static takeoff power. This same typical engine will deliver almost as much power at 20,000 ft and at high airspeed as it will produce under sea-level takeoff conditions.

Operating conditions and specific fuel consumption

The specific fuel consumption of a turboprop engine decreases with increasing airspeed and increasing altitude. Thus, the engine may have, at 20,000 ft and at high airspeeds, a specific fuel consumption approximately one-quarter less than that attained under sea-level static conditions. For these reasons, the performance of a turboprop engine should be evaluated under actual flight conditions as well as under sea-level static conditions.

Figure 17 · 4 Installation of turboprop engines in a Lockheed Navy transport aircraft.

Figure 17·5 Cutaway view of the Pratt & Whitney T34 turboprop engine.

THE PRATT & WHITNEY T34 TURBOPROP ENGINE

General description

The Pratt & Whitney T34 turboprop engine illustrated in Fig. 17·5 is an axial-flow gas-turbine engine which produces thrust by driving a propeller at the front and discharging high-velocity gases through a nozzle at the rear. This engine contains a 13-stage compressor, a can-annular combustion chamber with eight burner cones, and a 3-stage turbine. The compressor diameter compresses to a wasp waist at the forward end of the combustion chamber, thus permitting the installation of accessories at this point without increasing the overall diameter appreciably. The design of the combustion chamber and the 3-stage turbine also contribute to the small diameter of the engine.

The T34 is designed in several models, but no attempt will be made to point out the differences among them in this text, since the general arrangement and configuration of all models is quite similar.

T34 engines are manufactured to operate with either the Hamilton Standard Hydromatic reversing propeller or the Curtiss electric propeller. The T34-P-3 and YT34-P-5 engines incorporate two negative-torque-control switch actuator pushrods, which extend ¼ in. beyond the front case mount pad, and are adapted for use with a Curtiss electric propeller. In contrast the T34-P-6 engine incorporates a single negative-torque-control pushrod, which extends 3 in. beyond the front case mount pad, and is adapted for use with a Hamilton Standard propeller.

The T34 engine may be divided into three main sections, as shown in Fig. 17·6: (1) the

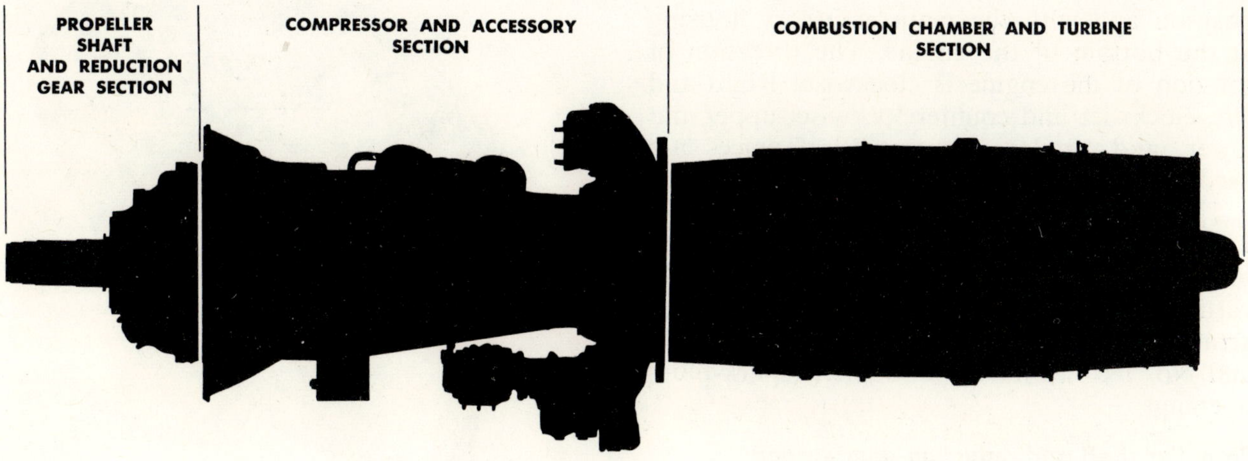

Figure 17·6 Main sections of the T34 engine.

propeller shaft and reduction-gear section, (2) the **compressor and accessory section,** and (3) the **combustion-chamber and turbine section.** These are major divisions and may be classified in smaller units and assemblies for convenience. In this text we shall discuss the principal units as they appear from front to rear.

Engine specifications and performance

The specifications of the T34 engine are as shown in the following list; however, variations may exist for different modifications of the engine.

Type of compressor	Axial-flow, single-spool
Compressor stages	13
Type of combustion chamber	Can-annular
Number of burner cones	8
Turbine stages	3
Compressor ratio	6.7:1
Propeller reduction-gear ratios:	
Low	0.354:1
High	0.257:1
Dry weight of engine, standard equipment	2,590 lb
Length	161.25 in.
Diameter	34.06 in.
Equivalent horsepower	5,700 hp, sea level, takeoff
Specific weight	0.46 (approx.) lb/hp
Specific fuel consumption	0.62 lb/hp/hr

Directional references

In discussing the various parts of a gas-turbine engine, it is necessary to follow a standard method for denoting position and direction. With the T34 we view the engine from the rear (exhaust duct) with the engine in a horizontal position and with the main accessory housing at the bottom of the engine. The direction of rotation of the engine is clockwise. Right and left, clockwise and counterclockwise, upper and lower, and similar directional references are used to indicate position and direction.

The combustion-chamber cones are numbered 1 to 8 in a clockwise direction, with the top combustion-chamber cone to the right of the engine vertical center line designated as No. 1. This arrangement is shown in Fig. 17 · 7. The No. 2 and No. 7 cones have sparkigniter (spark-plug) openings.

Propeller shaft and reduction-gear section

The **front case** of the engine, shown in Fig. 17 · 8, incorporates a recessed liner at its front end which supports the propeller-shaft thrust bearing. A boss is provided in the front case for the oil-transfer tube which carries to the **oil-transfer bearing.** A mounting pad is provided for a **torquemeter transmitter adapter** on the outside surface of the case. Five circular recesses are located in the front of the case to house the negative-torque-control (NTC) assemblies. An external pad located at the bottom of the front case is provided for the front scavenge oil pump.

The propeller shaft, which is provided with a

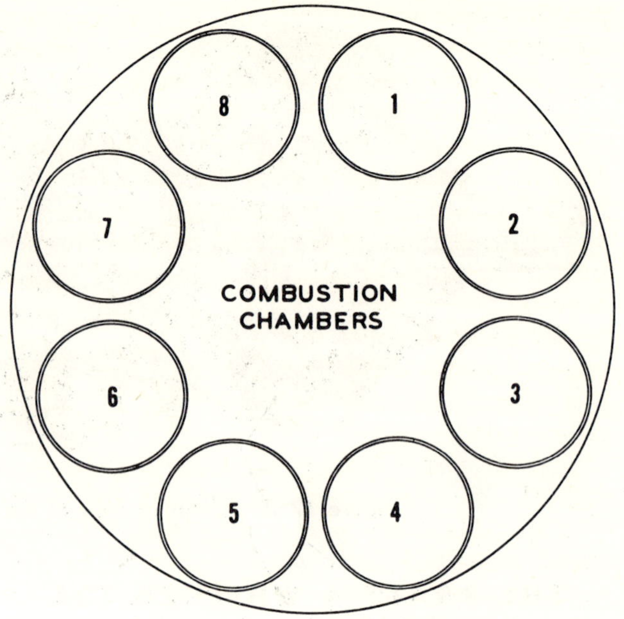

Figure 17 · 7 Combustion-chamber numbering.

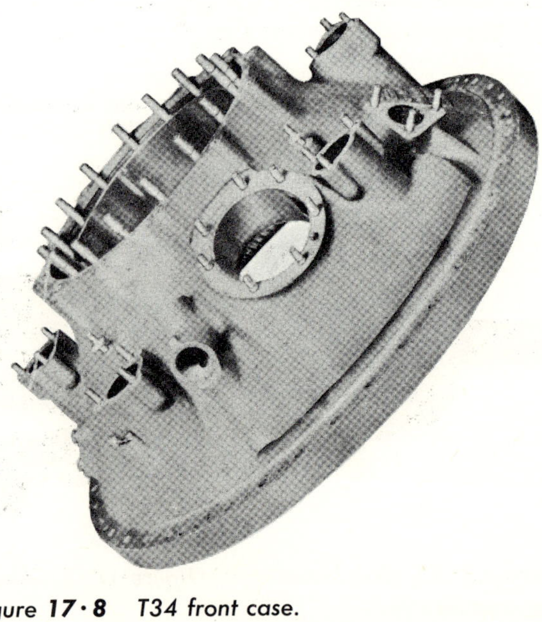

Figure 17 · 8 T34 front case.

No. 70 spline at the front end for the propeller, supports a two-stage planetary reduction assembly. The **low-speed reduction drive pinion cage** is integral with the shaft as shown in Fig. 17 • 9. The high-speed reduction drive gear assembly is held in place on the rear of the propeller shaft by the reduction-gear bearing support. The rear end of the shaft is supported by two steel-backed bronze bearings inside the high-speed pinion cage, and the main body of the shaft is supported by the ball and roller thrust bearings. The scavenge oil pump drive gear is bolted to the shaft just ahead of the low-speed reduction-gear pinion cage.

The propeller low-speed reduction ratio is 0.354:1. This reduction is obtained by means of a spur-planetary gear arrangement. The low-speed reduction fixed gear is a spur-ring gear with diagonal splines on its outside diameter which mesh with splines machined onto the rear of the front case. The fixed gear is held in the front case by the torquemeter piston which rides on the rear face of the gear and the five negative-torque-control pistons on the front face of the gear. A low-speed drive gear, splined and locked to the front end of the **high-speed pinion cage,** transmits power from the high-speed pinion cage to the nine **low-speed reduction drive pinions.** Each pinion is supported in the cage by a bearing, spacer, and a pinion bolt, and it meshes with teeth on the inside diameter of the fixed gear.

The **propeller-shaft oil-transfer bearing** is mounted on the propeller shaft in front of the scavenge-pump drive gear. An oil-transfer tube connects the bearing to the pressure oil inlet boss in the bottom of the front case. Engine pressure oil is transmitted through the oil-transfer bearing into the cavity between the propeller shaft and the propeller-shaft transfer liner, and thence through drilled passages in the high- and low-speed pinion cages to the reduction drive pinion assemblies.

Figure 17 • 9 Low-speed reduction-drive pinion cage and propeller shaft.

The propeller-shaft thrust bearings consist of a roller bearing and a ball bearing which has a split inner race. The outer race of the ball bearing seats on the front case liner, and the rear half of the split inner race seats against a shoulder of the propeller shaft. The thrust-bearing cover retains the outer race of the roller bearing, and the thrust-bearing nut retains the inner race. The thrust-bearing nut carries two oil seal rings which contact the liner in the thrust-bearing cover. The thrust-bearing cover spacer is installed between the thrust-bearing cover and the outer race of the roller bearing. An oil slinger is held between the thrust-bearing nut and the front face of the roller-bearing inner race. The bearings support the front of the propeller shaft and transmit the propeller thrust from the shaft to the engine-mounting brackets.

The propeller high-speed reduction gearing is of the spur planetary type and affords a gear reduction of 0.257:1. The high-speed reduction fixed gear is a spur-ring gear which is bolted to the front of the compressor front bearing support. The high-speed reduction drive gear is coupled to the compressor-shaft front coupling. Teeth on the outside diameter of the drive gear mesh with five reduction drive pinions housed in the high-speed pinion cage. Each pinion is supported in the cage by a bearing, a spacer, and a pinion bolt and meshes with the teeth on the inside diameter of the fixed gear.

Torquemeter

The use of torquemeters on aircraft engines has been a common practice for many years. The basic purpose of such a device is to enable the pilot of the airplane to determine the power output of the engine while the airplane is in flight. The torquemeter used in the T34 engine is similar in operation to those used in other turboprop engines and also in reciprocating engines.

As the low-speed reduction drive fixed gear moves rearward on the diagonal splines of its outside diameter in response to the torque applied to the propeller shaft by the compressor rotor shaft through the reduction drive gears, the rearward thrust of the reduction drive fixed gear is counterbalanced by pressure oil operating on the gear through the torquemeter piston. The piston carries two oil seal rings. A shoulder on the forward outside diameter of the piston maintains contact with the rear face of the fixed gear.

Engine oil from the pressure oil pump is carried through an external oil-transfer tube and a cored passage to the torquemeter oil-inlet connection on the lower right rear of the reduction-gear bearing support. From the oil-inlet connection, oil passes through the torquemeter valve to the rear of the torquemeter piston. The **torquemeter piston and valve** are illustrated in Fig. 17·10. Torquemeter oil pressure is maintained by the torquemeter valve, which is located in the oil-inlet connection boss. The pressure oil is received by the torquemeter piston which, by its own movement, opens or closes the torquemeter valve, thus metering the flow of oil to itself in direct proportion to the load applied to the piston through the reduction drive fixed gear. A torquemeter oil-bleed nozzle located on the upper rear side of the bearing support bleeds off any excess pressure. Thus, the reduction drive fixed gear is always balanced by oil pressure acting against the rear side of the piston. The oil pressure acting through the torquemeter transmitter-adapter transmits the pressure through tubing filled with air or a low-viscosity fluid to a pressure gauge in the airplane. By a calculation utilizing the oil pressure acting on the torquemeter piston, together with the compressor-rotor rpm and a previously determined torque constant, the pilot determines the horsepower being delivered to the propeller.

The various elements composing the torquemeter system of the engine are housed in the reduction-gear-bearing support. These units are the torquemeter piston, torquemeter valve, and torquemeter bleed nozzle.

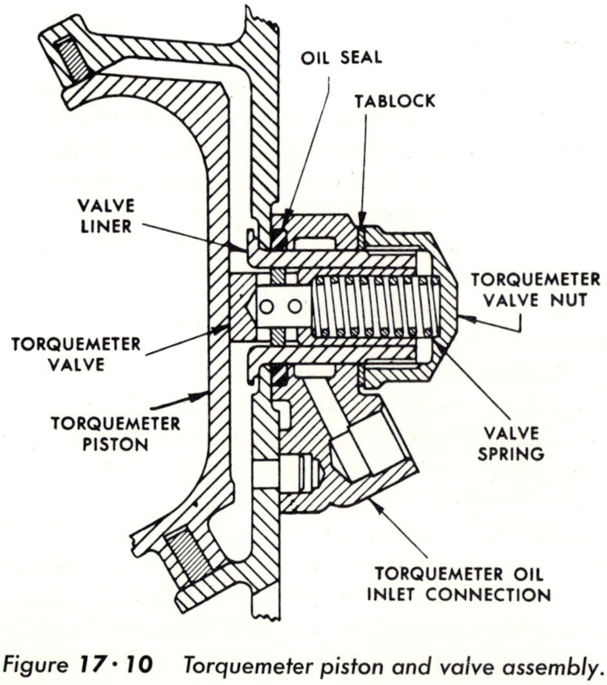

Figure 17·10 Torquemeter piston and valve assembly.

Compressor section

The compressor section of the T34 engine consists of the **compressor air-inlet case,** the **compressor front bearing support case,** the **compressor assembly,** and the **compressor front and rear cases.**

The compressor air-inlet case contains six hollow struts which project radially from the case hub across the air inlet passage to the outer rim of the case. These struts are structural members which serve to connect the inner cases and bearing supports to the outer case in order to give strength and rigidity to the entire front section of the engine. Compressor discharge air is circulated through the four upper inlet struts to prevent icing of the engine inlet. The two bottom struts circulate oil and do not require heated air. The struts located in the 11 o'clock, 1 o'clock, and 3 o'clock positions contain breather tubes. Oil baffles are bolted to the inlet case hub at the three breather-tube locations to prevent the loss of oil. The strut in the 11 o'clock position also contains a passage to the bearing support case for the transfer of water to the compressor section. Pressure oil to the front case and torquemeter system passes through the strut in the 5 o'clock position, and scavenge oil from the front scavenge pump passes through the strut in the 7 o'clock position.

Bosses for mounting a water-injection regulator, breather-pad covers, anti-icing air-outlet pad covers, thermocouples, and static pressure probes are located on the outer perimeter of the case. The compressor front bearing support case, together with the **high-speed reduction drive fixed gear,** is bolted to the inside of the air-inlet case, behind the hollow struts.

The **compressor front roller bearing** is contained in a liner supported in the support case hub. Located in the front of the bearing liner is a transfer tube elbow which transfers pressure oil from a hollow passage in the support case to a passage in the liner. An oil-metering plug, located in the inside shoulder of the liner, meters oil in a fine spray to the rear of the roller bearing. Oil is prevented from entering the compressor air system by a carbon seal housed in the rear of the front bearing liner. Nineteen hollow inlet guide vanes are radially supported between an inner shroud and an outer shroud, and the inner shroud is bolted to the rear of the support case hub. Four passages in the support case provide for anti-icing airflow from the hollow inlet guide vanes to the air inlet case struts. Bolted to the rear of the inlet-guide-vane inner shroud is a water-injection manifold spray ring which connects with a hollow passage in the support case to provide a fine spray of water to the inlet air system. The water spray is used to give added thrust to the engine when needed.

Compressor assembly

The compressor assembly is housed in the compressor front and rear cases. It consists of 13 rotor stages and 13 stator vane stages. The **compressor rotor** is illustrated in Fig. 17 · 11. The rotor assembly is composed of 13 rotor disks, 13 spacers, a front rotor shaft, and a rear rotor shaft. The **vane and shroud assemblies** are located between the successive rotor stages, and the inner shroud forms a sealing ring for the two air seals on the outside diameter of each rotor spacer. Dowels in the outer vane shrouds locate the vane and shroud assemblies. Mounted on the fourth-stage outer shroud is the **compressor air-bleed junction box** which transfers sixth-stage compressor discharge air to the front air-bleed valves. The front rotor shaft is supported by a roller bearing in the **compressor front bearing support liner.** Splined to the front rotor shaft is the **front shaft coupling.** An oil seal separates the front bearing void from the compressor air system. Two ball bearings support the rotor rear shaft and are housed in the bearing support in the accessory case. Splined to the rotor rear shaft is the **main accessory drive gear,** the spacer, and the **rear shaft coupling.** Two oil seals separate the rear bearing from the compressor air system.

The outer case surrounding the compressor is constructed in two sections called the **compressor front case** and the **compressor rear case.** The two sections are bolted together at a point between the sixth and seventh stages. The front flange of the front case is bolted to the air-inlet case, and the rear flange of the rear case is bolted to the accessory case. Two air-bleed valves located on the upper rear of the front case bleed sixth-stage compressor air overboard, and two bleed valves located on the upper front of the rear case permit bleeding overboard of seventh-stage air. The bleed valves on the right side of the engine are enclosed with a single-outlet air-bleed valve manifold, and the valves on the left are enclosed in a similar manner.

The purpose of the overboard air bleed is to make starting of the engine easier by reducing the load on the compressor. Air bleeds also aid in preventing compressor stall under certain critical conditions of rpm and pressure.

Accessory section

The accessory section is located between the compressor rear case and the combustion-chamber front case. The accessory case houses the two rear compressor shaft roller bearings and the bearing liner assembly. A carbon seal is incorporated at each end of the accessory drive pinion and bearing support housing to prevent oil from entering the compressor discharge airstream. Metered and screened oil from a groove in the center of the bearing liner lubricates the front and rear bearing mating faces and the rear bearing. Secured to the lower section of the accessory drive pinion support is a single-section scavenge oil pump.

Eight hollow struts which project radially from the case hub across the compressor discharge air passage to the outer rim of the case serve to diffuse the compressor discharge airflow and adapt it for entry into the combustion chambers.

Four large openings in the outside of the accessory case house the accessory drives which extend through the struts and engage the compressor rotor rear shaft accessory drive gear.

Figure 17 · 11 T34 compressor rotor assembly.

This gear is located immediately to the rear of the compressor rear bearing. A **generator drive adapter elbow** is bolted to the accessory case in the 12 o'clock position, an **engine brake and starter drive adapter** in the 3 o'clock position, and the **main accessory drive housing elbow** in the 6 o'clock position. Eight triangular bosses located at the rear outer diameter of the case house the fuel nozzle support assemblies, which extend rearward so that the dual fuel nozzles enter the forward ends of the burner cones. Bosses are also provided on the accessory case for anti-icing air connectors and a rear main bearing thermocouple.

The **lower accessory drive housing** and elbow are secured to the bottom of the accessory drive case. Housed in the bottom of the housing as viewed from the front are the **pressure oil pump,** the **scavenge oil strainer,** and the **scavenge oil pump.** Mounted on the front pads of the housing are the **fuel-control unit** on the left pad and the **dual fuel pump** on the right pad. Dual tachometer drives are mounted on the left side of the housing.

The **fire-seal ring and fuel-manifold assembly** is secured to the rear mating face of the accessory case at the front of the mating face flange. The fire-seal supports the primary and secondary fuel-manifold assembly on its forward face. The fuel-manifold assembly consists of a primary and secondary manifold tube and eight fuel-manifold connectors. Two steel tubes connect each fuel-manifold connector with the corresponding fuel-nozzle support. The two bottom connectors are also attached to the primary and secondary fuel lines from the fuel-pressurizing and dump valve.

Combustion-chamber section

The combustion chamber for the T34 engine is illustrated in Fig. 17 · 12. This assembly consists of eight burner cones which are supported at the front by the fuel nozzles and at the rear by the inner and outer burner liners. Thus the fuel combustion starts in eight separate chambers, and then the hot gases enter a single annular chamber formed by the inner and outer burner liners. The burner cones are interconnected by means of welded projections which form flame tubes for cross ignition of the fuel. Bosses for sparkigniters are provided in No. 2 and No. 7 combustion chambers. The entire assembly of burner cones and burner liners is locked together and secured to the combustion-chamber support by means of lockpins.

The forward ends of the burner cones contain

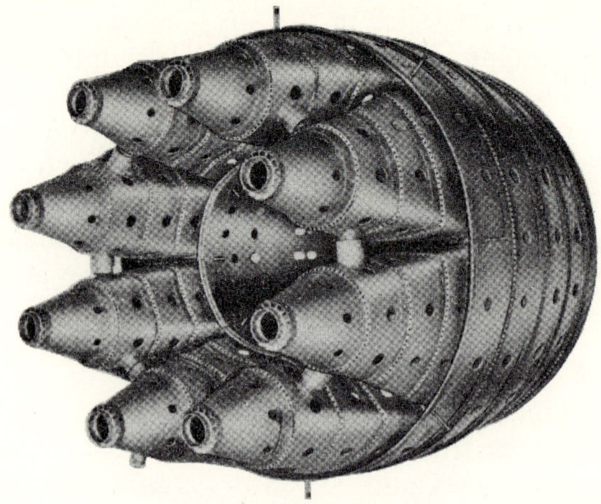

Figure ***17 · 12*** *T34 combustion-chamber assembly.*

swirl vanes which give the primary air a rotary motion as it mixes with the fuel being injected from the fuel nozzles. The swirling action helps to anchor the flame and assure complete combustion of the fuel. The burner cones and burner liners are perforated to permit the entrance of secondary air into the combustion area. The secondary air cools the hot gases to a safe level before they reach the turbine nozzle.

The combustion-chamber assembly is housed in the combustion-chamber outer case, which is a two-piece telescoping cylinder constructed of stainless steel. The front flange of the front section of the outer case is bolted to the rear flange of the accessory case, and the rear flange of the rear section is bolted to the turbine exhaust case. The two sections of the combustion-chamber outer case are secured together by the mating of an inner flange and an outer flange bolted together.

Turbine section

The turbine section includes the **turbine rotor assembly,** the **vane and shroud assemblies,** the **turbine exhaust case,** and the **turbine rear bearing and supports.**

The turbine rotor assembly, shown in Fig. 17 · 13, consists of the hollow turbine shaft, the 3-stage turbine disk and blade assemblies, and the turbine rotor hub. The turbine disks and spacers, together with the turbine hub, are secured to the turbine shaft with a tie rod. The turbine blades are of the shrouded type, and they are held in the outer rims of the turbine disks by means of fir-tree serrations and locked in place by means of rivets. The turbine shaft is secured at the front end to the compressor-shaft rear coupling with bolts and supported at the aft end

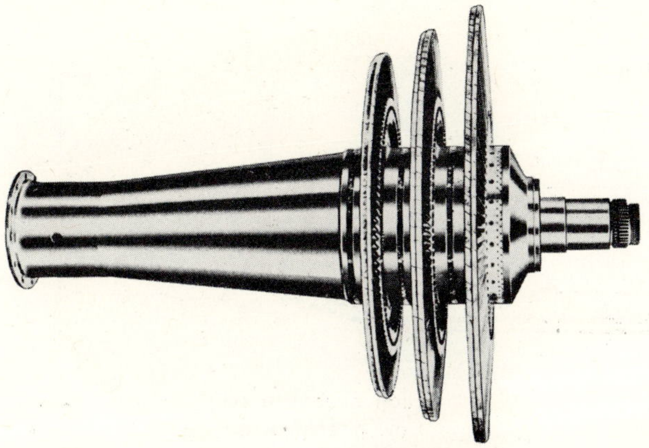

Figure 17·13 Turbine rotor assembly.

by the turbine rear bearing. The turbine bearing scavenge oil pump intermediate drive gear is coupled to the end of the rear shaft by means of a splined coupling and locknut.

The first-stage vane and shroud assembly, often called the "turbine nozzle," is bolted to the combustion-chamber cone and liner support ahead of the first-stage rotor disk. The second- and third-stage vane and shroud assemblies are located between the successive rotor stages. The outer rims of the vane and shroud assemblies are of sufficient width so that all the assemblies can be bolted together at the flanges and still provide proper spacing for the turbine blades between the stator vanes. Air seals are riveted to the second- and third-stage inner shrouds and to the rear of the cone and liner support. Dual air seals separate the outer vanes to form sealing rings with the turbine rotor blades.

The turbine exhaust case, which is secured to the **front gas seal bellows** and the rear combustion-chamber outer case on the forward end and to the turbine exhaust extension case at the rearward end, contains the turbine rotor and the turbine bearing support. Four struts are positioned radially around the bearing support and are held in place between the turbine exhaust inner and outer ducts by the bearing support arms and the strut support tubes. The lower left support arm contains the bearing thermocouple. The upper right support tube contains the **bearing breather tube,** and the two lower support tubes contain the scavenge oil and pressure oil tubes. Secured to the front of the bearing support housing is the **turbine rotor cooling air support.**

Attached to the rear of the bearing support are the **oil pump support,** the **oil sump,** and the **turbine exhaust inner duct support.** A single-section scavenge oil pump is bolted to the rear of the oil pump support. This pump is driven by the **intermediate drive gear** splined on the end of the turbine rotor shaft.

Housed in the center bore of the bearing support is the bearing liner. A spring-loaded carbon seal to provide a constant pressure on the oil seal face plate is located in the front of the liner. The roller bearing outer race seats on a shoulder in the liner. The turbine bearing void is sealed from the air system by the carbon seal and the bearing oil sump.

Turbine outer extension duct

The turbine exhaust outer extension duct is secured at the forward end to the turbine exhaust case. The exhaust outer extension duct houses the inner exhaust extension duct and cover. Bosses for four thermocouples are provided on the outside of the outer extension duct. The purpose of the exhaust outer extension duct is to carry the hot gases being ejected from the turbine section to the aircraft exhaust tailpipe. These high-velocity gases provide a substantial amount of thrust to the aircraft.

Lubrication system

The lubrication system for the T34 engine is shown schematically in Fig. 17·14. A **single-section pressure oil pump,** with its discharge pressure regulated by a **pressure oil relief valve,** furnishes high-pressure oil to the propeller shaft bearings, the reduction gearing, the torquemeter, and the compressor front bearing. The pressure oil pump also supplies low-pressure oil through a **pressure-regulating valve** to the compressor rear bearings, the accessory drive gear housings, and the turbine bearing. The torquemeter oil system is of a low-pressure design not requiring a booster pump; instead, it utilizes high-pressure oil directly from the pressure oil pump.

The scavenge oil system of the engine includes four individual gear-type pumps, which, working in parallel, scavenge the bearing voids and deliver the oil to the cooler. The first of these is located in the bottom of the front case and picks up oil from the thrust bearings, the reduction gearing, and the compressor front bearing. The second pump, located in the rear compressor bearing void, scavenges oil from the upper accessory drive housing and the compressor rear bearings. The third pump, located in the turbine bearing void, scavenges the oil from this bearing sump. The fourth pump, located in the bottom of the lower accessory drive housing, picks up the oil from the first three pumps and scavenges

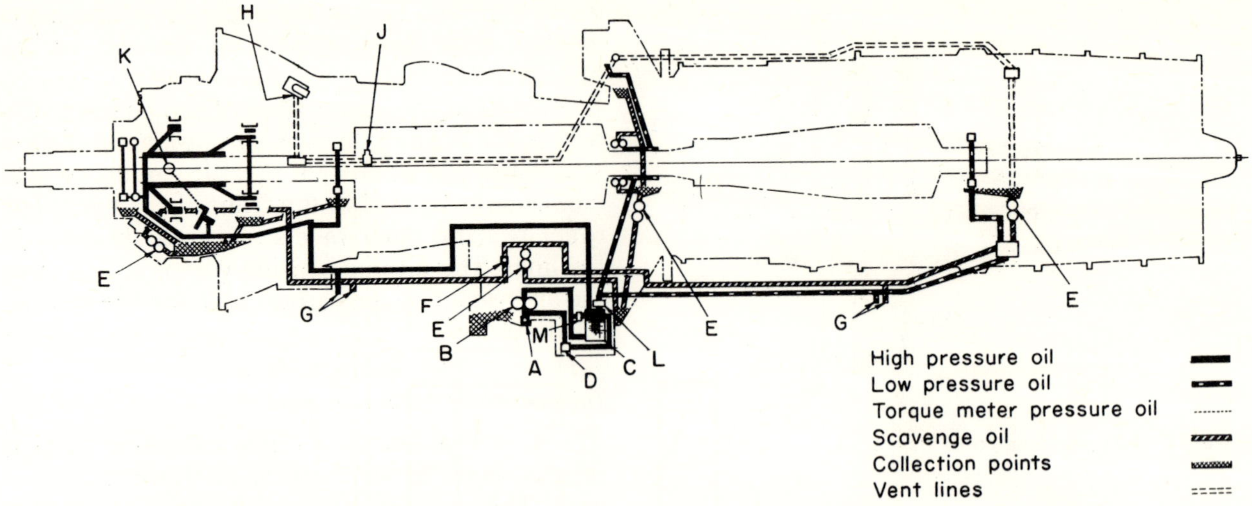

Figure 17·14 T34 turboprop lubrication system: (A) oil in connection; (B) oil pressure pump; (C) oil strainer; (D) oil pressure relief valve; (E) oil scavenge pumps; (F) oil out connection; (G) oil drain plugs; (H) breather connection; (J) oil-tank vent connection; (K) torquemeter pressure and vent connections; (L) low-pressure oil regulating valve; (M) oil strainer bypass valve.

the lower accessory housing. This pump also pumps the oil out to the engine oil cooler.

The pressure oil strainer is made up of a series of spacers and screens and is located in the bottom of the accessory drive housing between the pressure oil and scavenge oil pumps.

Breather connections are located at the top of the turbine exhaust case, the upper accessory drive elbow, and the three pads on the upper half of the air inlet case. External tubing connects the breather voids. An oil-tank breather connection is provided in the breather tubing located on the left-hand side near the engine horizontal center line and the forward compressor flange. It is important that the breathers operate properly to prevent a buildup of pressure which would interfere with lubrication and oil flow.

Ignition system

The T34 engine uses a high-energy capacitor-type ignition system. This system provides a very high voltage which makes possible starts with carbon-fouled sparkigniters and also facilitates air starts at high altitude.

A schematic diagram of the ignition system is shown in Fig. 17·15. The complete system consists of two identical circuits, each furnishing voltage to one sparkigniter. Both ignition exciter units are housed in one box at the bottom of the engine just aft of the compressor air inlet case. The ignition compositor units (transformers) are located one on either side of the engine just forward of the accessory section. The input voltage for the ignition system is usually 24 to 30 volts, but it will operate with as low as 16 volts. The input line of each of the two systems has a filter to prevent radio interference. The filter consists of induction coils in series and capacitors in parallel to eliminate fluctuations in the current, thus preventing the radiation of electromagnetic waves which would produce noise in radio circuits.

Since the circuits to both sparkigniters are identical, only one will be considered in detail. When the aircraft ignition switch is turned on, direct current is supplied through the filter to the motor operating the two cams controlling the opening and closing of the breaker points. Cam M, shown in the diagram of Fig. 17·15, has several lobes which open and close the breaker points quite rapidly, thereby delivering pulsating direct current to the autotransformer T_1. A capacitor C_2 is connected in parallel with the breaker points to prevent arcing. The voltage is built up in the autotransformer from the normal 24 volts to approximately 2,000 volts. It is then passed through the selenium rectifier V_1 to charge the storage capacitor C_1. Concurrently with the charging of this capacitor, the single-lobe cam S permits momentary closing of its breaker point, thus permitting one pulse of direct current through the primary coil of the high-tension transformer T_2 and to the capacitor C_3. The transformer increases this voltage to a value which is sufficient to break down the sparkigniter air gap and to provide for the discharge

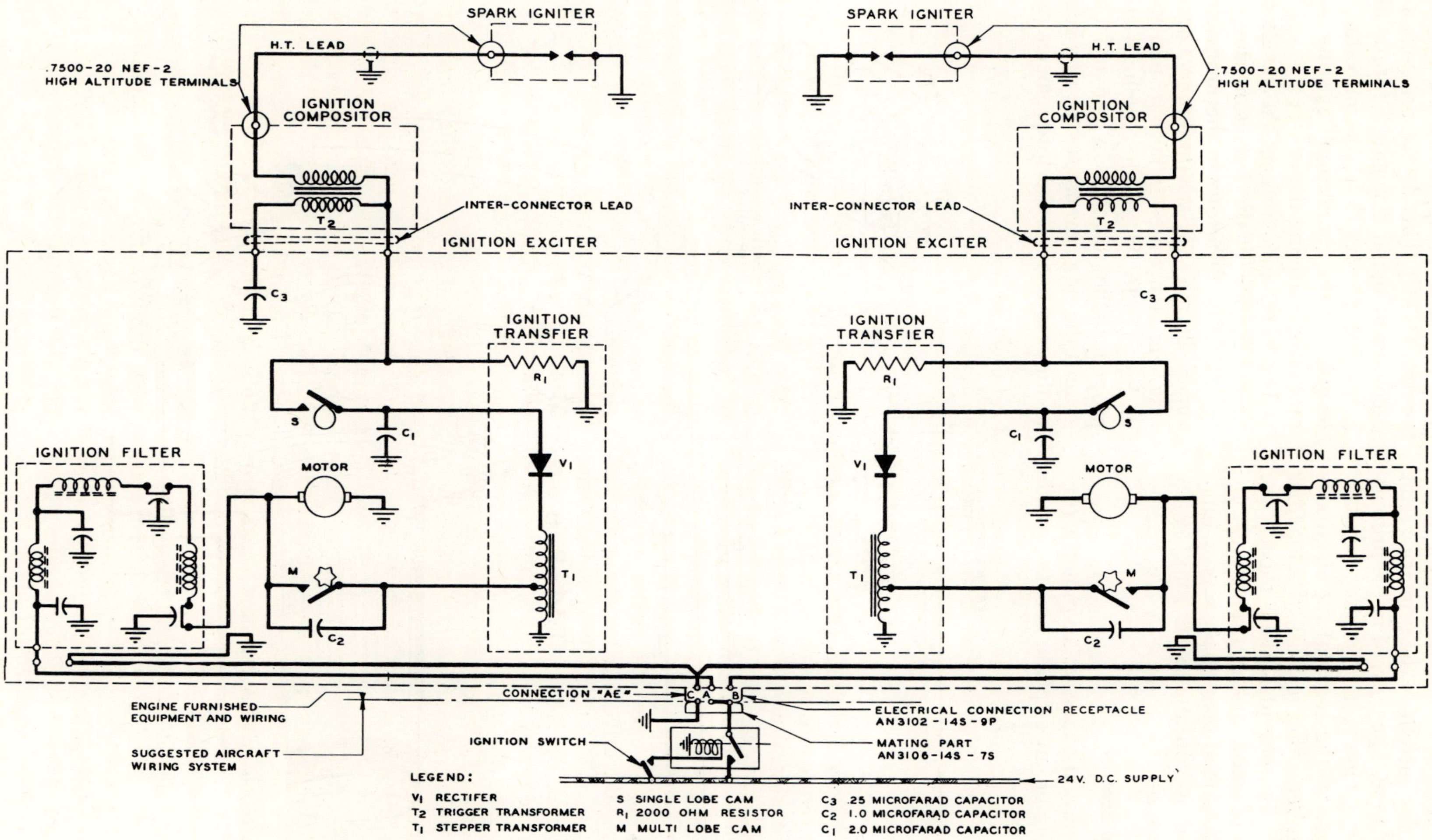

Figure 17·15 T34 ignition system.

of the capacitor C_1. When this capacitor discharges through the primary winding, the capacitor C_3 is charged. This results in high-voltage alternating current in the primary circuit for a very short time. The current in the primary coil induces a very high voltage in the secondary coil, which produces the spark at the electrodes of the sparkigniter. The resistor R_1, together with capacitor C_3, provides the proper time constant for oscillation in the primary circuit. The cycle just described takes place several times per second and is continuous as long as the ignition switch is on.

Negative torque control

In order to prevent excessive asymmetric (unbalanced) torque on an airplane and to provide for automatic reduction of windmilling drag to a tolerable value in the event of power loss, the T34 engine is equipped with a negative torque control (NTC). This device provides a powerful mechanical signal if negative torque at the propeller shaft exceeds a certain value. **Negative torque** is the condition existing when the propeller is driving the engine.

The NTC functions as the result of the reversed motion of the reduction ring gear when the direction of torque in the propeller shaft reverses because of windmilling. The reduction ring gear is mounted in helical splines, as shown in Fig. 17 · 16. When the engine is driving the propeller, the reaction of the reduction gears against the ring gear causes the ring gear to move to the rear against the torquemeter piston. The

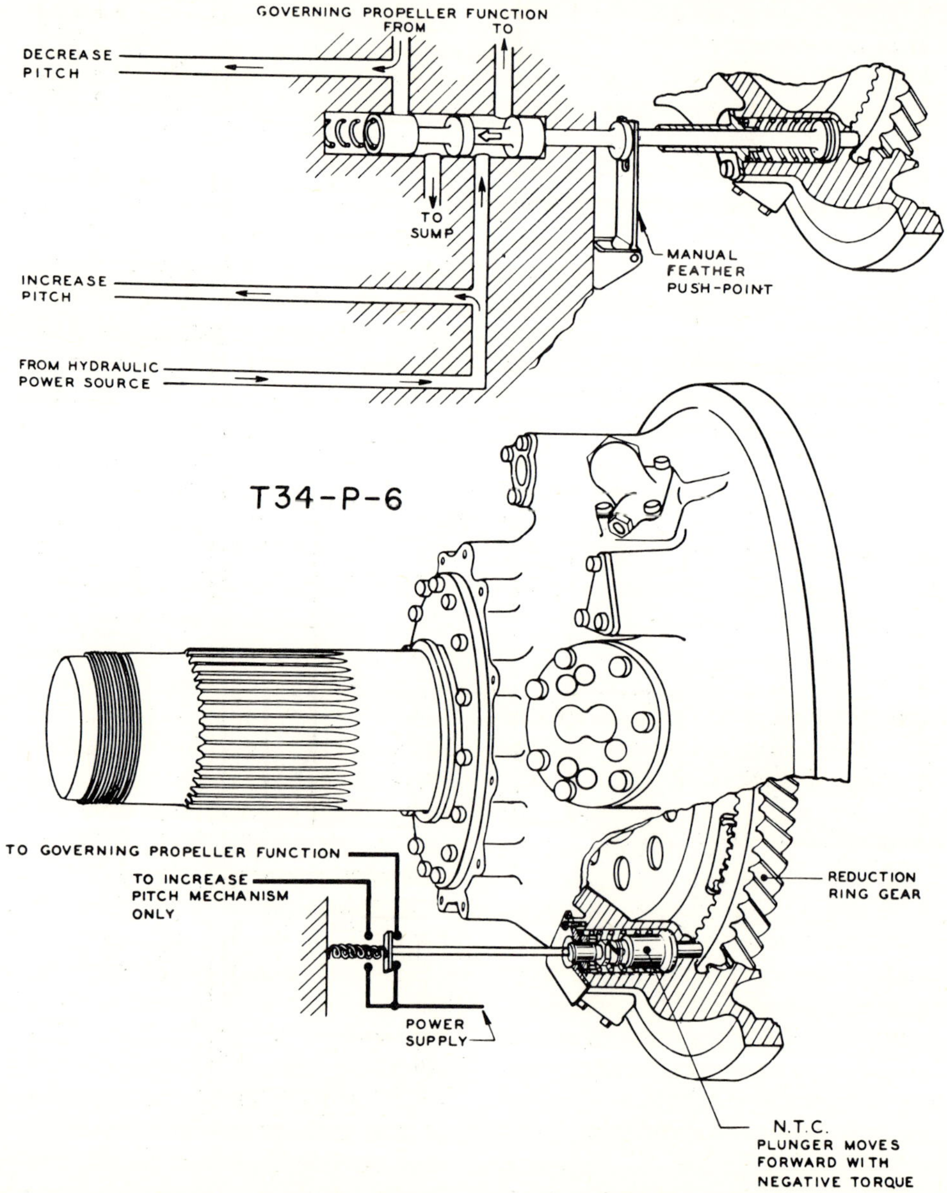

Figure 17 · 16 Operation of the negative-torque-control mechanisms.

oil pressure necessary to hold the gear in equilibrium provides the measure of torque for power-calculation purposes.

When negative torque develops, the forward motion of the ring gear provides the force that actuates the NTC. A ¼-in. movement is sufficient to give a positive signal. By opposing the travel of the ring gear with calibrated springs, it is possible to make the amount of forward motion proportional to negative torque. By selection of the proper spring rate, the NTC system produces the actuating signal when the required torque value is present.

Two of the spring assemblies are provided with switch-actuator shafts that move fore and aft with the motion of the ring gear. When the shafts have moved ¼ in. forward, the NTC signal is given, and the propeller mechanism starts turning the propeller blades toward the feathered position. If at any time the negative torque is reduced from the preselected value, the springs overcome the forward movement of the ring gear and return the gear to its normal position. Thus, if the NTC signal ceases, the propeller-pitch control is returned to the governing system.

The illustrations in Fig. 17 · 16 represent configurations for two different types of propellers. The drawing at the top shows the arrangement for the T34-P-6 engine which utilizes a Hamilton Standard turbo-hydromatic propeller. This is a hydraulically operated propeller; hence the NTC actuates a hydraulic valve. In the drawing at the bottom of Fig. 17 · 16, the NTC actuates a switch which controls two circuits in the Curtiss electric propeller system. When a negative torque signal is sensed, the "increase-pitch" circuit is energized and the propeller begins to feather.

Engine brake

The T34 engine is equipped with an engine brake which is mounted in conjunction with the starter drive gearing. The brake is used to prevent engine and propeller rotation when the engine is inoperative, regardless of whether the aircraft is in flight or is parked.

The brake is designed to provide static torque great enough to counteract the varying input torque that results when a propeller is feathered and the airplane travels at different airspeeds. The brake is not designed to overcome the input torque of an unfeathered propeller; hence, when it is used during flight to stop rotation of an inoperative engine, the propeller should be feathered (to a complete stop) before the brake is applied. To facilitate the stopping of engine rotation after shutdown on the ground, the brake may be applied at any engine speed below 5,000 rpm.

The brake consists of an **actuator,** a **brake,** a **brake disk,** a **brake housing,** a **coupling shaft,** and a **shaft gear.** The actuator is an electric control mechanism that operates from a standard 24- to 28-volt actuating motor. When the brake is turned on, the actuating mechanism drives a plunger in a hydraulically operated cylinder, thus raising the hydraulic pressure which applies the brake to the brake disk inside the brake housing. As pressure is applied in this manner, static torque is transmitted from the brake disk to the starter and brake coupling shaft to which the disk is bolted, and then to the starter and brake shaft gear which is splined into the coupling shaft. The braking action is then transmitted through the starter gear train to the compressor rotor shaft, thus tending to stop the rotor.

The brake remains in a "set" position because of a nonoverrunning screw thread in the actuator mechanism. When the hydraulic pressure reaches 340 to 350 psi (the absolute maximum is 450 psi, motor stall torque), an automatic switch opens the electric circuit to the actuator. As the hydraulic pressure drops, the pressure switch starts the actuator, again bringing the pressure up to the value required. As a safety precaution for the actuator, a cam-operated limit switch prevents the hydraulic plunger from reaching the end of its stroke. To release the brake, the actuator is reversed and is shut off at the end of its travel by a limit switch.

Anti-icing air system

In order to prevent dangerous icing of the inlet air surfaces, an anti-icing system is incorporated in the engine. High-pressure compressor-discharge air is piped forward to the air-inlet case by means of an external tube on the lower right side of the engine. An anti-icing air regulator and an anti-icing valve are incorporated in the tube. The regulator and valve are located just to the rear of the anti-icing tube mounting pad on the air-inlet case.

High-pressure compressor-discharge air, taken from the leading edge of the accessory case, is piped forward through the regulator and air valve to a well formed in the air-inlet case at the inlet-guide-vane location. From this well the air passes inward through the hollow inlet guide vanes into a space formed by the inside diameter of the inlet-guide-vane inner shroud and the rear outside diameter of the compressor-front-bearing support. From this chamber the air is forced forward through four passages in the compressor-front-bearing support into the four upper air-inlet struts. The two lower struts are not subject to icing, since oil circulates through these during engine operation. The airflow is discharged into the slipstream through ports located at the outer ends of the four inlet case struts.

An **anti-icing regulator** is provided with the anti-icing tube assembly to regulate automatically the flow of anti-icing air with changing temperature. A bimetallic coil secured to an inner valve expands at increased temperature, thereby moving the valve toward the closed position and restricting the flow of air through the regulator. At 70°F the slots in the valve are so aligned as to allow a maximum of air to pass. As the air temperature increases, the bimetallic coil rotates the valve, causing the slot openings to decrease until at 560°F the slots are so positioned as to allow only a minimum amount of anti-icing air to pass.

The anti-icing air valve is opened and closed by energizing an electric motor. When icing conditions are encountered, a switch is thrown to energize an electrically operated valve which allows anti-icing air to pass forward. The electric control system is shown in Fig. 17 · 17.

Water-injection system

The water-injection system consists of a water regulator and shutoff-valve assembly, water-spray ring, fuel-control water compensator, automatic air vent and drain valves, and external and internal tubing connecting the various components.

The maximum metered water flow for all operations is approximately 6,500 lb per hr. The flow control consists of a throttling-type water regulator and shutoff valve which holds a fixed pressure drop across a fixed orifice. A schematic diagram of this system is shown in Fig. 17 · 18.

Water is piped from the water tank through a 100-mesh screen to the inlet side of the hydraulically balanced shutoff and recirculating valve in the water regulator. When the valve is in the OPEN position, all the metered flow is directed to the engine, and when the valve is CLOSED, the water circulates back to the tank, thereby eliminating pressure surges because the system is never dead-ended.

Metered water from the water regulator and shutoff valve is piped through external tubing and cored passages in the air-inlet case and front-bearing-support housing, where it then enters the spray manifold ring and is sprayed into the compressor airstream. Connections are provided in the metered water chamber of the regulator for a cockpit water-injection indicator and a signal line to the fuel-control water compensator.

The fuel-control water compensator is mounted on the bottom of the fuel-control unit. Its function is to enrich the fuel flow when water is flowing to the engine. Metered water pressure

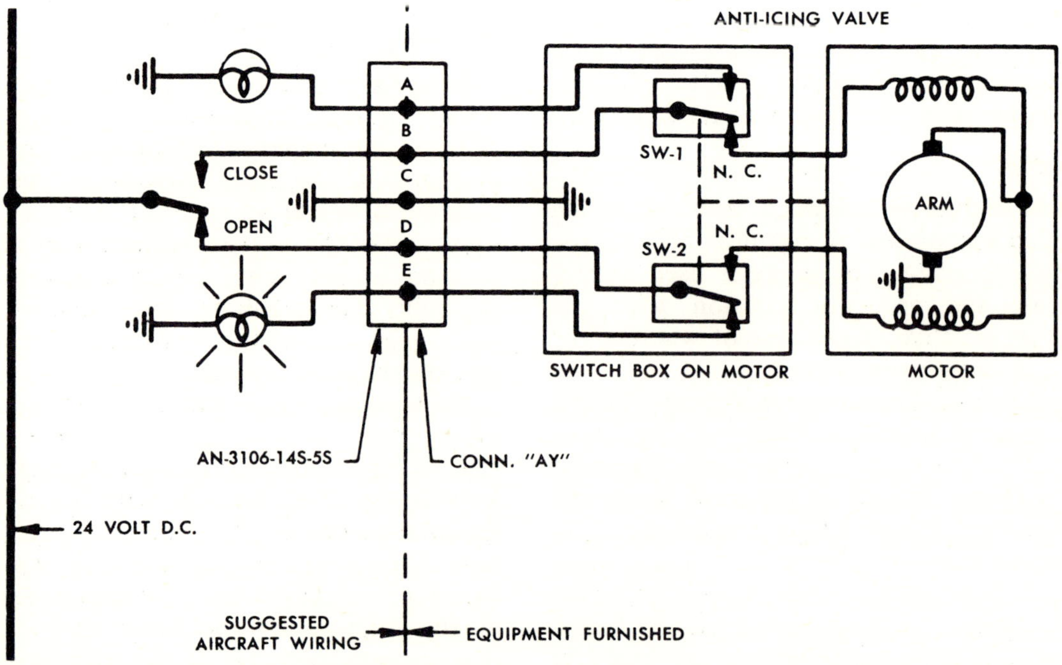

Figure 17 · 17 Anti-icing control circuit.

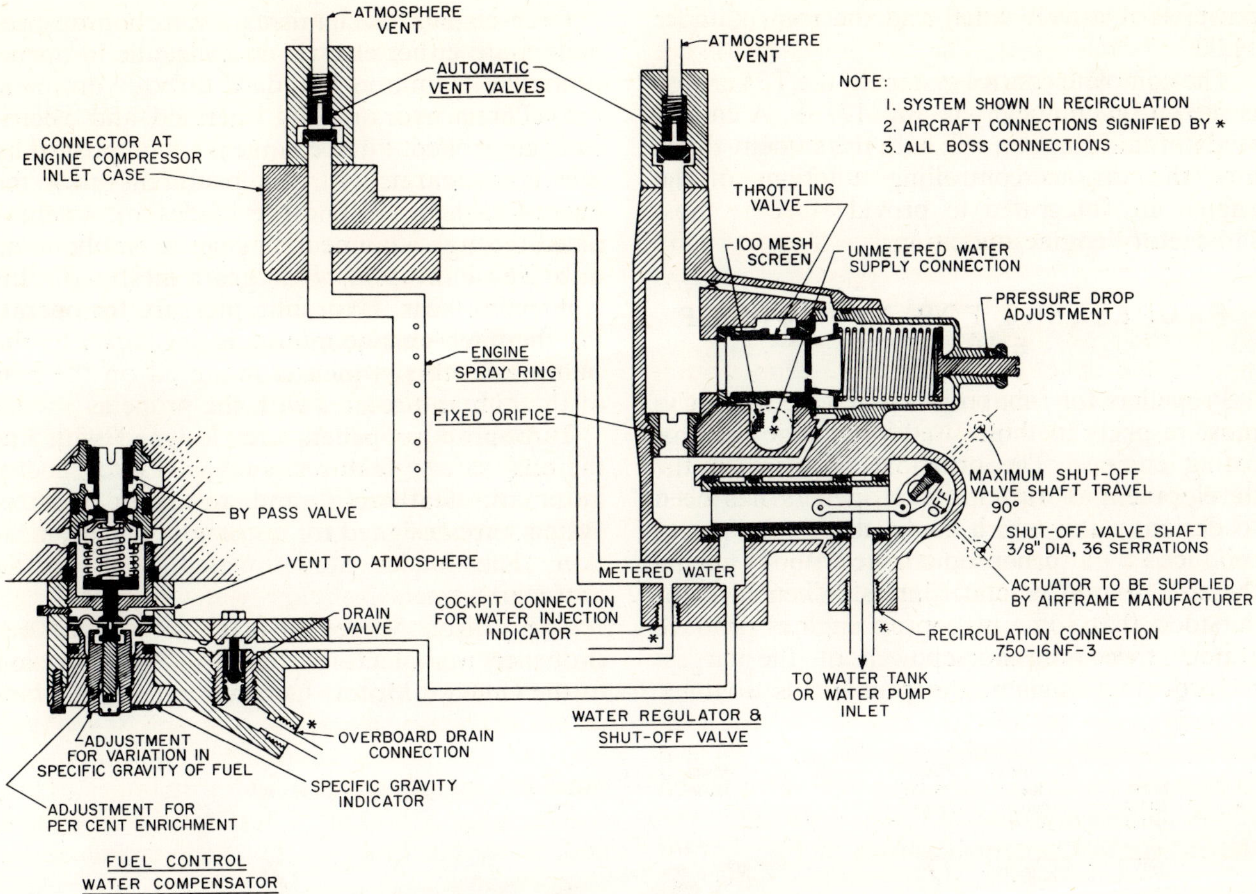

Figure 17·18 Water-injection system.

from the regulator acts on a diaphragm to change the tension on the fuel-control bypass valve spring, thus increasing the fuel flow.

Automatic vent valves and drain valves are incorporated in the water-injection system to drain the system completely after the completion of injection, to prevent damage to the components from freezing.

Fuel system and power control

Engine power is controlled by rotating a single shaft on the fuel-control unit. This shaft actuates both the fuel control and the propeller governor. Another shaft on the fuel-control unit controls a two-position on-off fuel shutoff valve.

The overall function of the fuel-control unit is to sense mass airflow and meter fuel to produce a turbine operating temperature sufficient to satisfy the selected power requirement. Mass airflow is not measured directly, but since it is a function of inlet-air temperature, engine rpm, and burner-case pressure, the combined sensing of these variables serves as the measure to which the fuel flow is metered. Fuel thus metered provides the desired turbine inlet temperature under all operating conditions, and it never falls below the minimum required to sustain combustion.

The fuel control meters both the maximum and minimum quantity of fuel which can be used by the engine in such a manner that the operating limits of the engine will not be exceeded. Thus, it is permissible to move the power-control shaft rapidly in either direction between the FLIGHT IDLE and TAKEOFF positions for any flight maneuvers. Likewise, for ground operation it is permissible to move the power shaft rapidly between the FULL REVERSE and FLIGHT IDLE positions. At 0°, or full clockwise position of the shaft, full reverse thrust is obtained. This occurs at 10,750 engine rpm with full reverse propeller pitch. **Ground idle** occurs at a control shaft position of 22½°.

When the power-control lever is set at FLIGHT IDLE position, the rpm is approximately 9,800. From flight idle to takeoff power, which is at the 90° position, the engine rpm is 11,000 and the speed is controlled by the propeller governor.

When the engine is operating at speeds below the propeller governing range, it is said to be in the "beta" range. In the beta range, the engine power is relatively small and the rpm is under 11,000.

The complete control system of the T34 engine is shown schematically in Fig. 17 · 19. A careful study of this diagram will help the student to see how the various controlling functions of the engine are integrated to provide for the most satisfactory engine operation.

PROPELLERS FOR TURBOPROP ENGINES

Propellers for tuboprop engines are similar in most respects to those used with large reciprocating engines. The principal problem in the development of turboprop propellers has been to design a unit which could absorb the power produced by a turboprop engine without having the size beyond practical limits. When it is understood that some turboprop engines produce almost twice the horsepower of the largest reciprocating engines, the problem is obvious.

Pitch-change mechanisms for turboprop propellers are either electric or hydraulic in operation. The Hamilton Standard turbo-hydromatic propeller is hydraulically operated and electrically controlled. Pitch change is accomplished by means of separate hydraulic piston motors in the butt of each blade, and the blades are synchronized through a connecting gear assembly in the hub. The individual blade gears mesh with this connecting gear. Hydraulic pressure for operating the pitch-change motors is developed by the pump assembly, which is mounted on the rear of the hub and rotates with the propeller.

Turboprop propellers are designed with numerous safety features such as pitch locks, automatic feathering, and reversing. Control systems are designed for automatic synchronization. The blades are provided with electric anti-icing systems.

The Allison Model A6441FN-606 and 606A propeller, manufactured by the Allison Division of the General Motors Corporation, is used with

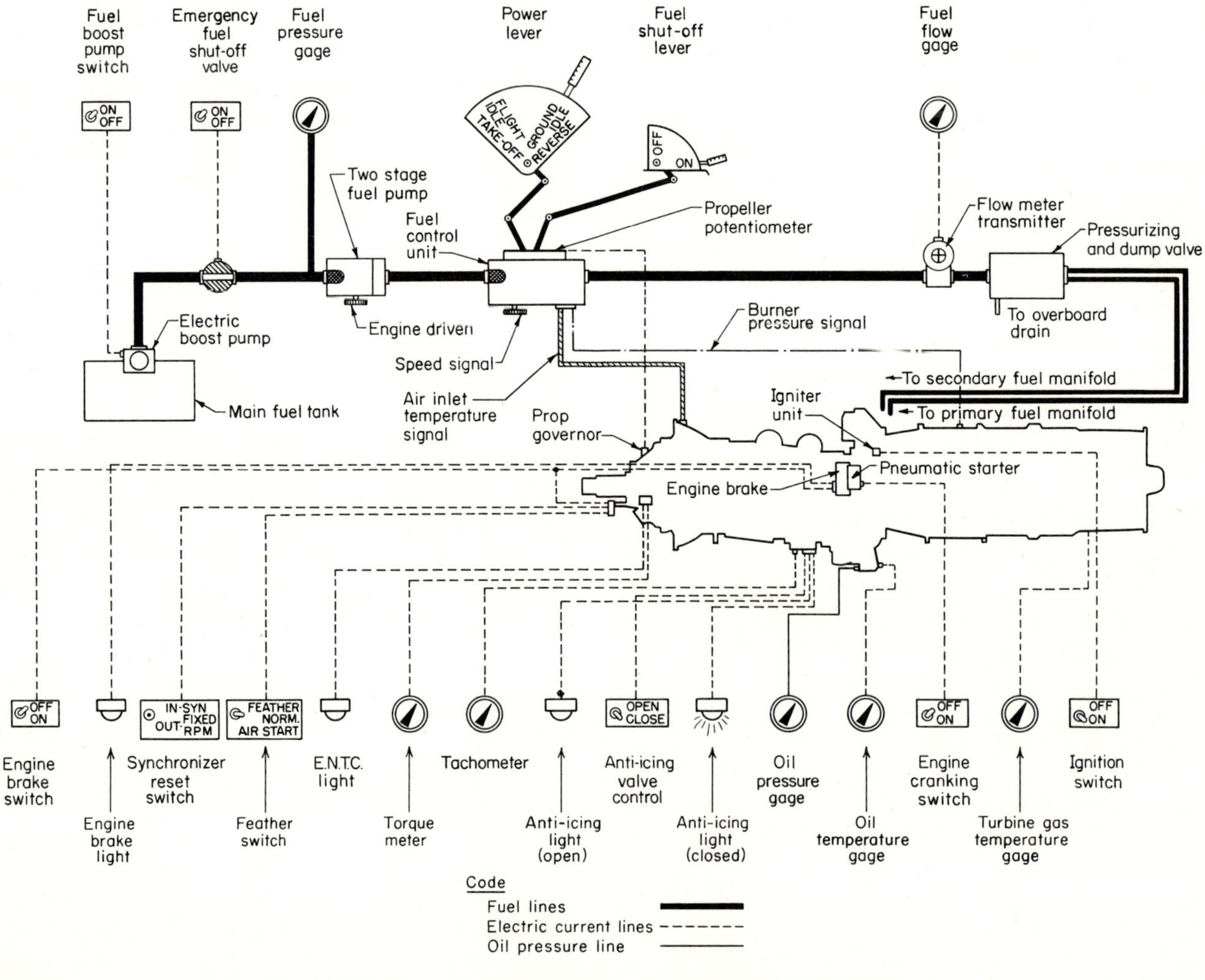

Figure 17 · 19 Schematic diagram of complete control system.

the Allison Model 501-D13 turboprop engine installed in the Lockheed Electra airliner. This propeller is hydraulically operated for normal conditions, feathering, and reversing, with power from four hydraulic pumps mounted within the rotating regulator section of the propeller. The pumps are driven by the rotation of the propeller, with the pump gears engaging an external fixed gear which is mounted on the adapter and surrounds the propeller shaft.

The speed-sensitive governor is mounted in the rotating regulator section and directs hydraulic fluid to four torque units, one unit being mounted on the root of each blade. The torque unit consists principally of a fixed spline, a piston, and a cylinder. The helical splines convert the linear motion of the piston to the rotation of the cylinder and blade. The blades are synchronized by means of a master gear to which all the blade gears are engaged.

The hydraulic fluid for the propeller operation is contained within the regulator assembly and the feathering tank at the front of the hub. An electrically driven feathering pump is mounted within the feathering reservoir, and the pump motor is mounted on the rear wall of the feathering tank.

The propeller is provided with an automatic feathering system which acts upon takeoff if the negative torque signal system indicates that an engine has failed. It also includes a negative torque control system to reduce propeller drag as explained previously.

STUDY QUESTIONS

1. Compare the turbine design for a turboprop engine with that of a turbojet engine.
2. Describe two different configurations for turboprop engines.
3. List the advantages of a turboprop engine.
4. How does airspeed affect the power output of a turboprop engine?
5. What are the three main sections of a Pratt & Whitney T34 turboprop engine?
6. Describe the compressor, combustion chamber, and turbine for a T34 engine.
7. Give the power output, specific weight, and specific fuel consumption for the T34 engine.
8. Describe the operation of the torquemeter.
9. Explain the operation of the negative torque control.
10. Explain the importance of the fuel-control unit.

CHAPTER 18

Commercial Turboprop Engines

The purpose of this chapter is to describe some commercial turboprop engines currently used in the United States. It is not intended to imply that these are the only such engines that will be used in the United States in the near future. As a matter of fact, there are many excellent turboprop engines in production in the United States, Great Britain, and other countries, and it is certain that turboprop engines other than those described in this chapter are now used or will be selected to power commercial aircraft in the United States.

The engines described in this chapter are the Allison Model 501, the Rolls-Royce Dart, the Lycoming T53, and the General Electric T58. The T53 and the T58 engines are suitable for use either as turboprop or turboshaft engines. The Allison 501 (military model T56) is an American-built axial-flow gas-turbine powerplant coupled to a reduction-gear assembly to drive a propeller. The Rolls-Royce Dart engine is a British- or Canadian-built 2-stage centrifugal-flow engine, with the propeller reduction gearing mounted directly on the front of the engine.

At present, the Allison Model 501 engine powers the Lockheed Electra airliner, shown in Fig. 18 · 1. The Rolls-Royce Dart is used in the Vickers Viscount airliner, shown in Fig. 18 · 2, the F-27 airliner, and the Grumman Gulfstream business aircraft. The Lycoming T53 powers the Grumman AO-1 Mohawk and certain helicopters, and the General Electric T58 is used primarily for helicopters at the present time.

Figure 18 · 1 Lockheed Electra airliner. (Lockheed California Co.)

Figure 18 · 2 Vickers Viscount airliner. (Continental Airlines.)

THE ALLISON MODEL 501 TURBOPROP ENGINE

General description

The Allison Model 501 turboprop engine is a conventional axial-flow gas-turbine engine consisting of a single-spool 14-stage compressor, a cannular combustion chamber, and a 4-stage turbine. The power unit is connected by an extension shaft and a supporting structure to a reduction-gear assembly having a single propeller shaft. Figure 18 · 3 shows an external view of the engine with the reduction-gear assembly and a cutaway view of the power section.

The military versions of the Model 501 engine are arranged with the extension shaft joined to the bottom of the reduction-gear case. This places the reduction-gear assembly above the center line of the engine. The commercial model, 501-D13, has the reduction-gear case inverted so that the extension shaft joins the top of the gear case, thus placing the gear case below the center line of the engine. This arrangement provides for the air inlet at the top of the nacelle, thus reducing the possibility of drawing foreign matter into the engine.

With the exception of the air-inlet housing and accessories housing, which are magnesium-alloy castings, the remainder of the power section is constructed entirely of steel. This, and employing the outer combustion shell as a structural load-carrying member, has resulted in a power section assembly of great ruggedness and minimum weight. Development testing has shown that in the event of compressor or turbine blade or vane failure, the case has sufficient strength to contain them. Also, the case is more corrosion-resistant than lighter-weight alloys.

The major components of the 501 engine are shown in Fig. 18 · 4. In our discussion of the engine, we shall consider these sections one at a time beginning at the front of the engine. The front of the engine is the propeller end. Rotation is clockwise and counterclockwise, looking at the engine from the exhaust end. Right and left of the engine are also considered as looking at the engine from the exhaust end.

Specifications and performance

The following specifications and performance ratings apply to the Allison 501 engine, and with some minor variations also apply to the military model T56.

Width	27.0 in.
Height	36.0 in.
Length	145.0 in.
Frontal area	5.4 sq ft
Weight (without propeller)	1,750 lb
Weight/power ratio	0.47 lb/eshp
Fuel consumption (max continuous)	0.55 lb/eshp/hr
Oil consumption (normal)	3.0 lb/hr
Fuel specification	JP-1 or JP-4
Oil specification	MIL-L-7808B
Power, takeoff, static	3,750 eshp or 3,460 shp plus 726 lb thrust

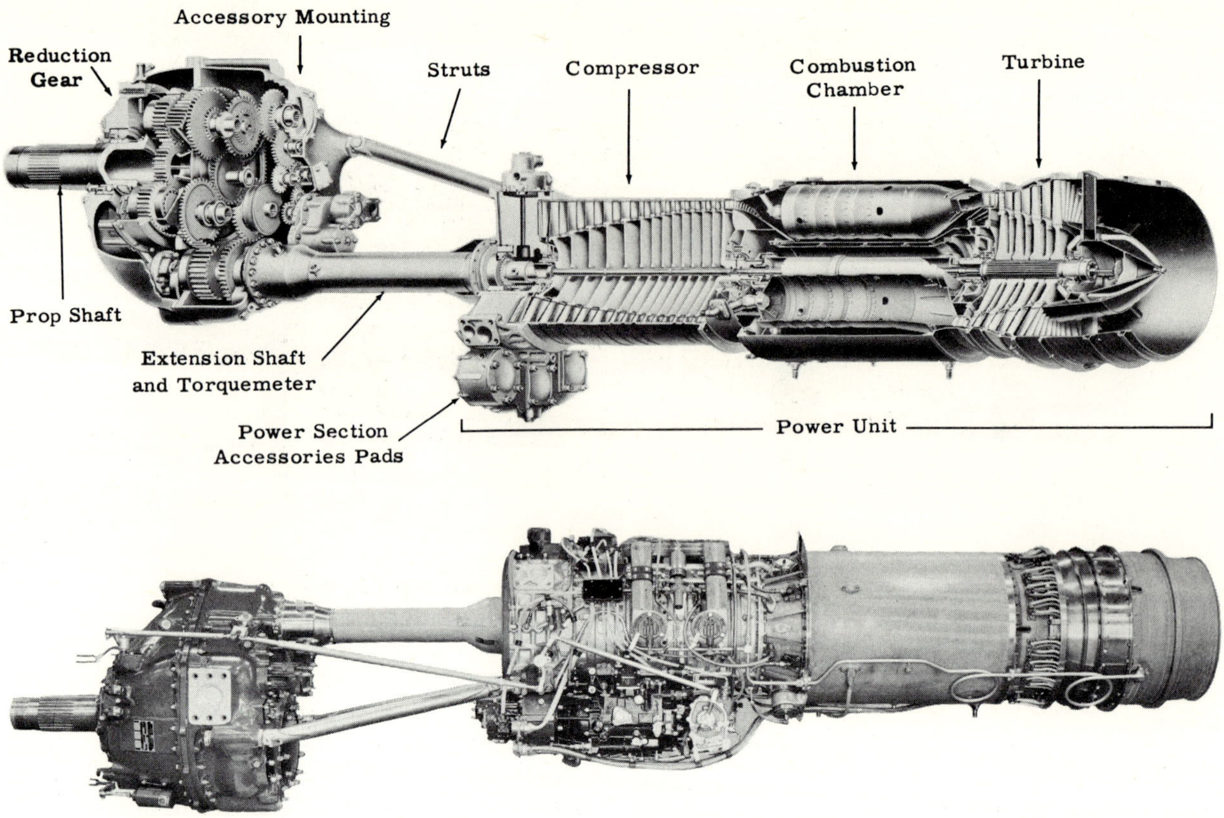

Figure 18 · 3 External and cutaway view of the Allison Model 501 turboprop engine. (Allison Division, General Motors Corp.)

Reduction-gear assembly

The reduction-gear assembly for the 501 engine is shown in Fig. 18 · 5. It is attached to the power section by means of the extension shaft and two struts.

The reduction-gear assembly incorporates a single **propeller drive shaft,** an independent **dry sump oil system,** a **safety coupling assembly,** a **propeller brake,** a **negative-torque-signal system,** and the necessary gearing arrangement.

The main reduction-gear train consists of two stages of reduction: the spur-gear-type first stage, with a reduction ratio of 3.125:1, and the planetary-type second stage with a reduction of 4.333:1. These two stages result in an overall

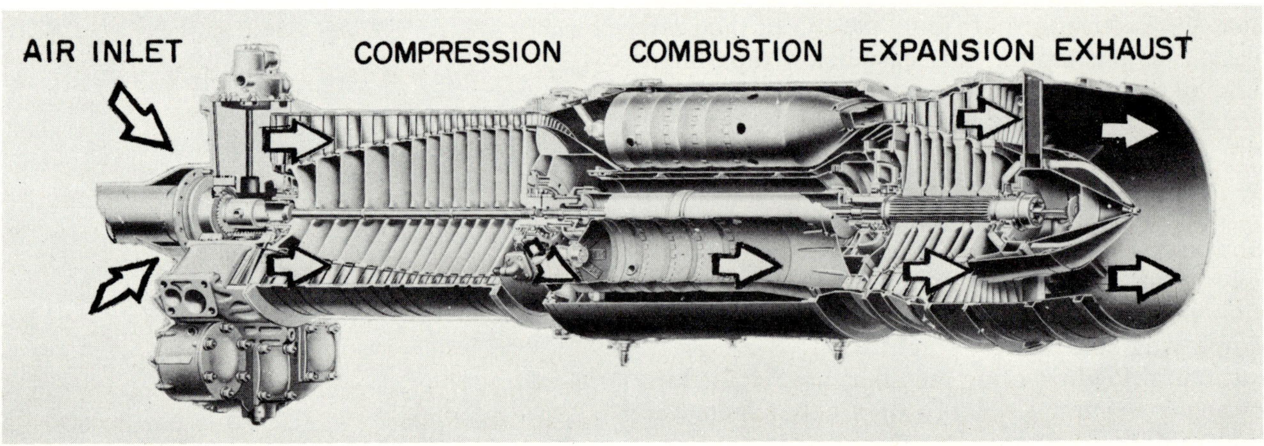

Figure 18 · 4 Major components of the Allison Model 501 engine.

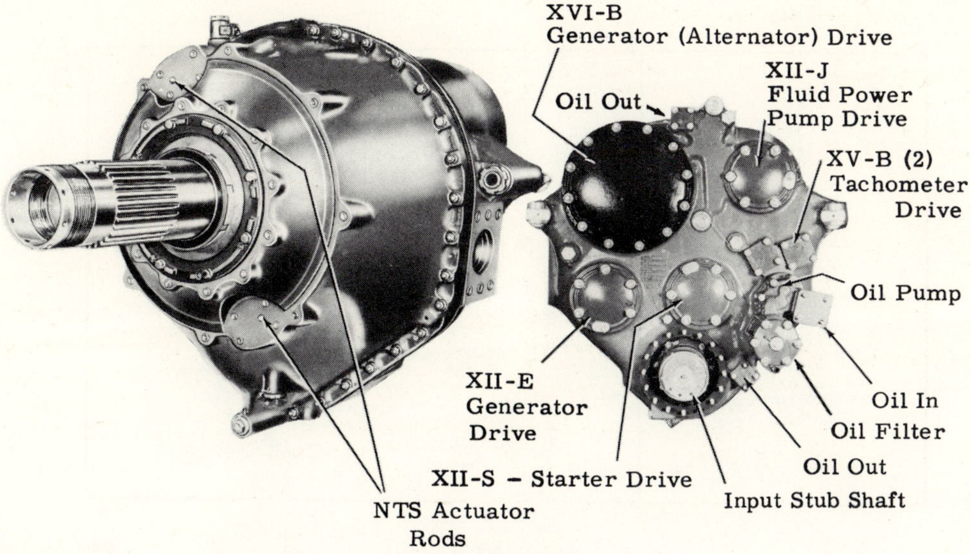

Figure 18 · 5 Reduction-gear assembly. (Allison.)

reduction of 13.540:1 and a propeller speed of 1,020 rpm at 13,820 engine rpm.

The **reduction-gear spur pinion,** which is driven counterclockwise by the power section through a safety coupling, drives the first-stage reduction gear. The **sun gear** is bolted to the first-stage reduction gear and meshes with the five **planet gears.** The planet gears mesh with the **stationary ring gear,** causing the **planet carrier** to rotate. The planet carrier is bolted to the propeller shaft. Therefore, the rotation of the planet carrier causes the propeller shaft to be driven clockwise.

The 501 engine is equipped with a negative torque signal (NTS) to prevent excessive negative torque and propeller drag. Negative torque develops when the engine power is reduced and the airflow drives the propeller, which turns the engine.

The NTS control system provides a signal which causes propeller pitch to increase as necessary to limit the negative torque. When a predetermined torque (105 ft-lb at the power section) is applied to the reduction gear, the stationary ring gear moves forward against spring force because of a torque reaction generated by helical splines. This movement causes two rods to be pushed forward through the reduction-gear nose. These rods actuate switches which direct current to the feathering motor. The feathering cycle continues until the reverse torque is relieved and the propeller returns to normal operation. External pads are provided at each of the actuating rod locations for mounting an electrical switch, hydraulic actuator, or mechanical linkage as required by the type of propeller. The engine components of the NTS system are shown in Fig. 18 · 6.

The Allison Model 501 engine employs a friction-cone-type propeller brake which acts through gears on the first-stage reduction pinion shaft. The brake is held in the OFF position by oil pressure acting on a piston. As speed is reduced and oil pressure drops, the braking surfaces are brought into contact, thus increasing the rate of propeller deceleration until the propeller stops. The brake is always ON when the engine is not running.

Helical splines are provided between the shaft and outer member, causing the brake to disengage when starter torque is applied during starting. With a propeller installed, starter torque is adequate to overcome the spring force and disengage the brake until oil pressure builds up enough to hold the brake in the OFF position. On shutdown, propeller-brake engagement begins at approximately 3,200 rpm.

The inner member of the brake operates on a helical spline. A small reverse movement of the propeller causes the brake to be self-energizing and thus prevents propeller rotation in the reverse direction. When reverse torque is removed from the brake, a spring forces the inner member to its normal brake-engaged position.

A **safety coupling** is provided on the Model 501 engine to disengage the reduction gear from the power unit in the event that the power unit is operating below a preset negative torque value considerably above that of the NTS. Details of the coupling are shown in Fig. 18 · 7. The cou-

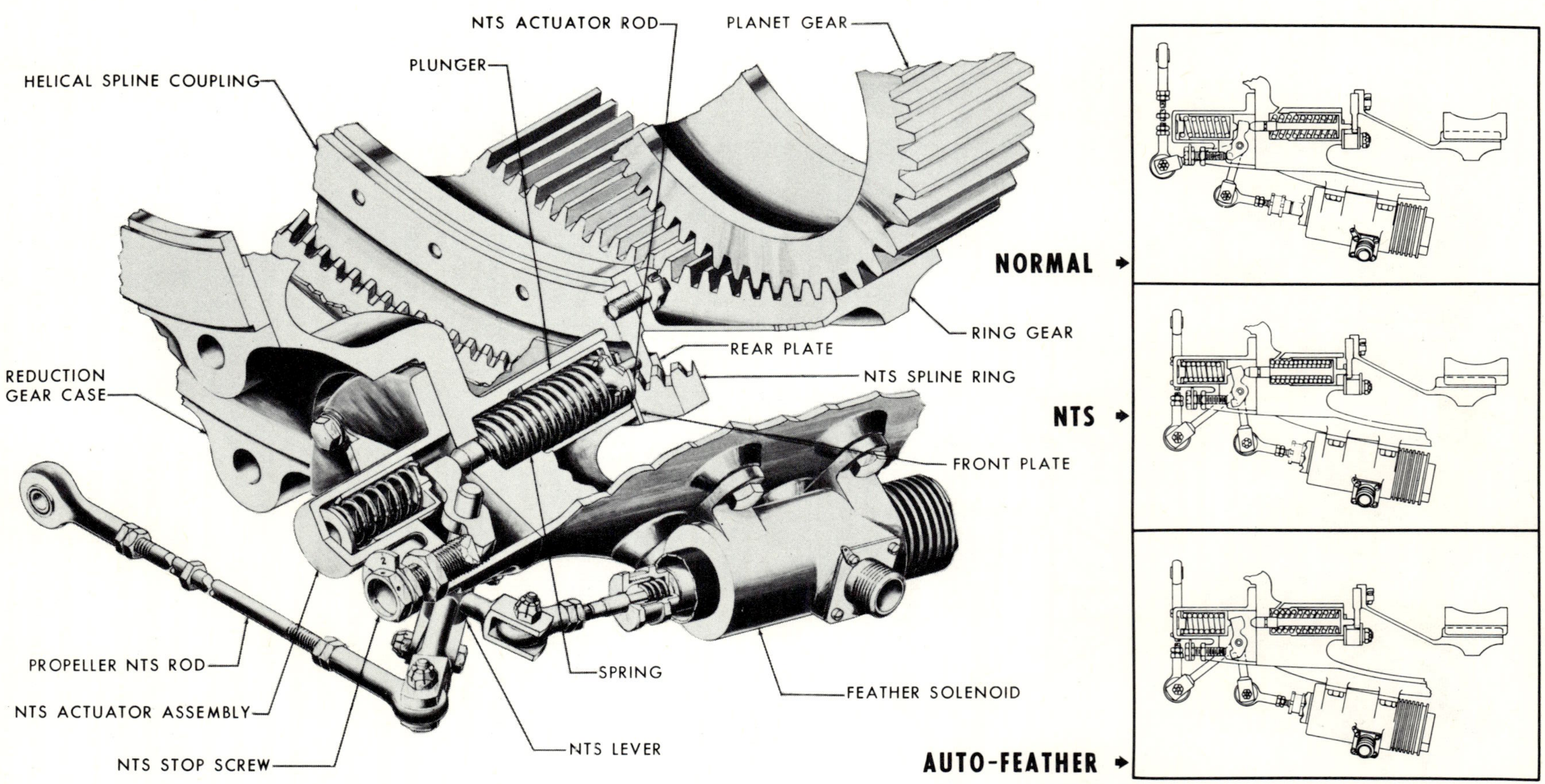

Figure 18·6 Negative torque signal device.

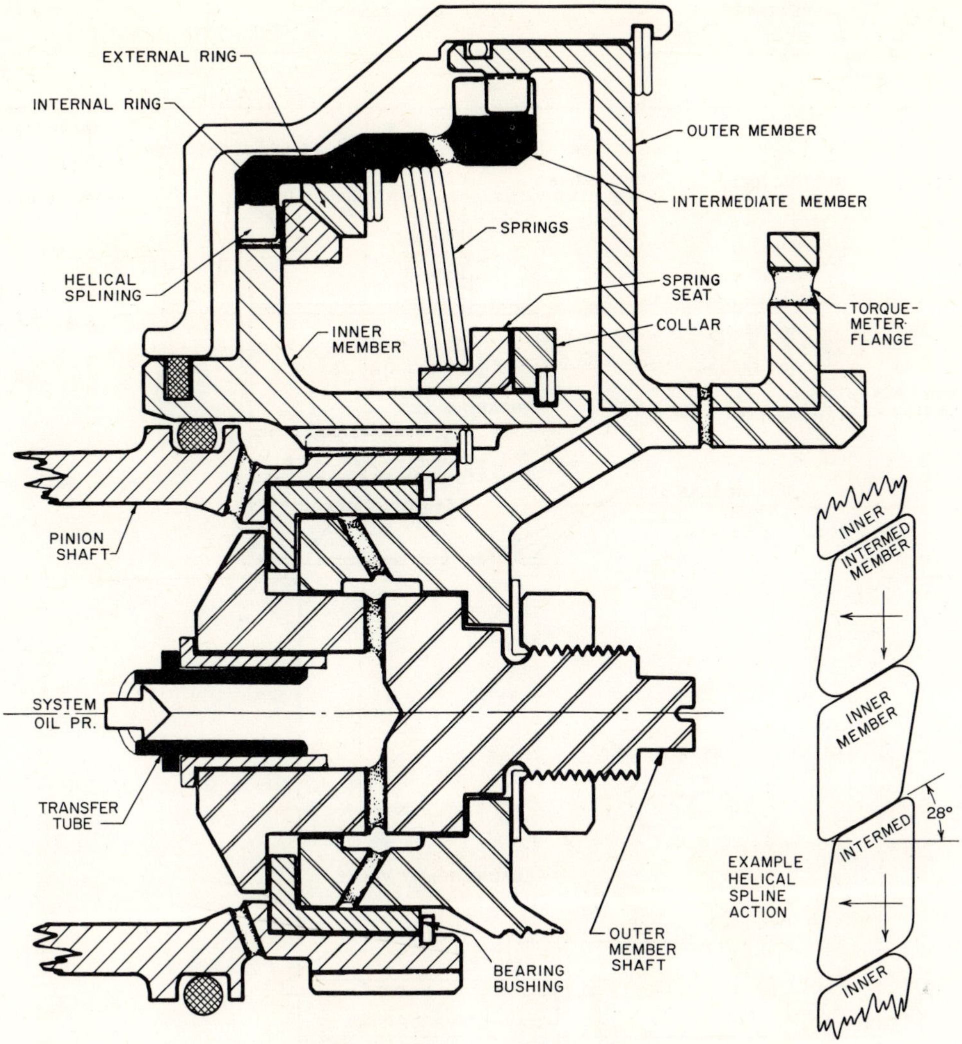

Figure 18 · 7 Safety coupling.

pling consists essentially of an inner member splined to the pinion shaft, an outer member bolted to the extension shaft, and an intermediate member which is connected to the inner member through helical teeth and to the outer member through straight teeth. The reaction of the helical teeth tends to move the intermediate member forward and into mesh when positive torque is applied and rearward and out of mesh when negative torque is applied. Thus, when a predetermined negative torque limit is exceeded, the coupling members disengage automatically. Reengagement is also automatic during feathering or power section shutdown. The safety coupling will come into play only in the event of a serious power-section failure, after which it must be inspected before returning to service.

In the event that the engine portion of the NTS system failed prior to engine failure, the safety coupling would become disengaged. This would result in a windmilling drag horsepower of approximately 35 propeller shaft horsepower after passing through a drag horsepower transient of approximately 1,200 shp for a fraction of a second.

The reduction-gear assembly is provided with an independent oil system which includes a main oil pump, a scavenge oil pump, and a nose scavenge oil pump. A diagram of the system is shown in Fig. 18 · 8. Oil for the system may be provided from the main oil tank which also supplies the power section of the engine.

Extension shafting and tie struts

The power section of the engine drives the reduction-gear assembly through an **extension shaft** and **torquemeter assembly,** which is shown in Fig. 18 · 9. The reduction-gear assembly is attached to the power unit by a pair of struts serving as the bottom supports and a torque-

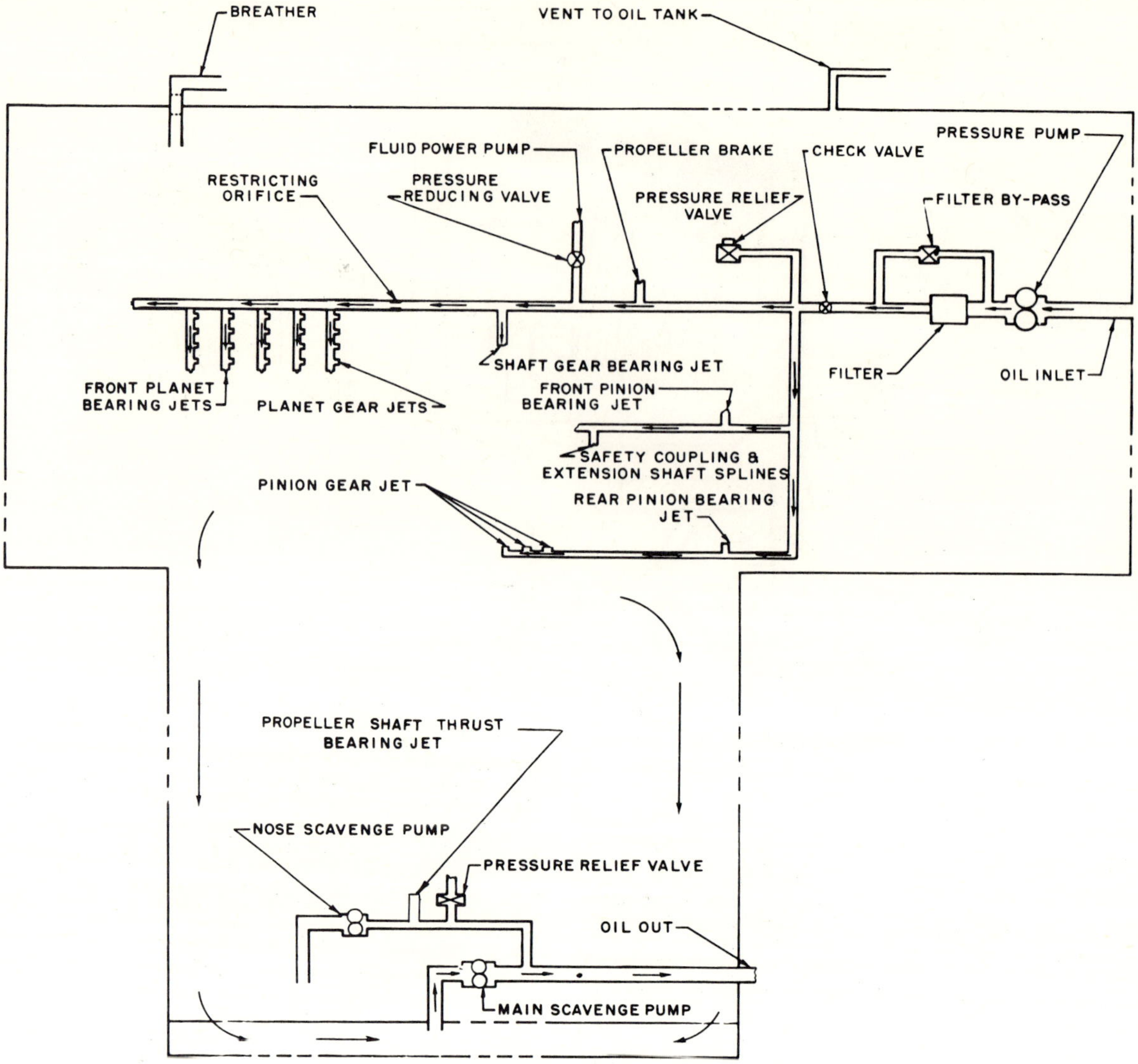

Figure 18 · 8 Reduction-gear oil system.

meter housing serving as the top support. On the military Model T56, this arrangement is inverted; the tie struts are at the top, and the extension shaft and torquemeter are at the bottom.

The **torquemeter** operates on the principle of accurate measurement of torsional deflection that occurs in any power-transmitting shaft. In this case, the torsional deflection occurs in the extension shaft. This torsional deflection is detected by magnetic pickups, measured electronically, and registered in terms of torque or horsepower, whichever is desired.

Two concentric shafts make up the shafting assembly. The inner shaft, or **torque shaft,** carries the load that produces the measured twist. The outer shaft, or **reference shaft,** is rigidly connected to the torque shaft at the drive input end only and provides a no-load reference used to compare the torsional deflection of the inner shaft.

There are separate flanges on both the torque and reference shafts at the reduction-gear-assembly end, as shown in Fig. 18 · 10. Rectangular exciter teeth are machined in line on each flange to enable the pickups to detect the relative displacement of the two flanges. The torquemeter housing assembly is composed of a **torque housing and pickup assembly.** The housing serves as a rigid lower support between the power unit and the reduction-gear assembly and provides mounting for the pickup limit assembly at the reduction-gear end.

The torquemeter signal is developed by means of magnetic pickups mounted over the teeth of the torque and reference shaft flanges. These pickups sense electrical impulses generated by the passage of each exciter tooth. Since the pick-

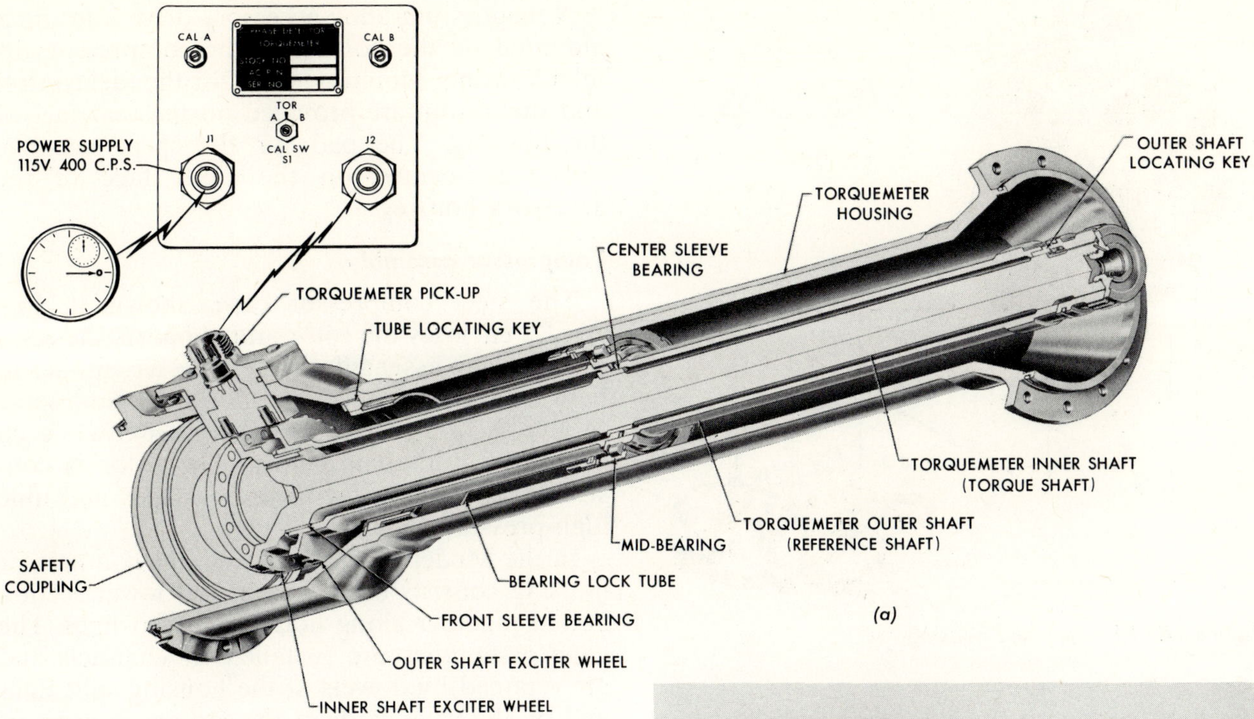

(a)

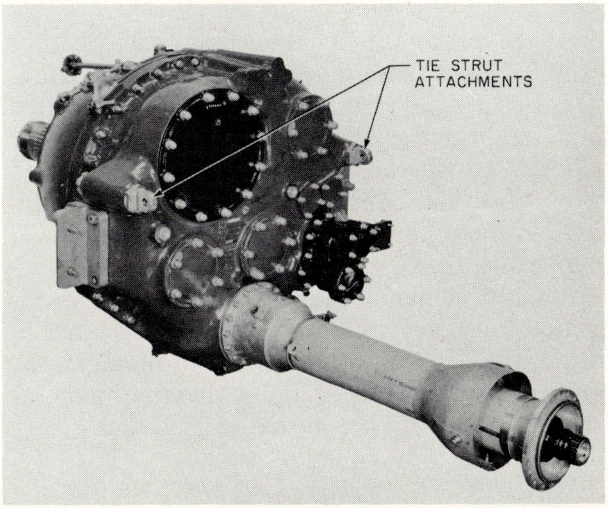

(b)

Figure 18 · 9 Extension shaft and torquemeter assembly.

ups are displaced circumferentially with respect to each other, the impulses from the pickup over the teeth in the reference flange and the impulses from the pickup over the teeth in the torque flange are slightly out of phase at zero load, thereby allowing measurement of both a positive and negative torque condition. As the torque shaft deflects under a load, the teeth on the torque flange are displaced with respect to the teeth on the reference flange, thus causing an increase in the phase displacement between

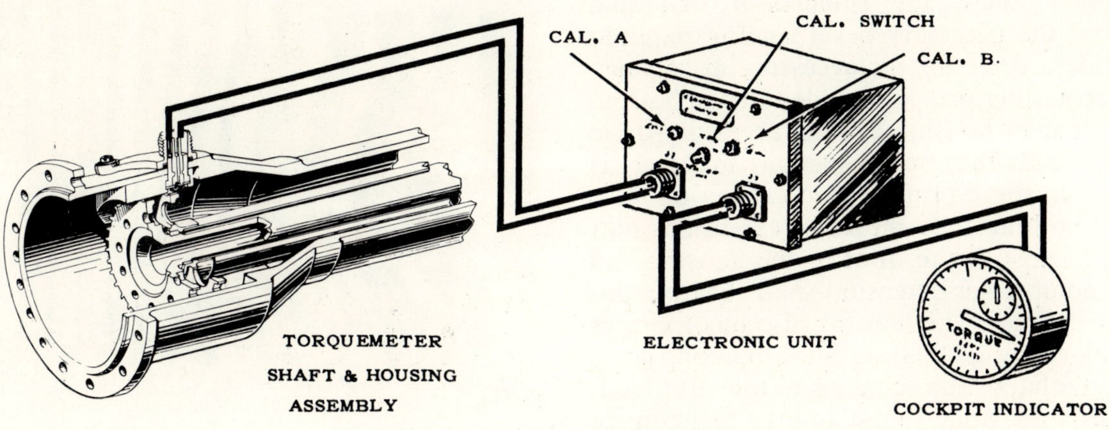

Figure 18 · 10 Torquemeter system.

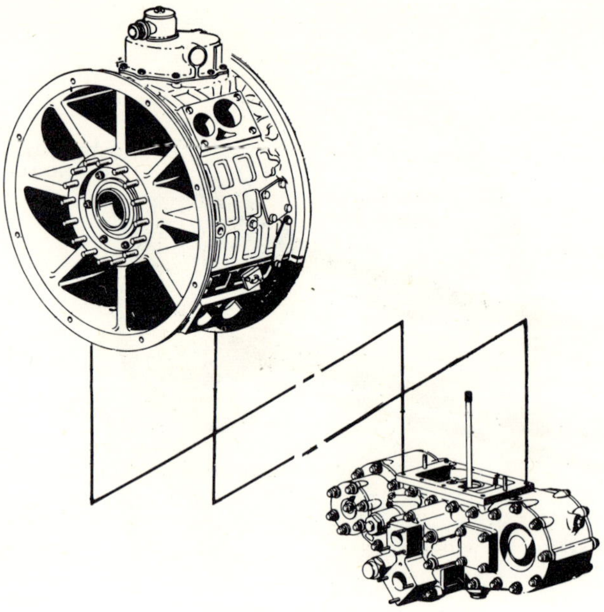

Figure **18 · 11** *Air-inlet housing.*

impulses produced by the magnetic pickups mounted over these flanges. This phase angle is measured by the phase detector and transmitted to a calibrated **synchro indicator.** The power required to operate the phase detector is 115 volts 400 cycles. The synchro indicators are located in the cockpit and require operating power of 26 volts at 400 cycles.

The torquemeter should operate with an indicated accuracy of ± 160 in.-lb from zero to maximum allowable power. All torquemeter assemblies are calibrated on a dynamometer or equivalent facility before being installed.

Air-inlet housing

The air-inlet housing, shown in Fig. 18 · 11, is a one-piece magnesium casting with eight supporting struts between the center hub and outer ring. It houses the **front compressor bearing,** the **inlet guide vanes,** the **compressor extension shaft,** and the necessary bevel gearing required to provide a drive for the **accessory drive housing.** A mounting pad is provided at the bottom of the air-inlet housing for the accessory-drive housing, and the power-section breather is mounted on top. Spring-loaded carbon oil seals are used both at the compressor-extension-shaft bearing located at the front of the housing and at the compressor-extension-shaft bearing located at the rear. The housing also incorporates anti-icing valves and passages for directing compressor-discharge anti-icing air to the strut leading edges, inlet guide vanes, and the fuel-control pressure- and temperature-sensing probes.

A magnesium-alloy accessory-drive housing is mounted on the bottom of the compressor air-inlet housing. Mounting pads for the fuel control and fuel pump are provided on the rear face of the housing. The pad for the speed-sensitive switch is located on the front face of the accessory housing.

Compressor assembly

The steel compressor case, shown in Fig. 18 · 12, encloses the rotor and supports the compressor **vane assemblies.** The vane assemblies fit between the rows of compressor blades and serve to give the air proper direction as it enters each stage of compressor blades. The rotor is constructed with five low-pressure stages and nine high-pressure stages.

In the Model 501-D10 engine, the compressor housing consists of an upper and lower section bolted together along horizontal split lines. The vane assemblies are installed in channels and are retained by dowels at the housing split lines and by the vane studs at the top and bottom of the housing. The compressor housing for the Model 501-D13 engine consists of four forged-steel quarters permanently bolted together in halves. The vane assemblies are installed in channels in each half and are held securely in place by five bolts.

The inner rings of the vane assemblies support the interstage air seals, which are made of soft aluminum. The seals thus formed prevent air from bleeding back to earlier stages.

The inside of the housing (compressor case) between the vane assemblies is coated with a special mixture of asbestos and aluminum polytherm paint to decrease the blade-tip clearances.

Figure **18 · 12** *Compressor case.*

This coating is sprayed on, baked, and then ground to the finished diameter.

The compressor **rotor assembly,** illustrated in Fig. 18 · 13, consists of 14 wheels secured to each other by serrations at the wheel rims. A continuous circle of compressor blades is pinned to the outer rim of each wheel. The wheels at each end of the compressor rotor provide the hubs for mounting of the bearings. The complete assembly of 14 wheels is held together as a unit by means of a through bolt which passes through the center of the assembly.

The compressor wheels, blades, casing, and vanes are made of steel. Over 30 lb per sec of air is handled by this 14-stage axial-flow compressor at a compression ratio in excess of 9:1 at rated power under standard sea-level static conditions.

Diffuser assembly

The compressor diffuser assembly is bolted to the rear of the compressor housing. It is of welded-steel construction, as shown in Fig. 18 · 14, and serves to diffuse the air from the compressor and direct it into the combustion chamber. In addition to its function of diffusing and directing the compressor-discharge air, the diffuser assembly supports the rear compressor ball bearing, the rear compressor bearing carbon seal, the rear compressor labyrinth seal outer member, the diffuser scavenge oil pump, and six fuel nozzles. The fuel nozzles supply fuel to the combustion-chamber liners and also serve as supports for the front end of the liners.

The diffuser assembly provides external two-bolt connections for the anti-icing air-supply tubes, the air-supply line for closing the compressor bleed valves, the oil sump vent line, the scavenge oil line, the oil sump drain cover, and the rear compressor labyrinth-seal vent lines. It also provides four ports for bleeding off compressor discharge air. These ports are provided with connecting flanges for use with "V-band" clamps. One diffuser bleed port is used as a starting bleed valve to improve engine starting characteristics. This leaves three ports available for aircraft use. Mounting brackets can be provided at either the top or the bottom of the diffuser depending upon installation requirements. A fire shield is provided at the rear diffuser split line. Each bleed port provides a maximum of 4 per cent bleed air. Two or more ports make available 8 per cent bleed air, which is the maximum permissible from each engine.

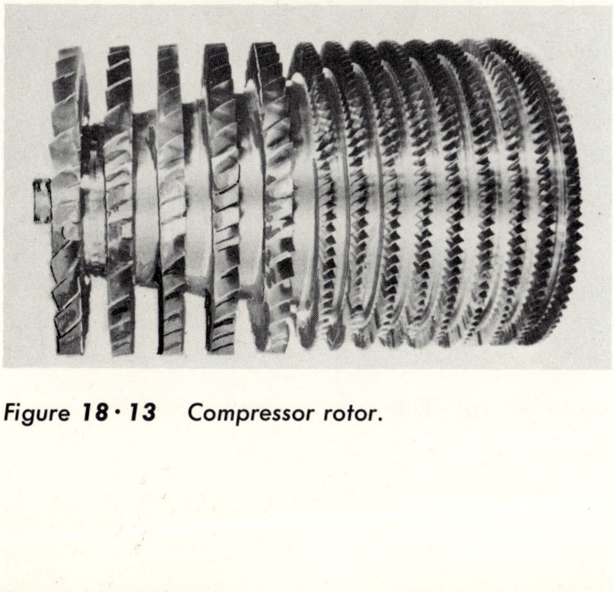

Figure 18 · 13 *Compressor rotor.*

Figure 18 · 14 *Diffuser assembly.*

Combustion-chamber assembly

The combustion-chamber assembly for the Allison Model 501 engine is shown in Fig. 18 · 15. The combustion chamber consists of six individual liners evenly spaced in one annular chamber formed by a one-piece outer casing and a one-piece inner casing. This arrangement is commonly called a cannular or can-annular combustion chamber.

The **combustion liners** are welded assemblies and are located radially by the fuel nozzles at the front and by the turbine-inlet-vane assemblies at the rear. Axial location is established by the igniter plugs in two liners and by "dummy" igniter plugs in the other four liners. Crossover tubes are located between the liners to carry the flame

Figure 18 · 15 Combustion-chamber assembly.

during starting from the two diametrically opposite liners which contain the igniter plugs to the other four liners. The fuel nozzles, which mount on the diffuser, extend into the center of each liner dome.

The combustion-chamber **outer casing** encloses the liners and serves as the supporting structure between the diffuser and turbine. It incorporates two burner drain valves which open when the pressure falls below 2 to 4 psig and drain any fuel which may have accumulated in the combustion chamber. The adapters for the igniter plugs are located in the forward end of the casing.

The combustion-chamber **inner casing** consists of an inner and outer shell which are separated by an air space and welded together to form a one-piece assembly. Both the inner and outer shells have bellows-type expansion joints to allow for differences in expansion and to make the assembly more flexible. The forward end of the inner casing is supported by a sleeve in the diffuser, while the aft end is bolted to the turbine inlet casing. This arrangement allows for differential expansion between the combustion-chamber inner and outer casings.

Practically all the compressed air which enters the combustion-chamber outer casing enters the combustion liners through holes provided for that purpose. A part of this air unites with the fuel and supports combustion. The balance of the air serves to lower the temperature of the hot gases to a safe temperature for the turbine inlet. Some air is directed through holes in the turbine front support to provide cooling for the first- and second-stage turbine wheels and wheel rims.

Turbine assembly

The turbine and exhaust assembly, shown in Fig. 18 · 16, consists of the **turbine inlet casing,** the **turbine vane casings** and **stator assemblies,** the **turbine rotor** and **rear bearing support,** and the **integral exhaust nozzle.**

The four stages of turbine wheels and intermediate spacer disks are spline mounted on a common shaft. All these parts are machined from heat-resistant stainless steel. The turbine blades in the first three stages are precision-cast GMR-235, a nickel-base alloy developed by the General Motors Research Laboratories Division. This alloy has superior high-temperature properties, making it exceptionally well suited for turbine blading; and at the same time it has a low critical-material index. The fourth-stage blades are forged S-816 stainless steel. In the area of the fourth-stage turbine wheel, the materials need not have extremely high temperature resistance. All four stages of blading are loosely fit to the turbine-wheel rims in broached serrations of the five-tooth fir-tree design. The blades are held in place axially by the rotor spacer disks and by wheel-rim lugs on the No. 1 and No. 4 wheels.

The turbine vanes or stators fit between the different rows of turbine blades. Their purpose is to give proper direction to the flow of the hot gases as they pass from one turbine stage to the next. The stator vanes are also constructed of GMR-235 precision castings. They are welded to inner and outer shroud casings. The inner rings form interstage labyrinth seals with the rotating knife edges on the rotor spacer disks. It

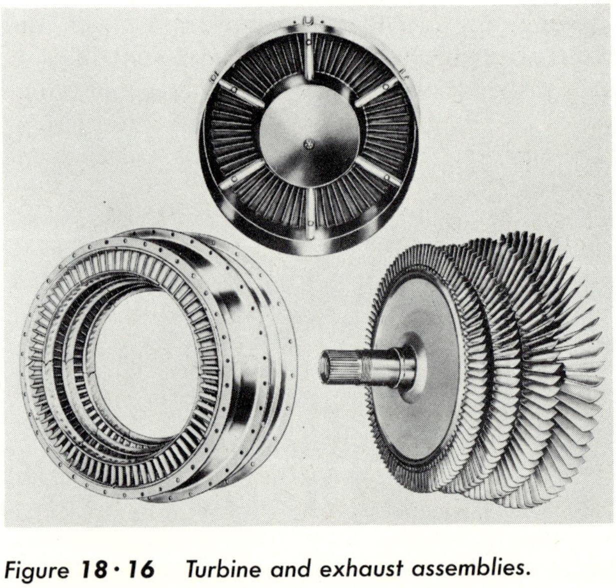

Figure 18 · 16 Turbine and exhaust assemblies.

will be noted that, with the exception of the fourth stage, extensions of the outer shroud rings together with the turbine casings form a double-walled construction. This arrangement minimizes distortion of the casings and appreciably reduces surrounding nacelle temperatures.

The turbine rotor is encased in the two-piece turbine casing assembly which includes the stator vane assemblies. The casing assemblies are bolted to the turbine inlet casing at the front and the turbine rear support at the rear.

The turbine front bearing support plate, which supports the turbine rotor assembly, is attached to the turbine inlet casing. The turbine rear bearing support is attached to the turbine rear vane casing and houses the rear turbine bearing. It also provides support for the turbine scavenge-oil-pump support assembly. The front and rear bearings for the turbine rotor are the roller type.

The turbine casing assembly is made up of the turbine inlet casing and two turbine vane casings, in which are located four turbine vane assemblies. Each turbine vane assembly consists of multiple vane segments which are formed into a full circle by an arrangement of keys and mating slots.

Lubrication system

The Model 501-D13 engine is provided with a low-pressure external oil system as shown in Fig. 18 · 17. This is an independent dry-sump system which includes one combination main pressure and scavenge-oil-pump assembly and three separate scavenge pumps and an oil filter.

The main oil pressure and scavenge pumps are located on the front face of the accessory drive housing. Oil from the aircraft oil tank is delivered to the inlet of the pressure pump, whence it is pumped through a filter and into the external oil lines to be delivered to the various points in the engine where lubrication is required. The pump system is provided with a pressure regulator which bypasses excess pressure to the inlet of the pump. The filter is equipped with a bypass valve to provide for oil flow in case the filter becomes clogged. A check valve in the line leading to the engine prevents the oil-tank pressure from forcing oil into the engine when the engine is shut down.

Pressure oil from the pump flows through the vertical strut into the compressor inlet assembly where it furnishes lubrication for the extension shaft bearing, the compressor front bearing, the compressor side gear, and the side gear bearings. Surplus oil returns by gravity to the accessory drive housing. Oil for the compressor rear bearing in the diffuser section and for the turbine front bearing flows through the external oil line and through a strut into the center of the diffuser section. Surplus oil drains to the diffuser sump and the turbine-front-bearing sump, where it is picked up and returned by a scavenge pump in each sump. The same external oil line which supplies the diffuser section and the turbine front bearing also carries oil to the turbine section for the lubrication of the turbine rear bearing. Sumps for collecting surplus oil are located at both the front and rear turbine bearings. Scavenge pumps at both locations return the oil to the main return line. The rear turbine scavenge pump returns the oil through the turbine coupling shaft and through the center of the compressor. Drilled holes in the shaft permit oil to flow out to lubricate the front turbine seal and the shaft couplings. Oil from the rear turbine scavenge pump flows entirely through the center of the turbine and compressor tie bolts and to the forward end of the extension shaft. At this point it drains into the accessory drive housing sump.

Engine control system

The system used to control a turboprop engine must be quite complex, since the engine is affected by many variables. Among the conditions which affect the engine operation are **rpm, required power output, air-inlet temperature, air pressure, turbine inlet temperature,** and **aircraft speed.** The control system must be designed to sense all these conditions and to regulate fuel flow and engine rpm so that the engine operating limits are not exceeded.

The system and units described in this chapter represent a typical system; however, when the Model 501 engine is installed in different types of aircraft, changes will be evident. A simplified diagram of the system is shown in Fig. 18 · 18a. The arrangement for the Model 501-D13 engine is shown in Fig. 18 · 18b.

The control system of the Model 501 engine is the single-lever type and includes the following major components: the **coordinator,** the **propeller,** the **control linkage,** the **speed-sensitive switch,** the **datum control valve,** the **electronic temperature datum control,** the **thermocouples,** the **relay box,** and the necessary wiring.

The propeller system includes a mechanical governor for the control of propeller speed in the flight range as a standby. A single synchronizer unit for all powerplants in multiengine aircraft is the main speed-governing device. Propeller blade angle in the taxi range is derived by

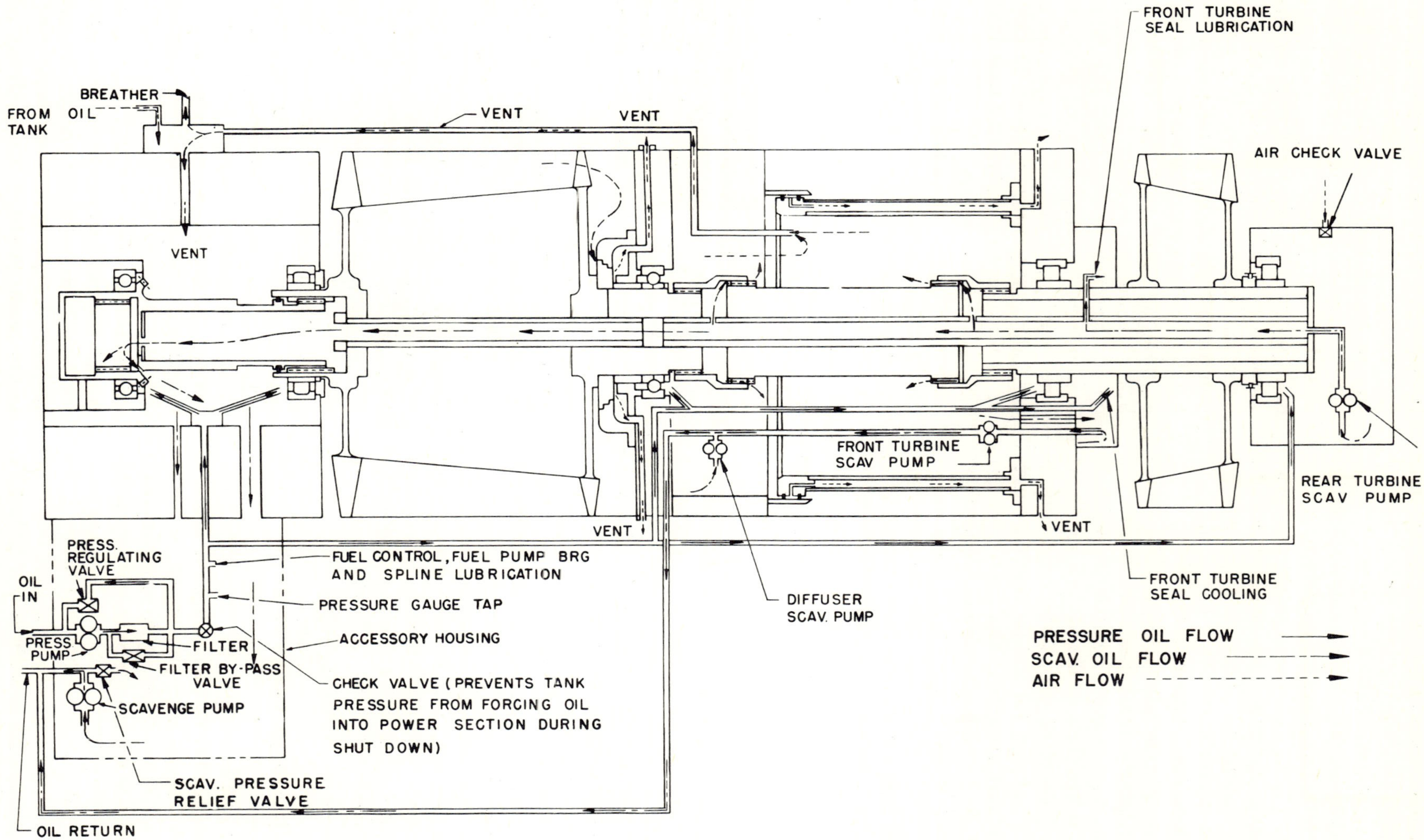

Figure **18·17** *Model 501-D13 oil system.*

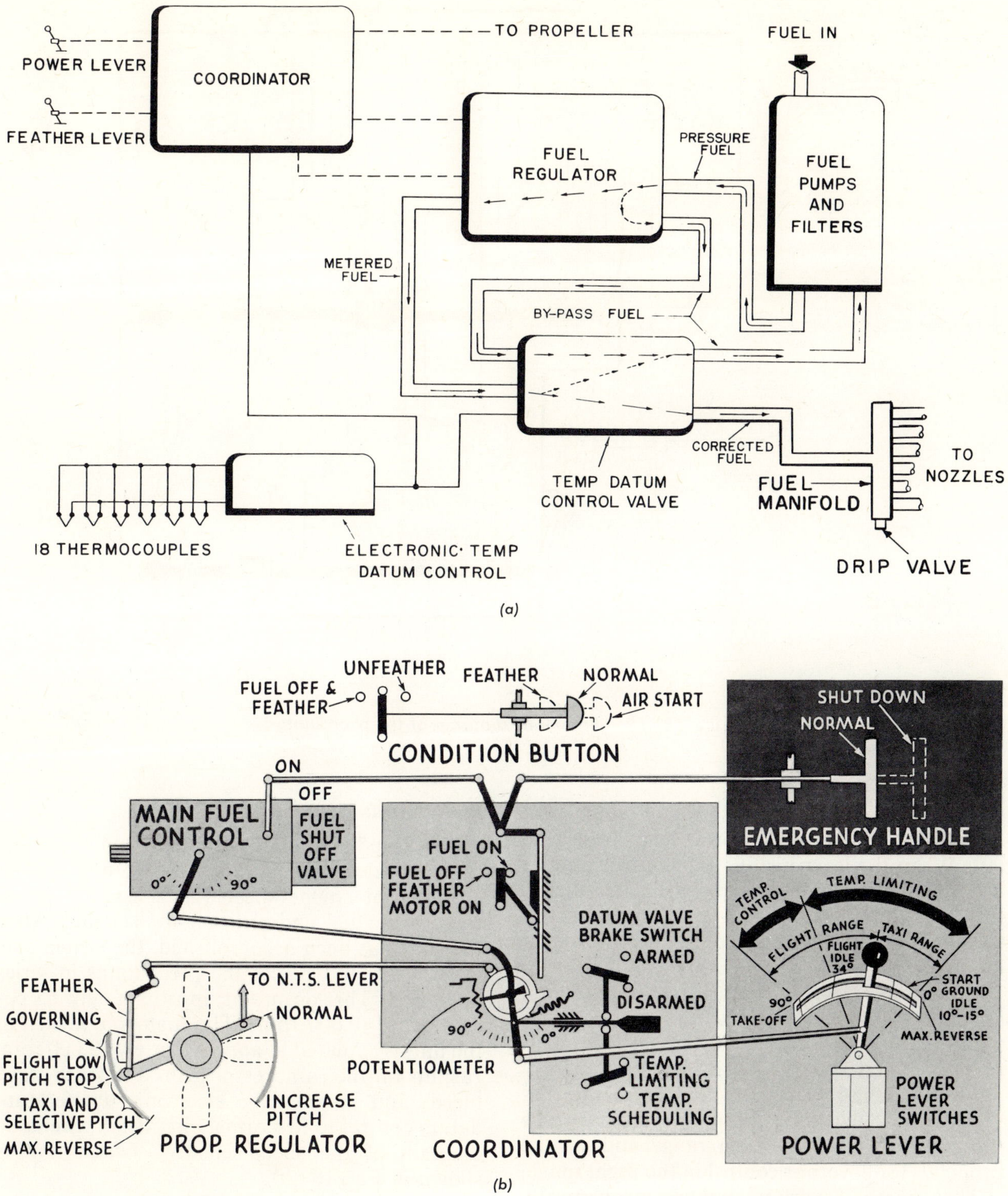

Figure 18 · 18 Engine control system: (a) simplified diagram; (b) Model 501-D13 control system.

mechanically positioning a rotating shaft on the propeller by the propeller-control linkage.

Coordinator

The function of the coordinator is to receive mechanical signals from the **power lever** and the **condition lever** and transmit these signals mechanically and electrically to the fuel-control unit and the propeller-control system. It is mounted on the rear face of the fuel-control unit. Figure 18 · 19 is a simplified schematic diagram of the coordinator.

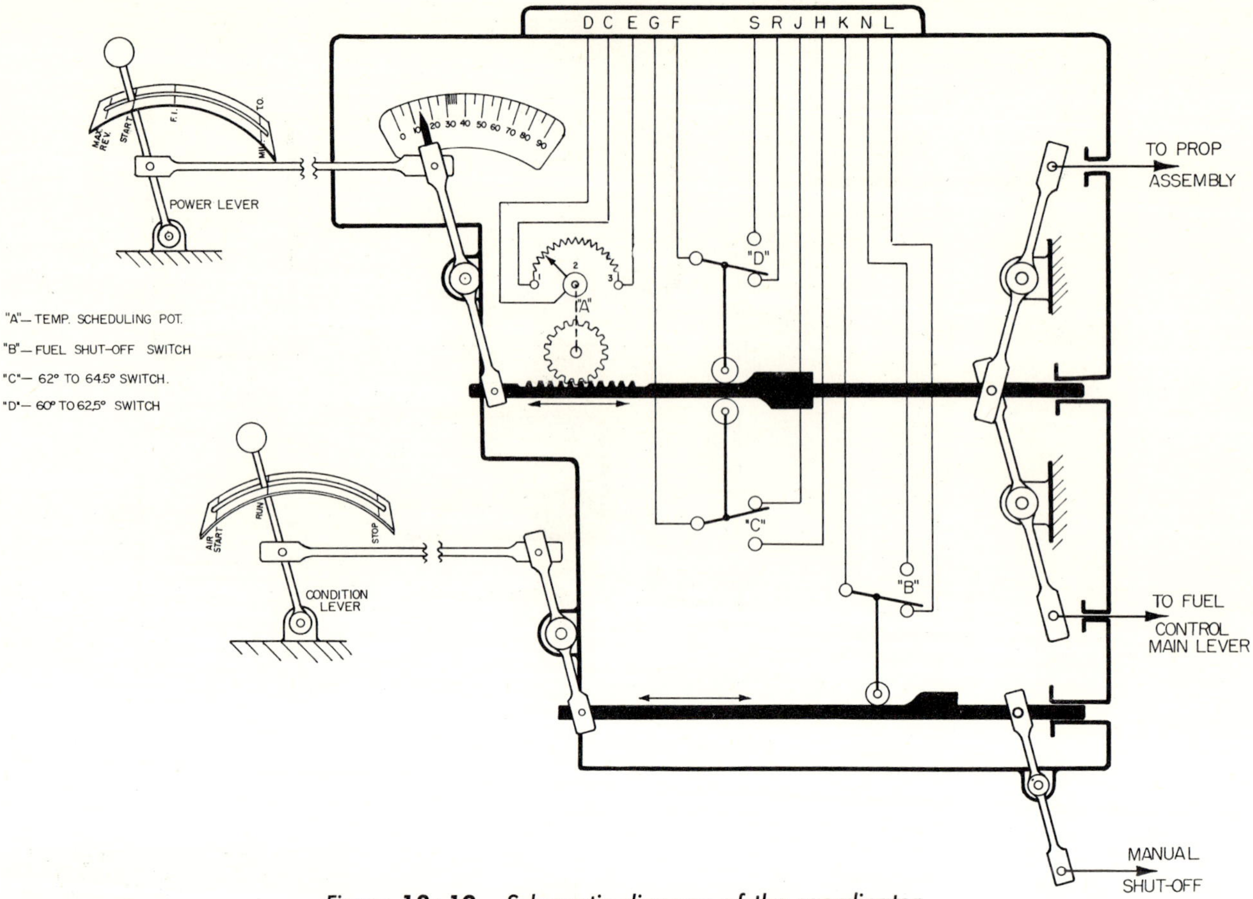

Figure 18 · 19 Schematic diagram of the coordinator.

On the coordinator is a protractor which indicates the position of the power lever from 0 to 90°. When the lever is at 0° the controls are set to provide maximum reverse propeller thrust. Positions for START, FLIGHT IDLE, and MILITARY TAKEOFF POWER are indicated on the power-lever quadrant.

The power lever actuates a potentiometer which establishes a variable turbine inlet temperature for the various power settings. This potentiometer is marked *A* in the diagram.

Two switches, marked *C* and *D* in the diagram, are actuated by the power lever to take the control out of the beta range and set the propeller for governor control in the flight range.

The power lever is linked mechanically through the coordinator to the fuel-control unit and the propeller control.

In the diagram of Fig. 18 · 19, the condition lever establishes conditions for AIR START, RUN, and STOP. This lever may also control propeller feathering. The switch marked *B* controls the electrical fuel-shutoff valve. In some installations, the condition lever is not used and the setting is fixed in the RUN condition.

Speed-sensitive switch

The speed-sensitive switch is mounted on the forward side of the accessory-drive housing. At 16 per cent engine speed (2,200 rpm) it serves to turn on fuel and ignition for starting. After the start has been accomplished, the switch cuts off ignition and places the fuel pumps in series operation. This occurs at approximately 65 per cent speed (9,000 rpm). The high-rpm switch in the unit is actuated at approximately 13,000 rpm. It shuts off the compressor fifth- and tenth-stage bleeds and restores the electronic temperature datum control to its normal circuits. A schematic diagram of the speed-sensitive switch unit is shown in Fig. 18 · 20.

Electronic temperature datum control

A schematic diagram of the control system of the Model 501 engine is shown in Fig. 18 · 21. This diagram shows the relative position of each unit in the system.

The electronic temperature datum control senses differences between the actual and desired turbine inlet temperatures at various power-

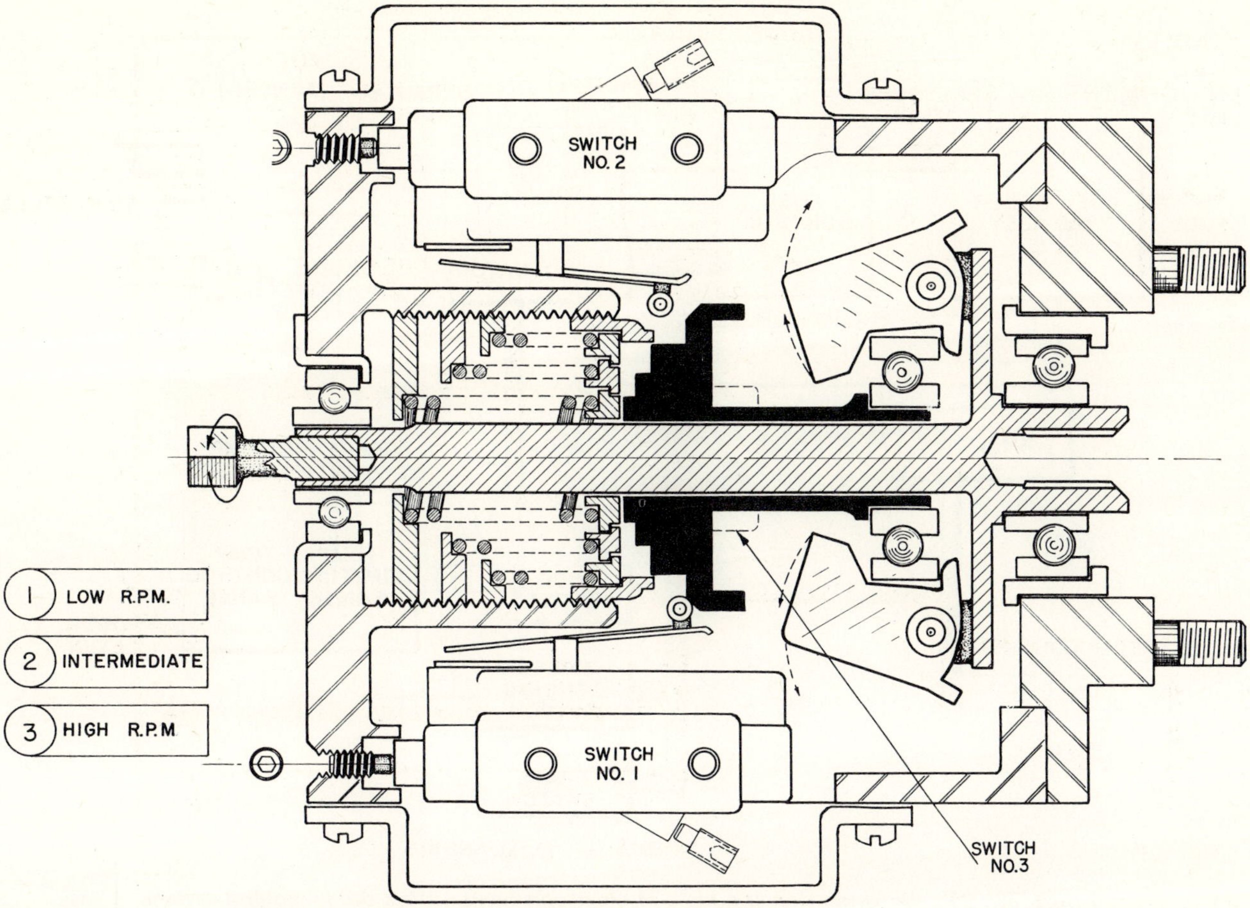

Figure 18 · 20 Schematic diagram of the speed-sensitive switch.

lever settings and corrects fuel flow to the engine. The desired temperature signal originates at the potentiometer in the coordinator and the actual temperature is sensed by the thermocouples located around the turbine inlet.

Determination of whether the turbine inlet temperature is greater or less than the desired signal is made by the **electronic temperature datum control amplifier.** If the temperature is not correct, the datum control signals the datum control valve-actuating motor, thus moving the datum control valve and adjusting the fuel flow to bring about the desired temperature. The datum control valve, shown in Fig. 18 · 22, changes the fuel flow by varying the amount of fuel bypassed to the fuel-pump inlet.

The datum control valve is so designed that it will bypass approximately 20 per cent of the fuel received from the fuel control when no signal is being received from the datum control amplifier. The fuel-control unit normally delivers approximately 20 per cent more fuel to the datum control valve than is required to maintain correct operating temperature. Thus, the datum control valve bypasses the excess fuel and the correct amount of fuel is delivered to the fuel nozzles.

If the datum control system senses excess temperature at the turbine inlet, the datum control valve is called upon to bypass more fuel. Likewise, when the turbine temperature is below that desired, the datum control valve will reduce the amount of fuel bypassed. Thus, on the basis of the desired turbine temperature, the valve can increase by 20 per cent or decrease by 50 per cent the nominal fuel flow to the engine at a particular setting.

For the landing condition and in order that the fuel can maintain a fixed shaft horsepower regardless of ambient temperatures, the valve is temporarily frozen in its corrective position by the operation of a cockpit-mounted, pilot-operated brake-lock switch.

Thermocouples

Thermocouples with two junctions and two sets of terminals in a single probe are used to provide two separate thermocouple circuits.

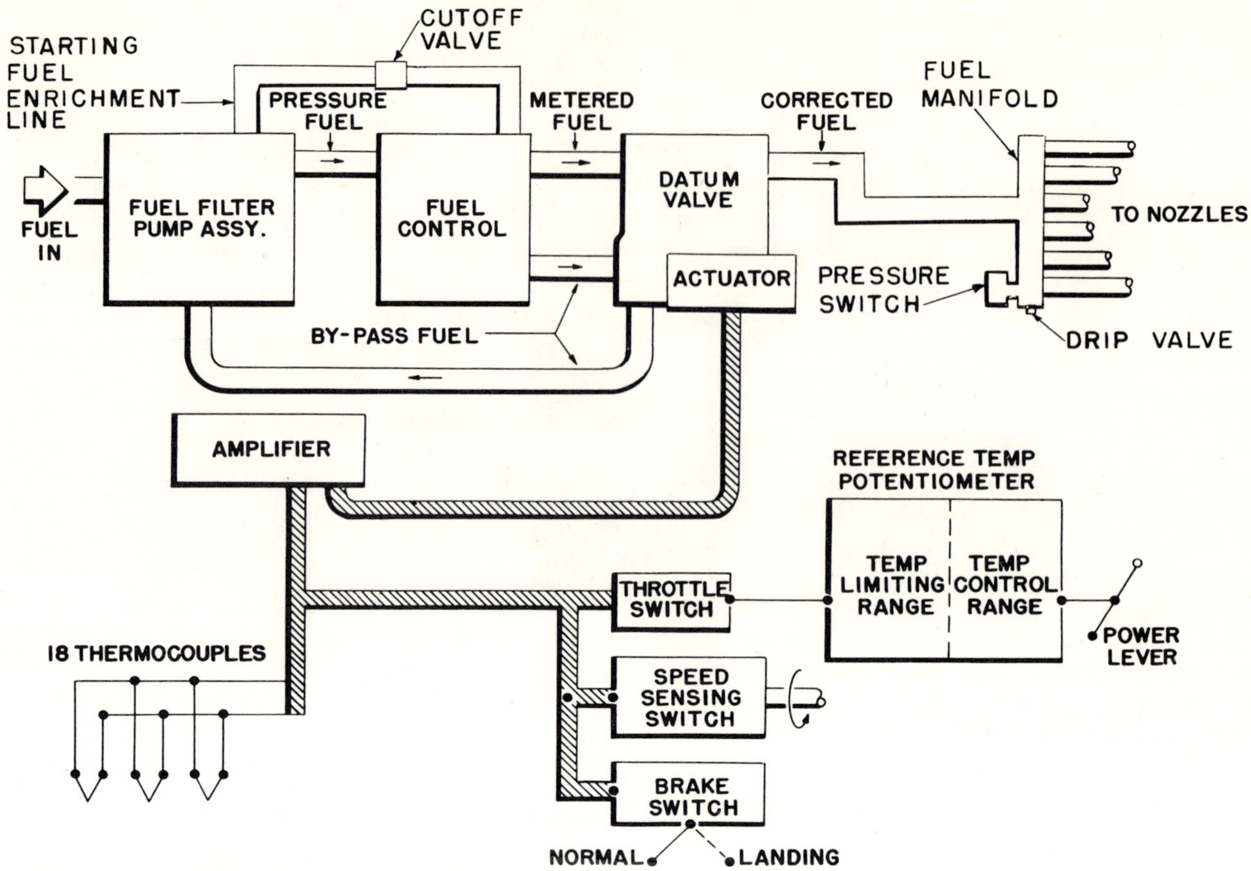

Figure 18 · 21 *Diagram showing units of electronic-temperature-datum control system.*

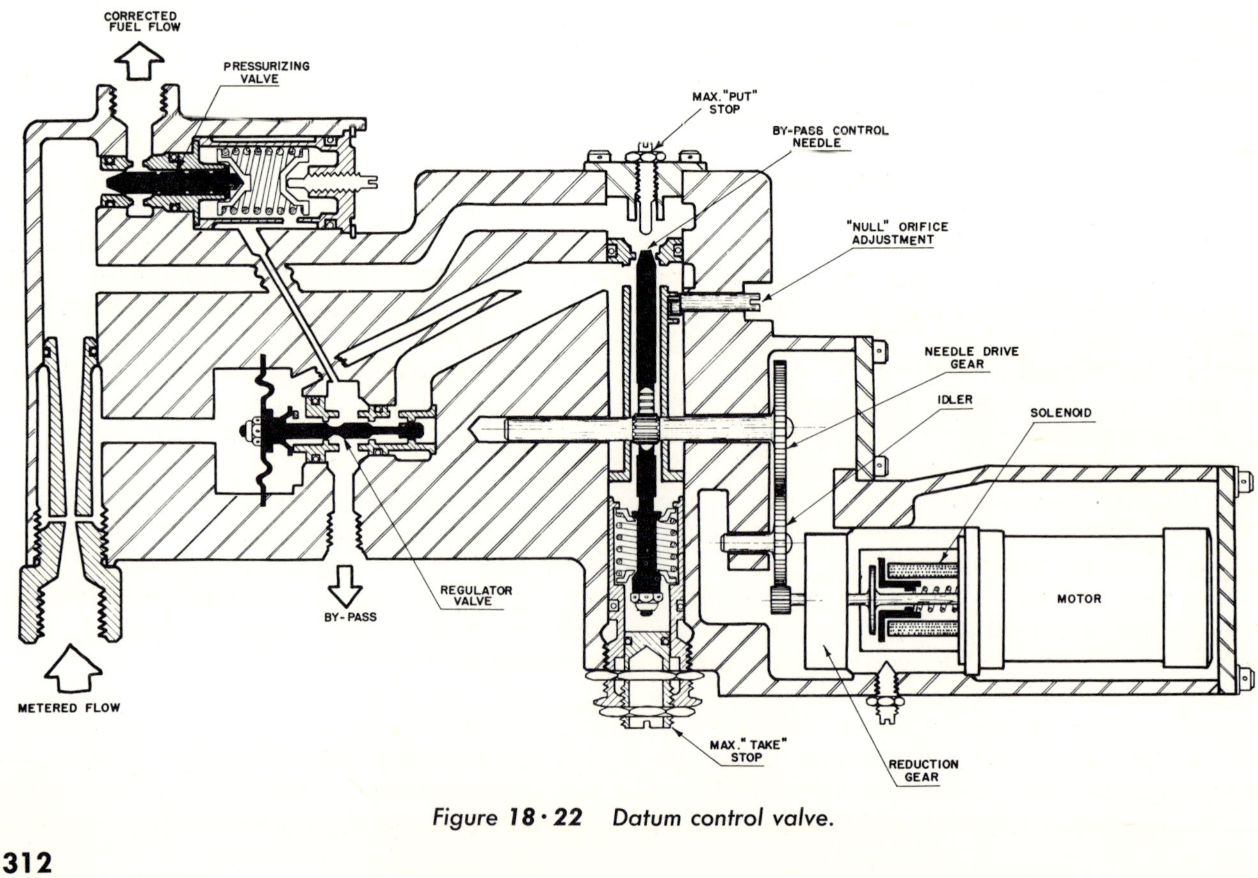

Figure 18 · 22 *Datum control valve.*

The thermocouples are located circumferentially around the turbine inlet immediately to the rear of the combustion chamber. They are connected in parallel so that the failure of one unit will not materially affect the signal.

One of the thermocouple circuits supplies the signal to the electronic temperature datum control, and the other circuit operates the cockpit temperature indicator. A failure in one of the circuits will not affect the temperature indications in the other circuit.

Relay box

The relay box contains a system of relays to receive signals from the speed-sensitive switch, coordinator, electronic datum control, and cockpit. The signals thus received are coordinated to give proper operation of the engine. Relays contained within the relay box are the **brake lock switch relay,** the **overtemperature brake lockout relay,** the **brake lock relay,** the **isolation relay,** and the **speed-sensitive relay.** Also contained in this assembly is the trim resistor network which establishes and permits some adjustment of the temperature schedule as called for by the coordinator.

Starting fuel-enrichment system

A starting fuel-enrichment system (Fig. 18 · 23) is included in the fuel system to provide an increased fuel flow during an engine start, thus ensuring an adequate fuel supply for starting. In addition to the normal fuel flow through the fuel-control unit, an enrichment bypass line allows pump discharge fuel to enter the system downstream of the metering portion of the fuel control but upstream of the fuel-control cutoff valve. The fuel is directed to the temperature datum valve and then to the fuel manifold and nozzles.

A normally closed solenoid enrichment valve is located in the enrichment bypass line. It is energized with the ignition system to allow pump discharge flow to parallel the flow from the fuel-control unit. The circuit to the solenoid which operates the enrichment valve is completed when the following events occur: the auto-feather, ignition, and enrichment relays are energized; the condition lever switch is in RUN position; the manual cutoff switch is ON; and the speed switch is energized at 2,200 engine rpm.

Attached to the fuel manifold is a normally closed pressure switch. In the closed position, the switch energizes the enrichment relay, which closes a pair of contact points, thus completing the circuit to the solenoid enrichment valve. When the fuel pressure in the manifold reaches 110 psig, the pressure switch opens, thus breaking the circuit to the enrichment valve.

Fuel pumps and filters

The fuel-pump assembly includes two gear-type pressure elements which are supplied by one centrifugal boost element. The design of the pump assembly is such that the capacity of the No. 1 gear-type element is 10 per cent less than that of the No. 2 element. This feature allows series operation of the elements without the need of a bleed valve to prevent the No. 1 element from taking over. In the event of failure of the No. 2 pump element, the No. 1 element is capa-

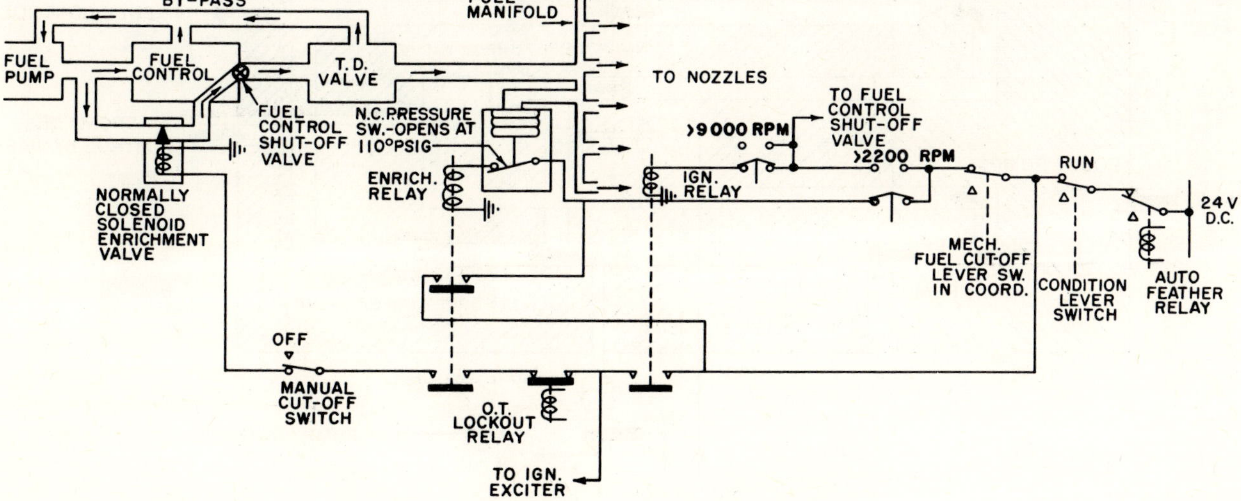

Figure 18 · 23 Starting fuel-enrichment system.

ble of providing sufficient fuel flow and pressure to operate the power section. The pumps are used in parallel for starting as shown in the schematic diagram of Fig. 18 · 24.

A pump warning light switch is located in the filter element. The function of the warning light is to notify the pilot when the parallel system is operating satisfactorily. Pressure in excess of 100 to 120 psig closes the pressure switch and turns on the light.

An electrically operated paralleling valve located in the fuel line between the two pump elements provides the paralleling feature for engine starting. When this valve is closed, fuel passes to both pump elements. Fuel from the booster pump goes to the No. 1 pump element, through the No. 1 filter element, through the No. 1 poppet check valve, and then to the fuel control. Fuel from the booster pump also goes through the No. 2 poppet check valve to the No. 2 pump element, through the No. 2 filter element, and to the fuel control. The warning light will be on during the parallel operation provided that the No. 1 pump element is operating. Should the No. 1 pump element fail, the pressure will drop, thus causing the pressure switch to turn off the warning light.

When engine speed reaches approximately 65 per cent, the speed-sensitive switch deenergizes the ignition relay, and this, in turn, deenergizes the paralleling solenoid valve. The valve opens and places the pump elements in series as shown in Fig. 18 · 25. At this time the warning light goes out.

During normal operation, the fuel goes from the booster pump through the No. 2 poppet check valve, the No. 2 pump element, and the No. 2 filter element to the fuel control. Fuel also flows through the No. 1 pump element, the paralleling valve, and to the No. 2 pump element. Bypass fuel from the fuel control and temperature datum valve flows through the No. 2 poppet check valve to the No. 2 pump inlet.

Each filter is equipped with a bypass valve so that fuel will be supplied to the engine even though the filter becomes clogged.

If the No. 2 pump element should fail during normal operation, the No. 1 element will assume the full load and the warning light will go on, indicating that the No. 2 element is not operating. When this occurs, the fuel flow is from the No. 1 pump element through the No. 1 poppet check valve and directly to the fuel control. Bypass fuel from the fuel control and the temperature datum valve will stop flowing to the No. 2 pump inlet.

Fuel-control unit

The fuel-control unit for the Allison Model 501 engine is a Bendix AP-B3 hydromechanical unit which operates in coordination with the propeller governing system and the temperature datum control. It is similar in construction to the fuel-control units used for many turbojet

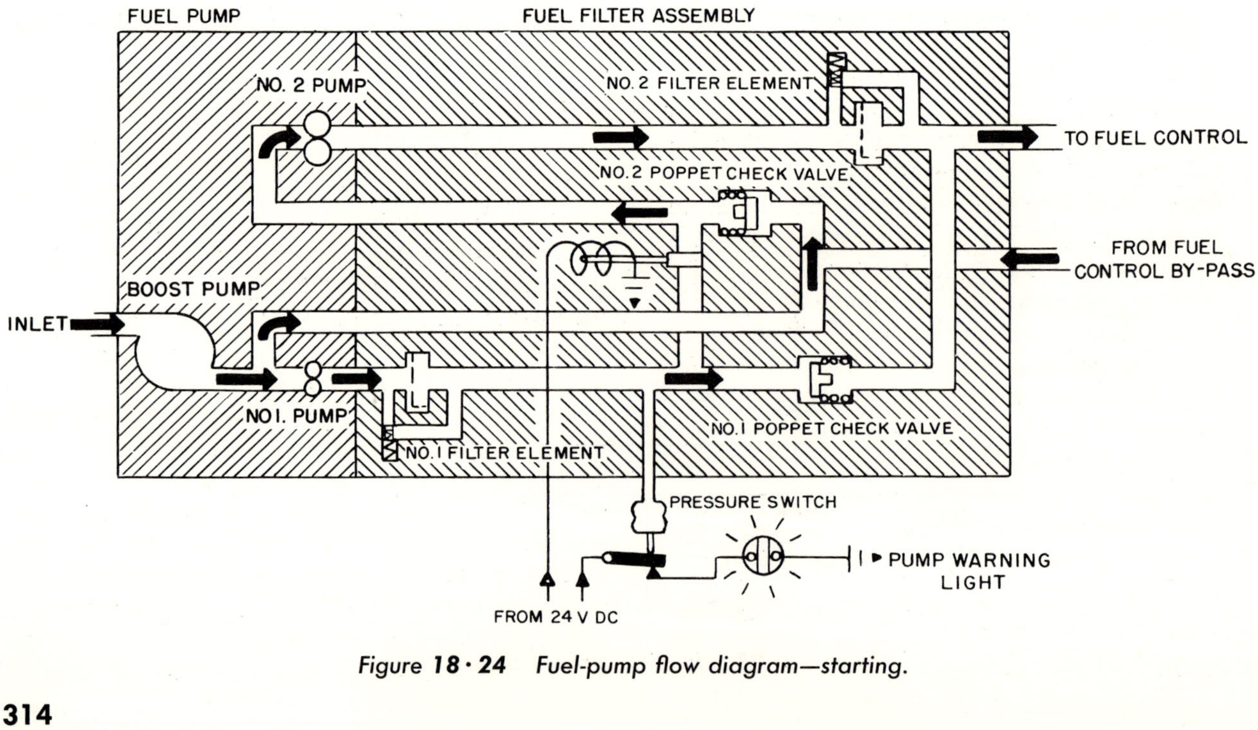

Figure 18 · 24 Fuel-pump flow diagram—starting.

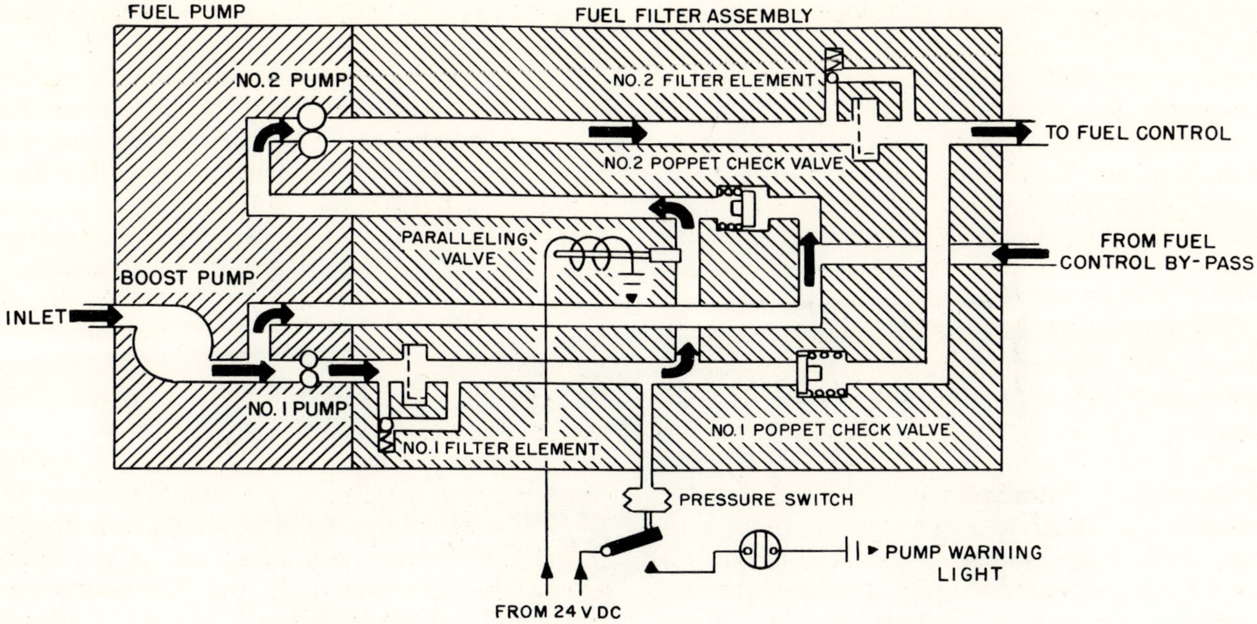

Figure 18 · 25 Fuel-pump flow diagram—normal operation.

engines; however, to operate satisfactorily with the constant-speed propeller, it must have certain additional features.

The fuel-control unit performs the following functions:

1. Supplies a controlled fuel flow to initiate an engine start
2. Supplies a controlled fuel flow during acceleration from start to the stabilized starting rpm (either low- or high-speed taxi rpm) to assist the fifth- and tenth-stage air bleed system in the prevention of compressor surge
3. Meters fuel flow in accordance with variations in air density caused by compressor inlet air temperature or pressure changes
4. Permits the pilot to vary fuel flow to the engine by movement of the power lever
5. Meters approximately 20 per cent more fuel than is required to operate the engine based upon rpm, air density, and power-lever setting, thus providing the temperature datum valve with a definite amount of fuel for trim
6. Provides the means of completely stopping fuel flow to the engine at shutdown
7. Limits the maximum possible fuel flow
8. Provides overspeed protection for the engine
9. Provides the means of selecting either low- or high-speed taxi operation
10. Controls power available in maximum reverse

The fuel-control unit is illustrated schematically in Fig. 18 · 26. As is true of other hydromechanical fuel-control units, this one utilizes a **bypass-valve assembly** which bypasses the excess fuel delivered to the control, and establishes a pressure differential of fuel pump discharge pressure P_1 minus metered fuel pressure P_2 across the metering valve. This differential, $P_1 - P_2$, will remain practically constant during all operations. The position of the bypass valve is determined by the differential forces acting on the valve's flexible diaphragm. When the bypass valve is stabilized, the opening force of P_1 equals the closing force of P_2 plus spring force. Therefore, $P_1 - P_2 =$ spring force. The length of the spring is a function of bypass-valve position, and spring force is a function of spring length. Thus, when the bypass valve moves to assume a new position, the spring length will vary, causing the spring force to change slightly.

The **relief valve** establishes the maximum fuel pressure within the fuel-control unit and thus establishes the maximum possible fuel flow from the control. It is set to open when P_1 exceeds P_o (low-pressure filtered fuel pressure) by a preset amount.

The **metering valve** meters all fuel flow to the temperature datum valve in accordance with variations in engine rpm, power-lever setting, and compressor air inlet temperature and pressure. The metering valve also provides protection from overspeed by reducing the fuel flow when a certain overspeed rpm is exceeded. The size of the metering valve orifice may be changed either by rotation or linear movement of the valve. The metering valve is rotated due to

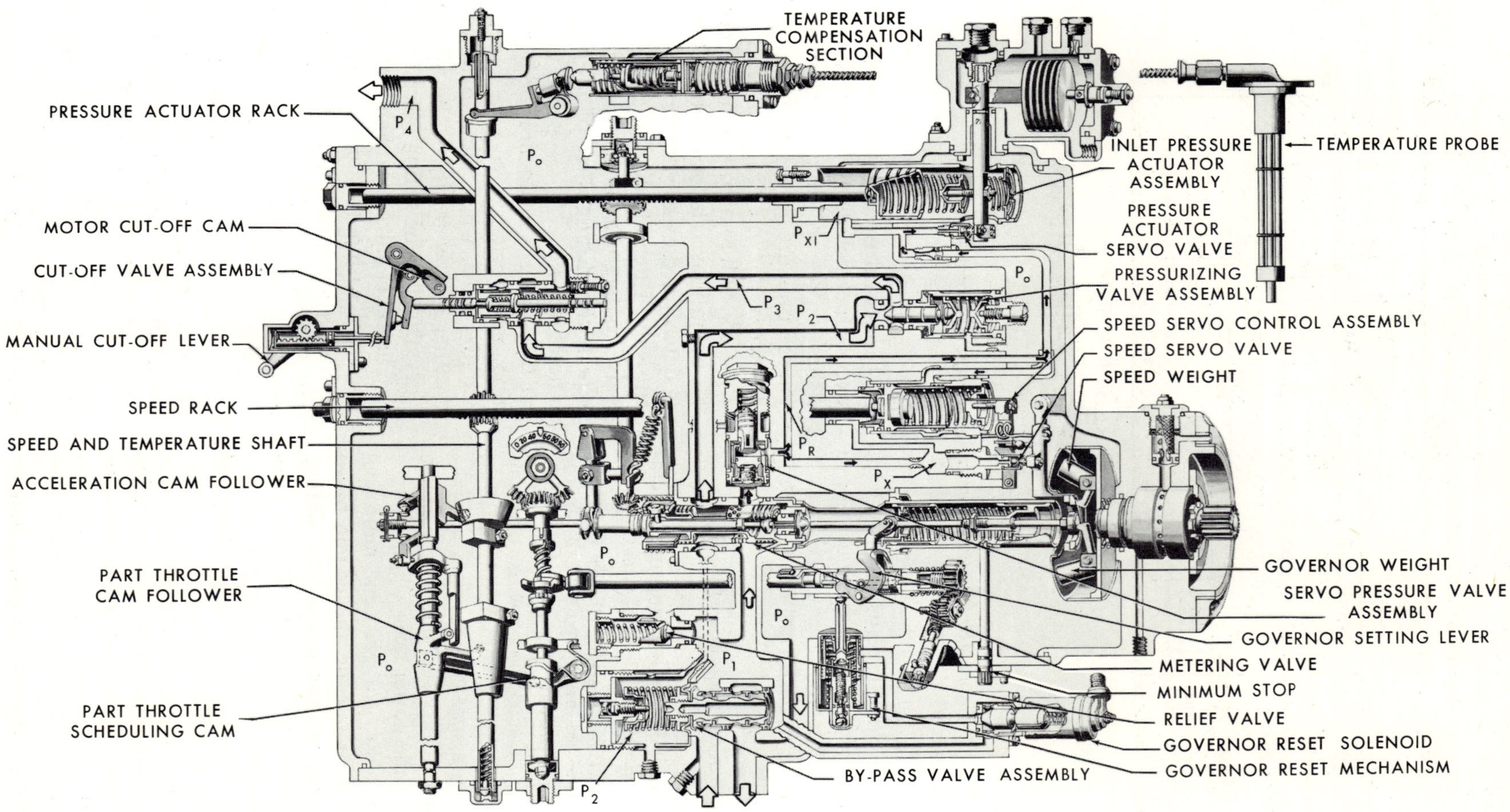

Figure **18·26** *Fuel-control unit, Bendix AP-B3.*

changes in compressor air inlet pressure, while changes in rpm, power-lever settings, and compressor air inlet temperature actuate a **cam assembly,** which results in linear movement of the valve. The metering valve linear opening force is a spring. The cam assembly establishes the linear position and thus the orifice size of the metering valve during normal operation. The **governor spring force** is established by the **governor setting lever,** which is controlled either by a cam and cam follower positioned by the throttle or by the solenoid-operated **governor reset mechanism.** When the solenoid is energized, the governor reset mechanism positions the governor setting lever in such a manner that the governor spring is set for 10,000, +300 to −100 rpm, low-speed taxi operation. When the solenoid of the governor reset solenoid assembly is deenergized, the reset mechanism positions the governor setting lever such that the governor spring is set for the overspeed rpm of the taxi and flight ranges. The solenoid of the reset solenoid assembly is controlled by the cockpit low-speed taxi switch. The force of the governor weights serves as a closing force for the metering valve. During low-speed taxi operation, the linear position of the metering valve is established by a balance of two forces, the governor spring force and the governor weight force. The cam assembly does not position the metering valve in low-speed taxi operation. During high-speed taxi and flight operation the governor spring force is greater than the governor weight force. Thus, the governor spring tends to move the metering valve fully open; however, the maximum linear opening of the metering valve is established by the cam assembly. In the event of overspeeding, the governor weight force increases with rpm; and when it overcomes the governor spring force, the metering valve moves to decrease its linear opening, thus reducing fuel flow from the control. This limits engine speed to a definite level above 13,820 rpm. During an overspeed condition, the cam assembly has no control over the linear position of the metering valve because the governor weight force moves the metering valve away from the cam assembly. When the overspeed is corrected, the governor weight force decreases, and the governor spring begins to move the metering valve toward the OPEN position. Then the cam assembly again determines the maximum linear opening of the valve.

The **inlet-pressure-actuator assembly** senses compressor air inlet pressure changes by means of a pressure probe in the left horizontal air inlet housing strut. The inlet pressure actuator initiates an action which causes the metering valve to rotate and provide a corrected fuel flow required by any air pressure variation. A partially evacuated bellows, sensitive to air pressure changes, repositions the **pressure actuator servo valve** by means of a lever action whenever compressor air inlet pressure changes. The position of the pressure actuator servo valve establishes servo pressure P_x', which in turn establishes the position of the **pressure piston** and the **pressure actuator rack.** When the air pressure changes, the pressure actuator servo valve causes P_x' to change. This moves the pressure piston and pressure actuator rack and results in moving the pressure actuator servo valve to a stabilized position. When the pressure actuator rack moves, the metering valve drive gear causes the metering valve to rotate by means of a bevel gear. Any change in air pressure results in the rotational movement of the metering valve.

The **speed servo control assembly** senses rpm changes and initiates an action which causes the metering valve to move linearly and monitor fuel flow to prevent compressor surges during engine accelerations. The position of the speed servo valve establishes servo pressure P_x which, in turn, establishes the position of a **speed piston** and **speed rack.** When rpm increases, the speed weights actuate linkage to move the speed servo valve toward a CLOSED position. This causes P_x to increase. Thus, the speed piston moves the speed rack and linkage to stabilize the speed servo valve. Movement of the speed rack rotates the speed and temperature shaft which has two cams: the **acceleration cam** and the **part-throttle cam.** These cams are contacted by the **acceleration cam follower** and the **part-throttle cam follower** respectively. The acceleration and part-throttle cams are designed so that during rpm changes the acceleration cam positions the acceleration cam follower; when there is no rpm change, the part-throttle cam positions the part-throttle cam follower, its shaft, and the acceleration cam follower. The acceleration cam follower's position establishes the metering valve's linear position and hence the flow of metered fuel from the fuel control. Any change in rpm results in linear movement of the metering valve.

The **temperature compensation section** which senses compressor air inlet temperature changes by means of a probe inserted through the air inlet housing beneath the left horizontal strut, initiates an action which causes the metering valve to move linearly to provide a corrected fuel flow as required by any air temperature variation. The probe and a bellows in this section are filled with alcohol and temperature variations cause the volume of the alcohol to change. Thus, the length of the bellows is dependent upon the sensed air temperature. When temperature changes occur, the mechanism causes the **speed** and **temperature shaft,** upon which are mounted the two cams, to move linearly. This arrangement can be seen clearly in the diagram of Fig. 18 • 26. Either the acceleration cam or the part-throttle cam, whichever is in control of the acceleration cam follower at the time of the air temperature change, will reposition the acceleration cam follower. This causes the metering valve to move linearly and correct the fuel flow. A return spring always retains the speed and temperature shaft in contact with the bellcrank of the temperature compensation section. Another bellows in this section is used to prevent any change in P_0 fuel temperature or pressure from moving the speed and temperature shaft. The fuel control only compensates for variations in air temperature and never for fuel temperature changes.

The **part-throttle scheduling cam** is positioned by the power lever; hence, when the power lever is moved, the part-throttle scheduling cam moves its follower. The **part-throttle scheduling cam follower** moves the part-throttle cam follower linearly on its shaft. This is brought about by the axial contour ground on the part-throttle scheduling cam, as shown in Fig. 18 • 26. The linear movement of the part-throttle cam follower changes its relative position with respect to the part throttle cam, and the contour of the part-throttle cam causes the part-throttle cam follower to pivot slightly. This pivoting action moves the acceleration cam follower, which, in turn, moves the metering valve linearly to schedule fuel flow as required by the power-lever movement.

The **servo pressure valve assembly** establishes and maintains the regulated servo pressure P_R at a predetermined value above P_0. The P_R is used by the **inlet pressure actuator assembly** and the **speed servo control assembly,** along with servo valves, to establish P_x and P_x' required by the assemblies. The position of the **servo pressure valve** is determined by P_R, the closing force, and P_0 plus a spring force, the opening force. The pressurizing valve causes metered pressure P_2 to build up to a predetermined value before it opens to allow metered fuel pressure P_3 to flow to the cutoff valve assembly. This causes P_1 to build up before fuel can flow from the fuel control and results in quicker stabilization of the fuel-control components during the initial phases of an engine start. The pressurizing valve is not designed to have any metering effect on the fuel, but there is a small decrease in pressure across the pressurizing valve. This is the reason for indicating a P_2 and P_3 metered fuel pressure. The opening force on the pressurizing valve is P_2, and the closing force is P_0 plus spring force.

The **cutoff valve assembly** provides the means by which fuel flow to the engine is started or stopped. Electrical actuation of the cutoff valve is desirable for automatic engine starts and normal shutdowns. Because the possibility of an electrical failure exists, mechanical actuation of the cutoff valve is required for emergency shutdowns. Therefore, the cutoff valve may be moved to the CLOSED position either electrically or mechanically. The cutoff valve must be permitted to open both electrically and mechanically during an engine start. The only time the cutoff valve will be mechanically held closed is when the emergency shutdown handle is pulled to the EMERGENCY position. Pulling the emergency handle causes the normal cutoff lever to move so that a bellcrank and lever push a plunger to compress a spring within the cutoff valve. When the force of this spring exceeds the force of an opposing spring, the cutoff valve moves against its seat to stop fuel flow. Energizing the cutoff valve actuator motor causes the cutoff cam to move the bellcrank, lever, and plunger to compress the spring and close the valve. When the cutoff valve is closed, P_3 is ported to P_0. The cutoff valve is not designed to have any metering action, but a small decrease in pressure does occur across the cutoff valve. This is the reason for indicating a P_3 and P_4 metered pressure.

Engine lubricating oil is used to lubricate the drive bearings of the fuel control. This oil is supplied and scavenged by the engine power section's lubricating system. Internal components of the fuel control are lubricated by the fuel which flows through the control.

To summarize the operation of the fuel-control unit, we must review the parameters by which the unit is influenced. These are engine speed, power-lever setting, compressor inlet temperature, and compressor inlet pressure. The unit also responds to acceleration or rate of change in rpm. Engine speed is sensed by the governor weights and the speed weights. The governor weights move the metering valve linearly in response to engine speed, thus providing constant correction for proper engine speed. The speed weights operate through the speed servo system and speed rack to control fuel flow in accordance with engine speed, thus preventing excessive fuel flow for any particular engine speed. Correction for air-inlet-temperature changes is accomplished by the temperature-compensation section, operating through the speed and temperature shaft. Changes in inlet air pressure are reflected through the inlet-pressure-actuator assembly, which moves the pressure actuator rack to rotate the pressure actuator cross shaft, and this in turn rotates the metering valve. Changes in engine power output are accomplished through the throttle (power lever) which compresses the governor spring when moved from low-speed and low-power settings to higher settings.

THE ROLLS-ROYCE DART TURBOPROP ENGINE

General description

The Rolls-Royce Dart engine, shown in Fig. 18 • 27, is a 2-stage, centrifugal-flow turboprop engine, used in the United States to power the four-engine Vickers Viscount, the Fairchild F-27, and the Grumman Gulfstream.

The principal components of the Dart engine are the **propeller reduction gear,** the **air-intake assembly,** the **2-stage compressor,** the **intermediate casing, seven can-type combustion chambers,** the **nozzle-box and turbine assembly,** the **heatshield,** and the **exhaust unit.**

The specifications and performance of the Dart engine are approximately as follows:

Equivalent shaft horsepower	2,020 (1,835 shp plus 485 lbt)
Specific fuel consumption (max)	0.57 lb/ehp/hr
Specific weight (max eshp)	0.66 lb/ehp
Oil consumption (cruising)	0.12 lb/hr
Maximum rpm	15,000
Diameter	38.0 in.
Length	98.0 in.
Frontal area	7.9 sq ft
Weight (without propeller)	1,335 lb
Fuel grade	JP-1, JP-4, or D.Eng.R.D. 2482 or 2486 kerosene

The main units of the Dart engine are shown in the exploded view of Fig. 18 • 28. This view shows the engine separated into subassemblies of the principal components mentioned previously.

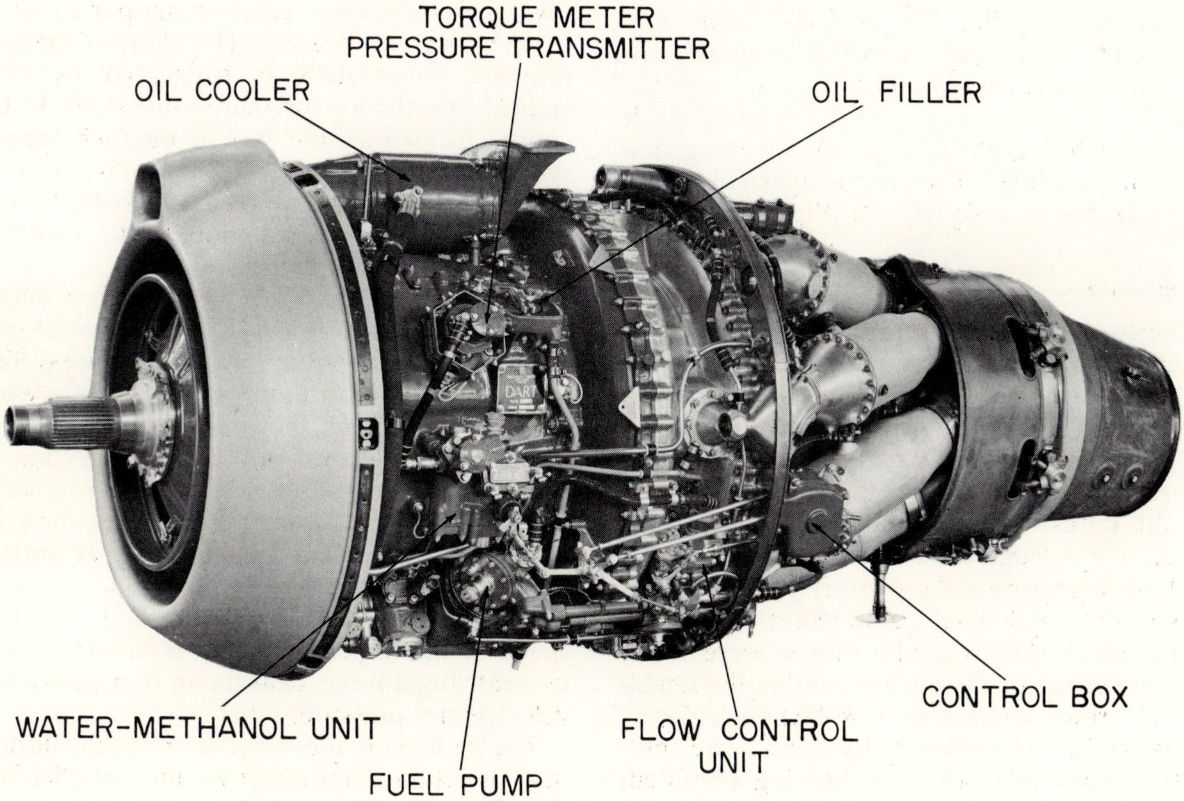

Figure ***18 • 27*** *Rolls-Royce Dart turboprop engine, left side view.*

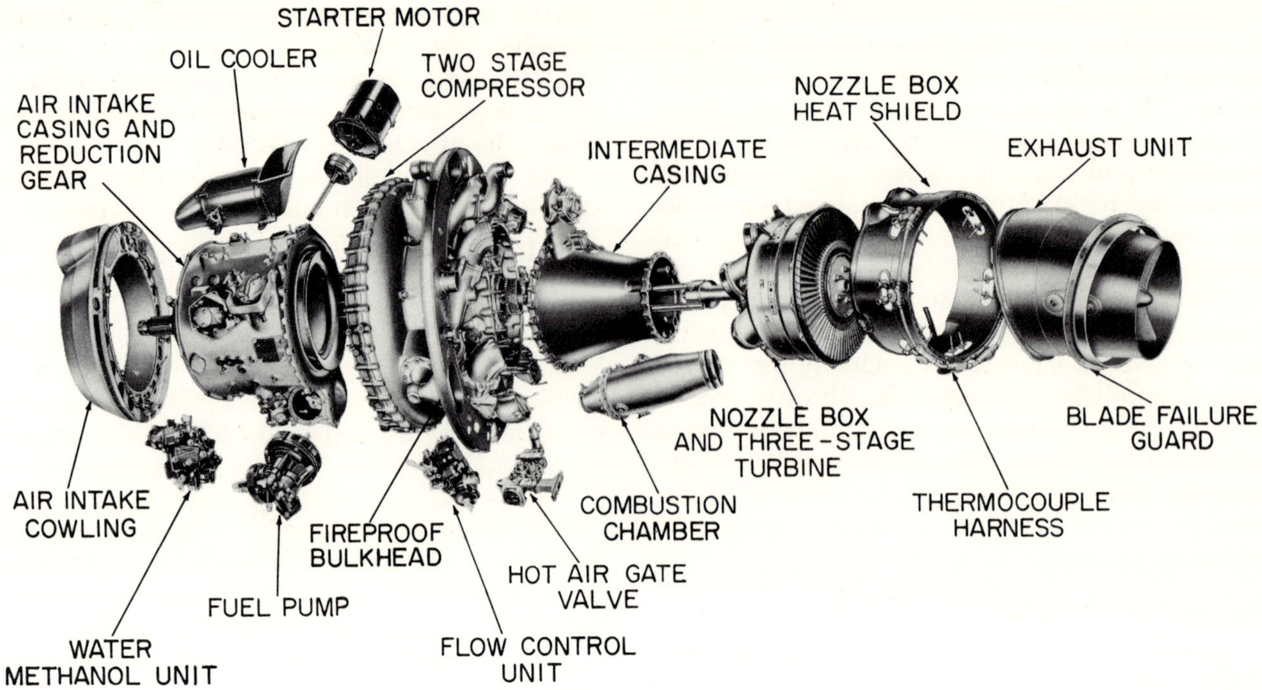

Figure 18 · 28 Exploded view of the Rolls-Royce Dart engine.

Air-intake cowling and casing

The air-intake cowling and casing are concentric units which form an annular air-inlet passage at the front of the engine immediately to the rear of the propeller. The propeller reduction-gear casing is located within the air-intake casing. Air flows rearward and inward through the passages within the casing to the rotating guide vanes attached to the front of the first-stage impeller at the center. The air passages are clearly shown in the cutaway view in Fig. 18 · 29.

Compressor

The two compressor rotors are housed within an assembly of three casings: the **front casing,** the **interstage casing,** and the **outlet casing.** These casings are bolted together at the outer periphery and form the structure which supports the rotor bearings. Inside the casing assembly are the **diffusers** and **guide vanes** which direct the airflow through the compressor passages.

The two compressor rotors are constructed of a light alloy with steel guide vanes. The guide vanes, which turn with the rotors, serve to reduce the shock of the air as it enters the impellers. The rotor assemblies are splined to separate shafts and are individually balanced. The split-shaft arrangement facilitates bearing alignment and makes it unnecessary to disturb the balance during engine buildup. The guide vanes and rotors are locked to the shafts by nuts and cup washers.

Both rotor shafts are supported by ball thrust bearings at the rear of the rotors. The low-pressure compressor rotor is supported at the front by a roller bearing. The front of the high-pressure rotor shaft is supported by being splined into the low-pressure rotor shaft. Helical splines formed on the rear of the high-pressure rotor shaft receive the drive coupling from the turbine. A separate spur gear is bolted to the front face of the helical coupling to transmit a drive to the **auxiliary gear box.**

Surrounding each rotor assembly is a magnesium **diffuser vane ring.** A set of **prewhirl guide vanes** is fitted between the low-pressure diffuser vane ring and the entry to the high-pressure assembly to reduce shock at this point.

Air enters the "eye" of the first-stage compressor from the intake casing. Here it encounters the rotating guide vanes, which are curved in such a manner as to reduce shock at entry to the impeller and cause a smooth flow into the radial passages. Because of the high rotational speed of the impeller, the air is driven outward by centrifugal force, causing an increase in both velocity and pressure.

The air leaving the impeller enters the diffuser, in a direction determined by the impeller rotational speed and the radial velocity the air has acquired. The divergent shape of the passages which form the diffuser will then convert a high percentage of the velocity energy into pressure energy. The airstream is then turned about and enters the convergent passages between the interstage guide vanes, whose object is to reduce shock at entry to the second-stage rotor by imparting whirl in the same direction as the spinning rotor. A small decrease in pressure will result from the convergent passage form.

The air now enters the eye of the second-stage impeller at an appreciably increased pressure and a velocity approximately equal to that at the entry to the first-stage eye. The sequence of energy changes through the second-stage compressor is the same as that through the first stage, and the pressure at the exit is about five times that in the intake with a corresponding temperature increase.

Combustion chambers

The seven combustion chambers of the Dart engine are arranged at an angle to the plane of the main rotor shaft; that is, they tend to form a left-hand spiral. This arrangement is shown in Fig. 18 • 27. Its purpose is to take advantage of the direction of the airflow as it leaves the second-stage compressor. The positioning of the combustion chambers shortens the engine and promotes a smooth airflow. The combustion chambers are numbered counterclockwise as viewed from the rear of the engine. No. 1 chamber is fitted to the top compressor outlet elbow on the right-hand side.

The construction of a combustion chamber may be seen in the cutaway engine illustration of Fig. 18 • 29. Each chamber assembly consists of an **expansion chamber,** an **air casing,** a **flame tube,** and **interconnectors.**

The expansion chambers are fitted to the compressor outlet elbow by two link bolts. The seating between the chamber and elbow is formed by a spherical joint ring. At the rear, the expansion chamber is attached to the air casing with a bolted flange. The expansion chamber is the forward portion of the outer casing and may be considered as a dome. The air casing is the main portion of the outer casing, and it extends rearward to the nozzle box.

The air casings are constructed of mild-steel sheet and are aluminized to prevent corrosion. They are bolted to the expansion chambers at the front and fitted into the combustion-chamber discharge nozzles with piston rings at the rear. Each air casing carries two interconnectors, three flame-tube suspension pins, and fuel-drain connections. Provision is made on No. 3 and

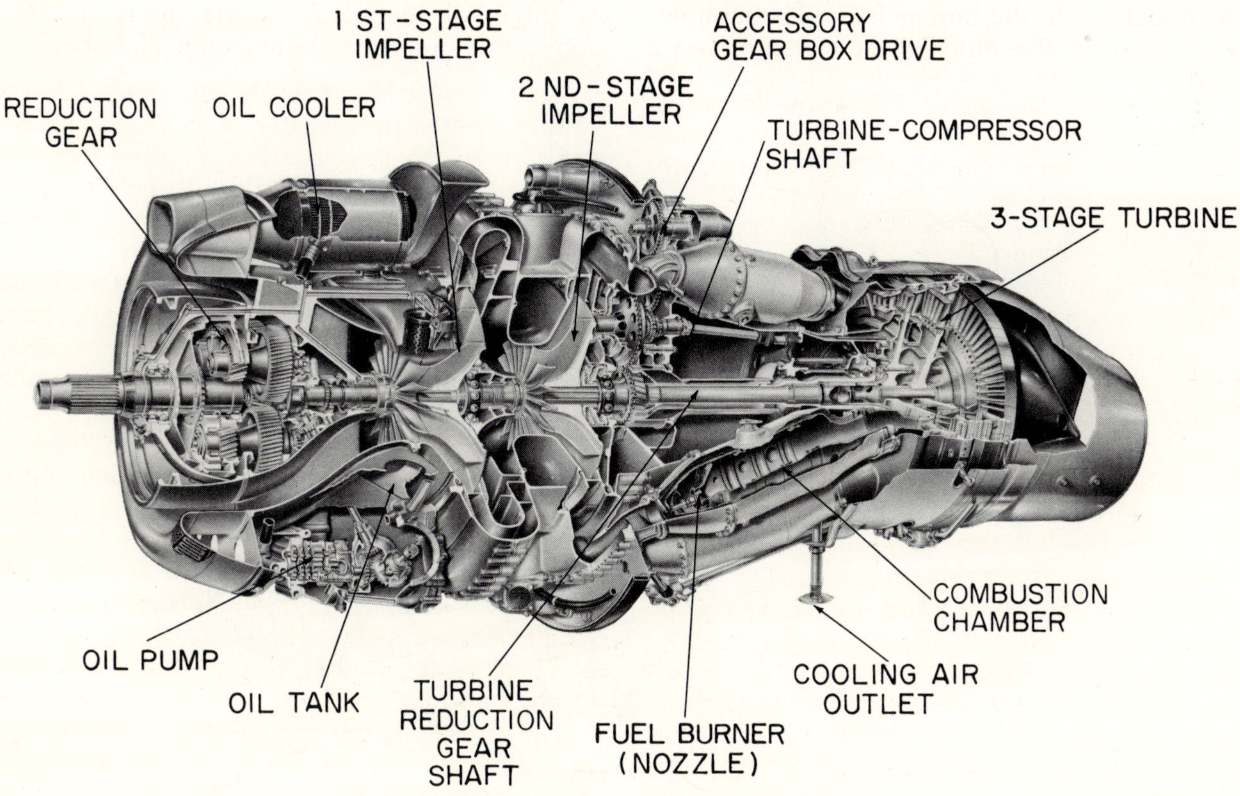

Figure 18 • 29 Cutaway view of the Dart engine.

No. 7 casings to carry the high-energy ignition plugs. Because of the various positions of interconnectors and fuel-drain connections, the casings are not interchangeable.

Inside the expansion chamber and air casing assembly is the flame tube. These tubes are fabricated in sections from Nimonic 75 steel alloy and the sections are welded together. The flame tube is located at the front in the air casing with three pins. To allow for expansion, it is supported at the rear by a spherical ring seating inside the discharge nozzle. The head (forward end) of each flame tube carries a set of fixed swirl vanes to assist in efficient mixing of fuel and air.

Fuel is supplied to the combustion chambers by means of **Simplex anticarbon burners.** These burners are mounted in the expansion chambers and extend forward so that they are surrounded by the swirl vanes in the forward end of the flame tubes. The burners (fuel nozzles) are fixed-orifice atomizers with no moving parts. Each burner contains a spring-loaded thread filter which protects the **swirl plug** and **swirl chamber** from foreign matter. The swirl plug has three flats on its periphery and two small tangential slots through the walls of a cylindrical chamber formed in its forward end. The swirl chamber surrounds the swirl plug and provides a 0.055-in.-diameter final orifice. The components are retained in position by a double outer shroud which locks onto the burner body. Radial holes in the rear of the outer shroud communicate with an annular space between inner and outer members. A drawing of the burner is shown in Fig. 18 · 30.

Fuel flows through the thread filter of the burner, enters the cylindrical chamber on the front face of the swirl plug through the tangential slots, and is directed around the walls of the swirl chamber. The high velocity causes the fuel to break up into small particles as it leaves the final orifice, producing a hollow conical spray. Air flows through the annular space in the double shroud and is directed across the burner nose to prevent the formation of carbon on the orifice.

A diagram of the arrangement of the Simplex anticarbon burner as it is installed in the expansion chamber is shown in Fig. 18 · 31. This diagram also shows the compressor outlet elbow with the cascade vanes which direct the air into the combustion chamber.

The interconnectors between the combustion chambers are necessary to equalize the gas pressures and to provide a means of passing on the "light-up" from the No. 3 and No. 7 chambers. Each interconnector consists of two concentric tubes, to connect the air casings and the flame tubes by independent passages. To provide an expansion joint between each interconnector, the outer tube carries two piston rings seating in a chromium-plated bore in the air casing. A three-bolt flange joins the interconnectors connecting adjacent combustion chambers.

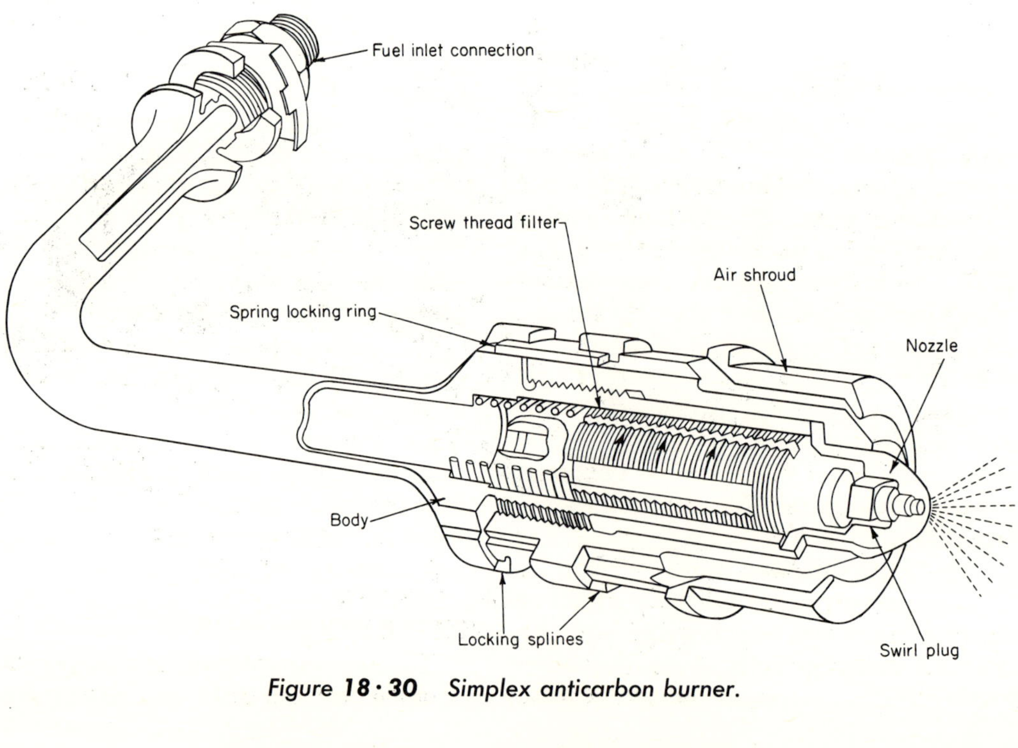

Figure 18 · 30 Simplex anticarbon burner.

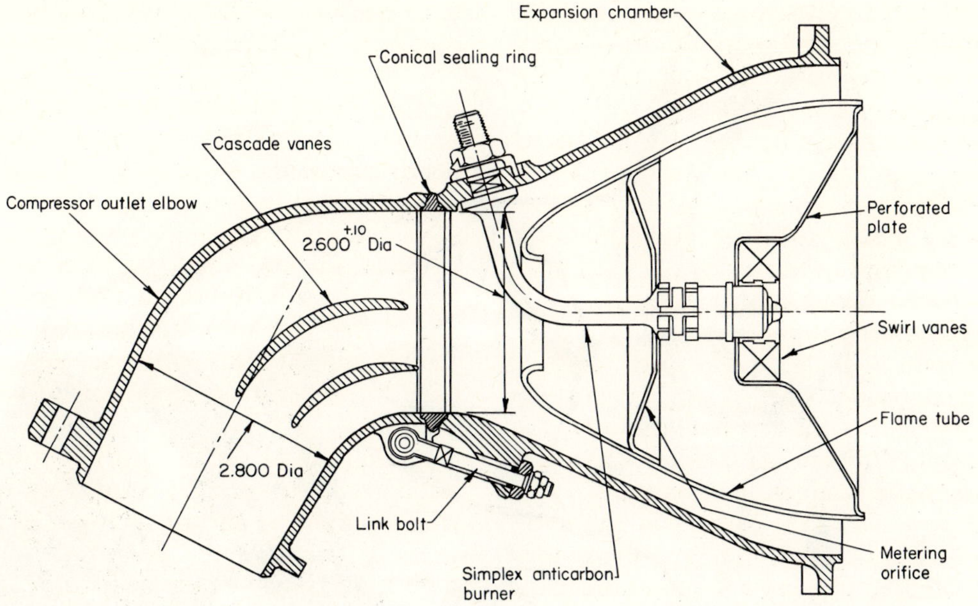

Figure 18 · 31 Installation of burner.

Turbine section

The **turbine assembly,** shown in Fig. 18 · 32, is bolted to the **intermediate casing** by means of the forward flange on the **nozzle box mounting drum.** This assembly consists of two principal subassemblies: the **nozzle box** and the 3-stage axial-flow **turbine.** The nozzles receive the hot gases from the combustion chambers and spread the flow of gases into the annular area formed by the **nozzle guide vanes** and turbine blades. The nozzle guide vanes impart to the gas stream a high whirl momentum which is then converted by the turbines into shaft horsepower to drive the compressor, auxiliary units, and propeller shaft.

The nozzle box consists of six main components: the **outer casing,** the **heatshield,** the **inner cone,** the **mounting drum,** the **turbine bearing housing,** and three sets of **nozzle guide vanes.** The outer casing carries the seven combustion-chamber discharge nozzles and the rear engine slinging attachment. On its front flange is attached the mounting drum, which, together with the turbine bearing housing, is bolted to the intermediate casing.

Machined on the inner periphery of the outer casing are flanges which provide the attachment for the three sets of nozzle guide vanes. The inner cone, which provides the inner location for the high-pressure (H-P) guide vanes, is attached to the mounting drum and rear bearing housing. It is positioned concentrically with the outer casing by adjusting washers on radial tie bolts.

The nozzle guide vanes form a series of nozzles in which the gas is accelerated. They are of airfoil section and are cast hollow to maintain as near as possible a constant sectional thickness to reduce thermal stress. There are 70 H-P vanes, which are hooked into flanges machined on the inner cone and nozzle box outer casing with 14 of the vanes used as locators. The inner location is provided by slots in the flange of the inner cone, and the outer location is provided by locating pegs fitted through the nozzle boss casing, engaging in the guide-vane outer platforms. The guide vanes are made of a high-temperature nickel-chrome alloy called Nimonic.

The intermediate-pressure (I-P) nozzle has 56 vanes, which are supported by an arrangement in which tongues on the outer platforms of the vanes are hooked into grooves in the nozzle box. The vanes are supported axially by two rings with the turbine interstage seal carried on their inner platforms. At the leading edge of 12 of the vanes, provision is made for mounting thermocouples. Twenty-eight of the vanes are used as locators to assure correct positioning.

The low-pressure (L-P) nozzle is similar to the intermediate-pressure nozzle in that there are 56 vanes with 28 used as locators. Their attachment in the nozzle box is the same as that of the I-P vanes.

Turbine rotating assembly

The turbine rotating assembly, which consists of three turbine disks and associated blading, is

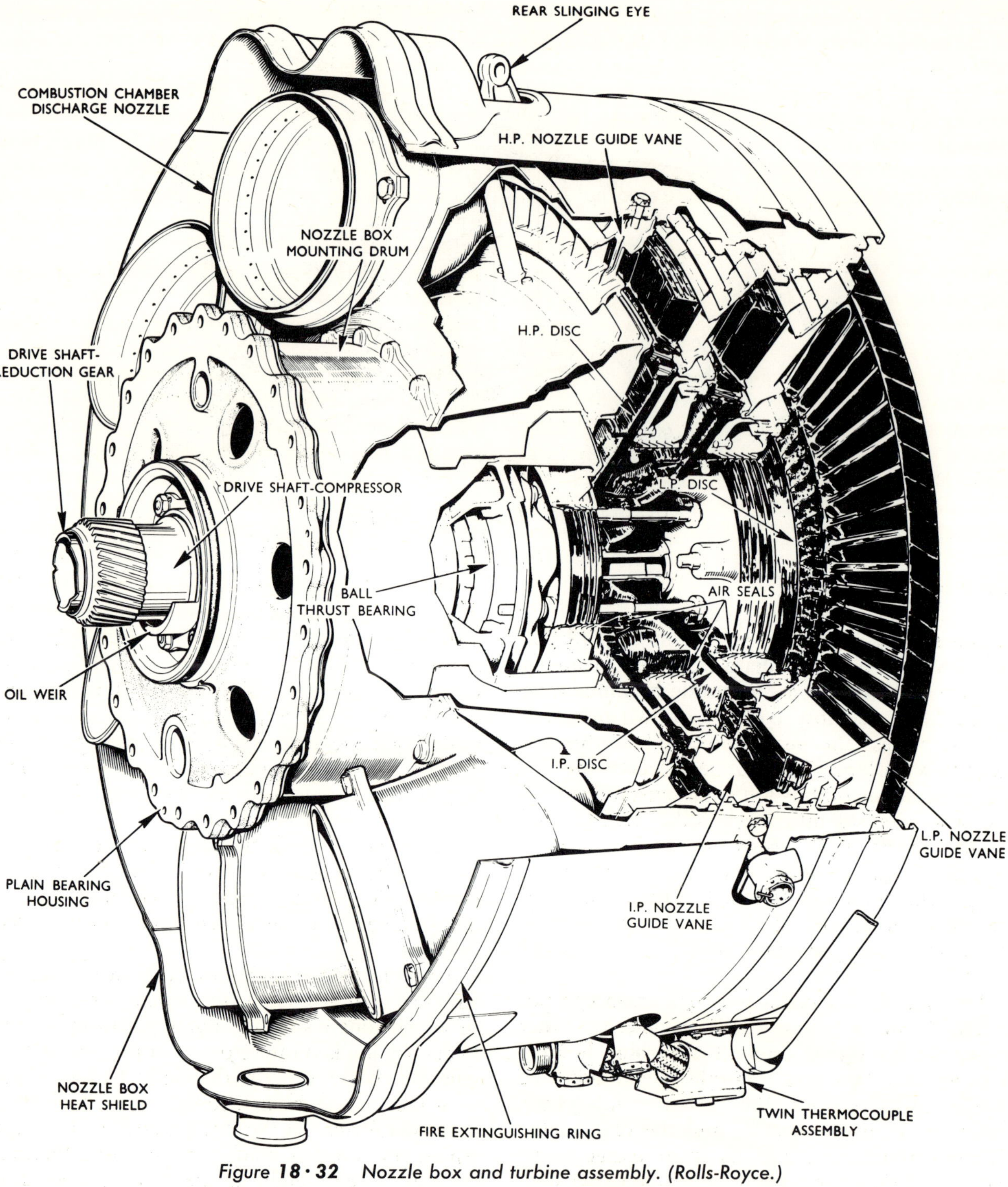

Figure 18 · 32 Nozzle box and turbine assembly. (Rolls-Royce.)

secured by ten bolts to the engine **main shaft assembly,** that is, the concentrically arranged compressor and propeller reduction-gear drive shafts. Each bolt has a tapered section on the shank to locate the turbine disks radially. Five of these bolts pass through the H-P and I-P disks and secure them to the flanged ends of the shaft assembly. Similarly, the remaining five bolts pass from the shaft assembly through all three disks and secure the L-P disk.

The entire assembly of disks and shafts is supported immediately forward of the H-P disk by a ball thrust bearing and a plain bearing, as shown in Fig. 18 · 33. The turbine blades are secured to their respective disks by fir-tree-shaped roots, which are fitted into a series of corresponding slots broached in the periphery of the turbine disks. Each blade is located axially between a tongue which projects from the root to lap over the turbine disk front face and a locking plate which engages with a slot in the rearward end of the blade root. Each locking plate secures a group of either two or three blades, depending upon circumferential location.

The labyrinth seals fitted to the turbine bearing housing and the turbine disks control the air pressure which is used for cooling. The air pressure also prevents oil loss because the air leakage through the seals is in a direction to keep the oil in the proper areas.

The flow of hot gases from the combustion chambers enters the nozzle box, where large nozzles convert the flow into annular form and direct it into the convergent spaces between the first-stage nozzle guide vanes. These vanes are at an angle which directs the flow of gases so that they enter the turbine at the correct angle. The guide vanes impart a high angular velocity to the gases, with a consequent drop in pressure and temperature. The flow is in the form of a free vortex, that is, a constant angular momentum at all radii, and in consequence will perform equal work at all radii.

To exploit the energy of the gases in its varying forms requires a changing turbine-blade section and an increasing amount of blade twist from root to tip. The turbine design is a combination of the two basic types and is known as an "impulse-reaction" turbine. It is impossible to state exactly the proportion of either type at any given point, but, broadly speaking, the larger portion of turbine work is accomplished by reaction, with a trend toward impulse at the blade roots.

On leaving the turbine assembly, the gases pass into the **exhaust unit,** where the velocity is reduced because of the increasing area. The reduction in velocity is accompanied by a slight rise in pressure and temperature. The exhaust cone blends the annular area of the turbine blades into a suitable circular form and protects the turbine disk from contact with hot combustion gases. The rear of the exhaust unit fits inside the front of the jet pipe, leaving a small annular clearance. The resulting expansion causes a drop

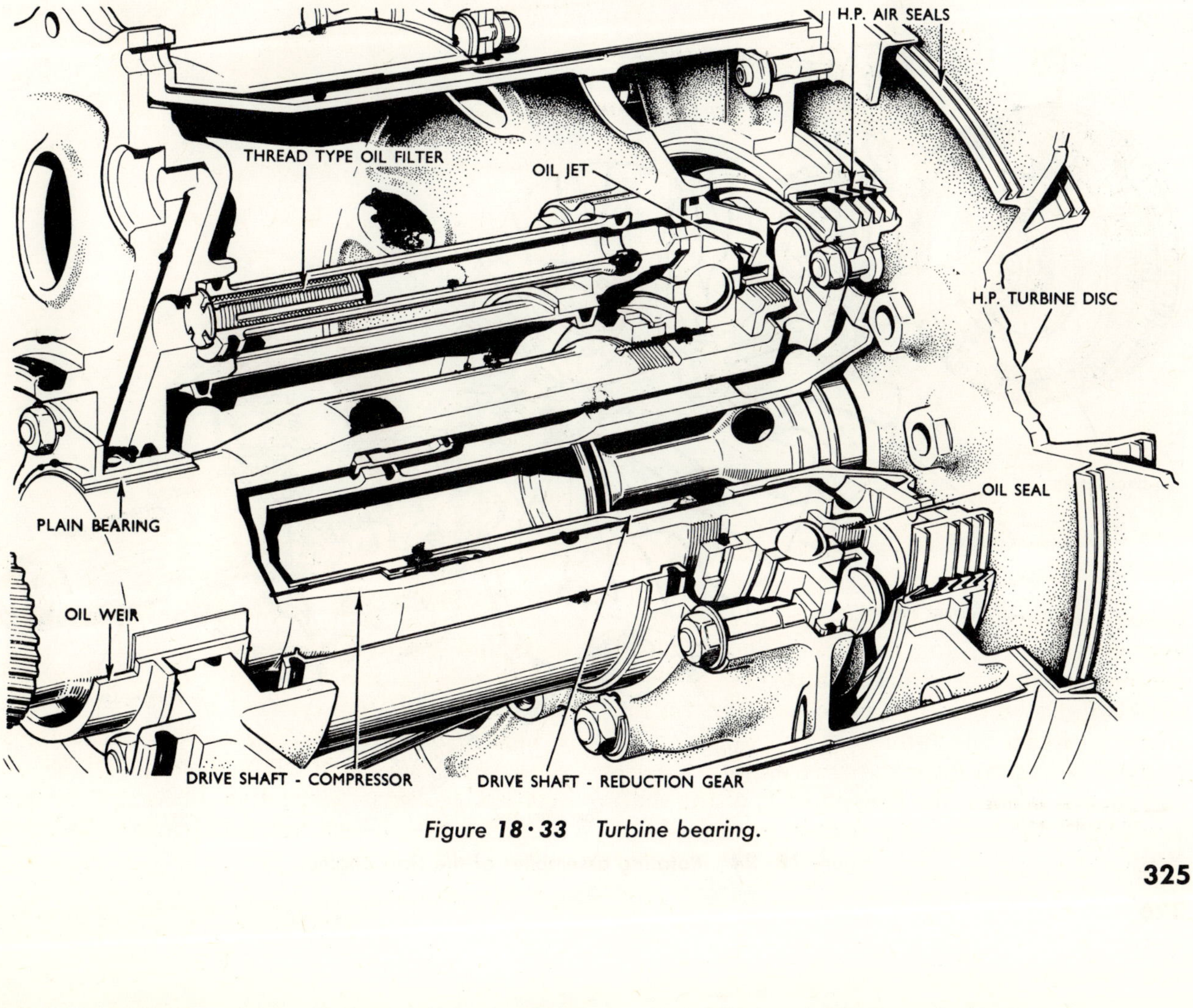

Figure 18·33 Turbine bearing.

in pressure which is used to induce a flow of cooling air through the engine bay and into the jet pipe.

Engine drives and gear layout

The rotating parts, including the gear drives, of the Dart engine are shown in Fig. 18 • 34. The 3-stage turbine drives the high-pressure compressor shaft through a helical coupling. A torsion shaft, splined into the front of the compressor shaft, carries the torque forward to the high-speed pinion at the rear of the reduction-gear assembly. The torsion shaft passes axially through the low-pressure compressor shaft; hence this shaft does not transmit any propeller torque.

The high-speed pinion, splined to the torsion shaft, drives three gears of the propeller reduction-gear assembly. These gears are integral with shafts called "layshafts." The forward ends of the layshafts form spur gears meshing with an internally toothed annulus gear to which the propeller shaft driving disk is splined. The reduction gears give a ratio of 0.093:1.

The **starter motor drive** is taken through a clutch assembly and bevel gears to a sleeve surrounding the high-speed pinion shaft. The motor rotates at twice engine speed. The sleeve carries nine ratchet teeth engaging with six pivoted pawls mounted on the high-speed pinion. The pawls are spring-loaded into engagement with the ratchets. Three pawls and three ratchets are engaged simultaneously, giving a 23° maximum rotation before engagement. The pawls are centrifugally disengaged at approximately 4,000 rpm, and they reengage at approximately 1,200 rpm after shutdown.

The clutch, spring-loaded to 900 lb-in., protects the starter motor in the event of excessive loading and reduces the initial shock when the starter is first energized.

Each reduction-gear layshaft contains a limit shaft internally splined to it and positioned axially by a ball thrust race. From the lower-limit shaft a drive is taken through bevel gears to the lower auxiliary drive housing, where a pack of five oil pumps is driven at one-fourth engine speed. Rearward of the oil pumps an additional bevel gear assembly forms a dual drive unit for the fuel pump on the left side and the rpm controller on the right-hand side, both being driven at a ratio of 0.192:1.

A spur gear is bolted to the front face of the turbine coupling at the rear of the high-pressure

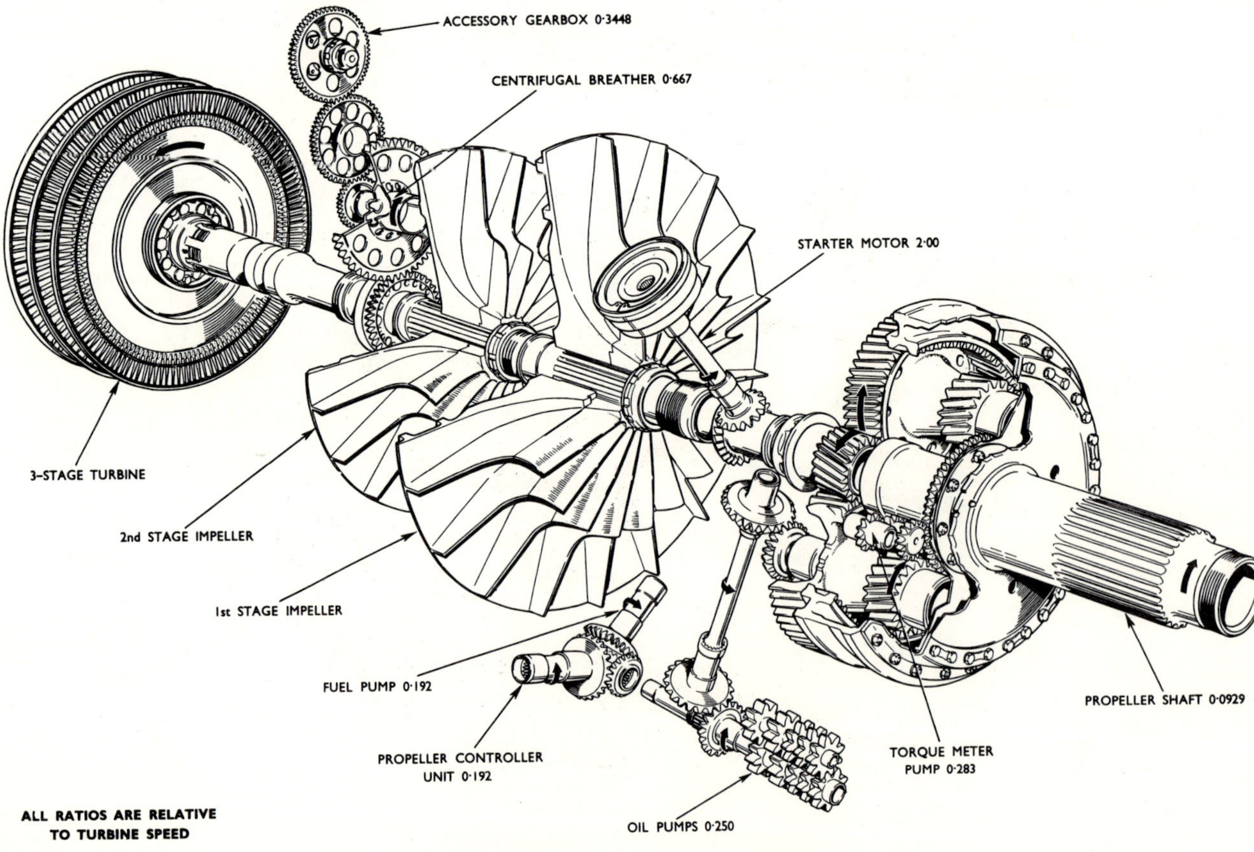

Figure ***18 • 34*** *Rotating assemblies of the Dart engine.*

compressor shaft. A vertical train of three spur gears provides the accessory-gear-box drive. The first gear of the train includes a centrifugal separator to allow air from the engine breather system to pass to atmosphere through the hollow shaft while retaining lubricating oil inside the engine.

Propeller reduction-gear assembly

A cutaway view of the propeller reduction-gear assembly is shown in Fig. 18 · 35. The reduction gear is a 2-stage spur-gear type employing three axial layshafts (countershafts). Each layshaft has a spur gear on each end. The housing is cast integral with the air-intake casing and is supported by webs in the center of the annular air intake. The drives, their associated bearings, and the torquemeter assemblies are supported in three magnesium panels and in the reduction-gear front cover. The propeller shaft is No. 4 SBAC standard.

The helical input gear (high-speed pinion) is located in the rear panel by a ball thrust bearing. It is supported at the rear by a bronze bushing inside the front of the first-stage compressor shaft. The three layshafts, driven by the input gear, are supported by roller bearings carried in the front and rear panels and are allowed a 0.100-in. axial movement. To limit this movement, each layshaft contains a limit shaft, positioned at the rear by a ball thrust race and supported at the front by splines in the layshaft.

The three layshafts, revolving at 0.488 engine speed, transmit the drive to an internally toothed annulus gear, thus forming the second-stage reduction of 0.190. The annulus gear is extended forward and attached by splines and three segment plates to a driving disk formed on the propeller shaft. The assembly overall reduction ratio is 0.093:1.

The propeller shaft is supported at the rear by a roller bearing housed in the center panel. It is

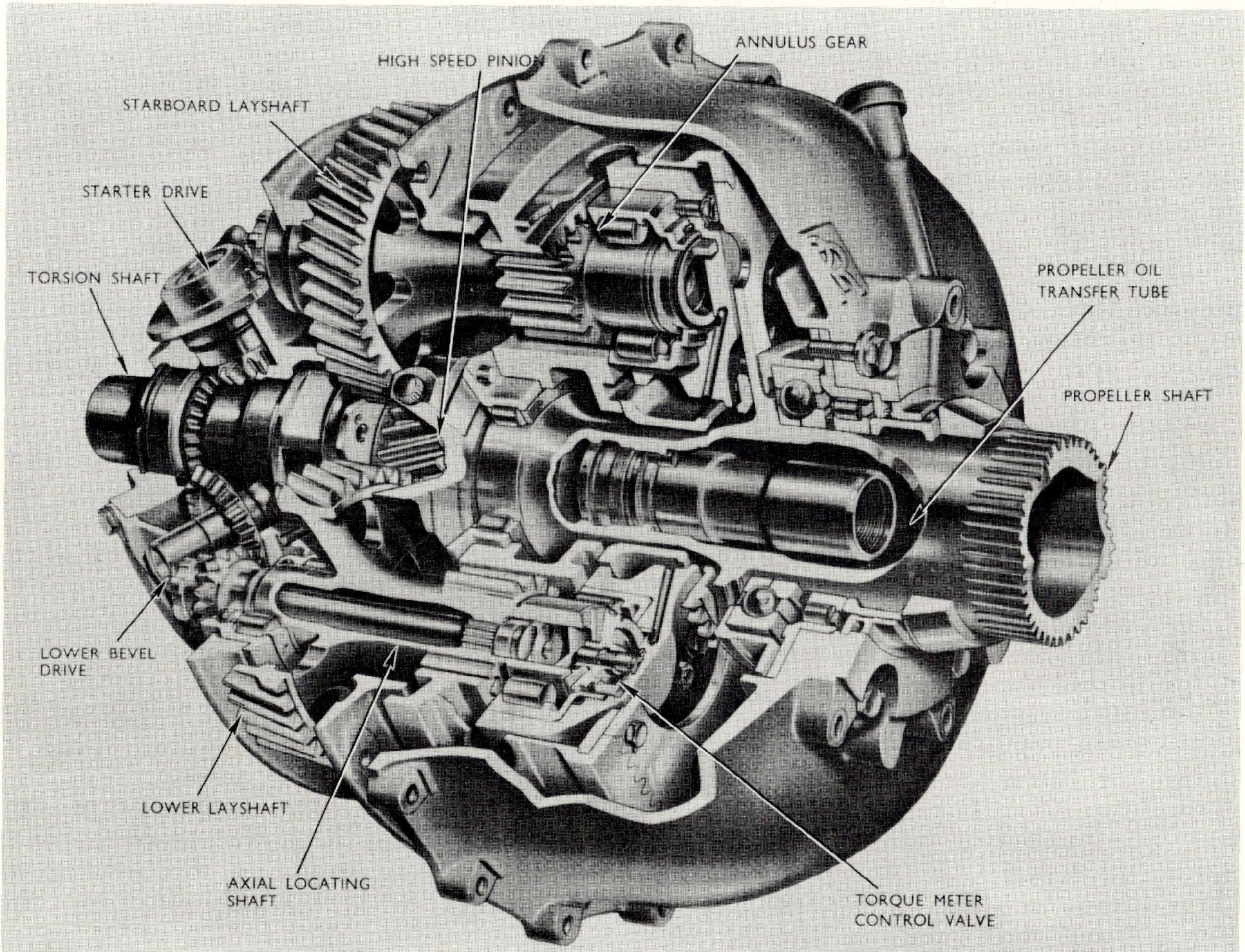

Figure 18 · 35 Propeller reduction-gear assembly.

located at the front by ball and roller bearings mounted in the reduction-gear front cover. A labyrinth-type seal assembly, pressurized by compressor discharge air, surrounds the propeller shaft where it passes through the front cover and prevents loss of lubricating oil to the atmosphere.

To permit propeller operating oil to be transferred from the stationary casing to the rotating shaft, a transfer seal assembly is bolted to the rear reduction-gear panel. The seal consists of white-metal lined bronze bushings fitted closely around the oil transfer tubes in the rear of the propeller shaft.

The **torquemeter** is an indicating system which receives a signal from the engine and provides a reading on an instrument in the cockpit. The engine signal is provided by the engine through the action of the helical drive gear on the layshafts. The helical teeth of the high-speed pinion produce a forward loading on each layshaft which is proportional to the torque transmitted. This forward loading is hydraulically balanced by a piston assembly fitted at the forward end of each layshaft. The pressure of the balanced fluid (engine oil) is recorded in the cockpit as an indication of the torque transmitted by the propeller shaft.

To provide sufficient balancing force with small piston areas, engine oil pressure is boosted by a small gear pump called the "torquemeter oil pump." This pump is housed in the front panel and driven by the rotating gear on the propeller shaft. A wire-mesh filter protects the pump gears from damage due to the ingress of foreign matter.

Lubrication system

The lubrication system of the Dart engine is shown in the diagram of Fig. 18 · 36. This is a dry-sump system with the oil supply and scavenge system completely self-contained within the engine. There are three main sections of the system: (1) the pressure system for engine lubrication, (2) the pressure system for propeller operation and the water-methanol operation, and (3) the scavenge system.

Two different oil pressures are encountered in the engine. The engine-lubrication system employs a pressure of 35 psi, while the propeller-control unit and water-methanol system use a pressure of 70 psi.

Compressor air is used with the system to provide seals against oil leakage from the bearings. The various sections, cavities, and sumps within the engine are connected with a vent system to prevent a build-up of pressure which could interfere with proper oil flow.

The oil tank is formed by an annular cavity in the rear portion of the air-intake casing. In this position it surrounds the eye of the low-pressure compressor when the engine is assembled. The oil tank is divided into upper and lower compartments by two horizontal webs. Oil passes to the lower compartment through gauze filters located in the center of each web. The oil supply to the pressure pump is through a short standpipe in the lower compartment. This ensures that 1¼ gallons of oil is retained for propeller feathering. The oil passes to the feathering pump through a drilled passage in the base of the tank.

From the standpipe in the tank, oil flows to the pressure pump which is the rearmost unit of the pump assembly. The pump is mounted in the lower auxiliary drive housing below the reduction gear. It circulates the oil through a Purolator oil filter located at the base of the auxiliary drive housing. From the filter the oil feed continues to a twin relief-valve assembly, the housing of which is integral with the propeller-control-unit mounting pad. The valves are in series, each having a relief pressure of 35 psi. The first valve is also loaded by the 35-psi oil pressure controlled by the second valve. This arrangement provides a total of 70 psi for the propeller controller and the water-methanol system.

Drawings of the oil pressure pump and filter are shown in Fig. 18 · 37. As shown in the drawing, the pressure pump contains five gear elements. One large element is the pressure pump, and the other four elements provide for scavenging of different sumps.

During operation, oil from between the two valves of the twin relief-valve assembly flows at 35 psi to the various points in the engine requiring lubrication. The reduction-gear assembly is fed through passages in the air-intake and reduction-gear casing, the oil being sprayed from 13 jets. The propeller-shaft front bearings are fed by splash, and the rear bearing is supplied by splash and leakage from the propeller oil transfer seal. Oil is fed through passages in the base of the reduction-gear casing to the plain bearings of the lower auxiliary drive assembly.

Apart from the front compressor bearing, which is fed by splash, the compressor interstage and rear bearings, auxiliary gearbox drive assembly, and the turbine bearing are fed from one common supply. This comes from the relief-valve assembly and passes through a rigid pipe inside the oil tank to an outlet at the top of the

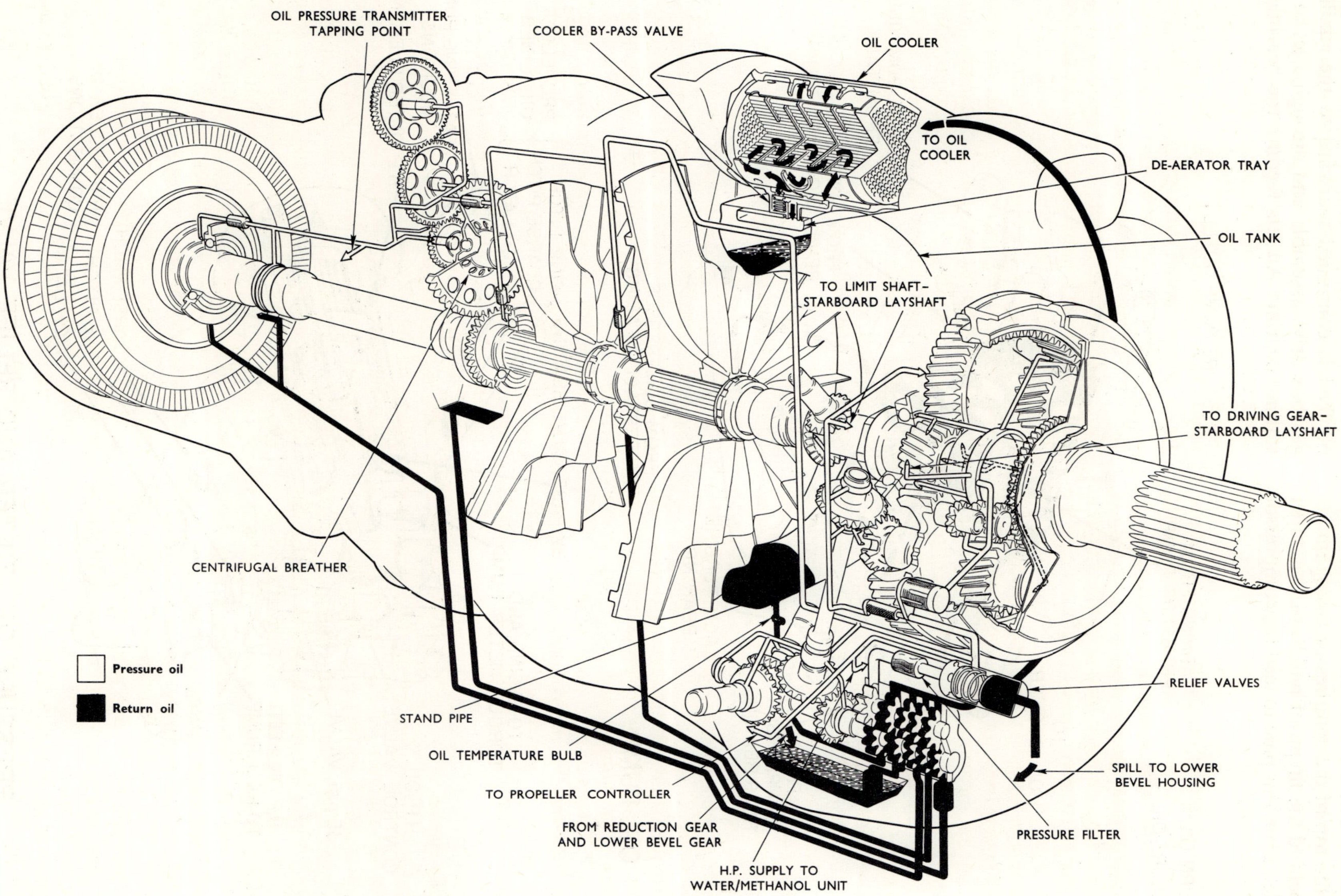

Figure 18·36 Lubrication system of the Dart engine.

air intake casing on the right-hand side of the oil cooler. An external pipe carries the oil rearward to the top of the compressor casing immediately behind the fireproof bulkhead. An outlet is provided from this pipe to feed the compressor interstage and rear bearing jets. From passages in the front face of the intermediate casing the oil feed is continued rearward to the auxiliary gearbox drive assembly and through a jet to the turbine bearing. All jets feeding the bearings of

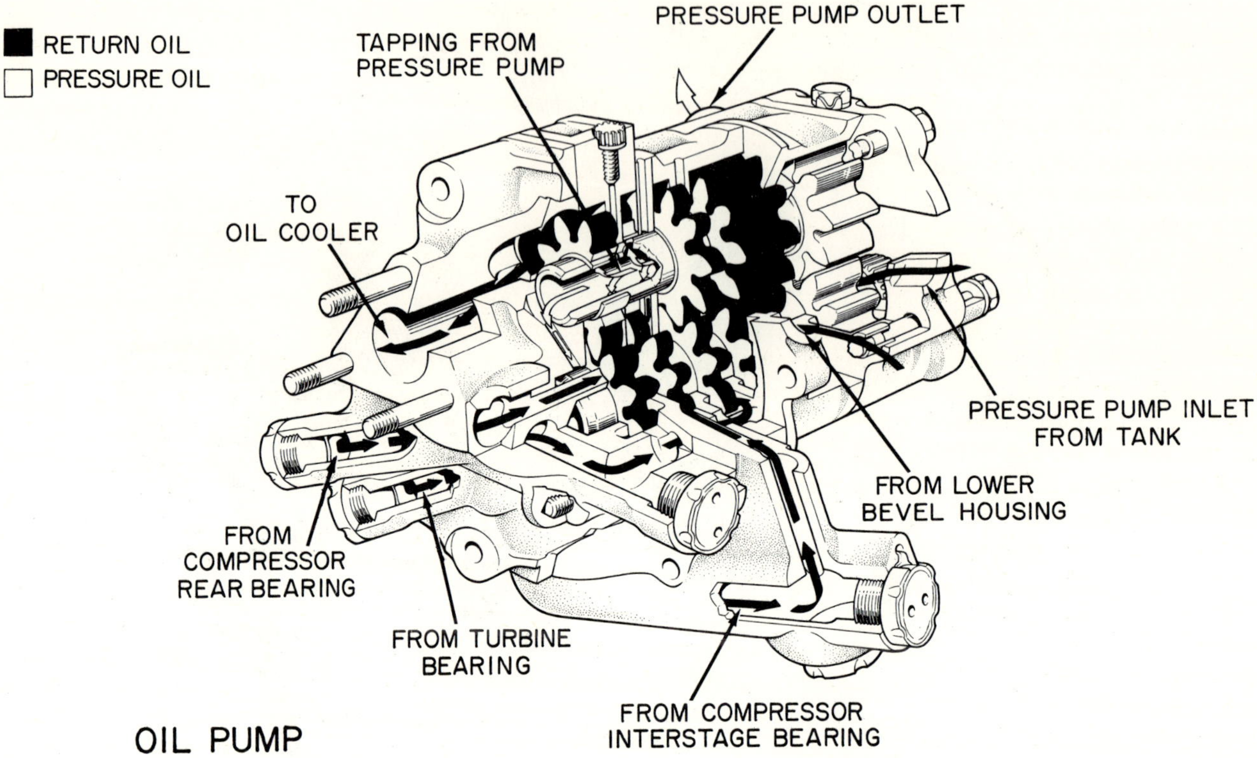

OIL PUMP

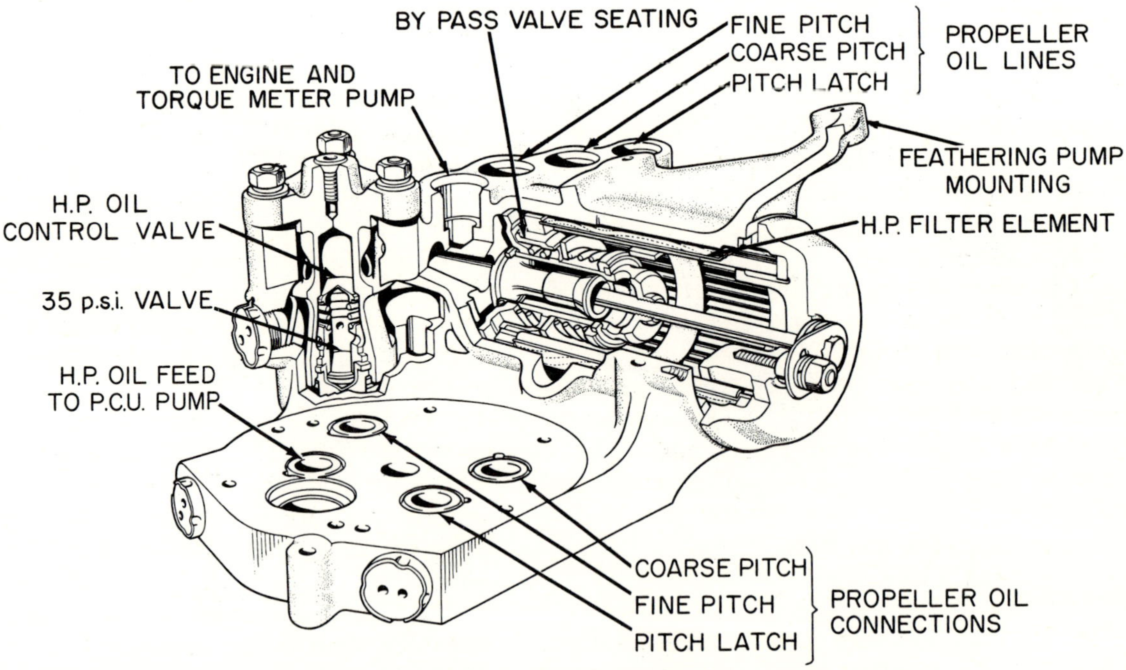

PRESSURE OIL FILTER AND RELIEF VALVE

Figure 18·37 Oil pumps and filters.

the compressor and turbine shafts are protected by thread-type filters. The bearings are provided with labyrinth seals to prevent the loss of oil. Similar seals are fitted to the propeller-shaft front bearing and the auxiliary gearbox drive assembly.

The four scavenge sections of the pump assembly receive scavenge oil flow from the reduction-gear assembly, the interstage bearing, the center bearing and auxiliary gearbox drive assembly, and the turbine bearing. The flow of the scavenge oil is shown in Fig. 18 · 36.

The oil-tank capacity of the Dart engine is 31.8 pints, with a 20 per cent air space. The level of the oil is indicated by a dipstick adjacent to the filler opening. An aperture in the top of the tank receives scavenge oil from the oil cooler. The scavenge oil passes over a de-aerator tray in the upper part of the tank before flowing on into the tank. A tank drain plug and spout are located on the lower left side of the tank.

A cylindrical oil cooler is mounted above the oil tank at the top of the air-intake casing. It is of the tubular type, with the oil flowing between and around the tubes. The cooler is protected against excessive pressures by a 30-psi bypass valve. Air for the cooler passes through an annular space between the air intake and the cowling and leaves by way of a rearward facing duct behind the cooler.

Air sealing and cooling

The Dart engine employs compressor air for sealing and cooling in much the same way as other turbine engines described in this text. Air is tapped from the low-pressure compressor to pressurize the threaded seals on the engine shafts and on the auxiliary gearbox drive shafts.

Air from the high-pressure compressor is used to reduce the working temperatures of the rear engine components by preventing hot gases from leaking away from the main gas stream. This high-pressure air flows to the area around the turbine hub and disks and out past the disk rims. The air which passes into the space between the two turbine disks is high-pressure air.

The air and gas seals are a self-clearing type, designed to expand away from the sealing surface in case of a seal rub.

Fuel system

The Dart engine fuel system includes three principal units: the **fuel pump,** the **flow-control unit,** and the **burners** (fuel nozzles). The fuel pump is an oscillating multiplunger type having a positive displacement of about 10 gallons per hr per 100 rpm of the pump at an outlet pressure of approximately 1,150 psi. The **pump casing** houses the **rotor,** the **cam plate,** the **servo mechanism,** and the **overspeed governor.** All the moving parts are lubricated by the fuel.

The aluminum-bronze rotor houses seven hollow steel plungers (pistons) spring-loaded against the cam plate. The plungers are formed with ball ends which fit into aluminum-bronze slipper pads to form the contact points between the cam plate and the plungers. The pads are surfaced by cadmium plating and are pierced by a small central hole to permit fuel lubrication. The seven slipper pads are retained in alignment with the cam plate by a spring-loaded plate. Between the plunger bores are seven radial passages which develop centrifugal pressure for operation of the overspeed governor. Fuel is fed to these passages through a center passage which is connected to low-pressure inlet fuel.

The governor consists of a diaphragm which actuates a half-ball valve controlling servo pressure to the servo mechanism. The governor diaphragm is spring-loaded so that it will permit required fuel flow to the engine before it starts to reduce fuel. On one side of the governor, pump inlet pressure is applied, and on the other side, fuel pump centrifugal pressure is applied. The pump inlet pressure remains constant, but the centrifugal pressure increases in proportion to engine speed. When engine speed is at the governed rate, the governor diaphragm moves against a lever which unseats the half-ball valve and permits servo pressure to bleed from the servo piston. The servo piston then moves in a direction to reduce the angle of the cam plate and shorten the strokes of the pump plungers. This shortening, of course, reduces pump output. When engine speed and fuel flow are in the correct proportion for governed speed, the governor diaphragm will be balanced to permit the bleed-off of servo pressure at a rate which will hold the cam plate at an angle to give correct fuel flow.

The fuel-flow-control unit, shown in Fig. 18 · 38, interacts with the fuel pump to provide proper fuel flow under all operating conditions. The principal controlling elements in the flow-control unit are (1) an intake pressure-sensing bellows, (2) a control piston, and (3) the throttle valve.

The intake pressure bellows consists of two capsules, one being connected to compressor intake-air pressure. The other capsule serves as a balancing bellows, and both are surrounded by fuel inlet pressure. The principal function of the bellows assembly is to sense air-intake ram pres-

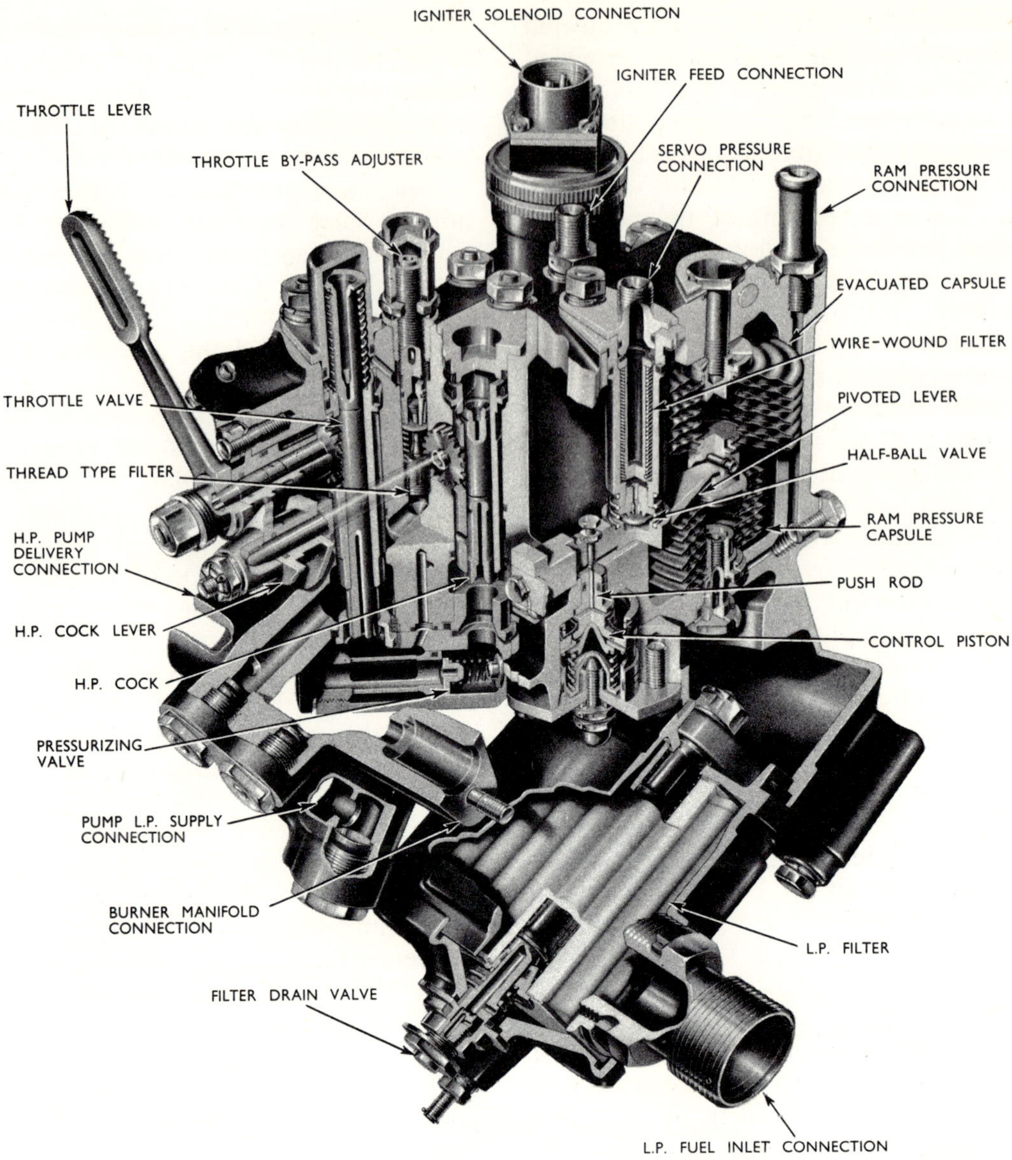

Figure 18 · 38 Fuel-flow-control unit for the Dart engine.

sure and control fuel flow in accordance with mass airflow. The bellows assembly is connected through a lever to a half-ball valve which controls servo bleed-off from the pump servo piston. Thus, when ram-air pressure decreases, the bellows moves to unseat the half-ball valve and allow a decrease in servo pressure. This causes the angle of the cam plate in the fuel pump to decrease and the fuel flow from the pump is reduced.

The purpose of the **control piston** in the flow-control unit is to maintain a correct pressure differential across the **throttle valve.** At sea level, the pressure differential is 100 psi. It will be noted in the schematic fuel-system diagram of Fig. 18 · 39 that the control piston is connected on one side to throttle valve upstream pressure, and on the other side to throttle valve downstream pressure. The spring on the control piston is adjusted to maintain a pressure differential of 100 psi. The control piston acts upon the same half-ball valve through which the intake pressure bellows controls fuel flow; hence the action of the intake pressure bellows will modify the action of the control piston. The result is that the pressure differential across the throttle will decrease as intake pressure decreases.

In a stabilized condition at any altitude, bellows (capsule) load plus pressure drop across the throttle valve equals control piston spring load. Therefore, an increase in bellows load (altitude) requires a lower pressure drop across the throttle valve to stabilize the half-ball valve. The fuel flow is reduced to maintain the turbine temperature within limits with decrease in mass airflow.

In order to maintain a controlled pressure differential across the throttle valve, it is necessary to stabilize the downstream pressure at a constant value. This is accomplished with the **back-pressure valve.** The back-pressure valve is spring-loaded so that it requires a pressure of 100 psi to open it. Thus, there is always a pressure of 100 psi existing between the throttle valve and the back-pressure valve during operation. Inlet pressure to the throttle valve will be 200 psi at sea level, thus producing the pressure differential of 100 psi which is required across the throttle valve.

Summarizing the action of the fuel-control system, the fuel pump delivers fuel to the engine according to the signal it receives from the controlling elements. The **speed control** is the governor within the pump itself. The **air-intake pressure** is sensed by bellows (capsules) within the flow-control unit. Fuel **pressure differential** across the throttle valve is maintained by the control piston in the flow-control unit. Fuel throttling and shutoff are controlled manually

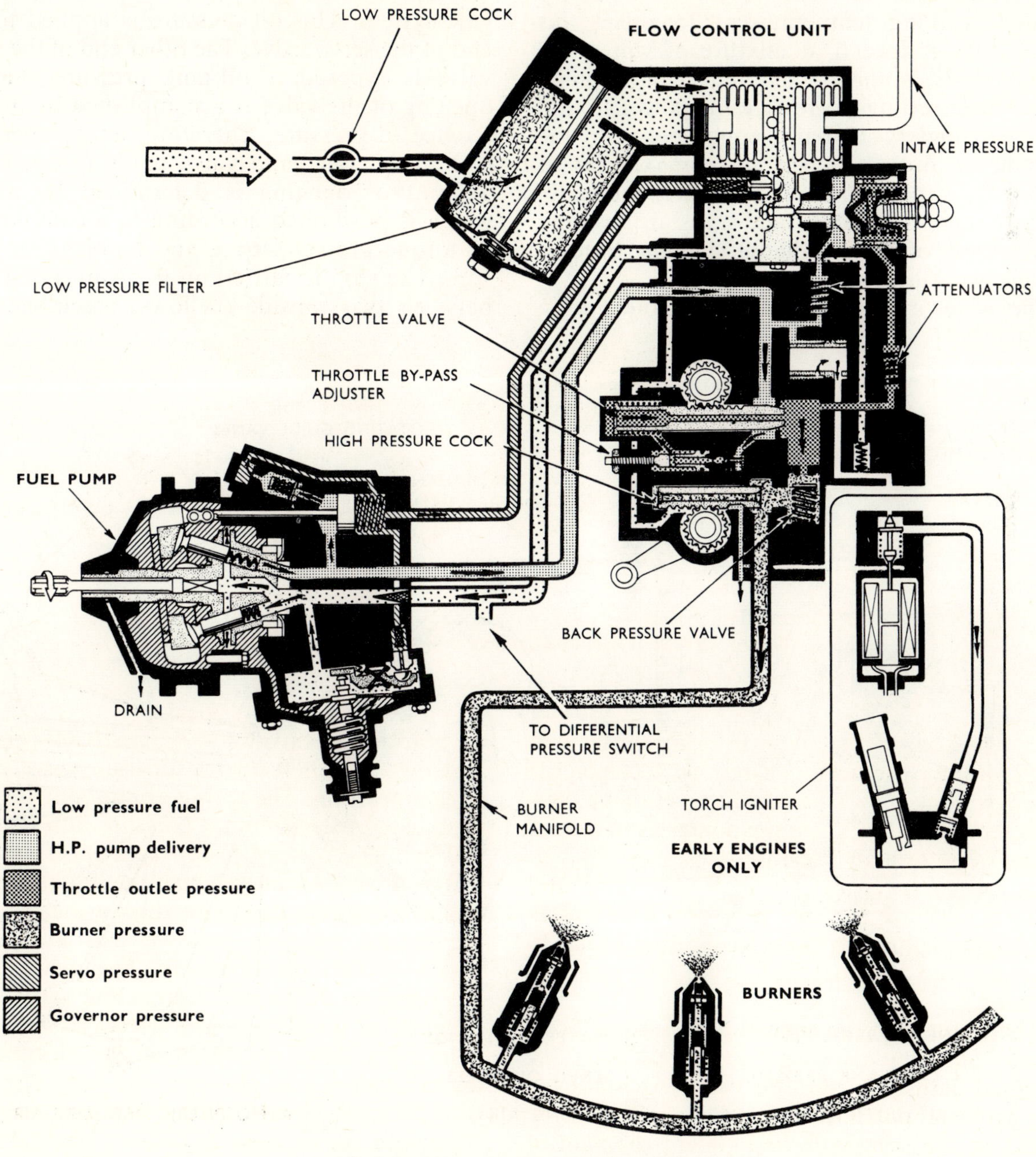

Figure ***18·39*** *Fuel-system diagram.*

from the cockpit. The throttle valve and the stop-cock are actuated through rack-and-pinion gears which are connected to the external control lever.

The engine is provided with a drain system so that all fuel remaining in the fuel manifold and burners will be drained when the engine is shut down.

Water-methanol injection

When operating under conditions of high ambient-air temperature and at airports above sea level, takeoff power is considerably reduced because of the low-mass airflow and the necessary reduction of fuel flow to prevent excessive turbine working temperatures. To offset this power loss at takeoff, a mixture of water and methanol is automatically injected at high power-lever settings to restore the power to that obtained under standard conditions for the altitude of operation, up to a maximum of 10,000 ft. Vaporization of the water-methanol mixture increases density and mass airflow, and combustion of the methanol content restores turbine working temperature.

The water-methanol flow is controlled by a unit mounted on the left-hand side of the air-intake casing, and it is introduced to the engine through passages in the first compressor rotor. The two major sections of the control unit are a **metering-valve assembly** operated by oil pressure and a **control assembly** regulating the oil pressure to the metering valve.

The control assembly, shown in Fig. 18 · 40, is sensitive to **torque pressure, engine rpm,** and **ambient air pressure.** During operation these conditions are used to position a servo valve in the control assembly. The servo valve controls oil pressure to a servo piston which operates the control valve.

Oil pressure is received at 70 psi from the engine system. This oil pressure is applied to one end of the servo valve. The other end of the servo valve is exposed to oil-tank pressure, and the opening of the valve is accomplished by the difference in pressure. Therefore, under operating conditions, the servo valve tends to open. The amount of opening is determined by a stop which is positioned according to a combination of torque pressure, rpm, and ambient air pressure. The stop is carried on the connecting stem between two capsule (bellows) assemblies, the

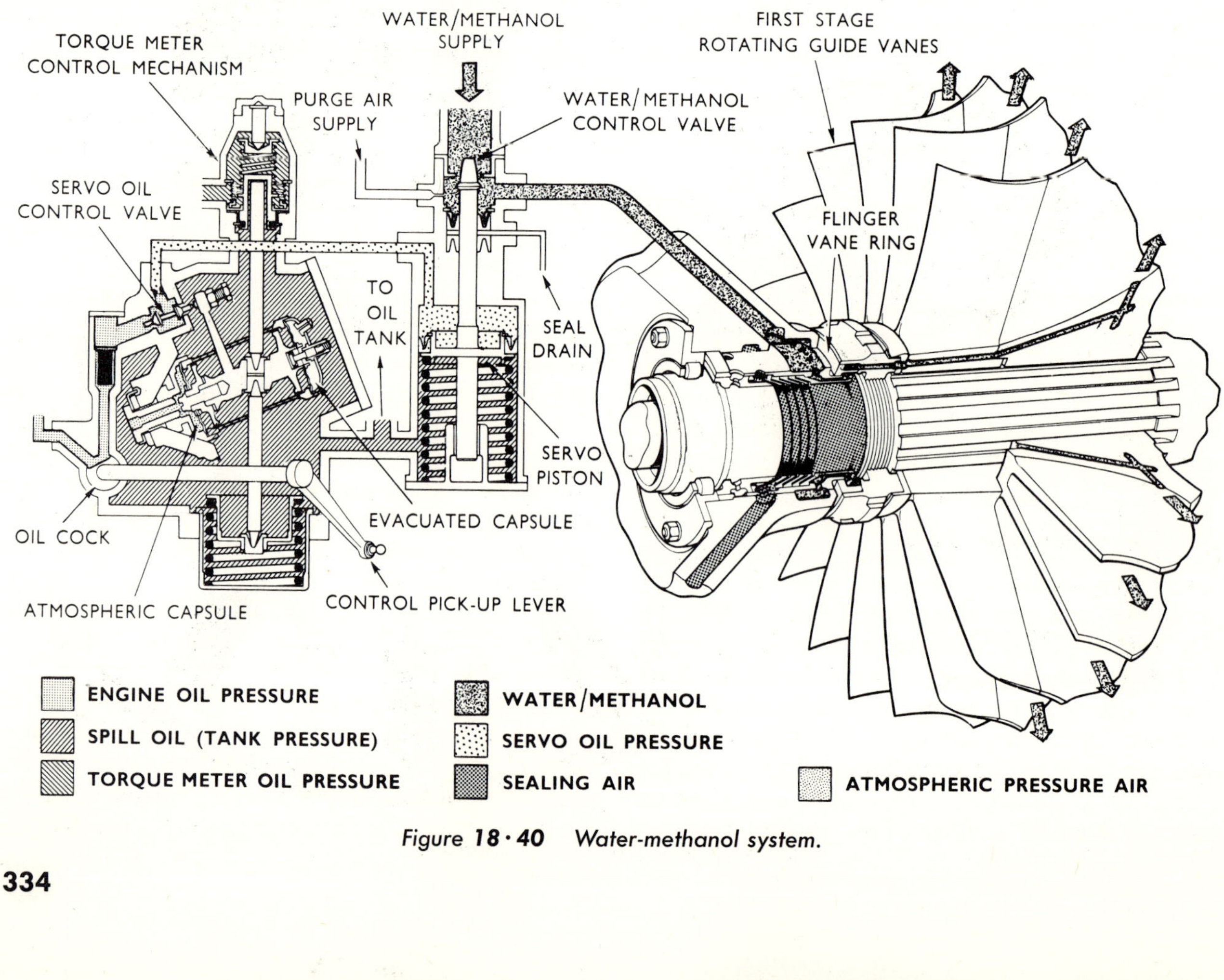

Figure ***18 · 40*** *Water-methanol system.*

left-hand capsule being evacuated and the right-hand capsule connected internally to nacelle pressure. The capsules are held stationary at their outer ends, and a reduction in nacelle pressure moves the connecting stem and the limiting stop to the right. This decreases the amount of valve opening and reduces servo oil pressure.

Bearing on the capsule stem are two vertical pushrods. The upper rod is loaded downward by torque oil pressure and tends to close the servo valve. The lower rod is loaded upward by a spring, and it tends to open the servo valve.

The control unit is equipped with a shutoff valve which is operated by an external lever. The lever is linked to the engine power-control lever, so that the shutoff valve cannot open until engine rpm is more than 14,500.

At maximum rpm on a standard day at sea level, the forces within the control assembly are in balance. Thus they hold the water-methanol metering valve in a sensitive position, where an increase in servo oil pressure would open the valve and allow injection to commence. At temperatures above standard, the forces in the control assembly are out of balance because of the reduced torque pressure. This unbalance allows the servo valve to open and causes injection to begin. Power increases until the rise in torque pressure restores the balance.

In the extreme case of a 45°C day, engine dry power would be 1,180 bhp. During acceleration to takeoff power, injection begins at 14,500 rpm and progressively increases until at 15,000 rpm approximately 350 gallons per min are being injected. This increase gives a 655-bhp increase in power, thus bringing the engine up to 1,835 bhp, which is standard.

Propeller system

The Dart engine is designed to operate with a four-bladed Rotol variable-pitch propeller. The propeller is controlled by an rpm controller which is fundamentally a constant-speed governor. The rpm selection mechanism is interconnected with the throttle valve (power lever) to ensure correct combinations of rpm and fuel flow, thus establishing correct turbine temperature. The feathering control is interconnected with the high-pressure cock (shutoff valve) so that fuel will be shut off before feathering begins.

To prevent excessive torque at low rpm (starting and idling), the propeller blades are allowed to move into "ground fine" (low) pitch by unlatching the flight fine-pitch stop. The unlatching operation is automatically carried out when the landing gear is loaded. This is accomplished through the action of microswitches fitted to the oleo struts. An automatic pitch-increasing device is incorporated in the system to prevent the propeller from moving to "ground fine" pitch and giving high windmilling drag in the event of engine failure during takeoff.

The operation of the propeller is similar to many other conventional constant-speed systems. The rpm controller (governor) utilizes engine-driven flyweights balanced by an adjustable spring to control rpm. As the flyweights move "in" by spring pressure or "out" by centrifugal force, they move a pilot valve which directs oil to and from the propeller piston.

The rpm is selected by means of the throttle and constant-speed unit control (power lever). Movement of this lever in a certain range will increase or decrease the spring pressure acting on the flyweights. An increase in spring pressure will cause an increase in rpm. In an overspeed condition, the governor weights move outward because of centrifugal force. This lifts the pilot valve and oil is directed to the front of the piston in the propeller dome. The piston moves rearward and increases the pitch of the propeller. This increases the load on the engine and the rpm decreases. An underspeed condition causes the pilot valve to move downward. This directs oil to the rear of the propeller piston and causes the propeller to change to a lower pitch, thus reducing the load on the engine and increasing rpm.

The propeller operating oil is taken from the engine oil system, and the pressure is boosted by means of a gear-type pump in the controller. This boosting is necessary because engine oil pressure is not high enough to operate the pitch-changing mechanism.

The propeller is feathered through the use of a feathering pump. The manual feathering-control lever is placed in the FEATHER position to place the governor pilot valve in the proper position and to shut off fuel to the engine. Then, by pushing the feathering button, the feathering pump is started and oil is directed under high pressure to the forward side of the propeller piston. This rotates the propeller blades to the feathered position.

To unfeather, the manual feathering-control lever is placed in the OPEN position, which returns the controller unit (governor) to the constant-speed position. Since the engine is stopped, the valve will be in "underspeed" and will direct oil pressure to the rear side of the propeller piston. The feathering switch must be closed to start the feathering pump, thus providing oil pressure to the propeller. When the propeller has returned to the normal range, the feathering switch is released.

When Dart engines are used in multiengined aircraft, the propeller speeds are automatically synchronized. The No. 3 engine of the aircraft is established as a "master" engine and the other "slave" engines are synchronized to the master engine.

Starter system

The electric circuit for the starting system is shown in Fig. 18 · 41. The starter is a 24–28-volt d-c motor which engages the main rotor shaft of the engine through a clutch and quill shaft. The starter drive may be seen in Fig. 18 · 34. The starter drive includes a pawl arrangement which is engaged at starting speeds. It disengages through the action of centrifugal force when engine speed is self-sustaining.

When it is desired to start the engine, the **starter master switch** is placed in the START position. This supplies power to the **starter button.** The **igniter test switch** is placed in the NORMAL position to route power to the **engine selector switch.** The engine selector switch is placed in the SELECT position for the engine being started.

With the switches' positions as indicated, the circuit is ready for starting the engine. When the starter button is depressed, it initiates current through the igniter test switch, the engine selector switch, and the **selector relay.** The selector relay provides a path for the power to the starter of the particular engine being started. There is also current directly from the starter button to the **main relay.** The main relay completes the main power circuit to the starter, and the engine begins to rotate.

Power from the engine selector switch flows through the **isolating relay** and the **relight time switch** to the high-energy ignition units. The ignition units supply the high-voltage high-energy current to the igniter plugs.

When the engine has reached approximately 1,500 rpm, fuel is supplied to the burners and combustion begins. As the engine accelerates, assisted by the starter motor, the load on the starter pawls decreases sufficiently to allow centrifugal force to disengage them. Also, as the starter load decreases, the starter-load current through the overspeed relay decreases. The relay opens, discontinuing the current to the starter-button holding coil, and the starter button releases, thus deenergizing all starter units.

Operation

Any turboprop or turbojet engine should be operated according to the instructions set forth

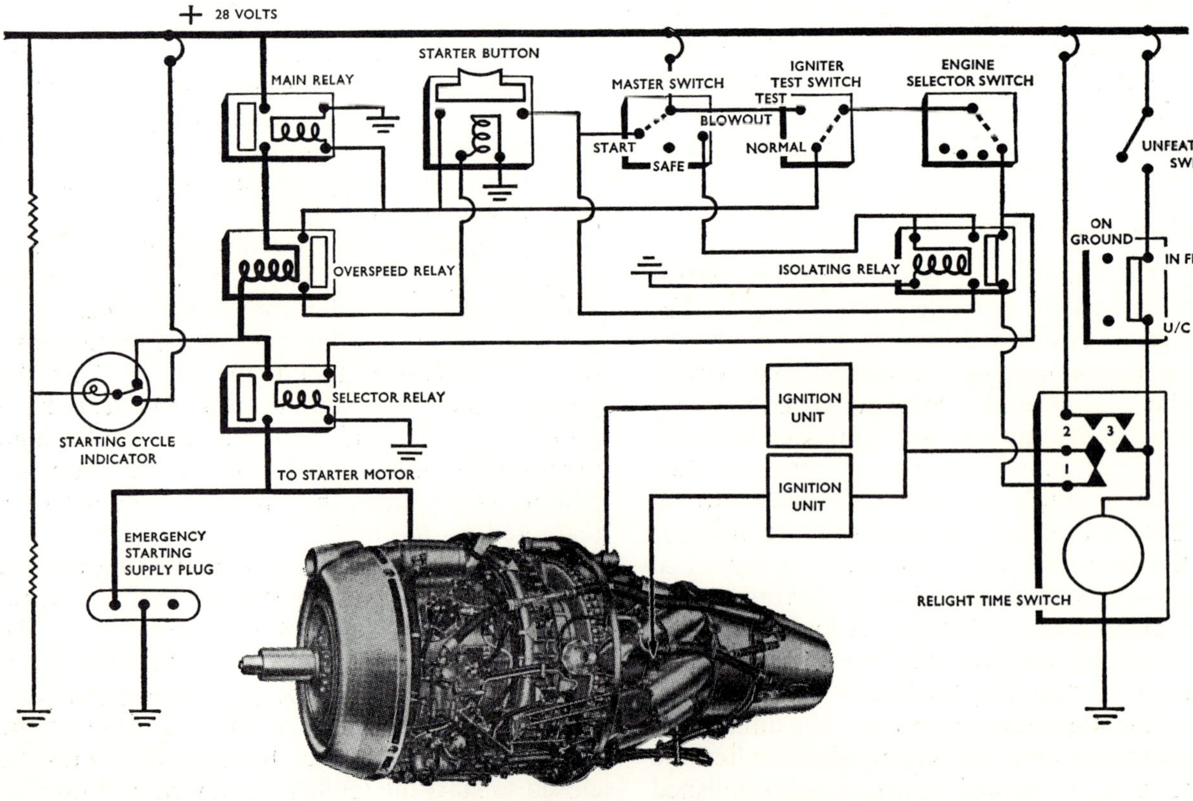

Figure 18 · 41 Starter system.

Table 18 · 1 Operating limitations

Condition	Engine rpm	Maximum Turbine gas temp, °C	Time limit
Ground idling	6,500–7,500	550	Unrestricted
Minimum approach	10,400	680	Unrestricted
Cruising	13,750–14,500	740	Unrestricted
Maximum continuous	14,500	775	Unrestricted
Takeoff			
Dry	15,000	795	5 min
Wet	15,000	825	5 min
Overspeed	17,000 (max)	. . .	20 sec

in the manufacturer's operation and maintenance manual.

As with all engines, there are certain maximum limits of temperature and speed beyond which the Dart engine should not be operated. These are listed in Table 18 · 1.

THE LYCOMING T53 TURBOPROP ENGINE

General description

The Lycoming T53 engine has been manufactured in several different models, some for use in aircraft as turboprop engines and others for use in helicopters as turboshaft engines. The basic engine is the same in either case; however, certain differences are necessary to accommodate the type of operation for which a particular engine is designed. For the purposes of this section we shall discuss the T53-L-3 turboprop engine, which is illustrated in Fig. 18 · 42.

The T53 turboprop engine is a reverse-flow gas turbine having both axial and centrifugal compressors. The performance and specifications are as follows:

Equivalent shaft horsepower	1,150 (1,100 shp plus 124 lbt)
Specific fuel consumption	0.64 lb/ehp/hr
Power/weight ratio	2.07 ehp/lb
Oil consumption	1.0 lb/hr
Maximum rpm	25,150
Diameter	23 in.
Length	58.4 in.
Frontal area	2.9 sq ft
Weight (without propeller)	555 lb
Oil specification	MIL-L-7808 or equivalent
Fuel specification	Kerosene, JP-1 or JP-4

ENGINE DESCRIPTION

The T53-L-3 gas-turbine engine is a free-turbine powerplant designed for conventional fixed-wing aircraft. The expression "free-turbine" means that the power turbine is separate from the turbine which drives the compressor. The engine consists of an **inlet housing,** a **reduction gear section,** an **axial-centrifugal compressor,** a **diffuser,** a **combustion chamber,** a **first-stage turbine,** a **second-stage turbine,** an **exhaust diffuser,** and a **power shaft.** The complete compressor rotor assembly consists of five axial stages, one centrifugal impeller, and the first-stage turbine rotor. The first-stage turbine rotor drives the compressor while the second-stage turbine rotor drives the power shaft, reduction gear and propeller.

Operation

The airflow through the T53 engine is similar to that described for other gas-turbine engines. It enters the annular air inlet surrounding the reduction gear and passes into the axial compressor section. From thence it flows through the centrifugal compressor, the diffuser housing, and into the combustion chamber, where it is scooped into the inner liner. When the engine is being started, fuel first is injected through five igniter nozzles. This fuel is ignited by two igniter plugs in the combustion chamber at the 4 and 8 o'clock positions. When the heat of the burning mixture is great enough to sustain combustion and engine speed has reached a preset value, main fuel begins to flow into the combustion chamber through 11 fuel vaporizers. The rotor speed increases as this additional fuel mixes with the compressed air and is burned; and when the speed is 23 to 27 per cent of full speed, the ignition system is deenergized, and the starting fuel is shut off. The engine then accelerates to

Figure **18·42** *The Lycoming T53-L-3 turboprop engine, left side view.*

ground idle speed. During this period, the starter deenergizes between 30 and 40 per cent of N_1 (compressor rotor maximum speed).

Combustion gases pass through the first-stage nozzle and impinge upon the blades of the first-stage turbine rotor, through the second-stage nozzle, and onto the second-stage turbine rotor blades. Approximately two-thirds of the gas energy passing through the first-stage turbine rotor is used to drive the compressor rotor, and the rest of the energy is used to drive the power shaft. The second-stage turbine rotor is splined to the power shaft and secured to it by the power-shaft bolt. The power shaft splines into the reduction sun gear in the inlet housing.

Inlet housing

The construction of the engine is clearly shown in the cutaway photograph of Fig. 18·43. The **inlet housing** is shown at 3 in the illustration. This assembly may be considered as embracing two principal areas. The **outer housing,** supported by six struts, forms the air inlet area and houses the de-icing manifold. The **inner housing** encloses the reduction gears, the propeller shaft rear bearing support, the accessory drive carrier, and the torquemeter. The overspeed governor and tachometer drive gearbox is mounted on the left side on the inlet housing, and the accessory drive gearbox is mounted on the bottom of the housing.

Interstage bleed, compressor, and impeller housing

The interstage bleed and centrifugal compressor housing includes the axial compressor housing and the impeller housing assemblies, the actuator, the controller, and the band and screw assembly. The axial compressor housing is machined to accommodate five compressor vane assemblies shown at 10 in the illustration of Fig. 18·43. The actuator, controller, and band and screw assembly form the **acceleration bleed valve.** The actuator and controller, mounted on the right side of the centrifugal compressor impeller housing, bleed air from a point between the axial and centrifugal compressors at speeds below 30 per cent normal rated power.

Compressor rotor

As previously explained, the **compressor rotor assembly** includes the axial compressor, the centrifugal compressor, and the first-stage turbine. The axial-flow section consists of five rotor disk and blade assemblies and five compressor rotor spacers, as shown in the illustration. The first-stage disk is machined from steel and has steel blades, while the other four disks are aluminum alloy with steel blades.

The centrifugal compressor impeller is made in two sections from titanium alloy. It is mounted on the same hollow shaft which supports the axial-flow compressor disks. This shaft is called the **compressor rotor sleeve.**

The first-stage turbine disk is mounted on the rear end of the hollow drive shaft which attaches to the compressor rotor sleeve. The drive shaft is supported by the No. 2 bearing shown at 15 in Fig. 18 · 43. The compressor rotor shaft encloses, but does not connect to, the power shaft.

Combustor turbine

Because the combustion chamber surrounds the two turbine disks, the combustion and turbine section of the engine is called the **combustor turbine.** This section consists of the combustion chamber, the second-stage turbine nozzle and cylinder, the second-stage turbine support, the fire shield, and the support cone. The second-

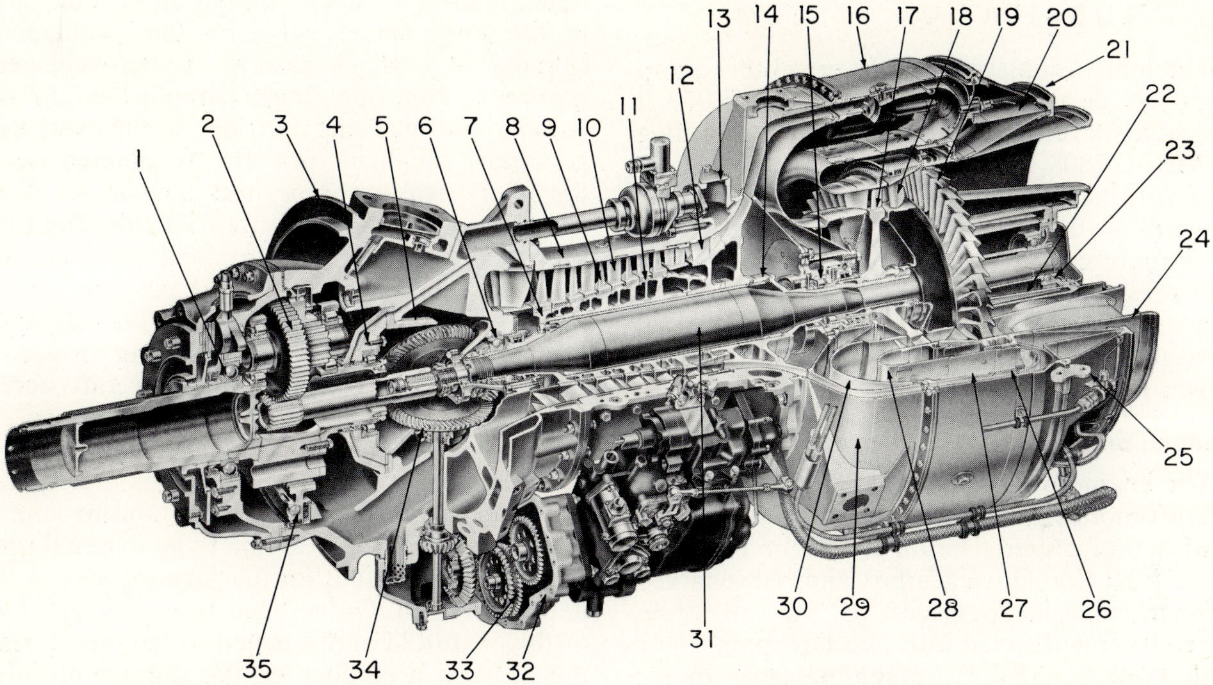

1. Propeller thrust bearing
2. Reduction gear
3. Inlet housing
4. Propeller shaft rear bearing support
5. Accessory drive carrier
6. No. 1 main bearing
7. Compressor rotor
8. Axial compressor housing
9. Compressor rotor disk and blades
10. Compressor vane
11. Compressor rotor spacer
12. Centrifugal compressor rotor impeller
13. Centrifugal compressor housing
14. Compressor rotor sleeve
15. No. 2 main bearing
16. Combustor
17. First-stage turbine rotor
18. Second-stage turbine nozzle
19. Second-stage turbine rotor
20. Fire shield
21. Support cone
22. No. 3 and No. 4 bearings housing
23. No. 4 bearing
24. Exhaust diffuser
25. V-band coupling
26. Scoop and shroud
27. Combustion chamber
28. Liner
29. Diffuser housing
30. Combustion chamber deflector
31. Power shaft
32. Accessory drive gearbox
33. Accessory gears
34. Accessory drive gear
35. Torquemeter

*Figure **18 · 43** Cutaway photograph of the Lycoming T53 engine.*

stage turbine support is made up of the second-stage turbine rotor and the exhaust diffuser. The second-stage turbine rotor consists of the second-stage turbine disk and blades, the bearings, and the No. 3 bearing seal. The welded exhaust diffuser contains hollow struts through which cooling air is supplied to the No. 3 bearing housing and the second-stage turbine disk.

The turbine blades are attached to the turbine disks by means of fir-tree blade roots inserted in the slots cut in the rims of the disks. The second-stage blades are shrouded.

LUBRICATION SYSTEM

The lubrication system for the T53 turboprop engine is designed to operate up to 25,000 ft above sea level with lubricating oil conforming to MIL-L-7808 specifications. This oil will provide satisfactory lubrication with an oil inlet temperature between −65 and 200°F.

The lubrication system consists of the manifold pressure and scavenge pump, the oil filter, and associated external oil lines and internal passages. A schematic diagram of the system is shown in Fig. 18 • 44.

Main oil pressure supply system

The engine oil supply is contained within an aircraft-mounted oil tank supplied by the aircraft manufacturer. The oil enters the oil pump on the accessory drive gearbox and is discharged through internal passages to the oil filter. Filtered oil is directed into two flow paths. One path flows internally through passages in the accessory drive gearbox and air inlet housing to the air inlet section to lubricate the reduction gears, torquemeter booster pump, torquemeter, accessory drive carrier, and the No. 1 bearing. The second oil flow path is external and leads to the Nos. 2, 3, and 4 bearings.

In the inlet section, oil is directed through the accessory drive carrier flanges into the rear-propeller-shaft bearing support. Internally drilled passages and an annulus in the rear-propeller-shaft bearing support serve as a common oil manifold. Oil from this manifold is directed to the accessory drive carrier assembly for forced-feed spray lubrication of the carrier gears, accessory drive pinion gear, and power-shaft support bearing. The No. 1 main bearing is lubricated by a transfer tube from the accessory drive carrier.

Pressurized oil from the support manifold constantly lubricates the propeller-shaft rear support bearing. A second oil transfer from this manifold is directed into the propeller shaft to spray the sun gear, propeller-shaft front support bearing, planet gears, and bearings. A third oil transfer from the manifold is directed through an air inlet strut to the mounting pad of the overspeed governor and tachometer drive. The oil is directed to a torquemeter booster pump mounted on and driven by the overspeed governor and tachometer drive. A pressure-regulating valve in the overspeed governor and tachometer drive controls booster pump output by circulating excess pressure oil back to the booster pump inlet port. The pressure oil from the booster pump is directed back through an air inlet strut to the torquemeter valve on the torquemeter cylinder. An offset passage in the overspeed governor mounting flange supplies oil to the strainer and metering cartridge in the overspeed governor, which in turn directs metered oil to the overspeed governor and tachometer drive gear train. An additional transfer tube from the propeller-shaft rear support bearing directs oil through inlet housing internal passages to the power takeoff mounting flange. This oil passes through a strainer and a metering orifice to supply oil to engine-mounted aircraft accessories.

External oil flow passes through a right-angle elbow and an adapter at approximately the 5 o'clock position on the inlet housing and is directed by a flexible oil line to an external rigid annular oil manifold on the forward side of the diffuser section. Oil is taken from the right side of this manifold and directed to the No. 2 bearing through a strainer on the diffuser housing. Oil is directed by means of a flexible hose from the left side of the manifold, through a strainer in the top of the oil supply tube in the exhaust diffuser to lubricate the Nos. 3 and 4 bearings.

Oil scavenge system

All internal scavenge oil from the air inlet housing section is drained into a hollow support strut in the inlet housing, through a strainer into the accessory drive gearbox. Scavenge oil from the reduction gears is fed by gravity into the hollow inlet strut. Scavenge oil from the No. 1 main bearing is pumped to the hollow inlet strut by an impeller mounted to the rear of the bearing. From the No. 2 main bearing, scavenge oil is pumped by two impellers, one on each side of the bearing, through a scavenge oil tube in the diffuser and to the accessory drive gearbox by an external scavenge oil line. Drain flow from the Nos. 3 and 4 bearings is aided by two impellers

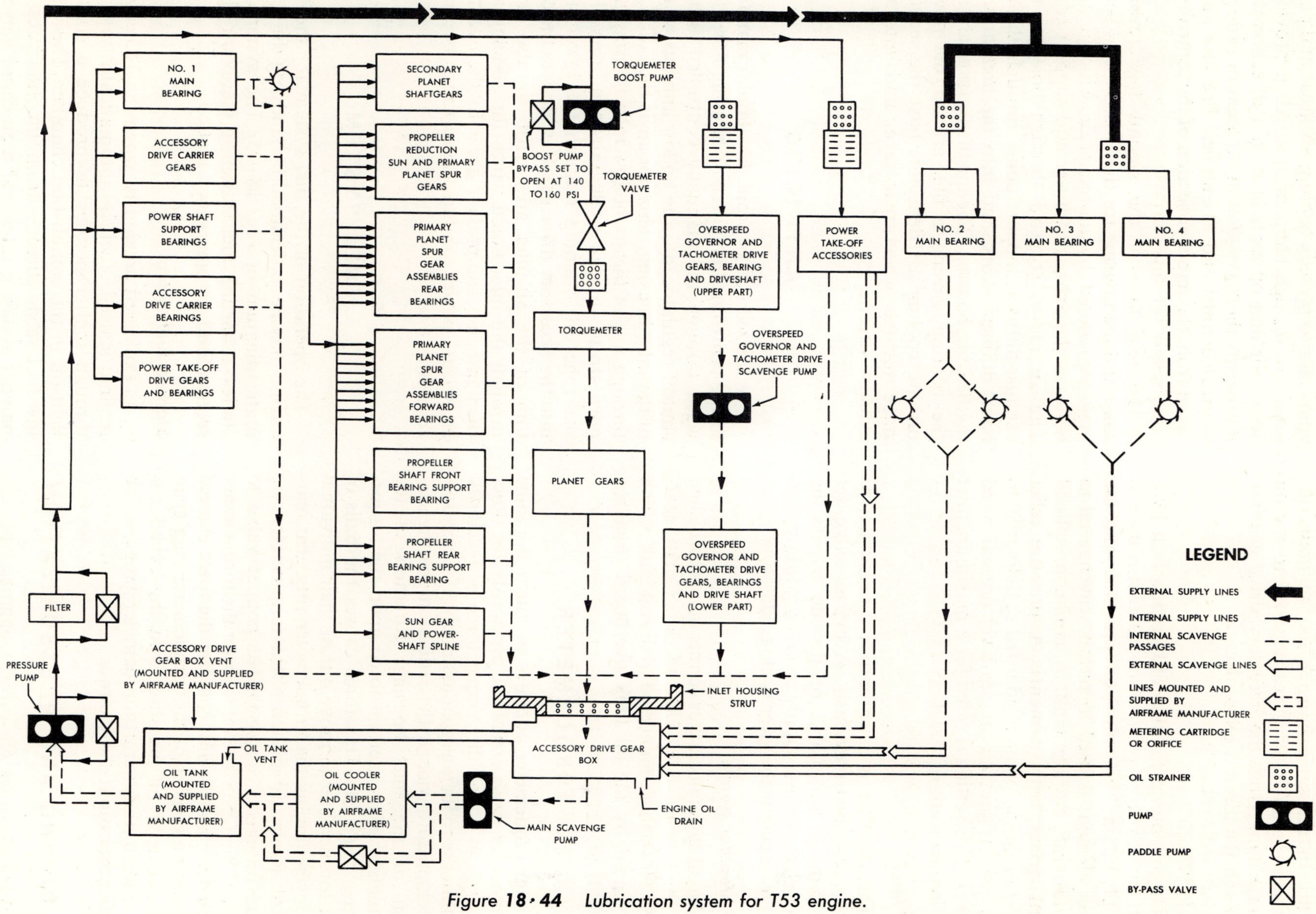

Figure 18·44 Lubrication system for T53 engine.

in the bearing housing. The gear-type scavenge pump, in the same housing as the main oil-pressure pump, returns scavenge oil to the oil storage tank from the accessory drive gearbox through the aircraft oil cooler.

Oil pump

The oil pump for the T53 engine is a two-element gear-type pump. One element is used to supply main lubricating oil pressure, and the other element is used to return scavenge oil to the aircraft storage tank. A common, splined shaft drives both elements. A pressure relief valve in the oil pump is adjusted to deliver 80 to 100 psi oil pressure at sea-level normal rated power. This setting is rated for a maximum inlet oil temperature of 200°F (93°C).

Oil filter

The oil filter is a wafer-disk type similar to those described previously. A bypass valve, set at 15 to 20 psi, allows oil flow to bypass the filter elements and supply emergency lubrication to the engine if the filter is clogged.

Oil breather system

An oil system breather connection is furnished on the forward side of the accessory drive gearbox. This vent is connected to the oil tank, which is vented by the oil tank overboard breather.

TORQUEMETER

The torquemeter operates in much the same manner as those explained previously for other engines in which the torque force is developed within the reduction-gear assembly. It is a hydromechanical device in the reduction-gearing section of the inlet housing. Rotation of the planetary gears exerts a force upon the torquemeter ring gear. This force causes the balls of the torquemeter to move out of their individual grooves, forcing the torquemeter ring gear rearward. Rearward motion of the torquemeter ring gear unseats a spring-loaded poppet valve to permit high-pressure oil to enter the torquemeter cylinder chamber and equalize the force exerted by the movement of the torquemeter ring gear. Torquemeter oil pressure from the cylinder is directed to the aircraft torquemeter gauge, which indicates output shaft torque.

COOLING AND PRESSURIZATION

As with other gas-turbine engines, the T53 engine requires controlled airflow through various parts of the engine to pressurize bearing voids and to provide cooling for areas and parts subjected to high temperatures. Air is provided across air and oil seals in such a direction that it prevents the leakage of oil to areas outside the lubricated units and assemblies. Pressure air is taken from the impeller section of the centrifugal compressor for this purpose.

The rear face of the second-stage turbine rotor and the housing for the Nos. 3 and 4 bearings are cooled by ambient air which enters between the exhaust diffuser and support cone. The air moves forward through holes in the exhaust-diffuser support cone, through the exhaust-diffuser struts, and into the area around the bearing housing. The air then moves around the bearing-housing baffle, around the power turbine cooling air deflector, past the second-stage turbine wheel rear face, and into the exhaust stream.

ANTI-ICING

Pressurized hot air from the annular manifold in the centrifugal compressor housing flows forward through the manually controlled airflow shutoff regulator into the hollow annulus on top of the inlet housing. This hot air is then directed through five of the six hollow inlet housing support struts to de-ice the air-inlet area. Hot scavenge oil, draining through the lower strut into the accessory drive gearbox, de-ices the bottom of the air-inlet area. Hot air also flows through the inlet guide vanes for de-icing and then into the compressor area. In the event of electrical power failure, anti-icing becomes continuous.

FUEL SYSTEM

The operation of the fuel system for the T53 engine may be understood by studying the schematic diagram of Fig. 18·45. The system includes a starting fuel system, a fuel-controlling section, and an emergency section.

Starting fuel system

As explained previously, when the T53 engine is first started, fuel is automatically injected through igniter nozzles to produce a starting light-off in the combustion chamber. This starting fuel feeds directly from the fuel pressure pump through a filter and a solenoid valve to the igniter nozzles.

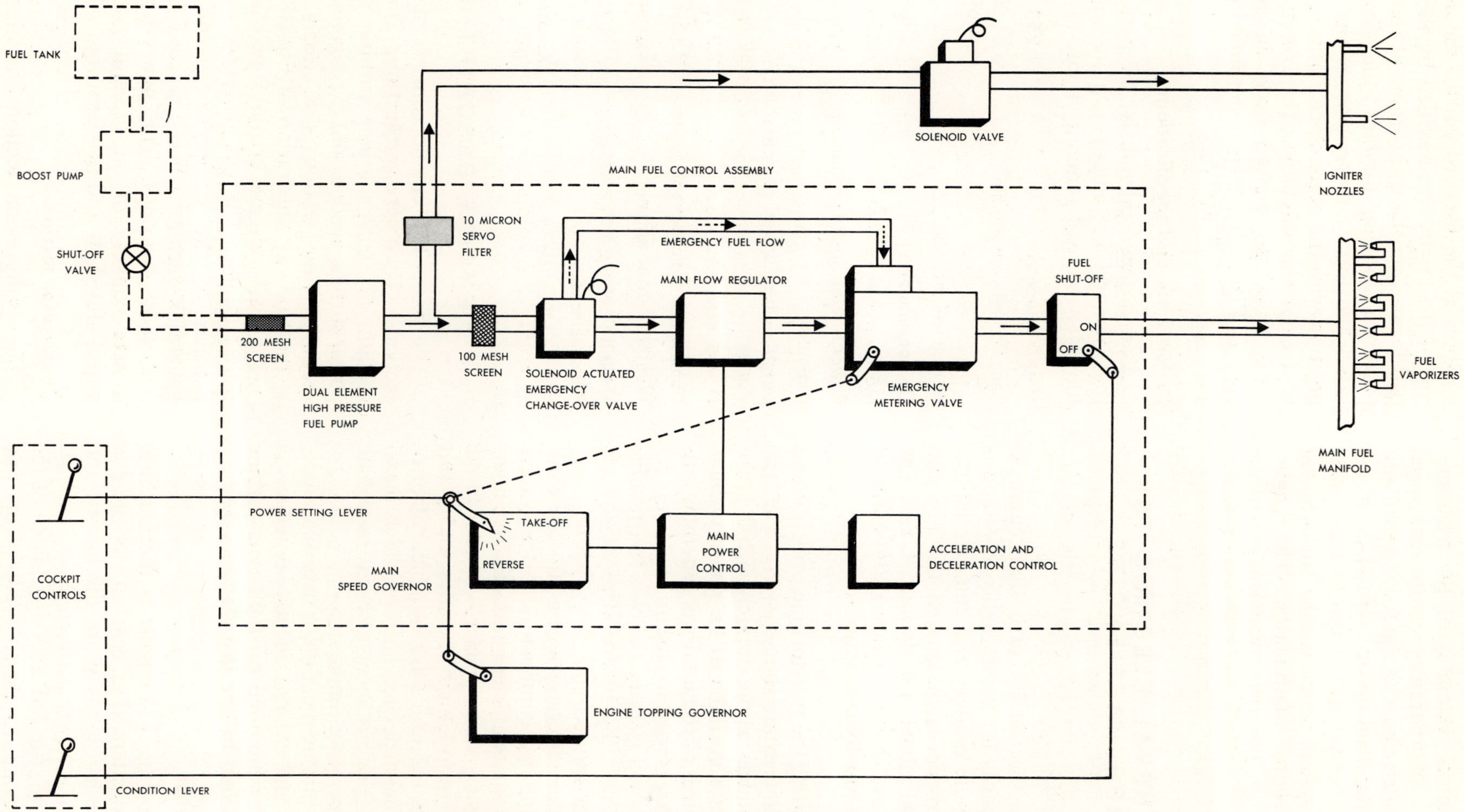

Figure 18·45 *Diagram of fuel and control system for the T53 engine.*

Main fuel-control assembly

The fuel-control system is of the hydromechanical type, incorporating centrifugal speed sensors. All the elements of the fuel-control unit are shown in the diagram of Fig. 18 • 45. It will be noted that the system incorporates an emergency bypass system which is solenoid-actuated in case of control malfunction. The engine topping governor prevents overspeeding of the engine.

THE GENERAL ELECTRIC T58 TURBOSHAFT ENGINE

General description

The General Electric T58 turboshaft engine is designed primarily for use in helicopters and is available for both military and commercial use. The military model is designated the T58 and the commercial model is designated CT58. The performance and characteristics of the different models are essentially the same.

The T58 engine is an axial-flow turboshaft engine incorporating the free-turbine principle and producing 1,250 shaft horsepower. The **gas generator** produces approximately 3,200 hp of which 1,250 hp is available at the power turbine. Nearly two-thirds of the total gas generator output, or 2,000 hp, is used to sustain engine operation. Output power is extracted by a free turbine which is mechanically independent of the gas generator rotor system.

The gas generator consists of a 10-stage compressor, an annular combustor, and a 2-stage turbine. To promote efficient and stall-free operation, the **variable-stator** principle is used in the compressor. The inlet guide vanes and stator vanes in stages 1, 2, and 3 are variable. A hydromechanical fuel-metering unit provides maximum engine performance without exceeding safe engine operating limits.

The power-turbine section consists of a single-stage turbine and exhaust section. A reduction gearbox can be mounted on the power-turbine section if required by the installation.

Engine orientation

The front of the T58 engine is the end into which the air enters. The rear end is the end from which exhaust gases are discharged and from which shaft power is taken. From the rear looking forward, the left and right sides of the engine correspond to the observer's left and right side. The bottom of the engine is identified by the accessories group located at the 6 o'clock position at the forward end of the engine. The top of the engine is identified by the mounting pad located on the front frame at the 12 o'clock position. Looking forward, the observer considers the top of the engine as the 12 o'clock position with the other clock positions following in a clockwise direction. The compressor rotor turns clockwise, viewed from the rear of the engine. Figure 18 • 46 is a simplified drawing to show engine orientation.

Performance and specifications

As mentioned previously, the performance of the T58 and CT58 engines is quite similar; hence the following will generally apply to both:

Shaft horsepower	1,250
Specific fuel consumption	0.61 lb/hp/hr (takeoff power)
Power/weight ratio	4.23 shp/lb
Oil consumption	1.3 lb/hr
Maximum rpm (gas generator)	26,300
Power turbine speed (optimum)	19,500
Diameter	16 in.
Length without speed decreaser or starter	59 in.
Frontal area	2.2 sq ft
Compressor ratio (takeoff power)	8.15:1
Weight of basic engine	295 lb
Weight of speed decreaser	106 lb (optional)
Oil specification	MIL-L-7808D
Fuel specification	Kerosene, JP-1, JP-4, JP-5

ENGINE DESCRIPTION

The basic T58 engine consists of five major sections: (1) **compressor section,** (2) **combustion section,** (3) **gas-generator turbine section,** (4) **power-turbine section,** and (5) **engine-accessories section.** These sections are clearly discernible in the cutaway illustration of Fig. 18 • 47.

The basic engine employs two independent rotor systems. The **gas-generator rotor system** consists of the compressor rotor, two gas-generator turbine wheels, a connecting shaft, and three bearings. The **power-turbine rotor system** consists of an integral power-turbine wheel and shaft and two bearings.

Compressor section

The compressor section includes the **front frame, compressor rotor, compressor stator,** and

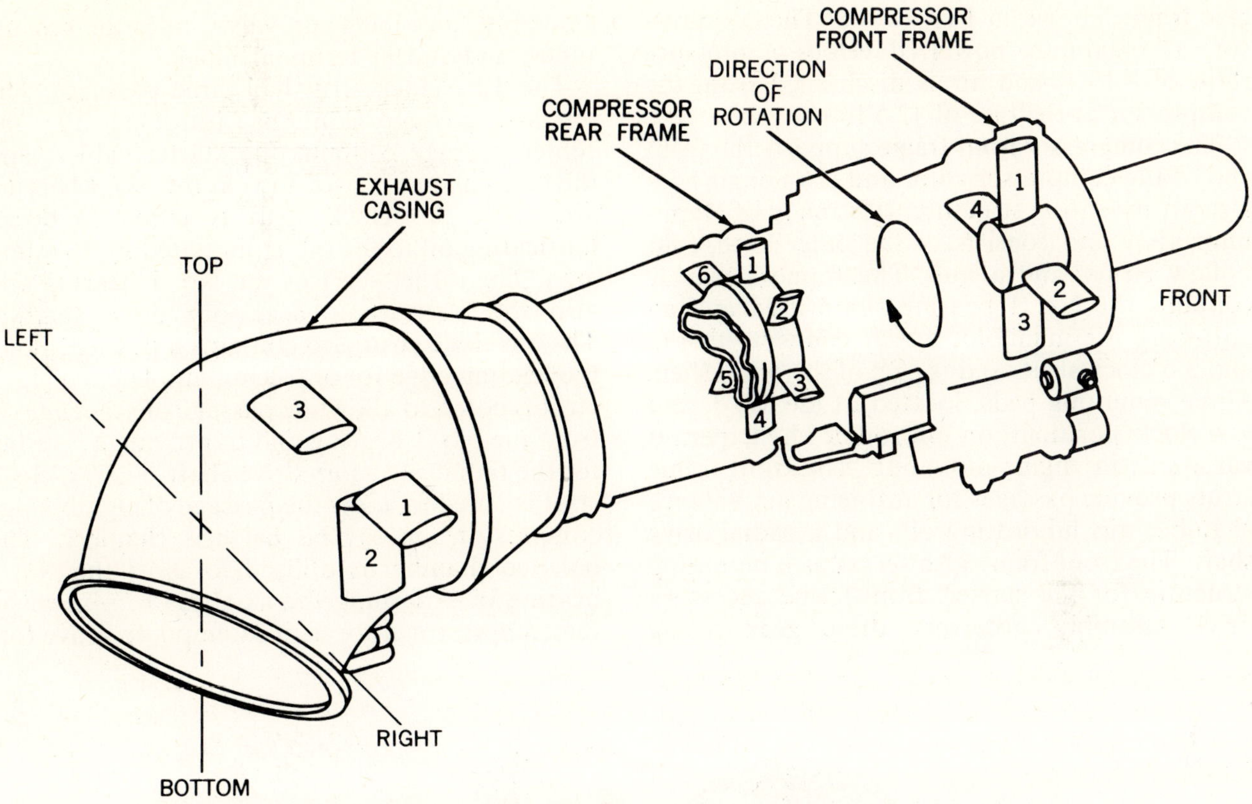

Figure **18·46** *Diagram showing orientation for the General Electric T58 engine.*

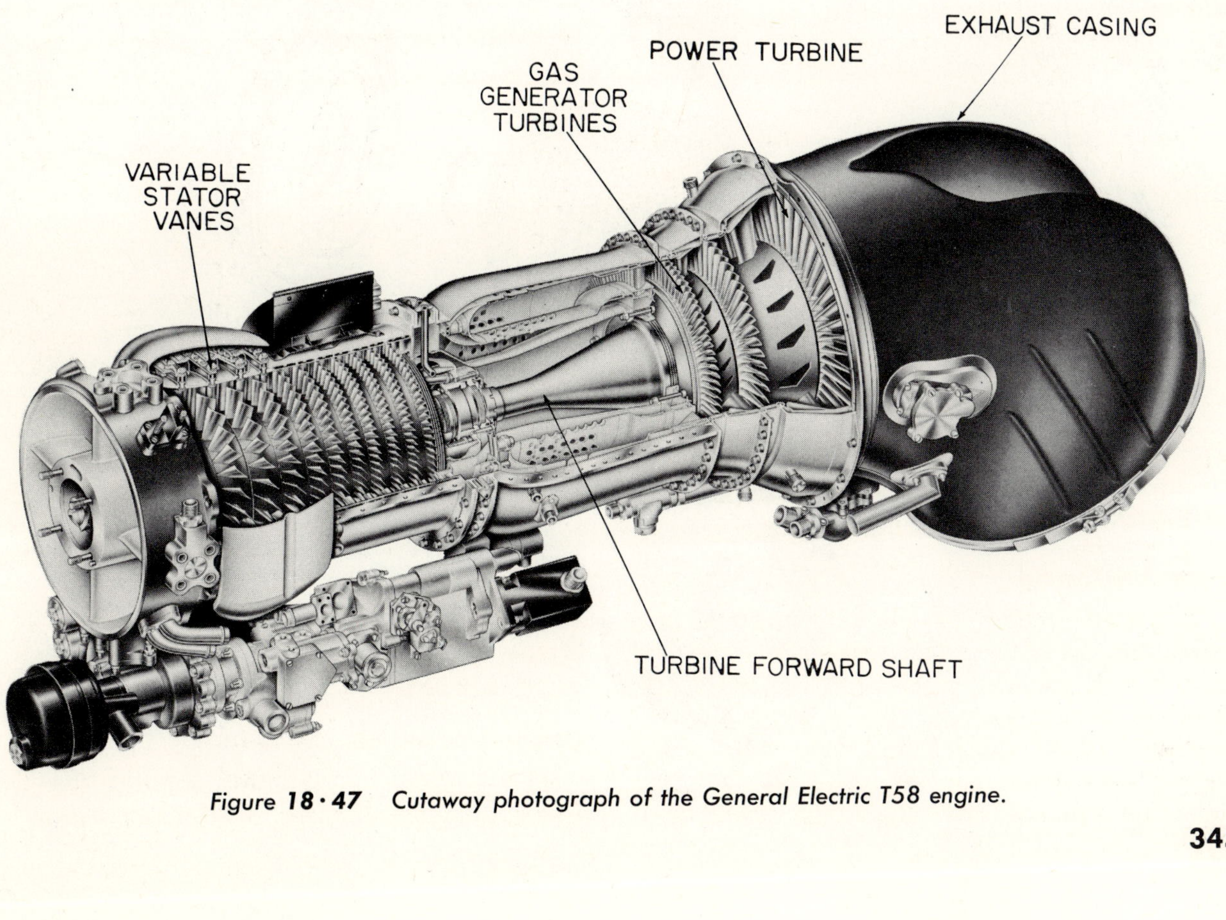

Figure **18·47** *Cutaway photograph of the General Electric T58 engine.*

rear frame, shown in Fig. 18 · 48. The compressor, at maximum power, develops a pressure ratio of 8.15:1, and air is discharged from the compressor at the rate of 12.5 lb per sec.

The **compressor front frame** supports the front end of the compressor rotor and is an engine-to-aircraft mounting structure. It is made of aluminum alloy and consists of the outer shell, four hollow struts, and a hub. The frame assembly contains the No. 1 bearing sump area and an anti-icing air manifold, which connects the 12 and 3 o'clock struts and is part of the outer shell. Three mounting pads, located at the 12, 3, and 9 o'clock positions on the outer shell, permit variations in engine mounting. The front frame struts provide passages for anti-icing air, balance chamber air, lubricating oil, and a radial drive shaft. The front frame also serves as a mounting structure for the starter, front-frame accessory drive assembly, accessory drive gear casing assembly, anti-icing air valve, inlet air sensing tubes, and starter terminal block.

The 12 o'clock strut has three passages. The forward passage contains anti-icing air; the center passage contains the starter cables; and the rear passage is used to vent the No. 1 bearing area. The 3 o'clock strut is cored to direct lubricating oil to the oil jet mounted in the sump area. The oil jet services the No. 1 bearing and also the front-frame accessory drive gearing. The 3 o'clock strut also contains a passage along the leading edge for anti-icing air. The 6 o'clock strut is cored to allow the passage of scavenge oil from the No. 1 bearing and to provide a housing for the radial accessory drive shaft. The 9 o'clock strut is cored to allow the passage of eighth-stage compressor air to the **balance chamber.** The balance chamber is utilized to assist the No. 2 bearing in absorbing the axial load imposed by the compressor rotor as it attempts to move for-

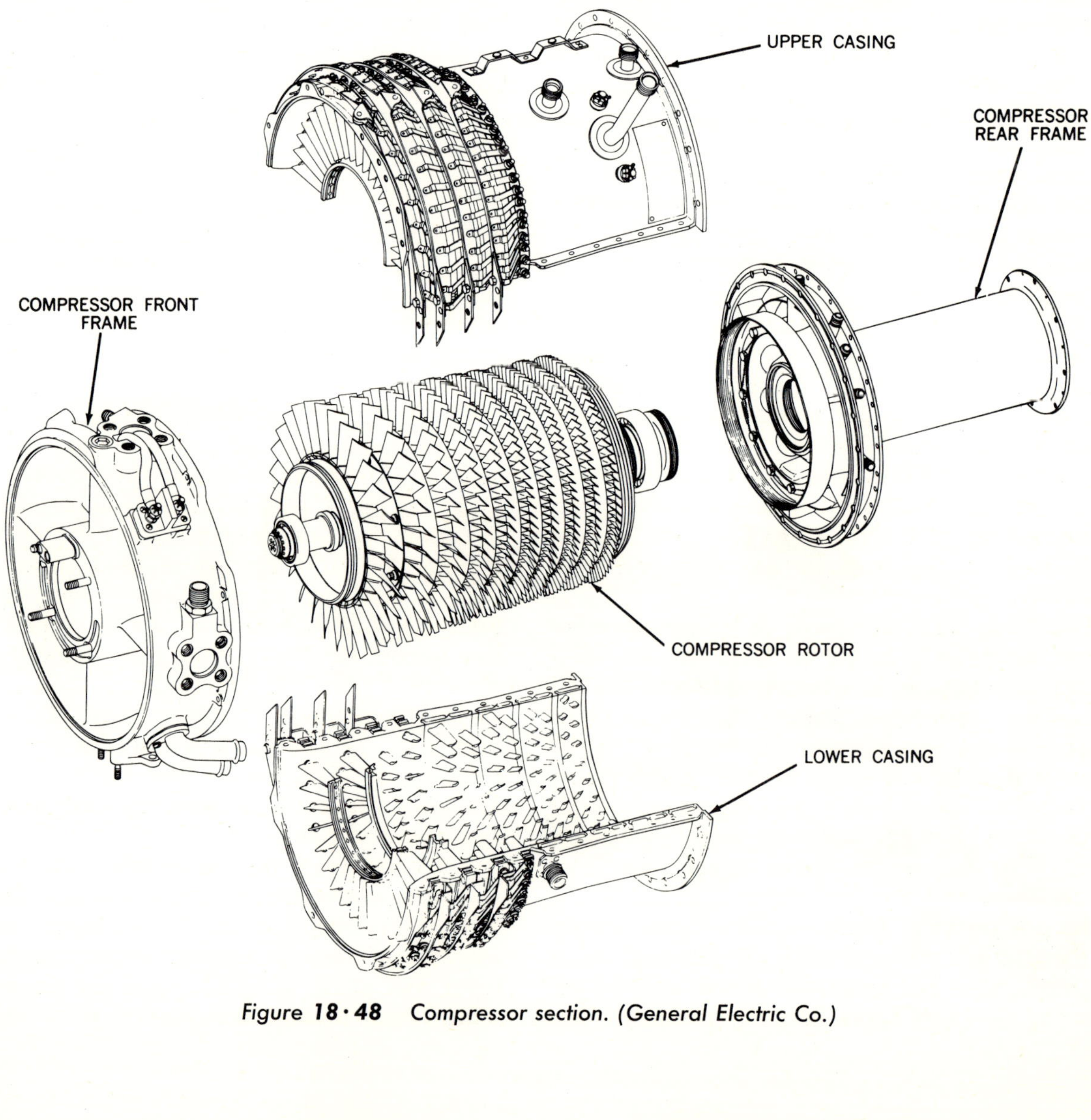

*Figure **18 · 48** Compressor section. (General Electric Co.)*

ward during operation. A starter terminal block is provided at the 11 o'clock position to allow electrical connection of the starter to an external power source. Two ports located at the 7 o'clock position are provided for the extraction and return of a small flow of compressor inlet air to be used for sensing compressor inlet temperature T_2 in the fuel-control unit. A seal ring is bolted to the aft face of the hub to provide a mating surface for the inlet-guide-vane shroud. Holes drilled in the outer circumference of the ring seal mate with a passage in each guide vane and lead to a chamber formed by the front frame hub on the forward side and by the seal ring on the aft side. This chamber contains tenth-stage compressor air, which is directed to the inlet guide vanes for anti-icing purposes.

The **No. 1 bearing air and oil seal** is a carbon rubbing seal mounted at the rear of an insert which is bolted to the inner surface of the hub. The seal rubbing surface is a steel ring mounted on the front disk shaft of the compressor rotor. The forward face of the steel ring rubs against the aft face of the carbon seal to prevent air from entering the No. 1 bearing area. Extreme caution should be exercised when handling these seals because any foreign matter will score the sealing surfaces, thus making the seal ineffective.

The **No. 1 bearing** is a roller bearing which transmits radial loads only, allowing the compressor to expand axially without conducting stress to the surrounding structures. It is a two-piece bearing, with the inner race removable from the outer race and rollers. The bearing outer race and rollers are housed in the forward end of the insert in the front frame, while the inner race is mounted on the front disk shaft of the compressor rotor. The No. 1 bearing sump area is sealed by carbon rubbing seals both forward and aft of the bearing. The forward seal is located in the front-frame accessory drive assembly.

The basic structural unit of the **compressor stator assembly,** shown in Fig. 18 · 49, is a cylindrical steel casing which is split and flanged along the horizontal center line for ease of assembly and maintenance. Mounted on its inner surface is one stage of inlet guide vanes, 10 stages of stator vanes, and one stage of exit guide vanes. For efficient compressor operation throughout the entire speed range, the inlet guide vanes and the first three stages of stator vanes are variable. These variable vanes remain in the CLOSED position until 60 per cent N_g is attained. N_g is gas generator speed, and 60 per cent N_g is 15,000 rpm. As N_g increases, the variable vanes rotate toward the fully OPEN position, which will be reached at 85 per cent N_g (23,000 rpm). Above 85 per cent N_g, the variable vanes are always in the OPEN position. These figures are based upon NASA standard sea-level static conditions. The total travel of each variable vane is 32°, with each vane stage rotating simultaneously and being controlled by the actuator assembly. The inlet guide vanes are supported at the interior of the engine by a shroud which mates with the seal ring mounted on the front frame. Each inlet guide vane has an internal air passage that uses tenth-stage compressor bleed air for anti-icing. The first-stage stator vanes are also supported by a shroud which, in addition to supporting the vanes, provides the stationary surface of the first-stage labyrinth air seal.

The outer shanks of the variable vanes extend outward through the compressor casings to individual stage support rings and vane actuating levers, as shown in Fig. 18 · 50. The pin in each lever engages in its respective slot in one of four actuating rings, which fit around the compressor casing and move circumferentially on bearings located in the support rings. The actuating rings are positioned by the actuator assembly, which pivots the vanes to the desired angle. The remaining seven stages of stator vanes and the exit guide vanes are stabbed into split rings and brazed (two stages per ring). These rings are dovetailed into the inner circumference of the compressor casing. All vanes are steel except the inlet guide vanes, which are made of copper alloy.

The compressor stator casings include mounting provisions for the variable vane actuator assembly, ignition unit, and lube cooler. Four bleed air ports, with fittings for external piping of compressor air, are located on the compressor casings. Fourth-stage air is bled from the 9 o'clock position to cool the power turbine. Sixth-stage air is bled from the 11 o'clock position to cool the No. 3 bearing area. Eighth-stage air is bled from the 10 o'clock position to pressurize the balance chamber. Tenth-stage air is bled from the 12 o'clock position to anti-ice the frontal area of the engine. The stator casings are not interchangeable because they are machined as a matched pair.

The **compressor rotor assembly,** shown in Fig. 18 · 51, is a 10-stage axial-flow unit and consists of 10 stages of compressor rotor blades, two disk assemblies, smooth tapered spool, compressor rotor rear shaft, air guide, and the tenth-stage seal. The first-stage compressor rotor disk as-

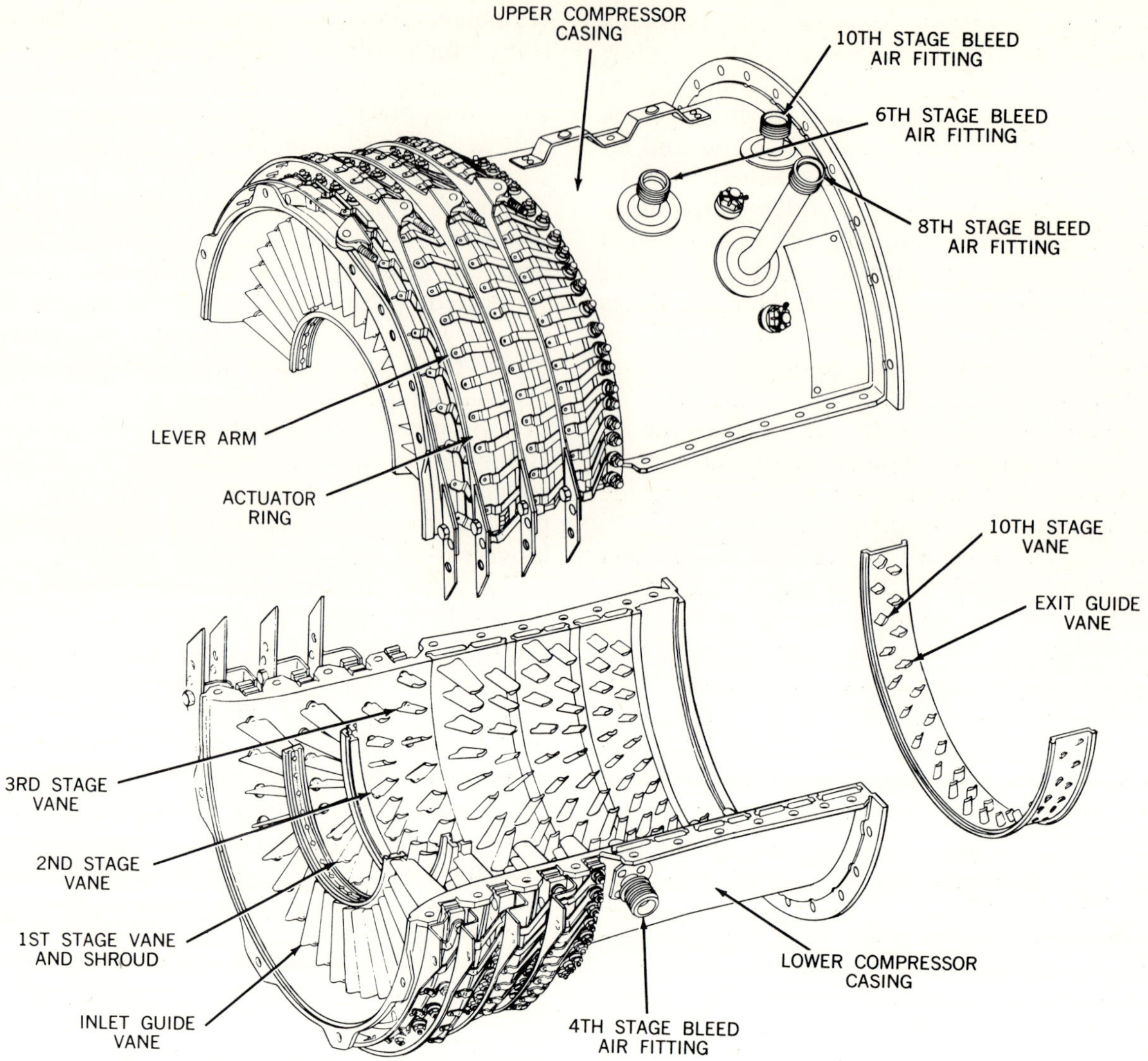

Figure 18·49 Compressor stator assembly.

sembly consists of a steel disk with steel blades axially dovetailed into the rim and held in place by a retaining ring.

The disk has two integral air seals: the **balance chamber air seal** and the **first-stage air leakage seal.** The front compressor rotor disk shaft has three functions: (1) it carries the second-stage rotor blades, which are axially dovetailed and held in place by a retaining ring; (2) it serves as a mount for the first-stage disk assembly; and (3) it is the front support for the compressor rotor assembly. On the shaft of the disk assembly are the sealing ring for the No. 1 bearing carbon oil seal, a spacer, the No. 1 bearing inner race, a lock washer, and the bearing lock nut. Both the compressor disk assembly and the compressor rotor disk shaft are bolted to the forward face of the one-piece spool. The remaining eight stages of compressor rotor blades are dovetailed circumferentially around the spool.

The compressor rear shaft, which contains the air guide, is bolted to the rear of the spool and is the rear support for the compressor. The rear shaft turns in the No. 2 (thrust) bearing, which is mounted in the compressor rear frame. The rear shaft is also splined and threaded internally to receive both the turbine shaft bolt and the turbine forward shaft. Slots machined in the flange of the rear shaft admit air to the inside of the spool where it flows aft to cool the gas generator turbine.

The compressor rotor carries 634 rotor blades and bears the serrated parts of three labyrinth-type seals: the balance chamber seal; the first-stage air seal, which prevents the trapping of inlet air and the loss of energy between the first-stage compressor disk assembly and the front compressor rotor disk assembly; and the tenth-stage air seal, which permits compressor air leakage for cooling the gas-generator first-stage turbine nozzle. The compressor rotor is a dynamically balanced assembly.

The **steel compressor rear frame** is bolted to the compressor stator casings at the front and to the combustion casings at the rear. It consists of an outer diffuser ring and an inner diffuser ring joined by six radial struts.

The top strut is hollow and contains the oil-in tube, which services the No. 2 (thrust) bearing housed in the interior of the inner diffuser ring. The bottom strut is hollow and contains the scavenge tube for draining the bearing oil sump, which the remaining four struts serve as structural supports.

The outer diffuser ring has seven connecting locations at the 12, 5, 6, 9, 10, 10:30, and 11 o'clock positions. An oil-in tube connection is located at the 12 o'clock position to provide lubrication to the No. 2 bearing. Compressor discharge air is taken from the compressor rear frame at the 5 o'clock position and piped to the bleed valve mounted on the ignition unit and to the main fuel control. A connection at the 6 o'clock position provides for the No. 2 bearing scavenge oil line. The 9 o'clock connection provides for the external piping of compressor dis-

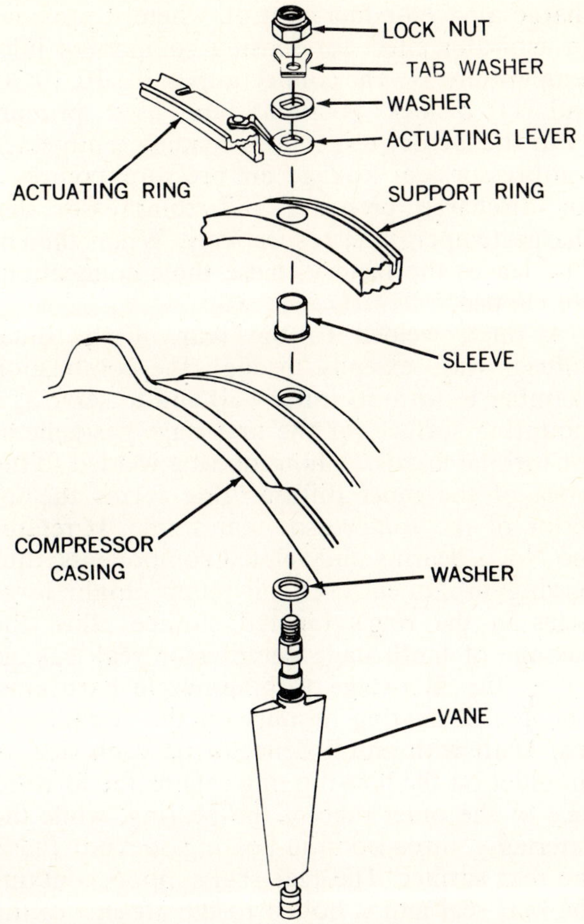

Figure 18 · 50 Installation of variable-stator vane.

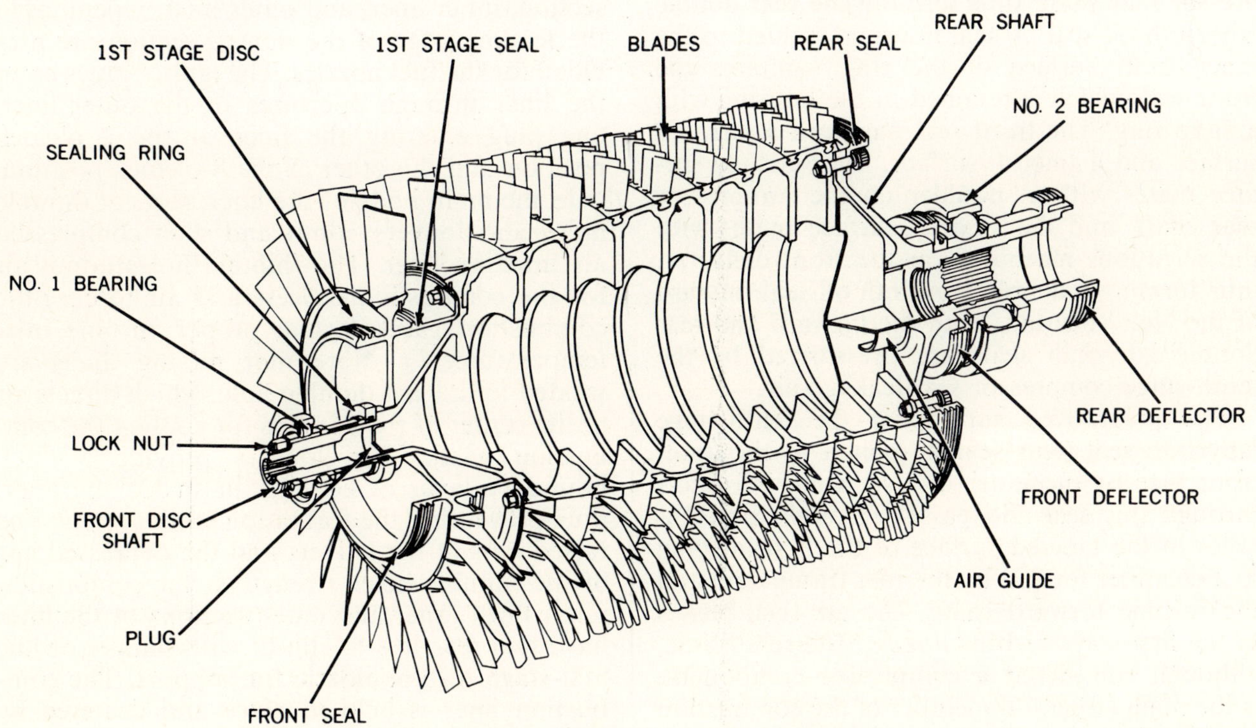

Figure 18 · 51 Compressor rotor assembly.

charge air to the fuel control, where it provides an aspirator effect for sensing compressor inlet temperature T_2. The connections at the 10, 10:30, and 11 o'clock positions are used primarily in the factory test cell for reading tenth-stage compressor seal leakage air pressure, compressor discharge pressure, and compressor discharge temperature, respectively. When the engine leaves the factory, these three connections are capped.

A duct, welded to the rear of the inner diffuser ring, extends through the combustion chamber to form its inner wall and to serve as a mounting surface for the first-stage gas-generator turbine nozzle. A housing ring welded to the front of the inner diffuser ring forms the interior of the compressor rear frame. It retains the No. 2 bearing and related components, thus serving as an enclosed oil sump. Eight large holes in the ring's tapered surface allow the passage of tenth-stage compressor seal leakage air to the first-stage turbine-nozzle partitions. The No. 2 bearing mounts on the compressor rear shaft with an oil deflector on each side. A shoulder on the housing ring retains the forward face of the outer race of the bearing, while the externally threaded mid-bearing locknut holds the rear surface. The rear seal support contains the rear seal and is bolted to the aft side of the housing ring. The rear seal has both a serrated surface and a smooth surface to mate with the rear deflector, which is mounted on the compressor rear shaft, thus forming the rear double labyrinth oil seal. A seal housing, welded to the inner front surface of the ring, contains the front seal, which is retained in the housing with a snap ring. The front seal has both a serrated surface and a smooth surface. The serrated surface mates with a shoulder on the compressor rear shaft, and the smooth surface mates with the serrations machined on the front deflector, thus forming a double labyrinth oil seal forward of the No. 2 bearing. Both the forward and rear double labyrinth seals are pressurized by the tenth-stage compressor seal leakage air.

The pregrooved surface of the tenth-stage labyrinth seal (rear seal) is bolted to the outer front face of the inner diffuser ring. Air leaks through this seal and passes through the eight holes in the tapered surface of the housing ring to a chamber formed by the rear frame duct and the turbine forward shaft. The air then passes to the first-stage turbine nozzle. The rear frame, although considered a compressor component, is the main structural member of the combustion section.

The No. 2 bearing is a one-piece ball bearing housed in the compressor rear frame. This bearing transmits both the radial and axial loads imposed by the compressor rotor and is referred to as the main engine bearing. The axial load on this bearing is relieved somewhat by the balance chamber.

Combustion section

The **combustion section,** shown in Fig. 18 · 52, is that portion of the engine in which fuel is added to the compressed air and ignited, thus causing a rapid expansion of gases toward the gas-generator turbine section. The combustion section of the T58 engine includes the outer casing, combustion liner, outer deflector, and various fuel system components.

The steel **combustion outer casing** is composed of two halves joined at flanges along the horizontal center line. This unit serves as a structural member of the basic engine, as it is bolted to the rear of the compressor rear frame and to the front of the first-stage turbine casing. The casings, machined as a matched pair, are noninterchangeable.

Mounted on the exterior of the combustion outer casing are the fuel flow divider, two igniter plugs, the combustion-chamber drain valve, two customer air bleed ports, and four locating pin bosses.

The **combustion liner** is a welded one-piece annular assembly of the straight-through type, consisting of a circular cowl section, inner dome section, inner liner, and outer liner. Openings in the leading edge of the domed section are provided for the fuel nozzles. The igniter plugs enter the liner through apertures in the outer liner, one plug entering the liner at the 2 o'clock position and the other at the 8 o'clock position.

In the main body of the liner, rows of thimble holes and louvers admit and duct compressed air into the liner. The thimble holes admit air for the combustion process and air to cool the combustion gases to the proper turbine inlet temperature T_4. Note that mixing ducts are welded to the last thimble hole, which directs air to the center of the annulus for cooling the combustion gases. The louvers provide a thick boundary layer of cooling air inside the liner, thus preventing the formation of hot spots. The raised tabs of the louvers and the depressed lips of the thimble holes strengthen the combustion liner. Both inner and outer sections of the liner have two rear lips to slip-fit with flanges on the first-stage turbine nozzle for support. The combustion liner is held in place and centered by four retaining pins which are inserted through bosses on the outer combustion casings into four

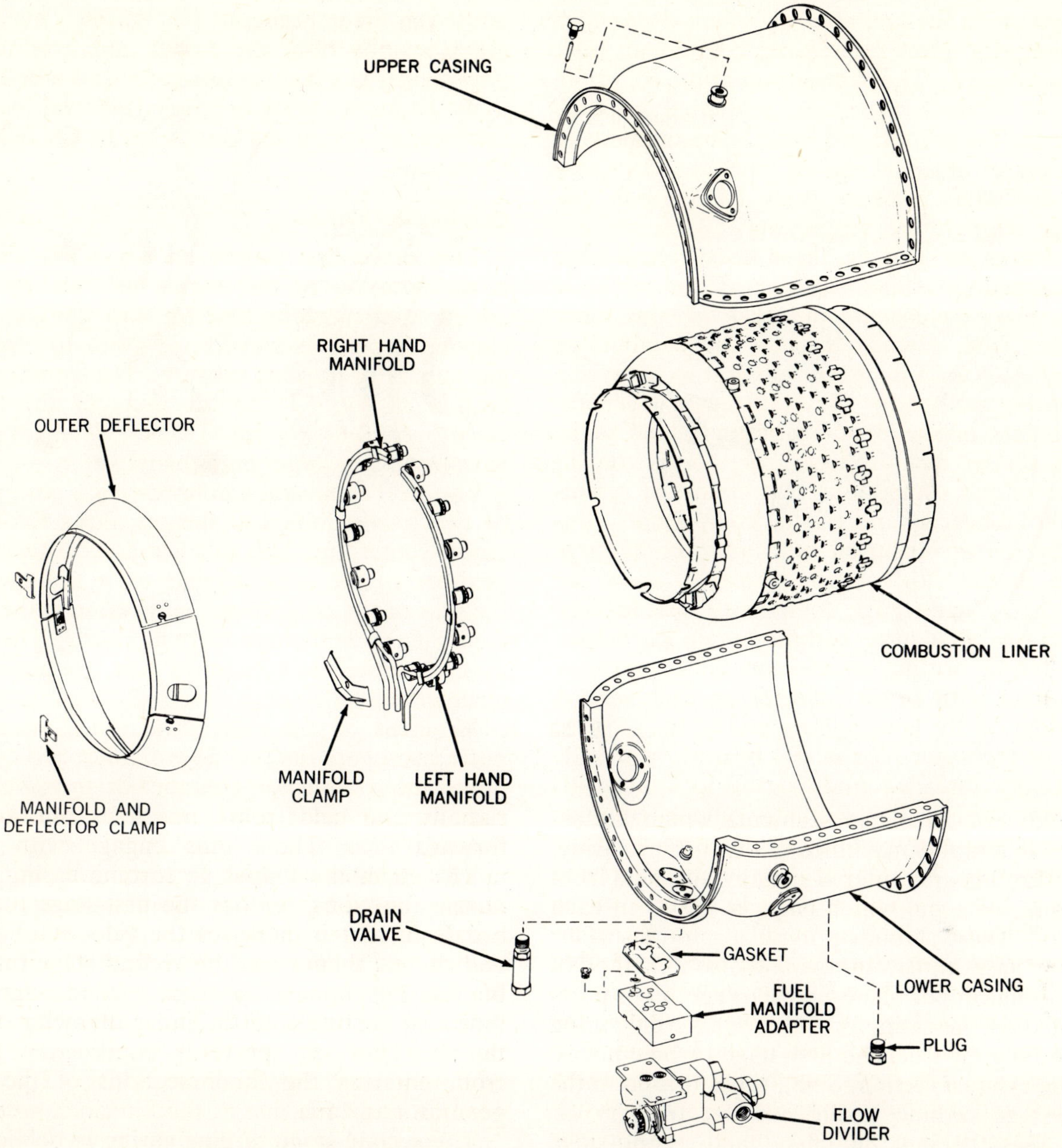

Figure 18·52 The combustion section.

bosses on the combustion liner. The two fuel manifolds are held in the combustion liner by the outer deflector.

The fuel flow divider and fuel drain valve are mounted on a boss at the bottom of the combustion lower casing. The two fuel manifolds extend through the outer casing into the flow divider, which controls the flow of fuel into either 8 or 16 nozzles.

Gas-generator turbine section

The turbine is of the 2-stage type, employing blades that utilize both the impulse and reaction principles to extract the required power from the exhaust gases for driving the compressor. The gas generator turbine section consists of the first-stage turbine casing, first-stage turbine nozzle, second-stage turbine nozzle, second-stage turbine casing, third-stage turbine nozzle, and the turbine rotor assembly. The No. 3 bearing area is enclosed within the third-stage turbine nozzle.

The **first-stage turbine casing,** shown in Fig. 18·53, is made of stainless steel and connects the outer combustion casing with the **second-stage turbine casing.** It houses the first-stage turbine shrouds, the first-stage turbine nozzle, and the second-stage turbine nozzle. The turbine

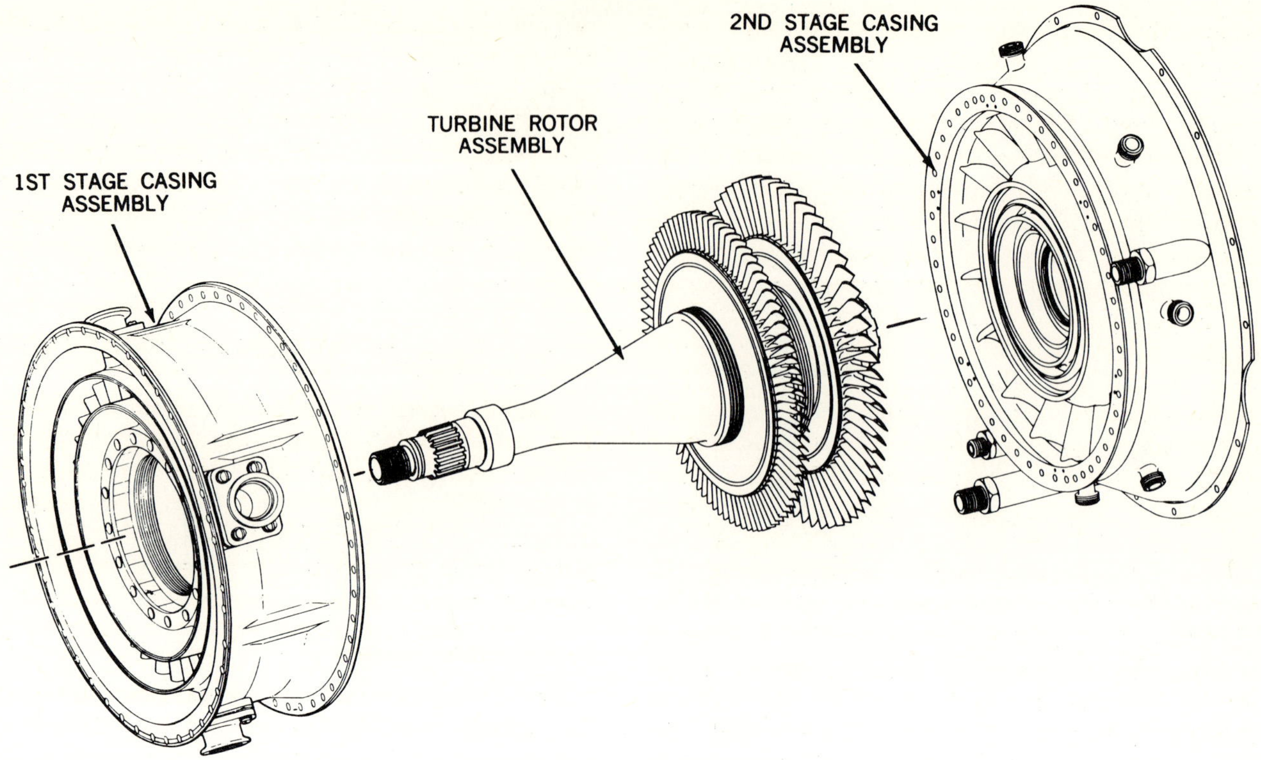

Figure 18·53 Turbine section.

shrouds consist of six segments which are secured in a rabbet machined in the turbine casing. The shrouds are centered radially and held from turning by a pin which engages a slot in each shroud. These shrouds provide a smooth surface for the gases to flow across and also a method of obtaining proper blade tip clearance. Four ports located on the casing provide for external piping of cooling air from the first-stage turbine nozzle to the exhaust section. Small tubes brazed to the first-stage turbine casing conduct primary air to areas behind the first- and second-stage turbine shrouds for shroud cooling.

The **first-stage turbine nozzle** consists of hollow partitions connected to an inner and outer band. The nozzle is bolted to the rear of the compressor rear frame duct and is enclosed by the first-stage turbine casing. The first-stage turbine-nozzle seal is riveted to the inner diameter of the nozzle. The function of this turbine nozzle is to increase the velocity of the combustion gases and direct them onto the first-stage turbine-wheel buckets.

The **second-stage turbine nozzle** is similar in basic design to the first-stage nozzle. It is located between the first- and second-stage turbine wheels and is housed within the first-stage turbine casing. It is held in place axially by the second-stage turbine casing and is centered radially and held from turning by pins on its forward face. These pins engage with slots machined in the first-stage turbine casing. The nozzle functions as does the first-stage turbine nozzle in that it increases the velocity of gases and directs them onto the second-stage turbine blades. The honeycomb-type second-stage turbine seal is riveted to the inner diameter of the nozzle. This seal prevents combustion gases from entering the inner sections of the gas-generator turbine.

The **second-stage turbine casing** is bolted forward of the first-stage turbine casing and aft of the power-turbine section. It houses the second-stage turbine shrouds and the third-stage turbine nozzle. The turbine shrouds are in six segments which are secured in a rabbet machined in the turbine casing. The shrouds are centered radially and held from turning by a pin which engages a slot in each shroud. The second-stage turbine casing contains eight bosses which admit the thermocouples for measuring power-turbine inlet temperature T_5, a lubricating-oil service boss, a scavenge-oil boss, and a boss to admit sixth-stage gas-generator cooling air.

The **third-stage turbine nozzle** is made of stainless steel and is similar in basic design and function to the first- and second-stage turbine nozzles. Located aft of the second-stage turbine wheel, the third-stage nozzle is housed within the second-stage turbine casing. The nozzle is held axially in place by the three service tubes, which extend through bosses on the casing. Lubricating service oil enters the interior of the turbine nozzle through one of the hollow partitions to provide lubrication for the No. 3 bearing. Scavenge oil is removed from the interior of the turbine nozzle assembly through another of the hollow partitions. Eight sleeves are provided in the turbine nozzle to permit the installation of eight hot-gas-indicating thermocouples. The inner structure has provisions for locating the outer race of the No. 3 roller bearing and the air and oil seals. Enclosing the No. 3 bearing sump area is an insulating material which is sealed into the nozzle and aids in keeping the sump cooled.

The **gas-generator turbine rotor assembly,** illustrated in the cutaway drawing of Fig. 18 · 54, consists of a turbine forward shaft, first-stage turbine wheel, coupling shaft, coupling heatshield, second-stage turbine wheel, turbine rear shaft, and the No. 3 bearing. All are coupled by curvic teeth and held together by the turbine shaft bolt except the bearing, which is mounted on the turbine rear shaft and held in place by a bearing locknut. Cooling plates are mounted and held in place on the fore and aft faces of both turbine wheels by the forward and rear turbine shafts and the coupling shaft. The cooling air plates, with the exception of the first-stage rear plate, contain gas seals, while the turbine rear shaft contains an oil seal. The forged turbine blades are held to the wheels by dovetail fastening. Each turbine wheel has a forward and aft blade seal that helps direct cooling air across the blade shanks and dampen blade vibration. The turbine rotor is dynamically balanced and matchmarked.

The **No. 3 bearing** is a two-piece roller bearing. The inner race and rollers are mounted on the turbine rear shaft, and the outer race is housed in the third-stage turbine nozzle. The bearing transmits radial loads only and allows the turbine rotor to expand axially without conducting stress to surrounding structures.

Power-turbine section

The power-turbine section is flange-bolted to the second-stage turbine casing and consists of the exhaust casing assembly, power-turbine assembly, power-turbine accessory drive system, and the reduction gearbox, if one is installed.

The **power-turbine exhaust casing assembly**

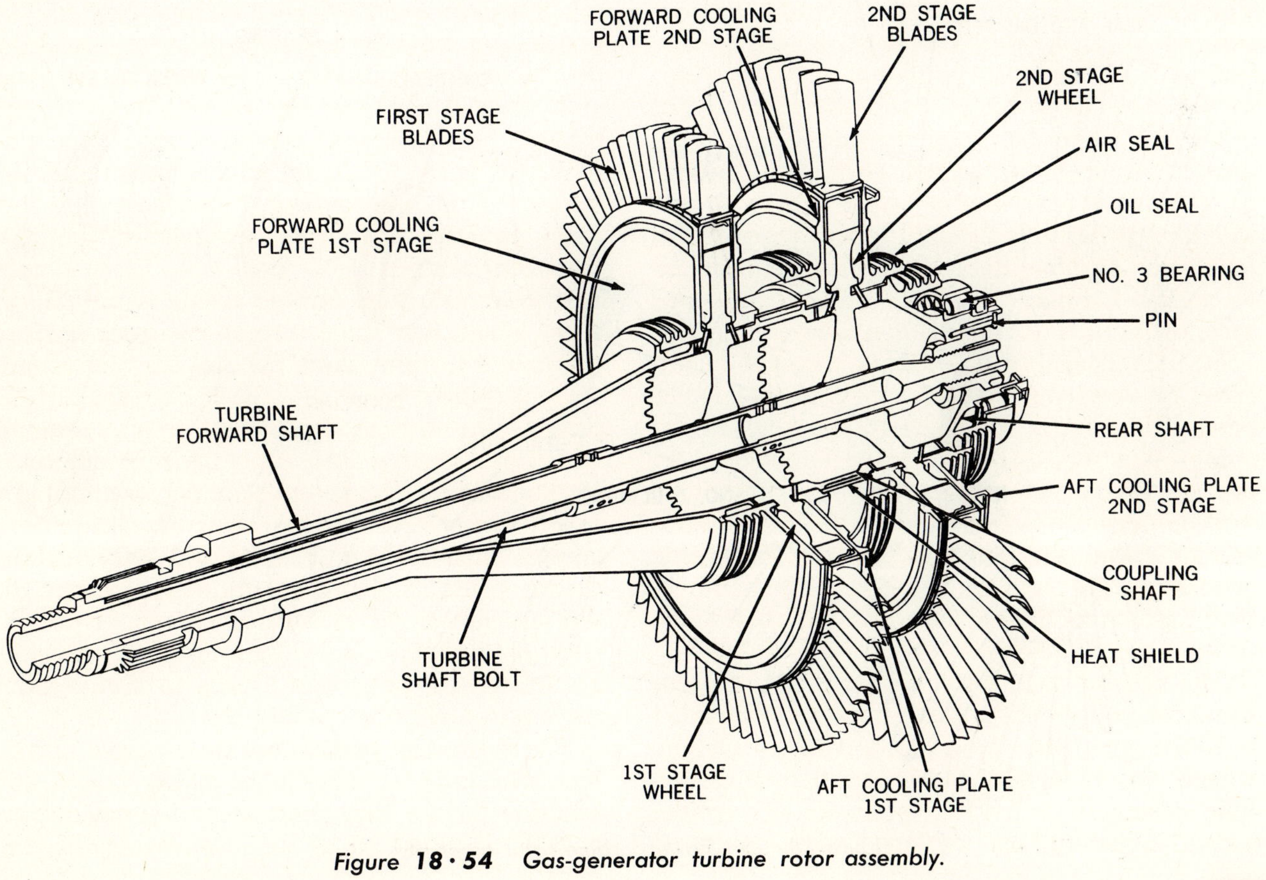

Figure 18 · 54 Gas-generator turbine rotor assembly.

consists of the exhaust casing, forward bearing housing, insulation blanket, air baffle, ring seal, accessory drive system, rear bearing housing, No. 5 bearing rubbing seal, and turbine shrouds. The exhaust casing is constructed to deflect the exhaust stream downward, to the left, or to the right, depending upon the installation requirements. The casing supports and exposes the rear flange of the inner casing and provides an external mounting structure for the reduction-gear assembly, if one is being used.

The outer casing includes three mounting bosses for the accessory drive system. Only one of these bosses is used, depending on whether the exhaust is down, to the right, or to the left. The accessory drive system must be mounted at the engine 6 o'clock position. Four ports on the outer casing duct tenth-stage compressor leakage air into the exhaust stream. The rear lip of the casing includes provisions for connecting exhaust ducting as necessary for the particular installation. The worm sleeve, right-angle drive assembly, Nos. 4 and 5 bearings, and the No. 5 bearing rubbing seal are contained within the inner exhaust casing. The forward and rear bearing housings comprise a conical oil sump which encloses these components. A one-piece insulation blanket surrounds the oil sump. Lubricating oil enters the engine 6 o'clock strut and flows to the oil jet, which is bolted to the inner surface of the forward bearing housing. The oil jet services the No. 4 bearing, the gears on the right angle drive assembly, and the No. 5 bearing. Scavenge oil is taken through the scavenge oil tube, which is also contained within the 6 o'clock strut. The radial drive shaft is housed within the oil scavenge tube. The power turbine shrouds consist of eight segments which are secured in a rabbet machined in the forward lip of the outer exhaust casing. It is held from rotating by pins in the exhaust casing. These pins engage with a slot in each shroud segment.

The **power-turbine rotor assembly,** shown in Fig. 18 · 55, consists of the power turbine wheel and integral shaft, which carries one stage of power turbine blades. The blades are axially dovetailed into the wheel with a seal and damper inserted between each pair of blades. This seal and damper is used to keep the hot gases off the blade shanks and to dampen out any blade

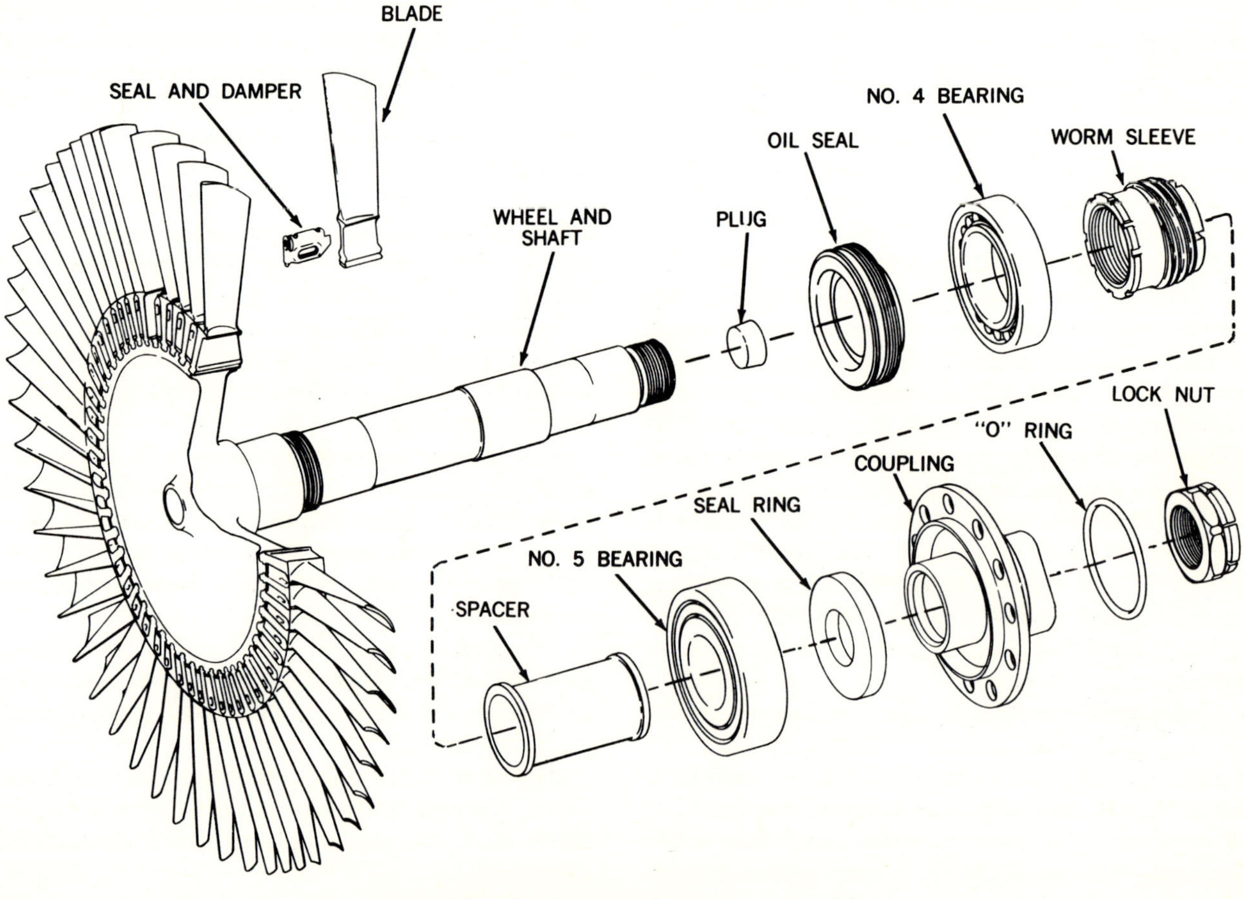

Figure 18 · 55 Power-turbine rotor assembly.

vibration. Mounted on the shaft is an air-oil seal, the No. 4 bearing, the worm sleeve, a spacer, the No. 5 bearing, the steel sealing ring, which mates with the carbon rubbing seal in the exhaust casing, and the flanged polygon coupling adapter. The worm sleeve is mounted on the shaft with a lefthand thread, and a locknut retains the entire wheel assembly. The power-turbine rotor assembly is dynamically balanced and matchmarked.

The worm sleeve on the shaft engages with the gearing on the right-angle drive assembly and, together with the radial drive shaft, drives the power-turbine accessory drive system. This generates a speed signal, which is transmitted to the fuel control by way of a flexible shaft. The fuel control will automatically stopcock the engine when power turbine speed N_f reaches 110 per cent. The flanged polygon coupling adapter, located at the end of the shaft, is the engine drive coupling, which fits directly into a connecting shaft provided by the customer.

The **No. 4 bearing** is a two-piece roller bearing located aft of the power turbine wheel. The outer race is housed in the power-turbine forward bearing housing, and the inner race and rollers are mounted on the shaft. This bearing transmits radial loads and allows the power turbine rotor to expand axially.

The No. 5 bearing is a one-piece ball bearing located aft of the power turbine wheel and housed in the power-turbine rear bearing housing. It is a thrust bearing and transmits both the radial and axial loads imposed by the power turbine rotor.

The **speed-decreaser gearbox** is provided as an optional piece of equipment with the T58 engine to reduce the relatively high output speed of the engine to a practical rpm for low-speed applications. An additional feature provided with the gearbox is a built-in torque sensor to provide remote indication of torque being transmitted by the gearbox. The gearbox is mounted on the power-turbine section of the engine at the exposed rear flange of the inner exhaust casing. It consists of a front and rear casing, internal gears, and an external oil and scavenge pump for lubrication and torque-sensor operation.

The magnesium front casing includes a flange for mounting the gearbox on the engine, two mounting pads for installation of the engine in the air frame, and a power-takeoff location. The magnesium rear casing includes a flange for mounting the oil scavenge pump and a power-takeoff location. The gears, of the herringbone type, produce a power-turbine to output-shaft-speed ratio of 3.25:1. The gear train consists of an input gear, an idler gear, and a split bull gear. The two halves of the bull gear are splined to each other through a ball spline which allows the two halves of the gear to separate from each other under load. Because this separating force would cause the idler gear and input gear to shift axially, a ball spline is incorporated at the gearbox input to ensure complete freedom of motion of the split bull gear.

Engine accessories section

The accessory drive system drives various engine components and is comprised of the front-frame accessory drive assembly, engine-accessories section, and power-turbine accessory drive system.

The **front-frame accessory drive assembly** is mounted at the front of, and extends through, the inner hub of the front-frame casing assembly. It provides a connection between the customer-furnished starter and the engine rotor assembly during initial starting. It also transfers power from the rotor shaft to the accessory drive gear casing assembly. The accessory drive assembly consists of a support assembly, accessory shaft, ball bearings, and gear train. The support assembly is a one-piece unit which provides rigid support for the right-angled transfer of power. The accessory shaft, externally splined at the aft end to mate with the rotor assembly, has starter jaws at the front which engage the starter during starting. This shaft turns in a ball bearing and drives a beveled pinion gear which in turn drives the beveled accessory gear. This gear is held in the support assembly by two ball bearings and is splined to mate internally with the radial accessory drive shaft. A carbon seal enclosed within the accessory drive assembly prevents oil leakage forward.

The engine-accessories section is composed of those parts that sustain the functions of the basic engine. They include the accessory drive gear casing assembly, main lubricating pump, fuel pump, dynamic fuel filter, fuel control, and two tachometer generators.

The accessory drive gear casing assembly is mounted at the bottom of the front-frame casing assembly and transmits power to drive the engine accessories. This power is taken from the compressor rotor shaft, through the accessory drive gear assembly, through the radial accessory drive shaft, to the accessory drive gear casing assembly.

The accessory drive gear casing assembly

houses the engine oil filter, relief valve, filter bypass valve, magnetic drain plug, bearings, and gears.

In the casing housing, two ball bearings support a beveled pinion that mates internally with the radial accessory drive shaft. The pinion drives a bevel gear directly below it, which is mounted on one end of a pinion shaft that holds a spur gear at its other end. The pinion shaft is mounted in two ball bearings which anchor it to the casing assembly housing. The spur gear acts as a pinion gear and meshes with two diametrically opposed spur gears mounted on individual shafts that are splined to mate with adapters from the engine accessories. The right shaft drives the main lubricating pumps while the left shaft drives the fuel pump and dynamic filter. Three accessories are mounted directly to the casings: the main lubricating pump and fuel pump on the aft face and the dynamic filter on the forward face. The gas-generator tachometer generator is mounted on and driven by the main lubricating pump. The fuel control is mounted on and driven by the fuel pump. The power-turbine tachometer generator is, in turn, mounted on the fuel control.

The power-turbine accessory drive system consists of the power-turbine right-angle drive assembly, radial drive shaft, power-turbine accessory drive assembly, and flexible shaft.

The power-turbine right-angle drive assembly consists of a circular support which mounts an offset pinion gear and a drive gear. Each gear turns in two ball bearings. The pinion gear mounts on its integral shaft a worm gear which meshes with the left side of the worm sleeve on the power turbine shaft. The pinion gear meshes with the drive gear at its right side, thus placing the drive gear directly below the power turbine shaft when the assembly is mounted. The integral shaft is splined to receive one end of the radial drive shaft, which extends down through the bottom strut to mate with splines in the power-turbine accessory drive assembly. This assembly has beveled gears which change the direction of power transfer.

A square-shaped tang on the flexible shaft fits into a similarly shaped receptacle in the accessory drive assembly. The flexible shaft is connected to a gear train in the fuel control. This, in turn, drives the power-turbine tachometer generator and internal fuel-control components.

Lubricating oil enters the power turbine through a boss on the accessory drive casing and is distributed by three oil jets. These jets supply oil directly to the No. 4 and No. 5 bearings and to the power-turbine accessory drive system gears. The oil exits through the oil drain tube surrounding the radial drive shaft in the 6 o'clock strut. The oil passes into the power-turbine accessory drive assembly from which it is then scavenged back to the main lube pump.

ENGINE AIRFLOW

The airflow system for the T58 engine is similar to that of other gas-turbine engines. It includes the main airflow through the engine plus the special airflow arrangements for cooling and pressurizing various parts of the engine. Air enters through the front frame and passes through the compressor where it is compressed at a ratio of 8.15:1 at maximum power. In the compression cycle, air pressure and temperature are increased, and volume is decreased. Compressor discharge air passes to the combustor at the rate of 12.5 lb per sec at maximum power. In the combustor, one-fourth of the air is burned with the fuel, and the remaining three-fourths is used to dilute the combustion gases to ensure safe-operating turbine inlet temperature and to cool the hot sections of the engine.

The hot combustion gases enter the gas-generator turbine where the velocity of the gases is increased and the turbine rotor extracts the necessary power to operate the compressor. Nearly two-thirds of the power generated is extracted from the gas flow to operate the compressor. The remaining power is extracted by the power turbine.

Air is bled at various stages of the compressor for cooling functions in addition to those performed by compressor discharge air. Additional air is required for anti-icing, pressurization, and sealing.

The compressor furnishes air through the fourth,- sixth,- eighth,- and tenth-stage bleeds for cooling, pressurization, anti-icing, balance chamber, and seal leakage. Tenth-stage air is used for anti-icing and for cooling the gas-generator turbines. Fourth-stage air is piped to the power turbine for cooling the rear face of the turbine wheel.

FUEL SYSTEM AND CONTROL

The fuel system of the T58 engine consists of a fuel reservoir, booster pump, filter screen, dynamic filter, 2-stage main fuel pump, hydromechanical fuel-control unit, static filter, flow divider, fuel manifolds, and fuel nozzles. The fuel reservoir, booster pump, and filter screen are furnished by the airframe manufacturer, while all other fuel-system components are furnished as engine-mounted units.

Operation

In operation of the T58 engine, fuel from the fuel tank is pressurized by the booster pump and passed through the filter screen and dynamic filter to the main fuel pump. The main fuel pump delivers a continuous flow of pressurized fuel through the fuel-control unit and a static filter to the flow divider. The flow divider feeds fuel to one or both manifolds, depending on engine operating conditions, and thence through the fuel nozzles to the combustor. The fuel system is shown in Fig. 18 • 56.

Dynamic filter

The dynamic filter removes fuel-system contaminants by centrifugal action. The filtering element spins at 4,200 rpm when the engine is operating at 100 per cent speed. The filter is mounted on and driven by the accessory drive gearbox. The filtering element is exposed for cleaning by disconnecting a clamp and removing the cover. A drain is provided in the cover to bleed off residual fuel before the cover is removed.

A casing, cover, and centrifuge element form the filter unit. The casing contains the drive shaft, bearing, and fuel IN and OUT ports. Fuel leaking by the shaft is ported overboard through a drain located in the casing. The cover contains a spindle and thrust bearing upon which the centrifuge element rotates. The centrifuge element is composed of solid shells, louvered shells, and vanes arranged concentrically in a two-piece construction to form the fuel chambers and contaminant traps. In order, from the center outward, are the primary fuel chamber, primary contaminant trap, secondary fuel chamber, and secondary contaminant trap.

Fuel enters the IN port in the casing and flows axially forward into the primary fuel chamber then radially through a series of holes into the secondary fuel chamber. After this it flows axially aft, exits through a series of holes into a helical passage in the casing, and discharges through the OUT port. Rotation of the centrifuge subjects the fuel to high centrifugal forces as it flows through the unit. Adjacent to each fuel chamber is a contaminant trap consisting of an inner louvered shell and solid outer wall. Under centrifugal force, contaminants travel from the fuel chamber outward through the louvers to the solid wall, where they are trapped. The contaminant trap is compartmented by a series of radial vanes which prevent centrifuge unbalance caused by contaminant shifting. The louver arrangement prevents contaminants from settling into the fuel chamber during engine shutdown.

Fuel pump

The fuel pump for the T58 engine is a gear-type unit having a worm-type centrifugal boost section. The boost section is gear-driven from the main pump drive shaft, as shown in the drawing of Fig. 18 • 57. Fuel from the inlet is

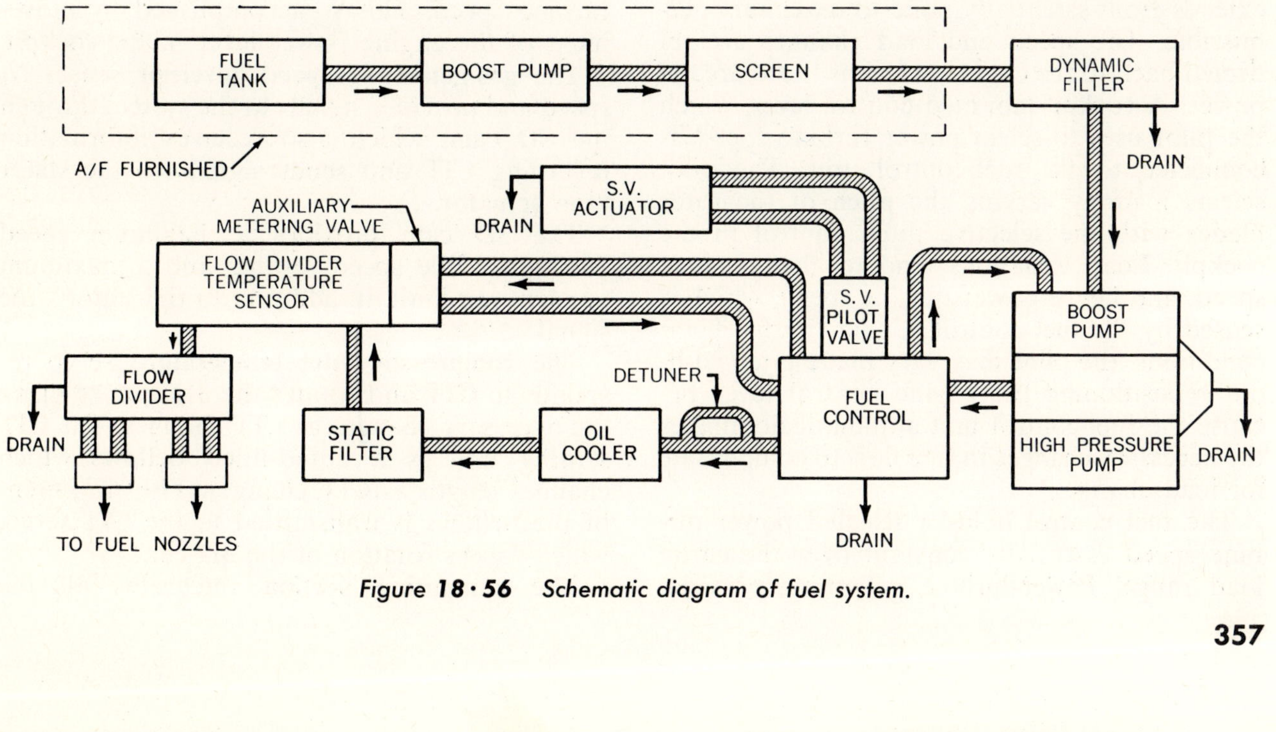

Figure 18 • 56 Schematic diagram of fuel system.

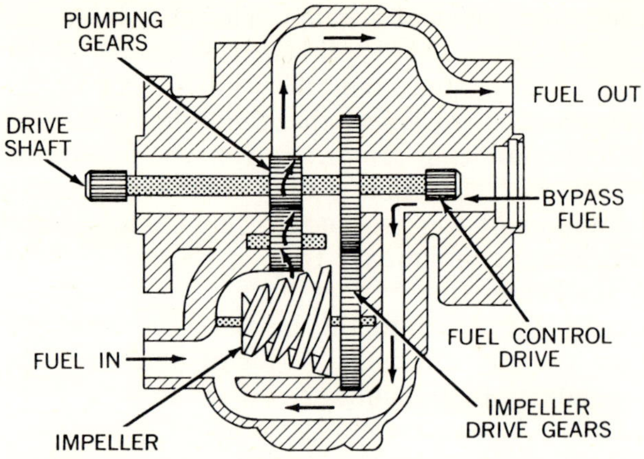

Figure 18 · 57 Fuel pumps.

carried along the axis of the pump by the rotating impeller and is discharged to the inlet of the gear section.

The gear pump consists of two closely machined spur gears, which operate in the pump housing. Fuel from the pump is delivered through an internal port to the fuel-control unit. Low-pressure bypass fuel from the fuel-control unit is returned to the boost element inlet through a port in the mating flange.

Fuel control

The power requirements of a helicopter rotor vary with speed and load. Rotor power is varied to meet the demands of flight conditions, hence the fuel control must adapt itself to situations somewhat different from those involved in driving a propeller. The speed range is approximately 15 per cent, that is, from 85 to 100 per cent of helicopter rotor speed. The load range extends from essentially none to maximum permissible. The speed and load changes are reflected back to the engine, which is the source of power. A cockpit-mounted control lever, which the pilot uses to select power turbine speed, is connected to the fuel-control unit. The pilot selects load by varying the pitch of the rotor blades with the selective pitch control in the cockpit. Load variations tend to change rotor speed, and hence power turbine speed, which is sensed by the fuel control. Under normal flight conditions, the pilot may vary blade pitch without repositioning the engine control lever because the fuel-control unit automatically makes the necessary changes in fuel flow to compensate for load changes.

The fuel control holds a selected power turbine speed essentially constant over the entire load range. Power-turbine governing provides maximum engine performance in every regime of helicopter flight operation. Safe engine operation is ensured by prevention of engine rotor overspeed, compressor stall, turbine overtemperature, and flameout. Engine efficiency is maintained by control of the variable stator vanes through a signal from the fuel-control unit.

The fuel-control unit senses five operating parameters of the engine in order to provide correct fuel flow for all conditions: (1) control-lever position, (2) gas-generator speed N_g, (3) power turbine speed N_f, (4) compressor discharge pressure CDP or P_3, and (5) compressor inlet temperature CIT or T_2.

The control has a fuel metering section and a computing section, similar to fuel controls previously described. The metering section selects the rate of fuel flow to the engine combustor in accordance with information received from the computing section. A simplified block diagram of fuel control is shown in Fig. 18 · 58.

The **pressure regulating valve** maintains a constant pressure drop of 60 psi across the **metering valve,** which results in fuel flow being proportional to metering valve orifice area. The orifice area of the metering valve depends upon the forces applied within the control in accordance with the parameters mentioned previously.

The **power turbine speed governor** maintains an essentially constant power turbine speed by correcting for any deviation between selected speed and actual power turbine speed as sensed by the flyweight governor. This unit is effective only in its governing range, which is 16,000 rpm or higher.

The **power-turbine-speed setting cam** and linkage permits the pilot to select a desired power turbine speed. This is accomplished by movement of the engine power lever in the cockpit.

The **gas-generator speed governor** senses N_g speed and provides signals to the system through the **3D cam,** which also receives information regarding CIT and sends signals to the **stator vane actuator.**

The 3D cam provides a maximum speed setting, an idle speed setting, and a maximum acceleration limit, in addition to the stator vane signal.

The **compressor inlet temperature** servo responds to CIT and rotates the 3D cam to make the necessary changes as CIT changes. The CIT sensing unit is a liquid-filled bellows which changes length as CIT changes. The movement of the bellows is transmitted to the CIT servo, which causes rotation of the 3D cam.

The **computing section** integrates all the

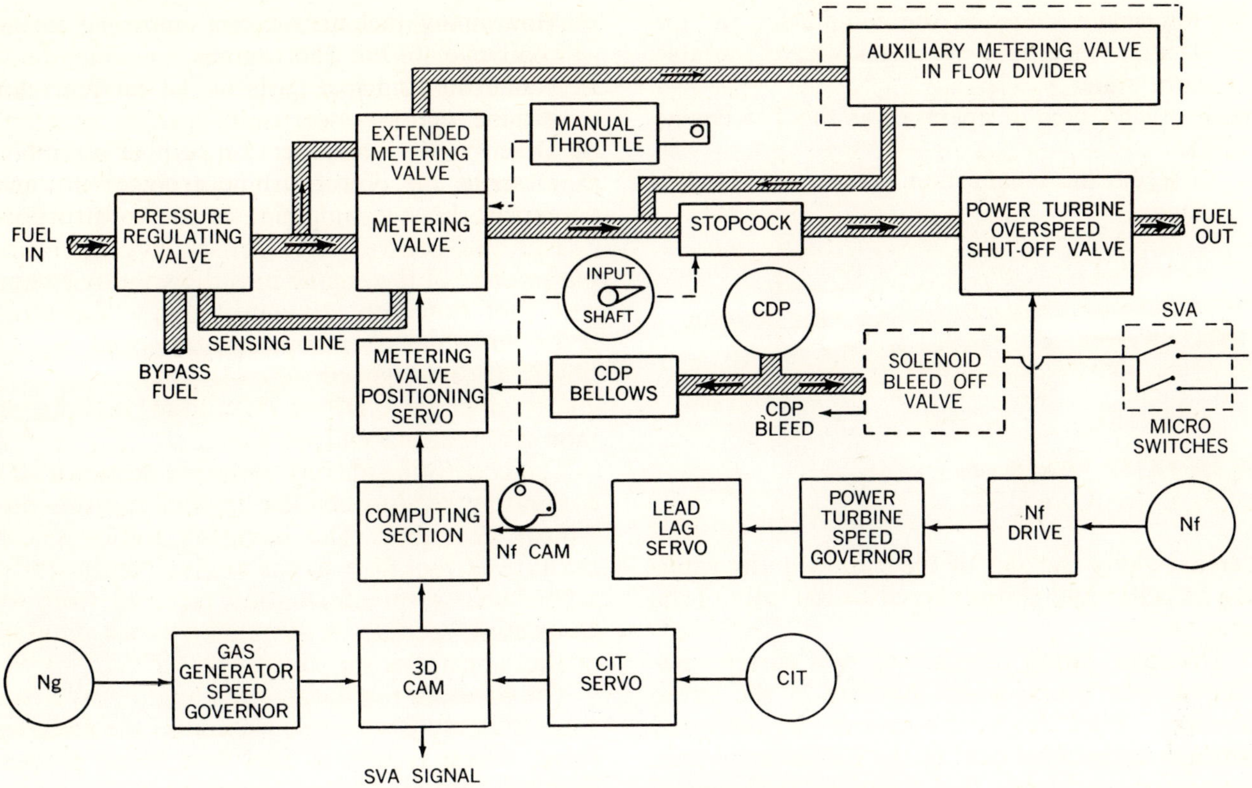

Figure 18·58 Block diagram of fuel-control system.

signals it receives and delivers a signal for metering-valve position to the metering-valve position servo. This signal is a ratio of fuel flow by weight to CDP or W_f/P_3.

The fuel-control system includes power-turbine overspeed protection in the event of a loss of the power-turbine speed-sensing signal. The overspeed governor activates at 106 to 108 per cent of maximum power turbine speed. The **overspeed shutoff valve** activates at 110 per cent of maximum power turbine speed in the event of a complete load loss in the helicopter transmission and power-shaft system.

OTHER ENGINE SYSTEMS

Because the lubrication system and electrical system for the T58 engine are similar in most respects to systems previously described, these systems will not be described in this section. If specific information about these systems is required, it may be obtained from the manufacturer's overhaul and maintenance manuals.

STUDY QUESTIONS

1. Describe the general configuration of the Allison Model 501 engine.
2. Explain the operation of the negative torque signal.
3. What is the function of the propeller brake?
4. Describe the operation of the safety coupling.
5. What is the location of the accessory drives on the Model 501 engine?
6. Describe the compressor rotor.
7. What type of combustion chamber is used in the Model 501 engine?
8. What is the function of the coordinator?
9. What is meant by "beta range"?
10. Describe the fuel-control unit.
11. Describe the construction of the Rolls-Royce Dart engine.
12. What type of lubrication system is used in the Dart engine?
13. Explain the function of the water-methanol system.
14. Describe the operation of the fuel-control unit for the Dart.
15. Briefly describe the propeller operation.
16. What type of starter is used with the Dart engine?
17. How does the compressor section of the Lycoming T53 engine differ from the conventional gas-turbine compressor?
18. What is unusual about the combustion and turbine sections of the T53?

19. How many stages of the compressor for the General Electric T58 engine have variable stator vanes?
20. For what type of operation is the T58 engine used?
21. Describe the combustion section for the T58 engine.
22. How many fuel nozzles are employed in the combustor of the T58 engine?
23. Name the principal parts of the gas-generator turbine section.
24. Describe the gas-generator turbine assembly.
25. Describe the power-turbine arrangement, and explain how it simplifies engine construction.

CHAPTER 19

Operation, Maintenance, and Overhaul

The purpose of this chapter is to familiarize the student with the general principles and practices utilized in the operation, maintenance, and overhaul of gas-turbine engines. It is not intended to give specific instruction on any particular engine, inasmuch as the instructions must be obtained from the manufacturers' manuals. The procedures used with a particular engine may be modified at any time, and in this case the manufacturer issues bulletins or directives which describe the change. Therefore, it must be emphasized that **the procedures to be followed with any specific engine must be obtained from the manufacturer's instructions.**

The first part of this chapter relates to operation of gas-turbine engines. Automatic devices have been developed to help prevent dangerous or damaging operation, but it is still quite possible to cause damage if proper care is not observed. In any case, it is essential that the operator understand the engine and the airplane with which he is working.

Maintenance procedures are also discussed in general because each engine model will have peculiarities which must be considered. Inspections and routine maintenance should follow the manufacturer's instructions. This is also true of overhaul, because the manufacturer is best qualified to establish the requirements for overhaul and the replacement of parts and assemblies.

OPERATION

Requirements for operation

Since the efficient operation and maintenance of a gas-turbine engine requires a thorough knowledge of the engine, it is essential that the operator be well acquainted with his equipment. He should know what the engine is designed to do and how it is going to do it. He should be familiar with the inspection procedures, starting procedures, fuel system, fuel-control operation, electrical system, and engine-operating limitations.

The operator should also know the engine-operating characteristics and specifications so that he will recognize any failure of the engine to meet operating requirements.

Ground safety precautions

The operating characteristics of gas-turbine engines have changed the ground safety rules for aircraft to some extent, because the turbine engine presents certain additional hazards which were not encountered with reciprocating engines. It has always been the practice to avoid the propeller on a conventional aircraft, and it is necessary to do so in the case of a turboprop aircraft. In addition, the rear of a turbine-powered airplane presents a hazard which must be avoided. Hot, high-velocity gases are discharged from the exhaust of the gas-turbine engine, and these may produce dangerous temperatures for more than 100 ft to the rear of the airplane.

The air intake of a gas-turbine engine draws in an extremely large volume of air, thus creating a very high-velocity stream near the inlet. This stream of air will pick up any loose material and may even draw a man's body into the inlet of the engine. Men have been killed or severely injured by being drawn into a jet-engine inlet. A turboprop engine provides its own warning to persons to keep their distance because of the whirling propeller. It is important, however, that all loose material be removed from the inlet area near the airplane, so that nothing will be drawn into the engine to cause damage.

Before starting any gas-turbine engine, the engine inlet should be inspected to make sure that the inlet duct is clear of any matter whatsoever.

This inspection must be very thorough, because it is easy to overlook such items as cotter pins, safety wire, washers, small nuts, screws, and other small items. Any of these can do severe damage to the engine.

When the gas-turbine engine is operated at full power, the jet exhaust may pick up and blow loose dirt, rocks, sand, bits of paving, and other debris for a considerable distance. For this reason, the airplane must be carefully parked in a location where no damage can result from flying particles which are driven to the rear.

The temperatures found to the rear of a gas-turbine engine operating at full power may be high enough to melt or ignite bituminous pavement; therefore it is recommended that concrete aprons be used for run-up of these engines. Sometimes excess fuel accumulates in the tail pipe of a gas-turbine engine, and when the engine starts, a long flame is blown out the tail pipe. Care should be taken that all flammable material is removed from the area to the rear of the engine so that this hazard may be avoided.

Tests have indicated that the carbon monoxide content of exhaust gases is relatively low, but other gases are present which have a disagreeable odor and cause eye irritation. The respiratory tract may also be adversely affected if these gases are breathed in too strong a concentration. In all cases, the breathing of the exhaust gases should be avoided.

After engine operation, no work should be performed on the hot section of the engine or aircraft until adequate time has been allowed for cooling. If it is necessary to work on a hot engine, asbestos gloves should be worn.

Gas-turbine engines produce noise capable of producing temporary and even permanent loss of hearing. Exposures to extreme noise may result in damage to the ear drum, and all persons working around operating engines should wear some form of ear protection.

Gas-turbine engines are normally equipped with high-intensity ignition systems capable of causing severe injury or death. No work should be performed on the ignition system or spark plugs when the engine is running, and care should be taken to avoid the possibility of a shock from the system at any time.

Fuels and lubricating oils should not be spilled on the skin. Fuels have a strong drying effect on the skin and may cause considerable irritation. Lubricating oils, especially the synthetic types, may have an irritating effect. If clothing should become saturated with either fuel or oil, it should be removed at once and the skin should be washed with soap and water.

Although jet fuels are not quite so flammable as aviation gasoline, they still have a flash point which is low, and the same precautions that govern the handling of fuel should be observed.

Starting a turboprop engine

In order to provide an example of the actual starting procedure for a typical turboprop engine, the following instructions are given as taken from the manufacturer's manual.

Preoiling of Engine. To assure ample lubrication of the propeller-shaft reduction-gearing assemblies in engines which have been idle for a period of 24 hr or longer, preoil the engine prior to starting as follows:

1. Fuel shutoff lever—OFF.
2. Emergency fuel shutoff valve—ON.
3. Fuel booster-pump switch—OFF.
4. Power lever—GROUND IDLE.
5. Ignition switch—OFF.
6. Propeller-feathering switch—NORMAL.
7. Release engine brake (check engine-brake light —OFF).
8. Energize starter and rotate the engine with the starter for 10 sec (check for oil-pressure rise).
9. Release the starter switch and allow the engine to rotate to zero rpm without the use of the engine brake.
10. Apply the engine brake (check engine-brake light—ON).

Ground Starts

1. Main power switch (as affects the engine)—ON.
2. Anti-icing valve control—OFF.
3. Anti-icing valve position light (for the closed position)—check ON.
4. Fuel shutoff lever—OFF.
5. Emergency fuel shutoff valve—OPEN.
6. Energize airplane fuel booster pump (check fuel inlet pressure 12 to 40 psi).
7. Set power lever in GROUND-IDLE position.
8. Release engine brake (check engine-brake light —OFF).
9. Set propeller pitch in NORMAL.
10. Engine cranking switch ON (check for oil pressure).
11. After tachometer indicates 9 per cent rpm, move the fuel shutoff lever to the ON position.
12. Fuel flow gauge should show increase of fuel flow.
13. Cranking switch—OFF.

Satisfactory Starts. The following statements define satisfactory start conditions:

1. Light-up will take place within 15 to 20 sec after fuel shutoff lever is placed in the ON position.
2. Engine will accelerate to 72.8 per cent rpm.
3. Exhaust gas temperature will not exceed 760°C during the transition to ground-idle.
4. Engine cranking switch, if not automatically cut out, will be turned off at 41 per cent rpm.
5. Ignition switch will be deenergized after engine reaches ground-idle rpm.
6. Exhaust gas temperature will settle to or below 430°C after ground-idle rpm is obtained.
7. Oil pressure should rise to 80 to 90 psi minimum at idle rpm.

Unsatisfactory Start. An unsatisfactory start is made when any of the following conditions occur:

1. HOT START. Exhaust gas temperature exceeds the temperature limit of 760°C (1400°F).
2. ABORTED START. The engine does not "light up" within 20 to 30 sec, providing the starter energy is adequate, after the fuel shutoff lever has been placed in the ON position. If the exhaust-gas temperature gauge does not indicate a temperature rise, a "light-up" has not occurred.
3. FALSE START. A "light-up" occurs, but the rpm does not increase to ground idle, remaining at some intermediate point, with the exhaust-gas temperature at or below 760°C.

Engine Clearing. In the event of an unsuccessful start, the engine should be cleared of trapped fuel as follows before additional starts are attempted:

1. Engine brake in the OFF position.
2. Fuel-shutoff lever in OFF position.
3. Ignition switch in OFF position.
4. Energize the starter and permit the engine to rotate for 20 to 30 sec.
5. Discontinue the "clearing procedure" by releasing the engine start switch.

Note. A fuel-drainage period of at least 30 sec should be allowed before attempting another start.

Starting a turbojet engine

The procedure for starting a turbojet engine is relatively simple; however, it should always be performed in accordance with the manufacturer's operating instructions for the particular engine.

The starting of a jet engine requires that certain events take place. First, the engine must be accelerated or rotated at a speed sufficient to draw adequate quantities of air through the engine. Ignition must be supplied before the fuel is turned on. The fuel is controlled with the power lever. Finally, the starter must be kept engaged until the engine attains self-sustaining speed.

Typical instructions for starting a turbojet engine are as follows:

1. Power control lever—OFF.
2. Main switch—ON.
3. Fuel control valve—ON or NORMAL.
4. Fuel booster pump switch—ON.
5. Starter or cranking switch—ON. Watch for increase in rpm and oil pressure.
6. Between 10 to 15 per cent rpm, advance power-control lever to IDLE position. (If a separate ignition switch is provided, it should be turned on before the power-control lever is advanced.)
7. Observe fuel pressure or flow, and watch tailpipe temperature. These conditions should not exceed specified limits.

Starting a turbofan engine with a pneumatic starter

To provide a good understanding of the starting procedures employed with a turbofan engine on a modern airliner, we shall describe here the recommended starting operations for the Pratt & Whitney JT8D engine installed in the Boeing 727 airliner.

The JT8D engine is a twin-spool turbofan engine, similar in many respects to the JT3D described in a previous chapter. The principal difference between the engines is the annular air passage, which carries the forward fan discharge air from the fan to the rear end of the engine where it mixes with the hot primary gases in the exhaust duct. The engine delivers 14,000 lb of thrust at takeoff power.

The starting air duct system for the three engines of the Boeing 727 airliner is shown schematically in the diagram of Fig. 19·1. This system supplies low-pressure air to any engine for starting.

The three control switches for the starters are located on the pilot's overhead panel and are marked ENGINE START. The switches have three positions: OFF, FLIGHT, and GROUND. This means that they may be used for starting either in flight or on the ground. Electric power for controlling the starters is received from the 28-volt

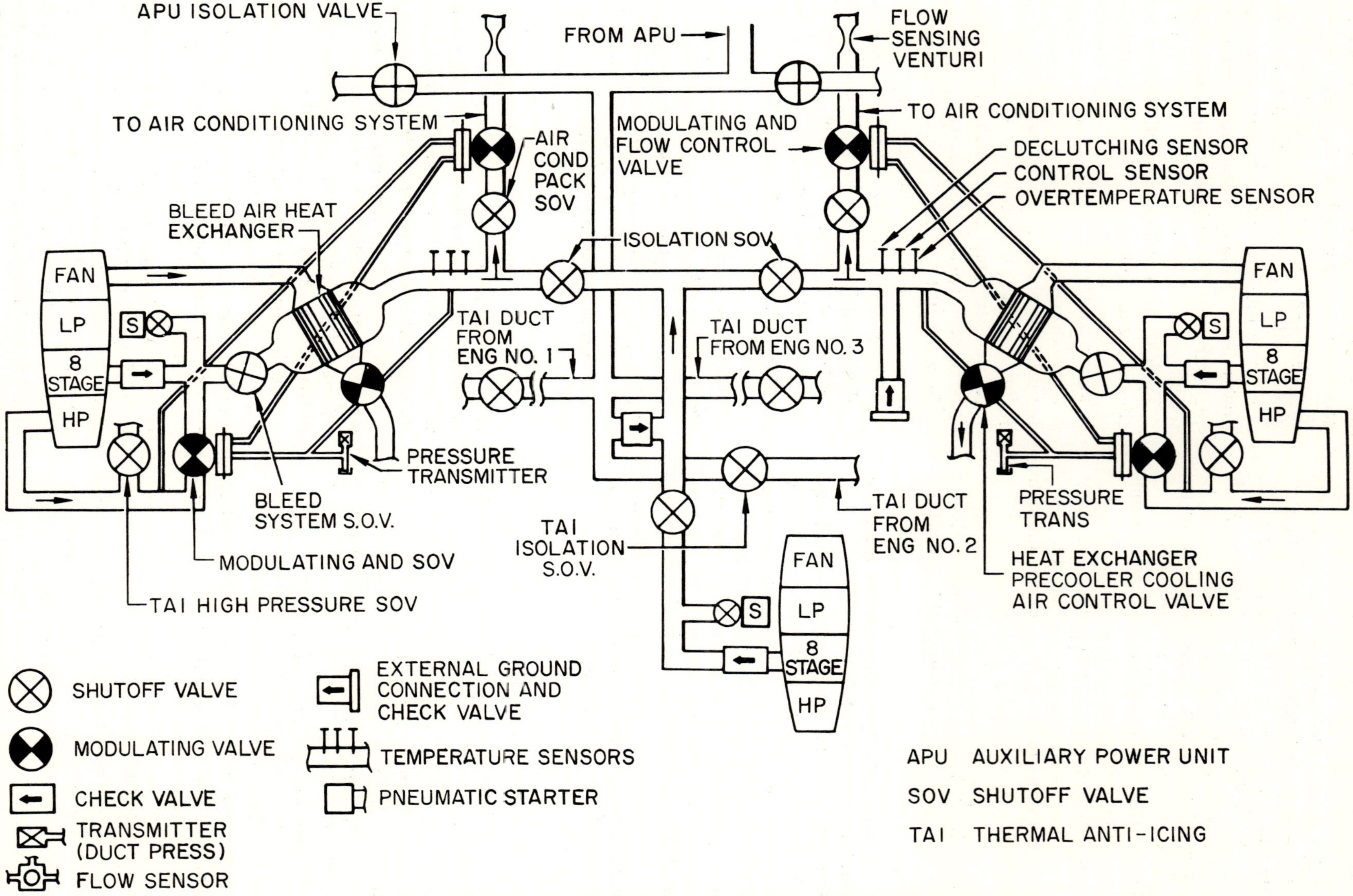

Figure 19·1 Starting air duct system for Boeing 727 airliner.

d-c battery bus in the main load control panel. Other controls used in conjunction with the cranking-system controls are the air-conditioning control switches, located on the flight engineer's upper panel, and the start levers on the pilot's pedestal.

A common method for starting the engine is to utilize an external air source which is supplied by a ground service unit connected to the ground air-supply fitting of the airplane. An electrically operated control valve permits air to flow to the pneumatic starter when the valve is actuated to the OPEN position by means of the starter switch.

When the left and right air-conditioning-pack shutoff valves are closed and the engine bleed air shutoff valves are open, low-pressure air is available at the starter valve. To start the engine, the start switch is held in the GROUND START position, thus supplying 28-volt d-c power through the normally closed **starter cutout switch** (in the starter) to open the **starter pressure regulating** and **shutoff valve.**

Because the engine start switch is of the momentary-contact type, care must be taken to hold the switch in the GROUND START position until the starter has reached cutout speed. This is 30 to 35 per cent on the N_2 (high-pressure compressor rotor speed) tachometer. If the starter switch is not held closed, a **hot start** may result.

With the engine start switch in the GROUND START position, the engine should accelerate; and when the N_2 tachometer indicates 12 to 14 per cent rpm, fuel and ignition are supplied by advancing the start lever to the START detent. The starter will continue to assist the engine rotation after light-up until the starter cutout switch opens upon attaining the proper speed. At cutout speed, electric power to the pressure regulating and shutoff valve will be interrupted by the centrifugally opened starter cutout switch, and the valve will close, thus ending the starter operation.

Engine light-up should occur within 20 sec after advancing the start lever to the start detent. After light-up the start lever is advanced to the IDLE position, the engine start switch is placed in the FLIGHT position to maintain ignition, and the engine bleed air shutoff valves are closed.

Considerable care must be exercised in the operation of the pneumatic starter. Because the starter is not designed for continuous operation, its duty cycle must not be exceeded. The normal duty cycle is 30 sec on and 1 min off. If the engine fails to start and accelerate normally, the operation should be stopped and no further attempt made to start until 15 sec after the N_2 tachometer indicates that the engine has stopped rotating.

Summary of operating procedures

The first requirement for jet-engine operation is that the operator be familiar with the engine and its operating instructions. A careful adherence to the instructions will result in long and satisfactory service from the engine.

Overheating of the turbine or other parts of an engine, either during starting or operation, will usually require that the engine be completely overhauled. Therefore, it is of the utmost importance that **hot starts** be avoided. Modern fuel-control units are designed to prevent hot starts; however, the operator may cause overheating damage by improper use of the equipment.

Engine ducts should be carefully inspected for foreign objects or material before starting the engine. Also, the airplane must be located so that it will not pick up dirt, rags, rocks, or other material. The area to the rear of the airplane must also be clear of anything which might be damaged by the hot exhaust gases.

MAINTENANCE

Daily or preflight inspection

The following items indicate typical inspections which should be performed daily or before flight on every turbine-engine installation. In any case, the manufacturer's inspection procedures must be followed.

The **engine compartment** or **nacelle** should be inspected for cleanliness and for security of all lines and fittings. It should also be inspected for fuel and oil leaks, loose parts, and any other undesirable condition.

Accessory mountings should be inspected for oil leakage, security of attachment, safetying, etc.

A complete external inspection of the engine should be accomplished to discover any loose or missing safety wire; loose nuts, bolts, and attachments; loose engine mounts; etc.

The engine and inlet ducts should be inspected for any loose parts, materials, tools, or other material.

The tailpipe or afterburner should be inspected for heat damage, loose parts, cracking, and any other undesirable condition. The turbine wheel should be inspected with a strong flashlight so that any damage may be observed.

Upon starting, the operator should listen carefully for noise. Any unusual sound which develops while cranking should be investigated.

Minor inspections

Minor inspections include many items not inspected during daily and preflight inspections but do not include those items which require engine disassembly.

During minor inspection, the following items should be checked in addition to those covered under preflight inspection.

The electrical system should be inspected for condition of harness, wiring terminals, thermocouples, spark plugs, and ignition units.

The fuel system units should be removed, cleaned, and reinstalled.

The oil strainers and drain plugs should be removed and cleaned. Any metal particles should be checked to determine whether excessive wear or other damage has occurred within the engine.

The engine controls should be operated to see that they are free and operate in the proper manner.

The air-inlet area should be inspected for damage caused by foreign matter. The area should be clean and all attachments secure.

The compressor area should be inspected for cracks, security of nuts and bolts, security of air-bleed units, and security of all case-mounted accessories.

Combustion chambers should be visually inspected for damage or deterioration. Drain or drip valves should be checked for proper operation.

The rear turbine and nozzle guide vanes should be inspected to the extent possible without disassembly. A strong flashlight will aid in this operation. The exhaust cone, afterburner, tailpipe, and similar units should be checked for warping, buckling, burning, and cracks. It must be emphasized that certain types of cracks are permissible in hot sections of a gas-turbine engine, according to manufacturers' instructions.

Major inspections

Major inspections include partial disassembly of the engine to see parts inaccessible during lesser inspections. A complete major inspection includes the previously described operations.

A major inspection requires an examination of the compressor blades and stator vanes, combustion chambers and liners, turbine blades, nozzles, fuel manifolds and nozzles, fuel system and units, and operating parts of the afterburner.

OVERHAUL

Requirements for overhaul

The overhaul of a gas-turbine engine is as exacting as any other mechanical operation involving precision equipment. For this reason, it is necessary that the proper tools, equipment, and facilities be available before starting an overhaul. Up to the present time, all overhaul of jet engines has been accomplished in Air Force or Navy depots or in large civilian shops. In these shops, the manufacturer's instructions are carried out in detail, and the overhauled engine usually performs almost as well as one straight from the factory.

Overhaul procedures

Briefly, the overhaul of a typical engine may utilize the following procedures:

1. Mount the engine on the proper type of overhaul stand.
2. Remove all dirt, grease, and foreign matter.
3. Following manufacturer's instructions, remove electrical harness, ignition units, case-mounted accessories, air-bleed units, fuel lines and units, control rods and mechanisms, and all other externally mounted parts.
4. Remove all parts to the rear of the turbine wheels. Place parts in suitable racks or stands.
5. Remove turbine wheels and nozzles in accordance with manufacturer's instructions. This procedure usually includes the removal of the turbine shaft.
6. Remove outer combustion-chamber case and liners.
7. Disassemble compressor section. Methods for accomplishing this operation will vary considerably, depending upon the type of casing which encloses the compressor. Great care must be taken in handling the compressor parts, because the compressor rotor and stator blades are easily damaged or knocked out of alignment. Also, the sharp edges of the blades can cause serious personal injury.
8. After complete disassembly, all parts should be cleaned as specified and prepared for inspection.
9. Inspection should include magnetic inspection of steel parts. Some alloy-steel parts cannot be magnetized and these must be inspected by dye-penetrant, fluorescent penetrant, X ray, or Sonigage. All fits and clearances must be checked with precision gauges to see that they are within the limits set forth in the manufacturer's table of limits.
10. Obtain all new gaskets and seals for assembly.

It is usually not advisable to reuse any gasket or seal, because the resilience of such materials is destroyed by pressure and heat.

11. Assembly procedures are usually the reverse of disassembly; however, the overhaul manual must be followed closely.

Overhaul precautions

A gas-turbine engine and its parts must be handled with great care. Parts which have been cleaned and oiled must not be touched with the bare hands, because perspiration can cause corrosion. Small parts such as bearings should be individually wrapped after inspection to prevent the accumulation of dirt and to prevent damage by contact with other parts. All parts should be placed in suitable parts racks or stands until they are assembled in the engine.

The proper overhaul tools and equipment should be used. Engine manufacturers design and manufacture special tools and fixtures for use with each engine model. The use of these tools greatly facilitates the assembly and disassembly of the engine.

The overhaul shop or area should be kept clean and orderly, to aid in keeping the engine and its parts clean and to provide for efficient operations.

During assembly the engine must be protected from loose bits of metal which might fall inside. If a nut, bolt, washer, cotter key, or any other item should be dropped, **the assembly of the engine must be stopped until the item is located.** Any small bit of metal can cause severe damage to a gas-turbine engine during operation.

Nuts, bolts, and studs must be safetied in accordance with approved practice. Safety wire must be used with studs, because pal nuts or lock nuts will not prevent a stud from unscrewing. Two or more studs should be safetied together. All safety wire should be installed so that its tension will tend to tighten the nut or bolt in which it is installed.

To summarize: the overhaul of a gas-turbine engine should always follow the procedure specified by the manufacturer, approved mechanical practices should be employed, the engine and its parts should be handled in the same manner as any fine precision equipment, and the overhaul mechanic should be familiar with the engine and its approved overhaul procedures.

ENGINE TRIMMING

For any particular turbojet engine, there are certain conditions under which the engine will produce maximum or rated thrust. These conditions include compressor inlet air temperature T_{t2}, engine inlet air pressure P_{t2}, engine rpm, turbine discharge pressure P_{t7} (for split-compressor engines) and P_{t5} (for single-compressor engines), and exhaust nozzle area.

For an engine such as the Pratt & Whitney JT8D, the exhaust nozzle area is constant and the rated thrust rpm of the engine is determined by test at the factory. After the test run, the rpm at which the engine produces rated thrust is stamped on the engine data plate. Because of the variations in manufacturers' tolerances and other slight differences in identical models of the engine, no two engines are likely to have the same rated-thrust rpm.

The thrust of these engines is measured by the **engine pressure ratio** (EPR) or the ratio of P_{t7} to P_{t2}, that is, $\frac{P_{t7}}{P_{t2}}$. The rpm for a particular engine must be set to provide the correct EPR to produce the rated thrust.

Adjusting the trim

To adjust the trim for a Pratt & Whitney JT8D engine, the fuel-control unit trimmers must be set in accordance with the directions furnished in the aircraft manufacturer's maintenance manual. There are two adjustments for trimming, one for MIL, or maximum thrust, and one for IDLE speed.

Before the trimming operation is started, the airplane is placed heading into the wind, and the barometric pressure and temperature of the ambient air are recorded. On commercial aircraft, the engine is trimmed at an rpm somewhat less than maximum, hence a **part power stop** on the fuel-control thrust lever is placed in a position to limit the movement of the lever. A test tachometer and a pressure gauge are connected to the engine test fittings to provide accurate readings of these values. The tachometer indicates the rpm of the high-pressure N_2 compressor, and the pressure gauge indicates turbine discharge pressure P_{t7}.

The engine is started and operated in accordance with the manufacturer's instructions. When the operation has stabilized as specified, the trim adjustments are made to provide the idle rpm and P_{t7} called for in the engine data. The idle speed is set to 57 per cent by means of the idle trim adjustment, and then the power lever is advanced against the part power stop. After stabilization, the MIL trim adjustment is made to provide the required P_{t7} and EPR. The engine is then shut down, and all data is recorded on the Engine Trim Check Sheet.

STUDY QUESTIONS

1. What rules apply to the operation of a gas-turbine engine?
2. What knowledge is essential to the proper operation of a gas-turbine engine?
3. What hazards are encountered in the ground operation of a gas-turbine engine?
4. What action should be taken to protect the engine during ground operations?
5. What protection should be used against the noise of jet engines?
6. What is meant by an "aborted start"?
7. What is the greatest hazard to the engine during starting?
8. List engine preflight and daily inspections.
9. What are "minor inspections"?
10. What equipment and information should be available before overhauling a gas-turbine engine?
11. What inspections are applied to nonmagnetic materials during overhaul?
12. What should be done if a nut, bolt, or other piece of metal should be dropped and lost during engine assembly?
13. What is the reason for *trimming* a turbojet engine?
14. Give a brief description of how trimming may be accomplished.

CHAPTER 20

Materials for Gas-turbine Engines

Because of the publicity given the aviation gas-turbine engine in various periodicals and newspapers, it is common knowledge that exceedingly high temperatures are attained when operating these engines. These temperatures provide the chief limiting factor in the performance of such engines, because the metals available for engine construction are limited in their capacity to withstand the high temperatures. This era has already been labeled the "high-temperature age" because of the impact that high temperatures have exerted upon modern life. The jet-power industry is consuming tons of hard-to-get alloys at an unbelievable rate; however, the great need for these materials has provided an incentive that is uncovering a wealth of knowledge about new metals and their combinations. These metals are known as **refractory metals.**

Until the sixteenth century, only seven metals were known and used by man: gold, silver, copper, tin, iron, lead, and mercury. These were found most freely in nature and were the easiest to obtain in a metallic state. During the next three centuries, seven more metals were discovered, but in the last 50 years, scientists have unearthed 69 additional metals to bring the total to 83, or more than three-quarters of the 100 elements known to exist. Now, the searchlight of science is being turned on the one-third of these metallic elements which have been little more than laboratory curiosities.

Impetus for this metallurgical research is the pressing need for materials which can withstand the consuming heat encountered in jet engines, rockets, and similar engines. One of the principal manufacturers engaged in the search for heat-resistant materials is the Ryan Aeronautical Company, which is a leading producer of "hot-end" components of ramjet, rocket, and turbojet engines.

It is not the purpose of this chapter to provide the reader with metallurgical education so that he can attain a complete understanding of the composition and manufacture of high-temperature steels and special alloys. Those who have the benefit of previous study or experience in the field of metallurgy will find the subject matter in this chapter quite easy to follow. For the benefit of those who have had no previous experience in the subject, certain definitions and explanations are included.

Definitions

yield strength: The load per square inch at which a material exhibits a specified limiting permanent set or a specified elongation under load.

creep strength: A continuing change in the deformation or deflection of metals and alloys, generally associated with a time rate of deformation continuing under stress intensities well within the yield point or the apparent elastic limit for the temperature in question.

critical temperature: Thermal points in the cooling or heating of the metal that are marked by the spontaneous evolution or absorption of heat. When steel passes through this critical range, its internal structure is altered, and rapid cooling of the metal through this range of temperature will prevent normal changes from taking place and the usual properties from being possessed by the material.

stress rupture strength: The sustained load which will bring about rupture of a material. The load is expressed in pounds per square inch and is applied for a period of 100 to 1000 hr.

ultimate tensile strength: The load under which a metal will continue to elongate and fracture in a short-term test.

ductility: Drawing out in length without fracture. Any ductile material can be easily deformed without breaking and resists failure under shock loads. As the term is employed in the testing of materials, ductility indicates the distortion a body undergoes before being ruptured.

ferritic alloys (stainless steel): Ferrite (carbonless iron) does not have any hardening properties, because of a lack of carbon in the alloy. Ferritic alloys are those alloys in the corrosion- and heat-resisting group which do not respond to heat-treatment.

austenitic steels (stainless steel): When steel is heated above the critical range, austenite is formed. As further cooling takes place, the critical range is reached and the austenite goes through a transition. This transition can be arrested by rapid cooling (cold water or oil quench) just above the critical range. Austenite contains enough nickel to make steel nonmagnetic. These steels have high creep strength up to 1200°F and are excellent where high temperature and corrosion are to be encountered.

martensitic steels: Straight chromium alloys which harden intensely if allowed to cool rapidly from high temperatures. If the chromium content is greater than 14 per cent, the hardening characteristic is reduced.

solution-treated: During heat treatment, the alloying constituents enter into solid solution. The alloying elements, which increase strength and hardness, are more soluble in solid metal at high temperatures. If the alloy is cooled quickly from this temperature, the particles will be retained in solid solution. This operation is performed in molten-type solutions.

precipitation hardening: After solution heat-treatment, when standing at room temperature, particles in the alloying elements of some alloys separate out of the solid solution. This is called "precipitation." In other alloys, these changes take place so slowly at room temperature that they must be heated for precipitation to occur. Precipitation imparts to the alloy a greater hardness, yield strength, and tensile strength. Where the precipitation of the alloys takes place at room temperature, the alloy is said to **age,** or **age-harden;** but if reheating is required, the heat-treatment is called a **precipitation heat-treatment** or an **artificial aging treatment.**

stress: A load acting on a material or the internal resistance to the load developed in the material; both expressed in pounds per square inch.

strain: The deformation of a material caused by a stress.

oxidation: The rust or corrosion formed when the metal combines with oxygen.

hot-worked: Rolled or forged when heated to a predetermined temperature.

cold-worked: Rolled or forged at atmospheric temperature. The process increases strength and hardness of a metal but reduces its ductility.

Chemical elements

The most commonly used elements in the manufacture of metals and alloys for gas-turbine engines are listed here with their symbols:

Carbon (C)	Tungsten (W)
Silicon (Si)	Columbium (Cb)
Manganese (Mn)	Titanium (Ti)
Chromium (Cr)	Nitrogen (N)
Nickel (Ni)	Copper (Cu)
Cobalt (Co)	Iron (Fe)
Molybdenum (Mo)	Aluminum (Al)

This list is not meant to include all possible elements used in high-temperature alloys, but it indicates the types of elements required.

Operating conditions for gas-turbine materials

With jet airplanes flying at supersonic speeds above altitudes of 50,000 ft in routine flights, the turbojet engine is already many times as powerful as the largest piston engine-propeller drive, and its installed weight is less than one-quarter that of a piston engine of comparable power at high speed. Nevertheless, it is safe to say that the full potential of the gas turbine has not been attained. Compact engines with power equivalent to more than 30,000 hp are in sight if temperatures and pressures can be increased, and that is the area in which the metallurgist is pacing performance. New materials with prescribed high-temperature properties must be synthesized to meet the challenge.

Highest readings of temperature are found in the combustion chambers and afterburners of jet engines. At these points, temperatures of more than 3700°F have been measured. Since this is 1000° hotter than the melting point of the alloys used in these structures, it is evident that wall temperatures are considerably lower than these figures. Actually the metals operate at temperatures within the range of 1200 to 2000°F. This disparity between flame and wall temperatures is accomplished by ingenious design innovations that ensure the centering of the combustion flames within the structures and provide a blanket of cooling (secondary) air around each component. As previously explained, the secondary air is fed through perforations in the combustion liners to form a blanket around the center core of combustion. This blanket insulates the walls of the liner from the high temperatures.

It may seem strange that metallurgists are having difficulty in finding suitable high-temperature materials when it is known that tungsten melts at 6200°F, molybdenum melts at 4800°F, and columbium melts at 4400°F. The problem emanates from the fact that melting point is not the important consideration in determining whether a metal can withstand the rigors of jet-engine environment.

The three characteristics that metallurgists search for in the metals for high-temperature applications are **oxidation resistance, high-temperature strength,** and **corrosion resistance.** Probably the most vital of these three characteristics is the capacity to resist oxidation, because structural strength is useless if a metal rapidly burns away. At jet-engine operating temperatures, ordinary steel burns like paper; and tungsten, molybdenum, and columbium oxidize quickly, as do most high-melting-point metals.

Oxidation resistance

Ability of a metal to resist oxidation at elevated temperatures is largely determined by the type of oxide scale it forms. Laboratory investigations show that many metals form a tough, impervious scale that retards progressive oxidation. Others form a porous type of scale which continues to build up and flake away, eating into the metal. However, over prolonged periods of exposure, almost any type of oxide film tends to break down because of compressive stresses that are set up. The effects of repetitive heating and cooling cycles, experienced in jet-engine operation, cause compressive stresses in the oxide film because the underlying metal tends to contract faster than the oxide scale. Alloys having high coefficients of thermal expansion aggravate this condition because they exhibit greater dimensional change under varying temperature.

High-temperature strength

The second vital quality that a metal must possess for jet-engine use is good strength at high temperatures. This involves all kinds of strength exhibited by alloys, such as ultimate tensile strength, yield tensile strength, creep strength, and stress rupture strength. Tensile strengths for various alloys are shown in Fig. 20 · 1.

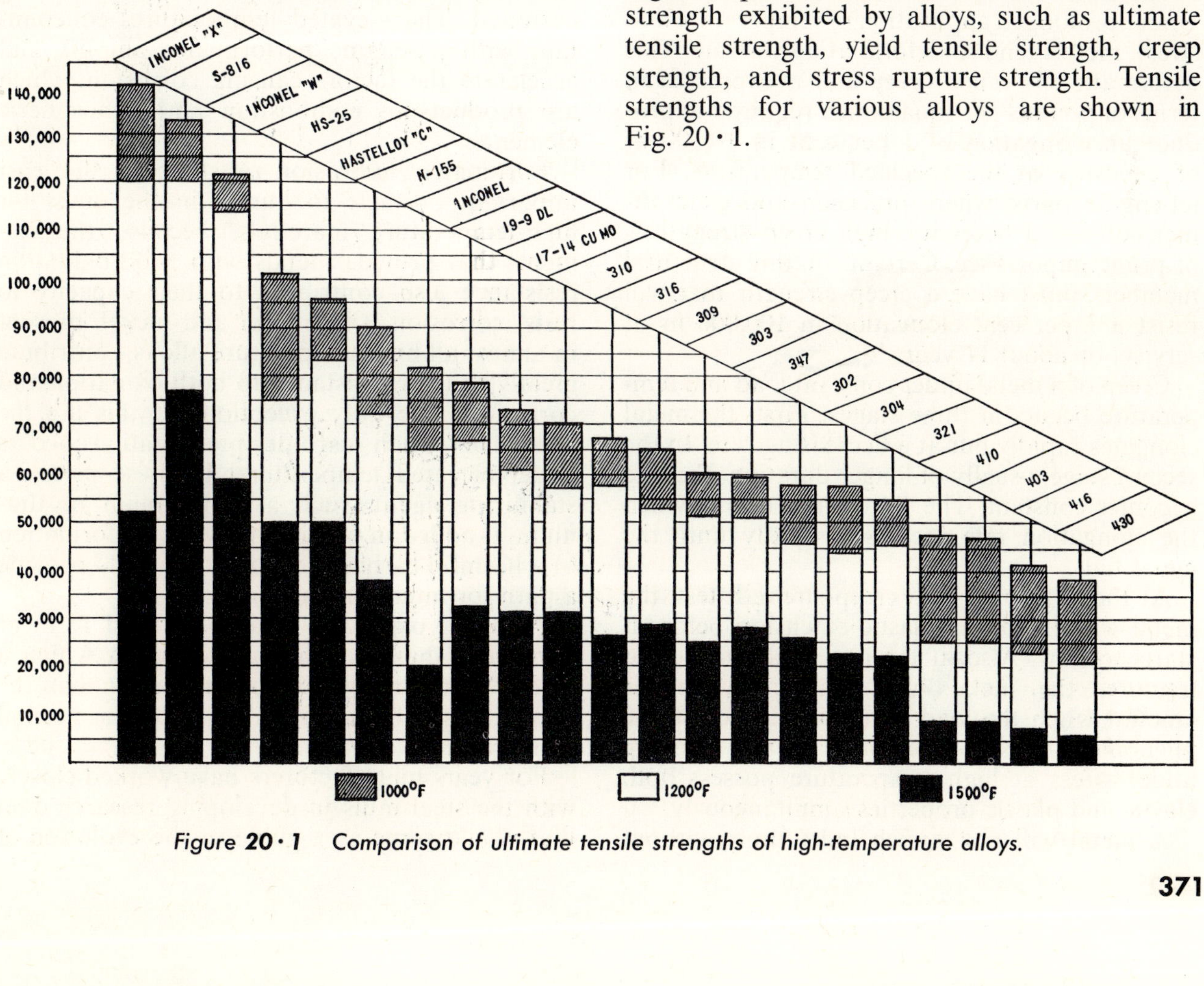

Figure **20 · 1** *Comparison of ultimate tensile strengths of high-temperature alloys.*

There is a striking difference in the behavior of metals at room temperature and in the environment experienced in jet powerplants. A good conception of this difference can be obtained when it is realized that structural carbon steel, with a room-temperature strength of 140,000 psi, would have an ultimate tensile strength of only 1,000 psi after exposure to 1400°F for 1,000 hr.

The design of load-bearing structures for service at atmospheric temperatures is generally based upon yield or tensile strength that is determined in ordinary room-temperature tensile tests. In service at ordinary temperatures, with a design stress determined in this way, a metal behaves essentially in an elastic manner. The structure undergoes an elastic deformation immediately upon application of the load, and no further deformation occurs with time. When the load is removed, the structure returns to its original dimensions.

At elevated temperatures the behavior is quite different. A structure will continue to deform with time after load application even though the design data may have been based upon tensile tests taken at the elevated temperature. This is because the tensile tests are relatively short-term operations, accomplished within a few minutes. Under prolonged application of temperature and stress, metals tend to deform at a slow rate. This deformation is called **creep** and is expressed in terms of pounds per square inch required to produce an elongation of 1 per cent in 10,000 hr of exposure at the specified temperature. For jet-engine parts, where precision and exact dimension are a necessity, high creep strength is of prime importance. Certain rotating structural members must have a creep strength that will resist a 1 per cent elongation in 100,000 hr of service, or about 11 years.

Creep of a metal under constant load and temperature occurs in three stages. First, the metal elongates rapidly but at a decreasing rate. In the second stage, usually of longer duration, the rate becomes constant. The third stage occurs when the elongation rate increases rapidly until the metal fails.

At the beginning of a creep-strength test, the metal will exhibit an elastic extension before it starts to flow. Also, if the test is not carried to fracture, the metal will contract a similar amount when the load is removed. From these phenomena it is apparent that metals creeping under stress at high temperature possess both elastic and plastic properties simultaneously.

A metal will not stretch indefinitely without fracturing. Therefore, it is important to the designer of high-temperature components to know when a metal will rupture under conditions that produce creep. Since creep tests are usually long-term undertakings (some have been run for 10 years), another means has been evolved to disclose the stress rupture point of metals. Similar to the creep test, this test utilizes a greater strain than the creep test in order to speed up the elongation process. By extrapolation of creep and stress-rupture test data, it is possible to determine when a metal will fail under prescribed conditions. This failure occurs at lower stresses than the ultimate tensile strength of the material. The chart of Fig. 20 · 2 shows the creep strength of various alloys used in jet engines. Tables 20 · 1 and 20 · 2 show yield strengths and stress-rupture strengths, respectively.

Corrosion resistance

The third characteristic that is of interest to the jet-engine metallurgist is corrosion resistance. Corrosion is the destructive alteration of a solid body by chemical or electrochemical reactions arising on the surface. Usually metals are corroded by chemical solutions in which some of the positively charged ions in the solution lose electrical charges to the metal being corroded. The elevated temperatures concomitant with jet-engine performance amplify and accelerate the factors causing corrosion, which are products of combustion and atmospheric elements.

Fortunately, corrosion resistance is the least important obstacle to hurdle in the quest for high-temperature materials, because the elements that provide metals with good oxidation resistance also contribute to their capacity to resist corrosion. Chromium and nickel, present in almost all high-temperature alloys, contribute greatly to their resistance to both oxidation and corrosion. The only exceptions to this are the alloys in which carbide precipitation occurs under elevated temperature. In these metals a stabilizing agent, such as columbium or titanium, is added in order to prevent the formation of a harmful carbide network that could provide a path for intergranular corrosion.

Actually, oxidation is a specialized form of corrosion, that is, a hot corrosion in which a black "rust" or oxide is formed rather than the familiar types of corrosion that occur at normal temperatures.

For years manufacturers have worked closely with the steel mills in developing research data that would serve as a guide to the evolution of

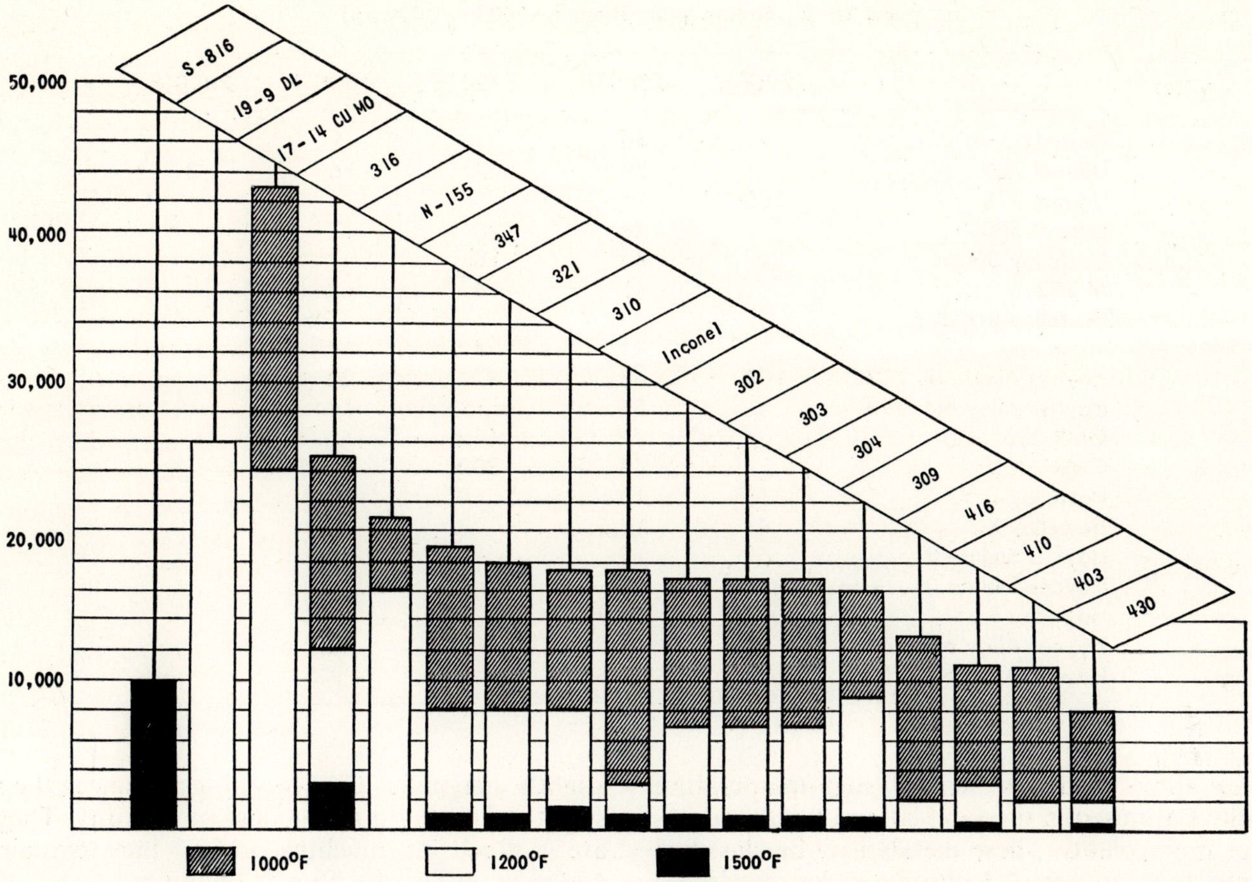

Figure **20·2** *Comparison of creep strengths in psi.*

Table 20 · 1 Yield strength—0.2% offset (1,000 psi)

	1200°F	*1500°F*	*1700°F*	*1800°F*	*2000°F*
Rene-41	145	118	. .	37	
Udimet 700	140	110	70	45	14
Udimet 718	140	85			
Udimet 500	125	95	60	35	
Hastelloy 500	125	95			
M-252	115	90			
Hastelloy R-235	111	85	. .	18	
Udimet 200	110	60			
Waspaloy	105	90	50		
Haynes Alloy No. 713C	102	105	. .	51	
Haynes Alloy No. 294	98	112	. .	49	
GMR-235	88	62	30		
GMR-235D	87	81	42		
Hastelloy C	43	40	. .	18	
Hastelloy X	40	37	. .	16	
Haynes Stellite 21	39				
Haynes Stellite 31	39				
Hastelloy B	38	37	. .	22	
Haynes Alloy No. 56	37	34	. .	20	
Haynes Alloy No. 25	35	36	. .	23	

Table 20 · 2 Stress-rupture strength—100 hr (1,000 psi)

	1200°F	1500°F	1700°F	1800°F	2000°F
Rene-41	92	40	. . .	9	
Udimet 700	. . .	59	27	16	3
Udimet 718	100	16			
Udimet 500	. . .	48	20	12	
Hastelloy 500	. . .	44	18		
M-252	. . .	41	14	8	
Hastelloy R-235	52	. . .	. . .	7.6	
Waspaloy	. . .	41	14	8	
Haynes Alloy No. 713C	101	68	. . .	20	
Haynes Alloy No. 294	. . .	55	. . .	18	
GMR-235	. . .	41	18.5		
GMR-235D	. . .	48	20.5		
Hastelloy C	55	18			
Hastelloy X	44	14.5	. . .	4.6	
Haynes Stellite 21	52	19	. . .	9.4	
Haynes Stellite 31	56	27	. . .	11.3	
Hastelloy B	50	16.2			
Haynes Alloy No. 56	49	17.5	. . .	5.2	
Haynes Alloy No. 25	. . .	22	. . .	7.5	

new alloys. Today, at least 27 high-temperature alloys are used in the fabrication of components for powerplants. These metals may be classified into four groups: (1) chrome-nickel steels, (2) chrome steels, (3) nickel-base alloys, and (4) cobalt-base alloys.

Chrome-nickel steels

The chrome-nickel steels contain these elements in approximately 18:8 ratios and are commonly referred to as "austenitic steels." This description refers to their crystalline structure, in which the ingredients are said to be in solid solution so that each crystal is composed of an intimate mixture of the constituents.

This group of steels is most widely used for high-temperature applications because of its good characteristics of oxidation and corrosion resistance and of high strength at elevated temperatures. These steels have about six times the electrical resistance of ordinary steels and respond readily to welding techniques. However, since they expand 50 per cent faster than plain steels and conduct heat only half as fast, they require special care during welding to prevent distortion. They do not need heat-treatment after welding to develop maximum physical strength.

These steels are nonmagnetic and cannot be hardened by heat-treatment. They may be hardened by cold-working, and then they become slightly magnetic. Cold-working enhances their structural strength at the cost of ductility. They are difficult to machine unless they contain added amounts of sulfur or selenium.

When subjected to temperatures above 800°F, they tend to precipitate carbides in their grain boundaries, thus lowering their resistance to corrosion. The corrosion resistance may be restored by heating them to 1850 to 1900°F and cooling them quickly, or they may be protected against the formation of harmful carbide precipitation by the addition of columbium or titanium to their formulas.

Many parts of gas-turbine engines are constructed from the chrome-nickel stainless steels. These steels are suitable for parts such as high-pressure compressor casings, combustion casings, turbine casings, and similar parts where extreme temperatures are not encountered. High-temperature stainless steels are also employed for compressor blading, compressor disks, turbine disks, exhaust cones, thrust reversers, afterburner casings, and miscellaneous sheet metal parts. The principal disadvantage of chrome-nickel stainless steels is the weight. For this reason, titanium alloys are often used where the temperature is not excessive for the material. Some gas-turbine engines, such as the Allison 501 turboprop, are made almost entirely of alloy and stainless steels, with the exception of the very hot parts and the compressor front frame.

Chromium steels

The straight chromium alloys, containing no nickel, are either martensitic or ferritic in crystalline structure, depending upon their hardening characteristics. Those containing up to 14 per cent chromium harden intensely if they are allowed to cool rapidly from high temperatures, forming a hard, relatively brittle substance called martensite. Because these alloys respond to heat-treatment, they are capable of being given a wide range of mechanical properties such as tensile strength and hardness.

Those chromium steels containing from 17 to 30 per cent chromium do not respond to heat-treatment or cooling and remain essentially ferritic at all times. The intermediate types, containing between 14 and 17 per cent chromium, behave in a manner that depends upon their exact chromium and carbon contents.

In the straight chrome group, corrosion resistance is a function of the amount of chromium dissolved in the iron. Chromium contents up to 4 per cent do not increase corrosion resistance. Amounts from 4 to 10 per cent improve resistance to corrosion appreciably. Chromium percentages from 10 to 30 per cent are designated when oxidation resistance at high temperatures is required.

The steels with chromium contents of 5 to 30 per cent are magnetic in all conditions. They exhibit distinctly inferior creep resistance compared with the chrome-nickel stainless steels. The 17 per cent chromium steels have excellent oxidation and corrosion resistance and may be used at temperatures up to 1500°F. The 25 per cent chromium content series is designed for severe heat and corrosion applications and is good for applications where temperatures reach 2000°F.

All the straight chrome steels are more difficult to weld than the chrome-nickel types because the welding heat leaves them in a brittle condition. They have about the same coefficients of expansion as ordinary steel, with several times the electrical resistance and slightly lower melting points. They tend to become brittle from welding heat in two ways: by slow cooling from around the 1200°F range, and by grain growth produced by holding them over 1650°F. The first type of embrittlement may be eliminated by quick cooling from above 1200°F. The second type cannot be remedied but may be avoided by careful welding techniques.

From straight chrome steel, jet-engine case weldments and bearing-housing assemblies are built for Pratt & Whitney turbojet engines. These are fabricated from welded alloys of Type 410. Additional jet-engine detail fittings and components are built from Type 416 and used in applications where temperature is not a critical factor.

Nickel-base alloys

The nickel-base alloys contain between 70 and 80 per cent nickel and cannot properly be called "steels" because of their small quantities of iron. In this group of alloys, many of which are called "superalloys," are the Nimonics, M-252, GMR-235, the Hastelloys, Nichrome, Rex 400, K 42 B, the Inconels, Waspaloy, Rene 41, Multi-alloy, Refractalloy 26, the Udimets, and Unitemp 1753. These alloys have been tailored for high-temperatures, and some offer excellent oxidation and corrosion resistance at temperatures of over 2100°F.

Inconel is a trade name for many of the high-temperature nickel-base alloys developed and manufactured by the International Nickel Company. These alloys can be cold-worked, hot-worked, or forged without difficulty and can also be welded and machined. Various types of the Inconel alloys are used for combustion chambers, transition liners, turbine nozzle parts, and turbine buckets.

The **Nimonic** alloys, numbered 75, 80, 80A and 90, are of British manufacture and contain a major percentage of nickel, with chromium as the next most important element. Nimonic 90 utilizes a large percentage of cobalt as an alloying element. The Nimonic alloys are largely used for the hot parts of gas-turbine engines, such as combustion chambers and turbine nozzles.

GMR-235 contains about 55 per cent nickel, 15.5 per cent chromium, and 10 per cent iron with smaller portions of molybdenum, aluminum, and titanium. It is of low strategic-alloy content, that is, it contains neither cobalt nor columbium; hence it can be manufactured even though these metals are not available. GMR-235 was developed by the General Motors Corporation primarily for gas-turbine wheels, buckets, and nozzle vanes operating at temperatures in excess of 1400°F. An improvement of this alloy, GMR-235D, contains increased amounts of aluminum and titanium, and less iron.

The **Hastelloys,** designated A, B, C, R235, W, and X, are nickel-base alloys developed and manufactured by the Haynes Stellite Company, a Division of the Union Carbide Company.

Alloys A, B, C, and W contain large percentages of molybdenum. These alloys are used in various applications where high temperature and corrosion resistance are desirable.

Haynes Alloy No. R-41, also known as GE Rene 41, is a vacuum-melted nickel-base alloy which has exceptionally high strength in the temperature range of 1600 to 1800°F. It is used for turbine buckets and other parts subjected to high stresses and temperatures.

Waspaloy was originally developed by the Pratt & Whitney Aircraft Division of United Aircraft Corporation for use in the manufacture of turbine buckets. It is a traditional superalloy suitable for use in the 1200 to 1600°F range although it retains good strength at higher temperatures. This alloy is also used for turbine disks and other heavy forgings in jet engines.

The **Udimet** alloys, manufactured by Special Metals, Incorporated, are superalloys of high strength, corrosion resistance, and high-temperature resistance. Udimet 500 is used for turbine buckets and small integral gas turbines. Udimet 700 is suitable for the 1400 to 1800°F range and can be used for a short time up to 2000°F. It is used for turbine buckets and for turbine disks in advanced jet engines. Udimet N-115 is one of the strongest wrought superalloys and is used for turbine buckets in applications requiring extended service up to 1850°F.

Cobalt-base alloys

Cobalt-base alloys, also called superalloys, are designed for use where high-temperature strength and corrosion resistance are important. These alloys usually contain substantial amounts of tungsten, nickel, chromium, and molybdenum in addition to cobalt. Some of the alloys contain either tungsten or molybdenum but not both.

Haynes Stellite No. 25 (HS-25) is used to build the eyelid assemblies for afterburners and afterburner flame holders. It has excellent strength and resistance to oxidation at high temperatures. It also exhibits good ductility and can be worked both hot and cold with a formability comparable to that of Type 347 stainless steel.

HS-25 has an ultimate tensile strength of 150,000 psi and a yield strength of 70,000 psi at room temperature. Its stress-rupture strength at 1800°F for 100 hr is 7,500 psi. It has good oxidation and carburization resistance for service up to 1900°F.

Haynes Stellite Alloy No. 21 (HS-21) is resistant to oxidizing and reducing atmospheres up to 2100°F. It has excellent strength which is maintained through a wide range of temperatures and thermal shock conditions.

Haynes Stellite No. 31 (HS-31) is a casting alloy, as is HS-21, and has good creep, endurance, and stress-rupture properties. It has a stress-rupture strength of 27,000 psi for 100 hr at a temperature of 1500°F and an as-cast tensile strength of 63,200 psi with an elongation of 15 per cent.

Another cobalt-base alloy manufactured by the Haynes Stellite Company is Alloy No. 151. This alloy has excellent properties at high temperatures and is suitable for turbine blade applications up to 1800°F.

Composition of typical high-temperature alloys

Even though the actual composition of a particular alloy may not have great significance to anyone other than a metallurgist or someone charged with the procurement of critical materials, it is believed that some knowledge of the elements included in various alloys will be of interest to the student of gas-turbine engines. Table 20 · 3 gives the chemical composition of a number of typical alloys in the iron-base, nickel-base, and cobalt-base classifications.

Variations in alloy properties

It must be emphasized that the properties of high-temperature alloys are not imparted merely by the mixing of various metallic elements. The elements are most important; however, the type of heat treatment and working applied to the alloy is also essential.

Among some of the treatments employed for developing desired characteristics in the superalloys are solution heat treating, precipitation hardening, strain hardening, cold-working, and hot-working. Some of these processes are time-consuming and complicated, hence they add to the cost of the material. It is important to note, however, that without the required treatment a particular alloy will not give the specified service.

Trends in alloy development

The development of new alloys which have superior characteristics is a continuing process. Extensive research has been done in the use of titanium alloys for various parts of gas-turbine engines in an effort to take advantage of the lighter weight of this metal. In many respects, titanium alloys compare favorably with alloy steels and today titanium is used extensively in both engines and aircraft.

Table 20 · 3 Chemical composition of superalloys

Alloy	*C*	*Mn*	*Si*	*Cr*	*Ni*	*Co*	*Mo*	*W*	*Cb*	*Ti*	*Al*	*Fe*	*Other*
Iron Base:													
19-9 DL	0.3	1.1	0.6	19.0	9.0	...	1.25	1.25	0.4	0.3	...	bal	
17-14 Cu, Mo	0.12	0.75	0.5	16.0	14.0	...	2.50	...	0.4	0.3	...	bal	3.0 Cu
A-286	0.08	1.35	0.5	15.0	26.0	...	1.25	...	...	2.0	0.25	bal	0.3 V
Discalloy	0.06	1.0	0.7	14.0	26.0	...	3.0	...	...	1.7	0.25	bal	
16-25-6	0.06	1.35	0.7	16.0	25.0	...	6.0	...	...	...	...	bal	0.15 N
Incoloy 901	0.05	0.5	0.35	13.0	40.0	...	6.0	...	...	2.5	0.2	bal	
D-979	0.05	0.5	0.5	15.0	45.0	...	4.0	4.0	...	3.0	1.0	bal	0.01 B
Nickel Base:													
Nimonic 75	0.12	1.0*	1.0*	20.0	bal	...	...	...	...	0.4	...	5.0*	
Nimonic 80	0.1*	1.0*	1.0*	20.0	bal	2.0*	...	...	...	2.25	1.25	5.0*	
Nimonic 90	0.1*	1.0*	1.0*	20.0	bal	18.0	...	...	...	2.50	1.50	5.0*	
M 252	0.15	0.5	0.5	19.0	bal	10.0	10.0	...	...	2.50	1.00	2.0	0.005 B
GMR 235D	0.15	...	...	15.5	bal	...	5.3	...	...	2.50	3.7	4.3	0.5
Hastelloy A	0.1	...	...	...	56.0	...	22.0	...	...	...	...	22.0	
Hastelloy B	0.1	...	...	...	62.0	...	32.0	...	...	...	...	6.0	0.4 V
Hastelloy C	0.15*	...	...	16.5	bal	...	17.0	4.5	...	...	...	6.0	
Hastelloy R235	0.15	...	...	15.5	bal	...	5.0	...	...	2.50	2.0	10.0	
Hastelloy W	0.12*	...	...	5.0	bal	...	25.0	...	...	...	...	5.5	0.6*V
Hastelloy X	0.15	...	...	22.0	bal	...	9.0	...	...	...	...	20.0	
Nichrome	0.25	...	...	18.0	80.0	...	...	...	...	...	...	1.0	
Rex 400	0.09	...	...	19.2	76.0	...	...	...	...	2.1	0.6	2.0	
K 42 B	0.06	...	...	18.0	42.0	22.0	...	...	...	2.6	0.6	13.0	
Inconel	0.04	0.35	0.2	15.0	78.0	...	...	...	...	...	...	7.0	
Inconel X	0.04	0.7	0.3	15.0	73.0	...	...	...	1.0	2.5	0.9	7.0	
Inconel X550	0.04	0.7	0.4	15.0	73.0	...	...	...	1.0	2.3	1.2	7.0	
Inconel 700	0.13	0.08	0.25	15.0	45.0	30.0	3.0	...	...	2.2	3.2	1.0	
Inconel 702	0.08	...	...	15.5	bal	...	...	...	...	0.6	3.2	1.0	
Inconel 703	0.08	...	...	10.0	80.0	...	2.6	...	...	2.4	2.95	0.4	
Inconel 709	0.10	...	...	9.0	80.0	...	3.0			2.75	3.75	0.4	
Inconel 710	0.10	...	...	8.25	77.0	...	3.0	...	...	2.7	3.65	0.4	
Inconel 718	0.05	0.2	0.3	19.0	53.0	...	3.0	...	5.3	1.0	0.7	bal	0.006 B, 0.03 Zr
Waspaloy	0.1	...	...	19.0	bal	14.0	4.0	...	...	3.0	1.3	1.0	
Rene 41	0.1	...	...	19.0	bal	11.0	10.0	...	...	3.0	1.5	3.0	
Multi-alloy	0.25	...	...	20.5	46.5	3.3	2.7	3.5	2.9	1.2	...	19.0	
Refractalloy 26	0.03	...	...	18.0	37.0	20.0	3.0	...	...	3.0	0.25	19.0	
Udimet 500	0.1	...	...	19.0	bal	19.0	4.0	...	...	3.0	2.9	4.0*	
Udimet 700	0.1	...	...	15.0	bal	19.0	5.0	...	...	3.5	4.5	4.0*	0.03 B
Unitemp 1723	0.24	0.05	0.1	16.0	bal	7.0	1.5	8.0	...	3.0	2.0	9.5	
Cobalt Base:													
S-816	0.4	1.2	0.4	20.0	20.0	bal	4.0	4.0	4.0	...	...	4.0	
V-36	0.25	1.0	0.4	25.0	20.0	bal	4.0	2.0	2.0	...	...	3.0	
S-590	0.4	1.2	0.4	20.0	20.0	20.0	4.0	4.0	4.0	...	...	bal	
HS-21	0.25	...	...	27.0	2.5	bal	5.5						
HS-31	0.5	...	...	25.0	10.0	bal	...	7.5					
HE-1049	0.4	...	...	26.0	10.0	bal	...	15.0	...	...	...	...	0.4 B
ML-1700	0.2	...	...	25.0	...	bal	...	15.0	...	...	...	...	0.4 B
N-155, HS-25	0.15	1.5	0.5	21.0	20.0	20.0	3.0	3.0	1.0	...	...	bal	0.15 N
WF-11	0.15	1.5	0.5	20.0	10.0	bal	...	15.0	...	...	...	2.0	

* Minimum

Titanium alloys are used to considerable advantage for compressor disks and blades where temperatures do not exceed 800°F. In this application and others, a large saving of weight is experienced. As explained previously, the principal advantages of titanium are low weight and resistance to corrosion.

The development of special coatings for high-temperature parts, such as turbine blades and nozzle vanes, has contributed to increased life for these critical components. For example, molybdenum provides great high-temperature strength, but it is subject to severe corrosion at high temperatures; hence a coating which resists corrosion makes it possible to use the material. It is common practice among engine manufacturers to take advantage of special coatings for turbine blades and disks because this makes it possible to employ higher exhaust-gas temperatures in the engine, thereby producing greater thrust.

Sintered-mesh porous media

A rather new material in its applications to jet engines is a sintered mesh sheet constructed of two or more layers of mesh manufactured from high-temperature alloys N-155 or L-605 and sinter bonded. This material is used primarily for **transpiration cooling** in high temperature zones.

Transpiration cooling, as the name implies, is the cooling which results from passing a fluid, either liquid or gas, through a porous media to remove heat from the porous surface. The porous media must have a large internal surface area for heat transfer to permit an efficient heat removal with a minimum of coolant.

Transpiration cooling is being used successfully in a number of applications in the aircraft, aerospace, and missile areas. The major applications in the aircraft field are found in turbojet engines. Afterburner liners as well as the liners of combustion chambers have been fabricated successfully out of the sintered-mesh porous media. In one application the coolant fluid is compressor bleed air at 600 to 900°F, and the porous media is known by the trade name Rigimesh, which is constructed as previously described. A typical material for this application consists of two or more layers of 12 × 64 mesh integrally sinter-bonded together.

Another important application in turbojet engines is the fabrication of transpiration-cooled turbine blades from porous media. For this application the material consists of sintered multilayers of 50 × 250 mesh in high-temperature alloy N-155.

Rigimesh plate is being used in the construction of most liquid hydrogen-oxygen engines as part of the fuel-injection system. Experience has demonstrated that, owing to the very high thermal conductivity of hydrogen, the heat from the combustion chamber of a rocket engine would back up into the hydrogen injection nozzles causing them to melt. With the development of the new system, a Rigimesh plate is used which utilizes transpiration cooling by means of cryogenic (supercooled) hydrogen. This system maintains its structural integrity through multiple firings. Porous injector plates for liquid hydrogen-oxygen rocket engines are being used by Pratt & Whitney Aircraft on the RL-10 rocket engine and by Rocketdyne on the J2 engine.

STUDY QUESTIONS

1. Give a common name for high-temperature metals.
2. Name and define the four "strengths" of high-temperature metals.
3. What is meant by **ductility?**
4. Compare **stress** with **strain.**
5. Explain **precipitation hardening.**
6. At what points in a gas-turbine engine are the highest temperatures found?
7. What are the three most important characteristics for high-temperature alloys?
8. What elements are most commonly found in stainless steels?
9. Name four alloys suitable for turbine blades and give their principal constituents.
10. What is the significance of the term "superalloy"?
11. Explain what is meant by **creep.**
12. Give an important function of chromium in high-temperature alloys.
13. Discuss the characteristics of Udimet 700, and tell how it is used.
14. What are the advantages of titanium?
15. How are gas-turbine blades and nozzle vanes protected from corrosion when used in extremely high-temperature conditions?
16. Discuss the use of sintered-mesh porous media in transpiration cooling.
17. Give some typical applications of porous media in the construction of jet engines and rockets.

CHAPTER 21

Modern Jet Transports

Ever since the first successful flight of a jet-powered airplane, air-minded people everywhere have dreamed of, and planned for, the day when commercial jet airplanes would become a reality. A number of jet-powered passenger-type airplanes have been designed and built, and some have been in operation for several years. A large measure of credit must go to Great Britain and its aircraft industry for pioneering commercial turbojet and turboprop aircraft. Notable among the jet transport aircraft built by British firms are the de Havilland Comet and the Avro Jetliner. Experiences with these and other jet aircraft have contributed a great amount of valuable information, with the result that manufacturers building new jet airliners today may avoid some of the problems encountered by the earlier aircraft.

It is the purpose of this chapter to describe some of the modern turbojet and turboprop aircraft now in operation or soon to be in operation. It is not our intent to emphasize the advantages of any particular airplane but only to describe each airplane in the light of information supplied by the manufacturer.

The airplanes selected for discussion in this chapter are those which are particularly active in the commercial airline industry. This does not mean, however, that others are not being developed. In addition to commercial jet airliners, there are several smaller jet airplanes in production for civilian use.

THE BOEING 707 JET AIRLINER

Since it was first announced in 1952 that the Boeing Airplane Company would build a commercial jet airliner, many changes have taken place in all phases of the air transport industry. From the original prototype model of the 707 airplane, numerous different models have evolved to serve the varying requirements of the industry. In addition to the family of 707 aircraft, which is primarily designed for long-range service, Boeing has developed and built the 720 and the 727 for intermediate and short-range operation. These airplanes will be described later in this chapter.

History of the 707 Jet Stratoliner

The design and development of the 707 Jet Stratoliner represent a private investment by the Boeing Airplane Company of more than 16 million dollars. After several years of research and development, the 707 prototype was rolled from the Boeing plant on May 14, 1954, and made its maiden flight on July 15, 1954. An advanced version of the 707, to be designated KC-135, was ordered into production for the Air Force a month and a half later. With its multipurpose capabilities, the military version of the airplane is employed as an aerial tanker for the Boeing B-52 heavy bombers and as a troop carrier and cargo transport.

On July 13, 1955, the Boeing Airplane Company was advised by the Air Force that there was no objection to building commercial jet aircraft concurrently with its production of military jet tanker-transports, the Air Force being satisfied that such commercial production would not interfere with or delay scheduled delivery of KC-135s.

The first commercial order for 707 Jet Stratoliners came from Pan American World Airways, which ordered 20 of the airplanes. On Nov. 9, American Airlines ordered 30, and since that time numerous other airlines throughout the world have placed Boeing 707 airplanes in passenger service.

In its first year, the 707 prototype was in the air for almost 200 hr testing and retesting all items of equipment, making intensive records of aerodynamic data, checking performance, and generally proving itself. In almost every case, original performance estimates were bettered. The airplane was flown at airspeeds of well over 600 mph and spent a substantial number of hours at altitudes above 42,000 ft. During one month when the main runway at Boeing Field was under repair, it operated from a 5,400-ft taxi strip with no difficulty.

Chronology of the Boeing 707

The time required to produce a new jet airplane of the 707 type is clearly revealed in the following chronology. Actually, this timetable does not include the planning and work which went on before the first date listed; however, it does indicate the time involved in the many steps to be taken between the first efforts and the finished product.

Aug. 30, 1952 First announcement that Boeing is engaged in a prototype jet-transport project.

May 14, 1954 Prototype is rolled out from plant in Renton, Wash., near Seattle.

July 15, 1954 First flight from Renton Municipal Airport; A. M. ("Tex") Johnson, Boeing chief of flight test, and R. L. ("Dix") Loesch, senior experimental test pilot, pilot and copilot, respectively. Landing made at Boeing Field after flight of 1 hr, 24 min.

July 17, 1954 Prototype flies above 42,000 ft and faster than 550 mph in third test flight.

Aug. 5, 1954 United States Air Force announces that it will purchase a number of advanced-design jet tanker-transports.

Sept. 1, 1954 KC-135 ordered into production.

July 13, 1955 Air Force advises Boeing that there is no objection to building commercial jet aircraft concurrently with military production.

July 28, 1955 Boeing moves to certificate 707.

Sept. 16, 1955 First of numerous airline pilots fly the 707 on demonstration flights.

Oct. 7, 1955 The 707 averages 550 mph on a 3,038-mile nonstop flight linking Seattle, Denver, Los Angeles, and back to Seattle.

Oct. 13, 1955 Pan American World Airways orders 20 Boeing 707 Jet Stratoliners.

Oct. 16, 1955 The 707 averages 592 mph in a nonstop flight from Seattle to Washington, D.C. The flight was made in 3 hr, 58 min. The return trip was made the same day in 4 hr, 8 min, an average of 567 mph.

Nov. 9, 1955 American Airlines orders 30 Jet Stratoliners.

March 11, 1957 Prototype 707, with members of the press aboard, flies nonstop from Seattle to Baltimore, Maryland, covering 2,350-mile distance in the record time of 3 hr, 48 min, for an average speed of 612 mph.

Oct. 28, 1957 First production 707 is rolled from factory.

Dec. 20, 1957 First production 707 makes maiden flight.

May 15, 1958 United States Air Force announces

Figure 21 · 1 *Boeing 707 airliner.*

order for three 707-120s, designated VC-137As, for military Air Transport Service special missions group for use by the President of the United States and other high-ranking officials.

Aug. 15, 1958 Pan American takes delivery of first U.S. commercial jet airliner, a 707-120.

Sept. 23, 1958 Civil Aeronautics Administration presents type certificate to Boeing for 707 jet airliners.

Oct. 26, 1958 Pan American World Airways begins first transatlantic service with the 707.

Dec. 11, 1958 First 707 jet airliners fly domestic routes.

Jan. 11, 1959 707 Intercontinental makes first flight.

Oct. 12, 1959 One millionth passenger carried by 707s.

April 30, 1960 First 720 is delivered to United Air Lines.

Oct. 26, 1962 Boeing 707 and 720 jetliners mark fourth anniversary in airline service with more than 30 million passengers carried.

The Boeing 707 airliner, in its various models, is one of the most widely used airliners in the world. The airplane is designed and built in a number of different configurations to meet the operational needs of the many different airlines throughout the free world. The Boeing 707-120 is the "basic" 707, designed principally for continental use but also capable of full-payload over-ocean operation. The 707-120B is the same configuration as the 707-120; however, it is equipped with the more powerful Pratt & Whitney JT3D turbofan engine, rather than the JT3C. The 707-120B Stratoliner is shown in the photograph of Fig. 21 · 1.

The specifications for the Boeing 707-120 Stratoliner are given below:

Dimensions	Wing: 130 ft, 10 in.; Length: 144 ft, 6 in.; Height: 42 ft; Sweepback: 35°
Gross weight	248,000 lb
Power	Four Pratt & Whitney JT3C-6 turbojet engines, rated at 13,000 lbt each, or four JT3D turbofan engines rated at 17,000 or 18,000 lbt each.
Speed	570 to 600 mph
Cruising altitude	25,000 to 40,000 ft
Range	Nonstop transcontinental or transatlantic
Payload	51,631 lb
Passenger capacity	179—High-density seating
Landing gear	Tricycle—Main undercarriage units are four-wheel bogie-type trucks; nose wheel, dual
Crew	Three—Pilot, copilot, flight engineer

The Intercontinental 707-320C

The Boeing Intercontinental 707-320C is the largest version of the 707 series airplane and is designed for greater range and payload than the −120. The wing span of this airplane is 145 ft, 9 in., and the length is 152 ft, 11 in. The gross weight is 328,000 lb with a passenger capacity of 189, a payload of 86,000 lb, and a range of 3,500 miles fully loaded. This airplane is designed to carry a much greater load of cargo than other versions.

The Boeing 720

The Boeing 720 airliner, shown in Fig. 21 · 2, was designed for economical operation on ranges as short as 250 miles. This does not mean, however, that it cannot operate on longer range routes, the maximum range of the airplane being 3,300 miles. It is the fastest and lightest of the four-engine Boeing airliners, cruising at more than 615 mph with a gross weight of 203,000 to 230,000 lb. The 720B is powered by four JT3D-3 turbofan engines rated at 18,000 lb of thrust each. The maximum passenger capacity of this airplane is 165.

The dimensions of the 720 airplane are almost the same as the original 707 prototype. The wing span is 130 ft, 10 in.; the length, 136 ft, 2 in.; and the height 41 ft, 6 in.

The Boeing 727

The Boeing 727 airliner, shown in Fig. 21 · 3, is a second-generation jetliner, featuring a three-engine powerplant system with one engine mounted in the rear part of the fuselage. The elevators are mounted on the top of the stabilizer to be free of any turbulence. The positioning of the engines at the rear of the airplane also permits an exceptionally clean wing design.

The 727 is a short-to-medium range jet especially designed to serve routes of 150 to 1,700 miles. It is powered by three Pratt & Whitney JT8D turbofan engines rated at 14,000 lb of thrust each. The 727 extends the advantages of jet flight to many areas where this type of service had not been available before and thereby replaces many of the older piston-engine airplanes.

The dimensions of the 727 are as follows:

Figure **21 · 2** *Boeing 720 jet airliner.*

Wing span, 108 ft, 7 in.; length, 134 ft, 1 in.; height, 33 ft, 9 in.; sweepback, 32°. The maximum gross weight is 142,000 to 152,000 lb, depending upon the configuration required. The airplane cruises at 550 to 600 mph at an altitude of 15,000 to 35,000 ft above sea level and carries a maximum of 114 passengers.

CONVAIR JET AIRLINERS

The Convair Division of the General Dynamics Corporation is the builder of two outstanding jet airliners, the 880 and the 990. The 880 is one of the fastest airliners in service with a cruising speed of approximately 615 mph, while the 990

Figure **21 · 3** *Boeing 727 jet airliner.*

is considered the fastest of the subsonic jet airliners with a maximum speed of approximately 640 mph.

The Convair 880

The Convair 880 and 880M jet airliners are slightly smaller and lighter than the Boeing 720 but they provide the same general type of service. The 880 airliner, shown in Fig. 21·4, includes all the refinements of the most modern jet airliners. The cabin configuration may be set up for 88 first-class or 110 coach passengers. The cabin is thoroughly insulated, soundproofed, and pressurized to provide a maximum of comfort for the passengers. The air conditioning system takes in fresh air from outside the airplane, adjusts the temperature to 75°F, and delivers the air to the cabin through two systems, either one of which can do the job by itself. The air in the cabin is completely changed once in every 2½ min.

The Convair 880M is equipped with four General Electric CJ805-3B pure-jet engines, each rated at 11,650 lb of thrust. Thus powered, the 880M is capable of operating from short runways, has a fast takeoff, and climbs at a comparatively high rate. The engines are equipped with thrust reversers to shorten the landing roll of the airplane.

The specifications and performance data for the 880M airliner are as follows:

Cruising speed	615 mph
Range, maximum	3,750 miles
Stalling speed	111 mph
Maximum cruise altitude	41,000 ft
Passenger capacity	88 to 110
Payload (coach)	26,780 lb
Fuel capacity	12,651 gallons (max)
Wing span	120 ft
Length	129 ft, 4 in.
Height over tail	36 ft, 4 in.
Maximum takeoff weight	193,000 lb

The Convair 990A jet airliner

The Convair 990A jet airliner, shown in Fig. 21·5, incorporates many advanced-design features not found in previously built jet airliners. As shown in the photograph, one of the distinguishing features of this airplane is the installation of the aerodynamic antishock bodies on the rear part of the wings. These devices reduce the formation of supersonic shock waves, which would develop at maximum speeds and increase the drag. The antishock bodies are also utilized for the storage of additional fuel to give the airplane extended range. The airplane can carry 15,562 gallons of fuel, thus providing a range of more than 4,500 miles.

The 990A airplane is powered by four General

Figure 21·4 Convair 880 jet airliner.

Figure 21 · 5 Convair 990 jet airliner.

Electric CJ805-23B aft-fan engines rated at 16,050 lb of thrust each. The airplane was specifically designed to take advantage of the performance of this type of engine. Each engine is housed in a pod almost 6 ft in diameter. Inside the pod, large bypass air ducts almost surround the basic powerplant and lead to the turbofan inlet. Access to the engine for maintenance is gained by raising the outer pod doors and then raising the hinged ducts on each side.

The 990A has an overall length of 139 ft, 9 in.; wingspread of 120 ft; height over tail of 39 ft, 6 in.; and fuselage width of 11 ft, 6 in. The plane has a maximum landing weight of 202,000 lb.

The normal cruise altitude of this airplane is 35,000 ft, and at this altitude the cabin interior is pressurized to the equivalent of 6,000 ft above sea level. At 20,000 ft altitude the system provides sea-level pressure within the cabin.

Easy maintenance is designed into the 990A airplane through the use of the systems package concept. Wherever possible, systems have been designed in units with a minimum of fastenings, connections, and ducts. These units are built for maximum accessibility to permit entire assemblies to be removed and replaced while the originals undergo maintenance work. Thus the airplane may be returned to service in a minimum of time when repairs are required.

The cabin of the airplane includes all the features possible for the comfort of passengers. The soundproofing is exceptionally effective, thus making it possible to carry on a normal conversation at any point in the cabin. Provisions are made to carry a maximum of 121 passengers in the coach configuration.

DOUGLAS JET AIRLINERS

The Douglas Aircraft Company, long the leader in the development and production of transport-type aircraft, continues to contribute to the air transportation industry with the production of a family of jet airliners. These are the DC-8 series, the DC-8F, and the DC-9. Among these aircraft, airlines are able to obtain suitable types for any conventional operation from extreme long-range to very short-range operation.

The Douglas DC-8 Jetliner

The DC-8 Jetliner was developed from its inception as a passenger-carrying airliner with a design embodying advanced aerodynamic and structural concepts as well as internal systems providing for maximum service reliability, convenience, and comfort. This successful transport is produced at a facility constructed expressly for the purpose at the Douglas Aircraft Division, Long Beach, California. The maiden flight of the first production model took place on May 30, 1958.

The DC-8, shown in Fig. 21 · 6, is built in a

Figure 21 · 6 Douglas DC-8.

series from the DC-8 Series 10 to the DC-8 Series 50. The specifications of these aircraft are given in Table 21 · 1. The airplane has a wingspan of 142 ft, 5 in.; an overall height of 42 ft, 4 in.; and a length of 150 ft, 6 in. The cabin is designed to accommodate from 105 to 173 passengers, depending upon the requirements of the carrier.

The wing sweepback of the DC-8 is 30°, which gives the airplane handling characteristics said to be superior to aircraft with larger amounts of sweepback. The leading edge slots provide an additional design feature to give the DC-8 superior low-speed characteristics. These are high-lift devices which open when the flaps are lowered. The effect is to reduce stalling speed and permit the airplane to take off and land on shorter fields than otherwise possible.

All DC-8s are capable of attaining speeds of 600 mph or more at maximum cruise thrust and of flying nonstop across the United States in either direction. Some models are capable of traveling nonstop for more than 7,000 miles under proper conditions.

Interior styling and utility are enhanced by the Douglas "Palomar" chairs, scientifically designed for cushioned comfort in a concept called **Unit-ized Seating.** Reading lights, call buttons, ash trays, dining or game tables, cold-air outlets, and oxygen outlets are built into the seats. This arrangement makes each passenger self-sufficient and eliminates all overhead and sidewall installations. It also gives the airline operator greater flexibility by permitting quick changes in seat spacing. Seats and bulkheads are attached to tracks for easy rearrangement. The DC-8 coach-type seating is shown in Fig. 21 · 7.

Special features, some of them unique, which have been incorporated in the DC-8 include:

1. A noise-suppressor which reduces the sound of the jet engines to satisfactory levels without losing takeoff performance.
2. Tandem landing-gear trucks, with the rear set of wheels castered to permit ground movement in smaller areas.
3. A thrust-brake system which reduces landing roll and also may be used for rapid deceleration in the air.
4. "Blowaway-Jet," an aerodynamic screen which prevents jet engines from sucking up debris during takeoff, taxiing, or landing roll.
5. "Skyview" windows, larger than on the DC-7.
6. A "Ski-Hi-Fi" cabin sound system for announcements and music during flight.
7. An advanced autopilot, designed especially for the DC-8.

The DC-8F Jet Trader

The DC-8F Jet Trader is a new concept in efficient air transportation, being the world's first convertible or combination cargo-passenger jet airliner. Taking its name from the early day sailing vessels which carried cargo in the holds and passengers on the upper decks, the Jet Trader was developed to help air carriers expand

Table 21 · 1 Douglas DC-8 specifications

	Series 10	Series 20	Series 30*	Series 40*	Series 50*
Engines	P&W JT3C-6	P&W JT4A-3	P&W JT4A-11	RCo 12	P&W JT3D-3
T.O. thrust, lb	13,500	15,800	17,500	17,500	18,000
Design gross wt., lb	273,000	276,000	318,000	318,000	318,000
Max. T.O. wt., lb	273,000	276,000	315,000	315,000	315,000
Design lndg. wt., lb	193,000	193,000	207,000	207,000	207,000
Operational wt. empty, lb	120,999	123,757	135,378	134,062	132,325
Fuel capacity, gal	17,600	17,600	23,397	23,397	23,397
Fuel capacity at 6.7 lb/gal, lb	117,900	117,900	156,760	156,760	156,760
Payload,† lb	34,030	34,030	34,360	36,175	34,360
Lev. fl. speed,‡ mph	542	579	592	586	579
Rate of climb,§ ft/min	1,280	2,900	2,850	2,100	2,380
FAA T.O. field at max T.O. wt., ft	9,380	8,240	9,650	9,650	10,380
FAA lndg. field at design lndg. wt., ft	6,400	6,400	6,800	6,800	6,550
Lndg. speed at design lndg. wt. at 1.3 VS, mph	148	148	153	153	153
Max. range,¶ stat. miles	4,120	4,792	5,968	6,480	7,090

* With 4 per cent leading edge on wing.
† Space limited.
‡ Max. cruise thrust at 220,000 lb (30,000 ft).
§ Enroute sea level at T.O. weight.
¶ Still air, zero payload, capacity fuel.

Figure 21 · 7 DC-8 coach seating.

the cargo markets by offering shippers the advantages of jet speed.

The interior design of the aircraft permits complete flexibility of operations, from all-cargo to all-passenger. A movable bulkhead divides the cabin into separate compartments when combination loads are carried.

The most noticeable characteristic and an outstanding feature of the Jet Trader is its huge forward cargo door. Measuring 85 × 140 in., it permits loading of long items of bulk cargo as well as standard pallets. The door swings upward to permit rapid entry and discharge of cargo, as shown in the drawing of Fig. 21 · 8 which illustrates the arrangement of the fuselage for all-cargo operation.

The DC-8F carries up to 95,282 lb of freight, the largest payload capacity for any commercial jet in operation at the time it went into service. The main cargo compartment has an overall length of more than 106 ft with a constant fuselage cross section measuring 127 by 86 in. As an all-passenger transport, the airplane accommodates 189 economy-class passengers.

The Douglas DC-9 Compact Jet

The Douglas DC-9, shown in Fig. 21 · 9, is designed to bring the efficiency of jet service to the smaller cities of the world where it is not economically feasible to operate the large jet transports. Designated the "Compact Jet," the twin-jet DC-9 will operate from airport runways of a mile in length and on route segments from 100 to 1,500 miles. Even though the maximum takeoff weight of the DC-9 is less than one-fourth that of the intercontinental DC-8, it carries up to 83 passengers at speeds up to 560 mph at no sacrifice in comfort or convenience. In normal operation it will take off on a 500-mile flight with 70 passengers and baggage from a 5,000-ft runway and will land in less than that distance at its maximum landing weight.

The DC-9 is powered by two Pratt & Whitney JT8D turbofan engines mounted on each side of the aft fuselage to permit an unusually clean wing design. Like its predecessor, the DC-9 wing has a varying airfoil from root to tip for additional stability and better performance at both

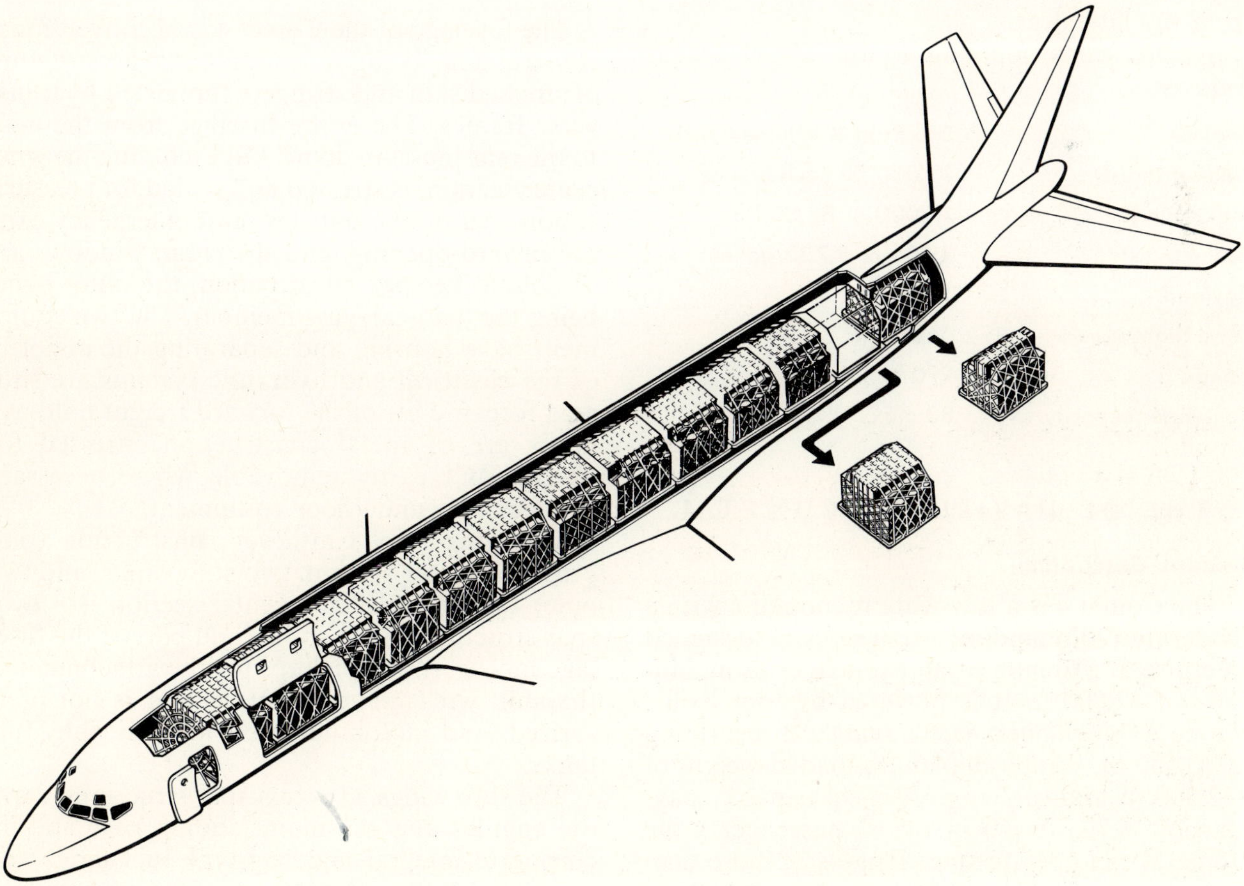

Figure 21 · 8 Fuselage arrangement of DC-8F Jet Trader.

Figure 21·9 Douglas DC-9 jet airliner.

high and low speeds. The wing sweeps back 24° from the lateral axis.

Specifications applying to the DC-9 are as follows:

Engines	Two Pratt & Whitney JT3D
Takeoff thrust	12,000 lbt each engine
Maximum takeoff weight	77,000 or 83,000 lb
Fuel capacity	2,806 or 3,722 gallons
Payload	19,700 lb
Level flight speed	560 mph
Range	990 to 1,940 miles
Passenger capacity	83 max

THE DE HAVILLAND COMET 4

General description

The Comet 4 is a low-wing monoplane with a moderate (20°) amount of sweepback designed to cruise at 520 mph at an average cruising altitude of 42,000 ft. It is powered by four Rolls-Royce Avon engines, each having a thrust rating of 10,500 lb. With a maximum loaded weight of 152,500 lb and carrying a typical capacity payload of 16,723 lb (60 first-class passengers), the Comet 4 can operate stage lengths of more than 2,900 miles with normal fuel reserves and allowances for head winds.

Construction

The fuselage of the Comet 4 is of conventional construction, being a semimonocoque structure of stressed skin and stringers supported by transverse frames. The entire fuselage from the nose to the rear pressure dome, but excluding the wing center section, is stressed and sealed for pressurization. All doors, hatches, and emergency exits are inward-opening, and the cabin windows are of double Perspex construction, the outer panel being the load-carrying member. The two equipment bays housing and separating the majority of the electrical and hydraulic systems are situated fore and aft of the forward freight hold. All floors are of metal construction, stressed for freight, and easily removable where necessary for access to underfloor equipment.

The wing is of cantilever construction comprising a center section, two stub wings, and two outer wing panels. The center section is a two-spar structure and is an integral part of the fuselage incorporating pickups for the attachment of the stub wings. The center section is not pressurized and accommodates four flexible fuel tanks.

The stub wings, also two-spar structures, carry the engines and the main landing gear and incorporate integral and bag-type fuel tanks and de-icing leading-edge ducts. Engine bays are separated by stainless-steel bulkheads.

The outer wing panels are attached to the stub wings by fore-and-aft bolt-plates extending between the spars. The outer panels carry the ailerons and the outboard section of the flaps, the flexible fuel tanks, nacelle-type fuel tanks, and de-icing ducts.

The tail section and the control surfaces are of conventional construction.

A good conception of the structural layout of the Comet 4 airplane may be obtained by studying the cutaway drawing shown in Fig. 21 · 10. A picture of a de Havilland Comet airplane in flight is shown in Fig. 21 · 11.

Accommodation

The passenger cabin of the Comet 4 is in two parts: a forward cabin about 25 ft in length, and a rear cabin of about 30 ft. The width is 9 ft throughout, and the height is 6½ ft.

Standard de Havilland chairs are mounted on permanent seat tracks, thus allowing different arrangements to be achieved quickly and easily in service. A typical arrangement is for 60 passengers (first class). Other layouts can accommodate up to 76 passengers (tourist class), and a mixed-class arrangement can also be made.

All seats of the Comet are forward of the jet pipes, which keeps noise at a minimum in the passenger compartments. Vibration is entirely absent.

A feature of the Comet 4 cabin is that the passenger door opens into a roomy and well-illuminated vestibule. Passenger lists can be checked with a minimum of delay, and light baggage and coats may be quickly stowed in the cloakroom leading off the vestibule. This area also opens on the well-equipped ladies' and gentlemen's dressing rooms and toilets.

The cabin furnishings are fireproof where possible. Wall upholstery is of plastic leather cloth, and soundproofing and heat insulation between upholstery and the outer skin are of glass wool. Cabin roof lighting is diffused evenly throughout both cabins, and an individual lamp is provided for each passenger.

The galley is designed for the compact but adequate accommodation of the ship's food stores and associated utensils. Features include two ovens, work-table tops, refrigerators, sink, and a seat for the steward. Stainless-steel fittings are used, and the floor is of heavy-duty plastic.

The cockpit, cabin, and luggage compartments are pressurized to a maximum differential of 8¾ psi, giving a cabin altitude of 8,000 ft when the aircraft is flying at 45,000 ft. Fresh warm air is bled from the compressors of each of the four engines, thus eliminating the need for mechanically driven blowers and combustion heaters. Adequate heat for warming the cabin is available from the compression of the air. Humidifiers and de Havilland cold-air units are incorporated in the circuit to regulate the humidity and the air temperature for maximum comfort.

Power plants

The Comet 4 airplane is equipped with four Rolls-Royce Avon RA. 29 engines, each developing 10,500 lb of thrust (dry).

The Avon RA. 29 engine is an axial-flow type having a 16-stage compressor, a two-position exhaust nozzle, and a silencer. Advanced models of the engine incorporate a 17-stage compressor providing a pressure ratio of 10.0 : 1. The turbine is of the 3-stage type with disks bolted to the drive shaft.

The experience gained in service with Avon engines has resulted in many improvements which have been incorporated in the RA. 29, the performance of which shows a marked advance both in regard to output and in specific fuel consumption over previous Avon engines.

The engines are mounted in the wing roots of the Comet 4 between the main spars. The mountings are specially designed for rapid engine changing and incorporate quickly detachable air-intake and tailpipe connections and special disconnect joints in the engine controls. Each engine is fitted with a hot-air anti-icing system, using air bled from the compressor, to prevent ice from forming in the air intakes during icing conditions. Protection of the air-frame structure in the vicinity of the turbine is provided by an armored shrouding around the engine in that region.

Aircraft systems

In connection with the brief descriptions which follow, it is important to note that the systems and equipment of the Comet 4 have been developed from earlier models over a period of years and may be considered fully proved.

Fuel System. Kerosene-type fuel is carried in fuel tanks installed in the wings and wing center section. The total fuel capacity is 8,760 imperial gallons or approximately 10,520 United States gallons.

Pressure-fueling, controlled from panels in the main-wheel wells, is accomplished from two connections under the wing at a rate of 200 imperial gallons per min through each point, with safety devices to prevent overfilling. The engines are

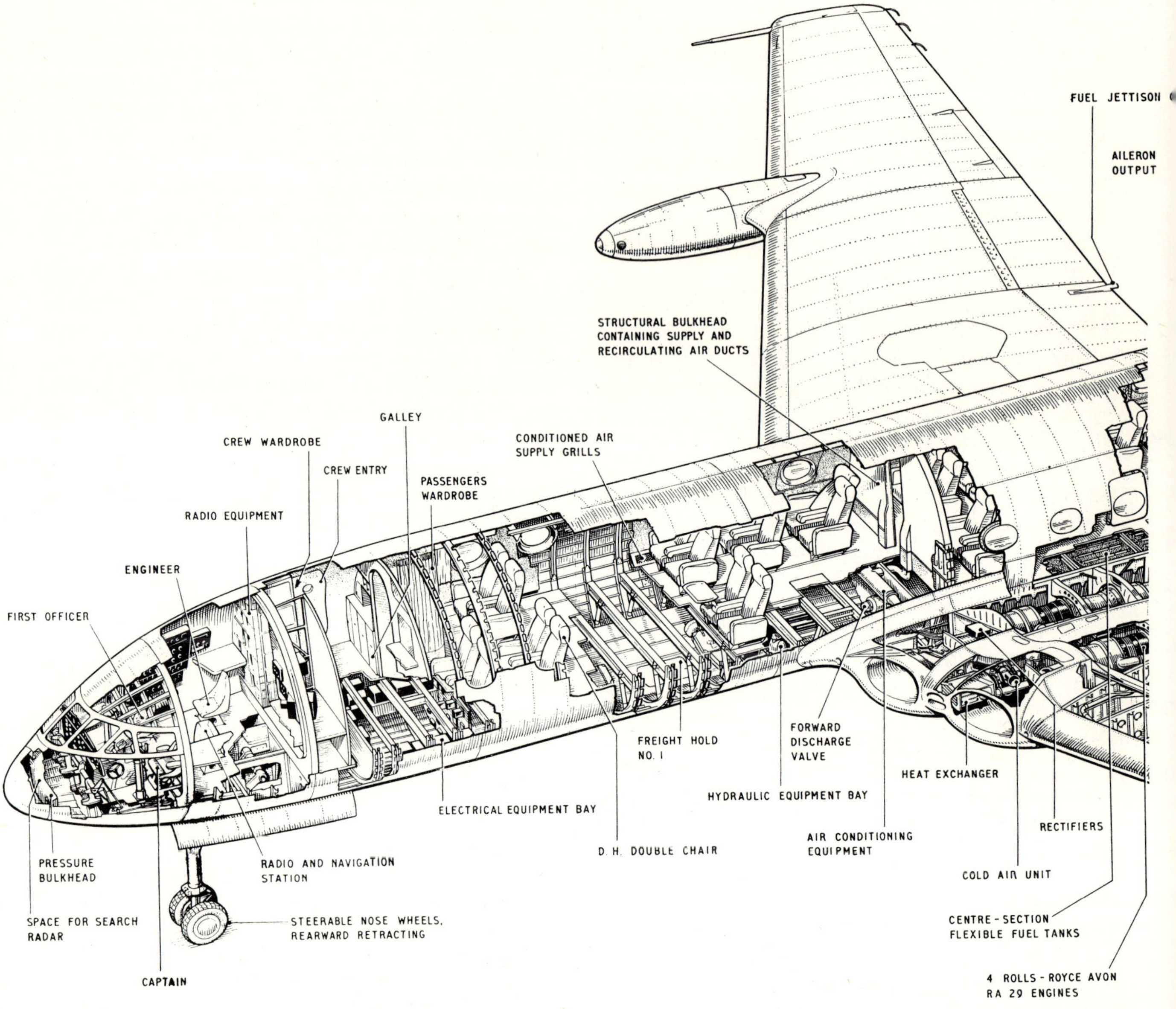

*Figure **21 · 10** Cutaway view*

supplied by two immersed pumps in each tank. Fuel quantity is shown by a capacity-type fuel gauge. Means are provided for dumping of fuel in case of emergency.

Flight Controls. The flight controls for the Comet 4 are power-operated by hydraulic means. Artificial feel proportional to airspeed is provided in the elevator circuit. Each control surface is moved by cables and pushrods from a duplicated hydraulic actuating cylinder. The cylinder is furnished with hydraulic pressure through duplicated valves which are controlled through cables connected to the rudder pedals and the control column. The primary hydraulic system for the flight controls is independent of the other hydraulic systems in the aircraft. There is a secondary system and also a stand-by system, so that power for the flight controls is always available. The airplane is also equipped with an automatic pilot.

In addition to the flight controls, the flaps, air

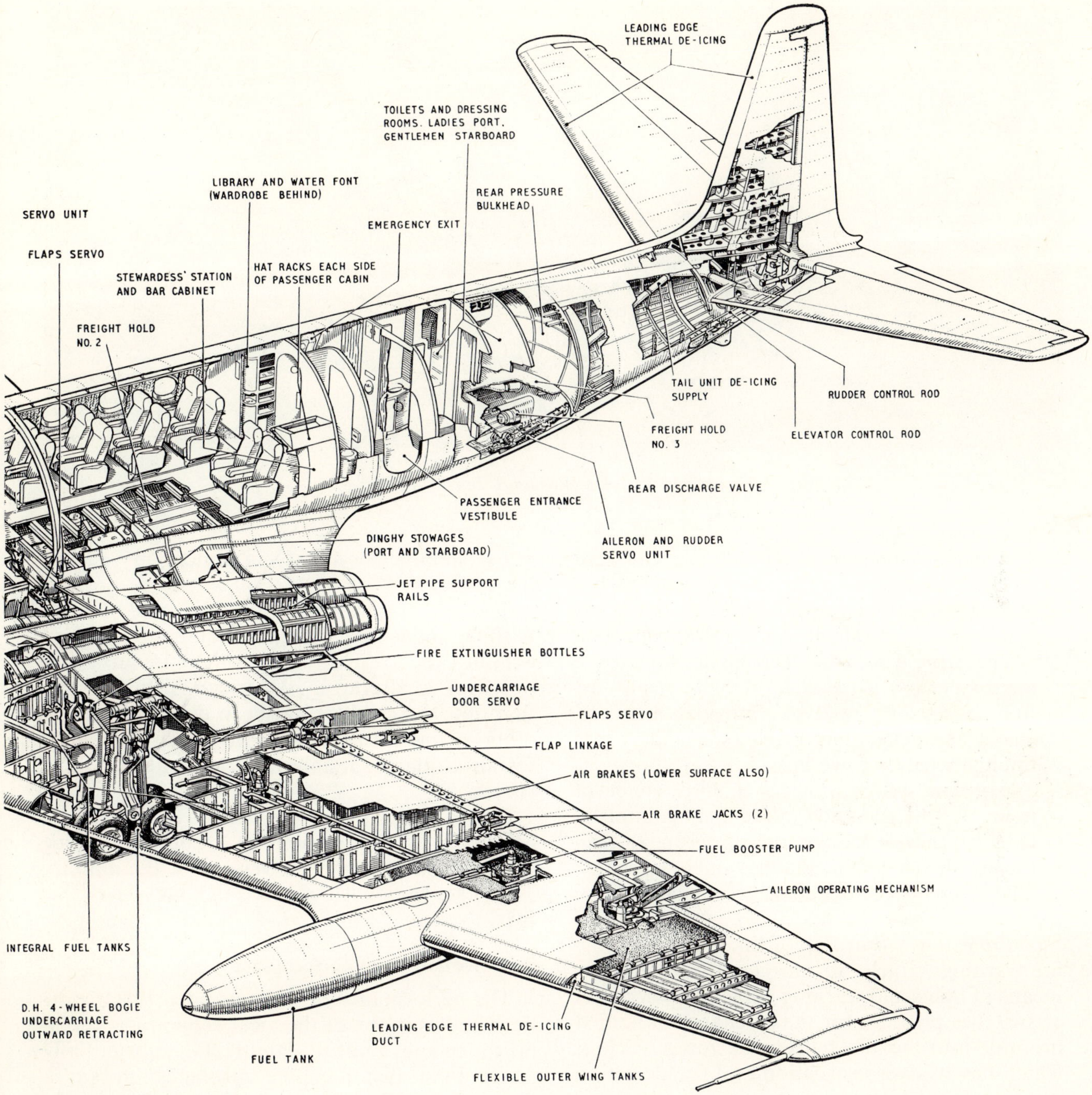

of the de Havilland Comet 4.

brakes, wheel brakes, landing gear, and nose-wheel steering are all powered through hydraulic systems.

Hydraulic Systems. There are four hydraulic systems in the Comet 4, which are used as follows:

1. The **green** system obtains pressure from pumps driven by the two outboard engines. It supplies, from an independent fluid reservoir and hydraulic accumulators, power to operate the landing-gear-retraction mechanisms, nose-wheel steering, flaps, air brakes, and wheel brakes. It also supplies power for the operation of the secondary flight controls in the event of failure of the **blue** primary system.
2. The **blue** system obtains pressure from pumps driven by the two inboard engines. It supplies, from an independent fluid reservoir and hydraulic accumulators, power to operate the primary flight controls.

Figure 21 · 11 De Havilland Comet in flight.

3. The stand-by **red** system receives pressure from an electrically driven pump. This system also has an independent fluid reservoir and accumulators and is used to operate the landing-gear-retraction mechanisms, flaps, and wheel brakes in case the main system is inoperative. A hand pump, fed from a separate reservoir, provides a second emergency method for lowering the landing gear through separate down lines.
4. The **yellow** system provides a third source of power for the operation of the flight controls in an emergency. It receives pressure from an electrically driven pump and has an independent fluid reservoir and hydraulic accumulator.

Fire Protection. Warning of fire is given on the fire-warning panels in the control cabin by means of lights and warning bells. The operation of switches provides for the discharge of methyl bromide into the affected engine zones, each of which has its own ventilation and fire-detection system. Smoke-detection devices and extinguishers safeguard the freight holds and equipment bays. Portable fire-extinguishing units are carried in the control cabin and passenger cabin.

THE LOCKHEED ELECTRA

General description

The Lockheed Electra "Prop-jet" airliner, shown in Fig. 21 · 12, is the first U.S.-built commercial airliner powered by turboprop engines. It was designed specifically for a variety of applications with airlines of the United States but is suitable for service in almost any part of the world.

The Electra was designed for an American Airlines-sponsored design competition, and winning the competition resulted in an initial order for 35 aircraft by American Airlines. Since that time the airplane has been utilized by many other airlines, especially those which have need for an airplane suitable for comparatively short routes.

The powerplant for the Electra is the Allison Model 501-D13 turboprop (prop-jet) engine which is the commercial version of the military T56. This engine is described in Chap. 18 of this text.

Specifications and performance

The performance of the Lockheed Electra airliner makes it one of the most economical transport airplanes built to date. It flies substantially faster than piston-engine airplanes and at a lower cost. The economical operation is the result of four principal factors: (1) clean aerodynamic design, partly the result of the small frontal area of the engines; (2) 60 to 100 mph greater speed; (3) low weight/power ratio of the engines, permitting a larger payload; and (4) lower cost of gas-turbine fuels.

Some of the specifications and performance data are as follows:

Dimensions	Wing: 99 ft; Length: 104 ft, 6.5 in.; Height: 33 ft
Gross weight	116,000 lb
Power	Four Allison 501-D13 engines

Speed	Over 400 mph
Cruising altitude	15,000 to 30,000 ft above sea level
Range	Over 2,300 miles
Passenger capacity	66 to 100
Payload	19,700 lb

Features and characteristics

The Electra was designed and built with the comfort and convenience of passengers in mind. The airplane is pressurized and air-conditioned in the same degree as the most modern jet transports and provides all the same safety and comfort features. It is well soundproofed so that propeller noise is reduced to a gentle rumble, and the noise of the jet exhaust is inaudible in flight.

Airlines operating the Electra find it easy to maintain and service and keep in flying condition. A full load of passengers can be discharged and a new load enplaned while the airplane is being serviced with fuel and supplies and baggage is loaded, all in a space of 15 min. This is typical in the frequent schedules maintained by these aircraft between large cities such as Los Angeles and San Francisco.

THE FAIRCHILD F-27 FRIENDSHIP

The Fairchild F-27 "Friendship" airliner, introduced in 1958, gave local-service airlines, for the first time, a brand new airplane tailored to their needs. During the time since its first appearance on the local scene, the airplane has continued to develop new interest and support among airline operators and has consistently grown in capability until it is almost literally a "new" aircraft. The gross takeoff weight has been gradually increased from 35,700 to 42,000 lb. The maximum landing weight has gone from 34,000 to 37,500 lb, and the payload has increased from 10,800 to 12,000 lb.

The F-27, shown in Fig. 21 · 13, is a high-wing, 40-passenger (44 in some versions), pressurized and air-conditioned transport aircraft, powered by two Rolls-Royce Dart 7 turboprop engines that give it a cruise speed in the 300 mph range. The high wing, which is an identifying feature, serves several important functions. From the aerodynamic standpoint, the high wing provides excellent inherent stability in all flight regimes, and it permits the use of larger diameter propellers with good ground clearance and improved takeoff and climb performance. Further-

*Figure **21 · 12** Lockheed Electra airliner.*

Figure 21 · 13 Fairchild F-27 Friendship airliner.

more, with the wing set high on the fuselage, it is above, rather than below, the windows. The result is that the F-27 offers an excellent view in the air. This view is enhanced by the large oval windows.

The high wing makes it possible to have a perfectly level cabin area that is only 43 in. above the ground. This eliminates the need for special passenger-loading ramps and cargo ground handling equipment. These advantages are translated into shorter turn-around time at route stops.

The F-27 is fully pressurized and air-conditioned to provide maximum passenger comfort. Regardless of whether the outside air temperature is over 100°F or many degrees below zero, the cabin temperature is always in the comfortable 70s.

The two Rolls-Royce Dart 7/2(528) or (529) engines provide the F-27 with more than 4,000 hp. This makes it possible to operate the aircraft at more than 300 mph, thus providing a substantial improvement over the piston-engine airplanes previously used.

CONCLUSION

Even though we have not had the space in this text to describe more than a few of the modern gas-turbine engines and aircraft, we have seen enough to know that modern air transportation has been revolutionized by the use of these highly efficient powerplants. The development of aviation gas-turbine engines has ushered in an era of undreamed-of speed and comfort in public transportation. It is now possible for a citizen of the United States to spend a week end in Paris, London, or Honolulu, and his traveling time is no longer than the time it would take to drive to a mountain resort a few hundred miles distant. During a two-week vacation, a person may visit almost any accessible part of the world—Japan, India, Palestine, South Africa, Australia, or any other country which suits his fancy. Thus, the gas-turbine engine has brought the world to our back door.

The comfort of modern jet airliners is a revelation to those who have never flown by "jet." We can hurtle through the sky, 8 miles above the earth at 10 miles per min, and be as comfortable as in our own living room. We are conscious of the gentle roar of the jet exhaust, and we can hear the slight sound of rushing air as our skyliner slips through the atmosphere; but these sounds are well muffled, and we can converse in a normal voice.

Every possible requirement for human comfort has been considered by the designers of "Jet-age" airliners. Medical experts have been consulted regarding the design of seats so that the body will be supported in just the right manner without undue pressure on any muscular or neural area. The air within the airplane is held at the proper pressure, temperature, and humidity and is changed completely at intervals of less than 3 min.

A substantial margin of safety is built into these aircraft. With two engines out of four stopped, the airplane will continue to fly at a safe altitude to any selected airport. The gas-turbine engine has proved remarkably free from failures, chiefly because there are no reciprocating parts and because the manufacturers have employed the most advanced types of metals and alloys in their construction. The service life between overhauls of a modern gas-turbine engine is more than double that of the best reciprocating engines. Additional safety is provided by radar, radio navigation systems, instrument landing systems, and all possible aids to assure safe and comfortable flight. Fire-detecting, warning, and extinguishing systems are provided so that a fire will be eliminated almost before it is started.

Flights in modern airliners are usually at altitudes far above storms and turbulence, so the passenger is hardly conscious that he is moving. The only time that turbulence is encountered is during climb and letdown, and even then the design and size of the airplanes reduce the effect of air "bumps" considerably.

The modern man may indeed consider himself fortunate to be able to enjoy the benefits of today's air transportation. To the uninitiated, there are few thrills which can compare with the first sensation as a mighty jet airliner whisks him off the earth and into a new experience which is truly "out of this world."

STUDY QUESTIONS

1. List the principal types of Boeing jet airliners, and tell what type of service each is best suited for.
2. What powerplant is employed in the Boeing 707-120B?
3. Describe the capabilities of the Boeing Intercontinental 707-320 airplane.
4. What is unusual about the design of the 727?
5. At what altitudes do jet airliners operate most satisfactorily?
6. What is the maximum range of a Series 50 DC-8?
7. What powerplant is employed for the DC-8-50?
8. What are some of the advantages of a jet airliner over a piston-engine airliner?
9. Describe the DC-9.
10. For what type of service is the DC-9 best suited?
11. What powerplant is used in the DC-9?
12. What is the range of the de Havilland Comet 4?
13. How is the cabin of the Comet 4 heated and pressurized?
14. What powerplant is used in the Lockheed Electra?
15. Discuss the advantages of the Electra as a modern airliner.
16. For what type of service is the Fairchild F-27 airplane best suited?
17. What are the advantages of the high-wing design of the F-27?
18. What type of engine is used in the F-27?

Glossary

absolute pressure: The sum of gauge pressure and atmospheric pressure.

absolute zero: The temperature at which all heat action ceases. It is 273.1° below zero centigrade and 459.7° below zero Fahrenheit.

acceleration: The rate of change of velocity of any object or material. It is expressed in feet per second per second.

acceleration of gravity: The acceleration of any body due entirely to gravity. In a vacuum a body will accelerate toward the earth's center at the rate of 32.2 ft per sec per sec.

adiabatic: Adiabatic changes in temperature are those changes which are due only to compression or expansion of a gas and during which no heat is added to or taken from the gas.

alloy: A solid solution consisting of two or more metallic constituents. Usually the alloy contains one predominant metal to which are added small amounts of other metals to improve strength and heat resistance.

alloy steel: A steel which contains metallic elements other than those found in carbon steels. These elements may consist of one or more of the following: molybdenum, chromium, nickel, vanadium, columbium, tungsten, cobalt, and manganese.

ambient: Surrounding conditions.

atmosphere: The envelope of air surrounding the earth, consisting of 78 per cent nitrogen, 21 per cent oxygen, and 1 per cent inert gases.

atom: The smallest possible part of an element.

augmentor: A device for augmenting the thrust of a turbojet engine. It may also refer to an arrangement of the exhaust outlet of a reciprocating engine which adds to the total thrust. An augmentor on a turbojet engine is usually called an "afterburner" or "reheat unit."

axial: Acting along an axis.

axis: An imaginary or real line around which a body rotates.

brake horsepower: The power produced by an engine and available for work through the propeller shaft. It is usually measured as a force on a brake drum or equivalent device, and is abbreviated bhp.

brake mean effective pressure: The portion of mean effective pressure on the piston which produces brake horsepower; abbreviated bmep.

bucket: A vane or blade of a turbine.

buzz bomb: The German V-1 pilotless bomb which was used to bomb Britain during World War II. The name was given because of the loud buzzing noise made by the pulsejet engine of the missile.

compressibility factor: The ratio of the pressure rise actually obtained from ram to that calculated from the simple hydraulic equation $\frac{1}{2}\rho V^2$.

console: Control panel in an airplane or a test cell. It is so named because it has some resemblance to the console of an organ.

contrarotating propellers: Two propellers mounted on coaxial shafts, one propeller being immediately behind the other and turning in the opposite direction. This type of arrangement is used with the Allison T40 engine.

density: The weight per unit volume of any substance, usually expressed in pounds per cubic foot or grams per cubic centimeter.

density altitude: Altitude expressed in terms of standard atmospheric conditions.

diffuser: That part of a turbine engine between the compressor and the burners. It is designed to reduce airspeed and increase pressure.

drone: A pilotless aircraft, remotely controlled, used for target work and reconnaissance.

equivalent shaft horsepower: The shaft horsepower plus the horsepower produced by the jet thrust in a turboprop engine; abbreviated eshp.

fatigue strength: The measured resistance of a body to failure caused by repeated applications of stress.

ferrous: The term describing metal which is derived from an iron base.

fuel consumption: The amount of fuel used by an engine. It is expressed in pounds per horsepower per hour or pounds per pound of thrust per hour.

fuel-flow meter: An instrument which indicates the amount of fuel flow in pounds per hour.

horsepower: The power required to raise 33,000 lb through a distance of 1 ft in 1 min. Thus 1 hp is 33,000 ft-lb per min or 550 ft-lb per sec.

hypergolic: Combinations of compounds or elements are hypergolic if the substances forming the combination burst into flame upon contact. This term is used to describe rocket fuels and oxidizers which require no ignition source.

indicated airspeed: The speed of the aircraft shown on the airspeed indicator and measured as the difference between static and dynamic pressures; abbreviated IAS.

indicated horsepower: The power produced within the engine cylinders and usually measured by obtaining indicator cards of the combustion process; abbreviated ihp. It includes brake, friction, supercharging, and accessory horsepower.

indicated mean effective pressure: The mean effective pressure on the piston, producing indicated horsepower; abbreviated imep.

jet propulsive efficiency: The ratio of the thrust horsepower obtained from the jet to the energy available in the jet velocity.

Mach number: The ratio of the speed of a body to the speed of sound under the same surrounding conditions.

mass: The actual quantity of material contained in a body. The weight of a body changes, depending upon its distance from the center of the earth, but the mass remains constant. Mass in **slugs** is equal to the weight divided by the acceleration of gravity, that is, 32.2.

mean effective pressure: The average pressure existing within the cylinder of an engine during the power stroke; abbreviated mep.

micron: The one-millionth part of a meter; hence, one-thousandth of a millimeter.

propeller efficiency: The ratio of the thrust horsepower obtained from the propeller to the brake horsepower put into it.

stratosphere: That part of the earth's atmosphere beginning about 7 miles out from the earth and continuing until there is no atmosphere. At the point where the stratosphere begins, there ceases to be any change in the temperature of the air as altitude is increased. The air temperature is 67° below zero Fahrenheit.

temperature, Kelvin: The absolute temperature scale in which there is 100° between the freezing point and boiling point of water at sea level; abbreviated K. The degree has the same value as a centigrade degree, and 0°K is 273.1° below zero centigrade.

temperature, Rankine: The absolute temperature scale in which there is 180° between the freezing point and boiling point of water at sea level; abbreviated R. The degree has the same value as a Fahrenheit degree, and 0°R is 459.7° below zero Fahrenheit.

thrust horsepower: The power available for pushing or pulling the aircraft; abbreviated thp.

troposphere: That part of the atmosphere between the earth's surface and about 7 miles above the earth. In this area the air temperature normally decreases as altitude increases.

true airspeed: Indicated airspeed corrected for air density; abbreviated TAS. This would be the ground speed of an aircraft flying near sea level with no wind.

Appendix

Standards

	English system	Metric system
Gravity (g)	32.174 ft per sec per sec	9.80665 m per sec per sec
Absolute zero	−459.7°F	−273.1°C
Horsepower	550 ft-lb per sec	76.04 kg-m per sec
π (pi)	3.14159	
Density (ρ_0)	0.002378 slug per cu ft	
Standard atmosphere at sea level:		
Temperature	59°F	15°C
Absolute temperature	518.4°R	288.2°K
Specific weight ($g\rho_0$)	0.07651 lb per cu ft	1.2255 kg per cu m
Pressure (P_0)	29.92 in. Hg	760 mm Hg
Pressure (P_0)	2,116 lb per sq ft	10,332 kg per sq m
Standard values at altitude:		
Isothermal level (Z_i)	35,332 ft	10,769 m
Isothermal temperature (t_i)	−67°F	−55°C
Temperature gradient (a)	0.003566°F per ft	0.0065°C per m

Airspeeds (true) vs. impact pressures (sea-level standard conditions)

Airspeed			Impact Pressure		
mph	*knots*	*ft per sec*	*in. H_2O*	*lb per sq ft*	*psi*
20	17.4	29.3	0.197	1.024	0.007
100	86.8	146.7	4.942	25.69	0.178
200	173.7	293.3	20.025	104.09	0.723
300	260.5	440.0	46.033	239.28	1.66
400	347.4	586.7	84.32	438.30	3.04
500	434.2	733.0	136.87	711.5	4.94
1,000	868.4	1,466.7	743.6	3,864.5	26.84
2,000	1,736.8	2,933.3	9,033.9	46,941.8	326.03

Standard atmosphere

Altitude, ft	Temperature t °F	°C	p, in. Hg	ρ	ρ/ρ_0
0	59.0	15.0	29.92	0.002378	1.0000
1,000	55.4	13.0	28.86	0.002309	0.9710
2,000	51.9	11.0	27.82	0.002242	0.9428
3,000	48.3	9.1	26.81	0.002176	0.9151
4,000	44.7	7.1	25.84	0.002112	0.8881
5,000	41.2	5.1	24.89	0.002049	0.8616
6,000	37.6	3.1	23.98	0.001988	0.8358
7,000	34.0	1.1	23.09	0.001928	0.8106
8,000	30.5	−0.8	22.22	0.001869	0.7859
9,000	26.9	−2.8	21.38	0.001812	0.7619
10,000	23.3	−4.8	20.58	0.001756	0.7384
11,000	19.8	−6.8	19.79	0.001702	0.7154
12,000	16.2	−8.8	19.03	0.001648	0.6931
13,000	12.6	−10.8	18.29	0.001596	0.6712
14,000	9.1	−12.7	17.57	0.001545	0.6499
15,000	5.5	−14.7	16.88	0.001496	0.6291
16,000	1.9	−16.7	16.21	0.001448	0.6088
17,000	−1.6	−18.7	15.56	0.001401	0.5891
18,000	−5.2	−20.7	14.94	0.001355	0.5698
19,000	−8.8	−22.6	14.33	0.001311	0.5509
20,000	−12.3	−24.6	13.75	0.001267	0.5327
21,000	−15.9	−26.6	13.18	0.001225	0.5148
22,000	−19.5	−28.6	12.63	0.001183	0.4974
23,000	−23.0	−30.6	12.10	0.001143	0.4805
24,000	−26.6	−32.5	11.59	0.001103	0.4640
25,000	−30.2	−34.5	11.10	0.001065	0.4480
26,000	−33.7	−36.5	10.62	0.001028	0.4323
27,000	−37.3	−38.5	10.16	0.000992	0.4171
28,000	−40.9	−40.5	9.720	0.000957	0.4023
29,000	−44.4	−42.5	9.293	0.000922	0.3879
30,000	−48.0	−44.4	8.880	0.000889	0.3740
31,000	−51.6	−46.4	8.483	0.000857	0.3603
32,000	−55.1	−48.4	8.101	0.000826	0.3472
33,000	−58.7	−50.4	7.732	0.000795	0.3343
34,000	−62.2	−52.4	7.377	0.000765	0.3218
35,000	−65.8	−54.3	7.036	0.000736	0.3098
36,000	−67.0	−55.0	6.708	0.000704	0.2962
37,000	−67.0	−55.0	6.395	0.000671	0.2824
38,000	−67.0	−55.0	6.096	0.000640	0.2692
39,000	−67.0	−55.0	5.812	0.000610	0.2566
40,000	−67.0	−55.0	5.541	0.000582	0.2447
41,000	−67.0	−55.0	5.283	0.000554	0.2332
42,000	−67.0	−55.0	5.036	0.000529	0.2224
43,000	−67.0	−55.0	4.802	0.000504	0.2120
44,000	−67.0	−55.0	4.578	0.000481	0.2021
45,000	−67.0	−55.0	4.364	0.000459	0.1926
46,000	−67.0	−55.0	4.160	0.000437	0.1837
47,000	−67.0	−55.0	3.966	0.000417	0.1751
48,000	−67.0	−55.0	3.781	0.000397	0.1669
49,000	−67.0	−55.0	3.604	0.000379	0.1591
50,000	−67.0	−55.0	3.436	0.000361	0.1517

Equations relating to engine power

Standard conditions for power measurement

At Sea Level

Barometric pressure—29.92 in. Hg
Humidity (water vapor pressure)—0.00 in. Hg
Temperature = 59.6°F

At Altitude

Pressure and temperature of standard altitude
Humidity zero at all altitudes

Propeller Load Curve

$$\text{Hp}_2 = \text{hp}_1\left(\frac{\text{rpm}_2}{\text{rpm}_1}\right)^3;\ \text{torque}_2 = T_1\left(\frac{\text{rpm}_2}{\text{rpm}_1}\right)^2$$

Brake Mean Effective Pressure

$$\text{Bmep} = \frac{792{,}000 \times \text{bhp}}{\text{displacement} \times \text{rpm}} \text{ lb per sq in.}$$

Torque

$$T = \frac{63{,}025 \times \text{hp}}{\text{rpm}} \text{ lb-in.} = \frac{5{,}250 \times \text{hp}}{\text{rpm}} \text{ lb-ft}$$

Indicated Horsepower

$$\text{Ihp} = \text{bhp} + \text{friction hp}$$

Mechanical Efficiency

$$\%\ \text{ME} = \frac{\text{bhp}}{\text{ihp}} \times 100$$

Thermal Efficiency

$$\%\ \text{TE} = \left(\frac{2{,}545}{\text{spec. fuel cons.} \times \text{Btu/lb fuel}}\right) 100$$

Weights of liquids

Liquid	Specific Gravity	Density, Lb per U.S. Gallon
Alcohol (methyl)	0.81	6.8
Benzine	0.69	5.8
Gasoline	0.72	6.0
Glycerine	1.26	10.5
Mercury	13.55	113.0
Oil	0.89	7.4
Ethylene glycol	1.12	9.3
Water	1.00	8.35

Weights of gases

Gas	Density,* Lb per Cu Ft
Air (at 59°F)	0.07651
Air	0.08071
Carbon dioxide	0.12341
Carbon monoxide	0.07807
Helium	0.01114
Hydrogen	0.005611
Nitrogen	0.07807
Oxygen	0.089212

* At atmospheric pressure and 0°C.

Laws of gases

Boyle's Law: When the temperature of a given mass of gas remains constant, the volume of the gas varies inversely at its pressure.

$$\frac{V_1}{V_2} = \frac{P_2}{P_1}$$

Charles's Law: When the pressure of a given mass of gas is kept constant, the volume of the gas is approximately proportional to its absolute temperature.

$$\frac{V_1}{V_2} = \frac{T_1}{T_2}$$

General Law

$$\frac{P_1V_1}{T_1} = \frac{P_2V_2}{T_2}$$

Centigrade-Fahrenheit conversion

$°F = \frac{9}{5}°C + 32 = 1.8(°C + 17.8)$
$°C = \frac{5}{9}(°F - 32)$

°C ←	°F ↓ °C	→ °F	°C ←	°F ↓ °C	→ °F
−62.2	−80	−112.0	54.4	130	266.0
−56.7	−70	−94.0	57.2	135	275.0
−51.1	−60	−76.0	60.0	140	284.0
−45.6	−50	−58.0	62.8	145	293.0
−40.0	−40	−40.0	65.6	150	302.0
−34.4	−30	−22.0	68.3	155	311.0
−31.7	−25	−13.0	71.1	160	320.0
−28.9	−20	−4.0	73.9	165	329.0
−26.1	−15	+5.0	76.7	170	338.0
−23.3	−10	14.0	79.4	175	347.0
−20.6	−5	23.0	82.2	180	356.0
−17.8	0	32.0	85.0	185	365.0
−15.0	5	41.0	87.8	190	374.0
−12.2	10	50.0	90.6	195	383.0
−9.4	15	59.0	93.3	200	392.0
−6.7	20	68.0	96.1	205	401.0
−3.9	25	77.0	98.9	210	410.0
−1.1	30	86.0	101.7	215	419.0
1.7	35	95.0	104.4	220	428.0
4.4	40	104.0	107.2	225	437.0
7.2	45	113.0	110.0	230	446.0
10.0	50	122.0	112.8	235	455.0
12.8	55	131.0	115.6	240	464.0
15.6	60	140.0	118.3	245	473.0
18.3	65	149.0	121.1	250	482.0
21.1	70	158.0	126.7	260	500.0
23.9	75	167.0	132.2	270	518.0
26.7	80	176.0	137.8	280	536.0
29.4	85	185.0	143.3	290	554.0
32.2	90	194.0	148.9	300	572.0
35.0	95	203.0	154.4	310	590.0
37.8	100	212.0	160.0	320	608.0
40.6	105	221.0	165.6	330	626.0
43.3	110	230.0	171.1	340	644.0
46.1	115	239.0	176.7	350	662.0
48.9	120	248.0	182.2	360	680.0
51.7	125	257.0	187.8	370	698.0

Centigrade-Fahrenheit conversion (*continued*)

°C ←	°F ↓ °C	→ °F	°C ←	°F ↓ °C	→ °F
193.3	380	716.0	287.8	550	1022.0
198.9	390	734.0	293.3	560	1040.0
204.4	400	752.0	298.9	570	1058.0
210.0	410	770.0	304.4	580	1076.0
215.6	420	788.0	310.0	590	1090.0
222.1	430	806.0	315.6	600	1112.0
226.7	440	824.0	326.7	620	1148.0
232.2	450	842.0	337.8	640	1184.0
237.8	460	860.0	348.9	660	1220.0
243.3	470	878.0	360.0	680	1256.0
248.9	480	896.0	371.1	700	1292.0
254.4	490	914.0	382.2	720	1328.0
260.0	500	932.0	393.3	740	1364.0
265.6	510	950.0	404.4	760	1400.0
271.1	520	968.0	415.6	780	1436.0
276.7	530	986.0	426.7	800	1472.0
282.2	540	1004.0	437.8	820	1508.0

Conversion factors

Multiply	*By*	*To Obtain*
Atm	76.0	Cm Hg
Atm	29.9212	In. Hg
Atm	33.8985	Ft H_2O
Atm	10,332.276	Kg per sq m
Atm	14.69601	Lb per sq in.
Atm	2,116.225	Lb per sq ft
Atm	1.0133	Bars
Cm	0.393700	In.
Cm	0.0328083	Ft
Cm Hg	5.352391	In. H_2O
Cm Hg	0.4460326	Ft H_2O
Cm Hg	0.193368	Lb per sq in.
Cm Hg	27.84507	Lb per sq ft
Cm Hg	135.9510	Kg per sq m
Cm per sec	0.0328083	Ft per sec
Cu cm	0.00099973	Liters
Cu cm	0.06102338	Cu in.
Cu ft	1,728.	Cu in.
Cu ft	$\frac{1}{27}$	Cu yards
Cu ft	7.480519	Gallons
Cu ft	28,317.017	Cu cm
Cu ft	28.31625	Liters
Cu ft	0.028317017	Cu m
Cu ft per min	0.471704	L per sec
Cu ft per min	0.028317	Cu m per min
Cu ft of water	62.42833	Lb
Cu in.	16.3871624	Cu cm
Cu in.	0.0163876	Liters
Cu in.	$\frac{1}{231}$	Gallons
Cu m	61,023.3753	Cu in.
Cu m	35.3144548	Cu ft
Cu m	264.170	Gallons

Conversion factors (*continued*)

Multiply	*By*	*To Obtain*
Degrees (arc)	0.017453292	Radians
Ft	30.4800613	Cm
Ft	0.3048006	Meters
Ft H_2O	0.029500	Atm
Ft H_2O	0.433530	Psi
Ft H_2O	62.3775	Lb per sq ft
Ft H_2O	304.8006	Kg per sq m
Ft H_2O	0.882671	In. Hg
Ft H_2O	0.24199	Cm Hg
Ft per min	0.0113636	Mph
Ft per min	0.018288	Km per hr
Ft per min	0.508001	Cm per sec
Ft per sec	0.681818	Mph
Ft per sec	1.09728220	Km per hr
Ft per sec	30.48006	Cm per sec
Ft per sec	0.3048006	M per sec
Ft per sec	0.5920858	Knots
Ft-lb	0.138255	M-kg
Ft-lb per min	1/33,000	Hp
Ft-lb per sec	1/550	Hp
Gallons	231.0	Cu in.
Gallons	0.133680	Cu ft
Gallons	3.785332	Liters
Gallons	0.832680	Imperial gallons
Gm	15.43236	Grains
Gm	0.0352739	Oz
Gm	0.002204623	Lb
Gm	1,000.0	Mg
Gm	0.001	Kg
Gm	980.665	Dynes
Gm-cal	0.0039685	Btu
Gm per cm	0.1	Kg per m
Gm per cm	0.06719702	Lb per ft
Gm per cm	0.0055914	Lb per in.
Gm per cu cm	1,000.0	Kg per cm
Gm per cu cm	62.42833	Lb per cu ft
Hp	33,000.0	Ft-lb per min
Hp	550.0	Ft-lb per sec
Hp	76.04039	Kg-m per sec
Hp	1.013872	Metric hp
Hp, metric	75.0	Kg-m per sec
Hp, metric	0.986318	Hp
Hp-hr	2,545.06	Btu
Hp-hr	1,980,000.0	Ft-lb
Hp-hr	273,745.4	Kg-m
In.	2.54000508	Cm
In. Hg	0.0334211	Atm
In. Hg	13.5951	In. H_2O
In. Hg	1.132925	Ft H_2O
In. Hg	0.4911570	Psi
In. Hg	70.72661	Lb per sq ft
In. Hg	345.3162	Kg per sq m
In. H_2O	0.0735559	In. Hg
In. H_2O	0.1868324	Cm Hg
In. H_2O	0.0361275	Psi
In. H_2O	5.1981	Lb per sq ft
In. H_2O	25.400051	Kg per sq m

Conversion factors (*continued*)

Multiply	*By*	*To Obtain*
Kg	2.20462234	Lb
Kg	35.273957	Oz
Kg	1,000.0	Gm
Kg-cal	3.9685	Btu
Kg-cal	3,087.4	Ft-lb
Kg-cal	426.85	Kg-m
Kg-m	7.2329983	Ft-lb
Kg-m	9.80665×10^7	Ergs
Kg per cu m	0.06242833	Lb per cu ft
Kg per cu m	0.001	Gm per cu cm
Kg per m	0.6719702	Lb per ft
Kg per sq m	0.00142234	Psi
Kg per sq m	0.2048169	Lb per sq ft
Kg per sq m	0.00289590	In. Hg
Kg per sq m	0.003280833	Ft H_2O
Kg per sq in.	0.1	Gm per sq cm
Km	3,280.833	Ft
Km	0.6213700	Miles
Km	0.539553	Nautical miles
Km per hr	0.9113426	Ft per sec
Km per hr	0.6213700	Mph
Km per hr	0.2777	M per sec
Km per hr	0.539553	Knots
Knots	1.0	Nautical mph
Knots	1.688944	Ft per sec
Knots	1.151553	Mph
Knots	1.853249	Km per hr
Knots	0.514791	M per sec
Liters	1,000.027	Cu cm
Liters	61.02503	Cu in.
Liters	0.035315411	Cu ft
Liters	0.264178	Gallons
Liters	0.219975	Imperial gallons
Meters	39.37	In.
Meters	3.280833	Ft
Meters	1.093611	Yards
M per sec	3.280833	Ft per sec
M per sec	2.2369317	Mph
M per sec	3.600	Km per hr
Miles	5,280.0	Ft
Miles	1.609347	Km
Miles	0.8683925	Nautical miles
Miles per hr	1.46666	Ft per sec
Miles per hr	0.4470409	M per sec
Miles per hr	1.609347	Km per hr
Miles per hr	0.8683925	Knots
Miles per hr per hr	2.151111	Ft per sec per sec
Oz	1/16	Lb
Oz	28.349527	Gm
Oz	437.5	Grains
Pounds	453.5924277	Gm
Pounds	0.45359243	Kg
Pounds	16.0	Oz
Pounds	32.174	Poundals
Pounds per cu ft	16.018369	Kg per cu m
Pounds per cu ft	0.016018369	Gm per cu cm
Pounds per cu in.	1,728.0	Lb per cu ft

Conversion factors (*continued*)

Multiply	*By*	*To Obtain*
Pounds per cu in.	27.6797424	Gm per cu cm
Pounds per sq in.	2.036009	In. Hg
Pounds per sq in.	2.306645	Ft H_2O
Pounds per sq in.	0.0680457	Atm
Pounds per sq in.	703.06687	Kg per sq m
Pounds per sq in.	0.07036	Kg per sq cm
Radians	57.29578	Degrees (arc)
Radians per sec	57.29578	Degrees per sec
Radians per sec	0.159155	Revolutions per sec
Radians per sec	9.84930	Rpm
Revolutions	6.283185	Radians
Rpm	0.104720	Radians per sec

Conversion factors (*continued*)

Multiply	*By*	*To Obtain*
Slugs	32.174	Lb
Sq cm	0.1549997	Sq in.
Sq cm	0.00107639	Sq ft
Sq ft	929.03412	Sq cm
Sq ft	0.092903412	Sq m
Sq in.	645.162581	Sq mm
Sq in.	6.45162581	Sq cm
Sq km	0.3861006	Sq miles
Sq m	10.76386736	Sq ft
Sq m	1.1959853	Sq yards
Sq miles	2.59000	Sq km
Sq yd	0.8361307	Sq m

Symbols and terms for gas-turbine engines

A	Area of jet or rocket nozzle
bsfc	Brake specific fuel consumption
C-D	Converging-diverging (nozzle)
CDP	Compressor discharge pressure
	P_{t3} for single-compressor engine
	P_{t4} for twin-spool engine
CDT	Compressor discharge temperature
	T_{t3} for single-compressor engines
	T_{t4} for twin-spool engines
CIP	Compressor inlet pressure
	P_{t2} for all engines
CIT	Compressor inlet temperature
	T_{t2} for all engines
CPR	Compressor pressure ratio
	P_{t3}/P_{t2} for single-compressor engines
	P_{t4}/P_{t2} for twin-spool engines
EGT	Exhaust gas temperature
	T_{t6} for single-compressor engines
	T_{t7} for twin-spool engines
EPR	Engine pressure ratio
	P_{t6}/P_{t2} for single-compressor engines
	P_{t7}/P_{t2} for twin-spool engines
eshp	Equivalent shaft horsepower
F	Thrust
F_g	Gross thrust
F_n	Net thrust
F_r	Ram drag of engine airflow
IGV	Inlet guide vanes
M	Mach number
N	Speed of rotation
NTC	Negative torque control
NTS	Negative torque signal
OAT	Outside air temperature
OGV	Outlet guide vanes
sfc	Specific fuel consumption
TIP	Turbine inlet pressure
	P_{t4} for single-compressor engines
	P_{t5} for twin-spool engines
TIT	Turbine inlet temperature
	T_{t4} for single-compressor engines
	T_{t5} for twin-spool engines
tsfc	Thrust specific fuel consumption
TSS	Thrust sensitive signal

Index